HEATH PHYSICAL SCIENCE

Authors **Louise Mary Nolan**
Science Teacher and Science Department Chairperson
John F. Kennedy Memorial Junior High School
Woburn, Massachusetts

Wallace Tucker
Science Consultant, Writer, and Lecturer
Harvard-Smithsonian Center for Astrophysics, and
University of California, Irvine

Consulting Author Zelda D. Tetenbaum
Science Teacher
Hinsdale Junior High School
Hinsdale, Illinois

Content Consultants Stephen S. Barshay CHEMIST
Assistant Professor of Chemistry
Rhode Island College
Providence, Rhode Island

Mario Iona PHYSICIST
Professor of Physics
University of Denver
Denver, Colorado

D.C. Heath and Company
Lexington, Massachusetts Toronto

Executive Editor: Roger R. Rogalin
Project Editor: Caroline N. Stevens
Production Coordinator: Donna Porter

Series Design: William Tenney
Book Designer: Cornelia L. Boynton

Cover Photo: Tom Magno

Teacher Consultants

Kathleen Carlig
Science Teacher
Dirksen Junior High School
Calumet City, Illinois

Barbara P. Groshens
Science Teacher
North Penn Junior High School
Lansdale, Pennsylvania

Barbara Hicks
Science Department Chairperson
Sylvan Hills High School
Atlanta, Georgia

Dr. John Ignacio
Science Supervisor
Irvington Public Schools
Irvington, New Jersey

Richard T. McKnight
Director of Science/Health
Waltham Public Schools
Waltham, Massachusetts

Nancy T. Watson
Assistant Professor
Burris Lab School
Ball State University
Muncie, Indiana

Series Consultants

Napoleon Bryant
Professor of Education
Xavier University
Cincinnati, Ohio

James H. McGill
Science Department Chairperson
Hill Country Middle School
Austin, Texas

Sister Rita Meredith, O.P.
Former Science Teacher
St. Agnes High School
Rockville Center, New York

Richard Merrill
Curriculum Specialist
Mt. Diablo Unified School District
Concord, California

William D. Thomas
Science Supervisor
Escambia County Schools
Pensacola, Florida

Reading Level Consultant

J & F
Milton D. Jacobson, Founder
Charlottesville, Virginia

Field Test Teachers

Grateful acknowledgement is given to the teachers and students who participated in field tests for this program.

ARIZONA
Harold Pruitt, Pima Middle School, Scottsdale

CALIFORNIA
L. Meyer, C. W. Tewinkle Middle School, Costa Mesa
Catherine Baird, Rancho San Joaquin Intermediate School, Irvine
Patrick Croner Venado Middle School, Irvine
Patrick Gleason, Vista Verde School, Irvine

ILLINOIS
Kristin Ciesemier, Naperville North High School, District 203, Naperville
David Klussendorf, Michael Sommers, Jefferson Junior High School, District 203, Naperville
Jean Kriebs, Butler Junior High School, Oakbrook
Kelly Douglas, Wilson Middle School, Rockford

INDIANA
Nancy Watson, Burris Lab School, Ball State University, Muncie

MASSACHUSETTS
Brian Flaherty, Ronald Gomes, Plymouth Carver Intermediate School, Plymouth
Warren Phillips, Nathaniel Morton Elementary School, Plymouth
Paula Garten, Russell Magee, Middle School East, Salem
Alan Chasse, North Intermediate School, Wilmington
Joseph Gilligan, West Intermediate School, Wilmington
Priscilla Lockwood, Wilmington High School, Wilmington

MISSOURI
Beverly Edwards, Visual and Performing Arts Center, St. Louis
Marie Globig, Investigate Learning Center, St. Louis

NEW JERSEY
Joan Camarigg, Eisenhower School, Freehold
Susanne Flannelly, Joseph Herman, Barkalow School, Freehold

NEW YORK
John Spring, Trinity Lutheran School, Hicksville
Sam Alaimo, Hoover Middle School, Kenmore
Elaine Johnson, Kenmore Middle School, Kenmore
James Morgan, Benjamin Franklin Middle School, Kenmore
Edith Newfield, Hendrick Hudson High School, Montrose
A. Charles Rossi, Blue Mountain School, Peekskill
Alan Goodman, Pamona Junior High School, Suffern

PENNSYLVANIA
Anna McCartney, Tidioute Junior/Senior High School, Tidioute

TENNESSEE
Jo Quarles, Jefferson Middle School, Jefferson City
Ted Frisby, Westwood Junior High School, Manchester
Peggy Mason, Gary Mullican, Central Middle School, Murfreesboro

TEXAS
Mary Jane Vasquez, Lake Travis Junior High School, Austin

Published simultaneously in Canada.
Printed in the United States of America.
International Standard Book Number: 0-669-05162-4

Excerpts from the position statement issued by the National Science Teachers Association in *NSTA Report*, November 1982:

"The goal of science education during the 1980s is to develop scientifically literate individuals who understand how science, technology, and society influence one another and who are able to use this knowledge in their everyday decision making. The scientifically literate person has a substantial knowledge base of facts, concepts, conceptual networks, and process skills which enable the individual to continue to learn and think logically. This individual both appreciates the value of science and technology in society and understands their limitations.

"The middle/junior high school science curriculum should be designed to accommodate the needs and learning styles of the early adolescent. Students should be provided with daily opportunities to explore science through reading, discussion, and direct learning experiences in the classroom, laboratory, and field."

We at D. C. Heath share these goals. In fact, the new science program Heath has developed for the middle/junior high school classroom—**Heath Life Science, Heath Earth Science,** and **Heath Physical Science**—fully address these concerns.

On the following 19 pages you will find an overview of **Heath Physical Science,** demonstrating the many ways in which this textbook meets the important goal of science education as defined by the NSTA.

How HEATH PHYSICAL SCIENCE meets its goal: an overview.

Heath Life Science. Heath Earth Science. Heath Physical Science. D. C. Heath's new science textbooks for junior high and middle school.

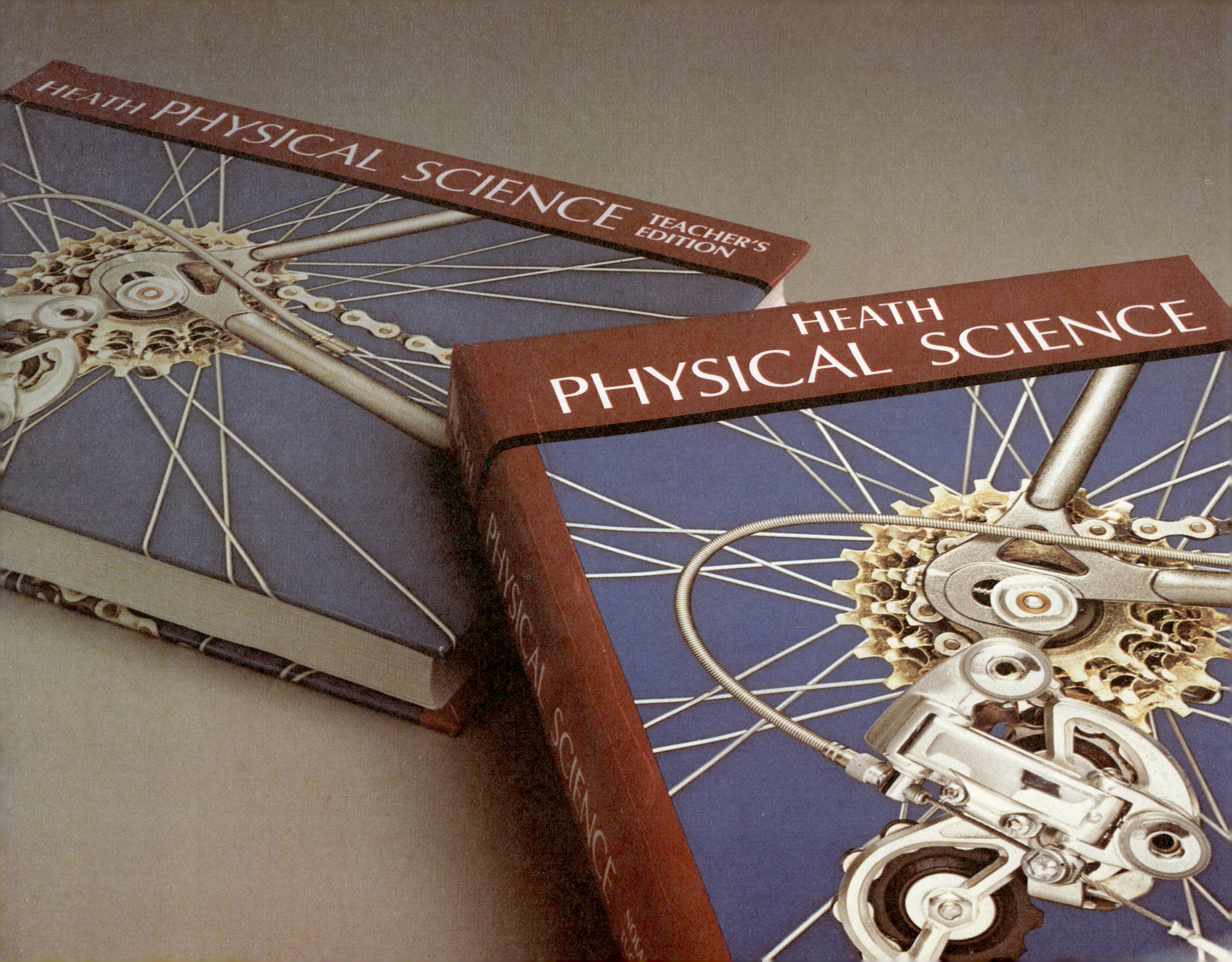

HEATH PHYSICAL SCIENCE TEACHER'S EDITION
HEATH PHYSICAL SCIENCE
PHYSICAL SCIENCE

INTRODUCING HEATH PHYSICAL SCIENCE. IT'S EVERYTHING YOU COULD ASK FOR. BECAUSE IT'S EVERYTHING YOU DID ASK FOR.

First, we talked to physical science teachers like you. Then we based our book on what you told us.

Teachers in junior high and middle school face some very tough challenges.

At D. C. Heath, we're fully aware of this fact. You see, while we were developing our new physical science book, we consulted hundreds of your colleagues in rural, urban, and suburban regions all over the United States.

The result is **Heath Physical Science**—custom-tailored to the concerns of physical science teachers at these critical grade levels.

You asked for strong content and clear presentation.

Whether we gathered responses from surveys or talked with teachers one on one, the outcome was always the same. Physical science teachers everywhere called for a book that was sound and accurate, but also clear and readable. They wanted a book that would motivate students—and foster scientific literacy.

Plus, you wanted dozens of teaching ideas (and you preferred them at your fingertips).

Then when it came to the Teacher's Edition, the teachers we talked to wanted something new. They wanted a practical tool—a no-nonsense, timesaving book—featuring plenty of valuable aids and ideas in a handy format, with everything easily accessible.

Heath Physical Science: responsive to the concerns of junior high/middle school physical science teachers.

Now there is good news for you and your colleagues. With the all-new **Heath Physical Science,** you get exactly what you asked for. You get a student text that offers all the strong, comprehensible content you want for your students. And, you get the kind of Teacher's Edition you've never had before: an efficient classroom tool, featuring step-by-step lesson plans and scores of imaginative ideas (all right where you need them—at your fingertips).

The following 17 pages give you a thorough overview of **Heath Physical Science**—including colorful sample pages and pictures.

GOOD, SOLID SCIENCE. IT'S WHAT YOU ASKED FOR—AND IT'S WHAT YOUR STUDENTS NEED.

The physical science teachers we talked to expressed great concern for the needs of their students. Above all, they were concerned that students at these grade levels get good, accurate information from their science books.

Heath Physical Science has that kind of scientific integrity. Here, at last, is a student edition that can effectively prepare your students both for high school science courses and for their changing world.

Comprehensive because it's based on teacher input.

This book covers all the topics traditionally required in the physical science curriculum—all the topics you expect to teach at your grade level. Just as important, it covers these topics completely. There's enough depth and enough detail to tell the whole story.

Checked by experts for accuracy.

A team of highly trained experts—specialists in various fields of physical science—have checked and double-checked every page of Heath's student book. Plus, our editors have carefully reviewed all illustrations to assure accuracy.

The result? You can rely on **Heath Physical Science** to give your students sound information—both about facts and concepts and about current scientific thought.

The latest information in a fast-changing field.

Heath Physical Science keeps abreast of recent scientific findings, theories, inventions, and discoveries. Throughout this book students read about such up-to-the-minute topics as electronics, optical fibers, holograms, alternate energy sources, recent developments in nuclear-fusion research, and the discovery and naming of elements 104 and beyond.

Career features present current information on the growing, changing world of physical science careers.

The text features functional and attractive illustrations to tell the complete story of physical science. Many photos present analogies to help students visualize abstract concepts in concrete terms.

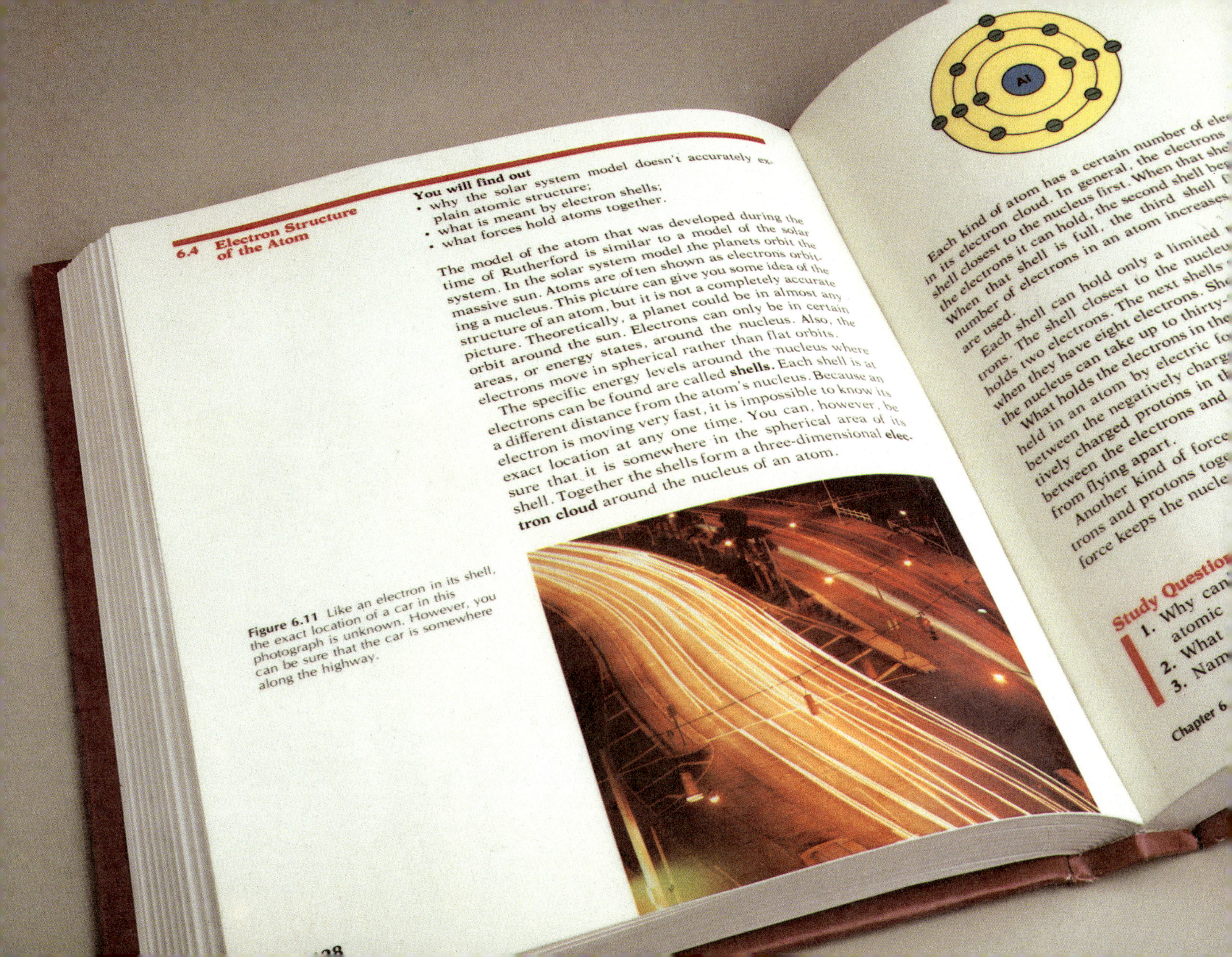

6.4 Electron Structure of the Atom
You will find out
• why the solar system model doesn't accurately explain atomic structure;
• what is meant by electron shells;
• what forces hold atoms together.
The model of the atom that was developed during the time of Rutherford is similar to a model of the solar system. In the solar system model the planets orbit the massive sun. Atoms are often shown as electrons orbiting a nucleus. This picture can give you some idea of the structure of an atom, but it is not a completely accurate picture. Theoretically, a planet could be in almost any orbit around the sun. Electrons can only be in certain areas, or energy states, around the nucleus. Also, the electrons move in spherical rather than flat orbits.
The specific energy levels around the nucleus where electrons can be found are called shells. Each shell is at a different distance from the atom's nucleus. Because an electron is moving very fast, it is impossible to know its exact location at any one time. You can, however, be sure that it is somewhere in the spherical area of its shell. Together the shells form a three-dimensional electron cloud around the nucleus of an atom.
Figure 6.11 Like an electron in its shell, the exact location of a car in this photograph is unknown. However, you can be sure that the car is somewhere along the highway.
Al
Study Question

GOOD SCIENCE DOES NO GOOD UNLESS STUDENTS UNDERSTAND IT.

Youngsters in junior high and middle school are at a crossroads. At no other time in their lives will they be growing and changing so rapidly. During these critical years, many things compete for their attention—and school subjects often lose out in the competition.

That's why physical science teachers face a unique challenge. How can you make sure your students fully understand the concepts of physical science? How can you sustain their interest and arouse their enthusiasm?

Heath Physical Science meets the challenge.

Heath Physical Science has been carefully developed in response to teachers' concerns about comprehension and motivation. Here is a book that translates facts and concepts into clear, colorful narrative—a book designed to keep students intrigued.

Clarity starts with the page and lesson format.

Every page in **Heath Physical Science** is clean, open, uncluttered: Students know where to start reading and where to find information.

Plus, each brief lesson starts with clearly stated objectives and ends with study questions that refer back, one to one, to these objectives. So students know what to look for as they read, and they can confidently test themselves when they're done.

The story of physical science is complete, not just a string of facts.

It's hard to learn from a lot of isolated facts and terms. So **Heath Physical Science** explains concepts fully, giving concrete examples and showing links between ideas. Students see how everything fits, and they're better able to understand.

Best of all, Heath's narrative text is so vivid it actually absorbs students' attention. Throughout, there's coverage of exciting topics that truly interest students: computers, lasers, making music, automotive engines, radioactivity, microwaves, black holes, space colonies, and much more. Plus, for every chapter, there's a fascinating spotlight on a science-related **Career** or a **Biography** of a well-known physical scientist, past or present. Youngsters learn that physical science is relevant and real.

Pictures work with text to make everything clearer.

The beautiful full-color pictures throughout **Heath Physical Science** reinforce text content. Each one is fully captioned and is placed right near the text item it helps explain. Students are more likely to understand each concept because they see it for themselves—in concrete, colorful terms.

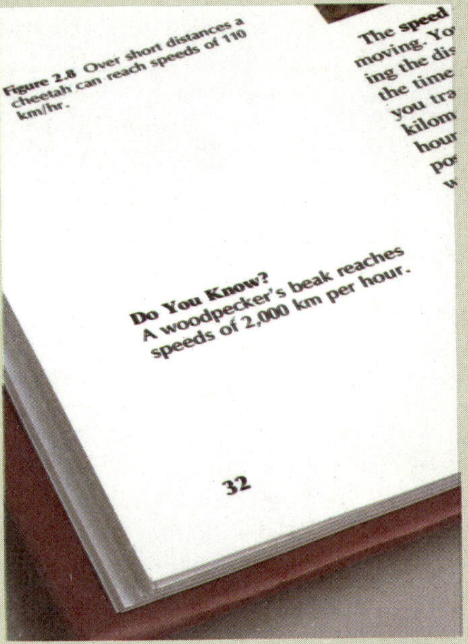

In the margins of the student book, high-interest *Do You Know?* features spotlight out-of-the-ordinary science facts.

Key concepts are illustrated with examples from everyday life to enhance student understanding. Students get a better sense of abstract concepts when they are presented in both pictures and words.

Falling objects all accelerate in the same
object falling near the surface of the earth wi
ate 9.8 m/s/s. This means the velocity of an o
increase by 9.8 meters per second every sec
falling. Take for example an apple dropped f
roof of a 20-story building. Before the apple is d
its velocity is 0 m/s. One second after the a
dropped, its velocity is 9.8 m/s. After two secon
apple has accelerated to 9.8 + 9.8, or 19.6 m/s.
another second the apple's velocity is 19.6 + 9.8, o
m/s. As the apple continues to fall, it continues to a
erate by 9.8 m/s every second. A falling brick w
accelerate the same way. So would a nickel, a per
and a sack of potatoes. All falling objects accelerate
m/s for each second they fall.

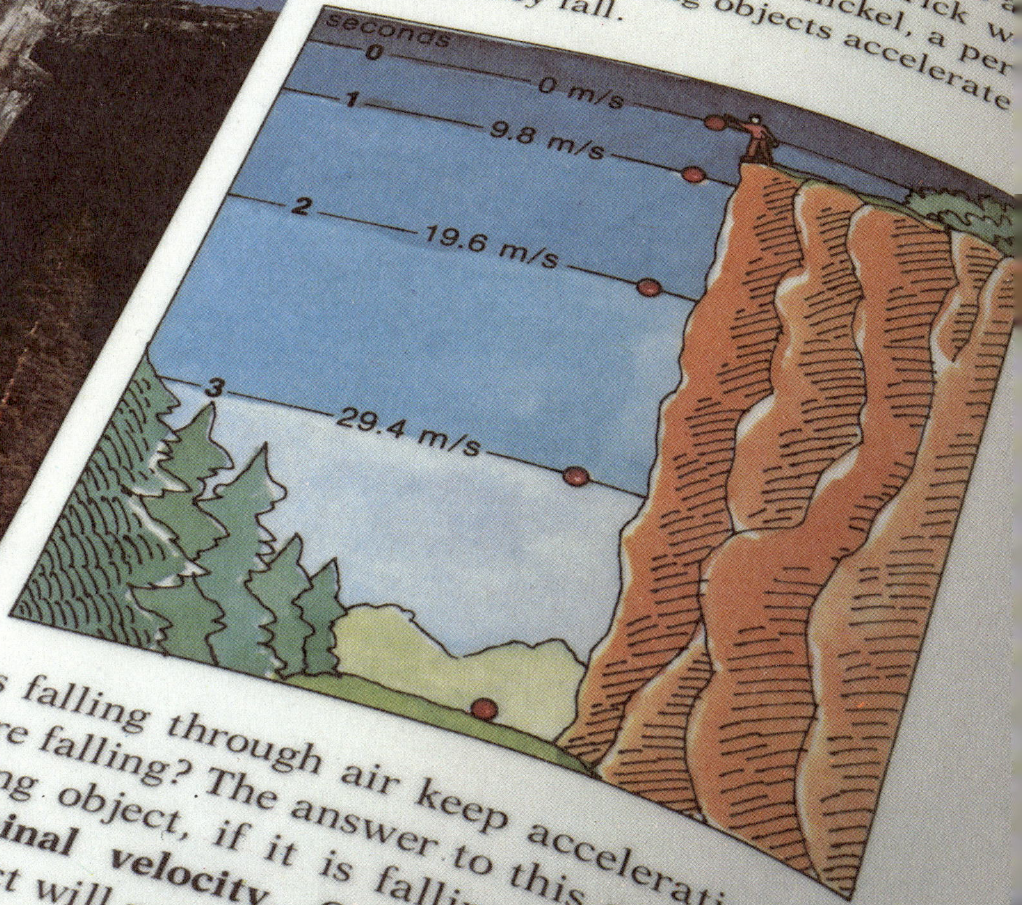

Figure 2.16 A falling object accelerates at a rate of 9.8 m/s/s.

Can objects falling through air keep accelerating as long as they are falling? The answer to this question is no. Every falling object, if it is falling long enough, reaches a **terminal velocity.** Once this velocity i
reached, the object will not accelerate any more. I
continue to fall at a constant rate, its termina

You will find out

- what is meant by the word *buoyancy*;
- what is stated in Archimedes' principle;
- how to find the density of an object;
- how to predict whether an object will float.

Have you ever noticed that it's easier to lift a person in water? This is due to the upward force, or **buoyancy** [BOY-uhn-see], of water. Buoyancy is the name given to the force that pushes an object up and makes it seem to lose weight in a fluid. In this case, the amount of weight a person seems to lose in the water is actually the weight of the water that the person moves aside by standing in it.

Why do some objects float in water while others sink? If you want to know which objects will float in water, you need to know more about the force exerted by the fluid. You need to know the strength of the buoyant force compared to the weight of an object.

Over 2,000 years ago the Greek scientist Archimedes figured out the strength of the buoyant force as he was taking a bath. Legend has it that Archimedes became so excited about having solved the buoyancy problem that he jumped out of the tub and ran naked down the main street of town, yelling "Eureka! Eureka!" which means "I have found it." What Archimedes discovered is that the buoyancy of any object is equal to the weight of the liquid that the object displaces, or pushes aside. This relationship between buoyancy and the weight of the fluid being displaced is known as **Archimedes' ...iple.**

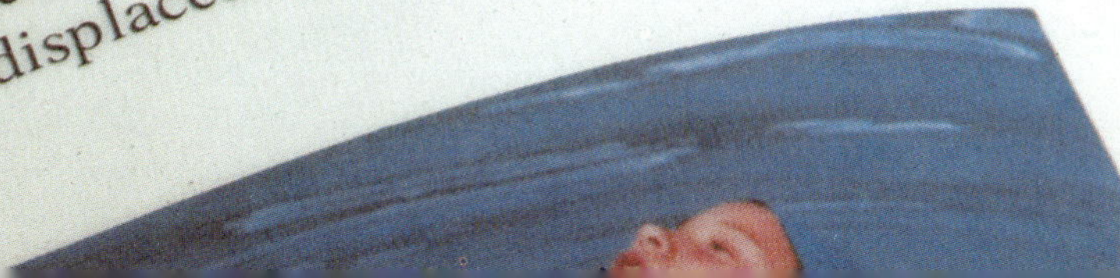

Figure 5.8 displaced ... of the bl...

Since the buoyant force acts upward, it opposes the downward force of the weight of the object. If the buoyant force is equal to the weight of the object, then the object will float. For example, a piece of wood having a volume of 100 cubic centimeters weighs only half as much as 100 cubic centimeters of water, so it will float on water. In other words, an object will float if it weighs less than the weight of the water it displaces.

On the other hand, a piece of steel having a volume of ten cubic centimeters weighs about eight times as much as the same volume of water. You can see that steel will not float. Why then are ships made out of steel? The answer lies in the ship's design. The ship is constructed so that it is made mostly of air. That is, the inside of the ship is hollow. Since the combination of steel plus air weighs less than the water the ship displaces, the ship will float.

HEATH HAS TEXT READABILITY DOWN TO A SCIENCE.

When it comes to reading, junior high and middle school students face a special challenge. Suddenly they are given more reading than they've ever had before—and less capable readers, or readers with poor concentration, often have trouble coping.

Heath meets the challenge—with a uniquely readable text.

Every single word of the student book has been computer-checked to make sure it is at the appropriate reading level for your students. The page layout and even the type size have been carefully designed for ease of reading.

Then there are the pictures. These attractive, instructive photos relate so closely to the surrounding text that they actually function as context clues.

BUILT-IN AIDS DEVELOP A RANGE OF SKILLS.

The **Heath Physical Science** student text is more than readable. It's also usable as a practical tool for building key skills such as content-area reading, using scientific processes, mathematical problem solving. In all, there are five categories of vital skills developed: reading comprehension, word, mathematics, laboratory, and reference/research.

How does it work?

Woven throughout the student text are opportunities for students to practice and apply the skills just mentioned. For instance, there are graphs students must analyze, new vocabulary words they must understand in context, and mathematical computations they must perform. The authors actually wrote the book with the development of these skills in mind, so learning opportunities are truly well integrated into the text. (The picture below shows a good example.)

For your convenience, all skills developed in the student text are identified in the Teacher's Edition with a blue bullet and a brief marginal description.

For more information on Heath's development of built-in skills, see the Skills Matrix and the Index of Skills on pages A-26–A-27.

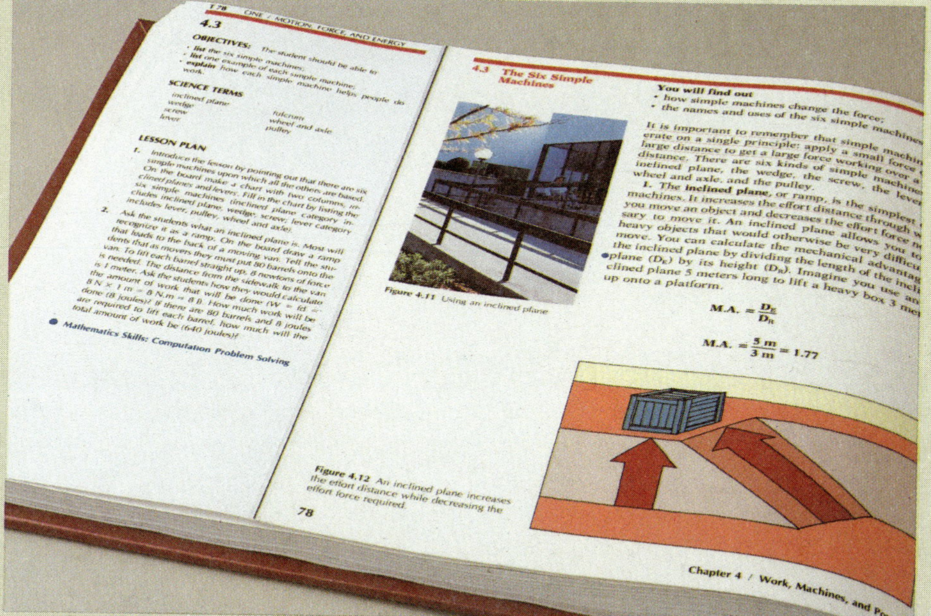

Key terms are boldfaced, phonetically respelled, and defined *in context*, where students can get a better sense of their meaning. Plus, they are included in both the Glossary and the Index.

ACTIVITY

How Do Pulleys Help Do Work?

Materials

pulleys
ring stand and ring
string
scissors
meterstick
book
spring scale

Procedure

1. Tie a string around the book and weigh it using the spring scale.
2. Attach the ring halfway up the ring stand.
3. Attach a pulley to the ring.
4. Tie a longer piece of string to the book and thread it through the pulley.
5. Place the book on the table and attach the spring scale to the free end of the string.
6. Have a friend hold the ring stand while you pull the spring scale and note the reading on the scale when the book is lifted off the table.

Questions

1. What was the weight of the book?
2. How much force was needed to lift the book?
3. Draw a diagram of the setup and label the fulcrum.
4. How is a pulley like a lever?
5. How do pulleys help do work?

Study Questions for 4.3

1. How do simple machines change the force?
2. [L]ist the names of the simple machines and briefly [...] how each is used.

HEATH ACTIVITIES: SET UP FOR STUDENT SUCCESS. . .

The teachers we talked to felt that activities could really be exciting tools for learning. But—they said—if these activities are hard to perform, students may well end up feeling confused and frustrated.

Heath's response: a step-by-step approach.
Each activity in **Heath Physical Science** has a numbered, step-by-step procedure, specially developed to fit in with the way students learn at these grade levels. This procedure makes everything clear and explicit. First it tells students exactly how to set up the materials—and gives them special cautions to ensure safety. Then it shows students what to look for and helps them organize their data, so they can answer the questions that come at the end.

The result? Students perform the activity with ease and confidence. And they learn a lot more in the process.

Heath helps students get started easily. The title poses the main question students will answer when they're done. It gives students focus and direction. Convenient **materials list** shows students just what they'll need. Clear, well-organized **steps** guide students all along the way so they can perform the activity independently. **Questions** prompt students to think about activity results—guide them in synthesizing data, help them draw conclusions.

. . . SET UP FOR YOUR CLASSROOM NEEDS.

Heath activities fit in with your time and budget.
Heath recognizes that physical science teachers must work within certain time, budget, and equipment constraints. So we've designed each activity to fit realistically with the amount of time devoted to an activity in a class period—and the facilities normally available.

Heath activities let you take a flexible approach.
The activities in **Heath Physical Science** give you all the options you want and need.

To begin with, there's a rich variety of these activities. They range from very simple ones involving only paper and pencil to much more complex ones that develop mathematical or laboratory skills. Then too, each activity—whether simple or complex—reinforces a concept that has come up earlier in the lesson.

It all adds up to flexibility. You can pick and choose among these activities on the basis of your teaching needs, time pressures, and the ability levels and interests of your class.

Heath activities give you more options. You may have every student do a given activity; you may have small groups or a few interested students do it; or you may prefer to do a demonstration, as the teacher is doing here.

RIGHT IN THE STUDENT BOOK: TOOLS FOR REVIEW AND INFORMAL TESTING.

Study questions at end of each lesson key back directly to lesson objectives.

Answers to study questions are clearly stated in the narrative text. Students can find them easily.

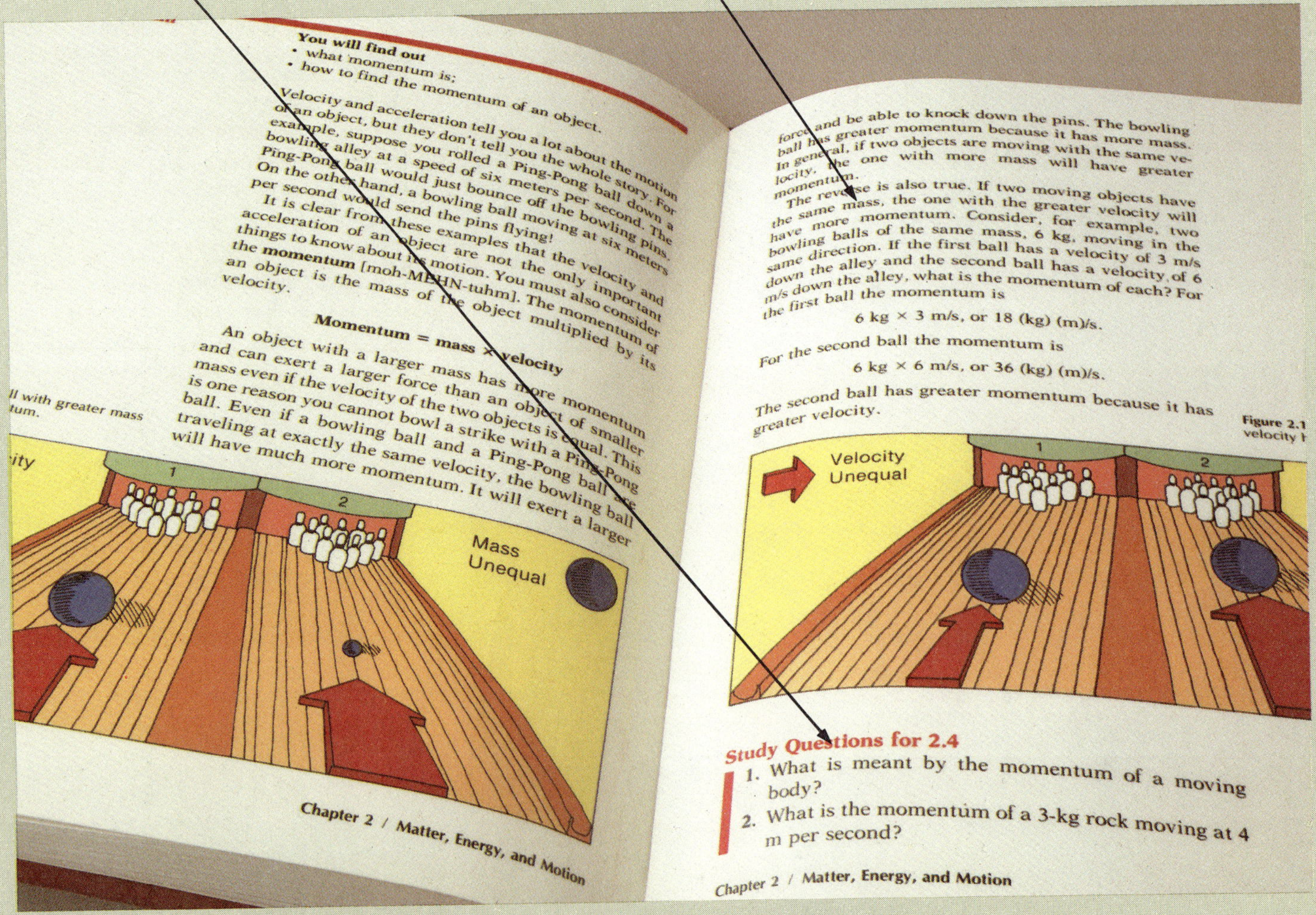

You will find out

- what momentum is;
- how to find the momentum of an object.

Velocity and acceleration tell you a lot about the motion of an object, but they don't tell you the whole story. For example, suppose you rolled a Ping-Pong ball down a bowling alley at a speed of six meters per second. The Ping-Pong ball would just bounce off the bowling pins. On the other hand, a bowling ball moving at six meters per second would send the pins flying!

It is clear from these examples that the velocity and acceleration of an object are not the only important things to know about its motion. You must also consider the **momentum** [moh-MEHN-tuhm]. The momentum of an object is the mass of the object multiplied by its velocity.

Momentum = mass × velocity

An object with a larger mass has more momentum and can exert a larger force than an object of smaller mass even if the velocity of the two objects is equal. This is one reason you cannot bowl a strike with a Ping-Pong ball. Even if a bowling ball and a Ping-Pong ball are traveling at exactly the same velocity, the bowling ball will have much more momentum. It will exert a larger force and be able to knock down the pins. The bowling ball has greater momentum because it has more mass. In general, if two objects are moving with the same velocity, the one with more mass will have greater momentum.

The reverse is also true. If two moving objects have the same mass, the one with the greater velocity will have more momentum. Consider, for example, two bowling balls of the same mass, 6 kg, moving in the same direction. If the first ball has a velocity of 3 m/s down the alley and the second ball has a velocity of 6 m/s down the alley, what is the momentum of each? For the first ball the momentum is

6 kg × 3 m/s, or 18 (kg) (m)/s.

For the second ball the momentum is

6 kg × 6 m/s, or 36 (kg) (m)/s.

The second ball has greater momentum because it has greater velocity.

Figure 2.1
velocity

Study Questions for 2.4

1. What is meant by the momentum of a moving body?
2. What is the momentum of a 3-kg rock moving at 4 m per second?

Chapter 2 / Matter, Energy, and Motion

Chapter 2 / Matter, Energy, and Motion

The physical science teachers we surveyed stressed the importance of thorough evaluation. That's why they wanted as many testing opportunities as possible—not just formal, end-of-instruction tests, but also informal reviews and quizzes and self-tests.

In response to this concern, **Heath Physical Science** provides—right in the student book—a collection of superb tools for informal, ongoing evaluation. Whether you use these tools alone or together with the formal tests described on page A-21, they'll give you a truly complete, balanced picture of each student's progress and performance.

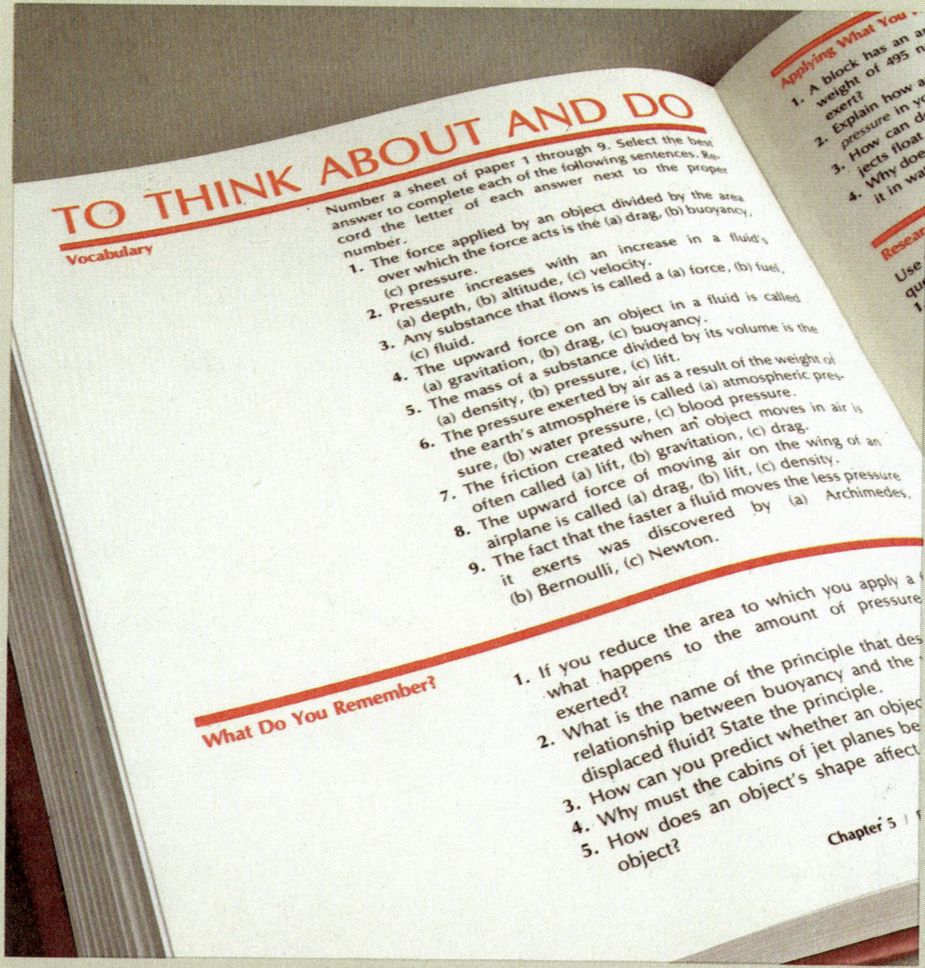

T 104 ONE / MOTION, FORCE, AND ENERGY

5.3

OBJECTIVES: The student should be able to
- **explain** how atmospheric pressure changes with altitude;
- **explain** how Archimedes' principle applies to air;
- **explain** one reason atmospheric pressure is important in daily life.

SCIENCE TERM

atmospheric pressure

ADVANCED PLANNING

For demonstrations you will need a tall heat-resistant glass jar, a piece of paper, a balloon that will fit over the mouth of the jar, a hard-boiled egg with the shell removed, and a liter (quart) glass milk bottle with an opening that the egg can rest in.

LESSON PLAN

1. Introduce the lesson by performing one or both of the following demonstrations:
   a. Crumble a small piece of paper and put it on the bottom of a tall heat-resistant glass jar. With a match light the paper and quickly put a rubber balloon over the mouth of the jar. Let the students observe what happens.
   b. Put a wad of paper in the bottom of the milk bottle. Light the paper and put the hard-boiled egg, pointed end down, in the neck of the bottle. Let the students observe what happens.
2. After you have completed the demonstrations, ask the students what they observed in each case (balloon was sucked into the jar, egg fell into the bottle). Ask why these things happened (oxygen in bottle was used up by fire, so there was more air pressure outside than inside and object was pushed in).

SCIENCE BACKGROUND

When weather reporters send up weather balloons or when people go for balloon rides, the balloons are only partially filled. As the balloons rise, they expand because there is less pressure outside the balloon than there is inside it.

5.3 Atmospheric Pressure

You will find out
- what is meant by atmospheric pressure;
- why atmospheric pressure decreases with height;
- how you use atmospheric pressure.

0 N/cm²

1 N/cm²

10 N/cm²

Figure 5.13 As altitude increases, atmospheric pressure decreases.

Jets fly at heights of ten kilometers above sea level. There the atmospheric pressure is less than a third of the pressure at sea level. Jet cabins must be pressurized for the comfort and safety of the passengers and crew.

Figure 5.14 Astronauts must wear pressurized suits because the pressure inside their bodies is much less than the pressure in space.

104

Study Questions for 5.3

1. What is atmospheric pressure?

Chapter 5 / Forces in Liquids and

AT LAST—A TEACHER'S EDITION THAT WORKS AS HARD AS YOU DO.

A Heath breakthrough: teaching support at your fingertips.

Introducing the Teacher's Edition for **Heath Physical Science.** If you've never used a teacher's edition before, this is the one you'll want to use.

Its secret? The *extended format.* There's a wide margin beside every student-book page—so we're able to place the well-organized aids and ideas right next to the material they refer to.

That means you don't have to flip back and forth from the section you're teaching to the front section of your Teacher's Edition. Instead, you can easily refer to the teaching support while you plan lessons or while you actually teach. The result? You plan more efficiently—and you teach without interruption.

Now you can do more for your students—without doing it all by yourself.

This book doesn't merely give you more accessible support. It gives you more support, period. There's a wealth of suggestions and imaginative ideas—something for everyone, from the first-year teacher to the seasoned pro.

Are you a new teacher who wants help in making lesson plans or in organizing your teaching strategy? A content specialist who has trained in an area other than physical science? Or a veteran physical science teacher looking for new ways to enhance your teaching? No matter what your situation, you'll want to see the time-saving **Lesson Organizer** that comes with every chapter—and the step-by-step guidelines for teaching each lesson. You'll also like the **Chapter Background** that opens each chapter and the informative beyond-the-book **Science Background** that's featured in every lesson.

You can pick and choose the teaching aids you want.

Best of all, the teaching support in **Heath Physical Science** is totally flexible. So, you can custom-tailor instruction to your own class situation, to your teaching style, and to the ability levels of your students.

HEATH SAVES YOU TIME IN PLANNING, SO YOU CAN FOCUS MORE ON TEACHING.

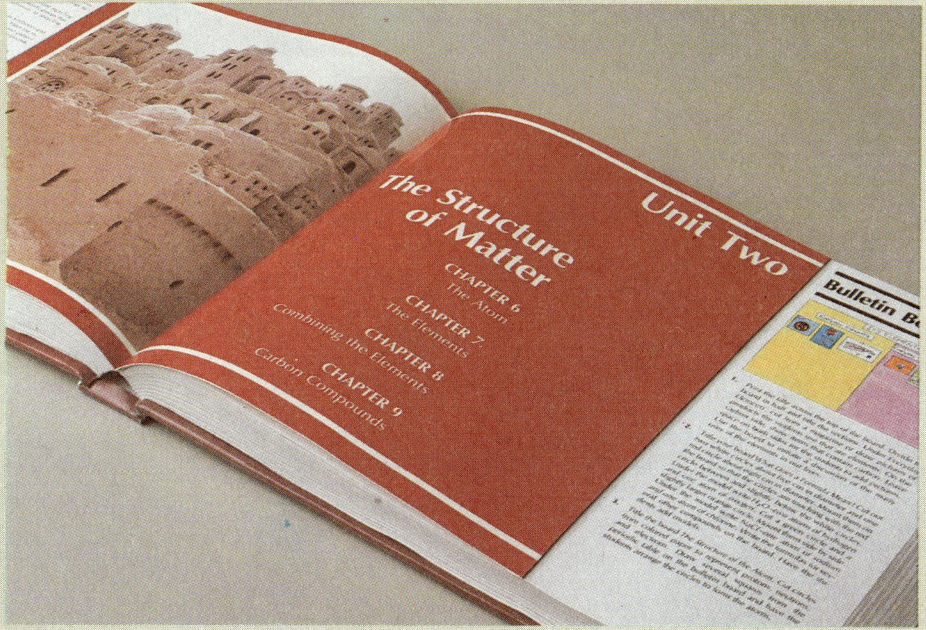

THE TEACHER'S EDITION LESSON PLAN: STEP-BY-STEP SUGGESTIONS RIGHT WHERE YOU NEED THEM.

Clearly stated **Objectives.**

Science Terms for this lesson listed for your convenience.

Advanced Planning—short- and long-term—helps you plan ahead to obtain needed materials, etc.

The core of the lesson support: numbered **Lesson Plan** that guides you every step of the way. Includes suggestions for class demonstrations.

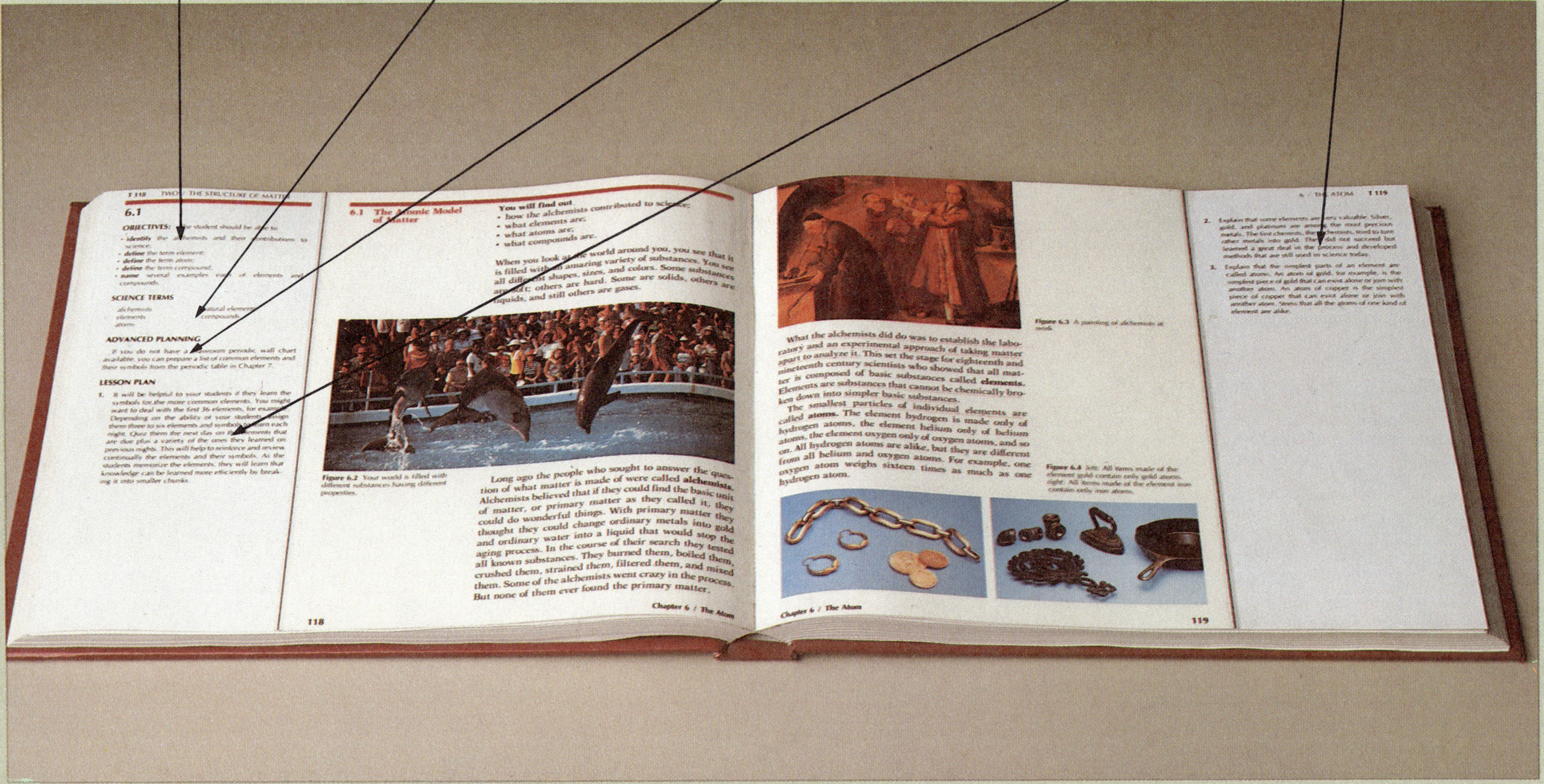

Important **Skills** developed in the student book pointed out here with a blue bullet, clearly identified in the margin, and cross-referenced to the Skills Matrix on page A-26.

Annotations always right next to the material they relate to.

Follow-up to customize the lesson for different students or groups of students, to fit different ability levels or interests.

Reinforcement—for students who need additional help in grasping the lesson concept, or for your whole class, if extra practice is needed.

Enrichment—for faster learners, or for students who have a special interest in the lesson topic.

Tie-ins to other subject areas—so students can see how broadly practical and useful physical science can be.

Science Background that goes beyond the book to give students even more insight into the lesson topic.

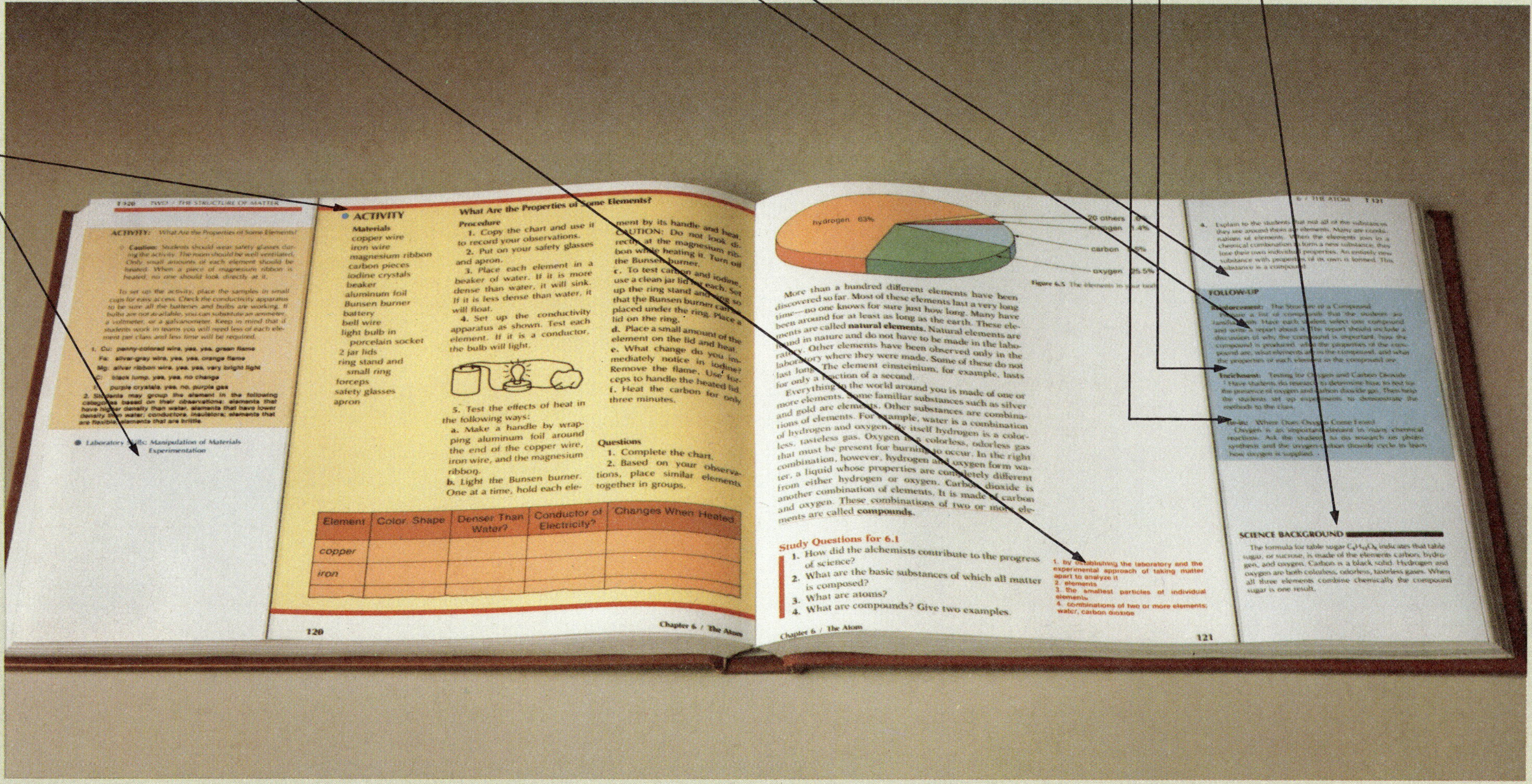

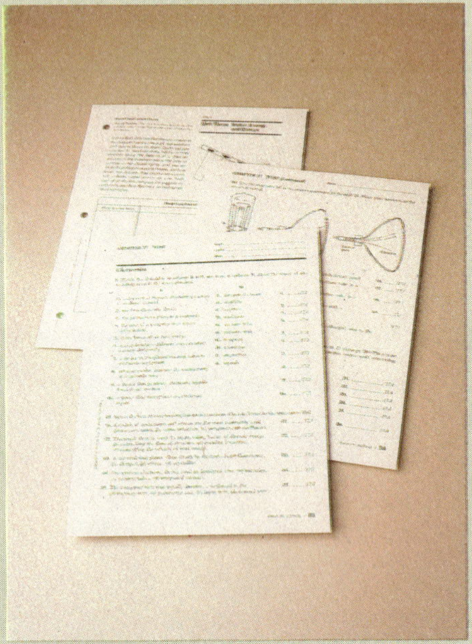

Chapter Tests evaluate on the literal, interpretive, and applied levels. **Unit Tests** give students valuable practice with a standardized-test format. Both kinds of formal tests are available in the **Copymasters** booklet as well as in the **Duplicating Masters** booklet. **Parent Involvement Sheets** are also available in both forms.

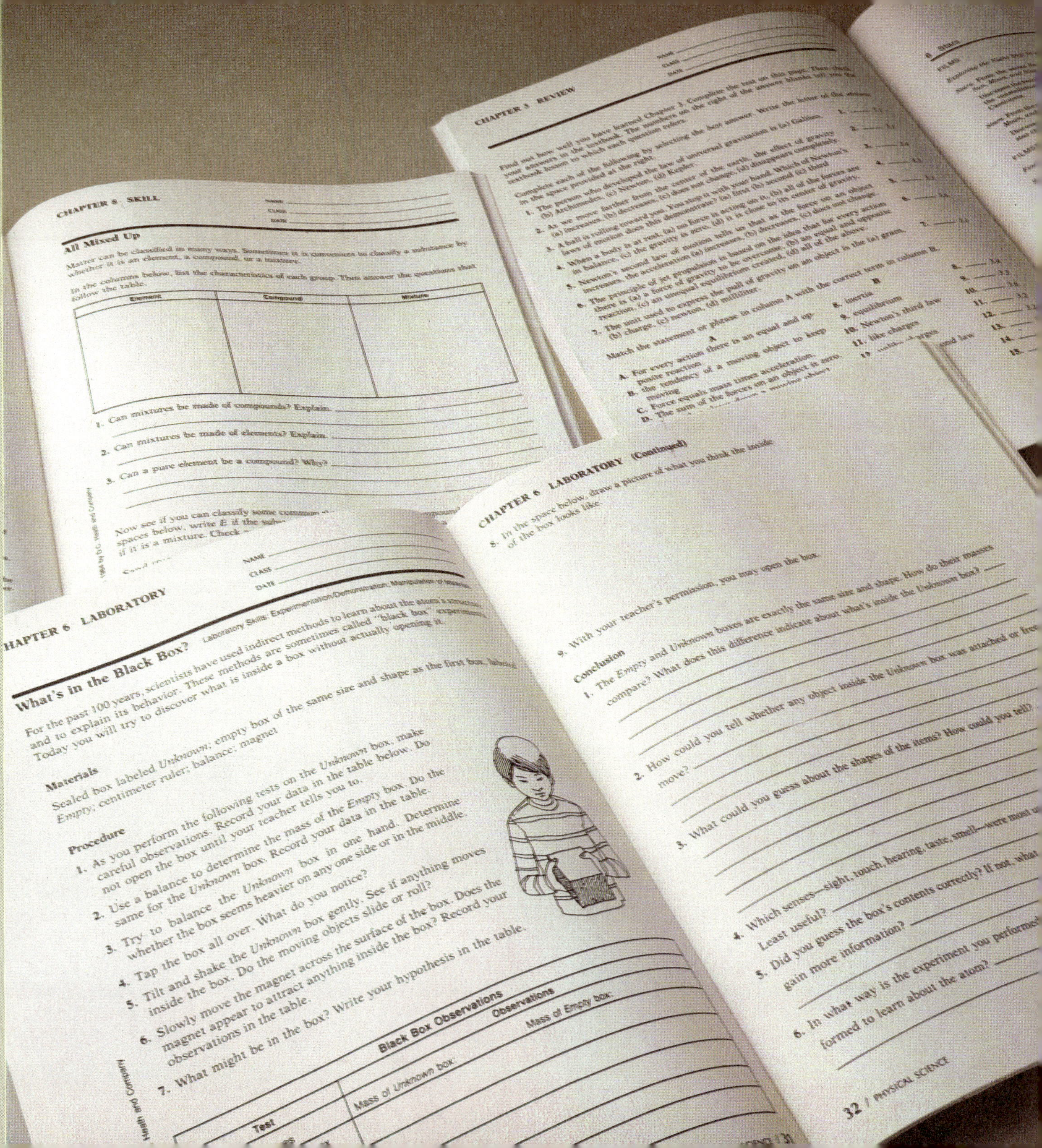

EXTRAS THAT SAVE YOU TIME—AND EVEN MONEY.

How can you fit helpful extras into your budget?

The teachers we consulted prefer to give their students additional opportunities for practice, review, and lab work. That's why so many of these teachers like to use workbooks and lab manuals.

Yet these extra items cost money—and, with today's tight budgets, that's a big concern.

Heath's answer: a workbook with built-in lab sections, so you won't have to buy a lab guide.

Heath responds to this challenge with **Heath Physical Science Activities**—a resource-filled workbook that does double duty as a lab manual. For every chapter in the student book you get

- one **Chapter Review page,** to help students prepare for the chapter test.
- one **Laboratory,** with many procedural steps—a full-fledged lab that truly involves the scientific method.
- a variety of **Skills Exercises,** all keyed to the Skills Matrix found on page A-26.

In addition, **Heath Physical Science Activities** comes with its own Teacher's Edition, containing all answers and solutions.

Also available: a Copymasters booklet with in-depth tests and much more.

To extend your options still further, Heath also offers a softbound, perforated **Copymasters** booklet, complete with an unusually thorough testing program, plus a set of special worksheets you can make into transparencies.

Chapter tests that meet your standards of thoroughness.

If you prefer to test students very thoroughly—on all levels, not just the literal level—you'll like the two-page **Chapter Tests** included in this **Copymasters** booklet. Here, at last, are commercially prepared tests that are truly comprehensive and reliable. They're the kind of tests you'd like to prepare yourself. But now Heath has saved you the time and the trouble.

What makes these tests so special? They don't merely test on the literal level; they test on the interpretive and applied levels too. The second page of each two-page test is devoted to these higher-level questions. So, for example, students may be asked to interpret a map, to explain a scientific process in their own words, or to state the practical applications of a concept covered in the textbook.

All copymaster **Chapter Tests** are also shown, reduced with answers plainly overprinted, at the back of the Teacher's Edition. And all are also available in a separate **Duplicating Masters** booklet.

Plus standardized Unit Tests. . .

Also available on copymasters in the **Copymasters** booklet are formal **Unit Tests** in standardized format. Each of these cumulative tests lets you evaluate a range of chapter objectives culled from the unit just completed.

Unit Tests are also shown, reduced with answers plainly overprinted, at the back of the Teacher's Edition. And each is also available separately in the **Duplicating Masters** booklet.

And special transparency masters. . .

If your class can benefit from visual aids, you'll want to see the handy transparency masters that are also included in the **Copymasters** and **Duplicating Masters** booklets. You can use these black-line masters to make transparencies for overhead projection, as well as copies for distribution to your class. While you fill in information on the diagram or chart on the projected transparency, your students follow along and fill in their own worksheets. It's an effective way to reinforce chapter content.

Parent Involvement Sheets too.

Finally, in both the **Copymasters** and **Duplicating Masters** booklets, you'll find six **Parent Involvement Sheets** (one per unit). These intriguing activities really get students and families working together on science, and let families know just what students are learning.

You can make transparencies from special copymasters provided in Heath's **Copymasters** booklet.

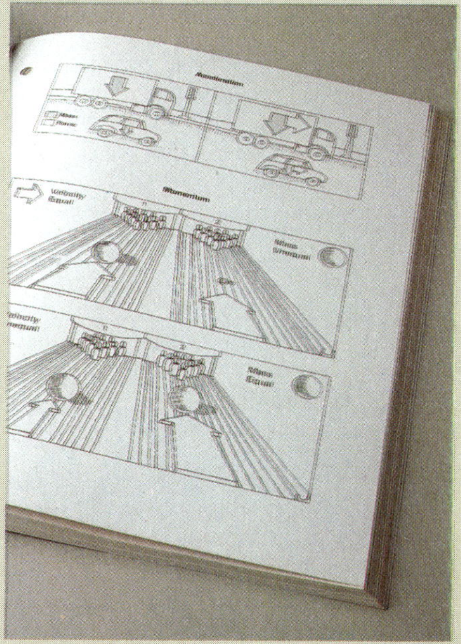

HEATH INTRODUCES THE TEACHER'S RESOURCE BOOK, WITH HUNDREDS OF IDEAS FOR MAKING SCIENCE COME ALIVE.

A treasury of lists and references for every possible resource.

"How can we make physical science really come alive for students?" Over and over, the teachers we talked to asked this question.

In response, we developed the **Teacher's Resource Book,** a valuable tool that will take you as far beyond the book as you may wish to go—with a wealth of aids and ideas for enhancing your instruction.

Throughout the **Teacher's Resource Book,** you'll find convenient, information-packed lists. You'll find lists of student literature, lists of audio-visual materials, lists of free and inexpensive materials, lists of teacher resources, and lists of commercially available computer software and related resources.

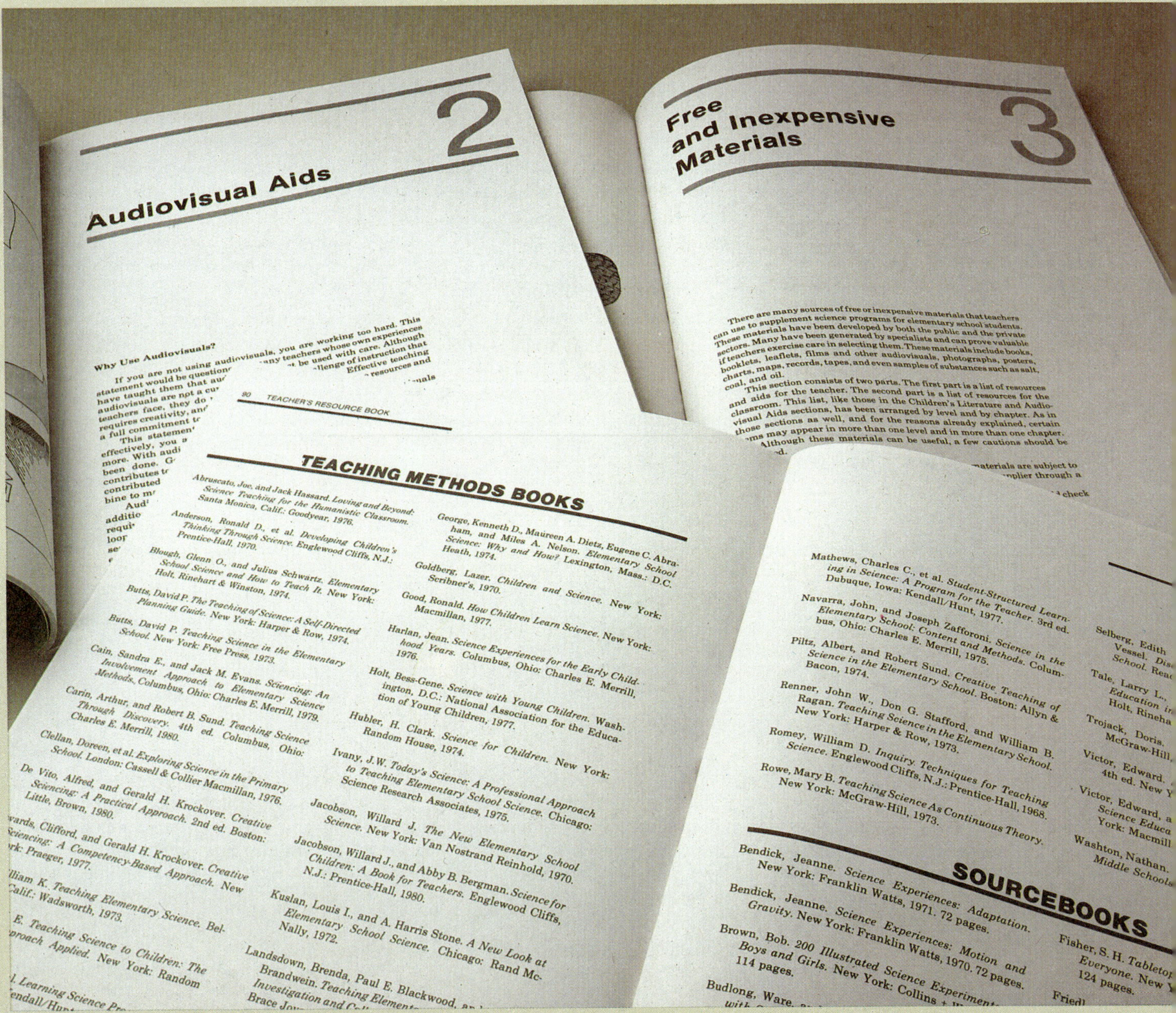

Lists of aids and resources in the **Teacher's Resource Book.**

CONTENT RATIONALE

Heath Physical Science presents a logical and comprehensive study of matter and energy. This study focuses on the basic principles of physics and chemistry. The students are introduced to physical science and the scientific method in Chapter One. The remaining twenty-three chapters are divided into six units.

Each of the six units concentrates on a major area of physical science. The units cover not only the traditional topics of physical science but also topics of current research and interest. As a result, the students gain a clear understanding of such fundamental concepts as Newton's three laws of motion, the action of magnetic and electric forces, and the behavior of waves. This understanding provides the students with a sound background that will help them to understand the world they live in and to make decisions about their lives as individuals and citizens. The students also learn about current technology in areas such as computers, lasers, alternative energy sources, and space exploration. This accent on current issues and technology captures the interest of students and shows them how the application of principles of physical science affects their lives.

SEMESTER COURSE OF STUDY

Chapter	Suggested Semester Courses				Full-Year Course
	Fundamental Physics Class Periods	Applications of Physical Science Class Periods	Energy Class Periods	Chemistry/ Energy Class Periods	Physical Science Class Periods
1	8	8	8	8	8
2	8	—	8	—	8
3	8	8	—	—	8
4	8	8	8	—	8
5	7	—	—	—	7
6	—	—	—	6	6
7	—	—	—	8	8
8	—	—	—	8	8
9	—	—	—	7	7
10	—	—	—	8	8
11	—	—	—	8	8
12	7	—	7	7	7
13	—	7	7	—	7
14	7	7	7	7	7
15	8	8	8	8	8
16	7	7	—	—	7
17	—	6	—	—	8
18	6	—	6	—	6
19	8	—	8	—	8
20	8	8	8	—	8
21	—	7	—	—	7
22	—	—	7	7	7
23	—	8	8	8	8
24	—	8	—	—	8
Total Class Periods	90	90	90	90	180

LESSON OBJECTIVES

Chapter 1 What Is Physical Science?
The students will find out
1.1 ▪what science is;
▪how facts and theories are used in science;
▪what physical science is;
1.2 ▪why the study of physical science is important;
1.3 ▪the four steps of the scientific method;
▪some qualities of a good scientist;
1.4 ▪what a laboratory is and its role in science;
▪the steps in a laboratory experiment;
1.5 ▪the standard units of measurement in the metric system.

Chapter 2 Matter, Energy, and Motion
The student will find out
2.1 ▪what matter is;
▪the three states of matter;
▪what energy is;
▪how energy changes from one form to another;
2.2 ▪how to get the average speed of an object;
▪the difference between speed and velocity;
▪how to combine two or more velocities;
2.3 ▪how to describe the changing velocity of an object;
▪why a body moving at a constant speed can be accelerating;
▪how falling objects accelerate;
▪what is meant by terminal velocity;
2.4 ▪what momentum is;
▪how to find the momentum of an object.

Chapter 3 Forces
The students will find out
3.1 ▪what is meant by the universal law of gravitation;
▪how gravity changes with mass and distance;
▪how weight and mass are different;
3.2 ▪how like and unlike charges behave toward each other;
▪how electric forces affect matter;
3.3 ▪when a state of equilibrium exists;
▪how to locate an object's center of gravity;
3.4 ▪how friction affects motion;
▪the first law of motion;
▪what is meant by equilibrium in motion;
3.5 ▪what happens when an object is not in a state of equilibrium;
▪the equation for Newton's second law of motion;
3.6 ▪Newton's third law of motion;
▪some examples of Newton's third law of motion.

Chapter 4 Work, Machines, and Power
The students will find out
4.1 ▪what a physical scientist means by the word *work*;
▪how to find the amount of work done by a force;
4.2 ▪how machines help you to do work;
▪the work principle on which all machines operate;
▪how to compute the mechanical advantage of a machine;
4.3 ▪how simple machines change the force;
▪the names and uses of the six simple machines;
4.4 ▪how to compute the rate at which work is done;
▪what is meant by efficiency.

Chapter 5 Forces in Liquids and Gases
The students will find out
5.1 ▪how to find the pressure exerted by an object;
▪how pressure varies with area;
▪how pressure varies with depth;
5.2 ▪what is meant by the word *buoyancy*;
▪what is stated in Archimedes' Principle;
▪how to find the density of an object;
▪how to predict whether an object will float;
5.3 ▪what is meant by atmospheric pressure;
▪why atmospheric pressure decreases with height;
▪how you use atmospheric pressure;
5.4 ▪what is meant by the term *drag*;
▪how the shape of an object affects drag;
▪what is meant by the term *lift*;
▪how the velocity of a fluid affects pressure.

Chapter 6 The Atom
The students will find out
6.1 ▪how the alchemists contributed to science;
▪what elements are;
▪what atoms are;
▪what compounds are;
6.2 ▪what ideas about the atom emerged from Rutherford's work;
6.3 ▪what types of particles make up the nucleus;
▪what is meant by atomic number and atomic mass number;
▪what is meant by isotopes;
▪what quarks are;
6.4 ▪why the solar system model doesn't accurately explain atomic structure;
▪what is meant by electron shells;
▪what forces hold atoms together.

Chapter 7 The Elements
The students will find out
7.1 ▪how the elements are organized on the periodic table;
▪what is meant by the term *valence electron*;
▪what determines families of elements;
7.2 ▪what some characteristics of elements in families IA, IIA, and the transition metals are;
▪how many valence electrons are in hydrogen, alkali metal and alkaline earth metal atoms;
▪what the transition metals are;
7.3 ▪how many valence electrons are in the atoms of elements in families IIIA through VIIA;
▪what some elements in families IIIA through VIIA are;
7.4 ▪why the noble gases are different from other elements;
7.5 ▪why elements within the lanthanide and antinide families are so similar;
▪how new elements are made in the laboratory.

Chapter 8 Combining the Elements
The students will find out
8.1 ▪how compounds and mixtures are different;
▪what is meant by a solution;
8.2 ▪what is meant by the term *molecule*;
▪how chemical bonds join atoms to form molecules;
8.3 ▪how an ionic bond is formed;
▪what ions are;
▪what crystals are;
8.4 ▪how a covalent bond is formed;
▪how electron dot diagrams illustrate covalent bonds;
▪how a diamond is formed;
8.5 ▪how metallic bonds are formed;
▪what explains many of the familiar properties of metals.

Chapter 9 Carbon Compounds
The students will find out
9.1 ▪why carbon makes so many compounds;
▪what organic compounds are;
9.2 ▪what hydrocarbons are;
▪what isomers are;
▪how crude oil is distilled;
9.3 ▪how alcohols are formed;
▪how organic acids are formed;
▪how some organic compounds are important to living things;
9.4 ▪what a polymer is;
▪how synthetic materials are made.

Chapter 10 Chemical Reactions
The students will find out
10.1 ▪how physical and chemical changes are different;
▪how to read chemical equations;
▪what the four main groups of chemical reactions are;
10.2 ▪how to balance chemical equations;
10.3 ▪how the temperature affects the rate of chemical reactions;
▪how the concentration of reactants affects the rate of chemical reactions;
▪what is meant by the term *catalyst*;
10.4 ▪what is meant by activation energy;
▪how exothermic and endothermic reactions are different;
10.5 ▪what acids and bases are;
▪what happens when an acid and base react;
▪how the strength of acids and bases is measured.

Chapter 11 Nuclear Reactions
The students will find out
11.1 ▪what happens in a nuclear reaction;
▪what holds the nucleus of an atom together;
11.2 ▪how radioactivity was discovered;
▪what happens in radioactive decay;
11.3 ▪what makes some isotopes radioactive;
▪what is meant by the term *half-life*;
▪how radioactive isotopes are useful;
11.4 ▪what is meant by the term *nuclear fission*;
▪how a nuclear chain reaction proceeds;
11.5 ▪what is meant by the term *nuclear fusion*.

Chapter 12 Heat
The students will find out
12.1 ▪what the kinetic theory states;
▪how heat energy and kinetic energy are related;
12.2 ▪how temperature and heat energy are different;
▪what is meant by specific heat;
12.3 ▪how substances are affected by temperature changes;
▪what happens when substances change state;
12.4 ▪how heat energy is transferred in conduction, convection, and radiation.

Chapter 13 Heat Energy in Your Life
The students will find out
13.1 ▪how heat engines do work;
▪where the heat for heat engines comes from;
13.2 ▪how steam engines developed;
▪how steam turbines work;
13.3 ▪how gasoline engines operate;
▪how diesel engines operate;
▪why heat engines cannot be perfectly efficient;
13.4 ▪how heat is transferred and moved in some cooling and heating systems.

Chapter 14 Magnetic and Electric Forces
The students will find out
14.1 ▪what magnetism is;
▪how magnetism and electricity are alike;
▪why some substances are magnetic;
14.2 ▪how the earth acts like a magnet;
▪how the north geographic pole and the north magnetic pole differ;
14.3 ▪how an object can build up a static electric charge;
▪how static electricity can be detected;
▪how an electric charge can be induced in an object;
14.4 ▪what moves an electric charge around;
▪how electric discharges occur;
▪how you can be protected from lightning.

Chapter 15 Electricity
The students will find out
15.1 ▪what is meant by the term *electric current;*
▪what causes electric current to flow;
▪how a dry cell produces voltage;
15.2 ▪how series and parallel circuits differ;
▪how alternating and direct currents differ;
▪what is meant by resistance;
15.3 ▪what is meant by electric power;
▪how electric companies measure and charge for electric energy;
15.4 ▪why fuses and circuit breakers are important;
▪what happens in a short circuit;
▪how you can avoid accidents with electricity.

Chapter 16 Electromagnetism
The students will find out
16.1 ▪how electric current and magnetism are related;
▪how electromagnets are made;
▪how electromagnets are used;
16.2 ▪what is meant by electromagnetic induction;
▪how electric generators produce electricity;
16.3 ▪how an electric motor changes electric energy into kinetic energy;
▪why transformers are important.

Chapter 17 Electronics
The students will find out
17.1 ▪what electronic devices do;
▪what vacuum tubes are;
▪how a picture is made on a television screen;
▪what semiconductors are;
17.2 ▪what rectifiers do;
▪what oscillators do;
▪what amplifiers do;
17.3 ▪how an integrated circuit is made;
▪what is meant by the term *microelectronics;*
17.4 ▪what the three basic functions of computers are;
▪what units are used to perform the three basic functions of computers;
▪how computers are used.

Chapter 18 Waves
The students will find out
18.1 ▪what a wave is;
▪how waves are grouped;
▪what some characteristics of waves are;
18.2 ▪how waves are changed by boundaries and barriers;
▪how waves are changed when the medium changes;
▪how waves are changed by other waves;
18.3 ▪how the frequency of waves can appear to change;
▪what the Doppler effect is;
▪how the Doppler effect can be used.

Chapter 19 Sound
The students will find out
19.1 ▪how sound waves are produced;
▪why sound waves must travel in matter;
▪what affects the speed of sound waves;
19.2 ▪what determines the loudness of a sound;
▪what factors affect pitch;
▪what ultrasonic sounds are;
▪how sound waves can be used to locate objects;
19.3 ▪what is meant by resonance;
▪how different instruments make music;
▪what is meant by overtones;
19.4 ▪what is meant by noise pollution;
▪what can happen when sound waves are reflected;
▪how echoes can be prevented.

Chapter 20 Light
The students will find out
20.1 ▪what electromagnetic waves are;
▪how sound waves and electromagnetic waves differ;
▪how some kinds of electromagnetic waves are used;
20.2 ▪what makes a rainbow;
▪why objects have the colors they do;
▪why the sky changes color;
20.3 ▪how mirrors reflect light waves;
▪how refraction of light waves can fool your eyes;
20.4 ▪how a diffraction grating separates the colors in white light;
▪how polarizing sunglasses reduce glare.

Chapter 21 The Uses of Light
The students will find out
21.1 ▪what a lens is;
▪what the two basic kinds of lenses are;
▪how to find the focal point and focal length of a convex lens;
▪how virtual images and real images differ;
21.2 ▪how your eyes see;
▪how a camera produces a photographic image;
▪how microscopes and telescopes use lenses;
21.3 ▪what laser light is;
▪how lasers produce laser light;
▪how lasers are used;
▪what optical fibers are.

Chapter 22 Energy and Society
The students will find out
22.1 ▪what fossil fuels are;
▪how fossil fuels are obtained;
▪what some of the problems associated with fossil fuels are;
22.2 ▪how fission is used to generate electricity;
▪what some of the problems associated with nuclear reactors are;
▪why fusion reactors are difficult to design and operate;
22.3 ▪why energy conservation is necessary;
▪what is being done to conserve energy.

Chapter 23 Energy Alternatives
The students will find out
23.1 ▪why people are interested in solar energy;
▪what devices are being used to collect solar energy;
▪how solar energy can be used;
23.2 ▪how hydroelectric power is produced;
▪how hydroelectric power is related to solar energy;
▪why hydroelectric power is not more widely used;
▪how tidal energy is used;
23.3 ▪how wind energy is used;
▪what is meant by geothermal energy;
▪what is meant by biomass.

Chapter 24 The Universe
The students will find out
24.1 ▪how astronomers study the universe;
▪what tools astronomers use;
24.2 ▪how astronomers have determined the life cycle of stars;
▪what some of the stages are in the life cycle of stars;
▪what astronomers believe about black holes and quasars;
24.3 ▪how satellites are being used;
▪what advantages there are to working in space;
▪what space colonies might be like.

MATRIX OF SKILLS DEVELOPMENT

I. Comprehension Skills

Literal: Recognition
Provides opportunities for recognizing main idea and related details; recognizing statements of conclusion; recognizing generalizations; detecting word meaning from explicit clues; and recognizing sequential relationships, cause-and-effect relationships, and compare-and-contrast relationships.

Interpretative: Inference
Provides opportunities for inferring main idea; inferring conclusions; inferring generalizations; inferring word meaning; and inferring sequential relationships, cause-and-effect relationships, and compare-and-contrast relationships.

Applied: Evaluation and Problem Solving
Provides opportunities for applying main idea and detail; evaluating conclusions; evaluating generalizations; applying sequential relationships; and using cause and effect relationships and compare and contrast relationships to interpret and evaluate information for solving problems and drawing conclusions.

II. Word Skills

Structure
Encourages word analysis through syllabication and identification of prefixes, suffixes, root words, and cognates.

Vocabulary
Reinforces vocabulary development through definitions, word exercises, puzzles, games, and synonyms.

Context Clues
Provide method of inferring definition or meaning of new term from the context in which it is used.

Decoding
Reinforces use of decoding skills in word recognition (pronunciation aids are provided in text for science terms that are difficult to pronounce).

III. Mathematics Skills

Computation
Reinforces basic arithmetic skills; develops skill of equation-solving.

Problem Solving
Develops skills of thinking through steps of a problem (analyses), finding similarities in real-life situations (applications), and selecting appropriate tools to work out problems (solutions).

Measurement
Develops skills to find linear measurements, mass, weight, density; provides opportunities for converting among metric units; encourages an appreciation of accuracy versus precision and of the importance of significant figures when recording measurements.

Estimation
Develops skills of checking mathematical operations, predicting rough measurements, and gleaning information from graphs.

Graphing
Develops methods for discovering relationship between two variable quantities.

IV. Laboratory Skills

Experimentation/Demonstration
Develops observation skills; encourages application of safety regulations.

Manipulation of Materials
Develops skills of constructing, calibrating, and reading instruments as well as techniques involved in making measurements and setting up equipment.

V. Reference/Research Skills

Locates and Reports Information
Develops skills of finding information in reference sources, such as dictionaries, atlases, encyclopedias, and original sources, organizes research into meaningful reports utilizing written materials, illustrations, charts, diagrams, and graphs.

INDEX OF SKILLS

Skill Category	Specific Skill	Development in Student's Edition
Comprehension	Literal: Recognition, Interpretative: Inference, Applied: Evaluation and Problem Solving	1.1, 1.3, 1.4, 1.5, 2.4, 3.1, 3.4, 3.6, 4.1, 4.4, 5.3, 5.4, 6.2, 6.4, 7.1, 7.5, 8.1, 8.2, 8.4, 8.5, 9.1, 9.2, 9.4, 10.1, 10.3, 10.5, 11.1, 11.4, 11.5, 12.4, 13.1, 13.3, 13.5, 14.2, 14.4, 15.1, 15.4, 16.2, 16.3, 17.2, 17.3, 17.4, 18.3, 19.1, 19.4, 20.1, 20.4, 21.2, 21.3, 22.1, 22.3, 23.3, 24.2, 24.3
Word	Structure, Vocabulary, Context Clues, Decoding	1.5, 2.4, 3.2, 3.6, 4.4, 5.4, 6.4, 7.3, 7.4, 7.5, 8.1, 8.5, 9.2, 9.4, 10.5, 11.5, 12.1, 12.4, 13.1, 13.3, 13.5, 14.1, 14.4, 15.1, 15.4, 16.1, 16.3, 17.1, 17.4, 18.2, 18.3, 19.2, 19.4, 20.1, 20.4, 21.1, 21.3, 22.2, 22.3, 23.1, 23.2, 23.3, 24.2, 24.3
Mathematics	Computation, Problem Solving, Measurement, Estimation, Graphing	2.2, 2.3, 2.4, 3.5, 4.1, 4.2, 4.3, 5.1, 5.2, 10.2, 12.2, 15.2, 15.3, 15.4, 18.1, 20.2, 22.3, 23.3, 24.1, 24.2
Laboratory	Experimentation/Demonstration, Manipulation of Materials	1.5, 2.1, 3.3, 3.6, 4.2, 4.3, 4.4, 5.1, 5.2, 5.4, 6.1, 6.3, 7.2, 8.1, 8.3, 9.3, 9.4, 10.2, 10.5, 11.2, 11.3, 12.3, 12.4, 13.2, 13.5, 14.3, 14.4, 15.2, 15.4, 16.1, 16.2, 17.4, 18.1, 18.2, 19.2, 19.3, 20.2, 20.3, 21.1, 21.2, 22.3, 23.1, 23.3, 24.1
Reference/Research	Locates and Reports Information	1.2, 1.5, 2.4, 3.6, 4.4, 5.4, 6.4, 7.5, 8.5, 9.4, 10.5, 11.5, 12.4, 13.2, 13.5, 14.4, 15.4, 16.3, 17.4, 18.3, 19.4, 20.4, 21.3, 22.3, 23.3, 24.3

SUGGESTIONS FOR A SAFE SCIENCE PROGRAM

Each science teacher should instruct his or her students in proper safety procedures. A set of safety rules should be available to students and should be posted in several places around the science classroom. These rules should be discussed frequently with the students. In order to assure that reasonable care will be taken, the science teacher must remind the students of the proper safety procedures at the beginning of each class where there might be any risk of personal injury.

To insure a safe environment in your classroom, the following safety suggestions have been prepared by the National Science Teachers Association:

1. Obtain a copy of the federal, state, and local regulations which relate to school safety, as well as a copy of your school district's policies and procedures.

2. Check your classroom on a regular basis to insure that all possible safety precautions are being taken. Equipment and materials should be properly stored; hazardous materials should not be left exposed in the classroom.

3. Storage facilities should be locked at all times when not under direct supervision of a responsible person.

4. Before handling equipment and materials, thoroughly familiarize yourself with their possible hazards.

5. Be extra cautious when dealing with fire and instruct your students to take appropriate precautions. Familiarize your students with the school's fire regulations, evacuation procedures, and the location and use of fire-fighting equipment.

6. The size of the group working on an experiment should be limited to a number that can safely perform the activity without causing confusion and accidents.

7. Student apparel should be appropriate for activity work. Remind your students to roll long sleeves above the wrists. Ties, coats, and bulky outerwear should be removed. Long hair should be tied back during an activity, especially when an open flame is nearby.

8. Instruct students to wear safety goggles when working with dangerous chemicals, hot liquids, and other potential sources of splashes, spills, or spatterings.

9. Students should be instructed never to taste or touch substances in the science classroom without first obtaining specific instruction from you. Students should also be instructed that it is unsafe to touch the face, mouth, eyes, and other parts of the body while they are working with plants, animals, or chemical substances until they have washed their hands.

10. Sufficient time should be planned for students to perform an activity. They should then clean up and properly store the equipment and materials after use.

11. You should dispose of dangerous waste chemicals and materials as prescribed by appropriate standards and the laws of your community. Provide separate waste receptacles for broken glass and waste paper.

12. In case of accidents, know your school's policy and procedures. Make accident reports promptly, accurately, and completely.

If you want more detailed safety information, order the booklet entitled *Safety in the Secondary Science Classroom*.
Mail request to:
National Science Teachers Association
1742 Connecticut Avenue, N.W.
Washington, D.C. 20009

SAFETY GUIDELINES FOR PHYSICAL SCIENCE

The following is a list of safety recommendations for specific activities frequently performed in physical science classrooms.

1. When doing an activity, always read the entire procedure before beginning the activity. Make special note of any cautions.

2. Always follow the procedure exactly. Do not leave out any steps. Do not add any steps.

3. Always clean up any spills immediately. Dispose of broken glass and wet papers properly.

4. Never place an eyedropper that has been used anywhere but in its proper container.

5. Never point any closed container that is being heated toward anyone.

6. Always treat any hot plate as if it were on. Never touch it or allow any clothing to touch it.

7. Always assume any container on or near a hot plate is hot. Handle it with tongs or a pot holder.

8. Always handle all chemicals with care. Never touch them with your hands.

9. Always light a Bunsen burner according to the directions your teacher provides.

10. Always dispose of matches by putting them in the proper container. Never leave matches on the lab table or throw them in the sink or wastebasket.

11. Never cover a test tube or beaker with your hand. Stopper it properly.

12. Always push away from your body when poking a hole through any substance. Never hold your hand beneath the substance.

13. To insert glass tubing through a stopper, always wet the tubing and the hole first. Then twist the tubing as you gently push it forward and away from you through the hole.

14. Always use thermometers very carefully. They break easily. If a thermometer breaks, notify your teacher immediately.

15. Never leave a wire attached to both terminals of a battery. Be sure all circuits are open before leaving them.

16. Always be sure that apparatus and equipment is kept in the center of the lab table.

17. Always clean up your work area after every activity. Put all materials back in their proper places.

18. Always report accidents, breakages, and spills to your teacher immediately.

MASTER MATERIALS LIST

Consumable Items	Chapters
Alcohol	
isopropyl	8, 9, 11, 13
methyl	9
Bags, large brown	7, 11
Baking soda	3, 8
Carbon pieces	6
Cardboard	
cylinder from paper towels	16
pieces of, or oaktag	3, 7
Cards	
index, small	5, 6, 24
white	21
Cellophane	
3 squares of, one color	20
Cooking oil	5
Copper sulfate	7
Corn syrup	5
Cups, clear plastic	20
Dry ice	11
Fingernail polish remover	9
Foil, aluminum	
pellets	5
sheets of	6
Food coloring	
blue	5, 8
green	8
red	5, 8
yellow	8
Gauze	13
Glue	6, 7, 23
Gumdrops	6
Hanger, wire	6
Hydrochloric acid, dilute	10
Ice cubes	12, 20
Iodine crystals	6
Insulating materials such as cloth, newspaper, Styrofoam chips	12
Lead nitrate solution	10
Lids	
can	9
jar	6
Magnesium ribbon	6
Matches, safety	5, 9, 18, 20
Milk cartons, small	2, 5
Newspaper	23
Paint, black spray	23
Paper	
blotting	11
carbon	18
chromotography	8
graph	11, 24
lined	8, 17, 22
white	1
Petroleum jelly	3
Phenolphthalein	10
Picture, colored	7
Ping-Pong balls	5, 23
Plastic wrap	23
Plates	
paper	16
tin pie	10
Protractors	23
Saccharin	9
Salicylic acid	9
Salt	8, 9, 12
Shoe boxes	19, 23
Sodium hydroxide, dilute	10
Sodium iodide solution	10
String	4, 5, 14
Styrofoam	
cups	5
pieces	5
Sugar	8, 9
Sulfuric acid	9
Tagboard	13
Tape	
adhesive	5
cellophane	5, 16, 18, 21, 23, 24
Thread	3, 6, 23
Toothpicks	5
Towel, paper	1, 2, 3
Vinegar	3, 5
Watches with luminous dial	11
Water, distilled	8
Wood	23

Nonconsumable Items	Chapters
Alligator clamps	15
Ammeters	15
Aprons	3, 5, 6, 8, 9, 10, 11, 13
Balances	5, 10
Batteries	
1.5 volt or 6 volt	6, 7, 8, 14, 15
6 volt	16
Beakers	
250 mL	6, 7, 8, 9, 10, 12, 13, 23
600 mL	12
Blocks	11
Boards	2, 4
Books	2, 4, 18
Bottles	
1 L clear plastic	3, 5
plastic, with vertical row of holes, or cans with vertical row of holes	5
Bunsen burners	6, 9, 13
Candles	18, 20
Cans, tin	12
Cars, toy	2
Clamps	3, 5, 13, 18
Clay	6
Clock	12, 20
Cloth, black	11
Combs, plastic	14
Compasses, magnetic	16
Containers, different-shaped	2
Copper bars	7
Dowels or round pencils	5
Eyedroppers	9, 10, 13
Flannel or fur	14
Flashlights	11
Forceps	6, 14
Funnels	23
Galvanometers	16
Glasses with straight sides	20
Graduated cylinders,	
100 mL	1, 2, 3, 5, 9, 10, 23
Hammers, tuning fork	18, 19
Hot plates	9, 12
Iron	
bar	7
filings	8, 16
Jars, glass, with lid	11
Lightbulbs, small, with sockets	6, 8, 14, 15
Magnets, bar	8, 16
Magnifying glasses	9, 10, 21
Marbles	1
Metersticks	4, 19, 22, 24
Nails	1, 3
Needles	
blunt	6
long	23
Pans, small baking	11
Paper clips or pebbles	3
Paper punches	3
Pebbles	1
Pens, felt	1
Pencils	
colored	24
long, with eraser	13
marking	9
number two	8, 11, 17, 21, 22
Pennies	5
Pins, straight	13, 18, 24
Pulleys	4
Rheostats	15
Ring clamps	4, 6, 9, 14, 16
Ring stands	3, 4, 5, 6, 9, 13, 14, 16, 18
Rubber bands of different thicknesses	13, 19, 20
Rulers, metric	3, 5, 17, 18, 24
Safety glasses	3, 6, 8, 9, 10, 12, 13, 18, 20
Scissors	3, 4, 7, 19
Screwdrivers	15
Silk	14
Slinkies	18
Spoons	8
Spring scales	4
Steel wool	7
Stirring rods	5
Stoppers	
one-hole rubber	13
rubber	3

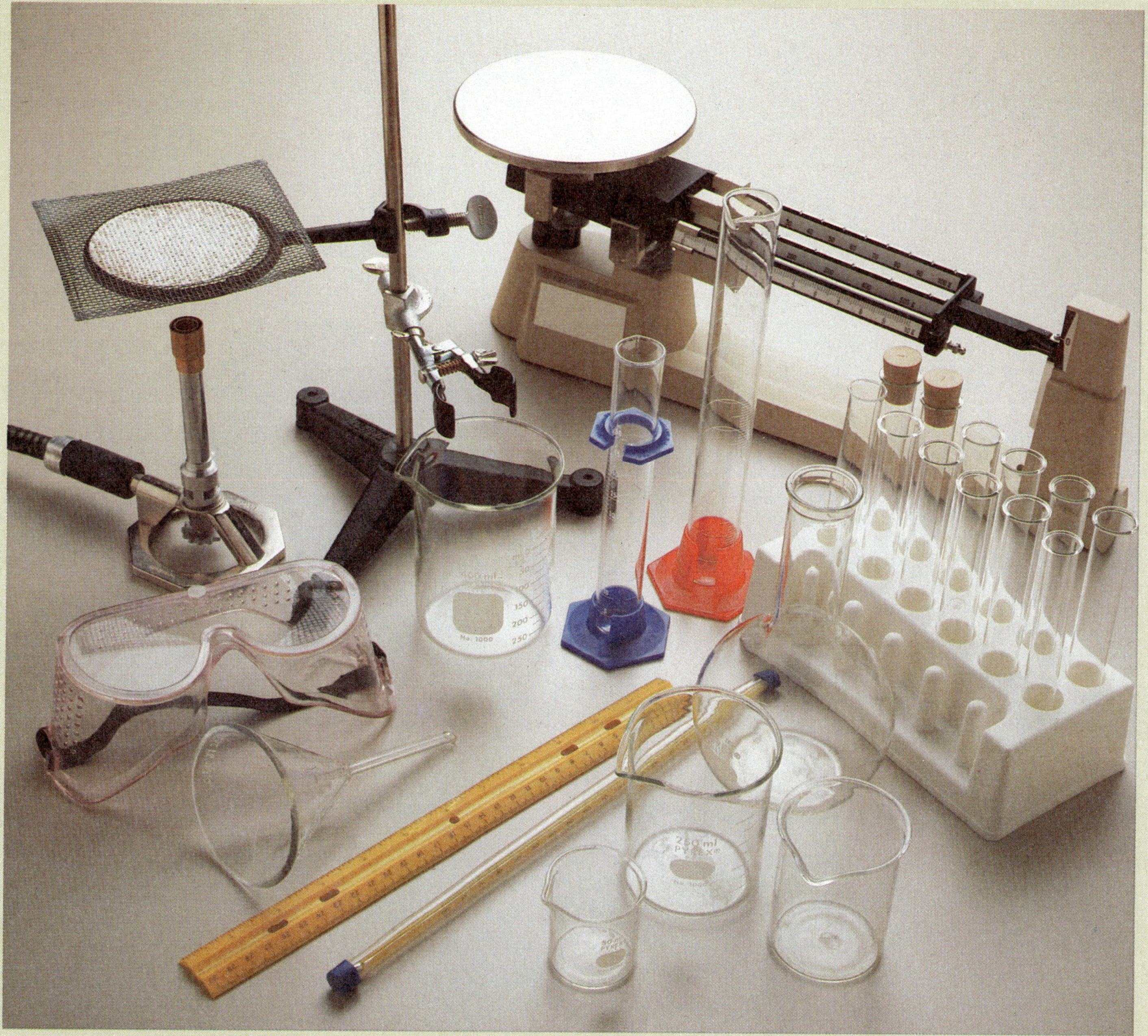

Using Heath Physical Science

OBJECTIVES: The student should be able to

- **locate** the table of contents, index, glossary, and appendices in *Heath Physical Science;*
- **use** the table of contents, index, glossary, and appendices;
- **describe** how the chapter features can be used as study aids.

LESSON PLAN

1. Begin the lesson by examining the table of contents with the students. Use Unit One as an example. Point out that the title of the unit is given first. On what page does Unit One begin (23)? What is the title of Unit One (Motion, Force, and Energy)? How many chapters are in the unit (4)? Have the students note that Chapter 2 is divided into four sections. Do all the chapters in Unit One have four sections (No, Chapter 3 has six sections)?

2. Have the students refer to the table of contents while you question them about each unit. For example, ask what unit deals with energy (6). How many chapters are in Unit Six (3)? Which chapter deals with wind (Chapter 23)?

3. Have the students turn to the index. Point out that the table of contents tells them in what chapter general topics are discussed while the index tells them on what page specific topics are covered. Have the students use the index to determine on which pages they would find information about specific topics such as the noble gases (page 150), density (page 100), and parallel circuits (page 319).

TABLE OF CONTENTS

4. Ask the students to turn to the glossary. Point out that the glossary is like a dictionary. It contains all the boldfaced terms from the chapters. The terms are listed alphabetically, and the definition of each term is given. For many terms, a pronunciation guide also appears.

5. Have the students page through the appendices. Point out that each appendix contains supplementary material. What are some of the topics covered by the appendices (metric information, using a balance, lighting a Bunsen burner)?

6. To familiarize the students with the chapter features, use Chapter 10 as an example. Have the students turn to page 205. Point out that the chapter begins with a photograph and an introduction.

7. Ask the students to turn to page 206. Tell them that the number **10.1** indicates that this is the first lesson in Chapter 10. Stress that the **You will find out** statement tells the students what they should understand when they finish the lesson. Point out that the **Study Questions** at the end of the lesson test the students' understanding of the statements listed at the beginning. Stress that each lesson begins with a **You will find out** statement and ends with **Study Questions.**

8. Look through Chapter 10 with the students. Point out that each figure has a caption. Stress that the students should study all figures and captions. This study will help the students to understand the new ideas in the lessons. Explain that any page with a yellow tint is an activity. The activities serve to reinforce chapter concepts and to develop laboratory skills. Call the students' attention to the **Do you know?** items. Point out that they supply interesting and often amazing facts related to chapter topics.

9. Have the students turn to page 223. Call their attention to the boldfaced terms *pH scale* and *indicator*. Explain that these terms are new science terms. Point out that whenever a new term is introduced, it is boldfaced and clearly defined in the same paragraph. Ask the students to turn to page 226. Explain that the list of **Key Terms** contains all the boldfaced terms in the chapter. Explain that the **What Did You Learn?** statements summarize the important facts in the chapter. These statements will help the students in reviewing the chapter.

10. Turn to the **Career** on page 227. Explain that at the end of each chapter either a **Career** or a **Biography** is featured. This feature helps to personalize and familiarize the students with people and jobs related to physical science.

11. Conclude the lesson by looking at the exercises on pages 228 and 229. Point out to the students that the **Vocabulary** exercise tests their knowledge of the new science terms introduced in the chapter. The **What Do You Remember?** questions refer to facts covered in the chapter. The **Applying What You Have Learned** questions require the students to do some thinking about the facts they have learned. The **Research and Investigation** section provides an opportunity for further study.

FOLLOW-UP

Reinforcement: Locating Information in *Heath Physical Science*

Prepare several work sheets that the students can complete by using the index, table of contents, glossary, and appendices of their texts.

Enrichment: Finding the Main Ideas

Explain to the students that it is important for them to be able to pick out the main ideas in a chapter. Point out that the **You will find out** statements and the **Study Questions** at the beginning and end of each lesson, respectively, are good indicators of the main ideas in each lesson. As the students read Chapter 1, have them list the main ideas for each lesson.

Tie-in: A Place to Start

During the year, the students may wish to do reports on topics related to physical science. Their text is the best place to start their research. Select a topic such as the noble gases and show the students how to use the table of contents, index, and chapters to locate facts and terms that could be used as starting points in a search for additional information on the topic. Assign everyone a topic. Each student should use the text to prepare a list of facts and terms for his or her topic. Keep the lists in a folder where they will be available for reference during the year.

CHAPTER BACKGROUND

Science is an organized body of knowledge. Through observation and experimentation, science offers explanations of the environment and makes predictions about it.

The organization of science into two broad fields, the biological and the physical sciences, and the further breakdown of these fields into more specialized branches such as genetics, geology, chemistry, physics, and meteorology make science easier to study for both the student and the scientist. However, a more thorough investigation of any one branch of science reveals that it is related to many of science's other branches. In today's world no scientists can be so specialized that they totally neglect those branches of science that are not their specialty.

A thread that unites both the biological and physical fields of science and their many branches is the scientist's use of the scientific method, an orderly process for solving problems and investigating nature. Scientists must first define what it is they are studying and specify the limits of their investigation. They must gather data about what they are studying by doing a thorough search of the literature and by using their senses to observe all that might be connected with the investigation. As scientists gather data, they will begin to formulate a basic hypothesis to explain the phenomena under investigation. As Frances Redi pointed out, however, "Belief would be in vain without the benefit of experimentation." Scientists must, therefore, test their hypotheses under carefully controlled conditions. This initial experiment, like all that scientists do, must be a controlled experiment. In this way a single variable can be easily identified as the cause of the results. Scientists must record all they do and observe so that others can verify their findings.

Analysis of the data allows scientists to organize their findings so that a pattern emerges. As scientists study this pattern, they will be able to begin the synthesis of a theory that, based upon data and experimentation, will offer a deeper understanding of the original problem and its solution.

Only when a theory has been tested under all possible conditions and found to be true can it be used to make predictions about what will happen in similar circumstances. The formulation and testing of a theory do not mark the end of scientists' work. Science is an ever-increasing body of knowledge.

1 Physical Science at First Glance

Have the students study the nine photographs very carefully and determine their correct sequence. Tell the students that the puzzle is not as easy as it first appears. It is very tricky. In order to determine the correct sequence, more than one object or item will have to be considered. Point out that in solving this puzzle, the students are thinking like scientists. They are looking at the photographs (observation), thinking about how the photographs are similar and how they are different (analysis), putting the photographs in what seems to be the correct sequence (synthesis), and testing that sequence by checking to see that each item in the sequence changes logically, by comparing answers with friends, and by comparing answers with the correct answer from the teacher's edition (test). The correct sequence is:

B, H, E, G, F, C, D, I, A

The trick to this puzzle is discussed in reference to the scientific method on pages 9 and 10. You may wish to refer students to those pages for help in solving the puzzle.

You live in exciting times. Frontiers are opening up all around you, from the depths of the ocean to the heights of space to the microworld of electronics. To get the most from these times, maybe even to survive them, you need to learn as much as you can about the world. What is it made of? How does it work? How can the matter and energy available to the planet earth be used to make life better for everyone?

The purpose of this book is to help you explore the answers to some of these questions. You will use a branch of knowledge called physical science.

In the next few pages you will learn a method of thinking about the world and everything in it. It is called the scientific method. It is used not only by scientists. Anyone at all who is curious about the world may find it a valuable method.

Figure 1.1 Look at the nine photographs very carefully. Can you put them in the correct order?

Lesson Organizer

1.1 What Is Physical Science?
Pages 2–5
Follow-up: Page 5
- **R** Scientific Data in the News
- **E** How Does Science Affect You?
- **T** Gathering Scientific Data

1.2 Why Is Physical Science Important to You?
Pages 6–7
Advanced Planning: Page 6
Follow-up: Page 7
- **R** Scientific Concepts Behind Inventions
- **E** How Do Inventions Help You?
- **T** Inventions and American History

1.3 Thinking as a Scientist
Pages 8–11
Advanced Planning: Page 8
Follow-up: Page 11
- **R** Using the Scientific Method
- **E** Applying the Scientific Method
- **T** A Theory Is like a Circus Tent

1.4 The Laboratory
Pages 12–13
Activity: How Can These Students Be More Careful? Page 13
Follow-up: Page 13
- **R** Charades—Danger in the Lab
- **E** Careers in the Laboratory
- **T** Speaking of Laboratory Safety

1.5 Measurement
Pages 14–17
Advanced Planning: Page 14
Activity: How Do You Find the Volume of an Irregular Object? Page 16
Follow-up: Page 17
- **R** Added Practice Using the Metric System
- **E** Scale Models of the Classroom
- **T** A Bulletin Board of Metric Measurement

See ACTIVITIES book pages 1–6 for laboratory, skill, and review exercises for Chapter 1.

1.1

OBJECTIVES: The student should be able to

- **define** the term *science;*
- **distinguish** between scientific data and nonscientific data;
- **explain** the relationship between facts and theories;
- **explain** how a hypothesis becomes a theory;
- **define** the term *physical science.*

SCIENCE TERMS

science	theories
facts	physical science

LESSON PLAN

1. Have the students brainstorm the meaning of the terms *science* and *scientist*. List all their ideas on the board. There are no correct or incorrect answers at this point. Guide the discussion so that the words *organized* and *curious* come up for *scientist* and the words *world, environment, scientific method,* and *future* or their synonyms are part of the list for *science*. Circle the words on the board that apply to the term *science*. Then ask the students to use the circled words to write a definition of *science*.

● **Comprehension Skill: Interpretative**

1.1 What Is Physical Science?

You will find out
- what science is;
- how facts and theories are used in science;
- what physical science is.

What is **science?** The answer is simpler than you might think. Science is a method of organizing curiosity. The scientific person asks questions and tries to understand the world in an organized, orderly way.

Science questions can be about almost anything: Why does an apple fall down from a tree and not up? What is the world made of? What causes measles? You may ask about anything at all as long as you are truly curious to know the answer and go about searching for the answer in an organized, orderly way.

To be scientific, you must follow some rules. Here is the main rule: If you gather some information to help answer a certain question, other people must be able to understand how you got your information. They must then be able to gather similar information on their own. That is, other people must be able to repeat what you have done and get the same results.

Do You Know?
According to George Bernard Shaw, "Science is always simple and always profound. It is only the half-truths that are dangerous."

Figure 1.2 Which statements are facts? Which are opinions?

The opinions are shaded in red.

For example, the observation that the average annual rainfall at the Honolulu, Hawaii, airport is greater than that at the Los Angeles, California, airport is scientific information. This is because the observation can be tested by someone else. On the other hand, suppose someone reports that little green creatures from outer space are moving more clouds over Honolulu and causing more rain there. This observation is impossible to check out (at least it has been so far!) and is therefore not scientific.

Observations that can be checked out, or scientific observations, go by a much shorter, simpler name: **facts.** Facts are the building blocks of science. These blocks are held together by **theories,** or ideas that explain groups of facts.

Figure 1.3 It is easy to understand why people years ago theorized that the earth was flat. A view such as the one above clearly suggests that. Later theories correctly suggested that the earth was round.

One of the most successful theories in the history of science is Isaac Newton's theory of gravitation. Using a few simple rules, he was able to explain how apples and other things fall to Earth, how the moon moves around the earth, and how the planets move around the sun.

2. Scientific observations are data that can also be gathered by other scientists. Give the students some examples of data that have been gathered and ask them which are scientific observations. Some examples you may use are the following:

a. Maple leaves fall off trees when the temperature falls below 6°C (42.8°F) for six consecutive days. (scientific data)

b. Collies shed some fur in the spring. (scientific data)

c. If there is a red sky in the morning, it will rain in five days.

d. An elephant's heart beats slower than a dog's. A dog's heart beats slower than a mouse's. (scientific data)

e. Thunder is caused by the angels in heaven bowling.

3. Stress that just because scientists observe something, it is not necessarily a fact. When something is first observed and scientists speculate on what caused it or what it may cause, their first idea is a hypothesis. A hypothesis is an untested but educated guess.

4. Ask the students to think back to the scientific events that occurred during the summer. As the students name the events, write them on the board. As you read over the list, ask the students how the events show that science is an incomplete body of knowledge.

Any good scientific theory not only explains and connects facts but also makes predictions. A theory makes predictions that should lead scientists to new ideas in the future. These predictions cannot be fuzzy ones of the kind found in fortune cookies, such as "you will meet an interesting friend tomorrow." They must be exact.

If a theory does not make predictions that can be tested, it is not a very good theory. Otherwise people have no way to agree on whether a theory is right or wrong. A good theory sets itself up as a target to be shot down if possible. In fact, as people's understanding of the world has deepened, almost all theories have turned out to be wrong in one way or another.

Some theories are so very wrong that they have to be thrown out completely. For example, today the theory that the earth is flat is known to be wrong. Other theories, such as Newton's theory of gravitation, are wrong only under certain extreme conditions—near black holes in space, for example. Such theories can continue to be used as long as people understand that they are not correct under all conditions.

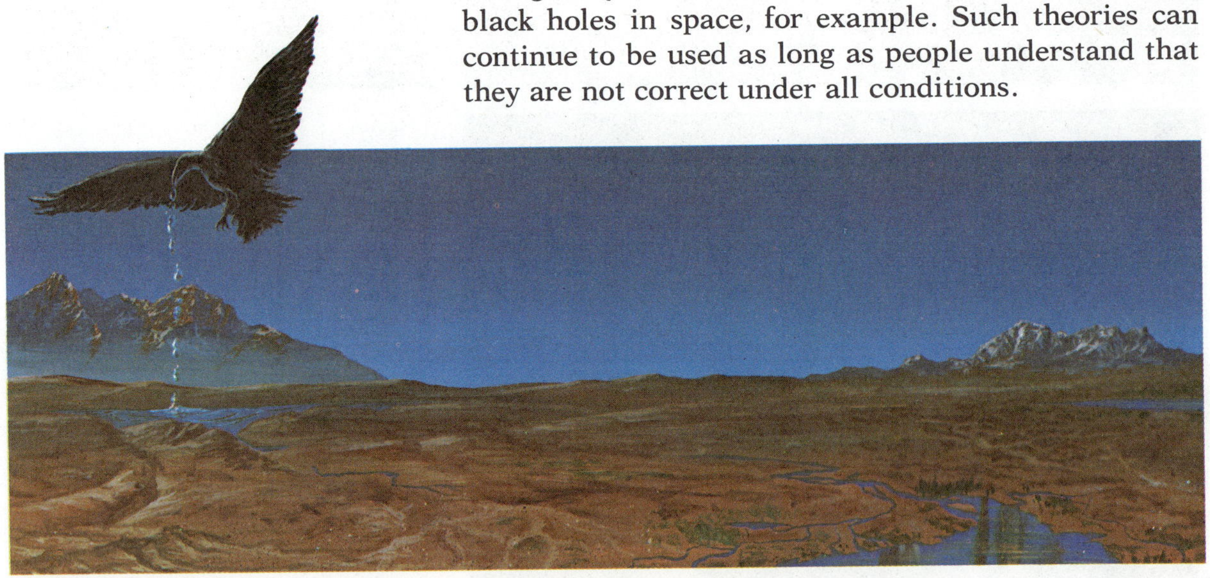

Figure 1.4 Many theories about the earth were developed in myths. An Eskimo myth offered a theory to explain the formation of rivers: A raven carrying water in its mouth spilled that water. It fell to the earth forming rivers.

Science is not a rigid body of laws carved in stone. Science is a way of doing things, a way of learning about the universe, a way of being curious. What really counts is the method of asking questions, making careful observations, coming up with answers, and testing these answers with more questions and observations.

Scientists do not have all the answers to all the scientific questions. They have not yet asked all the right questions either. They probably do not have any final answers. But today's scientists have better answers than they had in years gone by, and they will have still better answers in the years to come. Who knows? Maybe you will come up with some better answers yourself. Just by being curious about the world, you can take part in the adventure of search and discovery that makes science so exciting and important.

Figure 1.5 The same process that produces energy on the sun is being studied by physical scientists here on Earth.

Physical science is that branch of science that searches for answers to questions about matter and energy. Why is a fire hot? How is electricity generated? Where does the sun get its energy? All these questions involve matter and energy, and so they are questions for physical scientists.

Study Questions for 1.1

1. What is any method of organizing curiosity called?
2. What are the building blocks of science? How are they held together?
3. What is physical science?

1. science
2. facts; by theories
3. branch of science that searches for answers to questions about matter and energy

FOLLOW-UP

Reinforcement: Scientific Data in the News
Have the students collect articles from the newspaper. Which show scientific data? Which do not?

Enrichment: How Does Science Affect You?
Have the students research a scientific event that occurred during the past year. How do they predict that the event will change current thinking or life-style?

Tie-in: Gathering Scientific Data
Have the students gather some scientific data, e.g., annual rainfall, census statistics, crime rate, that could be used in a news article about their town. Have them use the data to write the article.

1.2

OBJECTIVES: The student should be able to

- **define** *technology;*
- **distinguish** between science and technology;
- **explain** why it is important to study physical science.

SCIENCE TERM

technology

ADVANCED PLANNING

You may wish to make an appointment for your class to do career research in the library.

LESSON PLAN

1. Introduce the lesson by reviewing the differences between our lives today and the lives of people 15 years ago.
2. Stress that physical science is important because it touches all aspects of students' lives. Point out each of the objects mentioned in the first paragraph of the lesson. Ask the students what each object's use is and how life would differ if that object had not been invented.
3. Help the students distinguish between science and technology. Point out that science is the process—the investigating, the experimenting, the discovering, the final statement of fact. Technology is the product—the use of the fact, science. Give the students a series of statements and ask them which express science and which express technology. Some examples are the following:
   a. Light rays in a vacuum travel in straight lines. (science)
   b. When molecules move quickly, the temperature increases. (science)
   c. Lenses bend light rays thereby forming clearer images. (technology)
   d. Ovens use heat to cook food. (technology)

1.2 Why Is Physical Science Important to You?

You will find out

- why the study of physical science is important.

Physical science is basic to understanding how things work. In fact, physical science is basic to understanding everything from simple machines, such as hammers and wheelbarrows, to computers and nuclear power plants. If you have ever wondered what causes rainbows or what causes stars to shine, you can expect to find an answer through physical science.

The information learned through physical science has helped to make life better for a great many of the people on Earth. This information has made it possible for scientists to do things that your ancestors never dreamed of. Scientists have put astronauts on the moon, landed a machine on Mars, and built a computer that will fit into a pocket.

Figure 1.6 Recent developments in physical science: *top left:* An ultrasound study of a baby before birth; *top right:* Scientists using lasers; *bottom left:* Photograph of Saturn's rings; *bottom right:* A semiconductor chip

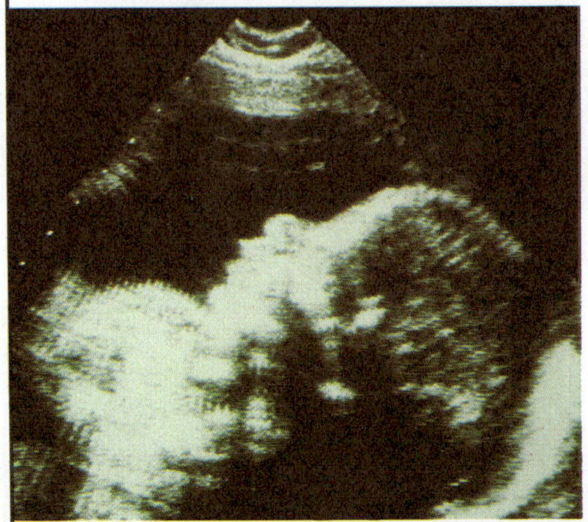

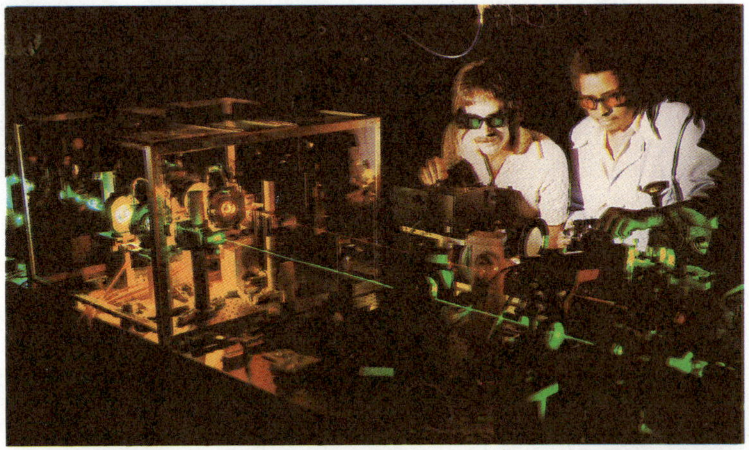

More than a billion people alive today owe their lives to agricultural and medical **technology** brought about by research in the physical and life sciences. Because of the tremendous progress that has been made, people sometimes forget that technology is just a tool, not a miraculous cure-all. Technology is simply a way of applying scientific discoveries to the real world.

The advances in science and technology are not entirely free of problems. People still have to worry about where to find enough energy to run machines, what to do about the pollution that these machines cause, and how to live with the challenges of nuclear power.

Science and technology may be in part responsible for many of the problems that face people today. But this is not the fault of science and technology themselves. Whether science and technology help or hurt people depends on how wisely people use them.

Some of you may want to do more than acquire a general knowledge of physical science. You may want to become an expert in one of the many fields of physical science and to make your living that way. If so, you will have a choice of any one of hundreds of interesting careers. Table 1-1 lists a few of these from A to Z.

Table 1-1 Physical Science Careers from A to Z

A	Aeronautical engineer, astronomer, audio engineer
B	Biomedical engineer
C	Crystallographer, chemist
D	Demolition specialist, draftperson
E	Environmental analyst, electrical engineer
F	Foundry metallurgist, flight engineer
G	Geologist, geophysical prospector
H	Health physicist, hydraulic engineer
I	Industrial designer, instrumentation technician
J	Jig and fixture builder, jet engine mechanic
K	Kettle operator, kiln operator
L	Laboratory supervisor, laser technician
M	Mechanical engineer, materials scientist
N	Nuclear physicist, nuclear reactor safety inspector
O	Optician, oceanographer
P	Pharmacist, physicist, petroleum engineer
Q	Quality control engineer, quarry supervisor
R	Rubber chemist, radar designer
S	Solar heating engineer, safety engineer
T	Teacher, transportation planner
U	Utilization engineer, ultrasound technologist
V	Vacuum technologist, volcanologist
W	Welding engineer
X	X-ray technician, X-ray astronomer
Y	Yeast culture developer
Z	Zanjero

Study Question for 1.2

List three ways that the discoveries of physical science have improved your life.

Answers will vary. Examples include pocket computers, medical technology, and advances in agricultural techniques.

4. Ask the students to name inventions or examples of technology they consider to be important. Write the list on the board. Ask the students to name a problem that each invention has caused and a use of the invention that they feel overshadows the problem.
5. Read with the students the list of careers that are related to physical science in Table 1.1. Ask the students how many careers they can identify. Help them to identify the others. You may use the career list to begin a variety of projects.

● **Reference/Research Skills**

FOLLOW-UP

Reinforcement: Scientific Concepts Behind Inventions

Ask the students to list ten inventions or technological advances. Beside each, have the students write one scientific concept the invention is based on.

Enrichment: How Do Inventions Help You?

Have the students read a book about an inventor or an invention. How was the invention developed from an idea? How did the invention help humankind?

Tie-in: Inventions and American History

Have the students write short stories in which they describe a day in contemporary America without mentioning a single scientific invention.

SCIENCE BACKGROUND

A kettle operator, or kettleman, may work in a brewery, lead refinery, or paint factory. He or she works with substances in a heated container. A zanjero oversees the distribution of water in irrigation ditches and canals.

1.3

OBJECTIVES: The student should be able to

- **list** and **explain** the four steps of the scientific method;
- **explain** how a problem can be solved by using the procedure for the scientific method;
- **list** three qualities of a good scientist.

SCIENCE TERMS

scientific method	synthesize
observe	hypothesis
experiments	test
analyze	

ADVANCED PLANNING

For the demonstration obtain a bottle of perfume and a bottle of ammonia. It will be helpful if the liquids are in atomizers, but you will have to be careful where you spray them!

LESSON PLAN

1. Begin teaching the steps in a scientific investigation by listing and defining them. Although the students will probably know what *observe* and *test* mean, they may find it interesting to learn the meanings of *analyze* and *synthesize* through their derivations. Analyze (analysis) comes from the Greek *analyein* meaning to dissolve. Synthesize (synthesis) comes from the Greek *syn* + *thithenal* meaning to put together.

1.3 Thinking as a Scientist

You will find out
- the four steps of the scientific method;
- some qualities of a good scientist.

If you want to think the way a scientist does, the most important thing you must do is to be curious about the world. Why is the sky blue at noon and red at sunset? What causes static on radios?

The next thing you must do is to believe that the world can be understood, that there are some rules that nature follows. These rules can be discovered and understood by anyone willing to make the effort.

Figure 1.7 The four steps of the scientific method

Finally, you must do something about your curiosity and your belief. What a scientist does can be summarized in these four words: *observe, analyze, synthesize,* and *test*. These four steps are what is meant by the **scientific method** of investigation.

For thousands of years people have had a sense of wonder and a belief in an order in nature. But it was less than four centuries ago that the scientific method was developed. The first person to make this method popular was Galileo Galilei, an Italian who has been called the father of physical science. Galileo developed the scientific method when he studied how the earth's gravity affects the way things move.

Figure 1.8 After you observed the photographs on the first page, you analyzed and synthesized your observations. You might have determined a logical order by studying only the milk. The milk makes sense in the order above. What other items suggest that the milk theory is wrong?

The first thing Galileo did was **observe;** that is, he recorded either mentally or in writing the things he noticed. The story goes that Galileo first became interested in the laws of motion when he was in a cathedral. While he was there, a small earthquake set in motion a chandelier attached to a long chain. It moved much like a giant pendulum. Galileo timed the swings with his pulse beat and noticed that each swing of the pendulum seemed to take the same amount of time. This happened even though the distance the chandelier swung got smaller with each swing.

Galileo was interested in far more than pendulums. He wanted to find out how things move under the influence of gravity. He built test tracks and rolled balls of different sizes down them. Time after time he rolled the balls. Time after time he measured and recorded the results. In this way he observed the motion of bodies under carefully controlled conditions. In other words, Galileo performed **experiments,** or measurements and tests under controlled conditions. As you will see, experiments are the backbone of science.

The next step was to **analyze** the observations, that is, to organize them in one way or another. For example, Galileo might have analyzed his observations by noting that the balls rolled down the tracks faster when the track was tilted at a steeper angle.

Do You Know?
Dr. Joseph Bell, who was supposed to have been the model used by Arthur Conan Doyle for Sherlock Holmes, once said, "Most people see, but they do not observe."

2. Show the students that *observe* means not just what they see with their eyes but also the information they gather with their other senses. (It can also mean information gathered through an extension of the senses with instruments such as microscopes and electrical meters.)

3. Have the students use their senses to observe by doing the following exercise. Ask them to close their eyes and tell you what you are doing. You should do the following things: Run the water; walk around with an open bottle of perfume; clap your hands. Ask the students which senses they used. Ask them which senses they did not use. Emphasize that you did not have them use touch and taste because it is often dangerous to use these senses in the lab.

4. Point out that when a scientist has completed the observe step, a great deal of preliminary work has been completed. The scientist has
   a. defined the problem and decided specifically what will be investigated.
   b. gathered data by using the senses and by reading books and periodicals.

5. In order to help the students examine the process by which a scientist completes the analyze step of the scientific method, have them think through some of the following problems.
   a. How does the size of a hole affect the rate of water flow from a can?
   b. Why does ice sink in alcohol but float in water?
   c. What color absorbs the most heat from the sun?

6. Explain that once the scientist has analyzed the data, the information can then be synthesized, or put together to form a new idea. If scientists base an idea only on data that they have read, then the data become a hypothesis. If the idea is based on data that they have read and experiments that they have performed, then the idea is more solid and is known as a theory. In synthesis a scientist makes what can be assumed, based on all the data, to be a true statement about what has been investigated.
7. On the board write two examples of hypotheses. You may use the following examples: If I put a rock in water, the water level will rise; If I put sugar in water, the water level will not rise. Ask the students what stage of scientific thought the statements are in (hypothesis). Ask the students what they would do to turn the hypotheses into theories.
8. The final step of a scientific investigation is to test. Stress that no theory, no matter how logical it sounds, can be taken as a fact until it has been checked under every possible condition. Only when this has been done is the theory accepted. Even after acceptance, however, new evidence can show that the theory is not true under all conditions.
9. Emphasize that the hypotheses and theories with which scientists work are not always completely true. By testing them, however, scientists develop a better understanding of the world and expand their knowledge base. The continual addition of knowledge through experimentation and observation means that science is not stagnant. With this in mind, students should realize that what is truly important is not that science is a body of knowledge but that science is an organized method of gaining new knowledge.

SCIENCE BACKGROUND

Scientists must test a hypothesis many times to be sure that it is valid. If it proves to be valid, the hypothesis becomes a theory. Scientists are reasonably sure that it is true. Only when the theory has been tested many times under as many conditions as possible and found to be true does it become a law.

Figure 1.9 You test the milk theory from page 9 by following other items such as the hamburger and the salad. By using all four steps of the scientific method, you conclude that the correct order is as shown above and a second glass of milk must have been involved.

The third step was to **synthesize** [SIN-thuh-syz], that is, to put the observations together in a new way and come up with a **hypothesis** [hy-PAHTH-uh-sis] that explains the observations. A hypothesis is really just an educated guess based on a few observations at hand. It is the first step in the synthesis. The next step in the synthesis is just a more educated guess, or theory. It is based on more work and more observations. Galileo's synthesis was the theory that objects of lighter weight fall just as fast as objects of heavier weight.

The last step of Galileo's scientific investigation was the **test,** or the check of the theory's predictions against new observations. This came, according to one story, when Galileo dropped a cannonball whose weight was relatively heavy and a musket ball of much lighter weight from the Leaning Tower of Pisa in the presence of a doubting crowd of onlookers. The crowd was amazed when both objects hit the ground at the same time.

These four steps—observe, analyze, synthesize, and test—are part of every scientific method. In most cases not all steps are performed by the same person. One group of scientists may do the experiments and make the observations. Another group may analyze the observations. Still another may synthesize the observations into a theory. Many times there is more than one theory, and it may take years to come up with a test to decide which theory is better. For example, there are several theories about how the earth and the other planets were formed, but until there are more observations of the other planets, scientists may not be able to decide between the different theories.

These four steps do not apply only to science. In fact, you go through them all the time, day in and day out. For example, suppose someone puts a cup of soup in front of you. You observe that steam is rising from the cup. You also observe that steam is not rising from a glass of cool milk. By analyzing this and other observations you put these facts together, or synthesize them, to conclude that the cup of soup is hot. Being a true scientist, you take a sip of the soup and find that it is hot, thereby proving your theory!

Good scientists have several qualities that set them apart. First, they ask good questions. These are often the simplest questions. But if Galileo had asked, as did many people in those days, how to turn lead to gold, he might have worked very hard with little result.

Second, good scientists are patient, careful, and honest in collecting observations. Incorrect observations are worse than useless. They can cause others to work very hard to explain observations that can be misleading. Scientists must report exactly what is observed even if it disproves their own theory.

Third, good scientists are able to make bold guesses about laws that bind the observations together. When Galileo put forth his theory, he was going against strongly held ideas. Since these ideas had been around for over a thousand years, almost everyone took them for granted. Yet he was right and the others were wrong.

Figure 1.10 An artist makes very careful observations of a painting's subject.

Study Questions for 1.3

1. List and describe the four steps of the scientific method.
2. List three qualities of a good scientist.

1. The four steps of the scientific method are observe, analyze, synthesize, and test. To observe is to record, mentally or in writing, the things that are noticed. To analyze is to organize the observations. To synthesize is to put observations together in such a way as to formulate a hypothesis and from the hypothesis, a theory. To test is to check the theory's predictions against new observations.
2. Answers will vary but may include the ability to ask good questions and have patience, honesty, and curiosity.

10. Have the students read the biography of Galileo on page 19. You may also want to have available other references that deal with Galileo's life and work. Ask which of the characteristics of a scientist would describe Galileo (curious, interested in world around him).

Comprehension Skill: Applied

11. On the board list the three important components of thinking like a scientist (curiosity, belief that the world can be understood, action/doing). Explain that in this context *action*, or *doing*, means to work in an orderly way in order to form a plan and then to work within that plan. Most discoveries are at best ten percent inspiration and ninety percent hard work.

12. Conclude the lesson by pointing out that good scientists have qualities other than the way they think that set them apart. Write the following list on the board:
Three qualities of good scientists:
1. They ask good questions.
2. They patiently and carefully collect observations.
3. They make bold guesses about the world.

FOLLOW-UP

Reinforcement: Using the Scientific Method
Ask the students to go to the library and find a book or an article about a scientist. Have them read the material and explain how the scientist used the scientific method.

Enrichment: Applying the Scientific Method
Present a series of topics that are appropriate for science projects. Ask the students to select one of the topics or a topic of their own choosing. Have the students prepare a basic plan showing how they will perform each step of the scientific method to complete the project.

Tie-in: A Theory Is like a Circus Tent
Theories are like circus tents. You may prove a theory or you may not. You may get your tent off the ground or you may not. Theories are difficult to devise, and once you have them you must sell them. A tent is sometimes difficult to put up, and once up you must support it correctly. Ask the students to think of other analogies. What can they compare a theory to?

1.4

OBJECTIVES: The student should be able to

- **explain** why extra care is needed when working in the laboratory;
- **list** several rules to follow when in the laboratory;
- **list** the steps that should be followed in a laboratory experiment.

SCIENCE TERM

laboratory

LESSON PLAN

1. Assign this lesson for students to read before you discuss it in class.
2. Introduce the lesson by asking the students what a laboratory is and what the most important rule of the laboratory is.
3. As a reminder of the proper behavior in the lab, write the words *Be careful* on the board.

1.4 The Laboratory

You will find out

- what a laboratory is and its role in science;
- the steps in a laboratory experiment.

The first business of science is to collect facts. Facts are the building blocks with which to construct theories, and they provide the final judgment as to whether those theories are suitable or not.

If you are an astronomer or a geologist, you collect most of your observations through a telescope or in the field. But if you are a physicist or a chemist or an engineer, you will collect observations in controlled tests or experiments. Most of the time these experiments are performed in a room where the testing equipment is set up and where conditions such as temperature, amount of light, and volume of gas can be carefully controlled. Such a room or building is called a **laboratory.**

Figure 1.11 A safe laboratory is necessary for collecting observations and testing theories.

The laboratory, or lab, is the workshop of science, the place where the action begins and ends. It is where observations that result in new theories are collected, tested, and retested. When you are doing an experiment in the lab, you should think of that lab as a workshop or testing ground. But your lab should not look like many workshops, with sawdust on the floor and tools scattered around. It should look more like an operating room with a place for everything and everything in its place.

The two most important words to remember when going to work in the lab are BE CAREFUL. Careless work in the lab is worthless. It could also cause serious injury to you or your fellow workers. Certain procedures or steps should be followed in every lab experiment.

1. Think what you are going to do in the lab before you go there. PLAN AHEAD. Read the procedure for the activity in your book. If you are doing an experiment that is not in your book, jot down what you plan to do. Then make a chart to record your observations.

2. Record in your notebook the date, the time, the place, the kinds of observations you intend to make, and the equipment you will use.

3. Organize your equipment.
4. Perform the experiment, observing the results.
5. Record your observations. If you don't keep a record, you may as well not have done the experiment.
6. Put the equipment back where you got it.

ACTIVITY

How Can These Students Be More Careful?

Materials

plain white paper
felt pens

Procedure

In the picture some students are doing something that is dangerous. What are they doing wrong?

Questions

1. For each mistake, what rule should the students have followed to have a safe lab?
2. Create a poster to illustrate one of the rules.

1. Just as a workshop is a place where things are made and tested, a laboratory is a testing ground for scientific observations.
2. plan; record date, time, place, kinds of observations to be made, equipment to be used; organize equipment; perform experiment; record observations; return equipment

Study Questions for 1.4

1. Why is the lab called the workshop of science?
2. List the steps you should follow in a lab experiment.

4. Set up a signal that you will use to get the attention of all students during lab time. The signal means they are to stop whatever they are doing, stand still, and listen to your directions.
5. Outline a procedure for students to follow in case an accident occurs. You may want to write it on a poster and display it where everyone can see it.

ACTIVITY: How Can These Students Be More Careful?

Ask the students what is dangerous in the picture and how the situation should be corrected. After you have discussed the picture, ask the students to list the rules they should follow in the laboratory. List these rules on the board under *Be careful*.

1. a. Wear safety glasses at all times.
   b. Do not heat a corked tube; do not point a tube at anyone while it is heating.
   c. Tie back long hair.
   d. Use a holder to pick up hot glassware.
   e. Don't fool around in the laboratory.
   f. Never taste any chemical.
2. Answers in the form of a poster should vary. Encourage students to illustrate a variety of the rules.

● Comprehension Skill: Applied

FOLLOW-UP

Reinforcement: Charades—Danger in the Lab

Have the students play a game of charades. The players must act out a troublesome laboratory situation, and the audience must guess what rule is being broken and explain how to remedy the situation.

Enrichment: Careers in the Laboratory

Invite a quality-control technician or a chemical analyst to describe what each does in the lab and to answer students' questions.

Tie-in: Speaking of Laboratory Safety

Have your students prepare oral presentations for younger students in which your students explain rules of laboratory safety.

1.5

OBJECTIVES: The student should be able to

- **define** the terms *mass* and *volume*;
- **identify** the kilogram, meter, second, and liter as units of mass, length, time, and volume respectively in the metric system;
- **define** the term *measurement* and **explain** what a standard of measurement is;
- **recognize** objects that have a length of about one meter, a mass of about one kilogram, and a volume of about one liter;
- **list** and **identify** the prefixes used with metric terms;
- **convert** metric terms from one unit to another.

SCIENCE TERMS

measurements	kilogram
mass	liter
volume	meniscus
metric system	second
meter	

ADVANCED PLANNING

Have metersticks and balances available for the students to use as they review measurement.

LESSON PLAN

1. Be sure the students understand that measurement is comparing something to a known standard. The standard may be anything, but it should be universal, unchanging, and easily reproduced. Students will be familiar with the words *time*, *temperature*, *length*, and *distance*. They may be unclear about *mass* and *volume*. Mass measures the amount of material in an object. Students may think that *mass* can be defined as the weight of an object. Point out that this is incorrect and stress the correct definition. (Mass and weight are compared in Chapter 3.) Volume is a measure of the amount of space an object takes up.

1.5 Measurement

You will find out
- the standard units of measurement in the metric system.

You learned earlier that the building blocks of science are scientific observations, or facts. If they are to be of any use to a scientist, the observations must be expressed in numbers. That is, they must be **measurements.** For example, it is not good enough to say that a large, heavy object moved over a long distance in a short time. Specific measurements give a more accurate picture of the size of the object than the words *large* and *heavy*. **Mass** is the measurement of the amount of material in an object. **Volume** is a measurement of the amount of space an object takes up. It is better to say that an object whose mass is 2,500 kilograms and volume is 1.2 cubic meters moved 510 meters in 5.4 seconds.

In making observations, words such as *long*, *hot*, *light*, and *big* are not useful. These words mean different things to different people. If mass, volume, length, temperature, and time are described in measurements instead of words, misunderstandings can be avoided. It is important that all measurements be taken by everyone in the same way. Such standards of measurement are necessary to make observations clear and understandable.

Figure 1.12 The word *far* means different things to different people.

Standards of measurement are so important to science that many countries have large laboratories whose main task is to build and maintain precise standards of measurement. In the United States there is a National Bureau of Standards with laboratories in Washington, D.C., and Boulder, Colorado.

The system of measurement used in laboratories in this country and around the world is the **metric system.** The metric system is a decimal system of measurement based on the **meter** as the unit of length, the **kilogram** as the unit of mass, and the **liter** as the unit of volume.

The metric system was first used in France. Here the meter was defined as one ten-millionth of the distance from the North Pole to the equator along a line passing through Paris. Today the meter is defined as 1,650,763.73 times the wavelength of orange-red light given off by the element krypton 86. In less precise units the meter is about equal to the height of a doorknob above the floor. For most measurements you don't have to have krypton 86 on hand. You can do quite well with a reasonably straight ruler, tape measure, or meterstick.

The standard for a kilogram is defined as the mass of a particular block made of platinum and iridium, which is preserved at Sèvres, France. Exact duplicates exist at the National Bureau of Standards and in other institutions around the world. A kilogram has a mass about equal to that of a brick. Mass can be measured by comparing the object you are measuring with carefully constructed masses.

Figure 1.13 Some metric measurements

2. The students should know that the kilogram, the meter, the second, and the liter are units of metric measurement, but they may need help reviewing exactly how large each of these is. Give them an approximate example of each:

1 meter = distance from the floor to the doorknob
1 kilogram = mass of 1 brick
1 second = time needed to say "one thousand"
1 liter = volume of 1 individual tonic bottle

3. Although most students will have used metersticks, some may need to review how to read them and how to express their answers with decimals. Provide practice by giving each student a ruler and a piece of paper. First have the students measure the length of the paper and express their answers in *millimeters* and *centimeters*. Have them measure the width of the paper using the same terms. Then have them fold the paper in different ways so that each fold results in a smaller size. Measure the paper after making each fold.

4. Some students may not have used balances or graduated cylinders. The use of the graduated cylinder will be taught in the activity. Demonstrate to the students how to use the kind of balance that you have available.

● **Mathematics Skill: Measurement**

ACTIVITY: How Do You Find the Volume of an Irregular Object?

Ask the students how they would find the volume of a cube. Then ask them why this method will not work with a nail or a pebble.

Give the students some time to review the experiment. Then point out the materials and remind the students of their responsibility for all materials and of the signal that you have arranged should you want their attention.

When the students have finished the experiment, ask them how they would find the volume of a ring.

● **Laboratory Skills: Experimentation**
Manipulation of Materials

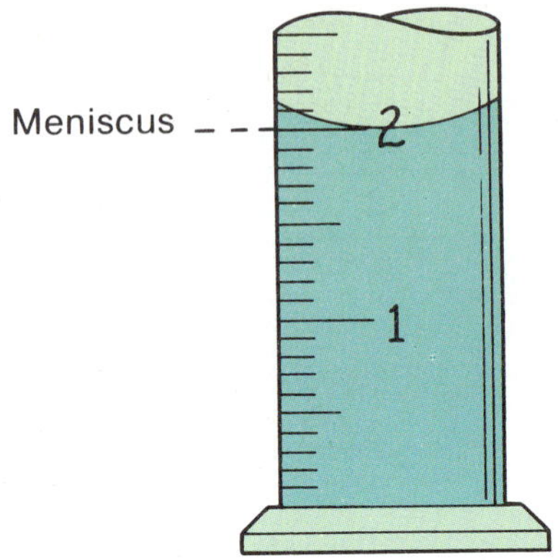

Figure 1.14 Observe a graduated cylinder at eye level. Read the volume at the bottom of the meniscus.

When you measure the volume of a substance, especially a liquid, you use a unit of measure called the liter. ● One liter is a cube 0.1 m on each side and contains one thousand cubic centimeters, or 1,000 cm^3, as they are abbreviated. Since 1 cm^3 is one thousandth of a liter, it is often called a milliliter.

Liquid volumes are measured with a graduated cylinder, which is much like a kitchen measuring cup. To find the volume, observe the level of the liquid and read the marks on the cylinder. For thin cylinders the liquid surface may be curved. Read the bottom of this curve, or the **meniscus** [me-NIS-kuhs], to get the volume.

You can measure the volume of solid objects by using this convenient fact: When the object is immersed in water, it will raise the water level by an amount equal to the volume of the immersed object.

● ACTIVITY How Do You Find the Volume of an Irregular Object?

Materials

water
100 mL graduated cylinder
nail
marble
pebble
paper towel

Procedure

1. Copy the chart.
2. Gather all the materials.
3. Fill your graduated cylinder with water to 50 mL.
4. Carefully drop the nail into the graduated cylinder.
5. Observe the level of the water. Record this level in your chart.
6. Repeat the procedure with the marble, and then the pebble.

Object	Volume of Water and Object	Volume of Water at Start	Volume of Object
nail		*50 mL*	
marble		*50 mL*	

Questions

1. What was the original level of the water in the cylinder?
2. What was the level of the water after you added the nail?
3. How much did the nail cause the water to rise?
4. What is the volume of the nail?
5. In your chart record and label the volume of the other two objects.

1. Answers will vary.
2. Answer should be greater than answer in question 1.
3. Subtract answer in question 1 from answer in question 2. Label this answer in milliliters.
4. answer 3 labeled in cubic centimeters
5. Answers will vary. Completion of chart is based on answers for questions 1, 2, 4, and 5.

The temperature scale in the metric system is based on the freezing and boiling points of water. The freezing point of water is given the value of zero degrees Celsius, or 0°C for short. The boiling point under standard atmospheric pressure is 100°C. A Celsius or Centigrade degree is one-hundredth the difference between 0° and 100°.

The metric unit of time is the **second.** The second was originally defined in terms of the earth's rotation as 1/86,400 of one day. Now the second is defined with atomic clocks that have an accuracy of about one second in 3000 years. Thanks to the marvels of solid state physics, split-second digital stopwatches with an accuracy of one second in about two weeks are available at reasonable prices.

The metric system has gained worldwide acceptance by scientists because it is easy to make larger or smaller units from the standard units. You simply multiply or divide by ten or powers of ten. Prefixes are used for each multiplication by ten, hundred, etc., and for each division by ten, hundred, etc.

Prefix	Abbreviation	Meaning
kilo	k	1000
hecto	h	100
deka	da	10
deci	d	.1 (1/10)
centi	c	.01 (1/100)
milli	m	.001 (1/1000)

Figure 1.15 Some common metric prefixes

Suppose, for example, you wanted to say one thousand meters in a simpler way. Just attach the prefix *kilo,* which stands for a thousand, to *meter,* and you have one kilometer, or a thousand meters. Likewise you could write one thousandth of a meter by putting the prefix for one thousandth *milli* in front of *meter* to get millimeter, one thousandth of a meter.

Do You Know?
The small slice of quartz crystal in an electronic digital watch vibrates over 56,000 times a second.

Study Question for 1.5

What are the metric units of measurement for mass, length, volume, and time?

mass—kilogram
length—meter
volume—liter
time—second

5. Review the fact that scientists and most of the nations of the world use the metric system. Mention its convenience as a decimal system with one basic unit for each measurement that all the other units are related to. Ask the students what a prefix is. Then ask them the names of the prefixes used in the metric system and the meaning of each prefix. As the students name the prefixes, write them and their meanings on the board in the proper order from largest to smallest.

FOLLOW-UP

Reinforcement: Added Practice Using the Metric System
Set up a learning corner devoted to the metric system. Have available objects that the students can measure such as balances, graduated cylinders, and metersticks. In an envelope, have a form the students can use to record their measurements. Students who need extra practice can be assigned to measure the objects in the learning corner.

Enrichment: Scale Models of the Classroom
Have the students draw scale models of the classroom using a scale in which 1 cm represents 1 m.

Tie-in: A Bulletin Board of Metric Measurement
Have the students create a bulletin board on which they display common items that are measured in metric terms.

SCIENCE BACKGROUND

The prefixes *kilo, deka,* and *deci* are taken from the Greek. They are reminders that ancient Greece and the early Roman Empire were two places where science originated. When the metric system was devised, all truly educated people spoke Latin. They preferred it because it was a descriptive but dead language whose word meanings would not change.

What Did You Learn?

1.1 • Science is a way of learning about the universe, a method of organizing your curiosity.

1.1 • The building blocks of science are facts. Theories are ideas that connect groups of facts.

1.1 • Physical science is that branch of science that deals with the study of matter and energy.

1.2 • Physical science is important because it is basic to understanding how things work.

1.3 • The four steps in the scientific method are observe, analyze, synthesize, and test.

1.4 • The two most important words for lab work are *be careful.*

1.4 • The six steps in performing an experiment are (1) plan ahead, (2) record date and conditions, (3) organize equipment, (4) perform experiment, (5) record observations, and (6) put equipment back where it belongs.

1.5 • Measurements are observations expressed as a certain number of units.

1.5 • The system of measure used by scientists is the metric system. In this system the primary units—mass, length, and volume—are measured in terms of kilograms, meters, and liters.

Key Terms

science
facts
theories
physical science
technology
scientific method
observe
experiments
analyze
synthesize
hypothesis
test
laboratory
measurements
mass
volume
metric system
meter
kilogram
liter
meniscus
second

Biography

Galileo Galilei (1564-1642)

Galileo was born in Pisa, Italy. He began training for a medical career but soon switched to mathematics and natural philosophy, which is what people called physical science in those days.

Galileo was a person who did many things well. He was a fair musician, an excellent writer, a brilliant lecturer, and one of the best scientists of all time. He was also good at making things.

In 1609 a man arrived in Italy with a new instrument said to be of great use for both military and shipping purposes. The instrument was called a telescope. He offered to sell it to the Grand Duke, but the Duke refused and sent a description of it to Galileo instead. A month later Galileo, who had never seen the instrument, presented the Grand Duke with a telescope three times more sensitive than the one offered for sale!

Galileo made over a hundred telescopes and was the first person to use a telescope to study the moon, the sun, the planets, and the stars. With it he discovered the moons of Jupiter, craters on the earth's moon, and sunspots. His work in astronomy proved beyond doubt that the earth moves around the sun.

Galileo's writings about his astronomical discoveries made him famous. He also faced a problem that many early scientists faced. Many powerful leaders refused to accept Galileo's new methods and ideas.

Because of his quick mind and his biting sarcasm, Galileo seldom lost an argument. But this time it was different. Because of his ideas, Galileo was sentenced to life in prison and spent the last eight years of his life under house arrest. It was there that he wrote the books on motion, which laid the foundation for the modern method of exploring science.

Vocabulary

1. facts
2. physical science
3. observe
4. analyze
5. hypothesis
6. liter

● Word Skill: Vocabulary

What Do You Remember?

1. Science is a method of organizing curiosity.
2. hypothesis—educated guess based on a few observations; first step in synthesis
theory—more educated guess than hypothesis; second step in synthesis; explains groups of facts
fact—scientific observation; event that can be checked out or tested by someone else
3. Answers will vary. Students may include machines, either simple or complex; oil refineries; nuclear power; rockets; and solutions to the problems of pollution.
4. Observe: Galileo observed a pendulum when he saw a chandelier swinging. He timed the swings with his pulse and noticed that each took the same amount of time. He built pendulums of different sizes and weights and observed their swings. He measured and recorded how the swing changed with the length of the pendulum, the weight attached, and the distance it was pulled back. He wanted to know how every moving object was influenced by gravity. He built test tracks, rolled balls of different sizes down them, and timed how long each ball took to roll down the track.
Analyze: Galileo may have noted that the balls rolled down the track faster when the track was steeper. Balls with greater mass were rolled at the same rate as balls whose mass was smaller.
Synthesize: If air friction is neglected, objects of smaller mass fall just as fast as those of greater mass.
Test: Galileo dropped a cannonball and a musket ball from the Leaning Tower of Pisa to test whether both objects would hit the ground at the same time.
5. asks good, simple questions; is patient, careful, and honest in making observations; able to make bold guesses
6. The laboratory is important as the place where the scientist tests observations that result in new theories.
7. Be careful.
8. kilogram
9. In order to change 45 kilograms to milligrams move the decimal to the right 6, or multiply by 1,000,000.

TO THINK ABOUT AND DO

Vocabulary ●

Select the term in parentheses that correctly completes each sentence.

1. The building blocks of science are called (facts, hypotheses, theories).
2. The branch of science that deals with matter and energy is (astronomy, meteorology, physical science).
3. When scientists record mentally or in writing the things they notice, they are performing the step of the scientific method called (analyze, synthesize, observe).
4. When scientists organize their observations in order to interpret them more easily, they are performing the step of the scientific method called (analyze, observe, test).
5. An educated scientific guess that is based on a few observations is a (hypothesis, theory, fact).
6. The basic unit used to measure volume in the metric system is the (second, liter, meter).

What Do You Remember?

1. What is science?
2. Distinguish among hypothesis, theory, and fact.
3. List three advances of physical science that have made life easier.
4. Describe how Galileo used each step of the scientific method.
5. What are the qualities of a good scientist?
6. Why is the laboratory so important to a scientist?
7. What are the two most important words to remember when working in the lab?
8. What is the basic unit used to measure mass in the metric system?
9. Explain how you would change 45 kilograms to milligrams.

Applying What You Have Learned

1. a. How long is the line in cm? In mm?

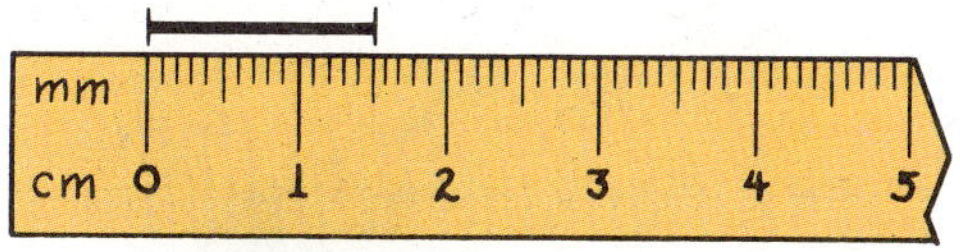

b. What is the diameter of the ball in cm?

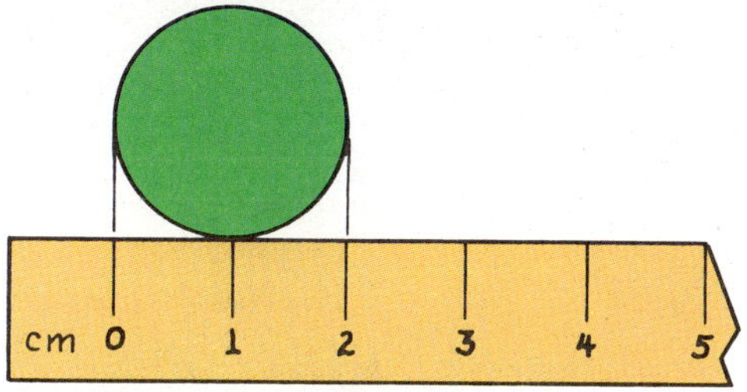

c. What is the volume of the rock in mL?

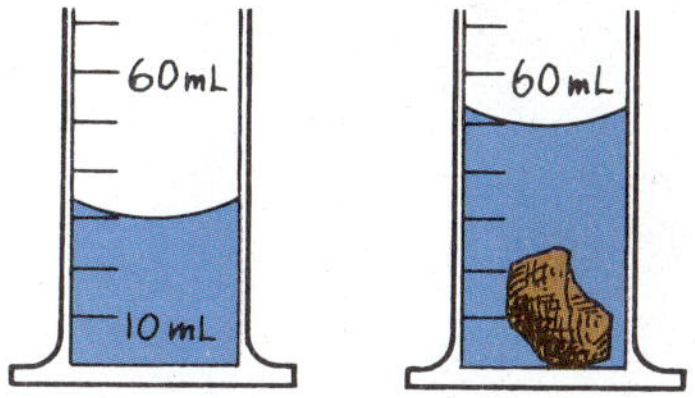

2. How would you use a balance to determine the mass of 1000 mL of water?
3. If you are given a meterstick, two wooden blocks, and a basketball, how would you find the diameter of the ball?
4. What would you do if you saw someone with long hair leaning near a lighted Bunsen burner?

Research and Investigation

Choose a scientific theory currently making the news. How did it evolve from a hypothesis? What experiments were done to prove it?

Applying What You Have Learned

1. a. 15 mm, 1.5 cm
   b. 2 cm
   c. 20 mL
2. a. Find the mass of the glass container.
   b. Find the mass of the container plus the water.
   c. Subtract the mass of the container from the mass of the container plus the water.
3. Select a principal mark or starting point and place inside edge of one block on it. Place ball against this block. Place other block on opposite side of ball. Find the difference between the two readings, i.e., inside edge of second block minus inside edge of first block.
4. Walk over and point out the danger to the student. Suggest that the long hair be tied back.

Comprehension Skill: Applied

Research and Investigation

Answers will vary. The student should note, however, that in order for a theory to evolve from a hypothesis, many controlled experiments must be done.

Reference/Research Skills

Long-Term Projects

1. Have each student make a scrapbook of at least 15 simple machines that they use. The students should label each simple machine. Under each picture, they should explain how the machine helps them to do work.

2. As you finish each chapter, brainstorm with the students devices that work on the principles discussed in the chapter. Have students select a device that interests them and write an illustrated report, perhaps including a model, to demonstrate how the device works. Two examples are air brakes and spray cans.

3. As a small-group activity, have students work together to prepare a *Museum of Flight*. The museum should include a time line of the important events in the history of flight, models of different types of planes with their parts labeled, and a wind tunnel students can use to demonstrate the forces that affect planes.

4. Obtain some blank filmstrip acetate from Bro-Dart Co., 1609 Memorial Ave., Williamsport, PA 17701. Have the students make filmstrips to identify the states of matter, their molecular arrangements, properties, and uses.

5. The Cottrell Precipitator works according to the laws of static electricity. Have students write reports to describe the device and its importance. Ask them to make models to demonstrate how the device works.

Unit One

Motion, Force, and Energy

Bulletin Boards

1. This board is designed to stress and reinforce Newton's contributions. Put the title at the top of the board. Divide it into four sections, and title each section. In the upper half of each section, draw or cut from a magazine and mount the indicated picture. Have the students explain how the drawings illustrate the desired concepts. Encourage the students to bring in and mount other pictures to illustrate the concepts.

2. Title your board *Kinetic and Potential Energy*. In the center of the board, draw a picture of a pendulum. Label the end points of its arc P.E. and the swing of the arc K.E. Ask the students how a pendulum illustrates energy conversions. To help reinforce the idea that potential energy can be converted to kinetic energy and vice versa, have the students bring in pictures to illustrate kinetic and potential energy. Mount them on the board.

3. To help illustrate the effects of electric forces, print the question *How Do Like Charges Act?* across the top of your board and the question *How Do Unlike Charges Act?* across the bottom. Illustrate the ideas with pictures of lightning striking a roof, rubbing a balloon on hair and then sticking it to a wall, and combing hair and then holding the comb near scraps of paper. Ask the students to bring in other pictures showing electricity affecting their lives. Add them to the board.

CHAPTER BACKGROUND

Matter is anything with mass and volume. Energy is the ability to move matter or cause changes in it. Everything can be classified as either matter or energy. Books, rulers, water, and air are examples of matter. Sunlight, heat, sound, and electricity are examples of energy. Energy affects matter by changing its state or by moving it to other positions.

Experimental evidence suggests that the atoms and molecules of which matter is composed are in constant motion. The rigidity of solids allows only a back-and-forth vibration of the particles. In gases and liquids the particles move more freely. The freer molecular movement allows gases and liquids to assume a variety of shapes. The idea that molecules are in a state of constant motion is the basis of the kinetic molecular theory.

The kinetic molecular theory is also based on energy. The theory not only implies that molecules of matter are in constant motion but also implies that the state of matter depends on this movement. As energy is added to matter, the molecules speed up. This increased vibration breaks the attractive forces that bind the molecules to each other, and a change of state results. (While the state is actually changing, the added energy does not cause the temperature of the substance or the speed of its molecules to increase. Both temperature and speed remain constant while the energy is used to free molecules from each other.) When energy is removed, the molecules slow down and become more attracted to each other. Again, a change in state occurs.

The rates at which energy changes matter are measurable. For example, it takes about 347 joules of heat energy to melt one gram of ice at 0°C. Our understanding of the factors that control these rates will allow us to put energy to better use in the future.

ACTIVITY

What Are the Characteristics of Liquids? ●

Materials

graduated cylinder
water
containers of different shapes
paper towels

Procedure

1. Mark the containers *A*, *B*, and *C* and observe the shape of each container.

2. Pour 100 mL of water into the graduated cylinder.

3. Pour the water from the cylinder into container A, then into B, and then into C. In each case note how the water changes.

4. Now pour the water back into the graduated cylinder. Observe the amount of water you have now.

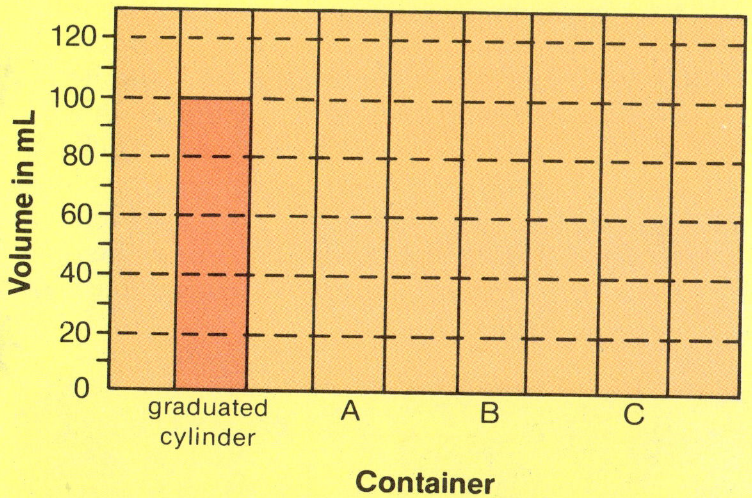

Questions

1. How do containers A, B, and C differ?

2. Was the volume of the water changed as you poured it from A to B to C? How do you know? Hint: Try pouring the water back into the graduated cylinder after you pour the water into each container.

3. Copy and complete the graph.

4. Explain how this experiment shows that liquids have a definite volume but can change their shapes.

1. different shapes
2. Volume was the same (100 mL), since it was still 100 mL at the end. No water was lost.
3. All bars on graph should be at 100 mL.
4. Water took the shape of container, but amount of water remained the same, which was apparent when water was poured back into graduated cylinder in procedure step 4.

A liquid, like a solid, does have a definite volume. A hundred milliliters of water take up the same volume whether the water is in a bottle, in a glass, or spilled on a tabletop. And if you add 50 mL of milk to the water, you'd better make sure that there is room for it. The milk will take up the same amount of space, 50 mL, whether it is mixed with the water or not.

3. Explain that the characteristics of an object can help us to place it in its proper category. Objects with a definite shape and volume are solids. Liquids have a definite volume, but their shapes change to fit the containers they are placed in. Gases have no definite shape or volume.

ACTIVITY: What Are the Characteristics of Liquids?

Remind the students of the proper way to read a graduated cylinder—from the bottom of the meniscus at eye level.

Stress the importance of not spilling any water, since spillage will interfere with their results. If the students do not have the same amount of water in the end as they had in the beginning, discuss with them the errors that could have caused the differences (reading the cylinder incorrectly, spilling some of the water, not getting all the water out).

● **Laboratory Skills: Experimentation/Demonstration**
Manipulation of Materials

4. Explain that if you pour a can of soup into a bowl, then into a cup, and finally back into the can without any spillage, you will have the same amount of soup as when you started. Yet in each container the soup took on a different shape. Like perfume as it evaporates, gases expand to fill all the space available to them.

5. Point out that although matter is commonly thought of as being a solid, a liquid, or a gas, there is a fourth state of matter called a plasma.

SCIENCE BACKGROUND

Plasmas are found in neon lamps, the exhaust of jets and rockets, the outermost layers of the earth's atmosphere, and the glowing gases of the sun and other stars. A plasma is like a superheated gas. In a plasma, however, charged particles move rapidly through large areas. The forces that act on them are acting over larger distances than they do in gases. Tell the students that at a million degrees Celsius, the approximate temperature of plasma making up the outer layers of the sun, all matter would be in the gas or plasma state.

Figure 2.3 The scent of a skunk expands to fill the space available.

A gas has no definite shape or volume. It spreads out to fill the space available to it. Imagine you are at one end of a room and someone spills a bottle of perfume at the other end. Some of the perfume will change to a gas, and pretty soon you will be smelling it at your end of the room. The perfume gas spread out and filled all the space available, the whole room.

Many scientists recognize a fourth state of matter that is quite different from most matter found on the earth. It is called the **plasma** state. It occurs when the temperature of a gas becomes very high. Some of the gas particles break apart into electrically charged pieces. The outer layers of stars are in the plasma state.

Most matter can exist in any one of the common states—solid, liquid, or gas. For example, think of a pail of water. Most of the water will be ice if the temperature is below 0°C. Above that temperature there will be only liquid water. At temperatures above 100°C the water will change to a gas, commonly called water vapor.

Other substances freeze to a solid, melt to a liquid and change to a gas at different temperatures. In other words, different substances change states at different temperatures. Many rocks, for example, do not melt until they are heated to temperatures of many hundreds or thousands of degrees Celsius. Helium, on the other hand, has never been cooled to a temperature low enough to turn it into a solid.

Figure 2.4 The three common states of water

Figure 2.5 Some solids with different properties

The temperature at which a particular substance changes states is a **property** of that form of matter. A property is a quality or characteristic of a substance and can be used to identify it. Other properties of matter include color, hardness, smell, whether it will burn, and whether it will mix with other substances.

As you move deeper into the study of physical science, you will keep coming back to two basic ideas: matter and energy. You know that matter is the stuff that the world is made of. But what is **energy?** Energy is the capacity to move matter from one place to another or to change matter from one substance to another. Put another way, energy can make things happen to matter.

One of the most important things to know about energy is that it never disappears. It simply changes form. A simple experiment can illustrate this changeability of energy. Try lifting a book about half a meter above the surface of your desk. It takes energy stored in the muscles of your body to lift the book. At first some of this energy is changed into the motion of the book. But now the book is not moving. So what happened to the energy? It is in the form of **potential energy,** or the energy of position. Potential energy is stored energy which can be released when matter moves. This energy can be released by simply letting go of the book.

When you let go of the book, it begins to move. The potential energy is changed into **kinetic energy,** or the energy of motion. In less than a second the book slams against the desk top. Where have the book's potential energy and kinetic energy gone? They have gone into the energy of sound waves and some heat energy.

Figure 2.6 The roller coaster has kinetic energy.

6. To introduce the concept of properties of matter, use the example in the text of the freezing and boiling points of water. Point out that some properties of matter, such as color and smell, may be relatively easy to observe. Others, such as whether the substance will burn, require more extensive observation and, often, experimentation.

7. Ask the students to recall the discussion in which they named examples of matter and energy shown in the opening photograph. Point out that the tidal energy of the wave and the sunlight are energy because they have the ability to move or change matter in some way. Ask what change or movement is brought about by the wave and the sunlight (carrying boats from one place to another, moving turbines, changing beaches, weathering rocks, heating houses).

8. Explain that energy can be classified. On the board create a two-column energy chart. Define each type of energy. Then ask the students to give you an example to illustrate each type. Your chart might look like this.

Kinetic	Potential
moving car	book held above a table
running boy	car on a hill before rolling down

ACTIVITY: What Is the Effect of Height on Potential Energy?

Explain to the students that some outside conditions can affect how much energy an object has. One of these is height. Tell the students that they will be determining how height affects potential energy in this activity.

In order to reduce the amount of equipment you will need, do this activity in groups. Be sure each member of every group has a specific job. Also be sure the books used are as close to the same thickness as possible.

After the activity you may want to discuss why this was a good controlled experiment (it had many factors that remained unchanged—the car, the board, and the cartons—and only one factor that was changed—the height of the boards). Because there was just one variable, you can be reasonably sure that the variable caused the results. Ask the students why they were told to do the experiment again at each height (the repeated trials helped to insure accuracy).

● **Laboratory Skills: Experimentation/Demonstration Manipulation of Materials**

● ACTIVITY

What Is the Effect of Height on Potential Energy?

Materials

toy car
6 books
1 board
3 milk cartons

Procedure

1. Slant your board by resting it against one book.

2. With the milk cartons, build a tower at the bottom of the board.

3. Place the car at the top of the board and allow the car to roll down it. Repeat this step.

4. Raise the board by placing three books under it and rebuild the tower.

5. Place the car at the top of the board and allow the car to roll down it. Repeat this step.

6. Repeat the experiment but increase the height of your hill by using six books to raise the board.

Questions

1. How did the height of the hill change in your experiment?

2. What kind of energy did the car have at the top of the hill?

3. What kind of energy did the car have as it was rolling down the hill?

4. When did the car have the least energy—at one book, three books, or six books? When did the car have the most energy?

5. Copy and complete the chart.

6. How does height affect the amount of potential energy in a body?

7. How did this experiment show that energy can be converted to other forms?

Height *(number of books)*	Amount Milk Cartons Moved *(least, most)*
1 book	

1. increased as more books were used
2. potential
3. kinetic, some potential
4. at one book; at six books
5. reading down right-hand column, a little change at one book; some change at three books; tower shattered at six books
6. The higher an object is, the more potential energy it has.
7. At the top of the hill the car was at rest. It had potential energy. When the car was rolling down the hill, it had kinetic energy and some remaining potential energy.

Figure 2.7 In the home, people benefit from different forms of energy.

The amazing thing about energy is that you can actually make it change from one form to another. You can light a fire and change the energy stored in wood into heat energy. Or you can flip a switch and change electrical energy into light. As a result you can do things that would put the magicians in the fairy tales to shame.

The key to technical progress has always been the ability to control and use energy. Prehistoric people started the progress when they learned to release energy in a controlled way by burning wood. They used the energy from the burning wood to keep warm, cook food, and make tools.

Today energy is used in many forms. People use energy to keep warm and to cook food. They use it also to change dark to light, to move themselves all over the planet, and to send astronauts and machines into space.

Do You Know?
The energy released per second by the space shuttle's three main engines is 23 times the Hoover Dam's output.

Study Questions for 2.1

1. What is matter?
2. List the three common states of matter and give an example of each.
3. What produces changes in matter?
4. Give an example of energy changing from one form to another.

1. anything that has mass and takes up space
2. solid, liquid, gas; ice, water, vapor
3. energy
4. When you drop a book, a change from potential to kinetic energy occurs.

9. Emphasize that just as the potential energy of a car at the top of a hill is converted to energy when the car rolls down the hill, other forms of energy are also easily converted. For example, in a light bulb, electric energy is converted to heat and light. Ask the students what forms of energy are released when dynamite is exploded (heat, light, sound). Give the students other examples of specific changes and ask what forms of energy result. Examples may include burning gasoline in a car engine and using a saw to cut wood.

FOLLOW-UP

Reinforcement: States of Matter
Have students divide a piece of art paper into four sections that represent solid, liquid, gas, and plasma. In each section the students should create a collage to show examples of that state of matter.

Enrichment: $E = mc^2$
Have the students do library research on Einstein's equation $E = mc^2$. They should find out the meaning of each letter and the correct unit for measuring each quantity. Then have the students determine how much energy could be produced from a 5-gram mass, given that $c = 300{,}000$ km/s $(1{,}500{,}000 \frac{g - km}{s})$.

Tie-in: Changing the Freezing and Boiling Points
In the winter it is helpful if antifreeze is added to the engine of a car. This will lower the temperature at which the water in the cooling system will freeze. Ask the students why antifreeze is sprayed on the leading edges of propellers as well as the wings of aircraft (prevents icing).

2.2

OBJECTIVES: The student should be able to

- **define** the term *speed* and **explain** how an average speed is determined;
- **define** the term *velocity* and **explain** how velocity is determined;
- **distinguish** between speed and velocity.

SCIENCE TERMS

speed
velocity

ADVANCED PLANNING

1. Prepare a work sheet of speed problems you can use in class. Use numbers the students can divide in their heads.

2. Prepare work sheets with problems that involve adding and subtracting velocities to find the resulting velocity. Use numbers the students can easily add and subtract in their heads.

LESSON PLAN

1. Introduce the lesson by explaining that whenever there is kinetic energy, movement is associated with it. When objects move, they do so at specific speeds. Mention that your speed coming to school today was 48 km per hour (30 mi per hour).
2. Explain that the average speed is found by dividing distance by time. If Paul drives 30 kilometers in 30 minutes, he drives $\frac{30 \text{ km}}{30 \text{ min}} = 1$ kilometer per minute. If Amy runs 4 kilometers in 16 minutes, she runs $\frac{4 \text{ km}}{16 \text{ min}} = \frac{1}{4}$ kilometer per minute.
3. To reinforce the concept that speed equals distance divided by time, give the students a work sheet of simple problems that they can do orally in class.

● Mathematics Skills: Computation
Problem Solving

2.2 Speed and Velocity

You will find out
- how to get the average speed of an object;
- the difference between speed and velocity;
- how to combine two or more velocities.

How do you describe the effects of energy on matter? To do this, scientists talk about matter that's in motion. They measure how fast a particular piece of matter is moving. For example, scientists need to know how fast a satellite must move in order to stay in a certain orbit. They also need to know how that motion is changing.

Figure 2.8 Over short distances a cheetah can reach speeds of 110 km/hr.

The **speed** of an object is a measure of how fast it is moving. You can find the speed of an object by measuring the distance the object has traveled and dividing by the time it took to travel that distance. For example, if you travel 80 kilometers in one hour, your speed is 80 kilometers per hour. If you travel 160 kilometers in two hours, your speed is still 80 kilometers per hour. Suppose you travel 40 kilometers in half an hour. What would your speed be?

$$\textbf{Speed} = \frac{\textbf{distance}}{\textbf{time}}$$

This formula gives the average speed. It does not tell the whole story of stops and starts, of slowdowns and speedups. Nor does it tell which direction the object is traveling in.

Do You Know?
A woodpecker's beak reaches speeds of 2,000 km per hour.

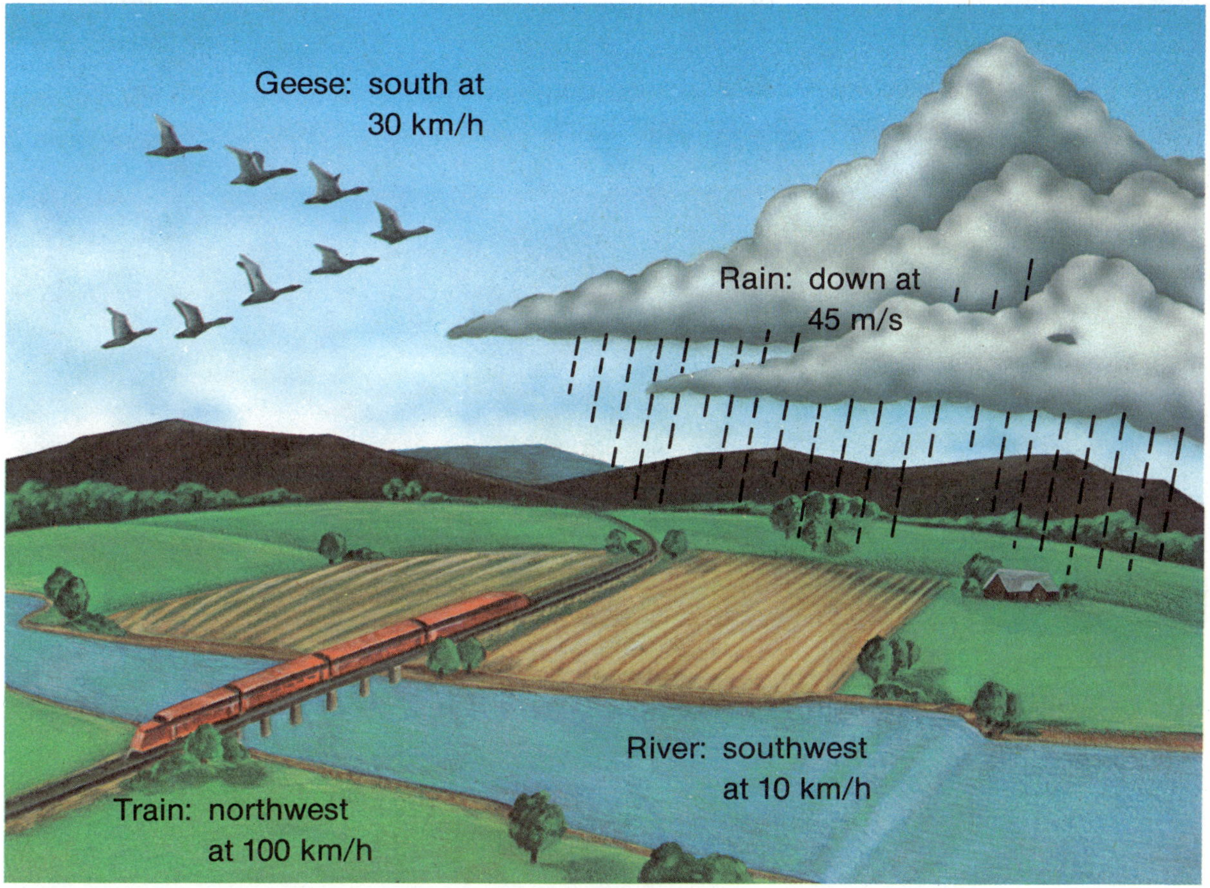

Figure 2.9 Velocities include both speed and direction.

The **velocity** [vuh-LAHS-uh-tee] of a body is a measure of both the speed of the body and its direction of motion. For example, suppose that in one hour's time an automobile travels 80 kilometers north along a highway. The speed of the automobile is 80 kilometers per hour. The velocity is 80 kilometers per hour due north. If the automobile had been traveling east, the speed would have been the same, but the velocity would have been 80 kilometers per hour due east.

It is especially important for air traffic controllers to know the velocities of the airplanes whose motion they control. Their job is to keep the airplanes from running into each other. For this reason they continually tell pilots the positions and velocities of other aircraft in the area.

4. Tell the students that if you say a car travels 32 km per hour (20 mi per hour), you are talking about speed. If you say the car travels 32 km per hour (20 mi per hour) north, you are talking about velocity. Then have the students tell you how speed and velocity differ (velocity gives both speed and direction).
5. Ask a student to give you an expression of speed. Then ask another student to make this into an expression of velocity (16 km per hour; 16 km per hour south). Repeat this five or six times until the students have the idea. Stress that the direction of velocity can be up or down, east or west, north or south.
6. Have the students read the biography at the end of the chapter. Emphasize that calculus is a type of mathematics that deals with the motion of objects. It can be used to calculate velocities.

7. Point out that not all objects move along stationary media the way a girl does as she runs between first and second base. When an object moves through air and water, the velocity of the air and the water must be considered in finding the resulting velocity of the object. If students are confused by this, tell them that all they have to be able to do to determine the resulting velocity is to add!

8. Ask the students to pretend that they are in a boat traveling 64 km per hour (40 mi per hour). The current of the river they are on has a velocity of 16 km per hour (10 mi per hour) south. Ask the students whether the river will help the boat move if the boat is also traveling south (yes). Explain that the velocity of the boat will be 64 km per hour (40 mi per hour) south plus 16 km per hour (10 mi per hour) south or 80 km per hour (50 mi per hour) south.

● Mathematics Skills: Computation
Problem Solving

Figure 2.10 The combined velocities of the boat and the river tell you how fast the boat is moving downriver.

Sometimes you may also need to know how to combine two or more velocities. Suppose, for example, you are on a boat moving down a river that flows at a speed of 10 km per hour. How fast will the boat go downstream? To find the answer, you add the velocity of the boat, 40 km per hour downstream, to the velocity of the river, 10 km per hour downstream, to get 40 + 10 = 50 km per hour downstream. In general, if you want to combine velocities that have the same direction, you simply add the speeds.

Figure 2.11 The combined velocities of the girl and the escalator tell you how fast the girl is moving upstairs.

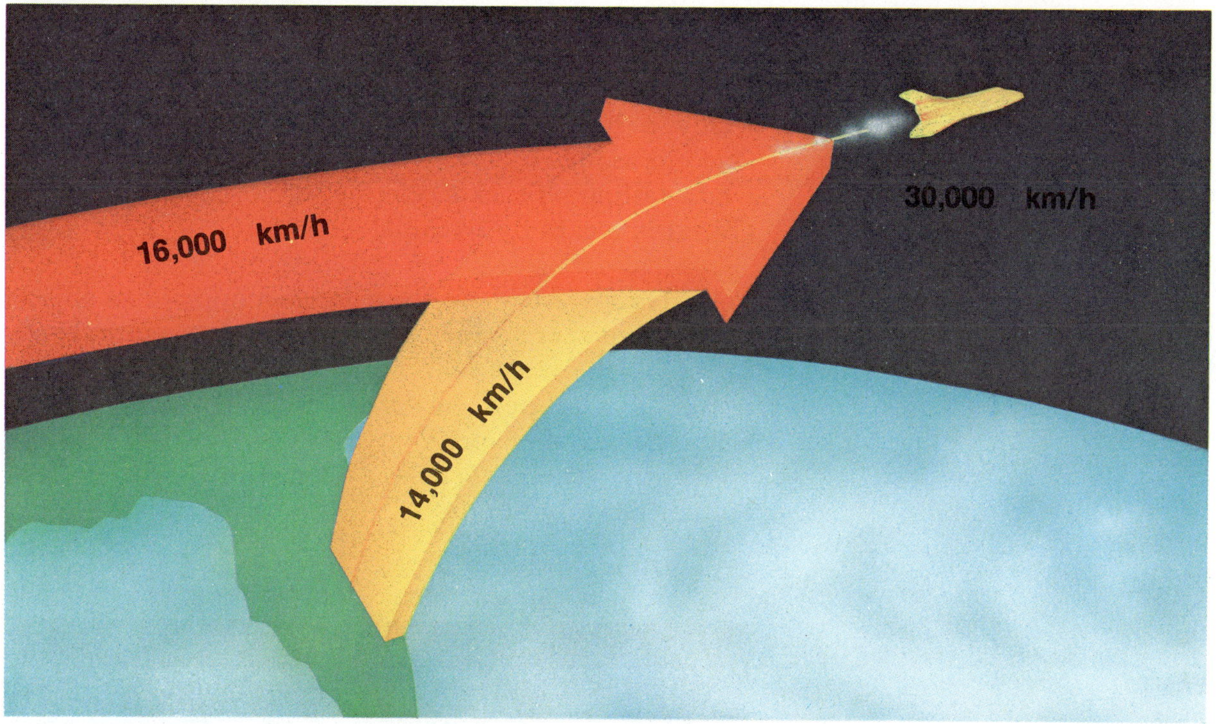

Figure 2.12 A rocket launched in the direction of the earth's rotation picks up an extra 16,000 km/hr.

Space scientists use this addition of velocities to their advantage when they launch rockets in the same direction as the rotation of the earth. By launching a rocket in the direction of the rotation of the earth, scientists pick up an extra 1,600 km per hour, the speed of the rotation of the earth at Cape Canaveral in Florida. This means that less fuel is needed to get up to the speed of about 30,000 km per hour, the velocity required to put a rocket on a course for the moon or beyond.

Study Questions for 2.2

1. What is the average speed of an ice puck that travels 10 m in 5 s?
2. What is the difference between speed and velocity?
3. A plane is moving 640 km per hour east in a wind current that is moving 80 km per hour east. What is the velocity of the plane?

1. Average speed = $\frac{\text{distance}}{\text{time}} = \frac{10\text{ m}}{5\text{ s}} = 2\text{ m/s}$

2. Velocity is also a measure of an object's direction of motion.

3. 640 + 80 = 720 km/h east

9. Distribute a sheet with several simple examples to reinforce the idea that if you are traveling in the same direction as a moving medium the medium will help you along. For example, what is the resulting velocity of a plane traveling 1,152 km per hour south if the wind is moving 160 km per hour south (1,312 km per hour)?

FOLLOW-UP

Reinforcement: Down the Mississippi
Have the students pretend that they are traveling down the Mississippi River. Each day they must travel 80 k in 5 hours. What must their speed be if the speed of the river is 4.8 km per hour (11.2 km per hour)?

Enrichment: For Race Buffs
Have interested students use an almanac to find the speeds of the fastest cars at the Indianapolis 500. Have them display this information on a graph.

Tie-in: Career—Air Traffic Controller
Have the students do library research on air traffic controllers. What is their job at the airport? Why is velocity important to them?

SCIENCE BACKGROUND

The ability to determine the correct velocity is important to airplane pilots and ship captains who must arrive at their destinations on time. Ability to determine correct velocity is also important to these people with respect to fuel conservation. By traveling with the current or the wind, fuel can be saved.

2.3

OBJECTIVES: The student should be able to

- **define** the term *acceleration*;
- **calculate** the acceleration of a given object;
- **define** the term *deceleration*;
- **calculate** the acceleration of a falling object;
- **define** the term *terminal velocity*;
- **explain** why falling objects reach a terminal velocity.

SCIENCE TERMS

acceleration
deceleration
terminal velocity

ADVANCED PLANNING

Prepare a work sheet of acceleration problems that you can use in class with the students. Use numbers that they can easily manipulate in their heads.

LESSON PLAN

1. Review the meanings of the terms *speed* (distance divided by time) and *velocity* (speed in a specific direction). Ask the students the speed of the car or bus that brought them to school. Does this mean that the bus was always traveling at that speed (no)? When might the bus have traveled faster (on straight roads, after it had picked up everyone)? When did the bus travel at a slower speed (in the busy center of town, slowing down to pick up students)? Stress that changes in velocity are spoken of as acceleration and deceleration. Acceleration is an increase in the velocity divided by the time.

2. As an example mention that when a certain train is in the station, its velocity is 0 km per hour (0 mi per hour) north. It will reach a velocity of 64 km per hour (40 mi per hour) north in just 5 seconds. Its acceleration is the change in velocity, 64 km per hour north-0 km per hour north, divided by the time, 5 seconds, or 12.8 km per hour per second (8 mi per hour per second) north. This means that every second the train travels 12.8 km per hour faster in the direction north.

● Mathematics Skills: Computation
Problem Solving

2.3 Acceleration

You will find out
- how to describe the changing velocity of an object;
- why a body moving at a constant speed can be accelerating;
- how falling objects accelerate;
- what is meant by terminal velocity.

Velocity tells you about the average motion of an object during a certain period of time. But it does not tell you about all the stops and starts that happen when the object picks up speed, slows down, or changes direction. This change of velocity divided by the time during which the change occurs is called the **acceleration** [ak-sehl-uh-RAY-shuhn].

Acceleration = change in velocity ÷ time

Figure 2.13 A bobsled accelerates every time it changes either speed or direction or both.

For example, suppose a car moves from a dead stop to twenty meters per second in five seconds. That is, the car went from zero meters per second to twenty meters per second in five seconds' time. If the car moves along a straight path, the average acceleration during this time interval is (20 m/s − 0 m/s) ÷ 5 s = meters per second each second or 4 m/s/s. That is, the car is picking up speed at a rate of four meters per second every second. Suppose now that the driver of the car brakes the car and slows it down. In this case, the car's speed has decreased. This change in velocity to a decreasing speed is called **deceleration.** You can find the deceleration of an object by using the same formula as that for acceleration.

Figure 2.14 Cars and trucks accelerate and decelerate frequently.

Since acceleration measures changes in velocity and velocity has direction as well as speed, a body can be accelerating even if its speed is constant! For example, even if the driver takes a corner at a constant speed, the automobile will be accelerating, since it is changing direction. The sides of racetracks are usually sloped upward to furnish an extra force to accelerate the cars as they go around a curve.

What do autumn leaves, sky divers, and hailstones have in common? They are falling toward the surface of the earth. There is a fact about all falling objects that may surprise you. It was proved by Galileo centuries ago. If you dropped a marble and a bowling ball at exactly the same instant from the top of the Washington Monument, both would hit the ground at the same time. This fact amazes many people who expect the heavy bowling ball to fall faster. Even though the bowling ball is much heavier, both objects fall through the air at the same velocity. Thus, they strike the ground together. That is what Galileo proved when he dropped a cannon ball and a musket ball from the Leaning Tower of Pisa.

Figure 2.15 Objects of different weight fall at the same rate.

3. To reinforce the concept that acceleration is equal to the change in velocity divided by the time, have available a work sheet of problems to be solved in class. For example, what is the acceleration of a motorcycle that increases its velocity from 54 km per hour south to 144 km per hour south in 15 seconds (6 km per hour per second)?
4. Recall with students that on the way to school their bus did not just accelerate. It often had to slow down, or decelerate. Emphasize that acceleration refers to a change in velocity in a certain period of time. Deceleration, on the other hand, is a specific term that refers to a decrease in velocity.
5. To introduce the concept that all falling objects accelerate at the same rate, hold up two objects of different masses (for example, a large dictionary and a small paperback book). Ask which would fall faster if both objects were dropped out of a seventh-floor window (although one object has a greater mass, both accelerate at the same rate 9.8 m/s/s (32 ft/s/s). You may want to refer to the discussion in Chapter 1 of Galileo's demonstration that objects of different masses fall at the same rate.

SCIENCE BACKGROUND

When moving objects travel around corners they are changing their direction. At high speeds this is not easy to do without deceleration. Cars traveling too fast around corners often roll over. Race drivers want neither to decelerate nor to roll over. The sides of racetracks are often built up at a slant to help make the change in direction at high speeds. This gives the cars the extra force to accelerate as they speed around a curve.

6. Remind the students that all falling objects accelerate 9.8 m/s (32 ft/s) each second they are falling. Then ask the students to imagine a baseball dropped from the top of a skyscraper. What is the velocity of the baseball before it is dropped (0 m/s [0 ft/s])? What will be the velocity one second after the baseball is dropped (9.8 m/s [32 ft/s])? What will be the velocity two seconds after it is dropped (19.6 m/s [64 ft/s])?

Falling objects all accelerate in the same way. Any object falling near the surface of the earth will accelerate 9.8 m/s/s. This means the velocity of an object will increase by 9.8 meters per second every second it is falling. Take for example an apple dropped from the roof of a 20-story building. Before the apple is dropped, its velocity is 0 m/s. One second after the apple is dropped, its velocity is 9.8 m/s. After two seconds the apple has accelerated to 9.8 + 9.8, or 19.6 m/s. After another second the apple's velocity is 19.6 + 9.8, or 29.4 m/s. As the apple continues to fall, it continues to accelerate by 9.8 m/s every second. A falling brick would accelerate the same way. So would a nickel, a person, and a sack of potatoes. All falling objects accelerate 9.8 m/s for each second they fall.

Figure 2.16 A falling object accelerates at a rate of 9.8 m/s/s.

Can objects falling through air keep accelerating as long as they are falling? The answer to this question is no. Every falling object, if it is falling long enough, reaches a **terminal velocity.** Once this velocity is reached, the object will not accelerate any more. It will continue to fall at a constant rate, its terminal velocity.

Figure 2.17 An open parachute encounters a large amount of air resistance.

What causes a falling object to reach a terminal velocity and to stop accelerating? To understand the answer to this question, imagine you have dropped a feather and a pebble. The pebble may be heavier than the feather, but all falling objects accelerate at 9.8 m/s/s. Both should reach the ground together. However, the pebble will probably fall faster. It will touch the ground first not because it is heavier but because of its shape. The air an object falls through provides resistance to the falling motion of the object. A feather encounters much more air resistance than a pebble. It falls more slowly. If the feather and pebble were falling in the absence of air, a vacuum, both would fall at the same velocity. There can be no air resistance when there is no air. It is air resistance that causes falling objects to stop accelerating and reach a terminal velocity.

Study Questions for 2.3

1. The velocity at which a car is moving changes from 20 m/s south to 30 m/s south in 10 seconds. Find the acceleration of the car.
2. How can a body moving at a constant speed be accelerating?
3. A package is dropped from a helicopter. Find the velocity of the package after 5 seconds.
4. What causes a falling object to stop accelerating?

1. $\text{Acceleration} = \frac{\text{change in velocity}}{\text{time}} = \frac{30\text{ m/s} - 20\text{ m/s}}{10\text{ s}} = \frac{10\text{ m/s}}{10\text{ s}} = \frac{1\text{ m/s}}{\text{s}}$
2. because acceleration measures changes in velocity, which has direction as well as speed
3. $5 \times 9.8\text{ m/s} = 49.0\text{ m/s}$
4. air resistance

7. To demonstrate air resistance, hold up a pencil and a tissue. Ask which will touch the floor first if you drop them altogether. Drop both objects. Point out that the reason the tissue falls more slowly is not because it is lighter. The shape of the tissue allows for much more air resistance than the shape of the pencil. Air resistance opposes the falling motion of an object. It causes falling objects to stop accelerating after a certain velocity, called terminal velocity, is reached.

FOLLOW-UP

Reinforcement: On the Way to School

Ask some students to have a parent drive them to school. Each student should take notes during the ride that includes the places where the car accelerated, where the speed remained constant, and where deceleration occurred. The student should also keep a record of the speeds traveled on different streets. Have the student use the information to prepare a map for getting to school.

Enrichment: Reading Acceleration and Deceleration Graphs

Have the students prepare a series of graphs to show a car that accelerates and comes to a sudden stop; a car that accelerates, decelerates and stops, and then accelerates again; a car that starts, accelerates, travels at a constant rate, and then decelerates to a stop; and a car that is moving in rush hour stop-and-go traffic.

Tie-in: 88 Kilometers per Hour Saves

Going 88 km per hour (55 mi per hour) on the highway saves fuel. Have a student contact the local EPA office or interview a gas station owner to find out why.

2.4

OBJECTIVES: The student should be able to

- **define** the term *momentum*;
- **calculate** the momentum of an object, given its mass and velocity.

SCIENCE TERM

momentum

ADVANCED PLANNING

Prepare a work sheet of momentum problems that the students can solve in class. Use numbers they can easily multiply in their heads.

LESSON PLAN

1. Introduce the lesson by asking the students to imagine that a tiny pebble and a baseball are thrown with the same velocity. Both the pebble and the baseball hit a plate-glass window. Which object is more likely to break the window (baseball)? Ask why the baseball is more likely to break the window, since both objects have the same velocity (baseball has more mass). Tell the students that when they are considering the mass and the velocity of an object they are speaking about momentum, or the mass times the velocity.

2. Ask the students to calculate the momentum of an object with a mass of 13.6 kg (30 lbs) and a velocity of 3 m per second (9.9 ft per second) (13.6 kg × 3 m per second = 40.8 (kg)(m) per second).

You will find out
- what momentum is;
- how to find the momentum of an object.

Velocity and acceleration tell you a lot about the motion of an object, but they don't tell you the whole story. For example, suppose you rolled a Ping-Pong ball down a bowling alley at a speed of six meters per second. The Ping-Pong ball would just bounce off the bowling pins. On the other hand, a bowling ball moving at six meters per second would send the pins flying!

It is clear from these examples that the velocity and acceleration of an object are not the only important things to know about its motion. You must also consider the **momentum** [moh-MEHN-tuhm]. The momentum of an object is the mass of the object multiplied by its velocity.

Momentum = mass × velocity

An object with a larger mass has more momentum and can exert a larger force than an object of smaller mass even if the velocity of the two objects is equal. This is one reason you cannot bowl a strike with a Ping-Pong ball. Even if a bowling ball and a Ping-Pong ball are traveling at exactly the same velocity, the bowling ball will have much more momentum. It will exert a larger

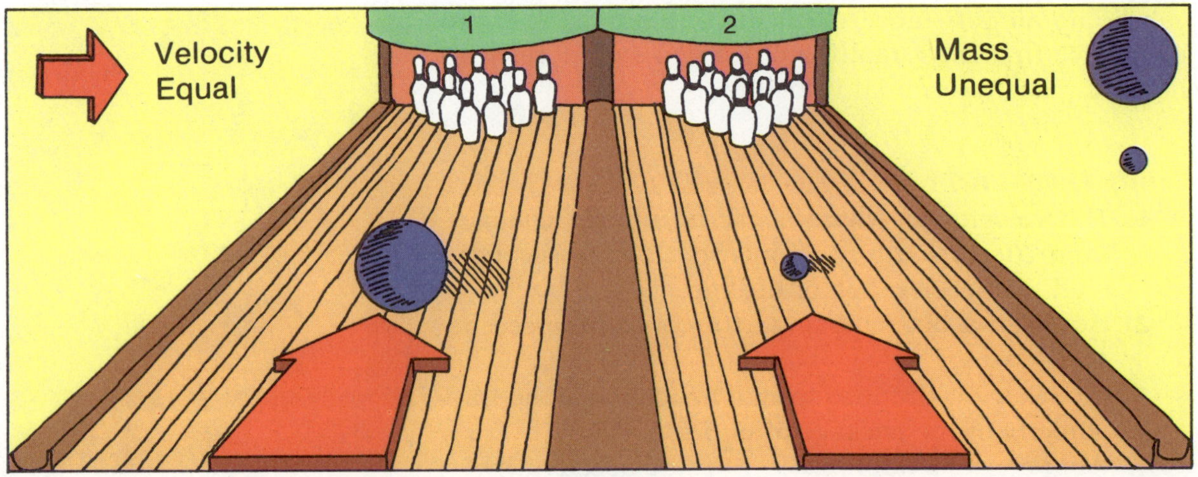

Figure 2.18 The ball with greater mass has greater momentum.

force and be able to knock down the pins. The bowling ball has greater momentum because it has more mass. In general, if two objects are moving with the same velocity, the one with more mass will have greater momentum.

The reverse is also true. If two moving objects have the same mass, the one with the greater velocity will have more momentum. Consider, for example, two bowling balls of the same mass, 6 kg, moving in the same direction. If the first ball has a velocity of 3 m/s down the alley and the second ball has a velocity of 6 m/s down the alley, what is the momentum of each? For the first ball the momentum is

$$6 \text{ kg} \times 3 \text{ m/s, or } 18 \text{ (kg) (m)/s.}$$

For the second ball the momentum is

$$6 \text{ kg} \times 6 \text{ m/s, or } 36 \text{ (kg) (m)/s.}$$

The second ball has greater momentum because it has greater velocity.

Figure 2.19 The ball with greater velocity has greater momentum.

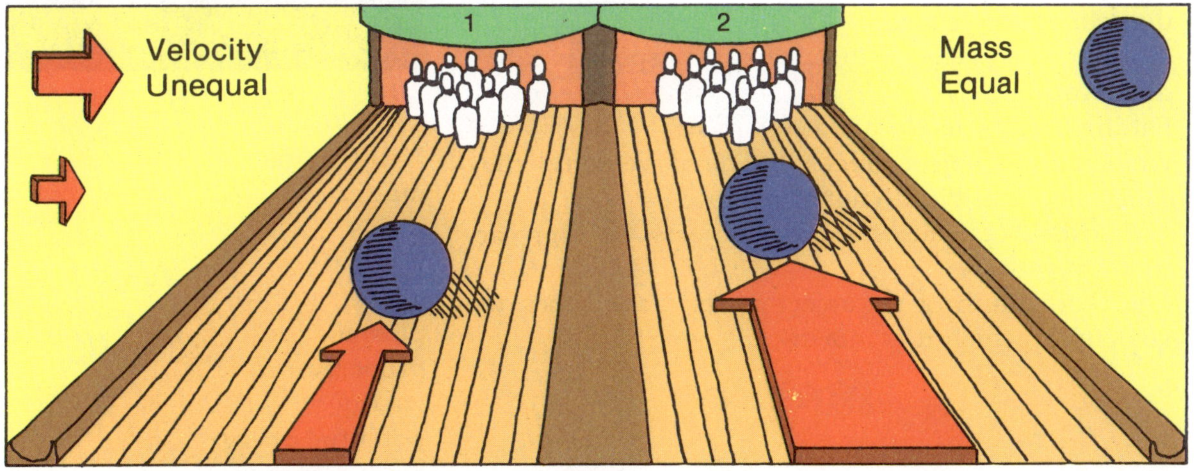

1. What is meant by the momentum of a moving body?
2. What is the momentum of a 3-kg rock moving at 4 m per second?

1. the mass of an object multiplied by its velocity
2. Momentum = mass × velocity = 3 kg × 4 m/s = 12 (kg) (m)/s

3. Provide the students with a work sheet of other examples. For example, what is the momentum of a 3-kg ball that has a velocity of 5 m per second down (15 (g)(m) per second down)?

● Mathematics Skills: Computation
Problem Solving

FOLLOW-UP

Reinforcement: Volkswagon versus Cadillac

Ask the students to solve this problem: If a Volkswagon and a Cadillac are both traveling 128 km per hour (80 mi per hour) and hit a brick wall, which car will do more damage (Cadillac)? Why (more mass)? Under what condition would each do the same amount of damage (Cadillac momentum must equal Volkswagon momentum)?

Enrichment: Car Safety in a Crash

Ask the students to contact local car dealers and find out the masses of cars commonly sold in the city. Students should use this information to determine the momentum of each car when it is traveling 64 km per hour (40 mi per hour). Have the students construct a graph to compare the cars.

Tie-in: Speed, Velocity, and Momentum

Ask the students to create posters to compare speed, velocity, and momentum. The posters should include pictures, definitions, and mathematical statements.

What Did You Learn?

2.1 • Anything that has mass and takes up space is matter.
2.1 • The three common states of matter are solids, liquids, and gases.
2.1 • The fourth state of matter is a high-temperature gas called a plasma.
2.1 • Energy is the ability to move matter and produce changes in it.
2.1 • Energy changes from one form to another but never disappears. For example, potential energy, or the energy due to position, may change to kinetic energy, or the energy of motion.
2.2 • The speed at which an object is moving is found by dividing the distance the object travels by the time it takes to travel that distance.
2.2 • Velocity gives both the speed of an object and the direction in which it is traveling.
2.2 • To combine two or more velocities that are in the same direction, you must add the speeds.
2.3 • An object's acceleration is found by dividing the change in its velocity by the time that is required for that change to occur.
2.3 • The change in velocity to a decreasing speed is called deceleration.
2.3 • Falling objects accelerate at 9.8 m/s/s.
2.3 • When falling objects reach terminal velocity, they continue to fall but do not accelerate.
2.3 • Air resistance causes falling objects to stop accelerating.
2.4 • The momentum of an object is its mass times the velocity at which it is moving.

Key Terms

matter
solids
liquids
gases
plasma
property
energy
potential energy
kinetic energy
speed
velocity
acceleration
deceleration
terminal velocity
momentum

Biography

Gottfried Wilhelm Leibniz (1646–1716)

Leibniz was a German mathematician noted for his idea of the kinetic energy of motion. Today he is even more famous for his invention of calculus. Calculus is a mathematical method of dealing with rates of change, such as accelerations, for complicated motions. Today calculus is used in every field of physical science to calculate velocities and accelerations and to deal with other changing quantities.

Leibniz was not the only inventor of calculus! Isaac Newton, an Englishman who lived at the same time as Leibniz, also invented a method for calculating rates of change that was almost identical to Leibniz's. It was apparently a coincidence that both men came up with the same idea at the same time. Neither man believed it was a coincidence, however. Each accused the other of stealing his ideas. The quarrel became so bitter that English and German scientists joined in the dispute on the sides of their countrymen. An international commission of scientists was then set up to investigate. The commission decided that both men had made their discoveries independently. Nevertheless, the bad feelings between Leibniz and Newton remained until they died.

Leibniz was involved in politics and philosophy as well as science. He got into many bitter arguments in these fields as well. As a result, he had few friends. When Leibniz died at the age of 70, only one person attended his funeral. This happened even though Leibniz was one of the most brilliant scientists in Germany.

Vocabulary

1. f.
2. i.
3. h.
4. b.
5. j.
6. c.
7. g.
8. e.
9. a.
10. d.

● Word Skill: Vocabulary

What Do You Remember?

1. solid—has definite shape and volume; liquid—takes shape of container, has definite volume; gas—has neither definite shape nor volume
2. plasma
3. Answers will vary. Examples: Burning fuel (heat and chemical energy) is used to produce steam to run a generator that produces electrical energy. Radiant energy from the sun can be converted to electrical energy. The explosion of an atomic bomb (nuclear energy) produces heat and radiant energy.
4. average speed $= \frac{\text{distance}}{\text{time}} = \frac{192 \text{ km}}{8 \text{ h}} = 24$ km/h
5. 48 km/h south + 16 km/h south = 64 km/h south
6. $(88 \text{ km/h} - 16 \text{ km/h}) \div 6 \text{ s} = 72 \text{ km/h} \div 6\text{s} = 12 \frac{\text{km/h}}{\text{s}}$
7. The truck has a greater mass than the car.

TO THINK ABOUT AND DO

Vocabulary ●

Match each term in column A to the phrase that describes it in column B.

A	B
1. acceleration	a. anything that has mass and takes up space
2. kinetic energy	b. matter that has a definite shape and volume
3. plasma	c. distance divided by time
4. solid	d. ability to move matter or to change it into another substance
5. gas	e. mass times velocity
6. speed	f. the change in velocity during a given time
7. velocity	g. property of motion that takes speed and direction into account
8. momentum	h. a fourth state of matter found in the outer layers of stars
9. matter	i. energy of object in motion
10. energy	j. state of matter with no definite shape or volume

What Do You Remember?

1. Name the three common states of matter. Describe the shape and volume of each state.
2. What is the name of a fourth state of matter making up the outer layers of stars?
3. Give an example of the way energy may change from one form to another.
4. A car travels 192 kilometers in 8 hours. What is its average speed?
5. A boat is sailing at 48 km per hour downstream on a river whose current is moving 16 km per hour downstream. What is the resulting velocity of the boat?
6. A car increases its velocity from 16 km per hour to 88 km per hour in six seconds. What is its acceleration?
7. Why does a truck traveling 88 km per hour have greater momentum than a car traveling at the same speed?

Applying What You Have Learned ●

1. A rock, a can of orange soda, and a helium balloon are all alike in some ways and different in others. How are they alike? How are they different?
2. Paul drove to the store with his mother. This graph shows the speeds at which the car moved. Use the graph to explain what happened in terms of acceleration, deceleration, and constant speed.

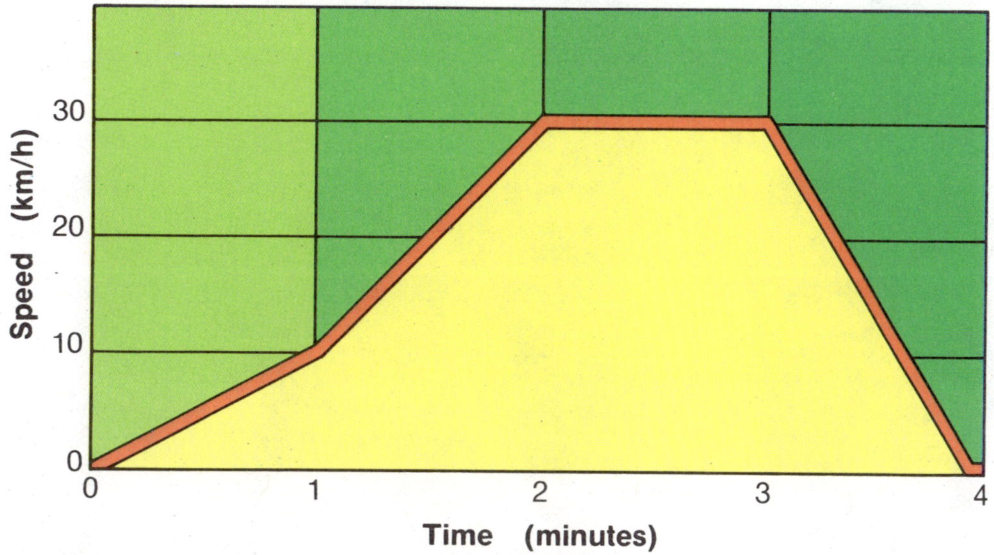

Research and Investigation ●

1. Parachutes are used to decrease the speed of falling objects. They can also be used to slow down race cars. How does a parachute reduce the speed of a fast-moving object? Will a larger parachute reduce the speed more than a smaller one?
2. If a balloon is put into a bell jar and a vacuum pump is used to remove the air from the bell jar, what will happen to the balloon? Why? What will happen if air is pumped into the bell jar? Why? What does this tell you about the volume of a gas?

Applying What You Have Learned

1. All three have mass and volume. They are all forms of matter. They differ in state of matter, and in many properties, such as color, taste, hardness, and smell.
2. The car was stopped. It accelerated to a constant velocity and then slowed down and stopped.

● Comprehension Skill: Applied

Research and Investigation

1. The increased surface area from the open parachute causes increased air resistance to motion. Yes, since a larger parachute has more surface area than a smaller one, there is greater air resistance.
2. As air is removed from the bell jar, the balloon will expand. This is because the removal of air from the jar results in a decrease in air pressure outside the balloon and an increase in pressure inside the balloon. As air is pumped into the bell jar, the balloon will decrease in size. Now air pressure is increased outside the balloon and decreased inside it.

● Reference/Research Skills

CHAPTER BACKGROUND

Although people generally think of motion as movement from one place to another, the scientific definition of motion is a change in the position of one object relative to another object. Acceleration is caused by an imbalance of forces, the acceleration itself occurring in the direction that the larger force is occurring.

In the seventeenth century Isaac Newton observed the motions of the planets, did countless experiments with falling objects, and studied the work of other scientists. The observations and conclusions that Newton drew were so accurate that scientists still refer to them as Newton's laws of motion.

Newton's first law of motion states that objects at rest remain at rest and those in motion remain in motion at constant velocity until an outside force acts on them. This property of matter is called inertia. People observe inertia daily when they are pushed back against the seat as the car starts and find themselves touching the dashboard when a car stops a bit too quickly. In the first case the people have remained at rest while the car moved. In the second they have continued to move after the car has slowed down abruptly.

Newton's first law explains what happens to an object if it is in a state of equilibrium, that is, when all the forces acting on the object are in balance, or add up to zero. Newton's second law explains what happens to an object when it is not in a state of equilibrium, that is, when the forces acting on the object are not in balance. This law states that an unbalanced force acting on an object will produce an acceleration of that object proportional to and in the same direction as the applied force. The law means that if you double the force with which you throw a baseball, for example, you will double the ball's acceleration. The law implies that as the mass increases the acceleration decreases if the force is kept constant.

Newton's third law, the law of action-reaction, states that for every action on one object there is an equal but opposite reaction on another object. The reaction is equal in magnitude but opposite in direction to the action. Jet planes, rockets, and rotary lawn sprinklers exemplify this law. Newton's laws describe motion on the earth and also help scientists to understand the movement of bodies in space.

3 Forces

Introduce the chapter by telling the students that there are forces all around them that influence their daily activities. Have the students imagine a baseball game, for example. Point out that many forces are involved in allowing the players to play ball. Define a force as any push or pull that affects matter in some way. In the baseball game examples of matter are the bases, players, bat, ball, gloves, and the baseline (ground). Ask the students to identify some of the forces (pushes or pulls) in the game (friction in sliding into safety, bat hitting ball, force of the arm's swing in pitching, swing of the bat, force of ball hitting glove). Tell the students that in this chapter they will study forces and the laws that govern the motion that forces cause.

What holds bridges up but can also cause bridges to collapse? What causes cars to speed up as well as slow down? What drives electric current through wires and water through pipes? What pushes you into the air when you jump and causes you to come down to earth again? Forces do! Forces give energy to matter and put it into motion.

The action of forces explain many circumstances that are familiar to you. When you ride up in an elevator, the elevator floor presses hard against the bottom of your feet. When the elevator goes down it feels as if the floor is pulling away from your feet. If the car you are riding in stops suddenly, your body lurches forward. If that same car rounds a corner at a relatively high speed, your body leans to the side. When you blow up a balloon and release it, the balloon shoots through the air as it empties and falls.

In this chapter you will find out how the idea of forces helps in understanding why things happen the way they do.

Figure 3.1 The Golden Gate Bridge in California

Lesson Organizer

3.1 Gravitational Forces **Pages** 48–51
Advanced Planning: Page 48
Follow-up: Page 51
- **R** Comparing Gravitational Force
- **E** Your Weight in Outer Space
- **T** How Do Astronauts Keep from Floating Away?

3.2 Electric Forces **Pages** 52–53
Advanced Planning: Page 52
Follow-up: Page 53
- **R** Electric Charges
- **T** Electric Storms
- **E** An Electroscope

3.3 Balance of Forces **Pages** 54–57
Advanced Planning: Page 54
Activity: How Is the Center of Gravity Found? Page 56
Follow-up: Page 57
- **R** A Great Tug-of-War
- **E** Center of Gravity
- **T** Circus Tightrope Walker

3.4 The First Law of Motion **Pages** 58–59
Advanced Planning: Page 58
Follow-up: Page 59
- **R** A Trick with a Penny
- **E** Car Safety
- **T** Kilometers per Liter

3.5 The Second Law of Motion **Pages** 60–61
Follow-up: Page 61
- **R** A Second Law Mobile
- **T** Rockets
- **E** Rube Goldberg

3.6 The Third Law of Motion **Pages** 62–63
Advanced Planning: Page 62
Activity: What Is Newton's Third Law? Page 62
Follow-up: Page 63
- **R** Rotary Sprinklers
- **T** Off to the Moon
- **E** Skiing

See ACTIVITIES book pages 13–18 for laboratory, skill, and review exercises for Chapter 3.

3.1

OBJECTIVES: The student should be able to

- **define** the term *force*;
- **define** the term *gravitational force*;
- **describe** how gravitational forces change with mass and distance;
- **state** Newton's universal law of gravitation;
- **define** the term *weight*;
- **state** one reason why the weight of the same object can vary at different locations on the earth;
- **explain** the difference between weight and mass.

SCIENCE TERMS

force	weight
gravitational force	newton
universal law of gravitation	

ADVANCED PLANNING

For class demonstration you will need a book, a pencil, and a piece of chalk.

LESSON PLAN

1. Introduce the lesson by asking the students to observe as you drop a book onto your desk. Then drop a pencil and then a piece of chalk. Ask the students to tell you what they observed (the objects fell). Ask them what made the objects stop (each hit something). Ask them why the objects fell (gravitation). Ask the students what gravitation is. They will probably answer that gravitation pulls all things to the earth.

3.1 Gravitational Forces

You will find out

- what is meant by the universal law of gravitation;
- how gravity changes with mass and distance;
- how weight and mass are different.

A **force** is a push or a pull that is exerted on matter. Whenever the car, plane, or elevator in which you are riding changes its rate of motion, you may feel this push or pull. You could say that force is the cause of motion. It speeds things up, slows things down, and pushes things around corners.

Physical scientists have identified several basic forces in nature. In this lesson you will learn about one of the most important forces—**gravitational force.** Gravitational force is the attraction between the earth and all the objects on it. When you drop a book, for example, gravitational force causes the book to fall to the ground rather than to remain floating in the air. When you jump off a diving board, gravitational force causes you to fall into the water.

Figure 3.2 Gravitational force moves skiers downhill and swimmers into the water.

Figure 3.3 Gravitational force keeps the moon in orbit around the earth.

Gravitational force acts between all objects in the universe as well. The gravitational force between the sun and the earth, for example, keeps the earth in its orbit. In the same way, the gravitational force between the earth and the moon keeps the moon in its orbit. It took the genius of the English mathematician Isaac Newton to realize that the force between the earth and the moon acts in the same way as the force between the earth and an apple. In fact, Newton is said to have come up with this idea while sitting under an apple tree and watching an apple fall to the ground.

2. Tell the students that although their definition of gravitation is correct it is not complete. Explain that the English scientist, Sir Isaac Newton, described gravitational pull, or gravitational attraction, in his universal law of gravitation. The law states that a gravitational pull or attraction exists between any two objects in the universe.

3. Mention the story that Newton discovered the universal law of gravitation while observing an apple as it fell to the ground. Tell the students that according to the universal law of gravitation the earth attracted the apple and the apple also attracted the earth. Ask why, if the apple is exerting a gravitational force on the earth, the earth doesn't jump up to meet the apple. Instead the apple falls down to the earth. Ask the students which has more mass (earth). Then point out that there is a direct relationship between mass and gravitational force. Objects with greater mass exert a greater gravitational pull. In the case you have just examined, the mass of the earth was millions of times greater than the mass of the apple.

SCIENCE BACKGROUND

Newton's universal law of gravitation states that every object in the universe attracts every other object with a gravitational force. This accounts for the fact that there is a gravitational interaction between the planets and the sun, which causes the planets to be continuously deflected from their otherwise straight-line paths.

Newton's **universal law of gravitation** states that the force of gravitation is present between any two objects in the universe. The amount of gravitational force that one object exerts on another depends on two things: (a) the masses of the objects and (b) how far apart they are.

If you cruised around the solar system, passing by each planet at the same distance, you would soon find that the gravitational pull of a massive planet such as Jupiter is far stronger than a less massive one such as Pluto. The strength of an object's gravitational pull depends on its mass.

As you travel away from the earth, you are affected less and less by the earth's gravitational pull. As your distance from earth increases, the gravitational pull decreases. If you travel far enough, you will not be affected

Figure 3.4 The amount of gravitational force increases as the mass of the objects involved increases.

by the earth's gravitational pull at all. In other words, gravitational pull between two objects depends on the distance between the objects. The closer the objects, the stronger is the gravitational pull.

Objects anywhere on the earth's surface are about the same distance from the center of the earth. For this reason, the earth's gravitational pull on an object depends only on the object's mass. The greater the object's mass, the greater is the gravitational pull. The measure of this pull on an object is the object's **weight.** When you stand on a scale, you are measuring the amount of gravitational pull on your mass. Scientists use the **newton,** abbreviated N, as the unit of weight.

It is important to remember that weight and mass are not the same. Mass is a measure of the amount of matter in an object. Mass is measured on a balance, not a scale, and the units are grams. The mass of an object is not changed by changes in gravitational force. Weight is affected by gravity. It increases when the gravitational force increases and decreases when the gravitational force decreases.

If you go on a diet, you say you want to lose weight. That's not really what you want. What you really want is to lose mass. You would lose a lot of weight very quickly if you could leave Earth and go to the moon. The moon's gravitational pull is weaker than the earth's so your weight would be less. However, you wouldn't look any thinner. Your mass would still be the same because the amount of matter in your body didn't change. Your clothes would fit just as they did on earth where you weighed more.

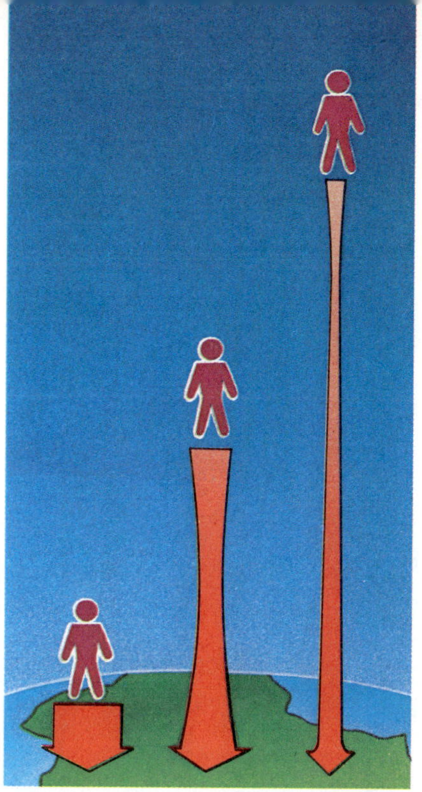

Figure 3.5 The amount of gravitational force decreases as the distance between the objects involved increases.

Do You Know?
A locust can jump more than 17 times the length of its body. In only a fiftieth of a second, the muscles in its two rear legs must exert a force more than a hundred times its weight.

Study Questions for 3.1

1. According to the universal law of gravitation, which objects have gravitational pull?
2. Explain why the gravitational pull at 12,000 km from the center of the earth is only a fraction as strong as at the surface of the earth.
3. How is weight different from mass?

1. all objects
2. As distance from the center of the earth increases, gravitational pull decreases.
3. Weight changes when gravity changes. Mass is not affected by gravity.

4. Emphasize to the students that just as they measure length in meters, volume in liters, and mass in kilograms there is a special unit for measuring force. Force is measured in newtons. The newton, abbreviated N, is the unit of force in the metric system. When a spring scale is used, gravitational force is being measured; thus force can be measured in newtons by using a spring scale.
5. Stress the differences between weight and mass. Weight is measured in newtons on a scale. Mass is measured in grams on a balance. Weight is affected by gravity. Mass is not.

● **Skill: Comprehension/Interpretative**

FOLLOW-UP

Reinforcement: Comparing Gravitational Force
Have students use a balance to find the mass of three or four objects, such as a paper clip, a book, a rock, and a roll of tape. After they have recorded each of the masses, ask the students to tell you which object, based on its mass, has the largest gravitational force between it and the earth and which has the smallest.

Enrichment: Your Weight in Outer Space
As the distance between objects increases, gravitational pull decreases. If the distance is doubled, the force of attraction between them is one fourth of what it was originally. If the distance is tripled, the gravitational force is one ninth what it was originally. Ask the students to calculate the weight of an astronaut as he travels farther from the earth. On the earth's surface (6,000 km from the center of the earth) he weighs 980 newtons. Hint: How much will he weigh when he is 1,300 km from the center of the earth?

Tie-in: How Do Astronauts Keep from Floating Away?
Ask the students to do research to learn how the astronauts performed such simple tasks as brushing their teeth and washing while in space. How did they move around inside the cabin?

3.2

OBJECTIVES: The student should be able to

- **describe** electric forces and the behavior of charged objects;
- **describe** electrically neutral objects in terms of their charge.

SCIENCE TERMS

electric forces	attract
charge	repel

ADVANCED PLANNING

For class demonstration you will need an inflated balloon, a rubber rod, a piece of wool, two pith balls or two pieces of paper attached to two pieces of thread, and a ring stand to suspend the pith balls. Before class attach the pith balls by silk or nylon strings that measure 20–30 centimeters (8–12 in.) to the ring stand.

LESSON PLAN

1. Rub an inflated balloon on your hair and stick it onto the wall. Then rub a hard rubber rod on a piece of wool for several seconds, put the wool aside, and bring the rod close to the pith balls (the balls should be attracted to the rod). Pull the rod out of the reach of the pith balls and again bring it toward the balls (they should move away from the rod). Ask the students what they have observed with respect to the wool, rod, and pith balls. Touch the balls with your fingers to discharge them. Tell the students that what they have seen is the result of electric forces. Explain that at first the rod had no charge. There was an equal number of positive and negative charges on the rod; therefore, the rod was neutral.

● Word Skill: Word Structure

SCIENCE BACKGROUND

Electrically charged bodies exert forces on each other. When the charged bodies are at rest, the forces are called electric forces, or coulomb forces. Scientists believe electric fields are responsible for these forces. When the charged bodies are moving, forces in addition to electric forces are present. These additional forces are magnetic forces.

3.2 Electric Forces

You will find out

- how like and unlike charges behave toward each other;
- how electric forces affect matter.

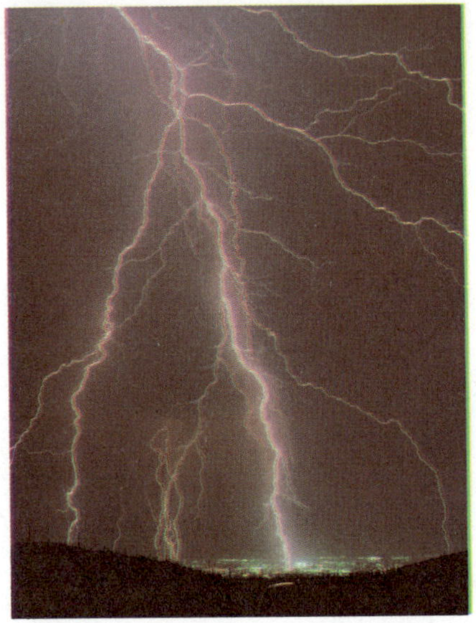

Figure 3.6 Lightning results from the action of electric forces.

Gravitational force is not the only force that can act between objects. To find out how another force behaves, run a comb through your hair or across a wool sweater. Then hold the comb just above some tiny pieces of torn newspaper. The paper is attracted to the comb. This attraction is caused by **electric forces.** The ancient Greeks observed the behavior of electric forces when they rubbed amber, a hard yellow-brown resin, with a cloth. The amber then attracted bits of straw. Our word ● *electricity* comes from the Greek word for amber *elektron*.

You have just seen that electric forces cause objects such as the comb and the paper to be attracted to each other. Whenever an object is affected by electric forces in this way, scientists say the object has acquired an **electric charge.**

In most cases objects are made up of particles with equal numbers of positive and negative electric charges. Scientists call objects with equal numbers of positive and negative charge **electrically neutral objects.** Sometimes, however, this balance between equal numbers of

Figure 3.7 Electric forces can affect your hair.

positive and negative charges is upset. This happened, for example, when you ran the comb through your hair. The comb got some negative charge from your hair. The comb, then, ended up with more negative than positive charge, or a net negative charge.

The electric force on an electrically neutral object is zero. If two objects have opposite net charges, the electric forces will cause them to be pulled toward each other. In other words, opposites **attract.** In the case of the comb and the paper the excess negative charge was attracted to the opposite, or positive, charge in the paper. On the other hand, two objects may have the same net charges. In this case, they will push each other apart, or **repel** each other.

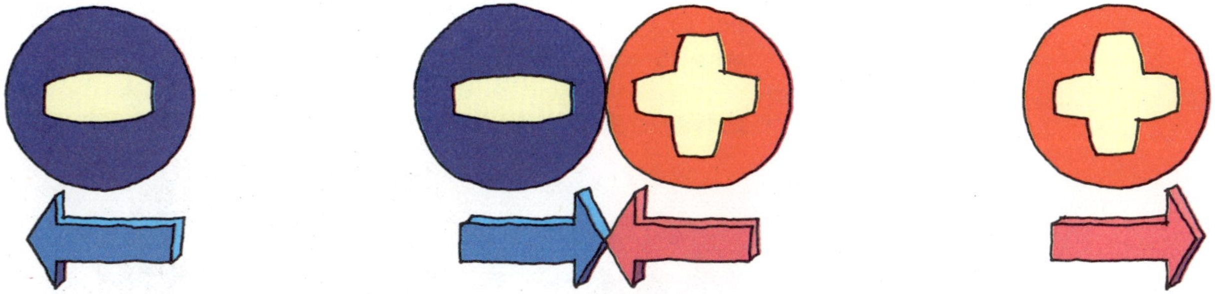

Figure 3.8 Like charges repel each other. Opposites attract.

In most objects, the number of positive and negative charges is about equal, so the object is almost electrically neutral. For this reason, the electric forces between two objects are usually small. Electric forces are still important, though, because they also determine the forces *inside* matter. For example, electric forces give steel its strength and rubber bands the ability to stretch. Electric forces cause glue to stick and muscles to tense. In fact, most of the forces that are not due to gravity can be traced to the electric forces between the charged particles that make up matter.

Study Questions for 3.2

1. Do like charges attract or repel? Unlike charges?
2. What are three examples of ways that electric forces affect matter?

1. repel; attract
2. cause muscles to tense, give rubber bands ability to stretch, cause glue to stick

2. Explain that when you rubbed the rod with the wool, you upset the balance of positive and negative charges and caused the rod to become charged. The balls were also neutral to start with. Their positive and negative charges were equal. When you put the rod near the balls, they were attracted to it because they were neutral and the rod was charged. While the rod and the balls were touching, some of the electric charge leaked off the rod onto the balls. Thus, the balls also became charged. After you removed the rod and brought it close to the balls again, the balls moved away because the two objects had the same charge. This demonstrates the fact that like charges repel each other.
3. Return to the balloon that you placed on the wall at the beginning of class and ask the students to explain how, in terms of electric charges, the balloon is stuck to the wall (the wall and balloon have unlike charges and are attracted to each other).

FOLLOW-UP

Reinforcement: Electric Charges

Have students make posters to illustrate the conditions under which electric attraction, repulsion, and neutrality occur. Students may want to use animated charges to show what happens, or they may want to construct a chart to illustrate the three categories.

Enrichment: An Electroscope

Scientists use electroscopes to investigate electric charges. Ask the students to look up the topic *electroscope* (*World Book Encyclopedia* is a good reference), build an electroscope, and demonstrate how it works.

Tie-in: Electric Storms

Ask the students to do research and write reports about electric storms. Students should include what causes such storms, what direction the electric current flows in, and how people can protect themselves and their homes from electric storms.

3.3

OBJECTIVES: The student should be able to

- **define** the term *net force;*
- **explain** the meaning of forces being in equilibrium;
- **explain** why it is important for forces to be in equilibrium;
- **determine** the center of gravity.

SCIENCE TERMS

net force	stable equilibrium states
equilibrium	center of gravity

ADVANCED PLANNING

You may want to obtain the movie, *St. Mary's Bridge*, from the Federal Highway Administration, National Highway Institute (HHI-3), Washington, D.C. 20590.

LESSON PLAN

1. Introduce the lesson by asking the students to imagine a tug-of-war between two boys who can each pull with a force of 300 newtons. Ask the students who would win and why (neither, they are evenly matched). Explain that when the forces are evenly matched you say that the boys are in a state of equilibrium. Point out that one boy is pulling with a force of +300 newtons while the other boy is pulling with a force of −300 newtons. Explain that you have called the forces + and − to indicate that they are in different directions. Explain that when you determine the sum of the forces you are finding the net forces. Here the net force is zero because +300 plus −300 is equal to 0.

3.3 Balance of Forces

You will find out
- when a state of equilibrium exists;
- how to locate an object's center of gravity.

Have you ever played tug-of-war? What happens when your team applies more force than the other team? The rope and the other team are pulled forward. Motion occurs because the pulling forces are no longer balanced. As long as your team and the other team pull with the same force but in opposite directions, neither the rope nor the teams move. These equal forces in opposite directions have canceled each other. In other words, the sum of these opposing forces is zero. The sum of the opposing forces is called the **net force.** When the net force is equal to zero, scientists say that a state of **equilibrium** [ee-kwuh-LIHB-ree-uhm] exists.

Figure 3.9 The opposing forces of the dam and the water cancel each other. A state of equilibrium exists.

Understanding how to get forces in equilibrium and how to keep them there is very important. An engineer who designs bridges must make sure that the bridge can provide enough force to balance the weight of the bridge and the weight of any vehicles that might cross over it. A dam must be strong enough to withstand the pressure of the water that the dam is holding back.

If the net force on an object is not zero, the object will start to move. In fact, the moving object will accelerate until another force comes into play and brings about an equilibrium. For example, a book resting on a desk top is in equilibrium because the gravitational force pushing down on the book is balanced by the desk top pushing up on the book. In other words, all the forces acting on the book add up to zero.

Suppose you exert another force on the book in order to lift it. The equilibrium will be destroyed, and the book will move.

Figure 3.10 *left:* A state of equilibrium exists. *right:* The state of equilibrium is destroyed.

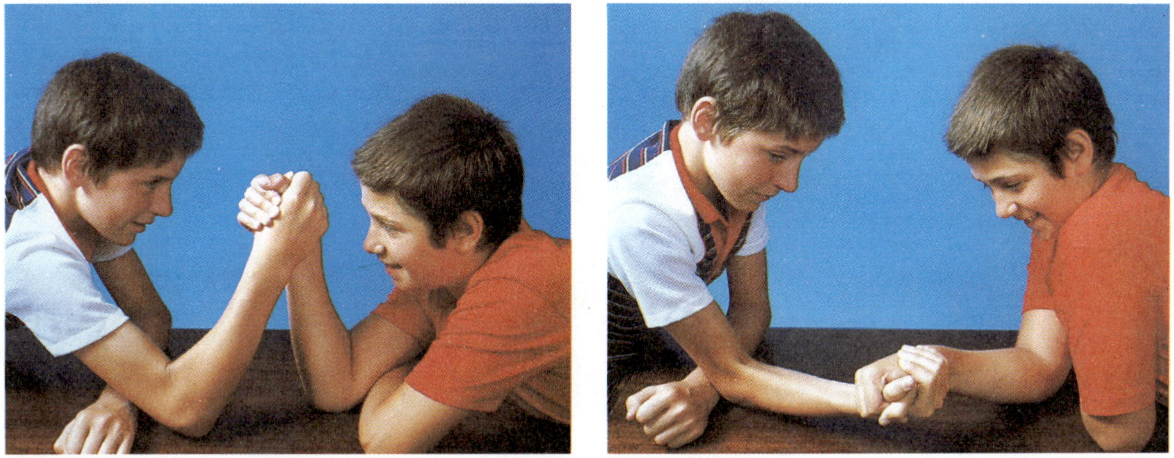

The earth's gravitational force pulls downward on every particle in an object. In order to find the forces necessary to keep an object in equilibrium, you do not need to consider every particle in that object. You can always assume that all the gravitational forces act on the object at a single point, the object's **center of gravity.** This is the point around which an object's weight is evenly placed.

2. To reinforce the concepts of net force and state of equilibrium, put other examples on the board, e.g., a. +890 + (−890), b. +456 + (−123), c. +234 + (−234), and d. +980 + (−762). Ask the students to determine the net force for each example (a. 0, b. 333, c. 0, d. 218). Which examples indicate a state of equilibrium (a and c)?

3. Tell the students that engineers who design and build bridges must be sure that the structure of the bridge will be able to exert enough force to balance the bridge so that it won't collapse. A nine-minute movie, *St. Mary's Bridge*, would be of interest at this point. The movie shows the tests that were performed to determine whether or not St. Mary's Bridge was safe after its sister bridge, Point Pleasant Bridge, collapsed because of unbalanced forces.

SCIENCE BACKGROUND

In order to construct bridges safely engineers must find stable equilibrium states, that is, states of equilibrium that cannot be destroyed by weak forces. To do this, engineers must be able to determine the center of gravity of the bridge. The center of gravity is the single point where the total weight of the object is considered to act.

ACTIVITY: How Is the Center of Gravity Found?

◇ **Caution:** Since the nail is a pointed object the students must be careful with it as with any other pointed object.

This may be the first experience students will have with clamps. Some students may need help adjusting them.

Emphasize to the students the importance of punching the holes away from the edge in order to avoid ripping the cardboard objects.

● **Laboratory Skills: Experimentation/Demonstration Manipulation of Materials**

● ACTIVITY

How Is the Center of Gravity Found?

Materials

cardboard
scissors
nail
ring stand
clamp
thread
weight such as a paper clip or pebble
paper punch
ruler

Procedure

1. Cut a piece of cardboard into a shape similar to the one on this page.

2. Around the edge of the cardboard, punch five holes, evenly spaced. Number the holes 1 to 5.

3. Try to balance the cardboard flat on your fingertip.

4. Cut a piece of thread 30 cm long.

5. Tie the weight to one end of the thread and make a loop in the other end.

6. Set up the ring stand, clamp, and nail as shown.

7. Place the cardboard on the nail at hole 1.

8. Place the loop of the thread in front of the cardboard on the nail.

9. Make a dot where the thread hits the bottom of the cardboard.

10. Remove the thread and cardboard and use a ruler to draw a straight line from the dot to hole 1.

11. Repeat steps 7 through 10 using each of the four remaining holes.

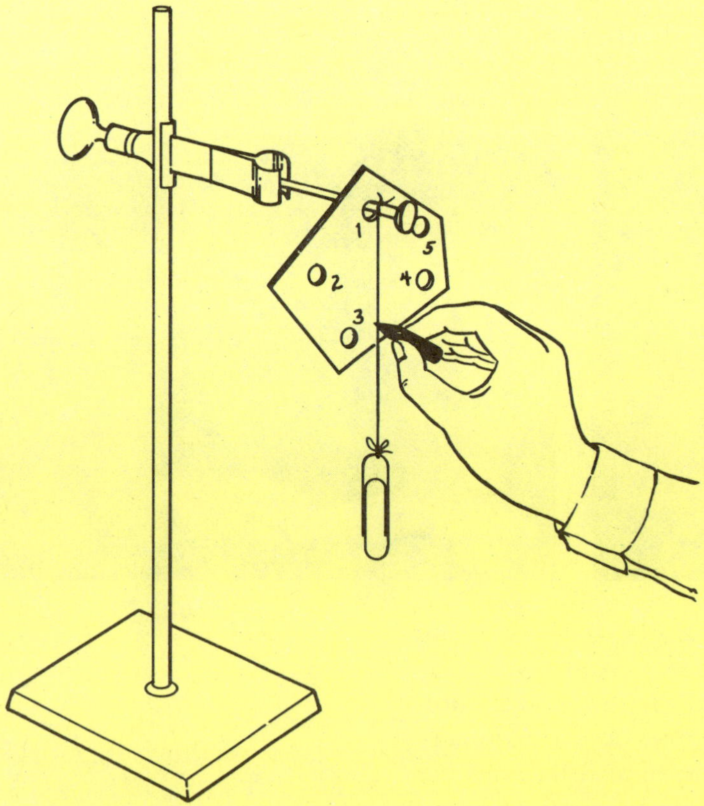

Questions

1. What do you notice about the five lines that you drew? Try to balance the cardboard on your fingertip by placing your fingertip at the place where the five lines cross.

2. What point have you found?

3. Why do objects balance at their center of gravity?

1. Lines all cross at a single point.
2. center of gravity
3. State of equilibrium is reached, and all forces equal zero.

The center of gravity of a regularly shaped object, such as a cube or a sphere, is easy to find. It is in the exact center of the object. The center of gravity of an irregularly shaped object is more difficult to find. It is the place on the object where the object can be balanced without any tendency to rotate.

In a stable equilibrium an object is usually supported with its center of gravity as low as possible. If the object is not supported in this way, a small force will cause the body to move to a new position where its center of gravity is lower. A top-heavy vase, for example, tips over because its center of gravity is high. More of its weight is near the top of the vase than is safe for it to be stable. The center of gravity of a weighted punching bag is low, because most of its weight is below the middle. It does not easily tip all the way over.

Figure 3.11 The athlete's center of gravity changes as her position changes.

Study Questions for 3.3

1. Does a state of equilibrium exist in a tug-of-war game in which one team pulls with a force of 150 N and the other team pulls with the same force in the opposite direction? How can you tell?
2. How do you know when you have found an object's center of gravity?

1. yes; balanced forces, no movement
2. You have found the place on an object where the object can be balanced without any tendency to rotate.

4. Conclude the lesson by asking the students to create two diagrams. One should illustrate forces in a state of equilibrium. The other should illustrate how the center of gravity can be determined and why it is important. If the diagrams are done for homework, have the students pass them in before school the next morning. Look through them and use the best ones to review the concepts of equilibrium, center of gravity, and state of equilibrium in class the next day.

FOLLOW-UP

Reinforcement: A Great Tug-of-War

Have students create two pictures of a tug-of-war. In each picture both teams should have ten members. In the first picture forces should be assigned to each member in a manner that will allow one side to win. In the second picture forces should be assigned in a manner that will not allow either side to win.

Enrichment: Center of Gravity

Set up a center of gravity corner in your room. In it place toys that balance on a single point. Challenge the students to design experiments to prove that the balancing point is the center of gravity. You may also want to use these toys as models in order to challenge your students to create their own balancing wonders.

Tie-in: Circus Tightrope Walker

Ask the students to pretend that they are tightrope walkers. Have them write reports to explain how they would use the concepts of center of gravity and equilibrium to prevent themselves from falling off the high wire.

3.4

OBJECTIVES: The student should be able to

- **name** a force that slows down or stops objects in motion;
- **explain** the meaning of the term *inertia;*
- **state** the law of inertia.

SCIENCE TERMS

friction
inertia
first law of motion

ADVANCED PLANNING

You will need a toy car and a book for the demonstration.

LESSON PLAN

1. Introduce the lesson by placing a toy car on a table. Ask the students why it isn't moving (balanced forces) and what must be done to make it move (unbalance the forces, or push it). Place a book at the end of the table and roll the car into the book. Ask the students why the car stopped (hit the book). Then roll the car on the floor. Push it gently enough so that it will stop before it hits a wall or a desk. Ask the students why it stopped (friction of tires on floor, lack of force to overcome other forces).

2. Go back and analyze each situation with the students. Explain to them that the car, like all other matter, has a property called inertia. This means that objects at rest remain at rest until an outside force acts on them, and objects in motion remain in motion until an outside force causes a change in their motion. The car was in motion (demonstrate again how the car moves until it hits the book). Ask the students what provided the force needed to stop the car (the book).

SCIENCE BACKGROUND

Frictional force is thought to be an electric interaction because the atoms on the surfaces of the objects that slide or rub against each other interact electrically.

3.4 The First Law of Motion

You will find out
- how friction affects motion;
- the first law of motion;
- what is meant by equilibrium in motion.

Figure 3.12 A hockey puck travels fast and far on smooth ice because friction is reduced.

Until about 350 years ago it was thought that the natural state of an object was that of rest. People believed that if a force were used to set an object in motion the object would come to rest when the force was removed. If you look around, your common sense seems to agree with this. A book pushed across the floor comes to rest after you stop pushing it. A ball thrown through the air takes longer to come to rest, but eventually it does.

Experiments by scientists such as Galileo showed that our common sense is wrong in many cases. Motion is, in fact, as natural a state as rest. Objects do not come to rest on their own. **Friction** is one force that slows objects down and stops them. Friction acts on a moving object in a direction opposite to the motion of the object. The more the friction can be reduced, the longer an object will stay in motion.

Newton gave a name to the tendency for objects to continue their state of motion or rest. He called this tendency **inertia** [ihn-UR-shuh]. Newton's **first law of motion,** or the law of inertia, states that every object stays in its equilibrium state unless it is acted on by outside forces. This equilibrium can exist when the object is at rest or as it moves in a straight line.

Objects will not change their motion if nothing acts on them to cause a change. For objects at rest, this is easy to understand. A book, for example, does not fly

Figure 3.13 The motion of an object in space does not change unless a force acts on it.

Figure 3.14 The law of inertia explains why you lurch forward when the car you are riding in stops suddenly.

across a room unless some outside force acts on it. What is sometimes difficult to understand is that motion with a constant velocity is also a state of equilibrium.

Consider, for example, a car moving along a straight, level road at a constant velocity. The car and its passengers are in a state of equilibrium. Now suppose that the driver of the car suddenly applies the brakes. Because of inertia, the passengers will keep moving at a constant velocity until some outside force acts to slow them down. If they are wearing seat belts, the seat belts will exert a force to cause the passengers to slow down to the same velocity as the car. If not, the dashboard or the windshield will slow them down, perhaps not so gently. Seat belts, then, protect people from the effects of the law of inertia.

According to the law of inertia, a car will have a tendency to keep moving in a straight line even if the road curves. However, friction between the tires and the road will cause the car to move in a curve along the road. As the car rounds the curve, the passengers that are not buckled in will keep moving in a straight line until they bump into the sides of the car. Next time you are riding in a car notice the ways in which inertia pulls on you as you round a curve, slow down, or speed up.

Do You Know?
The frictional force at any given joint in your body is less than a third of that encountered by the steel blade of a skate gliding on ice.

Study Questions for 3.4

1. What happens to motion when friction is reduced?
2. State the first law of motion.
3. Under what condition can a moving object be in a state of equilibrium?

1. Motion continues for longer time.
2. Every object stays in its equilibrium state unless acted on by outside forces.
3. moving with a constant velocity

3. Ask what force causes the toy car to stop (friction). If the students do not say friction, tell them that the floor and the wheels of the car are rubbing together. This rubbing slows the car down and eventually stops it. Mention that friction causes a car to stop when you apply the brakes. The brakes rub against tires and slow the car to a stop.

4. Conclude the lesson by asking the students to imagine that they are sitting in a car that is stopped. The driver suddenly steps on the gas and accelerates very fast. What will happen (they will be thrown against the seat)? Why (objects at rest tend to remain at rest until a force moves them, and here the car moves before they do)? Now have students imagine that they are in a car that stops short. Ask them what will happen (they may crash through the windshield). Why (objects in motion remain in motion until an outside force acts on them, and here the car stops before they do)?

Comprehension Skill: Applied

FOLLOW-UP

Reinforcement: A Trick with a Penny
Have the students place a cardboard square on top of a glass and put a penny in the center of the square. Have them flick their index fingers sharply against the cardboard. What happens to the cardboard? What happens to the penny? Why?

Enrichment: Car Safety
Have the students do research to find out how seat belts and air cushions prevent injury in car crashes. From their research which do they think is the better method of protecting passengers?

Tie-in: Kilometers per Liter
When car manufacturers are testing new cars to determine their mileage, they like to do so on clear, still days. Have the students list reasons why this is true.

3.5

OBJECTIVES: The student should be able to

- **explain** what happens when the net force on an object is not zero;
- **state** Newton's second law of motion;
- **state** the equation for Newton's second law of motion.

SCIENCE TERM

second law of motion

LESSON PLAN

1. Introduce the lesson by reviewing the terms *force, mass, acceleration, momentum,* and *velocity*.
2. Tell the students that Newton formulated three laws of motion. The first was the law of inertia, a law they are familiar with. Review this law by asking one student to state it and another to give an example of it. Point out that the second law concerns forces, mass, and acceleration.
3. To emphasize the first part of Newton's second law, write a Roman numeral one on the board. Under it draw two cars. Have a man pushing one car from the front and another man pushing the second car from the back. Point to each car and ask the students in what direction each car will move (backward and frontward). Point out that this is the first part of Newton's second law—objects move in the direction they are pushed or pulled.

3.5 The Second Law of Motion

You will find out
- what happens when an object is not in a state of equilibrium;
- the equation for Newton's second law of motion.

The law of inertia explains what happens when an object is in a state of equilibrium. When an object is not in a state of equilibrium, the object will accelerate. How will it accelerate? Newton's **second law of motion** answers this question. The second law of motion states that an object acted on by a constant force will move with a constant acceleration in the direction that the force is acting. This law can be written very simply in equation form.

Force = mass × acceleration

The second law of motion is one of the most important laws in physical science. If you know the mass of an object and the forces acting on it, you can use this law to tell exactly how an object will accelerate.

Figure 3.15 Rapid acceleration is important in a bicycle race.

The first part of the law is obvious. Objects accelerate in the direction that you push them. To get a feeling for the other parts of the law, imagine two vehicles at a stoplight. One is a subcompact car; the other, a dump truck.

Figure 3.16 The truck has greater mass so it requires a greater force to accelerate.

The mass of the truck is ten times the mass of the car. The drivers of the two vehicles want to accelerate away from the stoplight at the same rate. This means that the truck will need an engine that applies ten times more force than the subcompact car's engine. In other words, the greater the mass of an object the more force is needed to give that object the same acceleration.

Now imagine that two subcompact cars of the same mass are at the stoplight. One car has a normal four-cylinder engine. The other car has a high-powered turbocharged engine that can supply twice as much force as a normal engine. The car with the turbocharged engine will be able to move away from the stoplight with twice the acceleration as the car with the normal engine. The acceleration increases directly with the increased force.

Newton's second law applies to decelerations as well as accelerations. For example, it takes ten times as much force to stop a ten-ton dump truck traveling 80 km per hour as it takes to stop a one-ton car traveling the same distance at the same speed.

Figure 3.17 The car with the more forceful engine accelerates faster.

Study Questions for 3.5

1. What happens when the net force on an object is not zero?
2. Write Newton's second law in equation form. What happens to the force when either mass or acceleration is increased?

1. Object moves in the direction of the applied force.
2. force = mass × acceleration, increase in force

4. To emphasize the second part of this law, ask the students to imagine that an elephant and a mouse are to have a race. You and a friend have been hired to push the two animals so each will start at the same speed. Draw the animals on the board and label the drawing with a Roman numeral two. Ask the students which animal will have to be pushed harder if the two accelerate at the same rate (elephant). Why (more mass)?

5. Provide the students with other examples that will help to reinforce the second part of Newton's law. Ask them which of the following would require more force to accelerate the same amount: a Volkswagon or a truck (truck), a bicycle or a car (car). Ask which of the following objects would be given a greater acceleration if the same amount of force was applied to each: a bicycle or truck (bicycle), a Ping-Pong ball or a golf ball (Ping-Pong ball).

Mathematics Skill: Problem Solving

FOLLOW-UP

Reinforcement: A Second Law Mobile

Ask the students to create mobiles to illustrate Newton's second law. Each hanging mobile should have a picture of an object on each side. Each picture should also have a star showing where the force will be applied to the object and an arrow showing the direction in which it will move.

Enrichment: Rube Goldberg

Have the students look up Rube Goldberg. Tell them that Goldberg invented machines that did common tasks in outlandish ways. Ask the students to diagram one of Goldberg's machines and explain how it illustrates Newton's second law.

Tie-in: Rockets

Ask the students to do some research on rockets to determine the relationship between the weight of a rocket and the weight of its fuel. Ask the students what special adaptations engineers have made in rocket designs so that the rockets are very light when they are empty. Why do engineers want the empty rocket to weigh very little?

3.6

OBJECTIVES: The student should be able to

- **state** Newton's third law of motion;
- **list** some examples of Newton's third law.

SCIENCE TERM

third law of motion

ADVANCED PLANNING

You will need a deflated balloon, a pan of water, and a windup toy for demonstrations.

LESSON PLAN

1. Introduce the lesson by blowing up a balloon and letting it go. Ask the students what they observe and why they think this happens (the balloon flies through the air because air is coming out the back).

ACTIVITY: What Is Newton's Third Law?

Try the experiment before class to determine exactly how much vinegar and baking soda should be used for your bottles. Tell the students to imagine that the vinegar and baking soda are rocket fuel and the bottle is the rocket.

◇ **Caution:** The students must avoid a too-tight insertion of the stopper in the bottle and they must stand to the side of the bottle rather than behind or in front of it.

1. Mixture bubbles.
2. Gas is being released.
3. pops out of bottle; moves forward
4. They move in opposite directions.
5. The gas filled the bottle, then pushed the cork out as it continued to expand.
6. The third law states that for every action there is an equal and opposite reaction. The cork popped out, causing the bottle to be pushed backward.
7. No, the stopper moved farther. Since the stopper has less mass than the bottle, the same amount of force can move it farther.
8. to reduce friction and allow more movement

● Laboratory Skills: Experimentation/Demonstration
Manipulation of Materials

3.6 The Third Law of Motion

You will find out

- Newton's third law of motion;
- some examples of Newton's third law of motion.

Suppose you were to run into a tree. It would probably hurt. But it wouldn't hurt because you hit the tree. It would hurt because the tree hit you back. This observation is explained by Newton's **third law of motion.** This law states that for every action by one object there must be an equal and opposite reaction on the other object.

● ACTIVITY

What Is Newton's Third Law?

Materials

graduated cylinder
12 plastic straws, dowels, or round pencils
petroleum jelly
plastic gas collecting bottle or 1-L bottle
rubber stopper to fit bottle
100 mL vinegar
2 teaspoonfuls of baking soda
safety glasses
paper towels
apron

Procedure

1. Put on the safety glasses and apron.
2. Place the straws about 5 cm apart side by side on the table.
3. Pour 100 mL of vinegar into the gas collecting bottle.
4. Tilt the bottle on its side. Do not let the vinegar flow into the neck of the bottle.
5. Coat the rubber stopper with petroleum jelly.
6. Place the bottle horizontally on the straws.
7. Place the baking soda in the neck of the bottle.
8. Quickly push the stopper loosely into the bottle. As you do this, make sure that the baking soda is pushed into contact with the vinegar. CAUTION: Do not stand near the bottle.
9. Observe what happens.

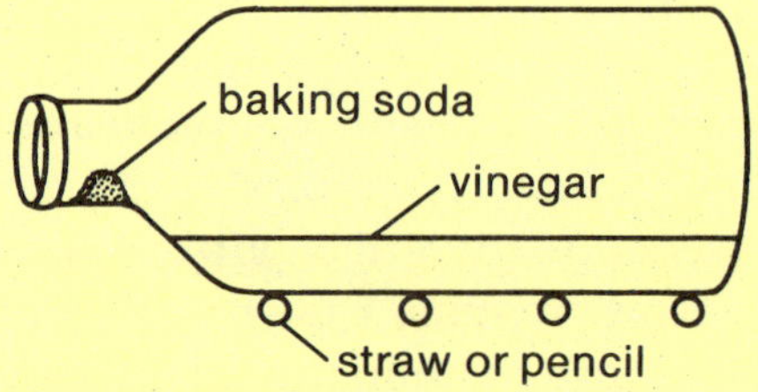

Questions

1. What happens when you mix vinegar and baking soda?
2. What does the foaming action indicate?
3. What happens to the stopper? The bottle?
4. How do the motions of the bottle and the stopper differ?
5. What caused the stopper to leave the bottle?
6. How can this experiment be used to illustrate Newton's third law of motion?
7. Did the stopper and the bottle move equal distances? Why or why not?
8. Why was the bottle put on straws?

Figure 3.18 According to Newton's third law, the oars push on the water and the water pushes on the oars.

The third law of motion completes the description of an object's motion. The first law describes how an object moves when there is no net force on it. The second law describes how an object moves when a force is applied. The third law tells what the effects are on the object that applied the force. For example, the third law explains why if you jump from a rowboat onto a dock, the boat will move backward as you move forward. When you jump, you push the boat backward, and the boat pushes you forward. In a similar way, a ship's propeller pushes on the water. The water pushes back on the ship's propeller which causes the ship to move forward.

A far more familiar example of the third law is walking. You push back on the ground with your feet and the ground pushes you forward. If the ground is icy, then you cannot push on it with enough force. The ground's push on you is therefore equally weak and you move forward slowly, if at all.

This law of action equals reaction is used in the design of jet engines and rockets for spacecraft. The rocket fuel burns producing hot gases. The gases are forced out of the rocket. This action produces an equal and opposite reaction that pushes the rocket up into space.

Figure 3.19 When you walk, you push on the ground and the ground pushes on you.

Study Questions for 3.6

1. What do scientists mean when they say that action equals reaction?
2. Give three examples of Newton's third law.

1. For every force (action) applied to an object there is an equal and opposite force (reaction).
2. pushing on the ground as you walk, jumping from a boat onto the dock, running into a tree

2. Wind up the toy and place it on the floor. Ask the students what makes it move across the floor (the toy pushes the floor back as the toy goes forward).
3. Explain that these two examples illustrate Newton's third law, i.e., for every action by one object there is an equal but opposite reaction on the other object. The air came out the back of the balloon so the balloon went forward. As the toy pushed against the floor, the toy went forward.
4. Conclude the lesson and review the chapter by giving the following examples to the students and asking them which of Newton's laws is represented by each.
   a. A girl dives out of a boat and the boat moves backward (3).
   b. A horse comes to a hedge and stops suddenly rather than jumping. The rider flies over the horse's head to the ground (1).
   c. A boy builds a go-cart that weighs less than his brother's. Both go-carts are pushed with the same force. The one that weighs less goes faster (2).

FOLLOW-UP

Reinforcement: Rotary Sprinklers
Ask the students to explain how, based on Newton's third law, rotary sprinklers work. Bring a rotary sprinkler to class for use in the explanation.

Enrichment: Skiing
Ask students to prepare short silent skits to show how skiers demonstrate Newton's third law as they ski down a mountain.

Tie-in: Off to the Moon
Have students do research on United States space flights. How do the rockets take off? How does this illustrate Newton's third law?

SCIENCE BACKGROUND

The action and reaction in Newton's third law coexist; one is impossible without the other. They are of the same magnitude but in opposite directions and they happen to two different objects. If, for example, the earth could not push back on an athlete's foot as she ran, the athlete could not push on the earth in the first place.

What Did You Learn?

- 3.1 • A force is a push or a pull that is applied to an object.
- 3.1 • The amount of gravitational force that one object exerts on another depends on the masses of the objects and how far apart their centers are.
- 3.1 • The universal law of gravitation states that the force of gravitation exists between any two objects in the universe.
- 3.1 • The measure of gravitational pull on an object is the object's weight in newtons.
- 3.1 • When gravitational force changes, mass remains the same but weight changes.
- 3.2 • Electric forces are the forces between electric charges.
- 3.2 • Electrically neutral objects have equal numbers of positive and negative charges.
- 3.2 • Two objects with like charges repel each other. Two objects with unlike charges attract each other.
- 3.3 • The sum of the forces acting on an object is the net force.
- 3.3 • A state of equilibrium is one in which the net force on an object is zero in all directions.
- 3.3 • The center of gravity is the point on an object where all the gravitational forces act.
- 3.4 • Friction is a force that slows down an object by acting in a direction opposite to the motion of the object.
- 3.4 • Inertia is the tendency for objects to continue their state of motion.
- 3.4 • Newton's first law, the law of inertia, states that every object stays in its equilibrium state unless acted on by an outside force.
- 3.5 • Newton's second law of motion states that force is equal to mass times acceleration.
- 3.6 • Newton's third law of motion states that for every force (action) there is an equal and opposite force (reaction).

Key Terms

force
gravitational force
universal law of gravitation
weight
newton
electric forces
electric charge
electrically neutral objects
attract
repel
net force
equilibrium
center of gravity
friction
inertia
first law of motion
second law of motion
third law of motion

Biography

Isaac Newton (1642–1727)

Isaac Newton, a giant of science, was so small when he was born that it was said he could fit in a quart mug. Isaac was born only a few months after his father died. When Isaac was three, his mother remarried, leaving Isaac to be raised by his grandmother. Although he considered himself a poor student at the grammar school he attended, Isaac showed a natural ability for making and inventing things. He invented a small windmill to grind wheat and corn, and he even created a sundial on the wall of his grandmother's house.

When Newton was 24, the bubonic plague struck London. In one summer 31,000 people died. Since schools had to close, Newton, who had been teaching mathematics at Cambridge University, went back to the family farm for two years. During those years, he set down the three laws of objects in motion. He also figured out the universal law of gravitation, which explains the conditions under which apples fall to the ground and the planets move around the sun.

Isaac Newton's discoveries changed the way people would look at the world forever after. Because Newton was so secretive, however, it would be almost 20 years before more than a very few people knew about his work. How was he able to do so much scientific work? "By thinking about it night and day," Newton himself said.

When thinking about a problem, Newton would forget to eat. Or if he did come to the table, he might wander off and forget to come back. According to one story, Newton got up soon after dinner was served and wandered off, leaving his food untouched. His companion took his plate, ate the food, and returned the empty plate to Newton's place. When Newton returned, he said, "I don't remember eating, but I see from my plate that I must have finished already."

Vocabulary

1. false - Force
2. false - unbalanced
3. false - mass
4. true
5. false - all
6. true
7. true
8. true
9. true
10. false - equal

● Word Skill: Vocabulary

What Do You Remember?

1. tendency of objects to continue their state of motion or of rest
2. masses of the objects, distance between objects' centers
3. measure of the gravitational pull on object; greater gravitational attraction on object the closer it is to earth
4. friction
5. First Law—Car is moving at constant speed, then brakes are applied. Car is slowed, but passengers continue moving unless restrained by seat belts.
Second Law—Two cars are stuck in snow. A greater force is required to push out the car of greater mass than the car of smaller mass.
Third Law—Rocket pushes fuel out of the thrusters, and fuel pushes back on the thrusters.
6. Multiply together the object's mass and acceleration.

TO THINK ABOUT AND DO

Vocabulary ●

Number your paper from 1 to 10. Write T beside the numbers of sentences that are true and F beside the numbers of sentences that are false. If you decide a sentence is false, change the word or words in italics to make the sentence true.

1. *Gravitation* is any push or pull that affects matter.
2. Motion is caused by *balanced* forces.
3. The quantity or amount of an object is the object's *weight*.
4. *Newton* is the unit used to measure force in the metric system.
5. A state of equilibrium is one in which the net force on an object is zero in *two* directions.
6. The attraction between two objects of opposite charges is caused by *electric* forces.
7. The point where all gravitational forces act on an object is the *center of gravity*.
8. *Inertia* causes all things at rest to remain at rest until an outside force moves them.
9. Force equals mass times *acceleration*, according to Newton's second law of motion.
10. For every action there is an *unequal* and opposite reaction.

What Do You Remember?

1. What is inertia?
2. What two factors affect gravitational force?
3. What is weight? Why does weight change as you move about the earth or the solar system?
4. What force slows down moving objects?
5. Give an example to illustrate each of Newton's laws of motion.
6. Explain how you would find the force of an object if you were given its mass and acceleration.

Applying What You Have Learned ●

1. If you double the net force on an object and keep the mass constant, how will its acceleration be affected?
2. What will happen to the acceleration if the mass of an object is tripled but the force is kept the same?
3. What must be done to overcome the inertia of an object?

Research and Investigation ●

Friction affects all moving objects. Design and perform experiments to answer these questions. Before you do the experiments, have your teacher check your procedures.

1. Is there more friction between two rough objects or two smooth objects?
2. Will ball bearings or oil reduce friction more?
3. How does an increase in mass affect the amount of friction?

Applying What You Have Learned

1. Acceleration will double.
2. Acceleration will decrease to one third of original acceleration.
3. Apply a force, e.g., friction.

● Comprehension Skill: Applied

Research and Investigation

1. Students might first attach a spring scale to a block of wood and measure how much force is needed to pull the block along a board. Then they should cover both the board and the block with sandpaper and measure how much force is required to pull the block.
2. Students could attach a spring scale to a metal block and pull it across a metal surface. They should record how much force is needed to pull the block. They could then put some ball bearings under the block and pull it across a metal surface. The amount of force needed to pull the block should be recorded. Finally, students could put oil on the metal surface and pull the block across it. The amount of force required this time should be recorded.
3. The amount of force required to pull a block across a surface can be determined by using a spring scale. A book could be placed on top of the block to increase its mass. The block and the book should be pulled with the spring scale in order to measure how much force is required.

● Reference/Research Skills

CHAPTER BACKGROUND

A person's daily activities are greatly influenced by machines. Machines aid people in work and during recreation. People use machines to do the work that they themselves had to do in the past. The chief source of power is no longer muscle power; it is the coal, oil, gas, or nuclear power that runs machines.

All machines, no matter how complex or simple, operate according to a single principle. The principle can be summarized by saying that work input is equal to work output. You cannot get more work from a machine (output) than you put into it (input). Rather than decreasing the amount of work that you must do, machines help you do that work more easily.

Machines help you by increasing your effort, or the force you apply to the machine. Machines can change the direction of the force so that you can simply pull down and have an object rise. Machines can allow you to work faster, and they can transform energy from one form to another.

The mechanical advantage, or M.A., of a machine describes exactly how much a machine helps people. The M.A. is simply a number that rates the performance of the machine. The M.A. of both ideal and actual machines can be calculated. Ideal machines are those that in theory are completely frictionless.

While frictionless machines do not actually exist, it is useful to speak of them because they serve as models of what should be the case under ideal conditions. The actual amount that an ideal machine would help is found by dividing the distance the effort moves by the distance the resistance moves. The calculation of the mechanical advantage of a real machine must include a correction for friction and other forces that hinder the efficiency of machines. This correction is made by determining the actual mechanical advantage, or by dividing the resistance by the effort. In all cases the actual mechanical advantage is less than the ideal mechanical advantage. Only through improved technology will a way be found to increase the actual mechanical advantage and thereby increase the efficiency of machines.

4

Work, Machines, and Power

Have available a hammer and a nail that is partly hammered into a block of wood, scissors and paper, a nutcracker and a nut in a shell, and a knife and an apple. Pull the nail from the wood with the hammer. Cut the paper with the scissors. Crack the nut with the nutcracker. Cut the apple with the knife. Open a door or turn on a faucet. As you perform each activity, ask what instrument is helping you. Stress that devices that help to do work are machines. The hammer, nutcracker, knife, and doorknob or faucet are all machines. Tell the students that in this chapter they will learn more about machines and how machines help people.

Henry Ford was a man who became famous and very wealthy by using machines to make automobiles. Very likely, the machines that help people today were beyond Ford's imagination. Since Ford's time, machines have helped change the face of the earth. They have even made it possible for people to leave the earth and travel into space.

Machines and the work they do are a vital part of your everyday life. You may think most machines are big, expensive, and complicated to operate. A crane certainly is. How many people on your block own a crane? If your neighbor did own one, would you know how to run it? You and your neighbors do own and use many kinds of machines, however. Some of them require fuel or electricity to operate. Most of the machines you use have no fuel tanks or plugs. They are very inexpensive to buy and easy to use. Yet without these simple machines you could not easily turn a faucet, open a door, or even swat a fly.

Lesson Organizer

4.1 Work
Pages 70–71
Advanced Planning: Page 70
Follow-up: Page 71
- **R** Identifying and Calculating Work
- **E** James Joule
- **T** Weight Lifting

4.2 Machines and Work
Pages 72–77
Advanced Planning: Page 72
Activity: How Does a Ramp Help Do Work? Page 76
Follow-up: Page 77
- **R** Input versus Output
- **E** Increasing Output
- **T** Career—Machine Operator

4.3 The Six Simple Machines
Pages 78–83
Activity: How Do Pulleys Help Do Work? Page 83
Follow-up: Page 83
- **R** Machine Collage
- **E** How Can Friction Be Reduced on an Inclined Plane?
- **T** Industrial Arts

4.4 Power and Efficiency
Pages 84–87
Advanced Planning: Page 84
Activity: How Much Power Do You Use to Walk up the Stairs? Page 86
Follow-up: Page 87
- **R** Power Problems
- **E** Horsepower and Mileage
- **T** James Watt

See ACTIVITIES book pages 19–24 for laboratory, skill, and review exercises for Chapter 4.

4.1

OBJECTIVES: The student should be able to

- **define** the term *work*;
- **calculate** the amount of work done.

SCIENCE TERMS

work
joule

ADVANCED PLANNING

1. To demonstrate the concept of energy, obtain a wind-up walking toy; two light objects that the toy will be able to move, e.g., a marble and a cork; and two heavy objects that the toy will not be able to move, e.g., a rock and a brick.

2. Prepare a work sheet of problems whose solutions require the use of the equation for work. Use numbers that the students can easily multiply in their heads.

LESSON PLAN

1. Introduce the lesson by reviewing the definition of *energy* (ability to move or change matter in some way).
2. Wind up the walking toy and hold it on your desk. Ask whether the toy has energy (yes). Point out that since the toy has energy it has the ability to do work. Energy is the ability to do work.
3. Place the light and heavy objects (marble, cork, rock, and brick) on the desk. In order for the students to see how much work the toy can do, ask the students to hypothesize which objects the toy will be able to move (answers will vary). On the board make a chart with headings *move* and *not move* to record the hypotheses of the students.
4. Wind up the toy and place it behind each object in turn. The lighter objects should move when hit by the toy, but the heavier objects should not. On the hypothesis chart rearrange the name of any object that was in the wrong column. Point out that when the toy moved an object, work was done.

● Comprehension Skill: Applied

4.1 Work

You will find out
- what a physical scientist means by the word *work*;
- how to find the amount of work done by a force.

Figure 4.2 No work is being done.

Many of the words used in physical science are familiar words that you use in your everyday life. Sometimes this causes a problem because physical scientists use the words in a slightly different way. Scientists don't use words differently just to be difficult. They are simply trying to communicate their observations as accurately and precisely as possible. The word *work* is a good ● example of this. Imagine that you spend two hours studying for a science test. Or suppose that you are a guard standing absolutely still at Buckingham Palace. In either case, there would be no doubt in your mind that you were working. But a physical scientist would say, "No. As far as physical science is concerned, you have done no work."

In physical science, **work** is done only when a force *moves* an object. The amount of work done in moving the object is equal to the strength of the force that you use times the distance the object moves.

Work = force × distance

Figure 4.3 Work is being done.

The unit of work is equal to the unit of force (newton) times the unit of distance (meter), or a newton-meter. In the metric system, the newton-meter is usually called the **joule,** abbreviated J. Therefore, one joule is the amount of work needed to maintain a force of one newton over a distance of one meter. The following example should help you understand how to find the number of joules of work that must be done.

EXAMPLE A student lifts a book that weighs 10 newtons a distance of 50 centimeters off the floor. How much work does the student do?

SOLUTION Here the force used to lift the book up must be equal to the gravitational force acting down on the book. Therefore an upward force of 10 newtons is needed. The distance moved is 50 centimeters, or 0.5 meters.

$$\text{Work} = \text{force} \times \text{distance}$$
$$\text{Work} = 10\text{ N} \times 0.5\text{ m}$$
$$\text{Work} = 5\text{ N.m, or } 5\text{ J}$$

Study Questions for 4.1

1. According to the physical scientist, what must happen in order for work to be done?
2. Tom uses 20 newtons of force to push a baby carriage a distance of 5 meters. How much work does he do?

1. A force must move an object a distance.
2. Work = force × distance moved in direction of force = 20 N × 5 m = 100 N.m = 100 J

5. On the board write the equation for work. From the equation review the term *force* (any push or pull that affects matter) and the unit used to measure it (newtons, pounds). Ask the students what the term *distance* means (straight-line length from one point to another) and what unit is used to measure it (meters, feet). Point out that according to the equation you can find the amount of work done by multiplying the force used by the distance moved. Then give the students a few examples like the following one:

   How much work will Paul do if he uses 5 newtons of force to move a box a distance of 8 meters (40 newton-meters, or 40 joules)?

● Mathematics Skills: Computation
Problem Solving

FOLLOW-UP

Reinforcement: Identifying and Calculating Work

Number each of the pictures on the bulletin board that was set up in this lesson. On each write f = ___ and d = ___. Fill in the blanks with appropriate distances and forces. Have the students identify the pictures that show work and determine how much work is being done.

Enrichment: James Joule

Ask the students to research the life of James Joule. What did he do that made other scientists honor him by naming a unit of measurement after him?

Tie-in: Weight Lifting

According to the physical science definition, weight lifting is work. Ask the students to visit a local body-building center. Have them observe and record some weight amounts that are lifted. For each weight amount that is lifted, students should also measure and record the height or distance the weight was moved. Students should then calculate the amount of work done for each lift.

SCIENCE BACKGROUND

Even though newtons times meters equals newton-meters, the more concise term is joules. This term was named after James Prescott Joule (1818–1889), a British physicist.

4.2

OBJECTIVES: The student should be able to

- **define** the term *machine*;
- **define** the term *simple machine*;
- **express** the relationship between effort force and resistance force and between effort distance and resistance distance;
- **explain** the meaning of the term *mechanical advantage*;
- **describe** the work principle on which machines operate;
- **compute** the mechanical advantage of a machine.

SCIENCE TERMS

machine	effort force
simple machines	effort distance
mechanical advantage	resistance distance
resistance force	

ADVANCED PLANNING

Prepare a work sheet in which students solve for mechanical advantage, given the effort and resistance forces or the effort and resistance distances. Use numbers that can be easily manipulated in the head.

LESSON PLAN

1. Introduce the lesson by asking the students what particular kinds of machines do. Use large machines like the crane, the plow, or the steam shovel so that students will come up with the idea that machines help people do work. Stress that a machine can be any device that helps work get accomplished. A machine can be as complex as a crane or as simple as a bottle opener, a broom, or a screwdriver.

4.2 Machines and Work

You will find out
- how machines help you to do work;
- the work principle on which all machines operate;
- how to compute the mechanical advantage of a machine.

A **machine** is a device that helps you to do work. If you remember that work is equal to force times distance, a machine helps you by changing the force or the distance or both. All machines are made up of one or more **simple machines.** Simple machines are devices that increase forces. In this way simple machines make it easier for you to do work.

No one would want a machine that makes work harder. There must be an advantage in using machines, or nobody would buy or use them. The advantage in using a machine is called a **mechanical advantage.** This mechanical advantage measures how much force you gain by using a machine rather than doing the work yourself. The bigger the mechanical advantage, the less effort you must supply.

Figure 4.4 How many of these simple machines have you used?

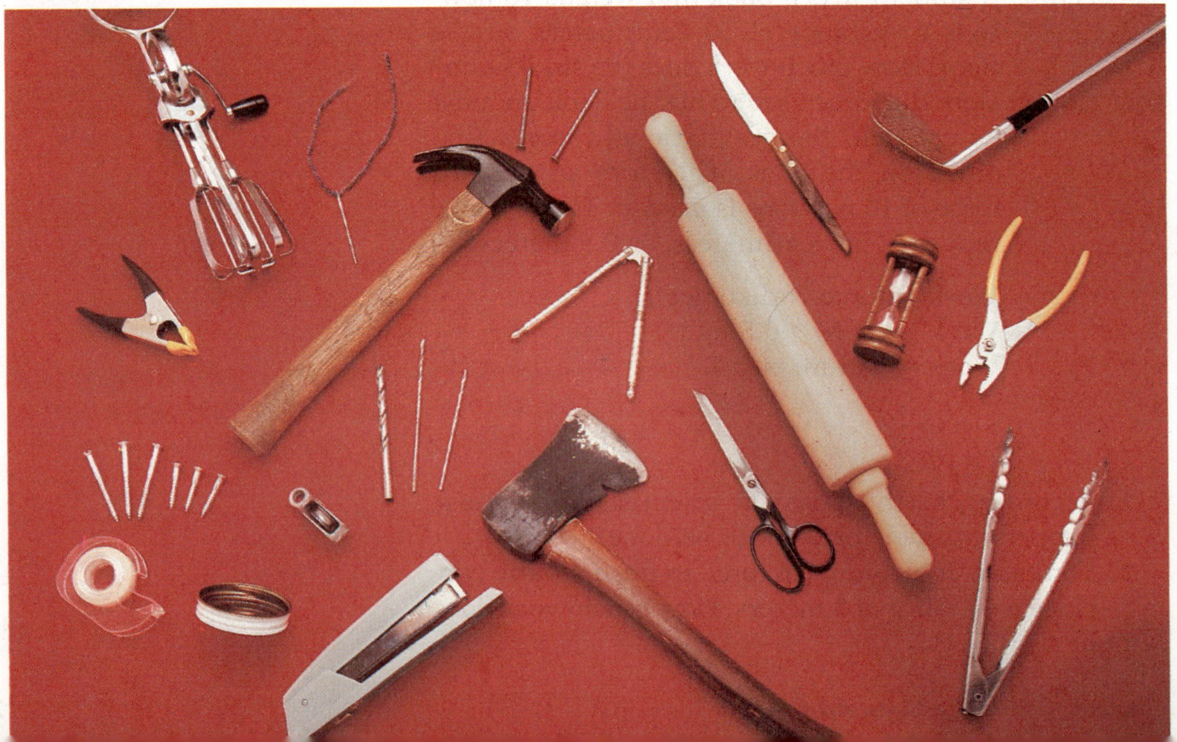

In using a machine to do any work (force times distance), there are really two forces involved. One force acts on the machine. It is called the **effort force.** When you use a machine, you provide this effort force. The second force is the force you and your machine are working against. This force is called the **resistance force.** Often the resistance force is the weight of the object to be moved.

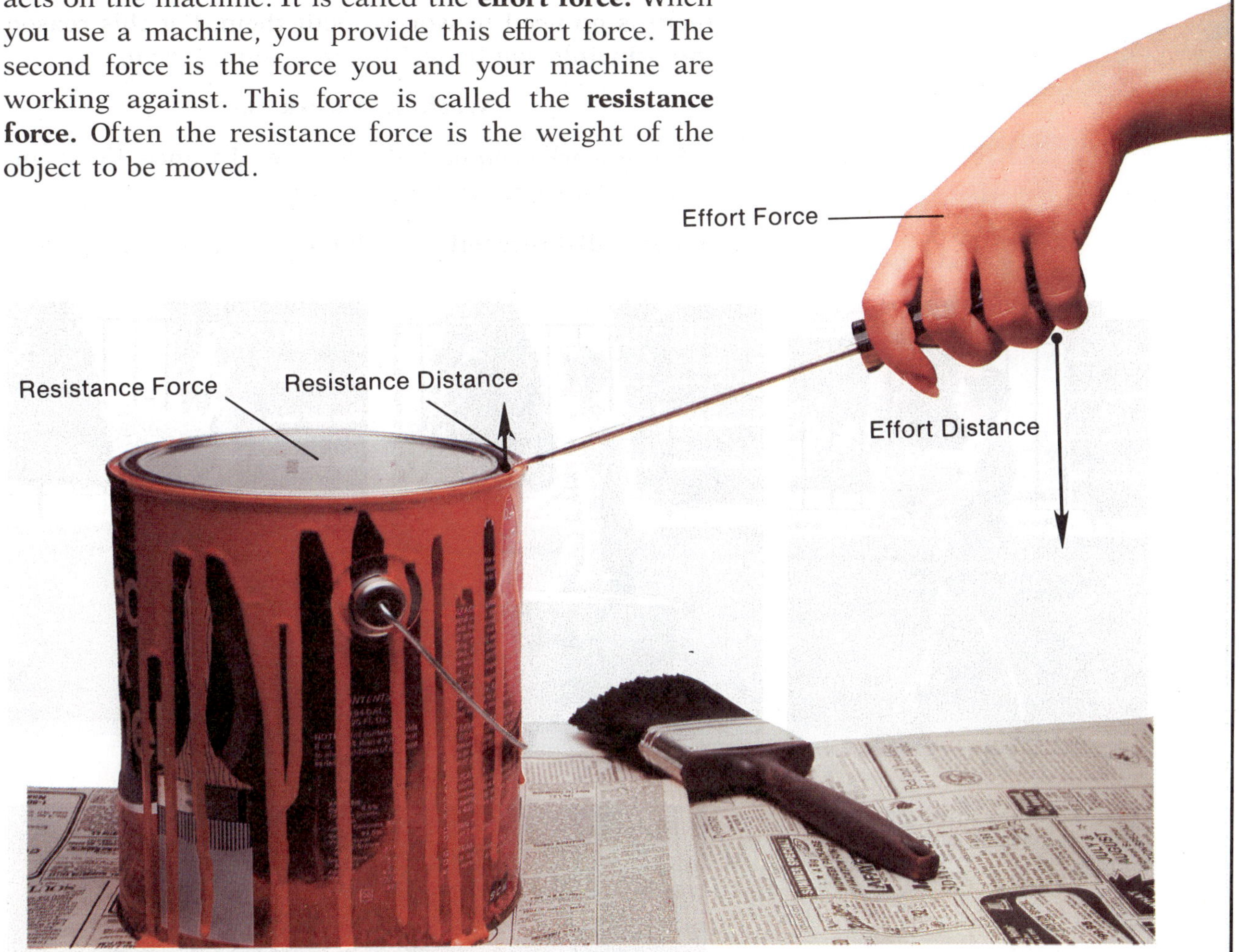

Figure 4.5 The forces and distances involved in doing work

There are also two distances involved in doing work. The distance over which your effort force acts is called the **effort distance.** The **resistance distance** is the distance the object you are doing work on is moved. Sometimes effort distance and resistance distance are the same.

All machines, no matter how simple or complex, operate on this one basic principle:

Work or energy in = work out + energy losses

2. Explain that machines help people in different ways. Machines may increase the force, e.g., a student alone could not lift a car, but a student could lift it with a jack. Machines may also increase the speed with which work is done, e.g., a bulldozer can move dirt faster than a person could move the same amount of dirt. Machines may change the direction of the force that is being used, e.g., you push down on a crowbar and a tree stump is lifted up.
3. Stress that no matter how simple or complex a machine is, the work input is equal to the work output plus the energy lost. You will never get more work out of a machine than you put into it. However, the machine can help you by increasing your force and by making the work easier for you to do.

4. Explain to the students that the degree to which the machine increases your effort is referred to as the mechanical advantage. The ability of machines to increase your effort allows you to do jobs that would otherwise be very difficult or impossible.

● Mathematics Skills: Computation Problem Solving

SCIENCE BACKGROUND

When you use a car jack, the force that you have to put in is small. This is because you move the bar of the jack up and down many times to achieve a large effort distance. Even though the jack moves a very small distance, the force it applies to the car is great enough to lift the car. The work done is the same; with the jack, however, the distance is less and the force is increased.

Keep in mind that the work that comes out of a machine can never exceed the work that goes in. For simple machines the energy losses are usually so small that there is no need to worry about them. For this reason the principle can be written in a simple form:

Work in = work out

Since work is equal to force times distance, this principle can also be written this way:

● **Force × distance (effort) = force × distance (resistance)**

Figure 4.6 Doing work without the help of a machine other than the body

Figure 4.7 Work finished

Figure 4.8 Doing work with the mechanical advantage of a machine other than the body

The mechanical advantage (M.A.) of any work can be calculated using either the two kinds of force mentioned or the two kinds of distance described. If you choose to use forces, you must divide the effort force (F_E) into the resistance force (F_R).

$$\text{M.A.} = \frac{F_R}{F_E}$$

Sometimes it is easier to measure the distances involved than it is to measure the forces. When that is the case, you can calculate the mechanical advantage using distances instead of forces. You divide the resistance distance (D_R) into the effort distance (D_E).

$$\text{M.A.} = \frac{D_E}{D_R}$$

Whichever way it is calculated, the mechanical advantage tells you the same thing. It tells you how much effort force was gained in using a specific machine to do a specific amount of work. Cartoon and movie superheroes don't have to worry about machines and mechanical advantages. They are strong enough to do anything without the help of machines. However, you are not! Some relatively easy work would be almost impossible for you to accomplish without the mechanical advantage of machines.

5. Solve several model mechanical advantage problems for the students before distributing the work sheet of these problems for class or home assignment. You may use examples like the following:

a. A bar is used to lift a weight of 2,000 newtons with an effort force of 50 newtons. What is the mechanical advantage of the bar (40)?

b. What is the mechanical advantage of a ramp 30 meters long and 2 meters high (15)?

ACTIVITY: How Does a Ramp Help Do Work?

In this activity the students will demonstrate that a ramp reduces the force needed to lift an object a certain distance. They will also calculate the ideal and actual mechanical advantage of a ramp. Before you begin the experiment, explain the difference between ideal and actual mechanical advantage (see Chapter Background). Then stress the importance of making precise measurements. The shop teacher may be able to provide the boards needed. Boards with smooth surfaces that reduce friction are best.

● **Laboratory Skills: Manipulation of Materials**
Experimentation/Demonstration

● ACTIVITY

How Does a Ramp Help Do Work?

Materials

smooth board
5 books
string and scissors
spring scale

1. Answers will vary.
2. Answers should be less than answers in number 2.
3. one
4. Answers will vary but should be greater than one.
5. Answers should not be same, since the ramp permits the use of less force resulting in greater mechanical advantage.

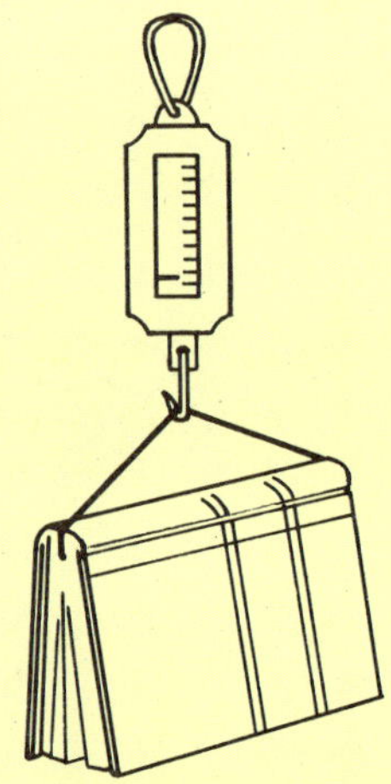

Procedure

1. Arrange 4 books and the board to form a ramp as shown in the diagram. Measure the length and height of your ramp.

2. Attach a string to the fifth book and use the spring scale to measure the effort force needed to lift the book straight up to a height equal to the height of your ramp.

3. Place the book on the very end of the ramp and use the spring scale to measure the effort force needed to pull the book up the ramp.

Questions

1. How much effort force was required to lift the book straight up? This amount equals the weight of the book.

2. How much effort force was required to pull the book up the ramp?

3. What is the mechanical advantage when you divide the weight of the book (resistance force) by the effort force needed to lift it straight up? (Hint: $\frac{F_R}{F_E}$ = M.A.)

4. What is the mechanical advantage when you divide the length of the board by the height to which you lifted the book? (Hint: $\frac{D_E}{D_R}$ = M.A.)

5. Were your two mechanical advantages the same? If not, can you think of a reason why?

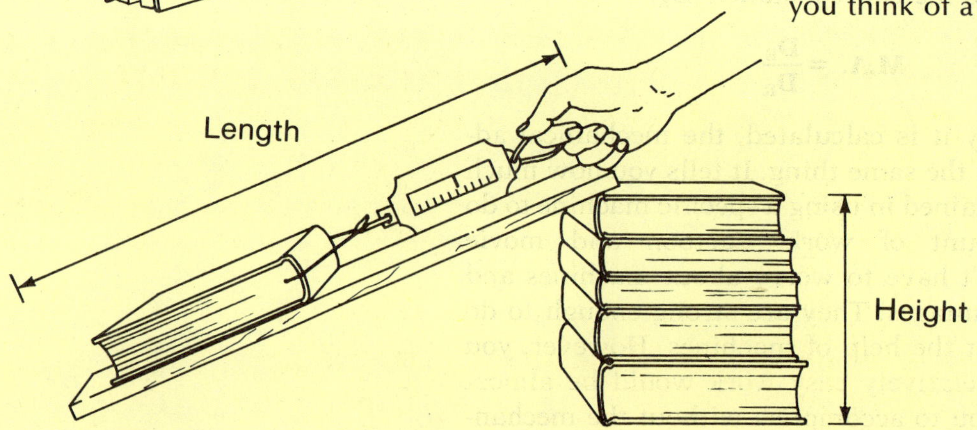

Figure 4.9 When you use a jack, the amount of effort force you must apply to lift a car is reduced.

Imagine that you have to change a flat tire. You have to do work. You have to use a force to lift an object, the car, up a distance off the road. You need the mechanical advantage of a machine to get this work done. A jack is a simple machine. It gives you a mechanical advantage by reducing the amount of effort force you must supply. It requires much less effort to operate a jack than it does to lock your hands under the bumper of a car and lift it up. Thus simple machines allow you to do the same amount of work with less effort.

Maybe you have never changed a tire. You are still an experienced simple machine operator. You're never out of practice. You use some types of simple machines nearly every day. You use them to open a bottle, open a door, or open a can. You use them to raise a flag, raise a sail, or raise a window shade. Simple machines help you land a fish, hit a home run, or split a log. You use them to shovel snow, rake leaves, and hammer nails. Without these simple machines many small tasks would become more difficult and time-consuming.

Figure 4.10 A simple machine you probably use often

Study Questions for 4.2

1. How do machines help you to do work?
2. What is the work principle on which all machines operate?
3. If a 100-newton force (F_E) is used to lift a 500-newton weight (F_R), what is the mechanical advantage?

1. Machines change forces or distances.
2. Work or energy in = work out + energy losses
3. $\frac{\text{Resistance force}}{\text{Effort force}} = \text{M.A.} = \frac{500\text{ N}}{100\text{ N}} = 5$

6. To conclude the lesson, make a chart on the board similar to the one below. The chart should list machines students use each day, how each machine helps them, and whether its mechanical advantage would be considered high or low.

MACHINE	HOW IT HELPS	MECHANICAL ADVANTAGE (relative amount effort is increased)
bottle opener	increases effort force	increase is high
bicycle	increases distance	increase is low
doorknob	increases effort force	increase is high

FOLLOW-UP

Reinforcement: Input versus Output

Set up an input-output corner of toy cars and trucks where students can use metersticks and spring scales to measure the input and output of these machines. This activity will help students to see that the input is equal to the output of a machine.

Enrichment: Increasing Output

Visit an appliance store and ask the clerk for the energy guides from two brands of dishwashers or similar objects. Have the students study the guides and determine which of the two machines is more efficient, i.e., which machine has an output that is closer to the energy input. Ask the students to think of ways to increase the output of a machine in order to make it more efficient.

Tie-in: Career—Machine Operator

Have students investigate the job of machine operator by interviewing people they know who hold this position or by writing to the International Association of Machinists and Aerospace Workers, 1300 Connecticut Avenue, N.W., Washington, D.C. 20036.

4.3

OBJECTIVES: The student should be able to

- **list** the six simple machines;
- **list** one example of each simple machine;
- **explain** how each simple machine helps people do work.

SCIENCE TERMS

inclined plane	fulcrum
wedge	wheel and axle
screw	pulley
lever	

LESSON PLAN

1. Introduce the lesson by pointing out that there are six simple machines upon which all the others are based. On the board make a chart with two columns, *inclined planes* and *levers*. Fill in the chart by listing the six simple machines (inclined plane category includes inclined plane, wedge, screw; lever category includes lever, pulley, wheel and axle).
2. Ask the students what an inclined plane is. Most will recognize it as a ramp. On the board draw a ramp that leads to the back of a moving van. Tell the students that as movers they must put 80 barrels onto the van. To lift each barrel straight up, 8 newtons of force is needed. The distance from the sidewalk to the van is 1 meter. Ask the students how they would calculate the amount of work that will be done ($W = fd = 8\ N \times 1\ m = 8\ N \cdot m = 8\ J$). How much work will be done (8 joules)? If there are 80 barrels and 8 joules are required to lift each barrel, how much will the total amount of work be (640 joules)?

● Mathematics Skills: Computation
Problem Solving

4.3 The Six Simple Machines

You will find out

- how simple machines change the force;
- the names and uses of the six simple machines.

Figure 4.11 Using an inclined plane

It is important to remember that simple machines operate on a single principle: apply a small force over a large distance to get a large force working over a small distance. There are six kinds of simple machines: the inclined plane, the wedge, the screw, the lever, the wheel and axle, and the pulley.

1. The **inclined plane,** or ramp, is the simplest of all machines. It increases the effort distance through which you move an object and decreases the effort force necessary to move it. An inclined plane allows you to lift heavy objects that would otherwise be very difficult to move. You can calculate the mechanical advantage of the inclined plane by dividing the length of the inclined plane (D_E) by its height (D_R). Imagine you use an inclined plane 5 meters long to lift a heavy box 3 meters up onto a platform.

$$M.A. = \frac{D_E}{D_R}$$

$$M.A. = \frac{5\ m}{3\ m} = 1.66$$

Figure 4.12 An inclined plane increases the effort distance while decreasing the effort force required.

2. The **wedge** is an inclined plane that moves. Most wedges—a woodcutter's wedge, an ax, and a nail—are made of two inclined planes. You move material *up* an inclined plane; you move a wedge *through* material. In both cases, material is moved across a sloping surface. In the case of the inclined plane, the sloping surface is held fixed and the material is pushed along it. In the case of the wedge, the material is held fixed and the sloping surface is pushed through it. For example, imagine a nail being hammered into a board. The board doesn't move. The nail, acting as a wedge, moves through it.

Figure 4.13 Using a wedge

3. The **screw** is just a combination of an inclined plane and a cylinder. Each time the screw makes a full turn, it is moved a definite distance up or down. The distance it is moved up or down in one turn depends on the distance between two threads, as shown in the diagram of the screw. In a finely threaded screw, the distance between the threads is quite small. If the distance between threads is small, you must turn the screw more times to move it into place. The more turns you use, the greater your effort distance will be. Thus the M.A., or $\frac{D_E}{D_R}$, of finely threaded screws is greater than that of coarsely threaded screws.

Figure 4.14 The finely threaded screw involves a greater effort distance but also requires less effort force.

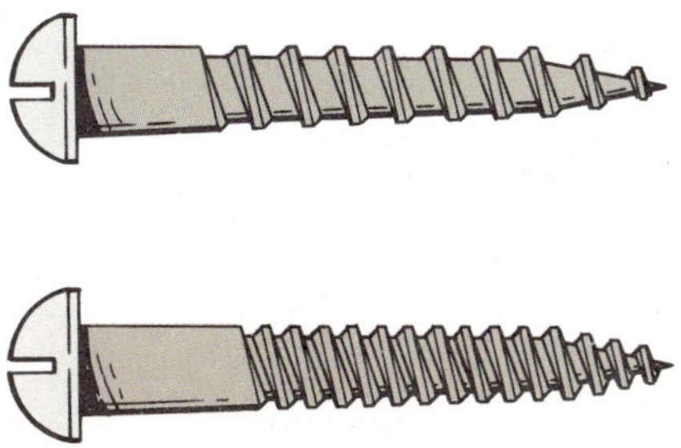

3. Explain that the wedge is a type of inclined plane. It can actually be two inclined planes arranged back to back. To show this, diagram a wedge on the board. Tell the students that the wedge increases the force used and helps the force to be brought to a single point. A log cannot be split with the hands alone, but with a wedge (or an ax, an example of a wedge), the log can be easily split. Wedges are often used for cutting. Ask the students to name a few wedges (knives, axes, nails).

4. The third type of inclined plane is the screw. Ask the students what part of the screw helps to do the work (ridges, or threads). Explain that if the threads on a screw are closer together the screw will help do work with less force. Screws with a small pitch (distance between the threads) have a high mechanical advantage. You may wish to direct some students to the tie-in follow-up at this time.

5. Explain that just as the first three simple machines were based on the inclined plane the last three are based on the lever.
6. Point out that levers are divided into three classes, the first, second, and third. A seesaw and a pair of pliers are in the first class. In first-class levers the fulcrum will always be between the two forces, the effort and resistance. In a second-class lever the fulcrum is located at one end. Where is the resistance (in the center, or between the fulcrum and the effort)? Mention the nutcracker as an example of a second-class lever. On the board draw a picture of a baseball bat with hands grasping it in the proper position. Ask the students which hand is the effort (top hand). What is the resistance (ball)? What is the bottom hand (fulcrum)? Tell the students that the baseball bat is a third-class lever. The effort is in the center.

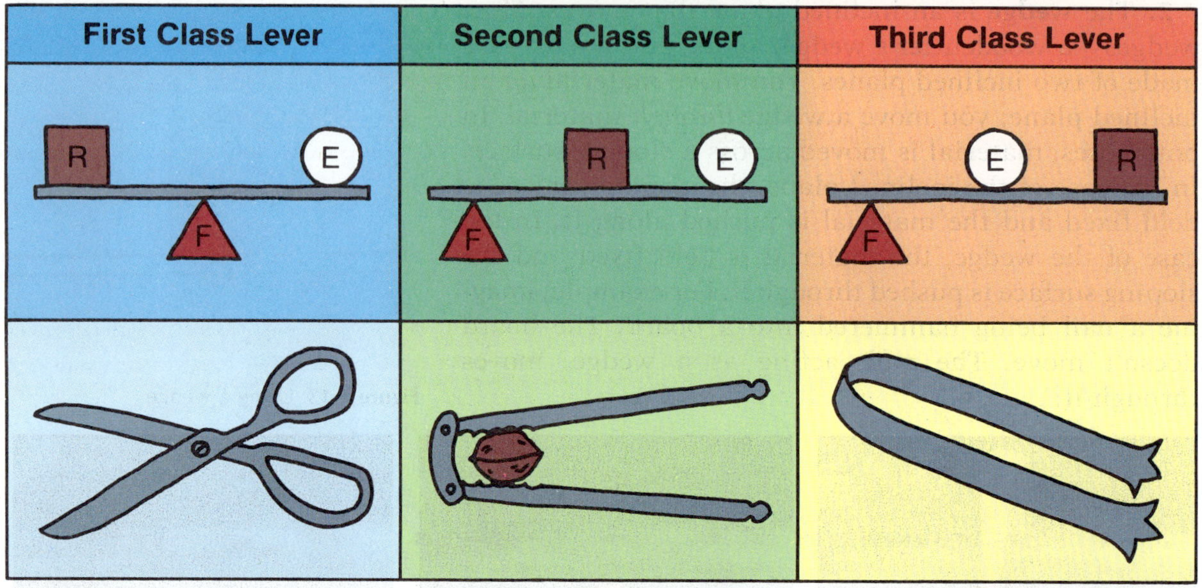

Figure 4.15 Notice the relative locations of the fulcrum, the effort force, and the resistance force.

4. The **lever** is probably the simple machine that you are most familiar with. Many household gadgets and tools are forms of levers. A lever is a bar or board that is free to pivot about a fixed point. The fixed point is called a **fulcrum.**

Levers are divided into three groups: first-class, second-class, and third-class. The classes are based on the relative locations of the fulcrum, the effort force, and the resistance force. Scissors and shovels are examples of first-class levers. Bottle openers and nutcrackers are second-class levers. Tweezers and fishing rods are third-class levers.

The mechanical advantage for any lever can be quite large. To calculate the mechanical advantage for a first-class lever, you divide the distance from the effort to the fulcrum (D_E) by the distance from the resistance to the fulcrum (D_R). Imagine you use a crowbar as a lever to lift up a window. The fulcrum is located very close to the bottom of the window. If the effort distance is 50 cm and the resistance distance is 5 cm, what is the mechanical advantage?

$$\text{M.A.} = \frac{D_E}{D_R} = \frac{50\text{ cm}}{5\text{ cm}} = 10$$

If you have a fixed point and a long and rigid enough lever, you can lift any weight. The great Greek scientist Archimedes said it best: "Give me a fixed point, and I will move the earth."

5. The wheel is often used much like a roller as a means to reduce friction. When used as an axle, on the other hand, it is considered to be a machine. The **wheel and axle** is a simple machine that is made up of two circles. The axle is the smaller circle. The wheel is the larger one. A screwdriver or a steering wheel is a wheel and axle. So is the crank shown in Figure 4.17.

From these examples you can see that the wheel and axle is really just a lever that can move around in a circle. Since the wheel has a larger radius than the axle, the distance over which the force is applied, or effort distance, is greater than the distance over which the weight is lifted, or resistance distance.

The mechanical advantage is calculated by dividing the radius of the wheel (D_E) by the radius of the axle (D_R). Imagine you are making ice cream in an ice-cream machine. The crank you are turning is the wheel part of your wheel and axle ice-cream machine. The radius of your wheel (length of the crank) is 25 cm. The radius of the axle is 5 cm. What is the mechanical advantage of using this ice-cream machine?

$$\text{M.A.} = \frac{D_E}{D_R} = \frac{25\text{ cm}}{5\text{ cm}} = 5$$

Figure 4.16 Using wheels and axles

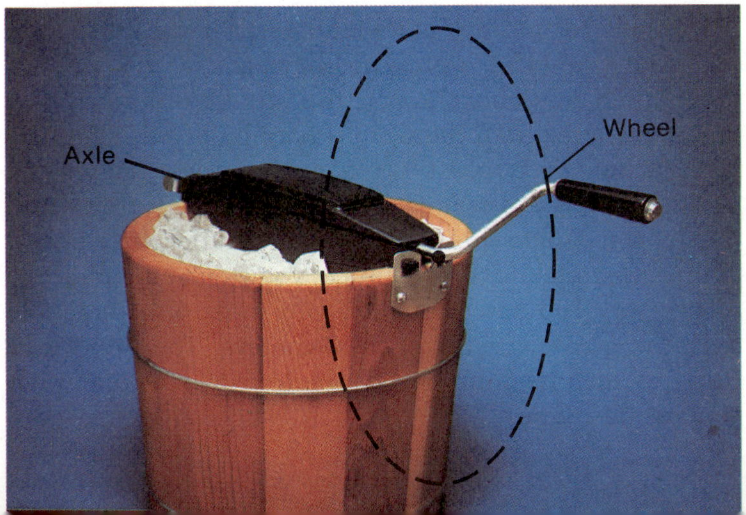

Figure 4.17 The crank is the wheel part of this wheel and axle machine.

7. Tell the students that the fifth simple machine is the wheel and axle. Define this machine as a kind of lever with a wheel that is attached to a rod. The wheel and the rod move together. Note that a doorknob, a steering wheel, and a windlass are all examples of a wheel and axle. Draw one of these on the board. Label its wheel and its axle. Point out that the axle has a smaller diameter than the wheel. Use a car steering wheel as an example to point out that when the driver turns the wheel a relatively small amount a great deal of movement is produced.

SCIENCE BACKGROUND

There are two basic differences between a pulley and a wheel and axle. A single pulley is a grooved wheel around which a rope is wound, and a wheel and axle is not; a single pulley moves while its fulcrum remains stationary, and the axle moves with the wheel in the wheel and axle.

8. Tell the students that another type of lever is a pulley. A pulley is a grooved wheel that rotates on a fixed point, or fulcrum. Ask the students to name some examples of pulleys (pulley on a flagpole, components of cranes and elevators). To illustrate the operation of a pulley, draw a flagpole on the board. Point out that you pull down on the rope and the flag goes up. In this case, the pulley changes the direction of the force. Note that you must pull down with the same amount of force as the object weighs. When the pulleys are used in combination, however, they allow you to use less effort.

9. Mention that the mechanical advantage of a pulley can be increased by using many pulleys that are joined together. This arrangement of pulleys is a block and tackle. You may want to refer some students to the tie-in follow-up at this time.

6. A pulley is a lever that rotates around a fixed point. It has two functions: to change the direction of the force and to provide a mechanical advantage. For the first function a simple pulley such as the one shown on the left will do. To achieve a mechanical advantage, you need a movable pulley such as the block and tackle shown on the right.

Figure 4.18 *left:* A simple pulley; *right:* A block and tackle

The rule for pulleys is the same as for the other machines. If you want to lift a heavy weight with a smaller force, then you must apply the smaller force over a larger distance than the weight is to be lifted. Suppose, for example, you must lift a car engine that weighs 1,000 newtons a distance of 1 meter. With a block and tackle, you can do this by exerting a smaller force over a greater distance. If, for instance, you exert enough force to lift only 100 newtons, then you must pull the cable over the greater distance of 10 meters. The important thing to keep in mind is that the total work done must remain the same.

ACTIVITY

How Do Pulleys Help Do Work?

Materials

pulleys
ring stand and ring
string
scissors
meterstick
book
spring scale

Procedure

1. Tie a string around the book and weigh it using the spring scale.
2. Attach the ring halfway up the ring stand.
3. Attach a pulley to the ring.
4. Tie a longer piece of string to the book and thread it through the pulley.
5. Place the book on the table and attach the spring scale to the free end of the string.
6. Have a friend hold the ring stand while you pull the spring scale and note the reading on the scale when the book is lifted off the table.

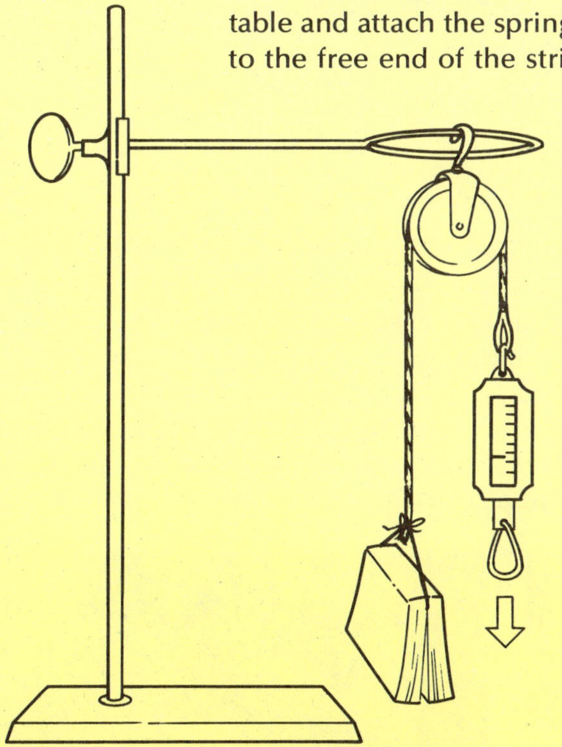

Questions

1. What was the weight of the book?
2. How much force was needed to lift the book?
3. Draw a diagram of the setup and label the fulcrum.
4. How is a pulley like a lever?
5. How do pulleys help do work?

1. Answers will vary.
2. Answers should be the same as in number 1.
3. See diagram in text. Fulcrum is the point where the string from the ring touches the pulley.
4. Like a lever, a pulley consists of a fulcrum, the resistance being moved, and the effort being applied to cause the motion.
5. Some increase force. Some change the direction of the force.

Study Questions
1. by applying a small force over a large distance
2. *inclined plane*—used to move heavy objects that would be impossible for you to lift directly from one level to another; *wedge*—applied through a material to cut it or drive another object into it; *screw*—used to move an object up or down much like an inclined plane does but in spiraling rotation; *lever*—used to pry up or pry open objects too heavy or awkward to move by hand; *wheel and axle*—used to gain force or to gain speed; *pulley*—used to pull down on an object in order to lift it more easily

Study Questions for 4.3

1. How do simple machines change the force?
2. List the names of the simple machines and briefly describe how each is used.

ACTIVITY: How Do Pulleys Help Do Work?

In this activity the students will demonstrate that pulleys can either change the direction of the force or the amount of force.

● Laboratory Skills: Manipulation of Materials
Experimentation/Demonstration

FOLLOW-UP

Reinforcement: Machine Collage

Have students cut out magazine pictures of machines that are based on the simple machines. Ask the students to circle the simple machine in each picture and then arrange the pictures in a collage.

Enrichment: How Can Friction Be Reduced on an Inclined Plane?

Point out that objects moving up inclined planes must overcome friction. Ask how this can be done (use of ball bearings, pencils, oil on surface of plane). Have students design experiments to prove their answers.

Tie-in: Industrial Arts

Ask the industrial arts teacher to demonstrate for students the relationship that exists between pitch and the mechanical advantage of a nail.

4.4

OBJECTIVES: The student should be able to

- **define** the term *power*;
- **define** the term *watt*;
- **calculate** power in a given problem;
- **define** the term *efficiency*.

SCIENCE TERMS

power
watt
efficiency

ADVANCED PLANNING

Prepare a work sheet of power problems the students can solve without a pencil and paper.

LESSON PLAN

1. Explain to the students that often the amount of work to be done is not the only factor to consider when a job must be done. It is also important to consider how long the job will take. When you consider the rate at which the work is done, you are speaking about power.
2. Define *power* as the amount of work done in joules divided by the time in seconds it takes to do that work. Explain that power is usually labeled in watts (or kilowatts), a term named after the Scottish engineer and inventor, James Watt. Refer some students to the tie-in follow-up at this time. Stress that a watt is equal to 1 joule per second of work.

4.4 Power and Efficiency

You will find out

- how to compute the rate at which work is done;
- what is meant by efficiency.

In many cases it is necessary to know more than the amount of work to be done. You must also know how long it takes to do the work. Suppose, for example, that you want to move a thousand books onto the storeroom shelves. You can use a forklift to raise all the books onto the shelves. Or you can place the books onto the shelves one at a time. In either case, the amount of work done against gravity will be the same. But the time it takes to do the work in the two cases will be greatly different. You need a quantity that measures how fast work is done. **Power** is the quantity that measures the rate at which work is done. In other words, power is the amount of work divided by the time it takes to do the work.

$$\textbf{Power} = \frac{\textbf{work}}{\textbf{time}}$$

The unit of power commonly used by physical scientists is the **watt,** named after the Scot, James Watt, who improved the steam engine. One watt is one joule of work done in one second. The watt is usually used in connection with electric power, as in a 60-watt light bulb, but a watt can describe any kind of power.

Figure 4.19 The electric mixer is more powerful than the spoon because the mixer can do the same amount of work in less time.

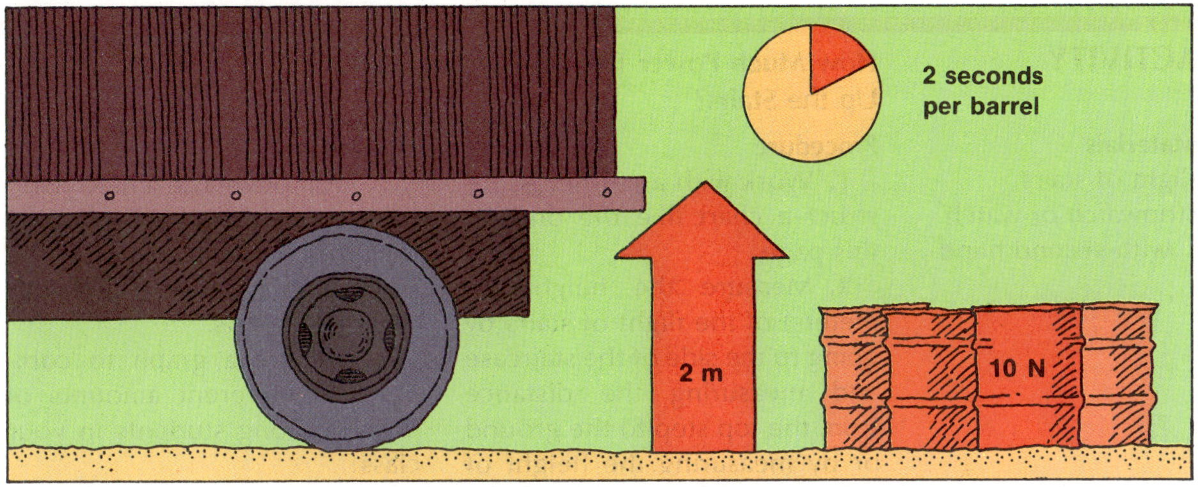

Figure 4.20 Calculating power

EXAMPLE One hundred barrels weighing 10 newtons each are lifted 2 meters onto a truck at a rate of 1 barrel every 2 seconds. How much power is involved in doing this work?

SOLUTION Since power $= \frac{\text{work}}{\text{time}}$, and you are not given the amount of work in the problem, you must first find the work in joules by using the formula, work = force × distance. Therefore, work = 10 N × 2 m = 20 N.m, or 20 J for each barrel lifted. For 100 barrels, 20 J/barrel × 100 barrels, or 2,000 J, are needed. The total time for all 100 barrels is 2 seconds per barrel times 100 barrels, or 200 seconds.

$$\text{Power} = \frac{2{,}000\ \text{J}}{200\ \text{s}} = 10\ \frac{\text{J}}{\text{s}},\ \text{or 10 watts}$$

How much power would be used if all the barrels are lifted onto the truck in 4 seconds? Since the work to lift 100 barrels remains the same, you use the 2,000 joules of work that you found above.

$$\text{Power} = \frac{2{,}000\ \text{J}}{4\ \text{s}} = 500\ \frac{\text{J}}{\text{s}},\ \text{or 500 watts}$$

(Since 1 kilowatt = 1,000 watts,
500 watts = 0.5 kilowatt.)

Do You Know?
A sleeping adult uses energy at a rate of about 80 watts.

3. On the board, write the formula power $= \frac{\text{work}}{\text{time}}$. Do some problems like the following to give your students the opportunity to use the formula.
   - **a.** How much power does a tractor have if it can do 8,400 joules of work in 10 seconds (840 watts)?
   - **b.** How much power does a locomotive have if it can do 7,500 joules of work in 5 seconds (1,500 watts)?
4. Have available a work sheet of power problems using numbers the students can easily manipulate. Have the students solve the problems with you in class. In order to be sure each student knows how to use the formula to determine power, provide enough problems so that you can call on every individual at least once.

● **Mathematics Skills: Computation**
Problem Solving

ACTIVITY: How Much Power Do You Use to Walk up the Stairs?

In the discussion before the activity, emphasize the importance of climbing the stairs at a normal brisk pace but not at breakneck speed.

Although students should work in pairs and time each other, each student should keep an individual data chart.

Have the students convert their weight to newtons by multiplying their weight in pounds by 9.8 and dividing the result by 2.2.

Laboratory Skill: Experimentation/Demonstration

ACTIVITY

How Much Power Do You Use to Walk Up the Stairs?

Materials

flight of stairs
stopwatch or watch with second hand

Procedure

1. Work with a partner. Construct a chart like the one on this page.

2. Measure the height (in meters) of the flight of stairs by going to the side of the staircase and measuring the distance from the top step to the ground or by measuring the height of one step and multiplying it by the number of stairs. Convert the height to meters.

3. Stand at the bottom of the stairs. When your partner says go, walk up the stairs.

4. Have your partner record the time in seconds that it took you to walk the distance.

5. Now time your partner and record the time on your partner's chart.

Questions

1. Complete the chart. Find the power you used by finding the work you did going up the stairs and dividing it by your time in seconds.

2. Prepare a graph to compare the different amounts of power among students in your class.

1. Chart will vary depending on weight of student and number of seconds taken to climb stairs.
2. Graph will vary depending on students sampled.

Name	Weight *(newtons)*	Distance *(meters)*	Work = Force (*N*) × Distance *(joules)*	Time *(seconds)*	Power = $\frac{\text{work}}{\text{time}}$ *(watts)*

Figure 4.21 The photographs show two machines used for snow removal. The power and efficiency of each machine is different.

A machine that could do a lot of work in a short time would be considered very powerful. When people buy and use machines, they are interested in power. They are also interested in the **efficiency** of their machines. Efficiency is a measure of how much work a machine puts out compared to the amount of work put into that machine. No machine can put out more work, output, than the work put in, input. If a machine's output equals its input, it is a very efficient machine. This almost never happens. The reason most machines have less output than input is that some of the effort force put into a machine is lost due to friction. Machines with the least amount of friction are the most efficient. High efficiency and power together are a winning combination.

Do You Know?
It has been estimated that all the machines in the United States do the work equivalent of over a thousand workers for every man, woman, and child in the country.

Study Questions for 4.4

1. Find the power delivered by a force of 10 newtons that acts through a distance of 4 meters in 8 seconds.
2. Why is the output of a machine almost always less than the input?

1. $\text{Power} = \frac{\text{work}}{\text{time}}$

Work = force × distance moved in direction of force = 10 N × 4 m = 40 N.m = 40 J

$\text{Power} = \frac{40\ \text{J}}{8\ \text{s}} = \frac{5\ \text{J}}{\text{s}} = 5\ \text{watts}$

2. Some of the input (effort force) is lost due to friction.

5. Tell the students that efficiency compares output and input. Explain that the ideal machine would be totally efficient. Work output would equal input. However, friction decreases efficiency.
6. Conclude the lesson by stressing that even though machines are important in our lives not even the most sophisticated computer can replace people. A machine is only as good as the person who is operating or programming it.

FOLLOW-UP

Reinforcement: Power Problems

Prepare an extra work sheet of power problems with one or two initial examples that are completely worked out. Students who are having difficulty can use the solutions as models in completing the problem sheet.

Enrichment: Horsepower and Mileage

The power of cars is measured in horsepower. Ask several students each to visit a different car dealer in the area. Have the students find out the horsepower and estimated mileage of the different models each dealer sells. List the information that the students gather on the board. Have all students organize this information into a chart and a graph. What relationship can they see between horsepower and fuel use?

Tie-in: James Watt

James Watt was one of the most prolific inventors of his time. Ask the students to do research on Watt and prepare a list of his inventions and ways they help us.

What Did You Learn?

4.1 • The amount of work done in moving an object is equal to the strength of the force that is used times the distance the object is moved in the direction of the force.

4.1 • The unit of work is the joule.

4.2 • A machine is any device that helps people do work.

4.2 • Simple machines are machines that do work by increasing forces.

4.2 • All machines operate on the principle that work or energy in equals work out plus energy losses.

4.2 • The effort force is the force applied to an object in order to move it; the resistance force is the weight of the object being moved.

4.2 • The distance over which the effort force acts is the effort distance; the distance over which the resistance force acts is the resistance distance.

4.2 • The mechanical advantage of a machine is the amount by which a machine increases the effort applied.

4.3 • To increase force, simple machines apply a small force over a large distance.

4.3 • The six simple machines are the inclined plane, the wedge, the screw, the lever, the wheel and axle, and the pulley.

4.3 • There are three classes of levers.

4.4 • The rate at which work is done, or the power, is the amount of work divided by the time it takes to do the work.

4.4 • The unit of power is the watt.

4.4 • Efficiency measures how much work a machine puts out compared to the amount of work put in.

Key Terms

work
joule
machine
simple machines
mechanical advantage
effort force
resistance force
effort distance
resistance distance
inclined plane
wedge
screw
lever
fulcrum
wheel and axle
pulley
power
watt
efficiency

Biography

Archimedes (287–212 B.C.)

Archimedes was perhaps the greatest in a long line of great Greek scientists. He was known far and wide as the best mathematician of his day, and his understanding of simple machines made him the one man in the world that the Roman army feared the most.

In the course of writing about the use of pulleys and levers to do work, Archimedes once boasted, "Give me a fixed point, and I will move the earth."

The king of Syracuse didn't want to move the earth, but he did have something heavy he wanted moved. It seems that his navy had built a very large ship. This ship was so large, in fact, that the navy could not push it into the water. The king heard of Archimedes' boast about moving the earth and took him up on it. "Move that boat into the water," he challenged Archimedes.

"I will," said Archimedes, and he constructed a system of rope pulleys and wheels and axles to do the job.

When the system was ready, Archimedes hooked one end of the rope to the ship and handed the other end to the king. As the king pulled the rope, the ship moved steadily down the beach into the water.

The last time the king of Syracuse came to Archimedes for help was when the Romans were planning an invasion. Almost single-handedly, Archimedes held off the Romans. He did this by building huge cranes that could seize the Roman ships, raise them into the air, and smash them to pieces against the rocks. In addition, he designed huge levers that hurled great rocks over the walls.

The Romans retreated in panic, terrified by "this mathematician, who plays pitch and toss with our ships and hurls a multitude of missiles at us."

Vocabulary

1. work
2. resistance force
3. power
4. mechanical advantage
5. wheel and axle
6. fulcrum
7. joule
8. pulley
9. watt
10. machine

● Word Skill: Vocabulary

What Do You Remember?

1. a force and a distance
2. any device that helps to do work
3. increase forces
4. Apply a small force over a large distance to get a large force working over a short distance.
5. increase force
6. inclined plane—slanted surface; ramp for the handicapped
wedge—inclined plane that moves; ax
screw—twisted wedge and bar that turns around fixed point; screwdriver
lever—bar that rotates on a fixed point; crowbar
wheel and axle—large wheel and small wheel that turn together; doorknob
pulley —grooved wheel rotating on a fixed point; arrangement on a flagpole
7. The effort force is used to move an object that otherwise resists being moved. The weight of the object to which the effort force is applied is caled the resistance force.
8. Mechanical advantage measures how much force you gain by using a machine.
9. Power is the rate at which work is done. It is important because the cost of having work done is determined in many cases by how long it takes to do the work. If this same work can be done at a faster rate, it will cost less.
10. $\text{Power} = \dfrac{\text{work in joules}}{\text{time in seconds}}$

TO THINK ABOUT AND DO

Vocabulary ●

Number a piece of paper from 1 to 10. Each numbered phrase below describes a term used in the chapter. On your paper write the term next to the number of the phrase describing it. Unscrambling the words listed may help you.

1. force times distance
2. weight of object you apply work on
3. work divided by time
4. amount by which a machine increases your effort
5. machine made up of two circles of different sizes
6. fixed point about which a lever turns
7. unit used to measure work done
8. a lever rotating about a fixed point
9. unit used to measure power
10. any device that helps to do work

wpeor
atwt
orkw
uoelj
kwro utnip
clufmur
kowr tuuopt
lyuelp
rvele
rscwe
seristcean recof
gdwee
nilnicde nealp
eewhl nda xela
hicenam
ehnclmcaia avnaedatg

What Do You Remember?

1. What two conditions must be met for work to occur?
2. What is a machine?
3. How do machines help you work?
4. What is the basic work principle on which simple machines operate?
5. How do all simple machines help do work?
6. Name six simple machines. Describe each. Give an example of each.
7. How are effort force and resistance force related?
8. What is meant by mechanical advantage?
9. What is power? Why is it important?
10. How is the power of a machine calculated?

Applying What You Have Learned ●

1. How much work is done if a box is moved 80 meters with a force of 5 newtons?
2. How much work is done if Sally spends 2 hours pushing a refrigerator. She uses 45 newtons of force, but the refrigerator doesn't move.
3. Explain how wedges and screws are like inclined planes and how pulleys and wheels and axles are like levers.

4. One of two trucks allows you to do 560 joules of work in 40 seconds. The other truck lets you do 999 joules of work in 45 seconds. Which is more powerful?

Research and Investigation ●

1. Construct a working model of each of the simple machines. Display them in such a way that others can see how they help to do work.
2. Many science-fiction stories deal with a world that is dominated by machines. Read a few of these stories in the library. Then write a story of your own to share with your classmates or to publish in the school journal.

Applying What You Have Learned

1. Work = force × distance moved in the direction of the force
   Work = 5 N × 80 m = 400 N.m. = 400 J
2. Work = force × distance moved in the direction of the force
   Work = 45 N × 0 m = 0 N.m. = 0 J
   (No work is done if object does not move.)
3. Wedges are inclined planes that move. A screw is a combination of an inclined plane and a cylinder. A pulley and a wheel and axle are like a lever that operates in a circle.
4. $\text{Power (truck 1)} = \frac{\text{work}}{\text{time}} = \frac{560\ \text{J}}{40\ \text{s}} = 14\frac{\text{J}}{\text{s}} = 14\ \text{watts}$

   $\text{Power (truck 2)} = \frac{\text{work}}{\text{time}} = \frac{999\text{J}}{45\ \text{s}} = 22.2\frac{\text{J}}{\text{s}} = 22.2\ \text{watts}$

   Truck 2 is more powerful.

● Comprehension Skill: Applied

Research and Investigation

1. In many cases plastic toys can be used.
2. A good model for the students to read is George Orwell's *Nineteen Eighty-Four*.

● Reference/Research Skills

CHAPTER BACKGROUND

Since both liquids and gases flow and take the shape of their containers, they are sometimes grouped into the single category fluids. Fluids exert pressure on the objects around them.

Certain characteristics distinguish the pressure that liquids exert from the pressure exerted by solids. Pressure from a liquid is applied in all directions. It depends on the depth of the liquid and not the shape of the container that is holding the liquid.

In the same way that pressure from liquids increases with depth, pressure from atmospheric gases decreases with altitude. As you move farther above the surface of the earth, the amount of air above you decreases. As a result, the air pressure decreases. This change in pressure can make you feel uncomfortable when you ride in an elevator.

Like liquids, gases exert their pressure in all directions. An increase in pressure on liquids and gases will cause a decrease in their volume.

The fact that people constantly experience air and moisture in their daily lives results in their ignoring the many ways that these two components of the environment affect them. Air, for example, is heated unequally above forests, cities, and bodies of water. This unequal heating causes up-and-down currents that affect wind direction. Such currents sometimes give rise to the turbulence that results in a bumpy airplane flight.

5

Forces in Liquids and Gases

Introduce this chapter by having the students observe the two demonstrations that follow. At this time students may not understand why certain things happen. Do not say that anyone is wrong in an observation or explanation. Rather, leave the students in a wondering mood.

1. Fill a drinking glass to the brim with water. Place a piece of cardboard over it, hold your hand against the cardboard, and invert the glass. Be sure that there aren't any air bubbles. Take your hand away from the cardboard. What has this demonstration shown about air (air pressure holds cardboard up so that water stays in glass)?
2. Hold your finger over the end of a straight glass tube or a clear soda straw. Lower it into a glass of colored water. Remove your finger and observe what happens (air pressure keeps water out of straw). Replace your finger on top of the tube and lift the tube from the jar. What happens (air pressure keeps water in straw)?

Although you are probably not aware of it, the forces in liquids and gases affect you every day of your life. Air (a gas) and water (a liquid) are important parts of your world. You experience the forces in liquids and gases every time you take a breath, turn on a faucet, or feel a soft breeze across your face. These same forces allow you to drink from a straw, float in a swimming pool, and vacuum a carpet.

Why do some objects sink in water while others float? How is it possible for a submarine to do both? What keeps an airplane up? What keeps a hot-air balloon up? What brings it back down? Why do astronauts have to wear special pressurized suits? The answers to these questions are simple when you understand the forces in liquids and gases.

Figure 5.1 Forces in liquids and gases keep a sailboat afloat and moving.

Lesson Organizer

5.1 Pressure in Liquids

Pages 94–97
Advanced Planning: Page 94
Activity: How Can You Measure Water Pressure? Page 96
Follow-up: Page 97
- **R** The Bends
- **E** A Hydraulic Ram
- **T** Water Pressure at a Garage

5.2 Buoyancy

Pages 98–103
Advanced Planning: Page 98
Activity: How Do the Densities of Some Solids and Liquids Compare? Page 102
Follow-up: Page 103
- **R** Density Diver
- **E** Floating Eggs
- **T** Why Does an Iron Ship Float?

5.3 Atmospheric Pressure

Pages 104–105
Advanced Planning: Page 104
Follow-up: Page 105
- **R** How Do People Breathe?
- **E** A Siphon Fountain
- **T** Forecasting Weather

5.4 Objects Moving in Air

Pages 106–109
Advanced Planning: Page 106
Activity: How Can You Demonstrate Differences in Air Pressure? Page 108
Follow-up: Page 109
- **R** Drag and Lift
- **E** A Paper Airplane Race
- **T** Airplanes

See ACTIVITIES book pages 25-30 for laboratory, skill, and review exercises for Chapter 5.

5.1

OBJECTIVES: The student should be able to

- **define** the term *pressure*;
- **calculate** pressure when given the force and the area;
- **explain** that pressure increases as the area to which the force is applied decreases;
- **list** an application of pressure in daily life;
- **explain** that pressure increases with depth in a liquid.

SCIENCE TERMS

pressure
fluids

ADVANCED PLANNING

1. Obtain tire hand pump, one-hole stopper that will fit into faucet, piece of glass tubing to fit into stopper, about six meters (20 feet) of rubber tubing that will fit glass tube, and wire that can be used to bind rubber tubing to pump and glass tube.

2. For use in demonstrating the relationship between water depth and air pressure, obtain a can with 5 holes punched in a circle 2 cm (¾ in.) from the bottom rim of the can.

LESSON PLAN

1. Introduce the lesson by telling the students that when you speak of pressure you are referring to the force divided by the area the force is acting on. Point out that you exert pressure on the floor when you stand on it and when you lie down on it. In both cases the force (your weight) is the same. Ask the students if the area the force is acting on is the same in both cases (no). In which case is the area smaller (standing)? In which case is the pressure greater and why (standing; force is exerted over a smaller area)?

● **Mathematics Skills: Computation**
Problem Solving

5.1 Pressure in Liquids

You will find out

- how to find the pressure exerted by an object;
- how pressure varies with area;
- how pressure varies with depth.

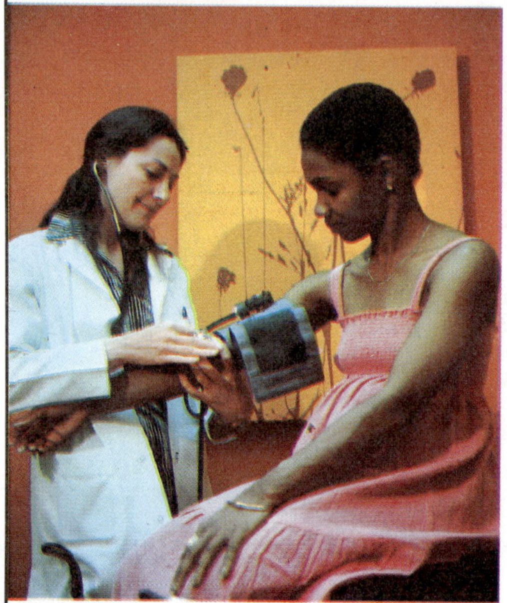

Figure 5.2 Measuring blood pressure

All of you have a rough idea of the meaning of the word *pressure*. Water pressure, air pressure, and blood pressure are three ways you may have heard the word used. To be precise, **pressure** is a force applied over a certain area. You can find pressure if you divide the force in newtons by the area in square centimeters.

$$\textbf{Pressure} = \frac{\textbf{force}}{\textbf{area}}$$

● EXAMPLE Consider an aquarium which has a bottom surface area of 1,000 square centimeters. The weight of the aquarium and the water it contains will exert a force of 300 newtons on the table supporting the aquarium. What is the pressure exerted on the tabletop by the aquarium and the water?

SOLUTION The pressure will depend on the area, 1,000 square centimeters, over which the force acts.

$$\text{Pressure} = \frac{\text{force}}{\text{area}} = \frac{300\ \text{N}}{1{,}000\ \text{cm}^2} = 0.3\ \text{N/cm}^2$$

Figure 5.3 The force is concentrated over the entire area of the bottom of the aquarium.

On the other hand, if instead of a table the aquarium was supported by narrow legs having a total area of 10 square centimeters, the force would be concentrated over a much smaller area. Now the pressure would be a hundred times greater.

$$\text{Pressure} = \frac{300 \text{ N}}{10 \text{ cm}^2} = 30 \text{ N/cm}^2$$

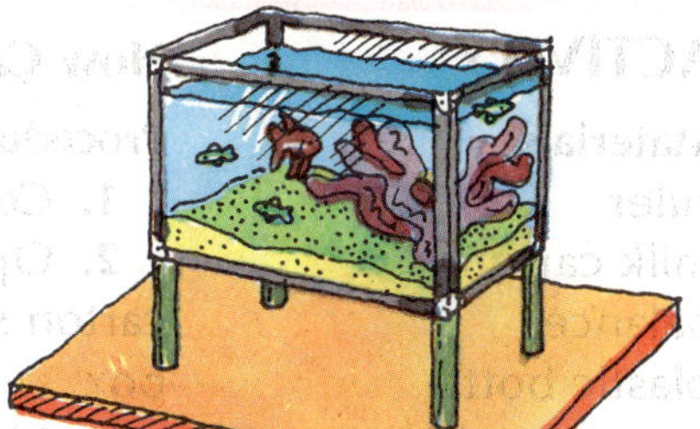

Figure 5.4 The force is concentrated over the relatively small area of the aquarium's narrow legs.

From the example you can see that the smaller the area to which a force is applied, the greater the resulting pressure. This explains why it is more comfortable to sleep on a mattress than on a bed of nails. On a mattress your weight is spread out evenly over a large area. On a bed of nails your weight is spread out over a small area—only the points of the nails.

Because a liquid has weight, it pushes down on the bottom of a swimming pool or an aquarium. In fact, the deeper the liquid the greater its pressure becomes. Consider, for example, two identical swimming pools, one full of water and the other only half full. The pressure at the bottom of the full pool is twice as great as the pressure at the bottom of the half-full pool.

Do You Know?

A person weighing 600 newtons in high-heeled shoes exerts a pressure on the floor of about 80 N/cm², more than three times that of a six-ton elephant.

Figure 5.5 The pressure at the water's surface in both pools is the same—zero. The pressure at the bottom of the half-full pool is only half as much as it is at the bottom of the filled pool.

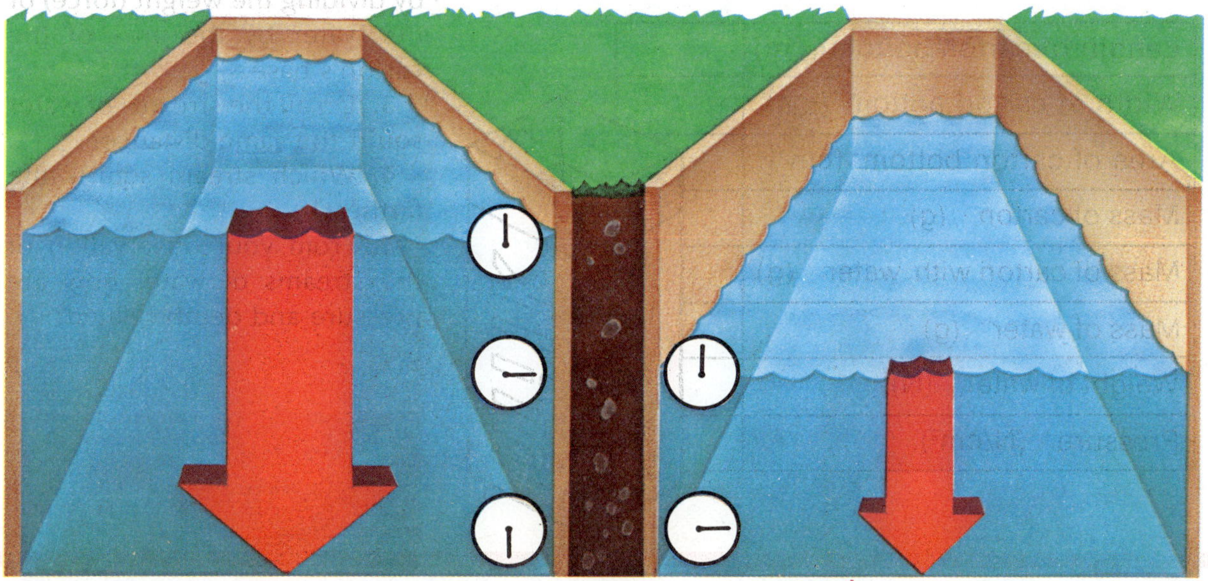

2. The following problems can be used as practice in reinforcing the concept that pressure is equal to force divided by area.

a. What is the pressure exerted by a 500-N force on an area of 10 cm² (50 N/cm²)? If you wish to use English terms, simply change the labels to pounds and square inches to retain easy-to-work-with numbers.

b. Is there more pressure when a force of 600 newtons is exerted over an area of 30 square centimeters or when a force of 800 newtons is exerted over an area of 40 cm square centimeters (same, or 20 N/cm²)?

3. Point out that both air and water can exert pressure. Ask the students if they think that the water coming from the faucet has enough pressure to lift one of them. Have them state the reason for their opinions. To test these opinions, construct a simple hydraulic elevator with a tire hand pump. Connect the tube from the pump to a 6-meter length of rubber tubing. Bind the connection with wire so that it will not blow out. Connect the tube to the water faucet with a glass tube and a one-hole rubber stopper. Bind the connection of the tube and the glass tube in the stopper. Ask one of your smaller students to assist. Have this student sit on the handle of the pump. Turn the water on slowly and watch the student rise as the handle rises. You may have to hold the stopper in the faucet.

SCIENCE BACKGROUND

The pump made for the demonstration is a hydraulic lift, a lift that operates because of a fluid that is under pressure. Explain that the fluid does not have to be water. It can be air.

ACTIVITY: How Can You Measure Water Pressure?

For this activity prepare a plastic bottle or can for each pair of students. Do this by punching holes of similar diameter in a vertical row from the bottom of the bottle or can to three fourths the way up the side. The holes should be covered with adhesive tape.

Have pans or cans whose diameters are large enough for the water to be directed easily into them when it spurts out of the bottle. Caution the students to work carefully so they will not get wet or get others wet. Review the method for finding the area of a rectangle.

Remind the students that their measurements should be accurate. Through observation the students should conclude that the pressure increases with depth.

● **Laboratory Skills: Experimentation/Demonstration Manipulation of Materials**

● ACTIVITY

How Can You Measure Water Pressure?

Materials

ruler
milk carton
balance
plastic bottle or can with a vertical row of holes
adhesive tape
collection tin

Procedure

1. Copy the chart below.

2. Open the top of your milk carton so that it is a rectangular box.

3. Measure the length and width of the base of the carton and multiply these numbers to find the area.

4. Use the balance to find the mass of the milk carton. Record this mass on your chart.

5. Fill the carton with water to a depth of 5 cm and use the balance to find the mass of the carton plus the water. Record this mass on your chart.

6. Subtract the mass of the carton from the mass of the carton plus the water to find the mass of the water. Record the result on your chart.

7. Think about how depth effects pressure.

8. Cover the holes in your container with tape.

9. Fill the container with water.

10. Place the container over your collection tin and remove the tape from all the holes.

11. Observe the streams of water to see whether they all squirt the same distance.

Length of carton bottom (cm)	
Width of carton bottom (cm)	
Area of carton bottom (cm^2)	
Mass of carton (g)	
Mass of carton with water (g)	
Mass of water (g)	
Weight of water (N)	
Pressure (N/cm^2)	

Questions

1. What is the weight of the water? Hint: 1000 g = 9.8 N.

2. Recall that weight is a measure of the pull of gravitational force. Then calculate the pressure on the base of the carton by dividing the weight (force) of the water by the area of the carton's base.

3. Do all the streams of water squirt the same distance?

4. Which stream squirts the farthest?

5. From your observation of the streams of water how are pressure and depth related?

1. Answers will vary.
2. Answers will vary but will equal the weight of the water in answer 1 divided by the area of the base of the carton.
3. no
4. stream from the bottom hole
5. Greater depth means a greater pressure.

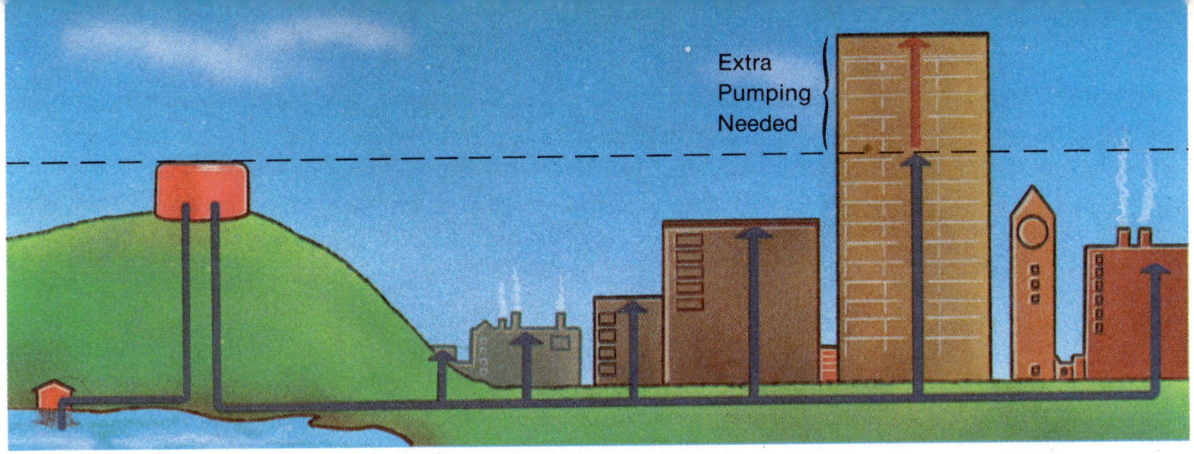

Figure 5.6 The weight of the water in the tower produces the pressure to drive water through pipes and into buildings.

One of the most important applications of the increase of pressure with depth is water towers. Water is pumped from a reservoir into a tower that is usually located on the highest hill in the area. The weight of the water in the tower produces a large amount of pressure at the bottom of the pipe. This pressure is used to drive the water through the pipes to houses and other buildings. In the case of houses on a hill higher than the water tower or in skyscrapers above the water tower, water pumps have to be used.

The increase of pressure with depth is true of all **fluids.** By fluids, scientists mean any substance that flows. Generally, liquids and gases are considered to be fluids, and solids are not. Even though liquids and gases flow readily, there is no perfectly fluid substance. All fluids do not flow at the same rate. Honey, water, and air, for example, flow at different rates.

Study Questions for 5.1

1. A container that weighs 120 newtons is placed on a tabletop. If the base of the container has an area of 3,600 square centimeters, how much pressure does the container exert on the tabletop?
2. Two blocks each weighing 100 newtons are placed on a table. If the base of block A has an area of 50 square centimeters and the base of block B has an area of 10 square centimeters, which block exerts the greater pressure?
3. What effect does increasing water depth have on scuba divers?

1. Pressure $= \frac{\text{force}}{\text{area}} = \frac{120\text{ N}}{3{,}600\text{ cm}^2} = 0.033\text{ N/cm}^2$
2. Since the two blocks act with the same force on different areas, the pressure is greater where the area is smaller. In this case, therefore, block B exerts the greater pressure.
3. Scuba divers are subjected to additional pressure as depth increases.

4. Ask the students to imagine a can that has holes in a row 2 centimeters (¾ inches) from the bottom of a can. Draw the can on the board. Ask the students whether the pressure pushing the water out of each hole will be the same or different. Take out a can like the one that you described. Tape the holes. Then fill the can with water. As you hold the can over a sink or collection tin, remove the tape and ask the students to observe (they should note that water squirts out the same distance from each hole). When the can is almost empty, ask the students for an observation (they should note that the water does not squirt out as far when the can is close to empty as it does when the can is full).

FOLLOW-UP

Reinforcement: The Bends
Ask the students to do library research to find out what happens to scuba divers who get the bends. Why do they get them? Can they be cured? If so, how?

Enrichment: A Hydraulic Ram
Hydraulic rams are sometimes used to raise water from a low level to a higher level. Have the students use library references to investigate the uses of hydraulic rams.

Tie-in: Water Pressure at a Garage
Have several students visit a garage near their homes. They should ask the garage owner how the hydraulic lifts are used to lift cars. Do the lifts work on air pressure or water pressure (water)?

SCIENCE BACKGROUND

When scientists speak about an increase of pressure with depth, they mean the depth below the free surface. This depth is measured straight down or perpendicular to the surface. The shape of the container does not influence the pressure.

5.2

OBJECTIVES: The student should be able to

- **define** the term *buoyancy*;
- **define** the term *density*;
- **calculate** density when given the mass and the volume;
- **explain** Archimedes' principle;
- **explain** why some objects float in water and others sink.

SCIENCE TERMS

buoyancy
Archimedes' principle
density

ADVANCED PLANNING

For demonstration you will need a cork, a beaker of water, a stone that will fit in the beaker, a piece of string, and a spring scale.

LESSON PLAN

1. Fill a glass with water and push the cork to the bottom of the glass. Let the cork go. As you do this, ask the students what they observe (the cork rises). Ask what is pushing the cork up (water).

5.2 Buoyancy

You will find out

- what is meant by the word *buoyancy;*
- what is stated in Archimedes' principle;
- how to find the density of an object;
- how to predict whether an object will float.

Have you ever noticed that it's easier to lift a person in water? This is due to the upward force, or **buoyancy** [BOY-uhn-see], of water. Buoyancy is the name given to the force that pushes an object up and makes it seem to lose weight in a fluid. In this case, the amount of weight a person seems to lose in the water is actually the weight of the water that the person moves aside by standing in it.

Why do some objects float in water while others sink? If you want to know which objects will float in water, you need to know more about the force exerted by the fluid. You need to know the strength of the buoyant force compared to the weight of an object.

Over 2,000 years ago the Greek scientist Archimedes figured out the strength of the buoyant force as he was taking a bath. Legend has it that Archimedes became so excited about having solved the buoyancy problem that he jumped out of the tub and ran naked down the main street of town, yelling "Eureka! Eureka!" which means "I have found it." What Archimedes discovered is that the buoyancy of any object is equal to the weight of the liquid that the object displaces, or pushes aside. This relationship between buoyancy and the weight of the fluid being displaced is known as **Archimedes' principle.**

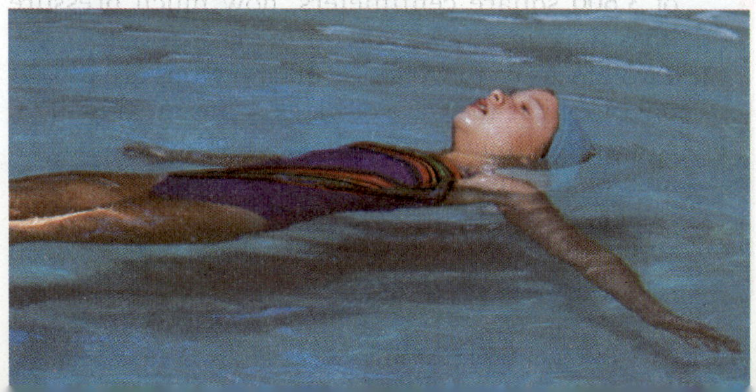

Figure 5.7 The buoyant force of water allows you to float on it.

Since the buoyant force acts upward, it opposes the downward force of the weight of the object. If the buoyant force is equal to the weight of the object, then the object will float. For example, a piece of wood having a volume of 100 cubic centimeters weighs only half as much as 100 cubic centimeters of water, so it will float on water. In other words, an object will float if it weighs less than the weight of the water it displaces.

On the other hand, a piece of steel having a volume of ten cubic centimeters weighs about eight times as much as the same volume of water. You can see that steel will not float. Why then are ships made out of steel? The answer lies in the ship's design. The ship is constructed so that it is made mostly of air. That is, the inside of the ship is hollow. Since the combination of steel plus air weighs less than the water the ship displaces, the ship will float.

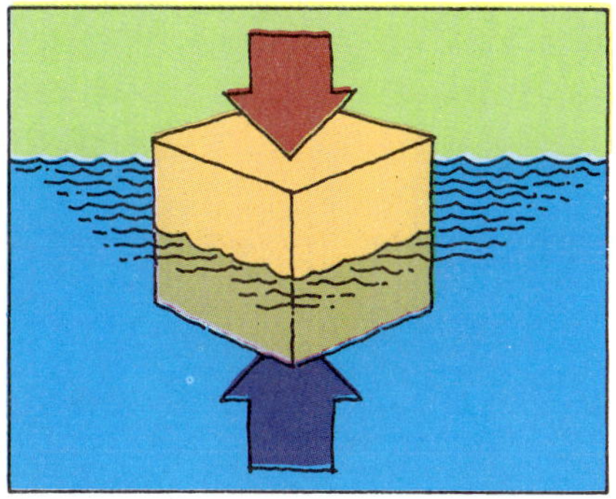

Figure 5.8 The weight of the water displaced by the block equals the weight of the block.

Figure 5.9 The inside of a ship is mostly air.

2. Use the string to suspend a rock from the spring scale. Have the students observe the weight of the stone. Lower the stone into the beaker of water and observe its apparent weight. How has the weight changed (stone's weight is apparently less in water)? Why does the stone seem to weigh less (the water is pushing it up)?

3. Tell the students that the force the water exerted to push the cork to the top of the water and to make the stone appear to weigh less is an upward force called buoyancy. In the cork example this means that the upward force exerted by the water is equal to the weight of the water that is pushed aside by the cork. If the displaced liquid's weight is greater than that of the object, as in the cork example, the object will float. If the displaced liquid's weight is less than the weight of the object, the object will sink.

4. Explain to the students that a property called density determines whether an object floats or sinks. Density is the relationship between the mass of an object and its volume. Density is a characteristic property, a property that can be used to distinguish one object from another. Refer the students to the density table. Have the students observe the label for density at the top of the density column. Ask what two properties must be known to determine the density (mass and volume).

5. Work out some sample density problems for the students. You may want to use the following examples:
   - **a.** What is the density of a liquid if its mass is 15 g and its volume is 10 mL (1.5 g/mL)?
   - **b.** What is the density of a solid whose mass is 140 g and volume is 7 cm^3 (20 g/cm^3)?

● **Mathematics Skills: Computation**
Problem Solving

Figure 5.10 Like submarines, the oceanographic research vessel *Alvin* sinks below the water's surface by filling its hollow tanks with water.

If a ship is built so that it can take water into tanks in its hollow interior, it can be made to sink. When the water is pumped out of these tanks, the ship will rise again. This is the principle behind the submarine.

Another way to express Archimedes' principle is in terms of the **density** of a substance. Density is the mass of a substance in grams divided by its volume in cubic centimeters.

Do You Know?
Not all wood floats. Ironwood has a density of 1.07 times that of water, so it sinks when put in water.

$$\textbf{Density} = \frac{\textbf{mass}}{\textbf{volume}}$$

● EXAMPLE The mass of 10 cubic centimeters of iron is 79 grams. The mass of the same volume of mercury is 136 grams. Which substance has the greater density?

SOLUTION

$$D(\text{iron}) = \frac{\text{mass (iron)}}{\text{volume (iron)}} = \frac{79\text{ g}}{10\text{ cm}^3} = 7.9\text{ g/cm}^3$$

$$D(\text{mercury}) = \frac{\text{mass (mercury)}}{\text{volume (mercury)}} = \frac{136\text{ g}}{10\text{ cm}^3} = 13.6\text{ g/cm}^3$$

Mercury, a liquid metal, has a greater density than iron, a solid metal.

Stated in terms of density, Archimedes' principle says that an object will float in a fluid if the density of that object is less then the density of the fluid. From Table 5-1, you can see that wood, ice, and gasoline will float in water, whereas most other substances will not.

Because it contains salt, sea water is more dense than fresh water. As a result, the same boat would float higher in sea water than in fresh water. The reason for this is that it takes a smaller amount of displaced salt water to equal the weight of the boat.

The water in the Great Salt Lake in Utah is very salty and much more dense than sea water. If you were to go swimming in the Great Salt Lake, you would immediately notice the amazing buoyancy of the very dense and salty water. In fact, you could almost float in a sitting position!

Table 5-1 Densities of Several Materials

Material	Density (g/cm^3)
hydrogen*	0.00009
helium*	0.00018
air*	0.0013
white pine	0.5
gasoline	0.7
oak	0.7
ice	0.92
human fat	0.92
water	1.00
sea water	1.02
bone	1.8
aluminum	2.7
iron	7.9
lead	11.3
mercury	13.6
gold	19.3

*gas densities measured at room temperature at sea level

Figure 5.11 The salt water of the Dead Sea is so dense that you can float effortlessly.

SCIENCE BACKGROUND

The same object will float differently in liquids of different densities because equal volumes of liquids of different densities will weigh different amounts. For example, one liter of fresh water weighs less than one liter of salt water because fresh water is less dense. Therefore, an object will float higher in salt water than fresh water.

6. Point out to the students that some liquids, such as oil and water, do not mix. If two liquids of different densities that do not mix are poured together, they will separate into layers. The less dense liquid will be on top. If the liquids are shaken or stirred together, they may appear to mix. However, once the stirring or shaking has stopped, the two liquids will return to their original layered positions.

ACTIVITY: How Do the Densities of Some Solids and Liquids Compare?

It would save time if you color the corn syrup and vinegar and break the materials into pieces of the appropriate size before the experiment. Baby oil may be substituted for cooking oil.

The students should note that the materials arrange themselves according to density with the most dense materials on the bottom and the least dense materials on the top. If you let the jar stand, the vinegar and corn syrup will eventually mix together.

Laboratory Skills: Experimentation/Demonstration Manipulation of Materials

You have seen how a solid can float in a liquid that is more dense. It is also possible for a liquid to float on another, more dense liquid. A very familiar example of this situation occurs with bottled salad dressing. What do you do to bottled dressing before pouring it on a salad? You shake the bottle to mix the dressing. Why won't the dressing stay mixed? Why must you shake the bottle every time you use it? Salad oil is less dense than vinegar. The oil floats in a layer on top of the vinegar.

ACTIVITY

How Do The Densities of Some Solids and Liquids Compare?

Materials

1 liter clear bottle
200 mL corn syrup
200 mL vinegar
200 mL cooking oil
red food coloring
blue food coloring
Styrofoam pieces about 1 cm × 1 cm
aluminum foil pellets the size of small pebbles
toothpick pieces 2 cm long
2 safety matches
penny
graduated cylinder
Styrofoam cup
stirring rod
apron

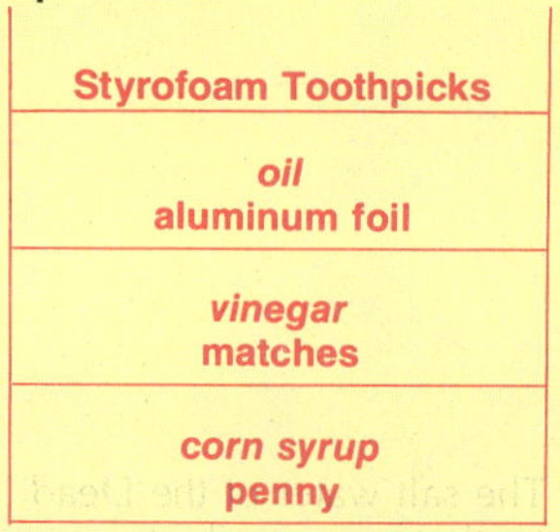

Procedure

1. Put on your apron.
2. Pour 200 mL of vinegar into your bottle.
3. Add a few drops of red food coloring. Stir.
4. Pour 200 mL of corn syrup into a Styrofoam cup.
5. Add a few drops of blue food coloring. Stir.
6. Pour the corn syrup into the bottle of vinegar and observe.
7. Pour 200 mL of cooking oil into the bottle and observe.
8. Drop the matches into the bottle and observe.
9. Drop the Styrofoam pieces into the bottle and observe.
10. Drop the penny into the bottle and observe.
11. Drop the aluminum pellets into the jar and observe.
12. Drop the toothpick pieces into the jar and observe.

1. The corn syrup sinks below the vinegar. Corn syrup is more dense than vinegar. They do not mix.

Questions

1. What happened when the corn syrup was added to the vinegar?
2. What happened when the cooking oil was added to the bottle?
3. Draw a picture with labels to show how the materials arranged themselves in the jar.
4. Which material is the most dense?
5. Which material is the least dense?
6. Name 2 materials more dense than vinegar.
7. Name 4 materials less dense than vinegar.
8. What determines whether a material will float or sink in a certain liquid?

2. The oil floats on the vinegar. The oil is less dense.
3. See diagram at left.
4. penny
5. Styrofoam and toothpicks
6. penny, corn syrup
7. oil, Styrofoam, toothpicks, aluminum foil
8. A less dense material will float on or in a denser liquid.

Figure 5.12 Because it is less dense, oil floats on top of water.

Shaking the bottle mixes the oil and vinegar only temporarily. After a short while the two liquids return to their original layered positions.

Another example of a liquid floating on a liquid occurs with crude oil and sea water. When a supertanker accidentally spills its cargo of crude oil into the sea, what happens? Crude oil is less dense than sea water so the oil floats on top of the water creating an oil slick. Oil slicks can cause serious ecological problems. Sea birds become coated with the oil and die. The oil sometimes floats to shore and fouls beaches. Sea life under the oil layer is threatened. Scientists and engineers are currently studying ways to clean up such oil slicks.

Study Questions for 5.2

1. What is the name of the upward force on an object in a fluid?
2. What is stated in Archimedes' principle?
3. The volume of a 20-gram bar is 5 cubic centimeters. What is the density of the bar?
4. If the density of mercury is 13.6 grams per cubic centimeter, is it possible for lead, whose density is 11.3 grams per cubic centimeter, to float in mercury?

1. buoyancy
2. The buoyancy on any object is equal to the weight of the fluid that the object pushes aside.
3. Density $= \frac{\text{mass}}{\text{volume}} = \frac{20\ \text{g}}{5\ \text{cm}^3} = 4\ \text{g/cm}^3$
4. Yes, this is possible because an object will float in a fluid if the density of that object (11.3 g/cm³) is less than the density of the fluid (13.6 g/cm³).

7. Conclude the lesson by asking the students to observe the following demonstration. Have two beakers on your desk. Fill one with water and the other with alcohol. Drop an ice cube in each. The students should observe that the ice floats in the water but sinks in the alcohol. Why is this the case (ice is less dense than water; alcohol is less dense than ice)?

FOLLOW-UP

Reinforcement: Density Diver

Have several students perform the following experiment. Obtain a tall glass jar with a wide mouth or a two-liter plastic bottle. Wrap a few turns of wire around the narrow part of a rubber bulb from a medicine dropper. Fill the jar to the brim with water. Put a little water in the bulb and float it in the jar of water. Since the bulb should contain enough water so that it is nearly sinking, almost all the rubber should be under water. If the bulb is not just at the point of sinking, adjust it by squeezing air from it a bubble at a time. When the bulb is adjusted, cover the top of the jar with a small piece of rubber sheet. Secure it with a rubber band or string. Press on the rubber sheet and make the diver sink. Release the rubber sheet, and the diver will rise. Ask the students to explain what is happening in terms of buoyancy and density.

Enrichment: Floating Eggs

Set up two glasses of clear liquid. One glass should contain water and the other salt water. Place an egg in each glass. Do not tell the students that the liquids are different. Ask students what they observe and what explanations they can offer. How could they prove their explanations? How does this experiment relate to the fact that ships ride higher in ocean water than in fresh water?

Tie-in: Why Does an Iron Ship Float?

Ask interested students to do research to determine why ships float. The students should first try to find out the mass of an ocean liner such as the *Queen Elizabeth II*. If the mass is known, what can the students conclude about the volume of the ship? Some students may want to do research on the *Monitor* and the *Merrimac*. What problems did their builders encounter? How did they solve them? Others may want to do research on submarines.

5.3

OBJECTIVES: The student should be able to

- **explain** how atmospheric pressure changes with altitude;
- **explain** how Archimedes' principle applies to air;
- **explain** one reason atmospheric pressure is important in daily life.

SCIENCE TERM

atmospheric pressure

ADVANCED PLANNING

For demonstrations you will need a tall heat-resistant glass jar, a piece of paper, a balloon that will fit over the mouth of the jar, a hard-boiled egg with the shell removed, and a liter (quart) glass milk bottle with an opening that the egg can rest in.

LESSON PLAN

1. Introduce the lesson by performing one or both of the following demonstrations:
   a. Crumble a small piece of paper and put it on the bottom of a tall heat-resistant glass jar. With a match light the paper and quickly put a rubber balloon over the mouth of the jar. Let the students observe what happens.
   b. Put a wad of paper in the bottom of the milk bottle. Light the paper and put the hard-boiled egg, pointed end down, in the neck of the bottle. Let the students observe what happens.
2. After you have completed the demonstrations, ask the students what they observed in each case (balloon was sucked into the jar; egg fell into the bottle). Ask why these things happened (oxygen in bottle was used up by fire, so there was more air pressure outside than inside and object was pushed in).

SCIENCE BACKGROUND

When weather reporters send up weather balloons or when people go for balloon rides, the balloons are only partially filled. As the balloons rise, they expand because there is less pressure outside the balloon than there is inside it.

5.3 Atmospheric Pressure

You will find out
- what is meant by atmospheric pressure;
- why atmospheric pressure decreases with height;
- how you use atmospheric pressure.

Figure 5.13 As altitude increases, atmospheric pressure decreases.

So far you have only seen examples of increasing pressure in liquids. This same effect also occurs in air, a gas. To be sure, it takes a huge column of air to produce a noticeable pressure. Keep in mind, however, that you are living at the bottom of a rather deep ocean of air. The weight of several kilometers of air produces what is known as the **atmospheric pressure,** or air pressure. Atmospheric pressure is therefore the pressure exerted by the weight of the earth's atmosphere.

At the earth's surface, the atmospheric pressure is equal to a little more than ten newtons per square centimeter. This pressure does not crush you because it is balanced by other pressures from inside your body.

As you go upward from sea level, the atmospheric pressure decreases. This happens because you're not as deep in the atmosphere. Therefore, the weight of the overlying layers of the atmosphere is less. In outer space there is little or no atmosphere. The pressure is zero. What would happen to an astronaut in space who did not have a pressurized suit? The astronaut would explode, since pressure inside the body would be greater than pressure outside the body.

Jets fly at heights of ten kilometers above sea level. There the atmospheric pressure is less than a third the pressure at sea level. Jet cabins must be pressurized for the comfort and safety of the passengers and crew.

Figure 5.14 Astronauts must wear pressurized suits.

Figure 5.15 The hot air in a hot-air balloon is less dense and, thus, weighs less than an equal volume of cooler air. This situation explains why hot-air balloons float in air.

The air exerts a buoyant force on objects, just as liquids do. According to Archimedes' principle, an object will float in air if it weighs less then the air it displaces. Thus, objects whose density is less than the density of air will float. Since air has such a low density, not many objects can float in it. But large balloons filled with hot air or helium are less dense than air and rise to great heights.

Back on the earth, you continually use the air pressure all around you to your advantage. For example, by expanding your lungs, you create a low-pressure region inside the lungs. The higher pressure outside your body will force air into your lungs. You do this thousands of times every day, each time you take a breath.

You also use air pressure to force liquids into your body. When you suck on a straw, for example, you lower the air pressure inside your mouth. This lowered pressure causes the outside air pressure to push the liquid in a glass up through the straw and into your mouth. This same principle of suction that causes a liquid to be forced into a straw is used in pumps to raise water from wells and in vacuum cleaners.

Do You Know?
To fill its lungs for the first time, a new baby must create a suction 50 times greater than a healthy adult creates to take an average breath.

Study Questions for 5.3

1. What is atmospheric pressure?
2. Where is atmospheric pressure greater, at sea level or on a mountaintop?
3. Describe two ways in which you use air pressure.

1. pressure exerted by air as a result of the weight of the earth's atmosphere
2. at sea level
3. to force air into your lungs for breathing, to force liquids into your body

3. Explain to the students that the air around them exerts pressure just as liquids do. Mention that the air pressure at sea level is about 10 newtons per square centimeter (14.7 pounds per square inch). On the board, draw a side view of a person. Draw arrows pointing to the person's back and front. Then point out that the arrows represent air and ask why the person has not been squashed flat (pressure on inside of person is pushing out and balances pressure on outside).

Comprehension Skill: Applied

FOLLOW-UP

Reinforcement: How Do People Breathe?
Ask the students to prepare research reports to explain how people inhale and exhale. Their reports should include the role that air pressure plays in the process.

Enrichment: A Siphon Fountain
Have interested students create siphon fountains by fitting a flask with a two-hole stopper. Through one hole place a jet tube which will extend about halfway to the top of the flask. Let about 2 centimeters (¾ inches) extend out of the flask. Through the other hole push a short length of glass tube so that it is just even with the bottom of the stopper. Let this tube also extend about 2 centimeters (¾ inches) outside the flask. Connect 20 centimeters (about 8 inches) of rubber tubing to the jet tube. Connect a 1-meter length (about 1 yard) of rubber tube to the other glass tube. Place some water in the flask and insert the stopper. Put the short tube in a beaker of water on a table and the long tube in an empty pail on the floor. Invert the flask. Ask the students to explain how the fountain operates.

Tie-in: Forecasting Weather
Ask students to construct a simple bottle barometer by filling a soda bottle halfway with water and inverting it in a pan of water. On a piece of paper they should mark off millimeters or fractions of an inch. This paper should then be glued to the side of the jar so that a rise or fall in the water level can be observed and measured. Have students observe the level of the water in the morning and in the late afternoon. What do they notice about the level of the water when the weather changes? Can the bottle barometer be used to predict the weather (it can indicate a trend)?

5.4

OBJECTIVES: The student should be able to

- **define** *drag;*
- **define** *lift;*
- **explain** how drag is produced;
- **explain** why engineers streamline planes;
- **explain** how the velocity of a fluid affects its pressure.

SCIENCE TERMS

drag
lift

ADVANCED PLANNING

To illustrate the concept of lift, you will need funnels and Ping-Pong balls.

LESSON PLAN

1. To introduce the lesson, ask the students to imagine themselves running through a tunnel. They are running as fast as they can and expect to be at the end of the tunnel in two minutes. Suddenly a strong wind comes up and blows against them. Will they still reach the end of the tunnel in two minutes (no)? Why (the wind will hold them back)?

2. Explain that even without the wind the students would experience a force acting against them as they ran through the tunnel. The force in still air is so much weaker that people are not aware of it. The force involved is friction. Students should recognize this term from Chapter 3. The term *drag* is often used to describe the friction created when an object, such as a car, a truck, or an airplane, moves through the air. Drag opposes motion and slows down an object.

SCIENCE BACKGROUND

When an object falls to the earth, drag is not the only force that acts on it. Gravitational force pulls falling objects to the earth while drag is a resistance to the fall. As the object falls faster and faster, the drag increases until it cancels out the gravitational pull. The falling object has now reached a constant, or terminal, velocity.

5.4 Objects Moving in Air

You will find out

- what is meant by the term *drag;*
- how the shape of an object affects drag;
- what is meant by the term *lift;*
- how the velocity of a fluid affects pressure.

Can a substance as light as air exert a force? Any time the wind blows, the force of air in motion becomes evident. It causes blades of grass and branches of trees to bend. It pushes kites up in the air. It can move a boat across a lake or even an ocean. By standing in front of a fan, you can get an idea of the strength of the forces of moving air.

Air, whether moving or still, causes friction with any object moving in it. Friction, you remember, is a force that acts on a moving object in a direction opposite to the motion of that object. Friction generally acts to decrease the speed of a moving object. When you are walking through calm air, you are not aware of the friction created between your moving body and the air. Even so, the force is there. When you throw or hit a baseball, friction is created between the moving ball and the air. A moving car or airplane also experiences friction with air. Engineers often refer to friction created when objects move in air as **drag**.

Figure 5.16 The force of air in motion is often evident during bad weather.

The amount of drag depends on the shape of the moving object. If the air has to make too sharp a turn in flowing around the wings of a plane, for example, the flow pattern becomes highly disordered. The flight becomes turbulent. In this case, there is too much drag. Engineers get around this by designing planes with streamlined bodies that have long tapered shapes. The air can then flow smoothly around the tapered bodies, and the drag is reduced.

The best shape for reducing drag depends on the speed of the object. For the low speeds of propeller-driven planes, the teardrop shape is the best. For supersonic aircraft, a slender body with a pointed nose is best.

Because of the increasing cost of fuel, engineers and scientists are doing a great deal of research on ways to reduce the drag of automobiles and trucks. They have found that changing the shape can change the air flow pattern and reduce the drag. Reducing the drag saves fuel.

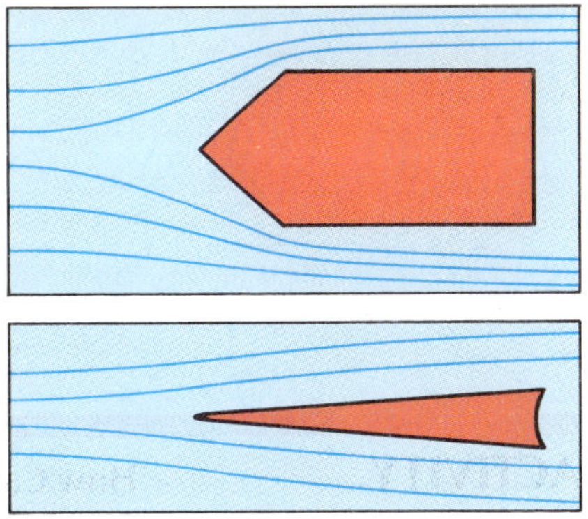

Figure 5.17 Air flows more smoothly around the tapered shape.

Figure 5.18 In this specially designed speed bicycle, the cyclists are positioned to minimize drag.

SCIENCE BACKGROUND

Fast moving objects experience more drag, or resistance. Larger objects also experience more drag. The drag is produced when the object bumps into the air. Since only one material can be in one place at one time, the air must be moved out of the way to allow the object to move forward.

Some race cars are designed with air tunnels on the sides. The tunnels are curved on the bottom and flat on the top. Air travels faster over the curved bottom creating an area of low pressure. This condition forces the car down onto the racetrack giving it better traction.

3. To illustrate the concept of lift, have the students perform the following short activity:

   Give some students funnels and Ping-Pong balls. Place the Ping-Pong balls in the funnel and ask the students to blow through the neck of the funnel in order to force the ball out of the funnel.

   After the students have performed the activity, diagram the situation on the board. Explain that fast moving air exerts less pressure than air that moves more slowly.

4. Ask the students to consider the Ping-Pong ball, which they probably only succeeded in causing to spin. Ask where the moving air was (under the ball). Ask where the low-pressure area was (under the ball) and where the high-pressure area was (above the ball). Why was the pressure greater above the ball (the air was moving more slowly)? Finally, why couldn't the light ball be easily blown out of the funnel (there was more pressure on top of the ball than under it)?

ACTIVITY: How Can You Demonstrate Differences in Air Pressure?

The experiments illustrate Bernoulli's principle. In Part I the Ping-Pong balls will move together because the air between them is moving and exerting less pressure than the air around them.

In Part II the cards will flip when students blow across them. The area of less pressure is on top of the cards. The air under the cards pushes up and flips them over. When students blow under the cards, the cards will bend down in the center because there is more pressure over the card than under it.

● **Laboratory Skills: Experimentation/Demonstration Manipulation of Materials**

Buoyant forces can lift objects into the air only if the objects are less dense than air. How do objects that are denser than air, such as airplanes, get off the ground and fly? **Lift** is used. Lift is the upward force of air moving around an object. In the case of an airplane, the lift is produced by a difference in air pressure on the top and bottom sides of the wing.

● ACTIVITY

How Can You Demonstrate Differences in Air Pressure?

Materials

2 pieces of string (30 cm long)
2 Ping-Pong balls
straw
ring stand and clamp
dowel
scotch tape
index card

Procedure—Part I

1. Place the dowel in the clamp so it is parallel to your desk.
2. Tape a piece of string to each Ping-Pong ball.
3. Suspend the balls from the dowel so they are the same distance above the desk and about 2 cm apart.
4. Use the straw to blow a stream of air between the balls. Observe.

Part II

1. Fold the ends of the index card on the imaginary lines shown in the diagram.
2. Place the index card on your desk so it is standing on the folded edges as shown in the diagram.
3. Use the straw to blow a stream of air under the index card. Observe.
4. Use the straw to blow a stream of air over the top of the index card. Observe.

Questions

1. How did the Ping-Pong balls act when you blew a stream of air between them?
2. Explain why they acted as they did.
3. When did the index card flip over?
4. Explain why it flipped over.

1. They move together.
2. There is an area of low pressure between them where the air is moving. The higher pressure on either side of the balls pushes them together.

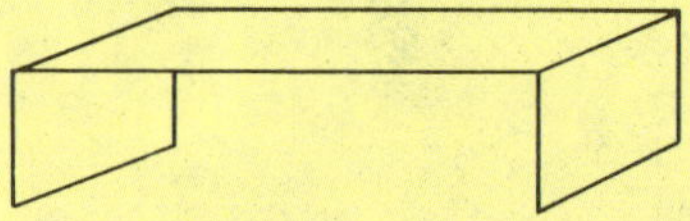

3. The card flipped over when you blew on top if it.
4. The air on top of the card was moving and therefore exerted less pressure. The higher air pressure under the card pushed up on the card and flipped it over.

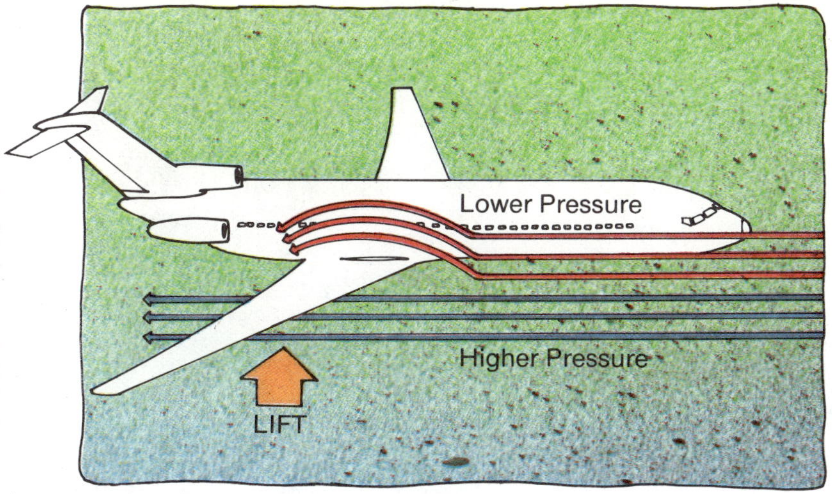

Figure 5.19 The lift of an airplane results from pressure differences below and above the wing.

Daniel Bernoulli, a Swiss scientist who lived in the eighteenth century, discovered that the faster a fluid moves the less pressure it exerts. You can see this effect by holding a thin slip of paper under your mouth and blowing across it. The pressure in the stream of air moving across the top of the paper will be less than the pressure in the still air underneath the paper. As a result, the force on the bottom side of the paper will be greater than that on the top, and the paper will be pushed upward, or lifted.

Airplane wings are designed so that the air will flow faster across the top of the wing than across the bottom. This makes the pressure lower on the top of the wing than on the bottom. The wing is pushed upward. The same principle lifts helicopters off the ground, except that in their case, the lift is produced by rotating wings.

Study Questions for 5.4

1. What is drag?
2. How can an object's shape reduce drag?
3. What is lift?
4. Why are airplane wings designed so that the airflow across the top of the wing is faster than the flow across the bottom?

1. Drag refers to the friction created when objects move in air.
2. by allowing air to flow smoothly around the object
3. the upward force of moving air on the wings of an airplane
4. The designer of the wings wants the air pressure lower on the top of the wings than on the bottom, so that the wings will be pushed upward. According to Bernoulli's principle, this will happen if the air on the top of the wings moves faster than the air on the bottom.

FOLLOW-UP

Reinforcement: Drag and Lift
Ask the students to do library research on the forces of drag and lift. They should prepare oral reports to share with the class.

Enrichment: A Paper Airplane Race
Have students design and construct paper airplanes. Which airplane will fly the greatest distance? Why?

Tie-in: Airplanes
Ask the students to do research to determine how each part of the airplane's structure helps it to fly.

What Did You Learn?

5.1 • Pressure is the force applied by an object divided by the area over which the force acts.
5.1 • The smaller the area to which a force is applied, the greater is the resulting pressure.
5.1 • Pressure increases as the depth of a fluid increases.
5.2 • Buoyancy is the upward force on an object in a fluid.
5.2 • Archimedes' principle states that the buoyancy of any object is equal to the weight of the fluid that the object pushes aside.
5.2 • Density is the mass of a substance divided by its volume.
5.2 • An object will float in a fluid if the density of that object is less than the density of the fluid.
5.3 • Atmospheric pressure is the pressure exerted by air as a result of the weight of the earth's atmosphere.
5.3 • Atmospheric pressure decreases as height increases because the weight of overlying air layers decreases with increasing altitude.
5.3 • Examples of the use of atmospheric pressure include forcing air into the lungs and liquids into the mouth, sucking dirt into a vacuum cleaner, and pumping water from wells.
5.4 • The friction that is created by an object moving in air is often called drag.
5.4 • The shape of an object can reduce drag by allowing air to flow smoothly around the object.
5.4 • Lift is the upward force of moving air on the wings of an airplane.
5.4 • The faster a fluid moves the less pressure it exerts.

Key Terms

pressure
fluids
buoyancy
Archimedes' principle
density
atmospheric pressure
drag
lift

Career

Aerospace Careers

Aerospace engineers, scientists, and technicians work as a team to design and develop airplanes and spacecraft. For example, scientists will do research on the drag and lift expected on a certain shape of aircraft. Engineers and technicians will then build a scale model of this design and test it in wind and heat tunnels. If this testing is successful, then a full-sized experimental model, or prototype, is built and thoroughly tested on the ground and in the air. Before the aircraft is put into production, the design will be changed many times until the test results are completely satisfactory.

The aerospace industry employs thousands of workers in many different companies as scientists, engineers, technicians, production workers, or administrators. The largest concentration of aerospace jobs in this country is in California, but New York, Washington, Connecticut, Texas, Missouri, and Kansas also have a large number of aerospace companies.

Vocabulary

1. c
2. a
3. c
4. c
5. a
6. a
7. c
8. b
9. b

● Word Skill: Vocabulary

What Do You Remember?

1. Pressure increases.
2. Archimedes' principle; The buoyancy on any object is equal to the weight of the fluid being displaced.
3. An object will float in a fluid if the density of the object is less than the density of the fluid.
4. At great altitudes atmospheric pressure has decreased from that at sea level. If cabins were not pressurized, there would be an imbalance between people's internal body pressure and the pressure exerted on people's bodies by the thinner atmosphere.
5. An object's shape can reduce drag by allowing air to flow smoothly around it; conversely, certain shapes can increase drag.

TO THINK ABOUT AND DO

Vocabulary ●

Number a sheet of paper 1 through 9. Select the best answer to complete each of the following sentences. Record the letter of each answer next to the proper number.

1. The force applied by an object divided by the area over which the force acts is the (a) drag, (b) buoyancy, (c) pressure.
2. Pressure increases with an increase in a fluid's (a) depth, (b) altitude, (c) velocity.
3. Any substance that flows is called a (a) force, (b) fuel, (c) fluid.
4. The upward force on an object in a fluid is called (a) gravitation, (b) drag, (c) buoyancy.
5. The mass of a substance divided by its volume is the (a) density, (b) pressure, (c) lift.
6. The pressure exerted by air as a result of the weight of the earth's atmosphere is called (a) atmospheric pressure, (b) water pressure, (c) blood pressure.
7. The friction created when an object moves in air is often called (a) lift, (b) gravitation, (c) drag.
8. The upward force of moving air on the wing of an airplane is called (a) drag, (b) lift, (c) density.
9. The fact that the faster a fluid moves the less pressure it exerts was discovered by (a) Archimedes, (b) Bernoulli, (c) Newton.

What Do You Remember?

1. If you reduce the area to which you apply a force, what happens to the amount of pressure being exerted?
2. What is the name of the principle that describes the relationship between buoyancy and the weight of a displaced fluid? State the principle.
3. How can you predict whether an object will float?
4. Why must the cabins of jet planes be pressurized?
5. How does an object's shape affect the drag on the object?

Applying What You Have Learned

1. A block has an area of 55 square centimeters and a weight of 495 newtons. How much pressure can it exert?
2. Explain how an eyedropper is used. Use the term *air pressure* in your explanation.
3. How can density be used to determine whether objects float or sink in liquids?
4. Why does an object seem to lose weight when you put it in water?

Research and Investigation

1. If you were a scuba diver, how would you protect yourself from increased water pressure at great depths?
2. How is specific gravity related to density?

Applying What You Have Learned

1. Pressure $= \frac{\text{force}}{\text{area}}$

Pressure $= \frac{495 \text{ N}}{55 \text{ cm}^2} = 9\text{N/cm}^2$

2. When the bulb of the eyedropper is squeezed, air is forced out. The air pressure in the eyedropper is less than the air pressure pushing on the water. When you release the bulb, air is forced into the eyedropper.
3. Density is the relationship between mass and volume. If equal volumes are used, the substance with the greater mass will sink into the substance with less mass. In order for an object to float, it must be less dense than the fluid it is in.
4. The water exerts an upward force called buoyancy that helps to counteract gravitational force.

● Comprehension Skill: Applied

Research and Investigation

1. Although answers will vary, students may find references to rebreathing devices for use in depths over 61 meters (200 feet). Rebreathing devices allow exhaled air to be repurified, so that it can be used again. Divers may also use decompression chambers at the surface in order to return to the pressure they experienced underwater. In such chambers the pressure is then reduced in stages.
2. Specific gravity is the ratio between the density of any object and the density of water, which is 1 g/cm^3.

● Reference/Research Skills

Long-Term Projects

1. Ask the students to create an extended comic strip to show the contributions of the alchemists to modern chemistry. They may wish to use their drawings to create shoe box televisions.
2. Help the students to initiate an aluminum recycling program. Have them contact community groups that will help to transport the aluminum to recycling centers.
3. Certain chemical compounds such as antibiotics and insecticides help to protect our health. Have the students work in small groups to prepare and present panel discussions to explain how such compounds protect us.
4. Have the students design and perform controlled experiments to investigate the differences and similarities that exist between natural and synthetic fibers. Check their experiment plans before allowing them to proceed.
5. Have the students work in groups to research the uses of elements. Ask each group to prepare a *Believe It Or Not* box to express their most unusual findings.
6. Ask each student to create a series of footsteps through time to trace the development of the atomic theory from the early ideas of Democritus to the present theories.
7. Chemistry can be used to improve natural products. Ask the students to work in groups to prepare editions of newspapers to show how advances in chemistry have helped to improve the manufacturing of rubber.
8. Ask the students to write a report explaining what "hard" water is, what damage it can do, and how the problems it causes can be prevented.

Unit Two

The Structure of Matter

CHAPTER 6
The Atom

CHAPTER 7
The Elements

CHAPTER 8
Combining the Elements

CHAPTER 9
Carbon Compounds

Bulletin Boards

1. Print the title across the top of the board. Divide the board in half and title the sections. Under *Everyday Elements*, cut from a magazine or draw pictures of products the students use that are elements. On the carbon side, show items that contain carbon. Leave space on both sides for the students to add pictures. Use the board to initiate a discussion of the many uses of the elements in our lives.

2. Title your board *What Does a Formula Mean*? Cut out two white circles about five cm in diameter and one red circle about eight cm in diameter. Mount them on the board so that the circles are touching with the red circle between and slightly below the white circles. Under the model write *H_2O–two atoms of hydrogen and one atom of oxygen*. Cut a green circle and a slightly larger orange circle. Mount them side by side. Under the model write *NaCl–one atom of sodium and one atom of chlorine*. Write the formulas for several other compounds on the board. Have the students add models.

3. Title the board *The Structure of the Atom*. Cut circles from colored paper to represent protons, neutrons, and electrons. Draw several squares from the periodic table on the bulletin board and have the students arrange the circles to form the atoms.

CHAPTER BACKGROUND

Chemistry is a study of matter and the changes it undergoes. People have always been curious about the basic structures that compose the world around them. Aristotle and Democritus, ancient Greek philosopher-scientists, recorded the first ideas about the structure of matter. Aristotle said that all things were made of four elements—earth, air, fire, and water—in different combinations. Democritus hypothesized that the elements were made of invisible, indivisible atoms. The ideas of the ancient Greeks were accepted until the alchemists, in their search for gold and the Philosopher's Stone, showed that earth, air, fire, and water were not elements and discovered a large number of the elements recognized today.

In his atomic theory, John Dalton confirmed, changed, and advanced Democritus' view of atoms.

This theory included the following four major ideas:

1. Matter is composed of atoms.
2. Atoms of one element are alike in size, shape, and mass but differ from the atoms of other elements in these properties.
3. During chemical change, atoms of different elements unite to form molecules of a new compound.
4. During chemical change the elements themselves do not change.

At the end of the nineteenth century J.J. Thomson proposed the plum-pudding model of the atom. In this model the atom was a positively charged sphere with negative electrons scattered in it like raisins in a plum pudding. Thomson's model pictured the atom as a solid.

In his experiments Sir Ernest Rutherford showed that atoms are not solids. He suggested instead that an atom consists of a dense nucleus surrounded by revolving electrons. Niels Bohr used Rutherford's data to suggest that the atom is like a solar system. In Bohr's model the sun is the nucleus. The planets are the electrons that revolve around it.

The Bohr model is no longer accepted as completely accurate. Today's model recognizes the wavelike nature of electrons and the influence of energy on their behavior.

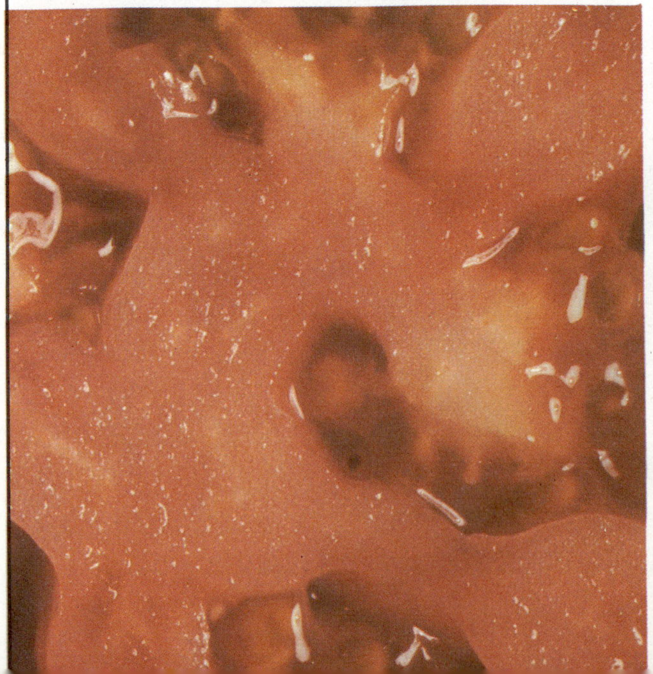

6

The Atom

Introduce the unit by explaining that chemistry is the study of the structure and changes of matter. Review the meaning of the term *matter* (anything with mass and volume) and examples of changes matter undergoes (changes of state, burning, etc.). If the students are not familiar with physical and chemical changes, you might demonstrate these changes in this manner. Hold up a piece of paper and ask the students to name some of its properties (white or other color, smooth, relatively small mass). Tear the paper into little pieces and point out that the pieces still have the same properties as the original paper. Place the pieces of paper in a beaker or evaporating dish and light them with a match. Point out that the ashes do not have the properties of the paper. In a physical change (torn page) the state or form changes but the properties are retained. In a chemical change (burned paper) the original properties are lost and new properties appear.

As you study the structure of matter, the simplest question you might ask is "What is matter?" You have already learned that matter occupies space and is the substance that the universe is made of. In this chapter you will probe more deeply into matter, take it apart, and find out what it is made of.

In probing matter, you will follow in the footsteps of people who lived over two thousand years ago in Egypt and Greece. For the Egyptians of that day the chemistry of matter was strongly connected with the embalming of the dead and religious ritual. The Greeks, on the other hand, were more concerned with changing one substance to another. If wood could be changed to fire and ashes, why couldn't iron be turned to gold?

Today scientists study matter in chemical laboratories and in giant atom smashers that tear apart the cores of atoms. The goal is the same, however. The scientists are trying to understand matter and to see if they can take it apart and rearrange it in more useful ways.

The items pictured are a sponge, a penny, a zipper, and a tomato.

Figure 6.1 Can you identify the four familiar items pictured? The photographs were taken at very close range. In this chapter you will examine matter at much closer range.

Lesson Organizer

6.1 The Atomic Model of Matter

Pages 118–121
Advanced Planning: Page 118
Activity: What Are the Properties of Some Elements? Page 120
Follow-up: Page 121
- **R** The Structure of a Compound
- **E** Testing for Oxygen and Carbon Dioxide
- **T** Where Does Oxygen Come From?

6.2 The Structure of Atoms

Pages 122–123
Advanced Planning: Page 122
Follow-up: Page 123
- **R** Rutherford's Probe of the Atom
- **E** Evolution of the Atomic Model
- **T** The Avogadro Number

6.3 The Atomic Nucleus

Pages 124–127
Activity: How Can You Build a Model of the Atom? Page 125
Follow-up: Page 127
- **R** Atomic Structure
- **E** Isotopes
- **T** Technology and Atomic Structure

6.4 Electron Structure of the Atom

Pages 128–129
Advanced Planning: Page 128
Follow-up: Page 129
- **R** The Electron Cloud
- **E** Theoretical Physicists
- **T** Locating an Electron

See ACTIVITIES book pages 31–36 for laboratory, skill, and review exercises for Chapter 6.

6.1

OBJECTIVES: The student should be able to

- **identify** the alchemists and their contributions to science;
- **define** the term *element*;
- **define** the term *atom*;
- **define** the term *compound*;
- **name** several examples each of elements and compounds.

SCIENCE TERMS

alchemists	natural elements
elements	compounds
atoms	

ADVANCED PLANNING

If you do not have a classroom periodic wall chart available, you can prepare a list of common elements and their symbols from the periodic table in Chapter 7.

LESSON PLAN

1. It will be helpful to your students if they learn the symbols for the more common elements. You might want to deal with the first 36 elements, for example. Depending on the ability of your students, assign them three to six elements and symbols to learn each night. Quiz them the next day on the elements that are due plus a variety of the ones they learned on previous nights. This will help to reinforce and review continually the elements and their symbols. As the students memorize the elements, they will learn that knowledge can be learned more efficiently by breaking it into smaller chunks.

6.1 The Atomic Model of Matter

You will find out
- how the alchemists contributed to science;
- what elements are;
- what atoms are;
- what compounds are.

When you look at the world around you, you see that it is filled with an amazing variety of substances. You see all different shapes, sizes, and colors. Some substances are soft; others are hard. Some are solids, others are liquids, and still others are gases.

Figure 6.2 Your world is filled with different substances having different properties.

Long ago the people who sought to answer the question of what matter is made of were called **alchemists.** Alchemists believed that if they could find the basic unit of matter, or primary matter as they called it, they could do wonderful things. With primary matter they thought they could change ordinary metals into gold and ordinary water into a liquid that would stop the aging process. In the course of their search they tested all known substances. They burned them, boiled them, crushed them, strained them, filtered them, and mixed them. Some of the alchemists went crazy in the process. But none of them ever found the primary matter.

Figure 6.3 A painting of alchemists at work

What the alchemists did do was to establish the laboratory and an experimental approach of taking matter apart to analyze it. This set the stage for eighteenth and nineteenth century scientists who showed that all matter is composed of basic substances called **elements.** Elements are substances that cannot be chemically broken down into simpler basic substances.

The smallest particles of individual elements are called **atoms.** The element hydrogen is made only of hydrogen atoms, the element helium only of helium atoms, the element oxygen only of oxygen atoms, and so on. All hydrogen atoms are alike, but they are different from all helium and oxygen atoms. For example, one oxygen atom weighs sixteen times as much as one hydrogen atom.

Figure 6.4 *left:* All items made of the element gold contain only gold atoms. *right:* All items made of the element iron contain only iron atoms.

2. Explain that some elements are very valuable. Silver, gold, and platinum are among the most precious metals. The first chemists, the alchemists, tried to turn other metals into gold. They did not succeed but learned a great deal in the process and developed methods that are still used in science today.

3. Explain that the simplest parts of an element are called atoms. An atom of gold, for example, is the simplest piece of gold that can exist alone or join with another atom. An atom of copper is the simplest piece of copper that can exist alone or join with another atom. Stress that all the atoms of one kind of element are alike.

ACTIVITY: What Are the Properties of Some Elements?

◇ **Caution:** Students should wear safety glasses during the activity. The room should be well ventilated. Only small amounts of each element should be heated. When a piece of magnesium ribbon is heated, no one should look directly at it.

To set up the activity, place the samples in small cups for easy access. Check the conductivity apparatus to be sure all the batteries and bulbs are working. If bulbs are not available, you can substitute an ammeter, a voltmeter, or a galvanometer. Keep in mind that if students work in teams you will need less of each element per class and less time will be required.

1. Cu: penny-colored wire, yes, yes, green flame
   Fe: silver-gray wire, yes, yes, orange flame
   Mg: silver ribbon wire, yes, yes, very bright light
   C: black lump, yes, yes, no change
   I: purple crystals, yes, no, purple gas
2. Students may group the elements in the following categories based on their observations: elements that have higher density than water, elements that have lower density than water; conductors, insulators; elements that are flexible, elements that are brittle.

● Laboratory Skills: Manipulation of Materials
Experimentation

● ACTIVITY What Are the Properties of Some Elements?

Materials

copper wire
iron wire
magnesium ribbon
carbon pieces
iodine crystals
beaker
aluminum foil
Bunsen burner
battery
bell wire
light bulb in porcelain socket
2 jar lids
ring stand and small ring
forceps
safety glasses
apron

Procedure

1. Copy the chart and use it to record your observations.

2. Put on your safety glasses and apron.

3. Place each element in a beaker of water. If it is more dense than water, it will sink. If it is less dense than water, it will float.

4. Set up the conductivity apparatus as shown. Test each element. If it is a conductor, the bulb will light.

5. Test the effects of heat in the following ways:

a. Make a handle by wrapping aluminum foil around the end of the copper wire, iron wire, and the magnesium ribbon.

b. Light the Bunsen burner. One at a time, hold each element by its handle and heat. CAUTION: Do not look directly at the magnesium ribbon while heating it. Turn off the Bunsen burner.

c. To test carbon and iodine, use a clean jar lid for each. Set up the ring stand and ring so that the Bunsen burner can be placed under the ring. Place a lid on the ring.

d. Place a small amount of the element on the lid and heat.

e. What change do you immediately notice in iodine? Remove the flame. Use forceps to handle the heated lid.

f. Heat the carbon for only three minutes.

Questions

1. Complete the chart.

2. Based on your observations, place similar elements together in groups.

Element	Color, Shape	Denser Than Water?	Conductor of Electricity?	Changes When Heated
copper				
iron				

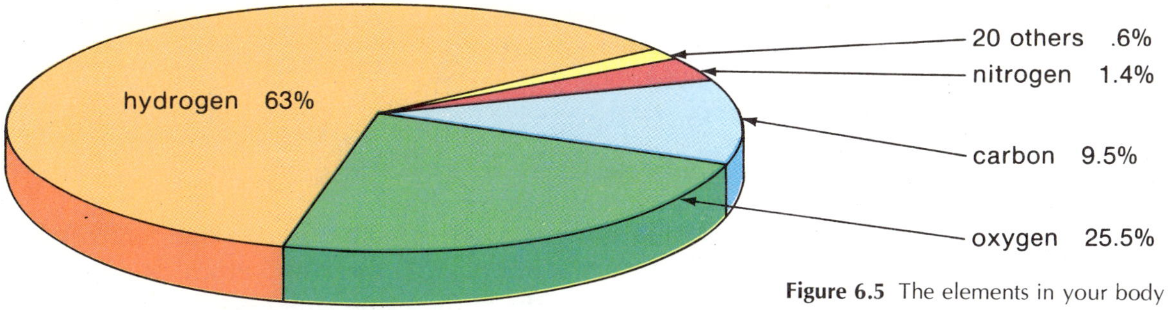

Figure 6.5 The elements in your body

More than a hundred different elements have been discovered so far. Most of these elements last a very long time—no one knows for sure just how long. Many have been around for at least as long as the earth. These elements are called **natural elements.** Natural elements are found in nature and do not have to be made in the laboratory. Other elements have been observed only in the laboratory where they were made. Some of these do not last long. The element einsteinium, for example, lasts for only a fraction of a second.

Everything in the world around you is made of one or more elements. Some familiar substances such as silver and gold are elements. Other substances are combinations of elements. For example, water is a combination of hydrogen and oxygen. By itself hydrogen is a colorless, tasteless gas. Oxygen is a colorless, odorless gas that must be present for burning to occur. In the right combination, however, hydrogen and oxygen form water, a liquid whose properties are completely different from either hydrogen or oxygen. Carbon dioxide is another combination of elements. It is made of carbon and oxygen. These combinations of two or more elements are called **compounds.**

Study Questions for 6.1

1. How did the alchemists contribute to the progress of science?
2. What are the basic substances of which all matter is composed?
3. What are atoms?
4. What are compounds? Give two examples.

1. by establishing the laboratory and the experimental approach of taking matter apart to analyze it
2. elements
3. the smallest particles of individual elements
4. combinations of two or more elements; water, carbon dioxide

4. Explain to the students that not all of the substances they see around them are elements. Many are combinations of elements. When the elements join in a chemical combination to form a new substance, they lose their own individual properties. An entirely new substance with properties of its own is formed. This substance is a compound.

FOLLOW-UP

Reinforcement: The Structure of a Compound

Prepare a list of compounds that the students are familiar with. Have each student select one compound and write a report about it. The report should include a discussion of why the compound is important, how the compound is produced, what the properties of the compound are, what elements are in the compound, and what the properties of each element in the compound are.

Enrichment: Testing for Oxygen and Carbon Dioxide

Have students do research to determine how to test for the presence of oxygen and carbon dioxide gas. Then help the students set up experiments to demonstrate the methods to the class.

Tie-in: Where Does Oxygen Come From?

Oxygen is an important element in many chemical reactions. Ask the students to do research on photosynthesis and the oxygen-carbon dioxide cycle to learn how oxygen is supplied.

SCIENCE BACKGROUND

The formula for table sugar $C_6H_{12}O_6$ indicates that table sugar, or sucrose, is made of the elements carbon, hydrogen, and oxygen. Carbon is a black solid. Hydrogen and oxygen are both colorless, odorless, tasteless gases. When all three elements combine chemically the compound sugar is one result.

6.2

OBJECTIVES: The student should be able to

- **describe** Rutherford's experiment;
- **state** the information about atomic structure that was learned from Rutherford's studies.

SCIENCE TERMS

radioactive
electrons
nucleus

ADVANCED PLANNING

Prepare transparency for the overhead projector to illustrate the nucleus of an atom surrounded by its electrons.

LESSON PLAN

1. Introduce the lesson by reviewing what elements are (simplest forms of matter that cannot be broken down by ordinary means) and what atoms are (smallest particles of individual elements).

● Comprehension Skill: Applied

SCIENCE BACKGROUND

Rutherford's idea of atomic structure became the basis for a model of the atom that was developed further in the early twentieth century by Niels Bohr. This model, the planetary or solar system model, is not used today, but it is sometimes useful in explaining atomic structure.

6.2 The Structure of Atoms

You will find out

- what ideas about the atom emerged from Rutherford's work.

Figure 6.6 The atoms in an enormous elephant are no bigger than the atoms in a tiny fly.

What are atoms made of? What makes one type of atom different from another? It is not the atom's size. Almost all atoms are about the same tiny size. About a million atoms stacked end to end would have a thickness equal to the thickness of only five pages in this book!

The clue to differences among atoms was found in 1911 in England by Ernest Rutherford and his co-workers. Rutherford was known to students as the alligator because of his ferocious personality. A short time before, it had been discovered that radium and certain other elements were **radioactive.** That is, certain elements give off high-energy particles. Rutherford saw that these high-energy particles could be used to probe the structure of atoms.

To understand how atomic structure was investigated, imagine you are an inspector at a secret physical science laboratory that is making a special metal. This metal is so strong that a sheet of it, no thicker than a page of this book, could cause bullets to bounce off it. A man shows up at the gate. He wants to take a truckload of old books and reports out of the laboratory to the dump. You suspect that he may be smuggling a sheet of this metal out of the laboratory in one of the books, but there is no way you can look through all the pages of all the books. What do you do?

After some thought you come upon a plan. You will shoot a bullet through each book. If the bullet passes through the book, it has none of the metal in it. If the bullet bounces off the book, it contains the metal. In this way you catch the smuggler and become a hero.

This is basically what Rutherford's group did. They shot high-energy particles into atoms and watched to see what happened. Some of the particles bounced back. A careful study of how the particles bounced back showed that the atom has a small but dense core called

Figure 6.7 Like a spider web, an atom is mostly open space.

the **nucleus.** At each point where a particle hit a nucleus, the particle bounced back. If the particle didn't strike a nucleus, it just kept going. The nucleus, which has a positive electric charge, is surrounded by negatively charged particles called **electrons.** The electrons have a much smaller mass than the nucleus has. Yet the actual size of the nucleus is very small. Its diameter is a hundred thousand times smaller than that of the whole atom. Nevertheless, the tiny nucleus has 99.9 percent of the mass of the atom.

A new idea of the atom emerged from these studies. In the center is the massive nucleus, with a positive charge. Orbiting around the nucleus at great distances from it are the negatively charged electrons.

The structure of the nucleus of an atom determines what kind of element you have. In the next lesson, you will see why.

Do You Know?
It takes about 300,000 atoms stacked end to end to get a thickness equal to a hair.

Study Question for 6.2

How did Rutherford's work contribute to knowledge of atomic structure?

discovered the positively charged nucleus when high-energy particles were shot into atoms; discovery of presence of negatively charged electrons emerged indirectly from this work.

2. Have the students read the paragraphs in Lesson 6.2 about searching for the special metal. Draw an analogy between Rutherford's experiment and the search for the metal. Point out that just as most of the bullets passed through the books they were shot at, most of Rutherford's bullets passed through the atoms. Only the bullet that hit the metal was stopped. Similarly only the bullets that hit the nucleus bounced back in Rutherford's experiment. Then stress that Rutherford's experiment showed that the atoms are mostly empty space with a small amount of solid material.
3. Use the transparency to point out the positively-charged nucleus and the negatively-charged electrons that surround it. Explain that because the nucleus and the electrons are oppositely charged they are attracted by electric forces that keep the atom from flying apart.

FOLLOW-UP

Reinforcement: Rutherford's Probe of the Atom
Have students set up a cardboard model of Rutherford's experiment. The model should have labeled parts that include fluorescent screen, gold foil, radioactive source, and alpha particles. A simple description of the experiment should be written on an index card and included in the display.

Enrichment: Evolution of the Atomic Model
Have students do library research on atomic structure from the ideas of Democritus to the present-day model. Then have the students prepare posters that can be used in explaining the different ideas.

Tie-in: The Avogadro Number
Scientists need to be able to compare the number of atoms in specific mass amounts of elements. To do this, they use the Avogadro number. Ask the students to research the Avogadro number and construct graphs to compare the mass amounts of hydrogen, chlorine, sulfur, magnesium, and neon that would contain the Avogadro number of atoms, or 6.02×10^{23}.

6.3

OBJECTIVES: The student should be able to

- **identify** the two types of particles in the nucleus of an atom;
- **explain** how to find the atomic mass number of an element;
- **define** the term *isotope*;
- **define** the term *quarks*.

SCIENCE TERMS

proton
neutron
atomic number
atomic mass number
isotopes
quarks

LESSON PLAN

1. Introduce the lesson by having the students read the first page of Lesson 6.3 and answer the following questions:
   a. What two particles are in the nucleus (protons and neutrons)?
   b. What is the charge of a proton (positive)?
   c. What is the charge of a neutron (neutral, none)?
   d. If a neutron has no charge what is the charge on a nucleus (positive)?
   e. What is the charge of an electron (negative)?

6.3 The Atomic Nucleus

You will find out

- what types of particles make up the nucleus;
- what is meant by atomic number and atomic mass number;
- what is meant by isotopes;
- what quarks are.

You have seen that most of the mass of an atom is concentrated in a very small region called the nucleus. Can the nucleus be made of smaller units, just as elements are made up of atoms? If the nucleus is hit with particles of high enough energy, it can actually be split apart. The answer then is yes; the nucleus is made of smaller units.

As a result of studies by Rutherford and others, it was learned that every nucleus consists of two types of particles. One type, the **proton,** has a positive charge of one unit. This is just the opposite charge of an electron, which has a negative charge of one unit. The other type of particle, the **neutron,** has no charge. It is electrically neutral as its name suggests. Neutrons have a mass that is about the same as that of the proton. Both neutrons and protons have a mass about 1,800 times larger than the electrons. This explains why most of the mass of an atom is concentrated in the nucleus.

Figure 6.8 The locations of the proton, the neutron, and the electron in an atom

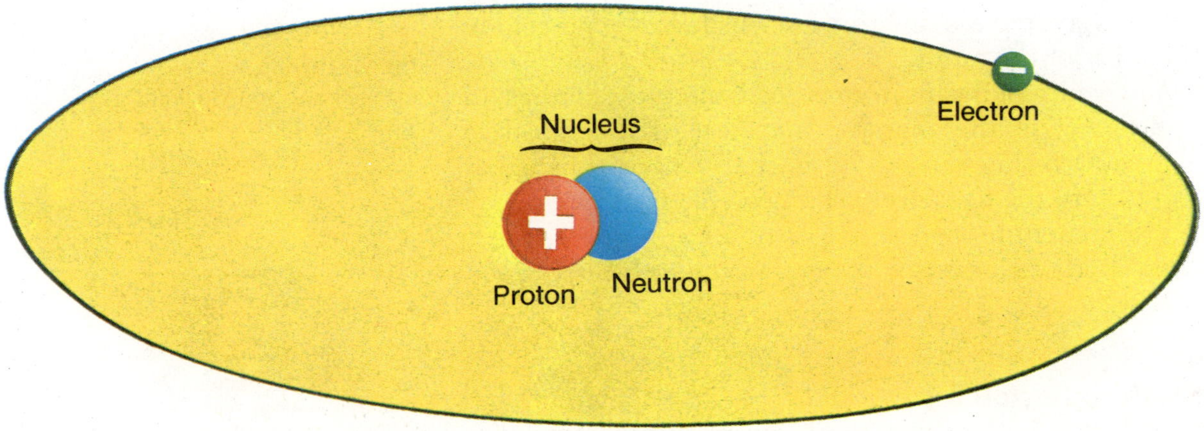

ACTIVITY

How Can You Build a Model of the Atom? ●

Materials

gumdrops
index cards
wire coat hanger
thread and blunt needle
clay
glue

Procedure

1. To create a model of helium, select three gumdrops of one color, three gumdrops of a second color, and three gumdrops of a third color.
2. Use the needle and thread to sew together two gumdrops of one color and two gumdrops of another color. This will be the nucleus.
3. Stretch out your coat hanger until it is as circular as possible. Then glue two gumdrops of a third color on the hanger.
4. Stand the hanger up in a lump of clay.
5. Use a piece of thread to suspend your nucleus in the center of the atom.
6. Glue a gumdrop of each color on an index card you have labeled Key to Helium Atom and write which particle of the atom each represents.

Questions

1. Why is the nucleus made up of two different colors of gumdrops?
2. Given a key for lithium with three protons and four neutrons in the nucleus, how would you create an atomic model of lithium?

1. The two different colors symbolize the two kinds of particles in the nucleus—the protons and the neutrons.
2. three red (or other color) gumdrops for the protons in the nucleus, four yellow (or other color different from proton color) gumdrops for the neutrons in the nucleus, three green (or other different color from proton and neutron colors) gumdrops for the electrons on the coat hanger

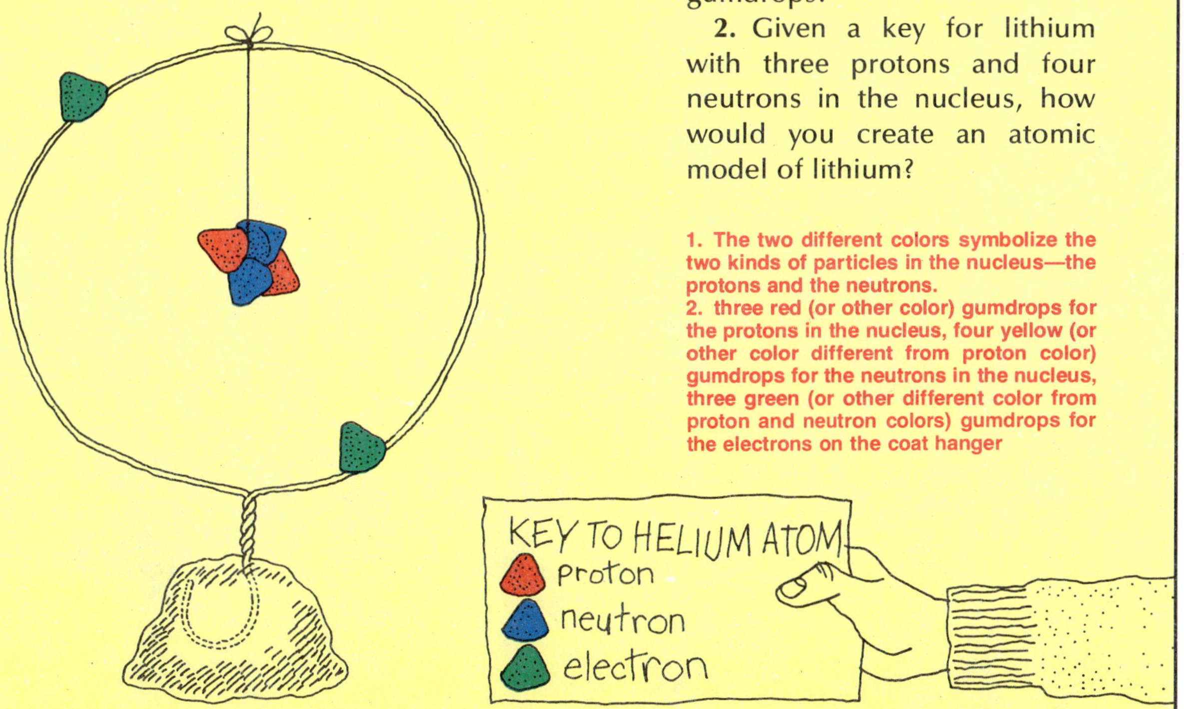

ACTIVITY: How Can You Build a Model of the Atom?

Remind the students to read the entire text of the activity before they begin. If you prefer having different students make different atomic models, assign only elements 1–18. Otherwise, the electron configurations may be difficult for students, and the materials requirements become excessive.

● **Laboratory Skill: Manipulation of Materials**

2. Have students differentiate between the atomic number and atomic mass number. Use practice problems to reinforce the distinction.
3. Explain that sometimes atoms with the same number of protons can have different numbers of neutrons. Since neutrons have no charge, this does not change the electric nature of the atoms. It does, however, make the atoms more or less massive. These atoms are called isotopes.
4. Explain that isotopes are often used as tracer elements; that is, they are used to help follow chemicals to see what paths they take in reactions. You might refer some of the students to the enrichment activity at this time.

SCIENCE BACKGROUND

The atomic mass number refers to the number of protons and neutrons in a specific atom or isotope. It is always a whole number. Another measure, the atomic weight, refers to the average number of protons and neutrons present in all the known isotopes of that element. Because it is an average, the atomic weight is often not a whole number.

Do You Know?
If you could collapse all the atoms in the Washington Monument down to the size of their nuclei, the Washington Monument would be smaller than the eraser of a pencil.

The total number of protons in a given nucleus determines the type of element you have. This number, which is also called the **atomic number** of the element, is different for each element. For example, hydrogen has one proton in its nucleus, or an atomic number of one. Helium has two, and lithium three.

In most nuclei, the number of neutrons is equal to or slightly greater than the number of protons. Most helium nuclei have two protons and two neutrons. Most carbon nuclei have six protons and six neutrons; most oxygen nuclei have eight protons and eight neutrons. As the elements get more massive, they tend to have an excess number of neutrons. For example, the nucleus of uranium, the heaviest natural element, has 146 neutrons but only 92 protons.

The sum of protons and neutrons in a nucleus is called the **atomic mass number** of that element. It is possible for different nuclei of the same type of element to have different atomic mass numbers. For example, rare carbon atoms have nuclei made up of eight neutrons and six protons, instead of the usual six neutrons and six protons. Atoms of the same element that differ only in the number of neutrons in their nuclei are called **isotopes.** Remember that the number of protons always determines which element the atom is. The number of neutrons may vary to form different isotopes of the same element.

Figure 6.9 The isotopes of hydrogen

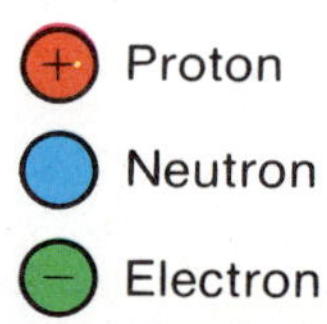

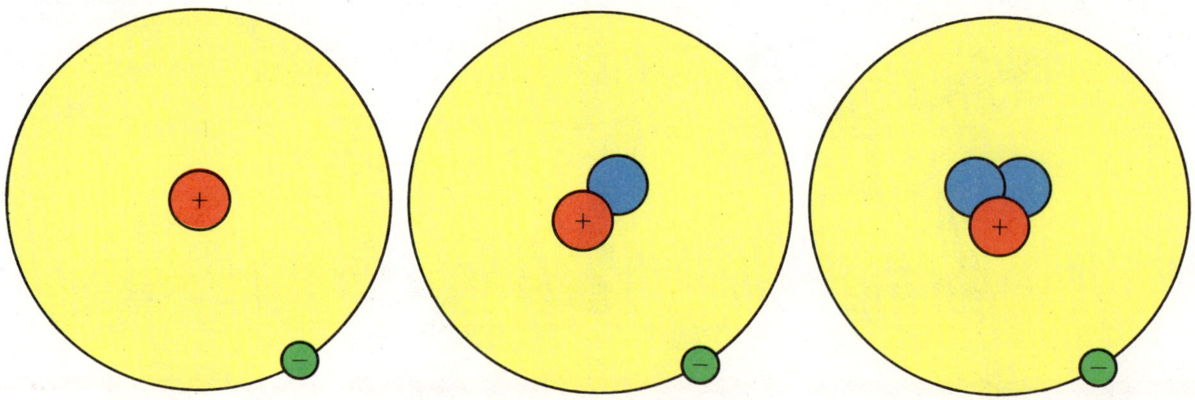

From what you have read, you may think that all matter is made up of only three particles—electrons, protons, and neutrons. It is true that different combinations of these particles result in all the matter known. Yet experiments with beams of very high-energy particles have shown that protons and neutrons may be made of even smaller particles called **quarks.** The evidence found so far seems to show that protons and neutrons are made of three quarks. There also seem to be several different types of quarks. These quarks combine in different ways to make protons and neutrons as well as short-lived particles that have been seen in high-energy experiments.

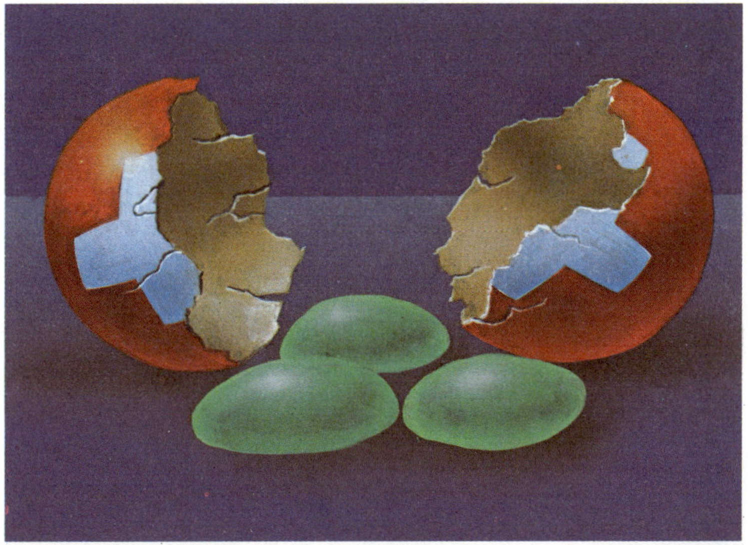

Figure 6.10 Recent studies suggest that protons are made of three quarks.

Study Questions for 6.3

1. What types of particles make up the nucleus?
2. What is the difference between the atomic number and atomic mass number?
3. What is the name of atoms of the same element that differ only in the number of neutrons in their nuclei?
4. Of what smaller particles are protons and neutrons composed?

1. proton, neutron
2. atomic number—number of protons; atomic mass number—sum of protons and neutrons in nucleus
3. isotopes
4. quarks

5. Point out that current research indicates that protons and neutrons may be made of smaller particles called quarks. Some of the research is being done at M.I.T. in Cambridge, Massachusetts, under the direction of Vera Kistiakowsky. Have the students read Kistiakowsky's biography at the end of the chapter.

FOLLOW-UP

Reinforcement: Atomic Structure

Prepare an atomic corner in your room. In the corner set up a bingo game based on the new terms in the first three lessons. The students can play the game during study periods and at lunch to reinforce the new terms.

Enrichment: Isotopes

Have the students do research to learn which elements have common isotopes and what the isotopes are used for.

Tie-in: Technology and Atomic Structure

In order for people's knowledge to advance beyond a certain point, technology must be able to extend their senses. Ask the students to find out about the cyclotron and other instruments currently being used to investigate the atom. The students should determine the purpose of each instrument, what the instrument does that a person could not do, and how the instrument is operated.

SCIENCE BACKGROUND

Scientists believe that quarks are held together by a very strong force called the strong interaction. This force holds the three quarks in a proton together so tightly that it appears to be impossible to separate one of the quarks from the other two.

6.4

OBJECTIVES: The student should be able to

- **explain** why the planetary model of the atom is not entirely satisfactory;
- **define** the term *shells*;
- **explain** how electrons fill their shells;
- **identify** the forces that hold an atom together.

SCIENCE TERMS

shells
electron cloud
nuclear force

ADVANCED PLANNING

You will need a piece of rope about 2 m (6 ft) in length for use in illustrating a wave pattern.

LESSON PLAN

1. Introduce the lesson by reviewing what an atom is as well as what its basic structure is like. Also, review what the atomic numbers and atomic mass numbers represent and how they can be used to identify the type of element you have.

2. Stress that since each of the elements' atoms has a different number of protons, each has a different atomic number and therefore a different attraction for its electrons. This difference in attraction results in having a different set of energy levels for each atom.

3. Explain that the energy levels are called shells. The shells are at different distances from the nucleus. The first shell (the shell closest to the nucleus) can hold two electrons. Other shells can hold eight electrons.

● **Comprehension Skill: Applied**

6.4 Electron Structure of the Atom

You will find out

- why the solar system model doesn't accurately explain atomic structure;
- what is meant by electron shells;
- what forces hold atoms together.

The model of the atom that was developed during the time of Rutherford is similar to a model of the solar system. In the solar system model the planets orbit the massive sun. Atoms are often shown as electrons orbiting a nucleus. This picture can give you some idea of the structure of an atom, but it is not a completely accurate picture. Theoretically, a planet could be in almost any orbit around the sun. Electrons can only be in certain areas, or energy states, around the nucleus. Also, the electrons move in spherical rather than flat orbits.

The specific energy levels around the nucleus where electrons can be found are called **shells.** Each shell is at a different distance from the atom's nucleus. Because an electron is moving very fast, it is impossible to know its exact location at any one time. You can, however, be sure that it is somewhere in the spherical area of its shell. Together the shells form a three-dimensional **electron cloud** around the nucleus of an atom.

Figure 6.11 Like an electron in its shell, the exact location of a car in this photograph is unknown. However, you can be sure that the car is somewhere along the highway.

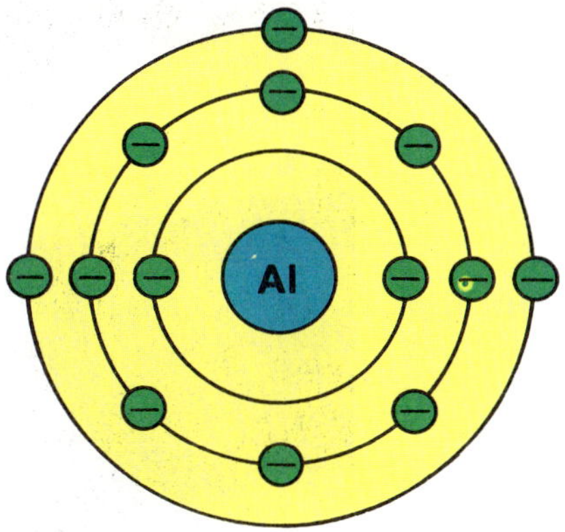

Figure 6.12 The electrons in an aluminum atom

Each kind of atom has a certain number of electrons in its electron cloud. In general, the electrons fill the shell closest to the nucleus first. When that shell has all the electrons it can hold, the second shell begins to fill. When that shell is full, the third shell fills. As the number of electrons in an atom increases, more shells are used.

Each shell can hold only a limited number of electrons. The shell closest to the nucleus, the first shell, holds two electrons. The next shells are considered full when they have eight electrons. Shells farther out from the nucleus can take up to thirty-two electrons.

What holds the electrons in their shells? Electrons are held in an atom by electric forces. Electric forces act between the negatively charged electrons and the positively charged protons in the nucleus. This attraction between the electrons and the protons keeps the atom from flying apart.

Another kind of force, **nuclear force,** holds the neutrons and protons together in the nucleus. The nuclear force keeps the nucleus from flying apart.

Study Questions for 6.4

1. Why can't the solar system be used to explain atomic structure?
2. What are electron shells?
3. Name the two forces that hold the atom together.

1. Electrons can only be in certain orbits. These orbits are spherical, not flat.
2. the energy states or spherical areas where electrons can be found
3. electric and nuclear forces

4. Illustrate how the shells are filled by drawing a diagram of a chlorine atom on the board. Ask how many protons chlorine has (17). How many electrons does chlorine have (17)? Draw a small circle to represent the chlorine nucleus and shade it in with chalk, preferably colored. Draw a circle around the nucleus to represent the first electron shell. Ask how many electrons can go in this shell (2). Draw in two dots representing electrons. Draw a circle for the second shell. Ask how many electrons can go in the second shell (8). Draw in eight dots. Ask how many more electrons are needed for chlorine (7). Ask where they should go (third shell). Draw the third shell with seven dots.
5. Point out that an atom's electrons are moving at very high speeds around the nucleus. There must be some force present to keep the electrons in the atom. Ask what charge electrons have (negative). Ask what charge protons have (positive). Ask what happens between opposite charges (they attract).

FOLLOW-UP

Reinforcement: The Electron Cloud

Reinforce the concept of a wavelike nature of electrons by asking students to imagine the electrons of an atom in a cloudlike area. Ask the students to explain how thinking of the electrons in a cloud rather than in orbits will change their picture of the atom.

Enrichment: Theoretical Physicists

Have the students do research on deBroglie, Heisenberg, and Schroedinger. The students should find out how each scientist contributed to the quantum model of the atom. They can present their research in the form of written interviews with the scientists.

Tie-in: Locating an Electron

Scientists indicate the electron's location in terms of probability. Ask the students to do research on probability. Before a World Series students could analyze possible combinations of games that might be played. (To win the Series one team must win four games.) Since there are 70 possible outcomes, the number of outcomes could be used as the basis for computing the probability of a given outcome.

What Did You Learn?

6.1 • The alchemists contributed to science by establishing the laboratory and the method of taking matter apart to analyze it.
6.1 • Elements are the basic substances of which all matter is composed.
6.1 • Atoms are the smallest particles of individual elements.
6.1 • Compounds are combinations of two or more elements.
6.2 • The atom is kept from flying apart by electric forces between the negative electrons and the positive protons.
6.3 • The positively charged proton and the electrically neutral neutron are the two types of particles that make up the nucleus.
6.3 • The type of element is determined by its atomic number, or the number of protons in an atom of that element.
6.3 • Most of the atomic mass is in the nucleus because neutrons and protons have a mass about 1,800 times larger than the electrons, which are outside the nucleus.
6.3 • The atomic mass number is the total number of protons and neutrons in the nucleus of an atom.
6.3 • Isotopes are forms of the same element whose atoms differ only in the number of neutrons in their nuclei.
6.3 • Quarks are the particles that make up protons and neutrons.
6.4 • Electrons orbit the nucleus at high speed within the spherical areas of their shells.
6.4 • Together all the electron shells of an atom form its electron cloud.
6.4 • Electric forces hold the electrons in their shells.
6.4 • Nuclear force holds the neutrons and protons together in the nucleus, thereby keeping the nucleus from flying apart.

Key Terms

alchemists
elements
atoms
natural elements
compounds
radioactive
nucleus
electrons
proton
neutron
atomic number
atomic mass number
isotopes
quarks
shells
electron cloud
nuclear force

Biography

Vera Kistiakowsky (1928-)

Vera Kistiakowsky is a professor of nuclear physics at Massachusetts Institute of Technology (M.I.T.). With a group of 30 other scientists, technicians, and engineers she studied the structure of the proton. By accelerating protons to very high energies and banging them together, the M.I.T. group and other groups of scientists around the world have shown that the proton is probably made up of three quarks.

When she is not busy probing the deepest structures of matter, Kistiakowsky works with students. She teaches them and directs their research projects. Since Kistiakowsky would like to see more women in science, she spends a large amount of her spare time giving talks to encourage young women to become scientists. Why does Kistiakowsky encourage women to enter the sciences? She does this because she herself likes science and she thinks other women would like it too.

"I like doing calculations; I like building equipment and making it work; I like thinking about explanations of the results. I enjoy being on the edge where you can really see things people haven't seen before and maybe contribute something to the knowledge of the universe."

Vocabulary

1. f
2. b
3. h
4. a
5. e
6. g
7. c
8. i
9. d

● Word Skill: Vocabulary

What Do You Remember?

1. basic substances of which all matter is composed
2. combination of two or more elements
3. more than a hundred
4.

Particle	Mass	Location	Charge
proton	1,800 × electron mass	nucleus	+
electron	1	outside nucleus	−
neutron	1,800 × electron mass	nucleus	0

5. atomic number—number of protons in the nucleus
mass number—total number of protons and neutrons in a nucleus
6. electric and nuclear
Electric forces exist between positive protons and negative electrons; they prevent the atom from flying apart. Nuclear force binds the neutrons and protons together; it keeps the nucleus from flying apart.

TO THINK ABOUT AND DO

Vocabulary ●

Number your paper from 1 to 9. Match the term in column A to the phrase that describes it in column B. Write the letter of the phrase next to the appropriate number.

A	B
1. compound	**a.** positively charged particle in the nucleus
2. nucleus	**b.** small but dense core of the atom
3. neutron	**c.** number of protons in the nucleus
4. proton	**d.** negatively charged particle outside the nucleus of an atom
5. shells	**e.** certain energy states in which the electrons of an atom can exist
6. quarks	**f.** combination of two or more elements
7. atomic number	**g.** particles of which protons and neutrons may be made
8. isotopes	**h.** particle in nucleus with no charge
9. electron	**i.** forms of the same element that differ only in the number of neutrons in their nuclei

What Do You Remember?

1. What is an element?
2. What is a compound?
3. How many different elements compose all matter?
4. Compare protons, electrons, and neutrons by making a chart with the electrical charge, location, and mass of each.
5. Distinguish between the atomic number and the atomic mass number.
6. What two forces exist in an atom? What is the function of each?

Applying What You Have Learned

1. Describe how an atom of chlorine and an atom of fluorine differ.
2. How would you determine how many neutrons an atom has?
3. Why are atoms electrically neutral?

Use the following chart to answer questions 4–6.

Symbol	Atomic Number	Atomic Mass Number
N	7	14
Al	13	27
Ge	32	73
Os	76	190
Mg	12	24
Be	4	9
K	19	39
Fe	26	56
Pb	82	207

4. How many protons are in an atom of each of the following: N, Al, Ge, and Os?
5. How many neutrons are in an atom of each of the following: Mg, Be, Os, and N?
6. How many electrons are in an atom of each of the following: Fe, K, Pb, and Ge?

Research and Investigation

1. Describe the contributions of the physicists J. J. Thomson and H. G. J. Moseley to modern chemistry.
2. Select one element. Write a report describing its properties and its uses. Then make a collage to illustrate its uses.

Applying What You Have Learned

1. An atom of chlorine and an atom of fluorine differ in their atomic mass number, derived from the number of protons and neutrons in the nucleus of each atom. Chlorine has 17 protons (therefore, 17 electrons) and 18 neutrons, whereas fluorine has 9 protons (therefore, 9 electrons) and 10 neutrons.
2. Subtract the number of protons (atomic number) from the atomic mass number for that atom.
3. because the number of protons (positive charge) is always equal to the number of electrons (negative charge)
4. N = 7; Al = 13; Ge = 32; Os = 76
5. Mg = 12; Be = 5; Os = 114; N = 7
6. Fe = 26; K = 19; Pb = 82; Ge = 32

Comprehension Skill: Applied

Research and Investigation

1. J. J. Thomson—English physicist who in 1898 provided model of the atom called plum-pudding atom in which electrons are sprinkled within a sphere of positive charge

H. G. J. Moseley—English physicist who determined the charges on the nuclei of most of the atoms by observing that each different element produces X rays with a characteristic wavelength

2. Answers will depend on the element selected.

Reference/Research Skills

CHAPTER BACKGROUND

As more and more elements were discovered, the need to arrange them into a pattern that would simplify their study became more evident.

Chemists in the early nineteenth century had enough knowledge to begin to group the elements. At this time Johann Dobereiner of the University of Jena in Germany observed that certain triads of elements existed. Each triad consisted of three elements that were similar in behavior.

Later in the nineteenth century, John Newlands reported the law of octaves. Newlands had observed that when the elements were arranged by increasing atomic mass number, every eighth element had similar properties (the noble gases had not been discovered at this time). The Royal Chemical Society in England awarded him its highest honor for his contribution to the development of the periodic law.

In 1869 Dimitri Mendeleev, a Russian chemist, and Lothar Meyer, a German chemist, independently published classifications of the elements. Like Newlands each had classified the elements by atomic mass number.

When Mendeleev published his periodic table, only 63 elements were known. Mendeleev left spaces on his table for elements that he predicted would be discovered. He even predicted some of their chemical and physical properties. This foresight was rewarded when three of the elements he predicted were discovered during his lifetime.

Mendeleev's original table was based on increasing atomic mass number. With the refinement of the technology used to investigate the elements and the discovery of new elements, several discrepancies were found.

The correct order of the elements was established by the English scientist Henry Moseley. Using X-ray emission frequencies, Moseley established that the elements should be arranged in order of increasing atomic number. This arrangement changed the placement of only six elements in Mendeleev's table, but it led to the current statement of the periodic law. The periodic law states that the properties of the elements are periodic functions of their atomic numbers. They repeat in a regular cycle. The discovery of isotopes made it more apparent that atomic number is the correct basis for arranging the elements.

7 The Elements

Introduce the chapter by placing the following data on the board in a chart: aluminum—D (density) = 2.7 g/cm^3, conductor, malleable, mp (melting point) = 660.2°C, color = silver; carbon—D = 1.8 g/cm^3, conductor, nonmalleable, mp = 3,350°C, black; copper—D = 4.93 g/cm^3, nonconductor, nonmalleable, mp = 1,083°C, orange; iodine—D =49.3 g/cm^3, nonconductor, nonmalleable, mp = 114°C, black; sulfur—D = 2.07 g/cm^3, nonconductor, nonmalleable, mp = 115°C, yellow. Ask the students to arrange these elements into categories based on these properties. Explain that classifying elements based on their properties is what the periodic table does. In this chapter the students will learn how the elements are categorized on the periodic table.

The element cadmium makes paint yellow. Cobalt makes porcelain and tile blue. Chromium makes rubies red. Other elements have familiar properties or uses. Uranium is radioactive. Mercury is a liquid metal. Barium produces green colors in fireworks. Strontium gives red colors to fireworks. Sulfur is a bright yellow solid. Chlorine is a yellowish-green poisonous gas. Osmium is very heavy. Lithium is explosive and very light.

In all there are 106 known elements on the earth. Are any of the elements similar? Is there some way they could be grouped? How could they be organized into groups? Heavy and light elements? Solids, liquids, and gases? Color groupings? Through observation and experimentation, scientists have agreed on a way to group the elements. Elements are grouped according to their properties.

Lesson Organizer

7.1 The Periodic Table of the Elements

Pages 136–141
Advanced Planning: Page 136
Activity: Can You Describe the Missing Piece? Page 141
Follow-up: Page 141
- **R** Creating a Periodic Table
- **E** Symbols
- **T** Elements in the Home

7.2 Hydrogen and the Metals

Pages 142–145
Advanced Planning: Page 142
Activity: What Is Copper Plating? Page 144
Follow-up: Page 145
- **R** Transition Metals
- **E** A Weather Detective
- **T** Alloys in Industry

7.3 From Metals to Nonmetals

Pages 146–149
Follow-up: Page 149
- **R** Families of Elements
- **E** Fluorine-Decay Prevention
- **T** Vulcanization

7.4 The Noble Gases

Pages 150–151
Follow-up: Page 151
- **R** Noble Gases
- **E** Discovery of Argon
- **T** Diving with Helium

7.5 The Lanthanides and the Actinides

Pages 152–153
Follow-up: Page 153
- **R** Lanthanides and Actinides
- **E** Naming Synthetic Elements
- **T** Fission

See ACTIVITIES book pages 37–42 for laboratory, skill, and review exercises for Chapter 7.

7.1

OBJECTIVES: The student should be able to

- **distinguish** between periods and families on the periodic table;
- **distinguish** between metals and nonmetals based on their properties;
- **define** the term *valence electrons*;
- **describe** the pattern that exists among the elements as one reads from left to right on the periodic table.

SCIENCE TERMS

periodic table	metals
periods	nonmetals
families	valence electrons

ADVANCED PLANNING

Prepare two transparencies of the periodic table. One should show the symbol, atomic number, atomic mass number, and electron structure of each element. The other should be blank. Obtain an overhead projector. If these are not available, an effective lesson can be taught using a large classroom periodic wall chart.

LESSON PLAN

1. Introduce the students to the periodic table by explaining that although many attempts were made to classify the elements, the most sensible was designed by Dimitri Mendeleev in the nineteenth century. Point out that the periodic table is an orderly classification of elements that shows the atomic number, the atomic mass number, and the symbol for each element.

7.1 The Periodic Table of the Elements

You will find out

- how the elements are organized on the periodic table;
- what is meant by the term *valence electron;*
- what determines families of elements.

For many years scientists tried to organize elements with similar properties into groups. The first person to sensibly group elements showing similar properties was a Russian scientist, Dimitri Mendeleev. In 1869 Mendeleev published a table of the elements arranged according to properties. An updated version of his table called the **periodic table** is shown on pages 138–139. The elements are arranged in order of increasing atomic number in horizontal rows called **periods.** Elements with similar properties appear lined up in vertical columns called **families.** For convenience, two subfamilies are separated out and displayed below the main body of the table.

If you look at the periodic table, you will see a key that explains the information in the boxes of the table. The dark line that looks like steps down the right side of the chart divides **metals** and **nonmetals.** Metals are on the left of the line, and nonmetals are on the right. Metals are good conductors of heat and electricity, are shiny, can be pounded into different shapes, and can be

Figure 7.2 In the periodic table, elements are grouped according to properties. In a band, players are grouped according to the instrument and part played.

Figure 7.3 Copper is a metal.

drawn out into wires. Nonmetals are poor conductors of heat and electricity, are dull, and break instead of bend. The division between metals and nonmetals is not so sharp as the dark line would show. The elements near the boundary line have properties of both metals and nonmetals.

Of the 106 known elements 85 are metals. However, this does not mean that the nonmetals are unimportant or even rare. In fact, 75 percent of the material in the earth's outer layer and atmosphere is made of nonmetals.

Each of the 106 known elements is made up of its own kind of atom. What is it about these atoms that causes elements to have similar properties? The answer lies with the tiny electrons.

Figure 7.4 Sulfur is a nonmetal.

SCIENCE BACKGROUND

When the elements are classified by properties, some elements seem to fit into both groups, metal and nonmetal. Elements that in some ways act like metals and in other ways act like nonmetals are called metalloids. The metalloids are the elements along the line that separates the metals and the nonmetals.

Periods are characterized by increasing atomic number and increasing atomic mass number as they are read from left to right. (Note that there are four exceptions to the increasing atomic mass numbers: 52 and 53, 18 and 19, 27 and 28, and 90 and 91.)

2. Ask the students to look at the key of the periodic table in their texts. Stress what each number and letter in the square means. This will give you a good opportunity to review some of the concepts that the students learned in Chapter 6. As you point to each number and symbol, ask the following questions:
   a. What does H stand for (hydrogen)?
   b. What do you call the letter or letters that represent an element (symbol)?
   c. When you write a symbol, what is always true of the first letter but never true of the second (upper case)?
   d. Tell the students that the number in the upper left-hand corner of each square is the atomic number. Ask them what it tells about an atom of the element (number of protons).
   e. Point out the atomic mass number. Ask the students what the atomic mass number tells them about the atom (protons plus neutrons).
   f. Ask the students how they can use the atomic mass number and the atomic number to find out how many neutrons an atom has (subtract the atomic number from the atomic mass number).

3. Reinforce the information in the key by having students study some squares on the chart. Ask questions such as the following:
   a. What is the element with atomic number 52 (tellurium)?
   b. What is the atomic mass number of element 63 (152)?
   c. What is the number of valence electrons in element 56 (2)?

Periodic Table

Families

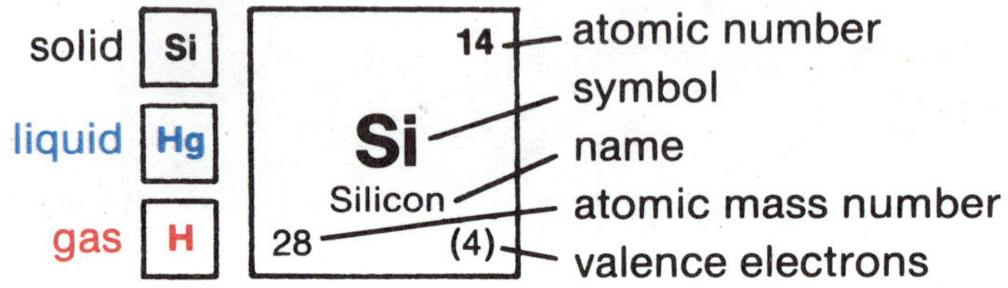

Transition

I A	II A							
1 H Hydrogen 1 (1)								
3 Li Lithium 7 (1)	4 Be Beryllium 9 (2)							
11 Na Sodium 23 (1)	12 Mg Magnesium 24 (2)							
19 K Potassium 39 (1)	20 Ca Calcium 40 (2)	21 Sc Scandium 45 (2)	22 Ti Titanium 48 (2)	23 V Vanadium 51 (2)	24 Cr Chromium 52 (1)	25 Mn Manganese 55 (2)	26 Fe Iron 56 (2)	27 Co Cobalt 59 (2)
37 Rb Rubidium 85 (1)	38 Sr Strontium 88 (2)	39 Y Yttrium 89 (2)	40 Zr Zirconium 91 (2)	41 Nb Niobium 93 (1)	42 Mo Molybdenum 96 (1)	43 Tc Technetium 97 (1)	44 Ru Ruthenium 101 (1)	45 Rh Rhodium 103 (1)
55 Cs Cesium 133 (1)	56 Ba Barium 137 (2)	57 to 71 Lanthanides	72 Hf Hafnium 178 (2)	73 Ta Tantalum 181 (2)	74 W Tungsten 184 (2)	75 Re Rhenium 186 (2)	76 Os Osmium 190 (2)	77 Ir Iridium 192 (2)
87 Fr Francium 223 (1)	88 Ra Radium 226 (2)	89 to 103 Actinides	104 Unq Unnilquadium 257 (2)	105 Unp Unnilpentium 260 (2)	106 Unh Unnilhexium 263 (2)			

57 La Lanthanum 139 (2)	58 Ce Cerium 140 (2)	59 Pr Praseodymium 141 (2)	60 Nd Neodymium 144 (2)	61 Pm Promethium 145 (2)	62 Sm Samarium 150 (2)
89 Ac Actinium 227 (2)	90 Th Thorium 232 (2)	91 Pa Protactinium 231 (2)	92 U Uranium 238 (2)	93 Np Neptunium 237 (2)	94 Pu Plutonium 244 (2)

of the Elements

Families

Metals

			III A	IV A	V A	VI A	VII A	VIII A
								2 He Helium 4 (2)
			5 B Boron 11 (3)	6 C Carbon 12 (4)	7 N Nitrogen 14 (5)	8 O Oxygen 16 (6)	9 F Fluorine 19 (7)	10 Ne Neon 20 (8)
			13 Al Aluminum 27 (3)	14 Si Silicon 28 (4)	15 P Phosphorus 31 (5)	16 S Sulfur 32 (6)	17 Cl Chlorine 35 (7)	18 Ar Argon 40 (8)
28 Ni Nickel 59 (2)	29 Cu Copper 64 (1)	30 Zn Zinc 65 (2)	31 Ga Gallium 70 (3)	32 Ge Germanium 73 (4)	33 As Arsenic 75 (5)	34 Se Selenium 79 (6)	35 Br Bromine 80 (7)	36 Kr Krypton 84 (8)
46 Pd Palladium 106 (18)	47 Ag Silver 108 (1)	48 Cd Cadmium 112 (2)	49 In Indium 115 (3)	50 Sn Tin 119 (4)	51 Sb Antimony 122 (5)	52 Te Tellurium 128 (6)	53 I Iodine 127 (7)	54 Xe Xenon 131 (8)
78 Pt Platinum 195 (1)	79 Au Gold 197 (1)	80 Hg Mercury 201 (2)	81 Tl Thallium 204 (3)	82 Pb Lead 207 (4)	83 Bi Bismuth 209 (5)	84 Po Polonium 209 (6)	85 At Astatine 210 (7)	86 Rn Radon 222 (8)

Metals | **Nonmetals**

63 Eu Europium 152 (2)	64 Gd Gadolinium 157 (2)	65 Tb Terbium 159 (2)	66 Dy Dysprosium 163 (2)	67 Ho Holmium 165 (2)	68 Er Erbium 167 (2)	69 Tm Thulium 169 (2)	70 Yb Ytterbium 173 (2)	71 Lu Lutetium 175 (2)
95 Am Americium 243 (2)	96 Cm Curium 247 (2)	97 Bk Berkelium 247 (2)	98 Cf Californium 251 (2)	99 Es Einsteinium 254 (2)	100 Fm Fermium 257 (2)	101 Md Mendelevium 258 (2)	102 No Nobelium 255 (2)	103 Lr Lawrencium 256 (2)

4. Point out that the periodic table also classifies elements as metals and nonmetals. Ask the students to name some metals (silver, gold, copper) and some nonmetals (oxygen, nitrogen, helium). Then ask the following questions about metallic and nonmetallic properties:
   - **a.** Which is shiny (metal; nonmetal is dull)?
   - **b.** Which can be pounded into shapes (metal; nonmetal breaks)?
   - **c.** Which can be made into wires (metal; nonmetal breaks)?
   - **d.** Which is a good conductor of heat (metal; nonmetal is a nonconductor)?
   - **e.** Which is a good conductor of electricity (metal; nonmetal is a nonconductor)?

5. Have the students look at the horizontal row of elements that begins with the element potassium. Ask them to look at the atomic numbers from left (K) to right (Kr). What do they notice (they increase)? Have the students look at other horizontal rows. Does the atomic number increase going from left to right (yes)? Now have the students look at the atomic mass numbers of the horizontal row that begins with potassium. As students read from left to right, what do they notice happens to the atomic mass number (increases in most cases)? Point out that the horizontal rows are called periods.

6. Point out the valence electron number and mention that this number tells students the number of electrons in the outermost electron shell. Emphasize that the outermost electrons are called the valence electrons.

7. Ask the students to look at the vertical columns, called families, on the periodic table. From the squares of a single column, or family, what similarity can students see among the elements in that column (same number of valence electrons)? Explain that because all the elements in a family have the same number of valence electrons, they have similar properties. Point out that some families have specific names other than the Roman numeral and letter designation.

8. Conclude the lesson by using your transparency or wall chart to guide the students as you ask them the following review questions:
   a. What do you call the elements to the left of the staircase (metals)? To the right of the staircase (nonmetals)?
   b. What are some properties of metals (are shiny, are conductors, can be pounded into different shapes)? Nonmetals (are dull, break easily, are nonconductors)?
   c. What are horizontal rows called (periods)?
   d. What happens to the atomic number as you go across a period from left to right (increases)?
   e. What happens to the atomic mass number as you go across a period from left to right (increases, with a few exceptions)?
   f. What are the vertical columns called (families)?
   g. How are the atoms of elements in a family the same (same valence electron number)?

● **Comprehension Skill: Interpretative**

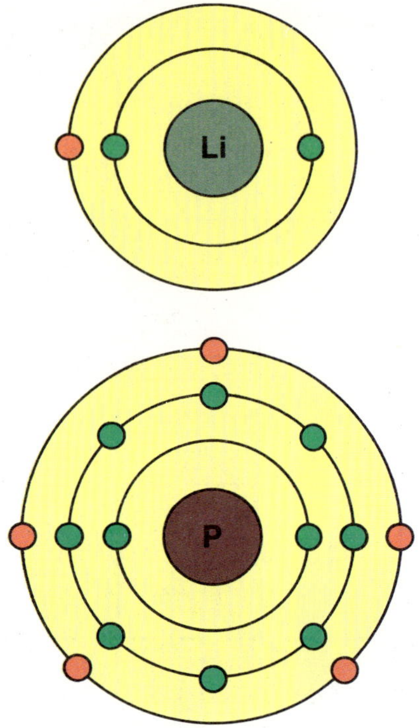

Figure 7.5 The valence electrons are in the shell farthest from the nucleus.

In Chapter 6 you learned that all electrons can be thought of as moving in different shells around the nucleus of an atom. The electrons in the shell farthest from the nucleus are called **valence** [VAY-luhns] **electrons.** It is the number of valence electrons that tells which family an element is in. For example, all the elements in family IA on the periodic table have one valence electron. All those in family IIA have two valence electrons, and so on. Elements with the same number of valence electrons have similar properties. It is important to remember that the number of valence electrons an element has is the key to its placement on the periodic table.

The periodic table has withstood the test of time and proved to be an accurate and useful way to group the elements. What is most amazing is that when Mendeleev first set up the periodic table there were only 66 known elements. Yet his arrangement still works today when 106 elements are known!

The idea behind Mendeleev's table was so accurate that he was able to predict correctly the existence of certain elements before they were discovered. He did this by describing the properties these elements would have. Mendeleev was so sure the elements germanium, gallium, and scandium existed that he left spaces for them on his periodic table. They were later discovered and proved that he was right!

Do You Know?
The oceans of the earth contain 20 million billion tons of sodium in the form of salt.

			Ti=50	Zr=90	?=180.
			V=51	Nb=94	Ta=182
			Cr=52	Mo=96	W=186.
			Mn=55	Rh=104,4	Pt=197,4
			Fe=56	Ru=104,4	Ir=198.
			Ni=Co=59	Pl=106,6,	Os=199.
H=1			Cu=63,4	Ag=108	Hg=200.
	Be=9,4	Mg=24	Zn=65,2	Cd=112	
	B=11	Al=27,4	?=68	Ur=116	Au=197?
	C=12	Si=28	?=70	Sn=118	
	N=14	P=31	As=75	Sb=122	Bi=210
	O=16	S=32	Se=79,4	Te=128?	
	F=19	Cl=35,5	Br=80	I=127	
Li=7	Na=23	K=39	Rb=85,4	Cs=133	Tl=204
		Ca=40	Sr=87,6	Ba=137	Pb=207.
		?=45	Ce=92		

Figure 7.6 In this section of Medeleev's periodic table, you can see the spaces he left for elements that were later discovered.

ACTIVITY

Can You Describe the Missing Piece?

Materials
colored picture
glue
cardboard or oaktag
scissors
brown bag

Procedure

1. Take your picture and glue it onto a piece of cardboard. Trim the cardboard to the size of the picture.
2. Carefully cut up the picture into approximately 25 puzzle-shaped pieces. You can draw a jigsaw outline on the back of the cardboard and follow it with your scissors.
3. Put all the pieces into a bag except one. Then exchange bags with a classmate.
4. Assemble the puzzle in the bag you received.

Questions

1. What gives you clues about the look of the missing piece?
2. Predict the size, shape, and colors of the missing piece.
3. Describe what the picture on the missing piece looks like.
4. Get the missing piece from your classmate. How accurate were your predictions about its appearance?
5. How do you think Mendeleev made predictions about the characteristics of the missing elements from his original table?
6. One of the elements missing from Mendeleev's original table was germanium (Ge). Which elements might give you some clues about the properties of germanium? Explain your reasoning. ●

Study Questions for 7.1

1. What are the vertical columns on the periodic table called?
2. What are the horizontal rows on the periodic table called?
3. What are valence electrons?
4. What do all elements of the same family have in common?

1. families
2. periods
3. the electrons in the shell farthest from the nucleus
4. They all have the same number of valence electrons.

ACTIVITY: Can You Describe the Missing Piece?

This activity simulates the methods Mendeleev used to predict the existence of the unknown elements for his table. The missing puzzle piece is analogous to the missing elements in his original work. The students can use their completed puzzle to help them describe the missing piece.

1. The surrounding pieces give clues to the identity of the missing piece.
2. The student should be able to give the piece's size, shape, and colors and may be able to predict exactly what is pictured on it.
3. The answer will vary depending on the student's puzzle.
4. Most students should have made very accurate predictions.
5. Mendeleev relied on the characteristics and properties of the elements surrounding the missing ones to make his predictions.
6. You might look at the elements that surround germanium on the periodic table: gallium, silicon, arsenic, and tin.

● Comprehension Skill: Literal

FOLLOW-UP

Reinforcement: Creating a Periodic Table

Have the students put an outline of the periodic table on the bulletin board or an empty wall. Have them label the metals and nonmetals. Then have them fill in the hydrogen square. They should include the symbol, atomic number, atomic mass number, and number of valence electrons for hydrogen. Students should label each family with its Roman numeral and letter. They should label each period with its number. Leave the table up so you can add to it after later lessons.

Enrichment: Symbols

The symbol of many of the elements is the first letter or the first two letters of the element. However, some of them are different. Ask the students to do library research to find out how the symbols of the elements were devised.

Tie-in: Elements in the Home

All household products are made of elements. Make a chart that lists the names of household substances and their chemical formulas. Ask the students to determine what elements are in each substance.

7.2

OBJECTIVES: The student should be able to

- **locate** the alkali metals on the periodic table;
- **identify** the alkali metals and the alkaline earth metals by the number of valence electrons in their atoms;
- **discuss** the importance of hydrogen;
- **identify** several properties the alkali metals have in common;
- **locate** the alkaline earth metals and the transition metals on the periodic table;
- **list** two reasons why calcium is important;
- **identify** the transition metals by the number of valence electrons in their atoms;
- **name** several transition metals and their uses.

SCIENCE TERMS

alkali metal family
alkaline earth metal family
transition metals

ADVANCED PLANNING

Obtain an alnico magnet and several items you can lift with it. Also obtain a magnet you know is weaker than the alnico.

LESSON PLAN

1. Introduce the lesson by referring the students to the periodic table. Ask what a vertical row of elements is called (family). How are the elements in a family similar (same number of valence electrons)? Using your transparency or wall chart, point out family IA. Tell the students that these elements are the alkali metals. Ask a student how many valence electrons the atoms of each of these elements have (1).

SCIENCE BACKGROUND

Even though hydrogen is a gas, it is placed with the solid alkali metals because, like the other alkali metals, it has one valence electron.

Some scientists believe that at the very high temperature and pressure found in the centers of Jupiter and Saturn, hydrogen may form a metallic solid similar to the alkali metals. On the earth, hydrogen is an important part of water, coal, oil, and natural gas.

7.2 Hydrogen and the Metals

You will find out

- what some characteristics of elements in families IA, IIA, and the transition metals are;
- how many valence electrons are in hydrogen, alkali metal, and alkaline earth metal atoms;
- what the transition metals are.

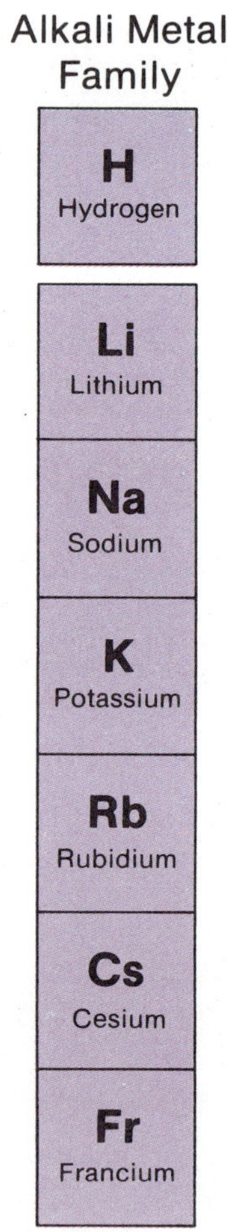

The atoms of elements in family IA of the periodic table have one valence electron. Hydrogen, although not a metal, is included in this family because it, too, has one valence electron. Hydrogen has a nucleus of one proton with a single electron around it. Although it is not the most common element on the earth, it is the most common element in the universe. The sun and the other stars are made mostly of hydrogen. Hydrogen is an essential part of many types of fuel. Petroleum and coal are compounds of hydrogen and carbon. Hydrogen is very reactive. It is almost never found alone on the earth. Hydrogen is in many important compounds such as water and alcohol.

Figure 7.7 Hydrogen is the most plentiful element in the universe.

The other elements in family IA are six very soft, silver-white metals. Family IA is called the **alkali** [AL-kuh-ly] **metal family.** All the alkali metals are very reactive. Even though they may be cool, they will burn your hand if you touch them. They will react quickly with oxygen. They are so reactive, in fact, that they are never found alone in nature.

Next to the very reactive alkali metals in the periodic table is a much calmer group of elements, IIA. These elements make up the **alkaline** [AL-kuh-lyn] **earth metal family.** Atoms of elements in this family have two valence electrons. The alkaline earth metals are slower than the alkali metals to react with other elements. The alkaline earth metal calcium is the fifth most common element in the earth's outer layer. It is found in limestone and gypsum. Your body has about one kilogram of calcium, which is found in your bones and teeth. Barium and magnesium, also in family IIA, have many different uses. Both elements are used in medicine and firecrackers.

Alkaline Earth Metal Family

Be	Beryllium
Mg	Magnesium
Ca	Calcium
Sr	Strontium
Ba	Barium
Ra	Radium

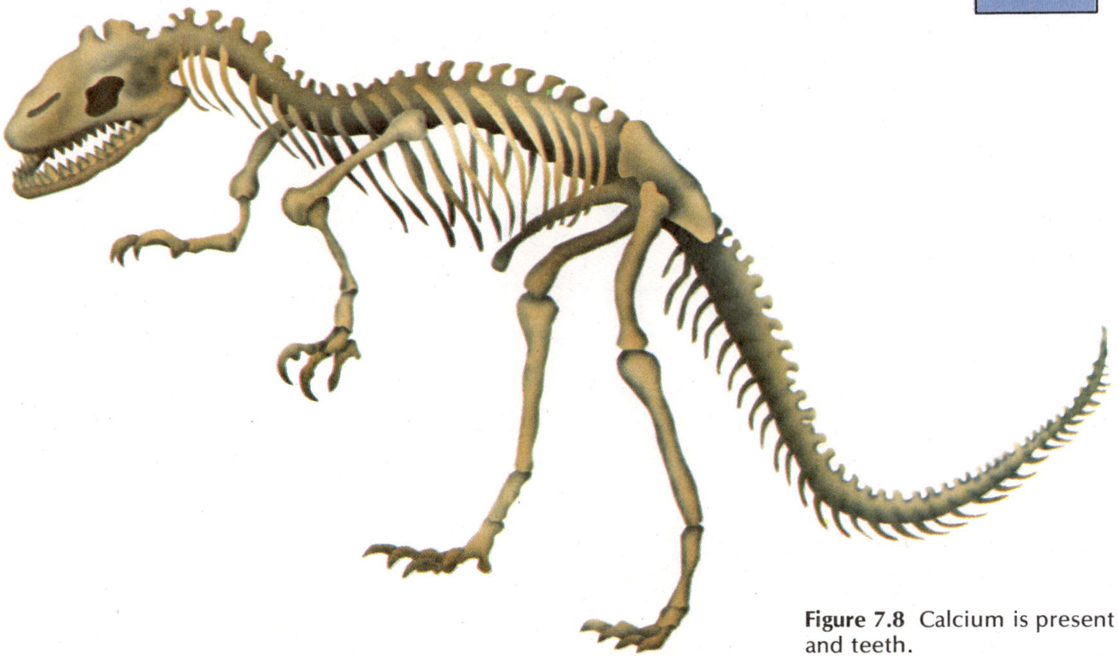

Figure 7.8 Calcium is present in bones and teeth.

2. Use your transparency or wall chart to point out the second family of elements, II A. Tell the students that these elements are called the alkaline earth metals. Ask the students to compare the number of valence electrons in this family to that of the alkali metals (alkaline earth metals = 2 valence electrons; alkali metals = 1 valence electron). Point out that since the electrons in II A are more tightly bound to the nucleus than those of the alkali metal family, the elements in group II A are less reactive than those in I A.

SCIENCE BACKGROUND

The single valence electron in the alkali metals is far from the nucleus. It is loosely bound to the nucleus and can be easily removed. As a result the alkali metals are very reactive. In air the alkali metals quickly react with oxygen. The elements are never found alone in nature. They are found in combination with other elements.

Calcium is the most common alkaline earth metal. It is found in limestone and gypsum. Gypsum is a white mineral used to make plaster of Paris, cement, and fertilizer. Calcium is also found in bones and teeth. Chalk and marble are made of a calcium compound, calcium carbonate.

3. Use your overhead transparency or wall chart to point out the shorter columns that exist between groups II A and III A. Tell the students that these elements are called the transition metals, which include some of the most familiar metals. Ask the students what properties the metals share (are shiny, can be pounded into shapes, are conductors).
4. Explain that the transition metals differ from other elements in the way that they fill their electron shells. In the atoms of most other elements, each new electron is added to the outermost shell until that shell is filled. Only when a shell is filled can an electron be added to the next shell. Explain that in atoms of the transition metals there are electrons in the outer shell even though the previous shell has not been completely filled. In these elements the last electron is added to the next to the last, not the last, shell.

ACTIVITY: What Is Copper Plating?

If copper and iron bars are not available, a penny and an iron nail will work. The students should be careful not to spill the copper sulfate solution. Copper can be poisonous. The materials that are copper plated should be handled with tweezers, as the copper is apt to rub off. Eye protection should always be used. Explain that this process is used to plate metals.

Step 7 calls for a half-hour wait. If the class period is too short to allow for 30 minutes, any time amount over 20 minutes should be sufficient.

◇ **Caution:** Students should wear safety glasses. They should never touch the copper sulfate.

● **Laboratory Skill: Manipulation of Materials Experimentation/Demonstration**

Between families IIA and IIIA is a large group of elements that do not exactly fit in any family. These elements are the **transition metals.** Nearly all the transition metals have two valence electrons. In these elements the next to last shell of electrons is filled as the number of electrons increases. The transition metals have similar properties.

● ACTIVITY

What Is Copper Plating?

Materials

copper sulfate
beaker
water
2 pieces of insulated copper wire
battery
iron bar
copper bar
steel wool
tweezers
safety glasses
apron
gloves

Procedure

1. Put on your safety glasses and apron. Get some copper sulfate in a beaker from your teacher. CAUTION: Do not touch the copper sulfate with your hands!
2. Rub a piece of copper and a piece of iron with steel wool until they shine.
3. Be sure the insulation has been removed from the wires near the ends.
4. Attach one end of each wire to the battery.
5. Look at the top of the battery. To the free end of the wire that is attached to the + pole, attach the copper bar. To the free end of the wire that is attached to the − pole, attach the iron bar.
6. Immerse the two strips in the copper sulfate solution but do not let them touch.
7. After one half hour remove the strips and examine them. CAUTION: Use a pair of tweezers. Do not touch the metals.

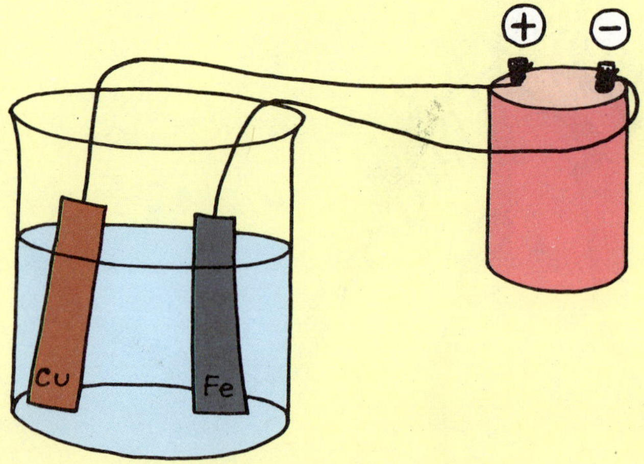

Questions

1. After one half hour what has happened to the iron bar?
2. Where did the copper on the bar come from?
3. How could you silver-plate a spoon?

1. covered with copper
2. the copper sulfate
3. Use a silver bar and silver sulfate rather than a copper bar and copper sulfate.

Transition Metals

Sc Scandium	Ti Titanium	V Vanadium	Cr Chromium	Mn Manganese	Fe Iron	Co Cobalt	Ni Nickel	Cu Copper	Zn Zinc
Y Yttrium	Zr Zirconium	Nb Niobium	Mo Molyb-denum	Tc Technetium	Ru Ruthenium	Rh Rhodium	Pd Palladium	Ag Silver	Cd Cadmium
	Hf Hafnium	Ta Tantalum	W Tungsten	Re Rhenium	Os Osmium	Ir Iridium	Pt Platinum	Au Gold	Hg Mercury

Most of the familiar metals—iron, nickel, copper, silver, and gold—are transition metals. Copper mixes with zinc to make brass. Copper and tin make bronze. Other less well-known transition metals have familiar uses. Chromium is used in stainless steel and chrome plate. Because it is almost perfectly rustproof, tantalum is used in skull plates and other surgical repairs to the human body. Yttrium is used in television tubes to make the color red.

Figure 7.9 The use of chrome plate was very popular in cars of the 1950's. Chromium is present in chrome plate.

Study Questions for 7.2

1. How many valence electrons are in hydrogen, alkali metal, and alkaline earth metal atoms?
2. What is different about the electron shell structure of transition metals?
3. What are some characteristics of elements in families IA and IIA?

1. one for hydrogen and the alkali metal atoms; two for alkaline earth metal atoms
2. The next to the last shell is not completely filled with electrons. It is gradually filled up as the atomic number increases.
3. Hydrogen and the alkali metals are very reactive. Alkaline earth metals are less reactive.

5. Refer students to the section on a career as a metallurgical engineer. If you have an alnico magnet, you can actually demonstrate a product of metallurgical research. Metallurgists developed alnico, which is a mixture of aluminum, iron, nickel, and copper. It is exceedingly hard and can be highly magnetized.

FOLLOW-UP

Reinforcement: Transition Metals

Ask students to create a mobile with ten hangings. Each hanging should describe a different transition metal. One side of the hanging should show the symbol, atomic number, and atomic mass number. The other side should illustrate some uses and properties of the element.

Enrichment: A Weather Detective

Dissolve some cobalt chloride crystals in water to form a dark red-violet solution. The ratio of $CoCl_2$ to water should be 1.5 g to 100 mL. Have the students draw a picture of their favorite personality. Some parts of the picture should be colored with felt pens, and other parts should be painted with the cobalt chloride solution. After the $CoCl_2$ dries, hang the picture near a window. Can the students tell you how it predicts the weather (blue in good weather, pink in wet weather)? CAUTION: Do not ingest the $CoCl_2$. Wash hands thoroughly after using.

Tie-in: Alloys in Industry

Ask students to contact some of the industries in the area. What alloys are important to them? What are the main uses of these alloys?

SCIENCE BACKGROUND

The following is a list of several of the transition metals and their uses:

Cr—chromium; car decoration
Fe—iron; magnets, construction
Ni—nickel; coins
Cu—copper; pennies, pipes, wire
W—tungsten; light bulbs
Pt—platinum; jewelry
Au—gold; jewelry
Ag—silver; jewelry
Hg—mercury; thermometers, barometers

7.3

OBJECTIVES: The student should be able to

- **locate** the boron, carbon, oxygen, nitrogen, and halogen families on the periodic table;
- **identify** families IIIA through VIIA by the number of valence electrons in the atoms of the elements;
- **describe** some uses of the elements in the carbon family;
- **describe** some uses of the members of the nitrogen family;
- **describe** some uses of the members of the oxygen family;
- **describe** some uses of members of the halogen family.

SCIENCE TERMS

boron family
carbon family
nitrogen family
halogen family
oxygen family

LESSON PLAN

1. Introduce the lesson by using your transparency or wall chart to help the students locate the boron, carbon, nitrogen, oxygen, and halogen families on their periodic tables.
2. Ask how many valence electrons are in atoms of each element of the boron family (3). Stress that each element has three electrons in its outer shell.
3. Ask how many valence electrons are in the elements of the carbon family (4). Point out that if one omits the transition metals the carbon family is in the center of the periodic table and has members whose valence shells are half empty and half full.

SCIENCE BACKGROUND

The protein, fat, and carbohydrate molecules of which the body is composed contain carbon. Many scientists believe that there is life on another planet and that it is not based on carbon. This life is probably based on silicon or one of the other members of the carbon family.

Carbon exists in many forms in people's daily lives. It is the black lumps that are burned as charcoal at a cookout. It is also the graphite lead in pencils and the hard diamonds that are worn as jewelry.

7.3 From Metals to Nonmetals

Boron Family	Carbon Family
B Boron	**C** Carbon
Al Aluminum	**Si** Silicon
Ga Gallium	**Ge** Germanium
In Indium	**Sn** Tin
Tl Thallium	**Pb** Lead

You will find out

- how many valence electrons are in the atoms of elements in families IIIA through VIIA;
- what some elements in families IIIA through VIIA are.

After the transition metals is the boundary between metals and nonmetals. This boundary runs through families IIIA, IVA, VA, and VIA. These four families—boron, carbon, nitrogen, and oxygen—are named after the first element in each family. All of these families have both metallic and nonmetallic elements. Atoms of elements in the **boron family**, IIIA, have three valence electrons. The elements include boron, a nonmetal, and aluminum, which is the most common metal in the earth's outer layer.

After the boron family is one of the most important families, the **carbon family,** IVA. Atoms of the elements in this family have four valence electrons. Because of the four valence electrons, the carbon family is halfway between the highly reactive metals and the highly reactive nonmetals. The first element in this family, carbon, forms an almost unbelievable number of different compounds. Many of these are necessary for life. Carbon is found in oil, plastics, and nylon. More than 5 million carbon compounds are known.

Do You Know?
The symbol for lead, Pb, comes from the word *plumbum,* the Latin word for *lead.* The word *plumbing* also comes from this same Latin word.

Figure 7.10 Among the elements present in emeralds are aluminum from the boron family and silicon from the carbon family.

Next to oxygen, silicon is the most common element in the earth's outer layer. Silicon, a member of the carbon family, is found in sand and many rocks. It is used to make glass and cement. Silicon, together with another member of the carbon family, germanium, forms the backbone of one of the most important modern industries, computers.

The heavier elements in the carbon family, tin and lead, show definite metallic properties. Tin is used to coat steel to make rustproof cans.

Atoms of elements in the **nitrogen family,** VA, have five valence electrons. Their outer shell is over half full. The air you breathe is 78 percent nitrogen. Nitrogen forms many compounds. It combines with carbon, oxygen, and hydrogen to form explosives such as TNT. Nitrogen compounds are a very important part of most fertilizers.

The next member of the nitrogen family, phosphorus, is important for life. It is a basic part of your organs and bones. Arsenic, another member of this family, is best known as a deadly poison although arsenic can also be used as a medicine. The heavier elements in the family, antimony and bismuth, have metallic properties.

Nitrogen Family	Oxygen Family
N Nitrogen	**O** Oxygen
P Phosphorus	**S** Sulfur
As Arsenic	**Se** Selenium
Sb Antimony	**Te** Tellurium
Bi Bismuth	**Po** Polonium

Figure 7.11 Nitrogen compounds are present in fertilizers.

4. Ask how many valence electrons members of the nitrogen family have (5). Point out that the valence shell is considered full with eight electrons. Therefore, in the nitrogen family the shell is more than half full.

SCIENCE BACKGROUND

In recent years lead compounds have been found to be poisonous. Research has shown that children who chewed on crib railings that were painted with a paint containing lead, people who ate from dishes improperly decorated with a paint containing lead, and people who drank water that had passed through lead pipes often suffered damage to their nervous systems. Analysis of artifacts dug from ancient Rome has shown that many of the materials used to make the aqueducts and Roman plates, cookware, and utensils had a very high lead content. Some historians believe that one reason for the decline of the Roman Empire was that many people were suffering from lead poisoning and were unable to function to the best of their abilities.

Nitrogen makes up 78 percent of the air. It is found in dynamite, fertilizer, and proteins. Nitrogen compounds such as NO_2 are air pollutants. Phosphorus is an important constituent of bones. Arsenic, antimony, and bismuth are all used in medicine. Each also has individual uses. Arsenic is a deadly poison often used to kill rats. Antimony is used to strengthen other metals. Bismuth is often used in fire alarms because some of its alloys have melting points as low as 50°C.

5. Ask how many valence electrons are found in each member of the oxygen family (6). Ask why oxygen is important (oxygen is necessary for survival; the body uses oxygen). Explain that oxygen is a very active element. It will join with many other elements.

● Word Skill: Vocabulary

SCIENCE BACKGROUND

Oxygen is present in the sugars that plants produce and in products that are considered destructive such as iron oxide or rust.

Sulfur is important because it is used to make matches and to vulcanize rubber. When rubber is taken from a tree, it is very sticky. When it is mixed with sulfur and other additives in vulcanization, it loses its stickiness and becomes stronger and more resilient. Sulfur is present in sulfuric acid, which is used to make paint, plastic, and fertilizer.

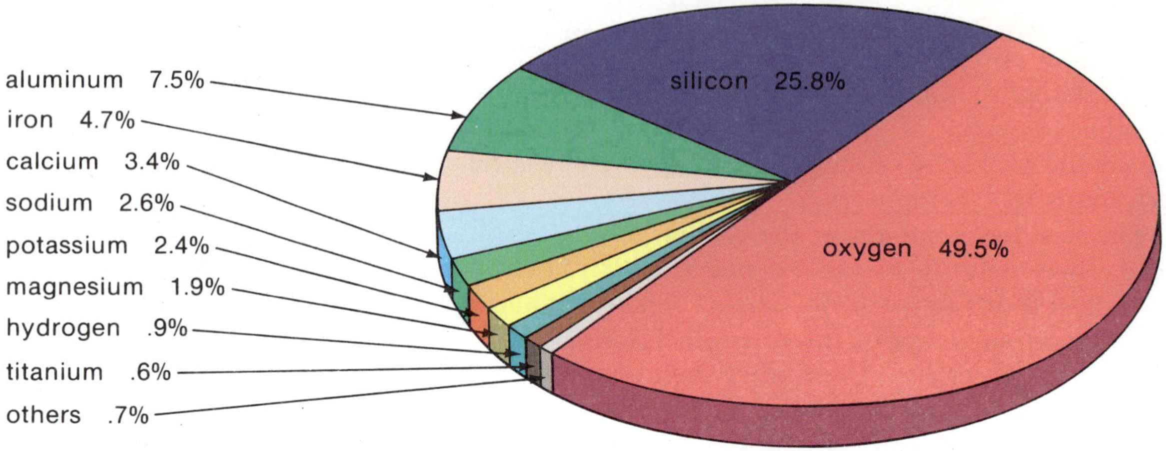

Figure 7.12 Elements in the earth's outer layer

Atoms of the elements in the **oxygen family,** VIA, have six valence electrons. Oxygen is the most common element in the earth's outer layer. It makes up about 50 percent of the total material on the surface of the earth. It is basic for life. Oxygen is very reactive, forming compounds with practically every other element.

Sulfur, another element in this family, is also a very reactive element. The Latin word for *sulfur* means ● "brimstone," or "burning stone." Sulfur is used to make matches and rubber tires and is used in practically every branch of modern industry.

Figure 7.13 Sulfur is used in the production of rubber.

The heavier members of this family—selenium, tellurium, and polonium—show a combination of metallic and nonmetallic properties. Selenium becomes more metallic if it is exposed to light. This makes it extremely useful for solar cells and light meters. Polonium is the least common natural element. It is used mainly in laboratories.

The family of elements next to oxygen in the periodic table is called the **halogen** [HAL-oh-jehn] **family,** VIIA. Atoms of the elements in this family have seven valence electrons. They are extremely reactive. They react most readily with the alkali metals. Halogen compounds are commonly used in disinfectants and cleaning agents. Compounds containing bromine are used in drugs, in photographic film, and in gasoline to prevent engine knocking. Fluorine compounds are used in toothpastes, in insecticides, and in the preparation of some plastics.

Halogen Family

Halogen Family
F Fluorine
Cl Chlorine
Br Bromine
I Iodine
At Astatine

Figure 7.14 Compounds containing bromine are used in photography.

Study Questions for 7.3

1. How many valence electrons are in atoms of elements in families IIIA, IVA, VA, VIA, and VIIA?
2. Name two elements in each of the following families: (a) boron family, (b) carbon family, (c) nitrogen family, (d) oxygen family, and (e) halogen family.

1. Family IIIA has three; IVA has four; VA has five; VIA has six; VIIA has seven valence electrons.
2. Use the periodic table to check answers.

6. Help the students to locate the halogen family (VII A) on their periodic tables. Ask how many valence electrons are in the halogen atoms (7).
7. Bonding is covered in Chapter 8. If you want to, you can expand here on the extreme reactivity of halogens. Explain that because the valence shell would be filled if it had eight electrons the halogens react with almost any other element to get the one electron that is needed to fill the outer electron shell.
8. Ask the students to name some uses of fluorine, chlorine, and iodine (fluorine—used to prevent tooth decay; chlorine—used in bleach and as a disinfectant; iodine—used in antiseptics).

FOLLOW-UP

Reinforcement: Families of Elements

Ask students to divide themselves into groups. Assign each group one of the families that are discussed in this lesson. Have each group prepare a series of posters (one per student) about its family of elements.

Enrichment: Fluorine-Decay Prevention

Ask the students to interview their dentists to learn how fluorine prevents tooth decay. Have them ask the dentist how the new fillings containing fluorine work. Students could record the interview on tape.

Tie-in: Vulcanization

Vulcanization is an important process to many industries that depend on rubber. Ask the students to create flow charts to show the steps of the process.

SCIENCE BACKGROUND

The halogens are often used as disinfectants. When halogens join with water, oxygen is released. Many bacteria that cause disease are anaerobic. This means that they cannot live in the presence of free oxygen. The oxygen released when the halogen and water react acts as a poison and kills the bacteria.

7.4

OBJECTIVES: The student should be able to

- **locate** the noble gases on the periodic table;
- **explain** why the noble gases are different from other elements;
- **identify** the noble gases by the number of valence electrons;
- **describe** some uses of the noble gases.

SCIENCE TERM

noble gas family

LESSON PLAN

1. Refer the students to the periodic table while you use your transparency or wall chart. Direct their attention to the last family of elements, the noble gases.

● Word Skill: Word Structure

SCIENCE BACKGROUND

Of the noble gases neon is used in signs, argon is used in welding, krypton is used in nuclear espionage, xenon is used in electronic flashbulbs, and radon is used in cancer therapy.

7.4 The Noble Gases

You will find out

- why the noble gases are different from other elements.

In the column on the far right of the periodic table are elements that are totally different from all the others. They are the members of the **noble gas family,** VIIIA. For years it was thought that the noble gases could not react with other elements. Just as the nobility in earlier times refused to mix with the common people, these elements were thought never to make compounds with other elements. That explains the name noble gases.

The noble gases mark the end of each period on the periodic table. They all have atoms with eight valence electrons except for helium. Helium atoms have two valence electrons. These electrons are in the shell closest to the nucleus, which holds only two electrons. The outer shells of all noble gas atoms are considered full. The noble gases are different from other elements because their electron shells are full.

Figure 7.15 *left:* The *Hindenburg* disaster occurred when the hydrogen used to inflate the blimp exploded and burned. *right:* Modern blimps use helium because it will not burn.

Helium is the lightest noble gas. Although it is the second most common element in the universe, helium is rare on the earth. It is found in natural gas wells. It is the lightest gas after hydrogen and does not burn. For these reasons, helium is used for inflating balloons and blimps.

The other noble gases are used in lighting and welding. In general, they do not react readily with other elements. Elements that burn are reacting with oxygen in the air. Because the noble gases will not react with oxygen and burn, they are much safer than other gases.

Noble Gas Family

He Helium
Ne Neon
Ar Argon
Kr Krypton
Xe Xenon
Rn Radon

Figure 7.16 Neon is often used in colorful signs.

Study Questions for 7.4

1. How many valence electrons are in a helium atom? A neon atom? An argon atom?
2. How are the noble gases different from the other elements?

1. eight for all but helium, which has two
2. Their electron shells are considered full. They do not react readily with other elements.

2. Ask a student how many valence electrons each noble gas after helium has (8). Explain that because the outer electron shell is full these elements do not easily join with other elements.
3. Stress that although the noble gases rarely join with other elements they are still useful. Ask students what helium is used for (blimps, balloons).

FOLLOW-UP

Reinforcement: Noble Gases

Ask the students to pretend that they have just discovered some of the noble gases. Have them write an article to publicize the newly discovered elements. The article should have a headline and description of the gases. It should also say why these elements will be nicknamed noble gases.

Enrichment: Discovery of Argon

The noble gases were not discovered when Mendeleev arranged his periodic table. Their low reactivity made them difficult to detect and study. Argon, the most plentiful noble gas, was discovered first. Have the students find out by whom, when, and how argon was discovered.

Tie-in: Diving with Helium

Scuba divers often carry and breathe a mixture of oxygen and helium in their tanks. Have the students find out why breathing this mixture is safer than breathing the normal air mixture when diving deep.

7.5

OBJECTIVES: The student should be able to

- **locate** the lanthanides and the actinides on the periodic table;
- **explain,** based on electron structure, why the lanthanides and the actinides are similar;
- **describe** the main uses of the lanthanides;
- **explain** why the nucleus of the actinides is important;
- **state** which elements are made in the laboratory;
- **explain** how the synthetic elements can be formed.

SCIENCE TERMS

lanthanide family
actinide family
synthetic

LESSON PLAN

1. Introduce the lesson by referring the students to the periodic table. Use your transparency or wall chart to help them locate the lanthanides and the actinides. Ask why the two families are found under the rest of the periodic table (all the lanthanides belong in the table at 57 but would not fit; the actinides belong at 89 but are below the rest of the table for the same reason).

2. Stress that all of the actinides and all of the lanthanides have very similar properties. Explain that in all the elements studied so far except the transition metals the last electron was added to the outer shell. In the transition metals the last electron was added to the second to the last shell. In the lanthanides and the actinides the electron is added to the third from the last shell. (Their two outer shells generally maintain the same number of electrons.) The addition of an electron to this third from the last shell has very little effect on the properties of the element.

SCIENCE BACKGROUND

Uranium is important not only because of its role as a nuclear fuel but also because other elements can be made from it. These other elements are made through the process of transmutation. Neutrons or nuclei of lighter elements are added to uranium to produce the synthetic, transuranium, or post World War II elements.

7.5 The Lanthanides and the Actinides

You will find out

- why elements within the lanthanide and actinide families are so similar;
- how new elements are made in the laboratory.

The **lanthanide** [LAN-thuh-nyd] **family** begins with the element lanthanium. The **actinide** [AK-tih-nyd] **family** begins with actinium. These are the families displayed below the main body of the periodic table. These families include elements with atoms in which the outer two shells of electrons are similar. The third shell from the outside is different. Since the third shell from the outside of an atom does not change the properties of the element much, it is not surprising that all the elements within these families are very similar.

The lanthanides are almost always found together. They are used in the production of steel and glass.

All the actinides are radioactive. The best-known actinide is uranium, the heaviest natural element. Uranium is the source of energy for nuclear power plants.

Any element heavier than uranium is made in nuclear laboratories. Samples of uranium are bombarded with neutrons or nuclei of light elements. This bombarding can change uranium into heavier elements that are called **synthetic** [sihn-THEHT-ihk]. The word *synthetic* means that the elements are produced in the laboratory.

Figure 7.17 Synthetic elements 93 through 105 were discovered here at the Lawrence Berkeley Laboratory in Berkeley, California.

La	Ce	Pr	Nd	Pm	Sm	Eu	Gd	Tb	Dy	Ho	Er	Tm	Yb	Lu
Ac	Th	Pa	U	Np	Pu	Am	Cm	Bk	Cf	Es	Fm	Md	No	Lr

The search for new synthetic elements continues. When elements 104 and 105 were discovered, they were placed in the periodic table without names. This happened because there were disagreements over who discovered them first. Normally the discoverer has the privilege of naming the new element. An international group of scientists approves the naming of all new elements. In this case, the chairperson of this group has met with all the scientists involved in the disagreement over elements 104 and 105. However, the disagreement is still not fully settled.

Element	Familiar Name or Place
Californium	California
Einsteinium	Albert Einstein
Nobelium	Alfred Nobel (Nobel prizes)
Neptunium	Neptune
Plutonium	Pluto
Americium	America
Berkelium	Berkeley, California
Curium	Marie and Pierre Curie
Francium	France
Scandium	Scandinavia
Polonium	Poland

Figure 7.18 Many elements were named after familiar people or places.

Study Questions for 7.5

1. Why are elements within the lanthanide and the actinide families so similar?
2. How can new elements be made in the laboratory?

1. because their two outer shells of electrons are similar
2. by bombarding the nuclei of uranium atoms with neutrons or nuclei of lighter elements

3. Explain that the lanthanides are heavy elements and are useful in the production of iron, steel, and glass.
4. Which is the best-known actinide (uranium, the heaviest natural element)? Which elements are made in the laboratory (any element heavier than uranium)?

Reference/Research Skills

FOLLOW-UP

Reinforcement: Lanthanides and Actinides

Ask the students to pretend that they are salespeople. Their job is to sell the lanthanides and the actinides. Ask them to create an advertisement that shows the properties and uses of at least three of these elements.

Enrichment: Naming Synthetic Elements

Ask several students each to select a transuranium element and find out why it was given its name.

Tie-in: Fission

Fission is a process that releases a great deal of energy. Ask the students to draw pictures of nuclear reactors that show how the energy released by the reaction is controlled.

SCIENCE BACKGROUND

Disagreement over the naming of elements 104 and beyond resulted in the suggestion by the International Union of Pure and Applied Chemistry that all new elements be named after their atomic numbers. Thus, element 104 would be unnilquadium; element 105, unnilpentium; element 106, unnilhexium; and so on.

Claims have been made for the synthesis of elements 106 and 107. These claims have been generally accepted. A claim by Peter Armbruster has also been made for the synthesis of element 109. Armbruster and his colleagues chose to work on the synthesis of element 109 and not 108 because they felt the creation of 109 would be easier to document.

What Did You Learn?

7.1 • Elements are arranged on the periodic table in rows called periods.

7.1 • Elements with similar properties are lined up on the periodic table in columns called families.

7.1 • Valence electrons are in the electron shell farthest out from the nucleus of an atom.

7.1 • Atoms of elements in the same family have the same number of valence electrons.

7.2 • Hydrogen and the alkali metals, family IA, have one valence electron.

7.2 • The alkaline earth metals, family IIA, have two valence electrons.

7.2 • Nearly all the transition metal atoms have two valence electrons. The next to last electron shell is filling up.

7.2, 7.3 • Some families of elements are very reactive.

7.3 • Atoms of elements in the boron family, IIIA, have three valence electrons.

7.3 • Atoms of elements in the carbon family, IVA, have four valence electrons.

7.3 • Atoms of elements in the nitrogen family, VA, have five valence electrons.

7.3 • Atoms of elements in the oxygen family, VIA, have six valence electrons.

7.3 • Atoms of elements in the halogen family, VIIA, have seven valence electrons.

7.4 • The noble gases mark the end of each period and all have eight valence electrons except for helium, which has two.

7.4 • The noble gases are different from other elements because their electron shells are considered full.

7.5 • Elements within the lanthanide and actinide families are very similar because their outer two electron shells are similar.

7.5 • Synthetic elements can be made in the laboratory by bombarding uranium atoms with neutrons or nuclei of light elements.

Key Terms

periodic table
periods
families
metals
nonmetals
valence electrons
alkali metal family
alkaline earth metal family
transition metals
boron family
carbon family
nitrogen family
oxygen family
halogen family
noble gas family
lanthanide family
actinide family
synthetic

Career

Metallurgical Engineer

Metallurgical engineers, or metallurgists, specialize in the study of metals. They probe metallic structure.

Some metallurgists work with ores, rocks containing metals that come out of mines. New and more efficient ways of getting metals from these ores are always needed. These new techniques are especially important because the supply of high-grade ores is being used up. Physical metallurgists study the physical properties of metals. They develop new types of metals to meet specific demands. For example, metals for spacecraft need to be heat resistant and very strong. Metals used in computers must be very pure. Mechanical metallurgists work on new ways to shape and form metals.

Many of the metal materials you encounter every day are the products of metallurgical research. These materials include tin cans, aluminum pots and pans, jewelry, braces and fillings for your teeth, fences, tools, and metals used in cars, bicycles, and other forms of transportation.

The iron and steel industries employ over half the metallurgical engineers. Metallurgical engineers also work in the aerospace, mining, and manufacturing industries. As technology increases and new types of materials are needed to meet new demands, more and more metallurgical engineers will be needed.

Vocabulary

1. true
2. alkali metals or halogens
3. true
4. true
5. true
6. actinides
7. bromine
8. true
9. metals
10. uranium

● Word Skill: Vocabulary

What Do You Remember?

1. 16 protons, 16 neutrons, 16 electrons, 6 valence electrons
2. Mendeleev
3. As you go from left to right, each element has atoms with one more proton than the previous element. In most cases the atomic mass also increases.
4. same number of valence electrons and similar chemical properties
5. electrons in the outer shell
6. Answers will vary. They may include 93—Np, 94—Pu, 95—Am, 96—Cm, 97—Bk, 98—Cf, 99—Es, 100—Fm, 101—Md, 102—No, 103—Lr.
7. metals—are conductors, are shiny, can be pounded into different shapes
   nonmetals—are nonconductors, are dull, can break instead of bend

TO THINK ABOUT AND DO

Vocabulary ●

Number your paper from 1 to 10. Read each of the following sentences. If the sentence is true, write *true* beside the appropriate number. If it is false, make it true by changing the underlined expression.

1. In a family of elements all the atoms have the same number of valence electrons.
2. The noble gases combine readily with other elements.
3. Elements that are made in the laboratory are called synthetic elements.
4. Rows on the periodic table are called periods.
5. All elements in the halogen family are nonmetals.
6. All the transition metals are radioactive.
7. According to the periodic table, the atomic number for chromium is 35.
8. All atoms in the alkali metal family have one valence electron.
9. Nonmetals are shiny and are good conductors of heat and electricity.
10. New elements are made in the laboratory from oxygen atoms.

What Do You Remember?

1. Look at this square. Tell at least three things about the atoms of the element sulfur.
2. Name the scientist who first developed the periodic table.
3. How are the elements in a period related?
4. How are the elements in a family similar?
5. What are valence electrons?
6. List the atomic numbers and symbols of five synthetic elements.
7. Make a list of properties of metals and a second list of properties of nonmetals.

8. Name the eight families of elements —IA, IIA, IIIA, IVA, VA, VIA, VIIA, VIIIA—and tell the number of valence electrons in each.
9. Name one element from each family in question 8 and explain why it is important.
10. Name three transition metals and explain why they are important.

Applying What You Have Learned ●

Give the letter from the diagram of the periodic table that corresponds to the family or elements described.

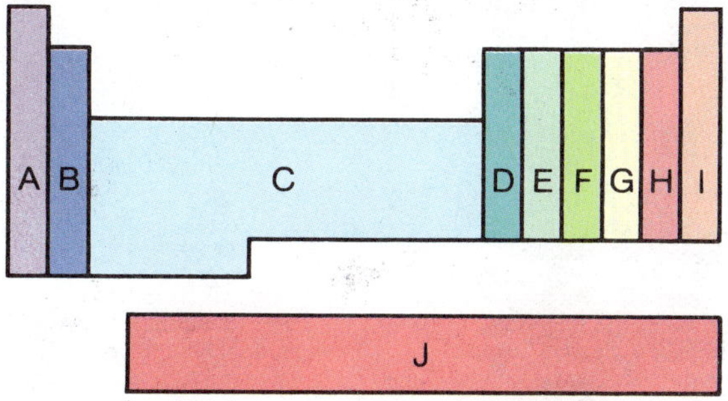

1. the alkali metals
2. the halogens
3. the alkaline earth metals
4. the lanthanides and the actinides
5. the noble gases
6. the transition metals
7. elements with atoms with one valence electron
8. elements with atoms with three valence electrons
9. elements with atoms with six valence electrons
10. elements with atoms with eight valence electrons

Research and Investigation ●

1. Silicon compounds are durable and hard. They are used in bone restructuring and grinding. Nitrogen has a low boiling point of 196°C. This makes it very useful in providing cold temperatures. Selenium, bromine, and cobalt are elements you do not often hear of. They have important uses. Find out what some of these uses are.
2. Fertilizers are labeled with three numbers. What element does each number refer to? How is each element important to plant growth?
3. Certain plants called legumes are important for their nitrogen-fixing ability. How do these plants make nitrogen available for people's use?

8. IA—alkali metal family; 1
   IIA—alkaline earth metal family; 2
   IIIA—boron family; 3
   IVA—carbon family; 4
   VA—nitrogen family; 5
   VIA—oxygen family; 6
   VIIA—halogen family; 7
   VIIIA—noble gas family; 8
9. Answers will vary. Some possibilities are the following:
   H—makes up sun and stars, is found in fuels
   Ca—is found in our bones and teeth
   C—is found in oil, nylon, and plastics
   N—is found in fertilizers and TNT
   O—is necessary for life
   Br—is found in drugs and gasoline
   Ne—is found in lights
10. Answers will vary. Some possibilities are the following:
   Au, Ag—are found in jewelry, money
   Cu—is found in wires and in brass and bronze
   Cr—is found in stainless steel and in chrome

Applying What You Have Learned

1. A	6. C
2. H	7. A
3. B	8. D
4. J	9. G
5. I	10. I

● Comprehension Skill: Applied

Research and Investigation

1. selenium—photoelectric cells, TV cameras, copying equipment
   bromine—ingredient in "antiknock" agents, sedatives, light-sensitive salt used in photography, disinfectants
   cobalt—alloyed with steel to make magnets, alloyed with tungsten to make cutting and machine steel
2. nitrogen—phosphorus—potash
   nitrogen—protein for plant growth
   phosphorus—protein for plant growth and ATP
   potash—a commercial name for compounds containing potassium, usually means potassium carbonate, used to increase crop production
3. The planting of legumes such as clover, peanuts, peas, and beans is one way to replenish essential soil nitrates. The plants have nitrogen-fixing bacteria in nodules on their roots. The bacteria convert the nitrogen in the atmosphere into nitrates that can be used by plants.

● Reference/Research Skills

CHAPTER BACKGROUND

Just as the words we know are composed of the letters of the alphabet, the world and all the things in it are composed of the elements. The elements join as mixtures and as compounds to form an endless variety of matter.

Chemical reactions occur when elements combine to form compounds. Each element that is involved loses its original properties, and the resulting compound has properties that are entirely its own. The valence electrons are of prime importance in the formation of compounds. Whether atoms transfer or share the electrons and how they share them determine what type of chemical bond will be formed. The bond, in turn, helps to determine the properties of the compound.

Mixtures are formed when elements or compounds mix mechanically. There is no bonding. Each substance keeps its own properties. The mixture can be easily separated into its parts.

Compounds and mixtures also differ in the ways their parts are arranged. A compound, like an element, is a pure substance. A compound is always composed of the same elements in a definite proportion by weight. This is the law of definite proportions. No such law exists for the formation of mixtures. The Great Salt Lake and the Atlantic Ocean differ in salinity, and yet both are mixtures we call salt water. Some mixtures are solutions, since their parts are evenly mixed.

Whether the matter around us exists as elements, as compounds, or as mixtures, the changes it undergoes naturally and the changes we cause it to undergo influence our daily lives in more ways than we realize.

8

Combining the Elements

Introduce the chapter by reviewing with the students their knowledge of the periodic table and the location of different elements on the table. Use your transparency or periodic wall chart to do this.

Have available on your desk a variety of materials that are not elements, such as sugar, salt, baking soda, paper, rubber, and glass. Ask the students if the materials are elements. Can any of them be found on the periodic table? Ask if they are made of elements. Tell the students that just as all the words they know are made of letters, all things in the world are made of elements. In this chapter they will learn about the different ways that atoms are joined to form more complex matter.

Suppose someone sprinkled a combination of two very dangerous chemicals on some food and ate it. Would you think that that person was crazy? Well, people do it every time they salt their food. Table salt is made of two dangerous chemicals, the elements sodium and chlorine. Yet when these elements are combined, they form a compound that is the most commonly used food seasoning.

Salt is just one example of a compound that looks and behaves very differently from the elements it is made of. A relatively small number of elements can make a world full of millions of different kinds of substances by forming compounds.

Why do some elements form compounds while others do not? How can compounds look and act so differently from the elements they are made of? Why do some materials break easily while others can be hammered into thin sheets? By understanding more about how elements combine, you will understand more about why the matter around you looks and acts the way it does.

Figure 8.1 Salt formations at Mono Lake in California

Lesson Organizer

8.1 Compounds and Mixtures

Pages: 160–163
Advanced Planning: Page 160
Activity: How Can You Separate a Mixture? Page 162
Follow-up: Page 163
R Mixtures
E Alloys **T** Miracle Mixtures

8.2 Forming Molecules

Pages: 164–167
Advanced Planning: Page 164
Activity: What Does a Molecular Formula Tell You About a Compound? Page 166
Follow-up: Page 167
R Compounds in Your Life
E A Slight Change **T** Allotropes

8.3 The Ionic Bond: Shifting Electrons

Pages: 168–171
Advanced Planning: Page 168
Activity: Which Solutions Contain Ions? Page 170
Follow-up: Page 171
R Potassium Bromide
E Svante Arrhenius **T** Salt and Your Health

8.4 The Covalent Bond: Sharing Electrons

Pages: 172–173
Advanced Planning: Page 172
Follow-up: Page 173
R Water
E Electronegativity **T** Hard Water

8.5 The Metallic Bond: Wandering Electrons

Pages: 174–175
Follow-up: Page 175
R Metallic Properties
E Comparison of Metals **T** Thermostats

See ACTIVITIES book pages 43–48 for laboratory, skill, and review exercises for Chapter 8.

8.1

OBJECTIVES: The student should be able to

- **define** the term *compound*;
- **define** the term *mixture*;
- **explain** how mixtures and compounds are different;
- **define** the term *solution*.

SCIENCE TERMS

mixture
solutions

ADVANCED PLANNING

For demonstrations you will need salt, pepper, a beaker, rocks that show clumps of different minerals, a can of vegetable soup, tonic, and sugar water.

LESSON PLAN

1. Introduce the lesson by pouring some salt and pepper into a beaker. Mix them together. Ask the students how many materials are in the beaker (2). Could they separate them (yes)? How (students may suggest by color; by taste; that salt will dissolve in water)?

8.1 Compounds and Mixtures

You will find out
- how compounds and mixtures are different;
- what is meant by a solution.

Substances made of two or more elements chemically combined are called compounds. A compound is always made of the same elements combined in the same proportions. For example, pure water is always 11.19 percent hydrogen and 88.81 percent oxygen.

A compound looks and acts very differently from the elements it is made of. Liquid water looks nothing at all like hydrogen gas or oxygen gas. Water freezes at 0°C. Both hydrogen and oxygen must be cooled to more than 200°C below zero before they turn into solids. Hydrogen will explode. Oxygen must be present for a fire to burn. Water puts a fire out. The properties of a compound are different from those of the elements that make it.

Figure 8.2 The element hydrogen burns. The element oxygen supports burning. Water, a compound of hydrogen and oxygen, is used to put out fires.

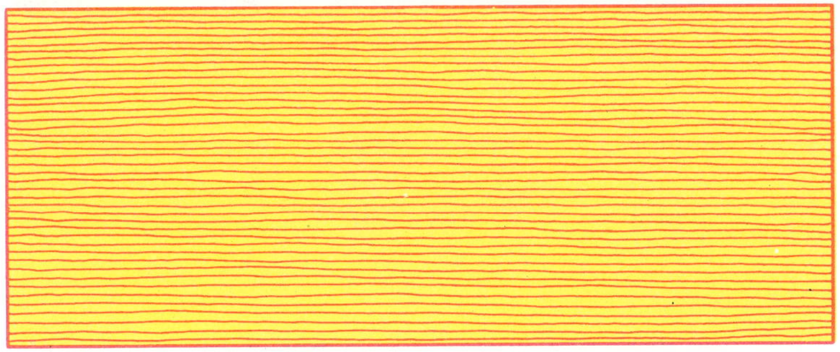

Figure 8.3 When you look at this mixture of colors from a distance, you see a new color. The original colors, however, remain unchanged.

Seawater, on the other hand, is not a compound. The amount of salt mixed in with the water can change. Seawater is an example of a **mixture.** Air, ice cream, brass, mayonnaise, orange juice, fog, and smog are also examples of mixtures. Mixtures are made of two or more elements or compounds that are not chemically combined. The amounts of each substance in a mixture are not fixed. For example, you can make a mixture of cereal and milk that has a little milk and a lot of cereal or a lot of milk and a little cereal or anything in between. This is an important difference between mixtures and compounds. The elements that form a compound are always combined in the same proportions. The proportions in a mixture can vary.

Do You Know?
Water tastes salty if there are 5 molecules of salt for every 10,000 molecules of water.

Figure 8.4 The proportions in a mixture of paint can vary.

2. Explain that you have made a mixture. A mixture is made of two or more substances physically combined. Each substance keeps its own properties. The substances are easy to separate. Review with the students the definition of a *compound*.

3. Ask the students to compare and contrast mixtures and compounds by asking the following questions about both:
   a. How are they joined (compounds—chemically combined; mixtures—physically combined)?
   b. Do the parts keep their properties (compounds—no; mixtures—yes)?
   c. Can they be separated (compounds—not easily; mixtures—yes)?

4. Point out that, unlike compounds, mixtures may contain particles that clump together in a certain area. Show the students the rock samples that you have and point out the clumping that has occurred. The rock is not made of only one material.

Comprehension Skill: Literal

ACTIVITY: How Can You Separate a Mixture?

Part I

Cut the chromatography paper into strips that are about 3 cm longer than your test tubes. The strips should be narrow enough to fit easily into the test tubes. Cut one end of the strips into a point.

To make the black ink, mix together in a beaker the red, blue, yellow, and green food coloring. Place a small drop of the resulting ink about 1 cm from the point of each strip. When it dries, add a second drop. When it dries, add a third drop. CAUTION: Do not let the drop spread to the tip of the paper.

After the students have separated the ink, challenge them to put it back together again. They can do this by cutting apart the colors and soaking the pieces in a test tube.

Part II

Students should note the physical differences between salt and iron filings. A mixture of the two substances can be separated based on solubility and magnetism.

To avoid ruining the test tubes, be sure to clean them thoroughly after they have had iron and water in them.

Laboratory Skill: Experimentation/Demonstration

ACTIVITY

How Can You Separate a Mixture?

Materials

test-tube rack
chromatography paper strip
iron filings
salt
4 test tubes
magnet

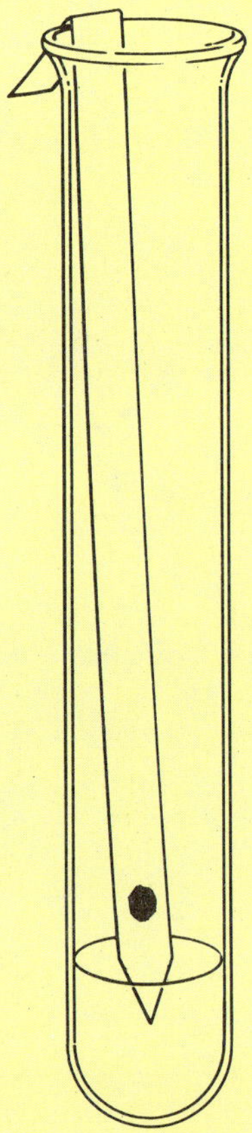

Procedure

Part I—Separating an Ink Mixture

1. Put 5 mL of water in a test tube.

2. Place a strip of chromatography paper stained with black ink into the test tube. The point of the paper should be in the water. Do not let the black spot touch the water. Fold the end of the strip over the test tube to hold it in place.

3. Allow the water to travel up the paper and observe what happens to the black spot.

4. Set your test tube in the test-tube rack. Check it periodically while you perform Part II. When the water has traveled to within 1 cm of the top of the test tube, remove the strip and allow it to dry.

Part II—Separating a Mixture of Iron and Salt

1. Place a sample of iron filings on a piece of paper. Place a sample of salt on a second piece of paper.

2. Fill two test tubes halfway with water. Add a small amount of iron to one and a small amount of salt to the other.

3. Move a magnet over your sample of iron filings. Then move the magnet over your sample of salt.

4. Mix your iron and salt samples together on a piece of paper. Divide the mixture into two parts. Place each part on a separate piece of paper.

5. Use a magnet to separate one sample of the mixture.

6. To separate the mixture by a different method, place a small amount of your second sample in a test tube and add 10 mL of water.

Questions

1. Describe what happens to the black spot as the water moves up the chromatography paper.

2. Can you use a magnet to separate a mixture of iron and salt? Explain your answer.

3. Can you use solubility to separate a mixture of iron and salt? Explain your answer.

4. Describe a mixture.

1. The black spot separates into different colors.
2. Yes, the iron is attracted to the magnet and the salt is not.
3. Yes, the salt dissolves in water and the iron does not.
4. A mixture is two or more substances joined in a physical combination. Each substance retains its properties so a mixture is easy to separate.

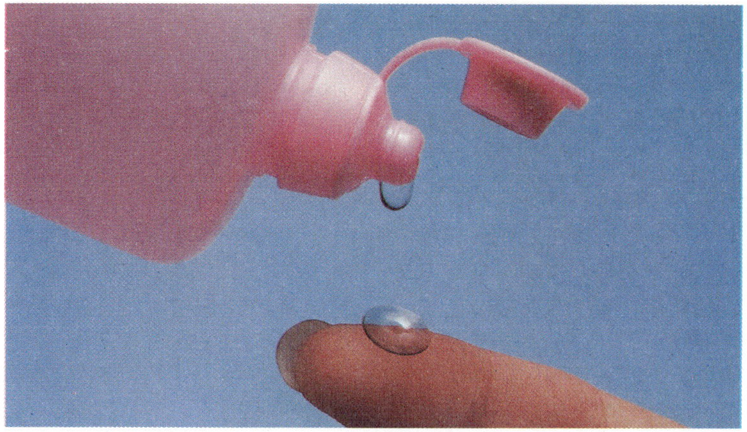

Figure 8.5 Many people are familiar with a liquid solution for contact lenses.

Some mixtures are called **solutions.** The substances in a solution are uniformly mixed. Sugar dissolved in water is a good example. If you stir sugar into water allowing enough time for all the sugar to dissolve, the resulting liquid will be the same throughout. One sample of the liquid would taste no sweeter or less sweet than any other sample. The sugar and water are uniformly mixed, forming a solution. The word *solution* is most often used with liquids. However, solids, such as brass, and gases, such as air, can also be solutions.

Figure 8.6 A liquid solution is being used to clean brass, a solid solution.

Study Questions for 8.1

1. How are compounds different from mixtures?
2. What is a solution?

1. The proportions in a compound are always the same. The proportions in a mixture vary. Substances in a compound are chemically combined. Substances in a mixture are not chemically combined.
2. a mixture that is the same throughout

5. Explain to the students that there are two types of mixtures—uniform mixtures (the same throughout) and mixtures that are not the same throughout, or nonuniform mixtures. The salt and pepper was a nonuniform mixture. So were the rocks. In one area of the beaker there could have been more salt than pepper. In the rocks the students could see the clumping of certain materials. Coffee is a uniform mixture. After the coffee has dissolved in the water, no matter where you take a sample from the cup, you will have coffee of the same composition. If you add cream, the cream will spread evenly throughout the coffee.
6. Place before the students the rocks with their mineral clumps, a cup of sugar in water, a bottle of tonic, and a can of vegetable soup. Ask which are uniform mixtures (tonic, sugar in water). Ask why the soup, like the rock, is a nonuniform mixture (it is not the same throughout).
7. Tell the students that uniform mixtures are called solutions. In a solution the particles are evenly distributed. The solution is the same throughout.

● **Word Skill: Vocabulary**

FOLLOW-UP

Reinforcement: Mixtures
Ask the students to create collages to illustrate different types of mixtures.

Enrichment: Alloys
A solid solution of metals is called an alloy. Brass, bronze, and alnico are examples. Have the students do research on alloys. How are they made? How are they used?

Tie-in: Miracle Mixtures
The *Gossamer Albatross*, the plane that was pedaled across the English Channel, was constructed with carbon fibers. The fibers are a nonuniform mixture that is like fiberglass. Fibers of carbon are lined up in the same direction, and epoxy plastic holds them together. The mixture is very strong and very light. Ask the students to investigate these fibers and the *Gossamer Albatross*. What information can they find to present to the class?

8.2

OBJECTIVES: The student should be able to

- **define** the term *molecule*;
- **interpret** a molecular formula;
- **define** the term *diatomic molecule*;
- **explain** the relationships between atoms and elements and molecules and compounds;
- **define** the term *chemical bond*;
- **identify** the atomic particle that is important in the formation of a molecule;
- **explain** the octet rule.

SCIENCE TERMS

molecule
molecular formulas
diatomic molecule
chemical bond
octet rule

ADVANCED PLANNING

You should have your transparency of the periodic table and an overhead projector or a periodic wall chart for use throughout the chapter.

LESSON PLAN

1. Explain that a molecule of a substance is the smallest particle that has the properties of the substance. As an example, mention a molecule of water. It is the smallest amount of water you can have and still have water. If you could cut a water molecule in half, however, you would have something that is *not* water.

2. Ask the students what information is given in a molecular formula that is not given by the name of the molecule of a compound (the number of atoms of each element combined in the molecule). How can the number of atoms of each element be determined (by the subscript immediately following the symbol for the element)? What is meant when there is no subscript following a symbol (there is one atom of that element)?

● **Comprehension Skill: Interpretative**

SCIENCE BACKGROUND

There are seven diatomic molecules: hydrogen, oxygen, bromine, nitrogen, iodine, chlorine, and fluorine.

8.2 Forming Molecules

You will find out
- what is meant by the term *molecule;*
- how chemical bonds join atoms to form molecules.

Sometimes the atoms of an element are found alone. At other times they are found joined together. Compounds are never made of single atoms. The different atoms in a compound are always found joined together. When two or more atoms, whether the same or different, combine, they form a **molecule** [MAHL-uh-kyool]. A hydrogen gas molecule is an example. Two hydrogen atoms combine to form a molecule of hydrogen gas. Just as elements ● can be represented by symbols, molecules can be represented by **molecular formulas.**

Figure 8.7 The air you breathe contains diatomic molecules.

The molecular formula for a hydrogen molecule is H_2. The H is the symbol for hydrogen. The number 2 beside and below the H, called a subscript, tells you that there are two hydrogen atoms in the molecule. This is an example of a **diatomic** [dy-uh-TAHM-ihk] **molecule.** Diatomic molecules are ones in which only two atoms join together. Another example of a diatomic molecule is O_2. Two oxygen atoms form an oxygen molecule.

Atoms	Molecules	Element or Compound?
I + I	I_2	element
O + O	O_2	element
C + O	CO	compound
C + O + O	CO_2	compound
H + H + O + O	H_2O_2	compound
H + H + S + O + O + O + O	H_2SO_4	compound
N + N	N_2	element

Figure 8.8 Elements are made of only one kind of atom. Compounds are made of two or more kinds of atoms.

Substances made of diatomic molecules such as H_2 and O_2 are considered elements, since they are made of only one kind of atom. In the molecules of compounds, atoms of two or more different kinds of elements join together. For example, a water molecule is a combination of hydrogen and oxygen atoms. The formula for the water molecule is H_2O. It contains two atoms of hydrogen and one atom of oxygen. Molecules are considered the basic units of compounds just as atoms are considered the basic units of elements.

The fitting together of atoms to form molecules can be compared to the fitting together of pieces in a jigsaw puzzle. If two atoms fit well together as pieces of a jigsaw puzzle do, a molecule is formed. When atoms fit together making a molecule, a **chemical bond** is formed.

Some atoms bond very readily with other atoms. Other atoms rarely form chemical bonds. The noble gases hardly ever form a bond with themselves or with any other kind of atom. On the other hand, hydrogen will very readily form bonds with many other kinds of atoms. So will fluorine. Some atoms readily form bonds, making molecules, while others do not.

3. Point out that molecules made up of two atoms are said to be diatomic. Suggest that the students remember the seven elements that occur naturally as diatomic gases by using the acronym *Brinclhof*.
4. Write the following analogy;

   atom : element as molecule : compound

   Ask the students to explain the analogy by defining each term. Here are some suggestions:

   atom = smallest complete part of an element that can stand alone or enter a chemical combination
   element = matter made of only one kind of atom
   molecule = two or more atoms chemically joined
   compound = elements chemically joined to form a new substance—each element has lost its original properties

● **Comprehension Skill: Literal**

SCIENCE BACKGROUND

When elements combine to form compounds they do so in specific whole number ratios. Water H_2O is made by joining two atoms of hydrogen with one atom of oxygen. Hydrogen peroxide H_2O_2, a material that is often used to clean cuts, is made by joining two atoms of hydrogen and two atoms of oxygen.

When a chemical bond forms, the atoms involved each lose their own properties. The compound they form has properties unique to itself. Table sugar $C_{12}H_{22}O_{11}$ is made of carbon (a black, tasteless solid), hydrogen (a colorless, odorless, tasteless gas), and oxygen (another colorless, odorless, tasteless gas). When they join, they form a white, crystalline solid that is sweet.

ACTIVITY: What Does a Molecular Formula Tell You About a Compound?

You may want to assign this activity as homework. If so, you could go over the examples and do the first formula from the list in class. Students could then confidently do the rest at home.

5. Ask the students to look at the periodic table. Explain that scientists have found that some elements combine together easily to form compounds while others do not. Point to the first vertical row, or family. How are the elements in a family similar (same number of valence electrons)? What are the electrons in the outer shell called (valence electrons)?

6. Point out to the students that the last group, the noble gases, is composed of members that each have 8 valence electrons except for helium, which has 2. Remind the students that the valence shell is considered stable when it has 8 electrons (2 in the case of helium).

ACTIVITY What Does a Molecular Formula Tell You About a Compound?

Materials
pencil
paper

1. 1 zinc, 1 oxygen
2. 1 sulfur, 2 oxygen
3. 1 lead, 1 sulfur, 4 oxygen
4. 1 sodium, 1 oxygen, 1 hydrogen
5. 3 carbon, 8 hydrogen, 3 oxygen
6. 2 carbon, 5 hydrogen, 1 oxygen, 1 sodium
7. 1 iron, 3 oxygen, 3 hydrogen
8. 1 calcium, 1 silicon, 3 oxygen
9. 3 hydrogen, 1 boron, 3 oxygen
10. 1 potassium, 1 manganese, 4 oxygen

Procedure

1. Read the following rules for interpreting molecular formulas and study all the examples.
2. Interpret the formulas by telling how many atoms of each element are in a molecule of the compounds.

Formulas for you to interpret

1. ZnO
2. SO_2
3. $PbSO_4$
4. $NaOH$
5. $C_3H_5(OH)_3$
6. C_2H_5ONa
7. $Fe(OH)_3$
8. $CaSiO_3$
9. H_3BO_3
10. $KMnO_4$

Rule	Examples	
Rule 1: The element symbol stands for one atom of that element in the molecule.	**CaO**	1 atom of calcium 1 atom of oxygen
	$NaOH$	1 atom of sodium 1 atom of oxygen 1 atom of hydrogen
Rule 2: If a subscript is after the symbol, there is more than one atom of that element in the molecule. The subscript tells you how many atoms of that element are present in the molecule.	**$MgCl_2$**	1 atom of magnesium 2 atoms of chlorine
	H_2CO_3	2 atoms of hydrogen 1 atom of carbon 3 atoms of oxygen
Rule 3: If a symbol or group of symbols is surrounded by parentheses with a subscript outside the parentheses, multiply the number of each of the atoms inside the parentheses by the subscript.	**$Ca(NO_3)_2$**	1 atom of calcium 2 atoms of nitrogen 6 atoms of oxygen
	$Mn(OH)_2$	1 atom of manganese 2 atoms of oxygen 2 atoms of hydrogen

Why do some atoms readily form bonds while others do not? The answer to this question can be found by looking at the valence electrons of an atom. Valence electrons are those found in the electron shell farthest out from the nucleus. An atom having eight valence electrons is considered very stable. The noble gases have eight valence electrons. They do not react readily with other elements and are considered very stable. Their outer electron shells are already stable.

In general, atoms will form chemical bonds with other atoms if the bonding will cause all atoms involved to have a stable outer electron shell of eight electrons. This rule is called the **octet** [ahk-TEHT] **rule.** Hydrogen and helium are exceptions. They each have only one electron shell. That shell is the one closest to the nucleus and is full and stable with two electrons instead of eight.

The octet rule states that atoms form chemical bonds to reach a more stable condition. That condition calls for complete electron shells. The complete shells are achieved in two general ways. One way is to shift electrons from one atom to another. The second way is to share electrons between atoms.

Figure 8.9 Which of these atoms would be most likely to form a chemical bond? F

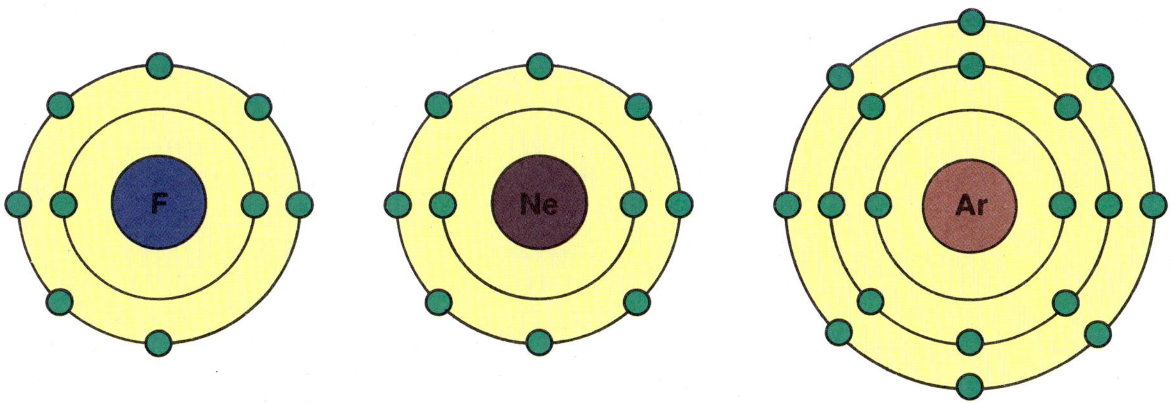

Study Questions for 8.2

1. What is a diatomic molecule?
2. What holds atoms together in a molecule?
3. What does the octet rule say?

1. a molecule that is made of two atoms
2. chemical bond
3. Most atoms form bonds to achieve a stable outer shell of eight electrons.

7. Point to the alkali metals and the halogens and explain that these two families are the most reactive because of their numbers of valence electrons. The alkali metals would have stable outer shells if they could lose their single valence electron, and the halogens would have stable shells if they could gain a single electron. When sodium (an alkali metal) and chlorine (a halogen) join, the sodium gives up an electron and the chlorine takes on an electron. As a result both atoms gain complete outer shells.

8. Remind the students that the rule that says atoms form compounds to achieve a stable valence shell of eight electrons is the octet rule.

FOLLOW-UP

Reinforcement: Compounds in Your Life

Ask the students to create mobiles in which they show pictures of compounds that they use in their daily lives. Displayed in the center of the mobile should be a paragraph explaining what a compound is.

Enrichment: A Slight Change

A slight change in chemical composition alters a compound greatly. Give the students examples of similar molecular formulas such as O_2 and O_3, CO_2 and CO, or H_2O and H_2O_2 and ask them to report on the differences that exist between the almost identical molecules.

Tie-in: Allotropes

Ask the students to read the biography of Shirley Ann Jackson at the end of the chapter. Explain to them that some nonmetallic elements such as carbon, sulfur, tin, and phosphorus are allotropes. That is, they have two or more forms in the same physical state. Ask the students to investigate one of these elements and report on its allotropic forms and their uses.

8.3

OBJECTIVES: The student should be able to

- **define** the term *ionic bond*;
- **explain** how an ionic bond is formed;
- **describe** how sodium chloride is formed;
- **explain** how an atom becomes a positive ion;
- **explain** how an atom becomes a negative ion;
- **define** the term *crystal*.

SCIENCE TERMS

ionic bond	negative ion
positive ion	crystals

ADVANCED PLANNING

1. Have available your transparency of the periodic table and an overhead projector or periodic wall chart.

2. Cut 28 red circles about the size of a quarter from a piece of colored paper. Mark each with a plus sign. Cut 28 blue circles about the size of a quarter from a piece of colored paper. Mark each with a minus sign. Cut 30 yellow circles about the size of a quarter from a piece of colored paper. Glue a tiny magnet (from a craft store or hobby shop) on the back of each. The red circles are protons. The blue are electrons. The yellow are neutrons. Most chalkboards have a metal sheet behind them and will hold a magnet. Test the magnets and make sure they will stick to your board. If they won't, ask the shop teacher if you can borrow a piece of metal that they will stick to.

LESSON PLAN

1. Using your colored-paper protons, electrons, and neutrons, have two students construct models of the atoms of sodium and chlorine (Na = 11 protons, 12 neutrons, and 11 electrons; Cl = 17 protons, 18 neutrons, and 17 electrons).
2. Ask how the number of protons compares to the number of electrons in each atom (they are equal; 11=11 and 17=17). Stress that atoms are neutral, not electrically charged, because the number of protons is equal to the number of electrons; the positive charges are equal in number to the negative charges.

8.3 The Ionic Bond: Shifting Electrons

You will find out
- how an ionic bond is formed;
- what positive ions and negative ions are;
- what crystals are.

An **ionic** [eye-AHN-ihk] **bond** is formed when one atom shifts an electron to another atom. This happens commonly when atoms with one valence electron, the alkali metals, are combined with atoms with seven valence electrons, the halogens. A good example is table salt, sodium chloride, which is formed by combining sodium and chlorine.

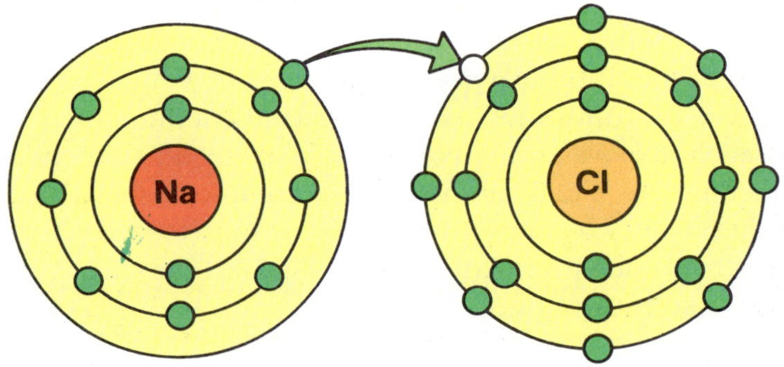

Figure 8.10 Forming an ionic bond

When an alkali metal atom such as sodium loses its one valence electron, a stable shell of eight electrons is left behind. If sodium's electron is taken by a chlorine atom, the chlorine atom will have 7 + 1, or 8, electrons in its outer shell. This gives it a stable outer shell. In forming an ionic bond both atoms have achieved a stable condition. Both have outer electron shells of eight electrons.

When the sodium atom loses an electron, the atom has one less electron than protons. The electrons each have a negative charge of one unit, and the protons each have a positive charge of one unit. The resulting atom has a net positive charge. Such an atom is called a **positive ion.** On the other hand, the chlorine atom has taken another electron. The chlorine now has an extra electron. As a result it has a net negative charge. Such an atom is called a **negative ion.**

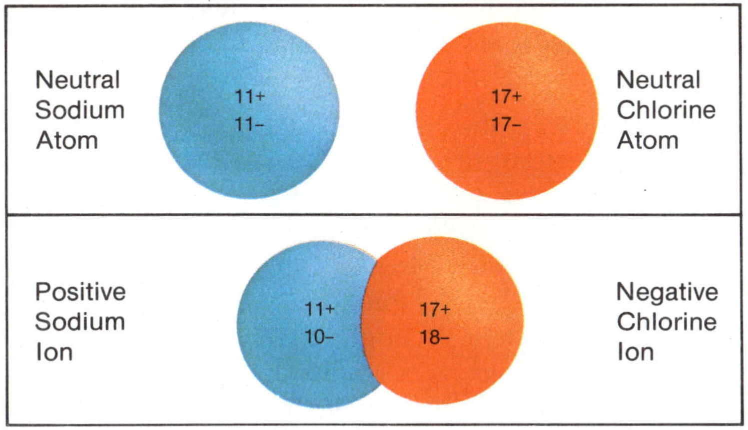

Figure 8.11 All atoms are neutral. All ions are charged.

Other members of the alkali metal family form positive ions in the same manner that sodium does. Similarly, other members of the halogen family form negative ions in the same manner that chlorine does. For example, the alkali metal potassium forms an ionic bond with the halogen bromine. Potassium shifts its one valence electron to bromine. Potassium becomes a positive ion. Bromine becomes a negative ion. The result is the ionic crystal potassium bromide.

Figure 8.12 The alkali metals form ionic bonds easily with the halogens.

3. Ask what must happen to the chlorine for its outer shell to be complete (gain an electron). Ask where that electron can come from (Na). Remove the outer electron from the sodium and place it in the outer shell of the chlorine. Tell the students that when sodium and chlorine combine, the sodium loses an electron and the chlorine gains the electron.

4. Point out that while sodium still has the 11 protons and 12 neutrons that it started with, it now has only 10 electrons. Ask why sodium is no longer electrically neutral (the numbers of positive charges and negative charges are no longer equal).

5. Tell the students that charged atoms are called ions. Because sodium has an excess of one positively charged proton, it is a positive ion.

6. Ask how many protons chlorine has (17). How many electrons (18)? Is the chlorine electrically neutral (no)? Why (it no longer has the same numbers of positive and negative charges)?

7. Ask what kind of ion has more negative charges than positive charges (a negative ion). What kind of ion is chlorine (a negative ion)?

8. Summarize by explaining that when ionic bonds form, one atom loses electrons. It becomes a positive ion. The other atom gains electrons. It becomes a negative ion.

ACTIVITY: Which Solutions Contain Ions?

The baking soda solutions and the salt solutions will be conductors. Salt and baking soda are ionic compounds. In water the ions will be free to move and thereby carry a current of electricity. You may wish to substitute ammeters or galvanometers for light bulbs.

◇ **Caution:** Students should wear safety glasses.

● **Laboratory Skill: Experimentation/Demonstration**

● ACTIVITY

Which Solutions Contain Ions?

Materials

beakers
copper wire
batteries
light bulbs in sockets
salt
distilled water
sugar
baking soda
alcohol
spoon
safety glasses
apron

Procedure

1. Make a chart like the one begun at the bottom of the page. Use this chart to record your observations.

2. Put on your safety glasses and apron.

3. Set up the battery, wires, and bulb as shown. To test each material for conductivity, place the free ends of the wires into the material. The wires should NEVER touch each other during the tests.

4. Fill a beaker about halfway with distilled water.

5. Test the distilled water for conductivity.

6. Put a spoonful of salt on a piece of paper. Test.

7. Dissolve a small amount of salt in the beaker of water. Test.

8. Add the rest of your salt sample to the salt solution. Test. Is the bulb brighter than before?

9. Empty and rinse the beaker. Repeat steps 4 and 6-8 with sugar and again with baking soda.

10. Pour a small amount of alcohol in the beaker. Test.

Test Material	Bulb Lights?	Ions Present?
water		
salt		
salt water		
sugar		
sugar water		

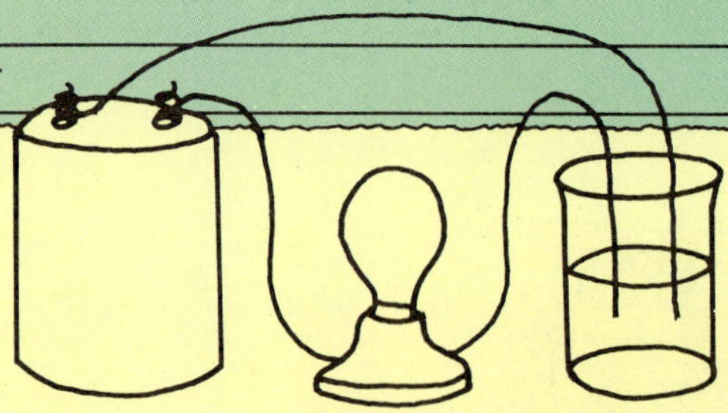

Questions

1. Solutions that allow electricity to pass through them contain ions. Complete the chart to show which contain ions.

2. Why do you think the bulb was brighter for the stronger solutions?

3. Why do you think solutions that contain ions are conductors while solutions that have molecules with covalent bonds are not?

4. Explain how you would determine whether or not a solution is a conductor of electricity.

1. The salt solutions and the baking soda solutions contain ions.
2. There were more ions present.
3. Ions are charged particles. They carry the current through the solution. Covalent molecules don't form ions.
4. Students should describe the apparatus and procedure used in this experiment.

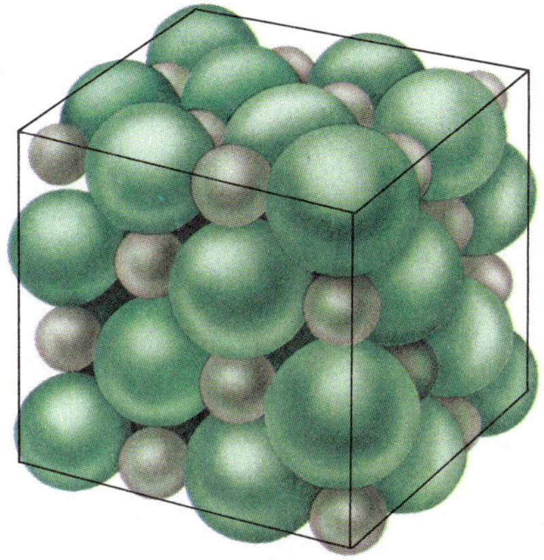

Figure 8.13 *left:* Crystals of sodium chloride; *right:* The crystalline structure of sodium chloride

The strong electrical attraction between the oppositely charged ions will pull them together in an ionic bond. Ionic bonds are strong. It takes a fairly large amount of energy to pull apart two ions of opposite charges. Because of this strong attraction, ionic bonds form materials called **crystals.** A crystal is a solid in which the atoms are arranged in a regular pattern. As a result, crystals tend to have flat surfaces. For example, salt (NaCl) forms crystals in which the positive sodium ions always have negative chlorine ions for neighbors. These salt crystals are cubes. Other crystals may have different shapes.

Study Questions for 8.3

1. How is an ionic bond formed?
2. What are negative and positive ions?
3. What is a crystal?

1. Electrons are shifted from one atom to another.
2. A negative ion is an atom with more electrons than protons. A positive ion has more protons than electrons.
3. a solid with atoms or ions arranged in a regular pattern

9. Explain that as a result of the transfer of electrons in a salt crystal and other ionic compounds the ions are very strongly attracted to each other. This attraction is electrical in nature, since the negative and positive charges are strongly attracted to each other.
10. Ask the students what a crystal is (a solid whose atoms are arranged in a regular pattern). Point out that even though the units of which crystals are built are too small to be seen there are similar examples in everyday life. Ask the students to name a few (table salt, or NaCl; salt substitute, or KCl; quartz crystals).

FOLLOW-UP

Reinforcement: Potassium Bromide
Ask the students to create a model of the atoms potassium and bromine joined with an ionic bond.

Enrichment: Svante Arrhenius
Arrhenius proposed an ionic theory to explain why some types of matter acted differently from other types. Ask the students to look up Arrhenius at the library. They should each pretend that he or she is the scientist and write an autobiography to describe Arrhenius' life and work on ionic theory.

Tie-in: Salt and Your Health
Common table salt, an ionic compound, is thought to be a contributing factor to heart disease. Ask the students to find out why salt is thought to be harmful and what foods are high in salt. They should create nutrition charts to show the salt content of common foods.

SCIENCE BACKGROUND

The strong electrical attraction between negative and positive charges gives ionic compounds specific properties. They have high melting points because it takes a great deal of energy to break the ionic bonds. They are also hard because of the strong attraction.

The repulsion between like electrical charges is what makes ionic compounds shatter when you hit them. You cannot stretch or bend an ionic substance. It shatters.

8.4

OBJECTIVES: The student should be able to

- **define** the term *covalent bond;*
- **describe** how two hydrogen atoms join to form a hydrogen molecule;
- **draw** electron dot diagrams of given molecules.

SCIENCE TERMS

covalent bond
electron dot diagram

ADVANCED PLANNING

Keep available the colored-paper protons, electrons, and neutrons that you used in 8.3.

LESSON PLAN

1. Introduce the lesson by reviewing the octet rule. Ask the students to recall that in order to achieve stability atoms must have valence shells with eight electrons. Hydrogen is an exception because it is stable with only two electrons.
2. Explain that when an alkali metal and a halogen form a bond, the atom with one valence electron gives it to the atom with seven valence electrons so that that atom will have enough. This is what happened with sodium and chlorine. Point out that the atom that gives up the electron (in this example, sodium) also achieves a full valence shell of eight electrons. This is because once the valence electrons are lost, the shell next to the outer shell becomes the valence shell. That shell is full and stable.
3. As an example of covalent bonding, ask students to construct electron shell diagrams for hydrogen and nitrogen atoms (H = 1 proton, 1 electron; N = 7 protons, 7 electrons). Point out that the hydrogen shell is half full. To fill it, hydrogen can share its electron. Nitrogen's outer shell has five electrons. Two are closely joined in an electron pair (move two electrons together on the model) and three are unpaired.

● **Comprehension Skill: Literal**

8.4 The Covalent Bond: Sharing Electrons

You will find out

- how a covalent bond is formed;
- how electron dot diagrams illustrate covalent bonds;
- how a diamond is formed.

● Sometimes atoms form bonds in which they share two electrons. This is called a **covalent** [koh-VAY-luhnt] **bond.** The simplest kind of covalent bond is formed when two hydrogen atoms combine to form a hydrogen molecule, H_2. Two electrons, one from each atom, merge into one pattern, or shell, about the two hydrogen nuclei. The two shared electrons fill each hydrogen atom's electron shell. The strength of the covalent bond comes from the attraction of the positively charged nuclei for the negative electrons.

Figure 8.14 Atoms of hydrogen joined by a covalent bond

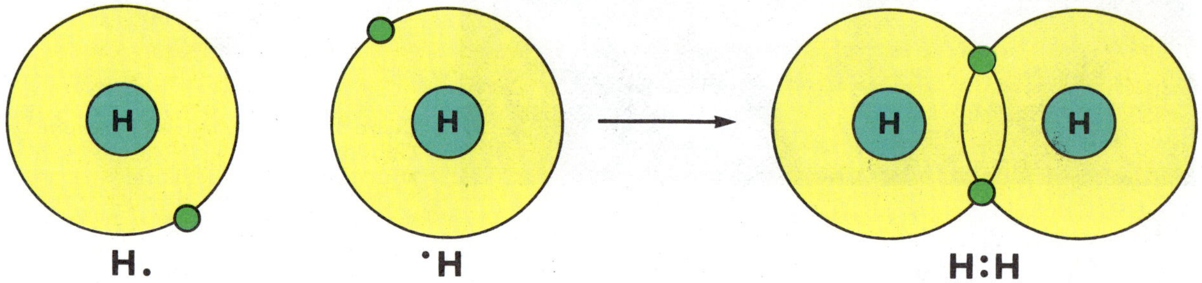

The hydrogen molecule is shown above using an **electron dot diagram.** In this kind of diagram the valence electrons of the atoms involved in the covalent bond are shown as dots. The pairs of dots placed between the element symbols represent shared electrons.

One atom can form more than one covalent bond. That is, it can share more than one pair of electrons. For example, a nitrogen atom shares three pairs of electrons with three hydrogen atoms to make a covalent ammonia molecule, NH_3. Ammonia is a strong-smelling compound used in cleaning solutions. Since the nitrogen atom has five valence electrons, sharing electrons with three hydrogen atoms gives it 5 + 3, or 8, electrons in its outer shell. At the same time, the sharing of electrons gives each hydrogen atom 1 + 1, or 2, electrons in its only electron shell.

Figure 8.15 Ammonia, a covalent compound, is used in window-washing solutions.

Another example of a covalent molecule is carbon dioxide, CO_2. Carbon has four valence electrons. It can complete its outer shell by sharing two pairs of electrons with one oxygen atom and two pairs with another one. Notice that in this case two pairs of electrons are shared between two atoms. In covalent bonding more than one pair of electrons can be shared between two atoms.

Carbon atoms also form covalent bonds with other carbon atoms. The results can be very beautiful. Diamonds are made of carbon atoms. In a diamond, each valence electron in the carbon atom is shared with another carbon atom. As a result, each atom is bonded to four other atoms. A diamond has a very high melting point, around 3,600°C, and is extremely hard. A large number of covalent bonds would have to be broken to destroy the crystal structure.

Do You Know?
Scientists have even produced diamonds from peanut butter! Unfortunately, to get gem-quality stones, it costs more to do this than to dig up the diamonds.

Figure 8.16 A diamond crystal

Study Questions for 8.4

1. How is a covalent bond formed?
2. Draw the electron dot diagram for ammonia.
3. How is a diamond formed?

1. Pairs of electrons are shared between atoms.

2. H : N̈ : H
   H

3. Each carbon atom forms four covalent bonds with four other carbon atoms.

4. Stress that there are three covalent bonds in an ammonia molecule. The nitrogen and the hydrogen are sharing electrons and so each has a complete valence shell.

5. Explain that when chemists are showing bonds they draw electron dot diagrams. Define an electron dot diagram as a picture of an atom that identifies the atom by its symbol and shows only the valence electrons. Illustrate electron dot formulas and stress that the members of a family have the same number of valence electrons. Illustrate this point by asking the students to look at the alkali metal family IA on their periodic tables. Ask how many valence electrons the elements have (1). Therefore, the electron dot formulas would be the following:

   H· Li· Na· K· Rb· Cs· Fr·

6. Ask what the valence of the alkaline earth metals is (2). How many dots should be on the diagrams (2)? You may wish to explain that the dots are placed side by side because the electrons are usually paired.

   Be: Mg: Ca: Sr: Ba: Ra:

FOLLOW-UP

Reinforcement: Water
Ask the students to create a poster or a model of a water molecule showing the sharing of electrons that exists.

Enrichment: Electronegativity
Electronegativity refers to the attraction of atoms for the electrons they share. When one atom has a greater attraction than the other, a polar bond forms. Have students do research on the polar nature of the water molecule and explain why the polarity of water is important.

Tie-in: Hard Water
Although water is a covalent molecule, it often has ions dissolved in it. Some of the ions give water a characteristic that we refer to as hardness. Ask the students to investigate hard water to determine how it can damage pipes. Ask students to make a model of a damaged pipe and explain how the condition can be corrected.

8.5

OBJECTIVES: The student should be able to

- **define** the term *metallic bond*;
- **describe** what happens when metallic atoms are brought together;
- **explain** how metallic bonds are similar to and different from covalent bonds;
- **explain** how metallic bonds give metals their properties.

SCIENCE TERM

metallic bond

LESSON PLAN

1. Introduce the lesson by asking the students what a covalent bond is (electrons are shared).
2. Explain that the metallic bond is similar to the covalent bond. In a covalent bond electrons are shared by two atoms. However, a metallic bond is a bond in which the electrons are shared among many atoms. Point out to students that when many metallic atoms come together a few of the electrons break away from each atom. The electrons are very light and move around through the metal, being attracted to the many positive metal ions left behind. Stress that in a metallic bond the electrons are free to wander because they are shared by many metal ions.
3. Help the students to realize that the wandering electrons are responsible for many of the properties of the metals. Explain that metals are good conductors of electricity. The freely moving electrons carry a current. They are good conductors of heat because the freely moving electrons carry heat through the metal. Metals can be hammered into shapes and can be drawn out into wires because the electrons can easily adapt to any new position.

● Comprehension Skill: Interpretative

8.5 The Metallic Bond: Wandering Electrons

You will find out

- how metallic bonds are formed;
- what explains many properties of metals.

Figure 8.17 Metallic bonds are present in this suit of armor.

There is another kind of chemical bond that involves sharing electrons. In a **metallic** [muh-TAL-ihk] **bond** many electrons are shared between many metal atoms. This sharing of electrons is different from covalent bonds. In covalent bonds electrons are shared between two specific atoms.

In the atoms that make up metals, the valence electrons are very loosely bound to the nucleus. When metal atoms are close together, the valence electrons break free, leaving behind positive metal ions. These ions are much heavier than the tiny electrons. The ions stay put, forming a crystal structure. The electrons, being lighter, wander through the crystal. The electrons are not attracted to any one, or even two ions, but to many metal ions.

The freely moving electrons of metallic crystals explain many familiar properties of metals. When you add heat to one part of a metal, the freely moving electrons will quickly carry heat to all parts of the metal. If you ● touch the metal handle of a heated pan without using a pot holder, you may burn your hand. Even though the handle of the pan is not directly over the burner, the electrons in the metal will very quickly transfer the heat from the stove to the handle and to your hand. This is the reason many pans and kitchen and barbecue utensils have wooden or plastic handles. Wood and plastic do not transfer heat as metals do.

In diamonds and salt crystals, all the electrons are bound to atoms or ions and cannot be easily moved. This makes these crystals hard but easy to break. If you hammer the crystal, the electrons cannot move to adjust. The crystal breaks into pieces. In metallic crystals, on the other hand, the electrons move easily. It is possible to change the shape of metals without breaking them. If a hammer pounds one layer of atoms over

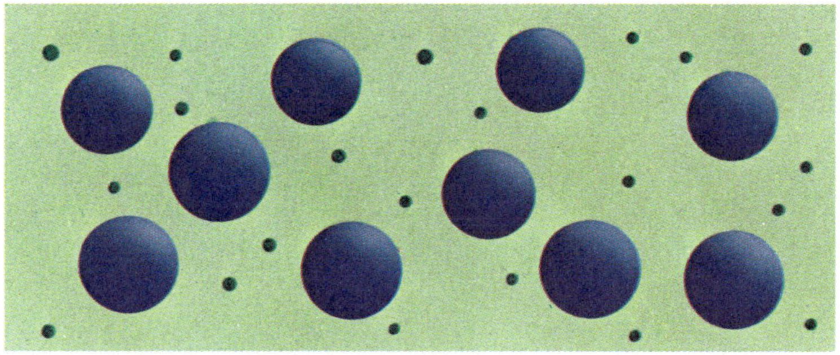

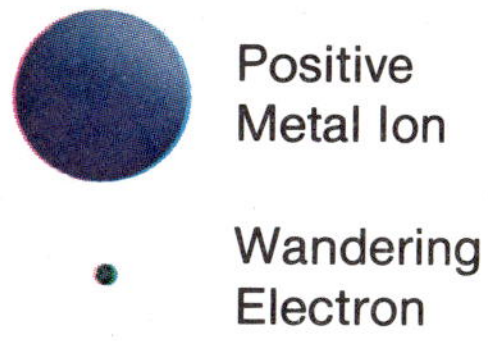

Figure 8.18 In metallic bonds, the wandering ele-trons are attracted to many metal ions.

another, the electrons move quickly and adjust to the new position. This makes it possible to hammer and stretch metals into many different shapes.

The properties that metals exhibit as a result of their bond structure make them very useful. Copper, tin, aluminum, and iron are used in cooking utensils. These metals conduct heat quickly and evenly. Copper is also used in pipes and electrical wiring. Gold and silver are pressed and pounded into many shapes to form jewelry and coins. Aluminum in aluminum foil can be folded and pressed tightly to cover and protect foods.

Figure 8.19 Metallic bonds make it possible to pound metals into different shapes.

Do You Know?
Only 28.4 g of gold can be drawn into a thin wire over 60 km long.

Study Questions for 8.5

1. How is a metallic bond formed?
2. What explains many of the familiar properties of metals?

1. Electrons are shared by many metal atoms.
2. Freely moving electrons transfer heat and allow metals to be pounded into shapes.

4. Conclude the lesson with a chart called Compounds. Make three columns—Ionic, Covalent, and Metallic. Ask the students the following questions and use the answers to complete the chart.
   a. What is an ionic bond (shifted electrons)?
   b. What are charged atoms called (ions)?
   c. What ion has more protons than electrons (positive ion)?
   d. What ion has more electrons than protons (negative ion)?
   e. Give an example of a compound with an ionic bond (salt).
   f. What is a covalent bond (shared electrons)?
   g. How many atoms are involved in a covalent bond (two)?
   h. What is an example of a compound with covalent bonds (water)?
   i. What is a metallic bond (many shared electrons)?
   j. How many atoms are involved in a metallic bond (many atoms)?
   k. Give an example of a metallic bond (copper).

FOLLOW-UP

Reinforcement: Metallic Properties

The electrons in metals are like a sea of electrons. Have students write a report that explains how the sea of electrons helps to carry electricity and heat and how it allows metals to be pounded into shapes and drawn into wires.

Enrichment: Comparison of Metals

Ask the students to obtain two samples of metals. They should compare them with respect to their malleability, ability to conduct heat, and ability to conduct electricity. On the basis of their comparison, which metal do students think has a freer sea of electrons? How can students support their answer?

Tie-in: Thermostats

Thermostats are made of bimetallic strips. Ask the students to find out how the strips work. You might have a bimetallic strip the students can use to demonstrate what they learn. Bimetallic jumping disks are available from many supply houses and science museums.

What Did You Learn?

8.1 • The amounts of substances chemically combined in a compound are fixed.

8.1 • Compounds have different properties from the elements that form them.

8.1 • Mixtures are made of two or more elements or compounds that are not chemically combined.

8.1 • The amounts of substances in a mixture can change.

8.1 • A solution is a mixture that is the same throughout.

8.2 • Molecules are formed when two or more atoms join together.

8.2 • In diatomic molecules two atoms are combined.

8.2 • Molecules are the basic units of compounds.

8.2 • Chemical bonds join atoms together in molecules.

8.2 • Most atoms form bonds to achieve a stable outer shell of eight electrons.

8.3 • In ionic bonds electrons are shifted from one atom to another.

8.3 • Ions have unequal numbers of protons and electrons.

8.3 • A crystal is a solid in which the atoms or ions are arranged in a regular pattern.

8.4 • In covalent bonds pairs of electrons are shared between atoms.

8.4 • Electron dot diagrams illustrate covalent molecules.

8.4 • Diamonds are made of carbon atoms joined by covalent bonds.

8.5 • In metallic bonds electrons are shared by many metal atoms.

8.5 • The freely moving electrons of metallic crystals explain many familiar properties of metals.

Key Terms

mixture
solutions
molecule
molecular formulas
diatomic molecule
chemical bond
octet rule
ionic bond
positive ion
negative ion
crystals
covalent bond
electron dot diagram
metallic bond

Biography

Shirley Ann Jackson (1946-)

Shirley Ann Jackson is a physical scientist who specializes in the study of matter in the solid state. She works at the research and development branch of Bell Laboratories. Dr. Jackson seeks to describe and explain what happens to a solid, such as a metal, when outside conditions are changed. Two outside conditions she is concentrating on are temperature and pressure. Dr. Jackson hopes to explain at the level of atoms and molecules the effects of changes in temperature and pressure on solids. To do this, she uses computers, mathematics, physical laws, and imagination. The results of her work are useful in electronics, communications, and the development of new materials.

Dr. Jackson graduated valedictorian of her class at Roosevelt High School in Washington, D.C. From there she went to the Massachusetts Institute of Technology, M.I.T. After receiving a bachelor's degree in physics, she remained at M.I.T. and continued to study physics. She later received a doctor's degree in theoretical high-energy physics. Dr. Jackson was the first American black woman to receive a doctorate from M.I.T.

Dr. Jackson's involvement with M.I.T. goes beyond her degrees. She has worked successfully to encourage the enrollment of black students there. She is also a member of the M.I.T. Corporation, the school's board of trustees.

Vocabulary

1. crystal
2. ion
3. mixture
4. metallic
5. octet rule
6. solution
7. covalent
8. molecule
9. diatomic
10. diamond

● Word Skill: Vocabulary

What Do You Remember?

1. A molecule is the basic unit of a compound.
2. valence electrons
3. Hydrogen and oxygen are colorless, odorless, tasteless gases. Hydrogen burns. Oxygen supports burning. Water is a colorless, odorless, tasteless liquid that neither burns nor supports burning. In water both hydrogen and oxygen have complete valence shells.
4. An atom loses one or more electrons.
5. An atom gains one or more electrons.
6. ionic = electrons are shifted from one atom to another
covalent = electrons are shared by two atoms
metallic = electrons are shared by many atoms
7. H:H

```
  ••
H:N:H
  ••
  H
```

8. Mixtures are combinations of elements and compounds. Each element or compound keeps its own properties. The proportions of substances in a mixture can vary. The elements that make up compounds are in fixed proportions. A compound has different properties from those of the elements that make it up.
9. Metals have metallic bonds. The electrons move about and are shared by many metal ions. The wandering electrons allow you to bend the metal because the electrons can easily adapt to a new position. They allow the metal to be a conductor because they will pick up heat and carry it.

TO THINK ABOUT AND DO

Vocabulary ●

Unscramble the following words and then use them to complete the sentences.

1. In a salt ________ sodium and chloride ions are arranged in a definite pattern.
2. A charged atom is a(n) ________.
3. Seawater is a(n) ________ because the proportions of salt and water are not fixed.
4. The bond formed when electrons are shared by many atoms is a(n) ________ bond.
5. The rule that says atoms are stable when their valence shell has eight electrons is the ________.
6. A mixture that is the same throughout is called a(n) ________.
7. A bond in which the electrons are shared between two atoms is a(n) ________ bond.
8. When two or more atoms are joined together by a chemical bond, they form a(n) ________.
9. H_2 and O_2 are examples of ________ molecules.
10. Carbon atoms can bond together to form a very hard and very beautiful ________ crystal.

lecmoule
catmiido
ctote erlu
tasylrc
niico
vatcnoel
xuemrit
middoan
tlmlicea
nio
ntlsouio
erehenoegusto

What Do You Remember?

1. What is the relationship between a molecule and a compound?
2. When atoms join to form molecules, what part of the atoms is involved?
3. How is water different from the hydrogen and oxygen that compose it?
4. Explain how a positive ion is formed.
5. Explain how a negative ion is formed.
6. Tell how ionic, covalent, and metallic bonds differ.
7. Draw electron dot formulas for H_2 and NH_3.
8. How do mixtures and compounds differ?
9. Explain why metals can be pounded into shapes and why they are good conductors of heat.

Applying What You Have Learned ●

1. Rubidium chloride is an ionic compound. Explain how it is formed.
2. How would potassium become an ion?
3. How would fluorine become an ion?
4. Describe, in terms of electrons, how copper and oxygen join to form copper oxide (Cu_2O).
5. For each of the following elements tell whether you would expect it to form an ionic bond, a covalent bond, a metallic bond, or to remain stable: Li, Kr, I, Si, Au, Cs, S, Cr, At, He.

Research and Investigation ●

1. Oxidation is a process that we see all around us. It results in the formation of compounds that are useful, but it also results in the formation of compounds we would rather not have. Oxidation is useful when food is broken down to release energy. It is not useful when rust is the result. What is oxidation? How does it occur?
2. Some covalent bonds are said to be polar. What does this mean?

Applying What You Have Learned

1. Rubidium has one valence electron, chlorine has seven. Rubidium loses its electron to chlorine.
2. Potassium has one valence electron. It loses the electron and becomes a positive ion.
3. Fluorine has seven valence electrons. It gains an electron and becomes a negative ion.
4. Oxygen has six electrons in its outer shell. Eight electrons would complete the shell. Copper has one electron in its outer shell. The electrons from two copper atoms join to the oxygen to complete the valence shell, forming the compound copper oxide.
5. Li forms ionic bond, Kr remains stable. I forms ionic bond (or forms covalent bond as diatomic molecule), Si forms covalent bond, Au forms metallic bond, Cs forms metallic bond, Si forms covalent bond, Cr forms metallic bond, At forms ionic bond, and He remains stable.

● Comprehension Skill: Applied

Research and Investigation

1. Oxidation refers to any chemical reaction that involves the loss of one or more electrons by an atom or ion.
2. A polar covalent bond is a bond in which the electrons are not shared equally.

● Reference/Research Skills

CHAPTER BACKGROUND

Organic chemistry is the study of compounds that contain the element carbon. The foundations of organic chemistry can be traced to 1685 when Nicolas Lémery published a book classifying substances based on their origins. This was the first attempt to distinguish between materials that originated in plants and animals and those that were from minerals. A few years after Lémery's book was published, scientists began to classify matter as organic (living) or inorganic (nonliving). Lémery and the scientists who used his system of grouping matter believed that the main difference between the organic and the inorganic classifications was the presence of a vital force that gave life to the organic compounds. Their belief was strengthened by their inability to create organic material. They believed that they lacked the ability to infuse a vital force into compounds.

One hundred and fifty years after Lémery proposed his ideas, they were proved to be incorrect. Friedrich Wöhler, a professor of chemistry, created an organic compound from a mixture of two inorganic salts. Wöhler's discovery brought about a change in the concept of organic compounds. Today all compounds that contain carbon (with a few exceptions) are called organic compounds.

Today advanced knowledge and technology allow scientists to create organic compounds that Lémery and Wöhler never dreamed of. Many of the compounds scientists create make our lives fuller and more enjoyable.

9 Carbon Compounds

As the class begins, have on your desk the following materials: cloth, different types of food, water, different types of rocks, sand, and an assortment of chemicals that the students will recognize. Ask the students to group the materials into two separate categories. Perhaps their groups will be things they use such as food, cloth, and water versus things they don't use. Whatever pattern the students select, the food materials will probably be in the same group. Tell the students that scientists would also put the foods in a group. They would do this because the foods contain carbon. Add to the food group the cloth and any chemicals that are organic. Explain to the students that these materials are grouped together because they contain carbon. The others have no carbon. Tell the students that in this chapter they will be studying the compounds that contain carbon.

Every morning you wake up, climb from between the sheets, and dress for the day. The sheets you sleep on and the clothes you wear contain carbon compounds. After washing your face with a carbon compound, you sit down to breakfast. Your chair contains carbon compounds. You eat carbon compounds. After breakfast you brush your teeth with a toothbrush made of carbon compounds. The bus you take to school has rubber tires made of carbon compounds. The fuel the bus runs on is a carbon compound.

Carbon compounds are absolutely necessary for you to survive. Your body contains carbon compounds. Other living things, plants and animals, contain carbon compounds. Your food contains carbon compounds. So do the fuels that heat your home and power the cars, buses, trains, and planes you ride in. This book you are reading contains carbon compounds. By now you should be getting the idea. Carbon compounds are everywhere!

Figure 9.1 Look for items in the photograph that contain carbon compounds.

Lesson Organizer

9.1 Organic Compounds

Pages 182–183
Advanced Planning: Page 182
Follow-up: Page 183
- **R** The Bonds of Carbon
- **E** Saturated and Unsaturated Compounds
- **T** Classification and the Vital Force

9.2 Hydrocarbons

Pages 184–189
Advanced Planning: Page 184
Follow-up: Page 189
- **R** Hydrocarbons
- **E** Structural Formulas
- **T** Engine Knock

9.3 Adding Other Atoms to Hydrocarbons

Pages 190–193
Advanced Planning: Page 190
Activity: How Can You Prepare an Organic Molecule? Page 192
Follow-up: Page 193
- **R** New Compounds
- **E** Preparing Ethyl Alcohol
- **T** Food Chemists

9.4 Synthetic Materials

Pages 194–197
Activity: How Do Three Substances Compare? Page 196
Follow-up: Page 197
- **R** Polymers
- **E** Nylon
- **T** Rubber Industry

See ACTIVITIES book pages 49–54 for laboratory, skill, and review exercises for Chapter 9.

9.1

OBJECTIVES: The student should be able to

- **describe** the electronic structure of a carbon atom;
- **describe** the types of bonds carbon atoms may form;
- **explain** why carbon atoms can form so many different compounds;
- **define** the term *organic compounds;*
- **identify** organic chemistry.

SCIENCE TERMS

double bonds
triple bonds
organic compounds

ADVANCED PLANNING

Have available an overhead projector and your transparency of the periodic table or periodic wall chart for use throughout the chapter.

LESSON PLAN

1. Ask the students to locate the square on the periodic table that represents carbon. What is its symbol (C)? What is its atomic number (6)? How many protons does a carbon atom have (6)? How many neutrons (6)? How many electrons (6)? How are the electrons arranged (2 in the first shell; 4 in the second)? Diagram a carbon atom on the board.

● **Comprehension Skill: Literal**

2. Ask the students which part of the atom is involved in forming compounds (valence electrons). How many valence electrons does carbon have (4)? Explain that because the valence shell is half full, carbon shares electrons when it forms compounds. Ask what kind of bond is formed when pairs of electrons are shared (covalent).
3. Explain that carbon can form three different types of covalent bonds. Have three students illustrate one each of the three types on the board. The students should be told to show the carbon symbol with the appropriate number of dots to indicate shared electron pairs (single bond C:, double bond C::, triple bond C:::).

9.1 Organic Compounds

You will find out
- why carbon makes so many compounds;
- what organic compounds are.

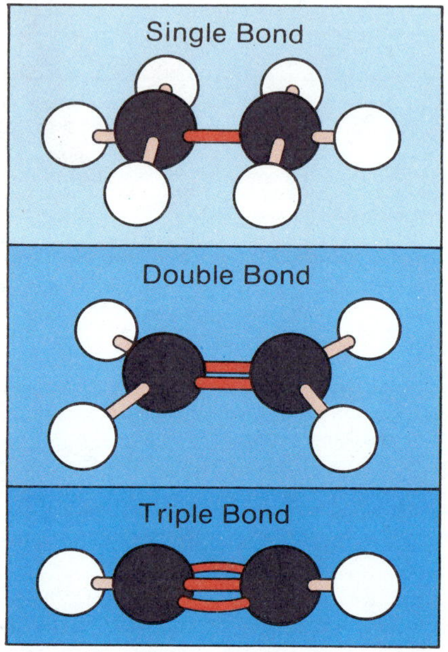

Figure 9.2 Carbon atoms form single, double, and triple bonds with other carbon atoms.

The element carbon is the world champion when it comes to forming compounds. Carbon is unique because its atomic structure allows it to form many different molecules. A carbon atom has four valence electrons. It can form four single covalent bonds with other atoms to make a full, stable outer shell. It can also form double and triple bonds. In **double bonds,** two pairs of electrons are shared. Similarly, in **triple bonds,** three pairs of electrons are shared. A carbon atom can form single, double, and triple bonds with other carbon atoms as well as with other kinds of atoms. As a result, there are over five million different molecules that can be formed with carbon atoms.

You might think that other members of the carbon family would also form millions of different compounds. These other elements also have four valence electrons. However, atoms of other elements in the carbon family all have more shells of electrons than carbon atoms. These extra shells make it more difficult for the atoms to form bonds. Carbon, therefore, is truly unique.

Figure 9.3 Most of the substances in this picture contain carbon.

Figure 9.4 Seawater, rocks, and sand contain inorganic compounds.

Compounds that contain carbon are called **organic** [ohr-GAN-ihk] **compounds.** The study of organic compounds is called organic chemistry. The word *organic* was first used many years ago. At that time it was believed that all matter could be divided into three categories; animal, vegetable, and mineral. The animal and vegetable, or living part of the world, was called organic. Because organic material came from living things, it was thought to be made of substances different from those in nonliving, mineral material. The nonliving material was called inorganic.

Scientists no longer use the word *organic* to describe only living matter. It is now known that living things are made of the same substances as nonliving things. Both living and nonliving things are made of elements. Today, the word *organic* is used to describe any carbon compound whether or not it is found in living matter. Therefore, things such as toothbrushes, which contain carbon, are said to be made of organic compounds, even though toothbrushes are not alive, never have been alive, and never will be alive.

Study Questions for 9.1

1. What kinds of bond do carbon atoms form?
2. What is an organic compound?

1. A carbon atom has four valence electrons. It forms single, double, and triple covalent bonds to achieve an outer shell of eight electrons. Carbon atoms bond with other carbon atoms as well as other kinds of atoms.
2. a compound that contains carbon

4. Ask why carbon forms so many compounds (because carbon can form different types of bonds).
5. Conclude the lesson by asking students to explain why they think that carbon is such an important element (carbon is unique in its ability to form bonds; as a result it can form more different types of compounds than any other atom; we rely heavily on organic compounds in our lives).

FOLLOW-UP

Reinforcement: The Bonds of Carbon

Ask the students to create a poster which explains the different types of bonds that carbon can form. On the back of the poster students should write a paragraph to explain why carbon's ability to form different types of bonds is important.

Enrichment: Saturated and Unsaturated Compounds

Ask the students to do library research to determine the difference between saturated and unsaturated organic compounds. Why do doctors believe that a diet rich in saturated fats is harmful? The students should write reports to present their research.

Tie-in: Classification and the Vital Force

Ask the students to do research on the early attempts to classify matter as organic and inorganic. What is the vital force theory? How was it shown to be incorrect? Ask the students to prepare reports that include the works of Lémery, Berzelius, Wöhler, and Kolbe.

SCIENCE BACKGROUND

Carbon atoms have the unique ability to form covalent bonds with other carbon atoms as readily as they do with other kinds of atoms. In theory, there is no limit to the number of carbon compounds that can be formed.

Some compounds that contain carbon are not considered to be organic. Examples include CO_2 and CO.

9.2

OBJECTIVES: The student should be able to

- **define** the term *hydrocarbon;*
- **identify** methane;
- **describe** a structural formula;
- **define** the term *isomer;*
- **compare** long and short chain hydrocarbons;
- **describe** how crude oil is distilled;
- **explain** how fractional distillation occurs.

SCIENCE TERMS

hydrocarbons
structural formula
straight chain hydrocarbons
isomers
distillation
fractional distillation

ADVANCED PLANNING

Obtain 60 Styrofoam balls and a box of toothpicks. Paint 15 balls black to represent carbon atoms. Paint 40 balls white to represent hydrogen atoms. Paint 4 balls red to represent oxygen atoms. Paint 1 ball orange to represent a nitrogen atom. You will need toothpicks to represent shared pairs of electrons as you construct models of molecules.

LESSON PLAN

1. Introduce the lesson by telling the students that the simplest organic compounds are hydrocarbons. These compounds are composed of only hydrogen and carbon.
2. Ask what kind of bond is formed when carbon atoms form compounds (covalent). Explain that the simplest hydrocarbon is methane, or CH_4. What kinds and numbers of atoms are in a methane molecule (1 carbon atom and 4 hydrogen atoms)?

9.2 Hydrocarbons

You will find out
- what hydrocarbons are;
- what isomers are;
- how crude oil is distilled.

Hydrocarbons [hy-droh-KAHR-buhnz] are compounds made of only carbon and hydrogen. Hydrocarbons are the basic parts of all organic compounds. The simplest hydrocarbon is methane. Methane, CH_4, is made of one carbon atom and four hydrogen atoms. The hydrogen atoms are joined to the carbon atom by single covalent bonds.

Covalent molecules can be illustrated by using electron dot diagrams. Two dots represent a single covalent bond. When dashes are used in place of the pairs of dots, the result is called a **structural formula.** Figure 9.5 shows the electron dot diagram and the structural formula for a methane molecule. Structural formulas are a little neater way to represent molecules with many atoms. For this reason they are used to illustrate hydrocarbons.

Figure 9.5 *left:* A model of methane; *right:* Formulas for methane

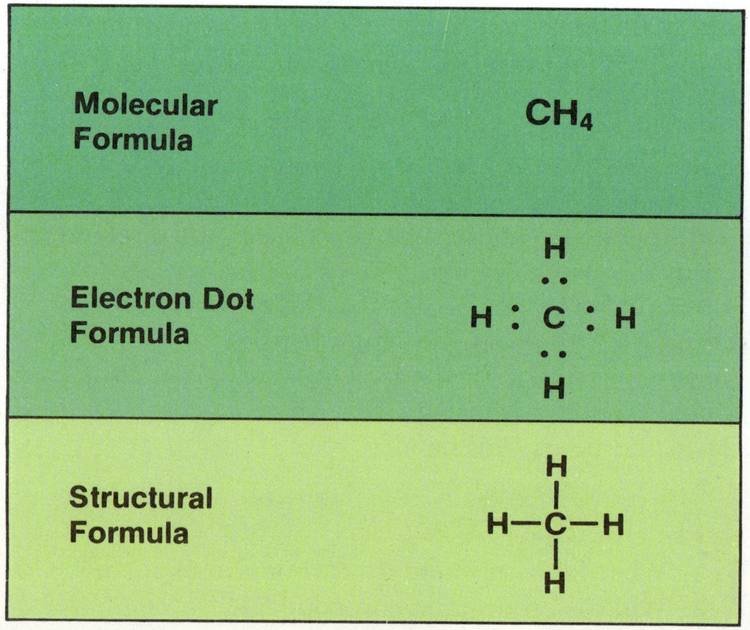

Figure 9.6 A model of ethane

Structural formulas give an idea of the structure of the molecule. However, you should always remember that they do not give the exact structure. They are drawn on a flat sheet of paper, and the molecules are three-dimensional. For example, the actual shape of a methane molecule is a pyramid with four faces. The carbon atom is in the center, and the hydrogen atoms are on the points of the pyramid.

Other hydrocarbon molecules similar to methane are made by linking together one or more carbon atoms in a chain. For example, in the ethane molecule, C_2H_6, two carbon atoms are bonded together and surrounded by hydrogen atoms. In propane, C_3H_8, three carbon atoms are bonded together. Each time another carbon atom is added to the chain, a new molecule is formed. These compounds are called **straight chain hydrocarbons.** The carbon atoms are bonded together in a single line.

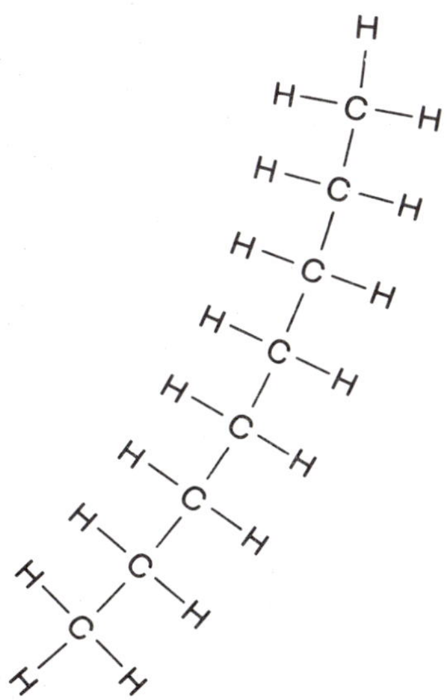

Figure 9.7 Gasoline contains the hydrocarbon octane C_8H_{18}.

3. Hold a black ball in your hand. Tell the students that it is a carbon atom. In it stick four toothpicks in a pyramidal arrangement, not in a flat square. Each toothpick represents a single covalent bond. To each toothpick attach a white ball to represent hydrogen atoms. Point out that in your model of methane there is one carbon atom and four hydrogen atoms just as the molecular formula CH_4 indicates. Point out that each of the four toothpicks represents a pair of shared electrons.

4. Explain that when you write the structural formula for methane the lines represent shared electron pairs. Point out that a structural formula shows how the atoms are arranged to form the molecule. Ask why the structural formula is limited (it is a flat representation on paper while a real molecule has three dimensions).

Comprehension Skill: Interpretative

5. Explain that ethane is made when two carbon atoms join with six hydrogen atoms. Join two black balls with a toothpick. Ask the students how many more bonds each carbon atom can make (three each). Stress that since each carbon atom can form four bonds and one was made when they joined, each atom has three more electrons ready to share. Stick three toothpicks in each carbon atom so that they are equidistant. Tell the students that ethane is a hydrocarbon and then ask what other element is needed (hydrogen). Add a white hydrogen atom to each toothpick. Point out that each carbon atom has formed four bonds while each hydrogen has formed one bond.

SCIENCE BACKGROUND

The so-called straight chain hydrocarbons are kinked rather than straight. However, this arrangement of atoms is still referred to as a straight chain.

6. Ask some students to create ball-and-toothpick models of propane, butane, and pentane molecules while other students draw the structural formulas on the board. Remind the students that the carbon atoms each form four bonds while the hydrogens form one bond.

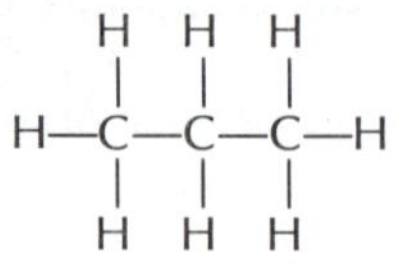

propane

C_3H_8

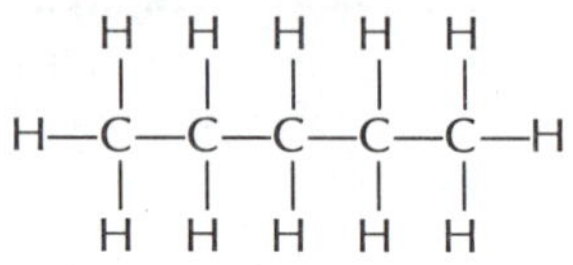

pentane

C_5H_{12}

7. After explaining that compounds with the same numbers of carbon and hydrogen atoms can have different arrangements called isomers, have students make ball-and-toothpick models of isobutane from the butane models.

Sometimes the hydrogen and carbon atoms do not form single chains. Instead they form branched or looped chains. Molecules formed with branched or looped chains have properties different from straight chain hydrocarbons. Two hydrocarbon molecules may have the same numbers of carbon and hydrogen atoms bonded together in different ways. That is, the molecular formulas are the same, but the structural formulas are different. For example, butane, C_4H_{10}, is a straight chain hydrocarbon. Isobutane, C_4H_{10}, is a hydrocarbon with a branched chain. Butane and isobutane have the same molecular formulas but different structural formulas and different properties. Two compounds with the same molecular formulas but different structural formulas are called **isomers** [EYE-suh-muhrz].

Figure 9.8 The carbon atoms in butane form a straight chain. In isobutane, they form a branched chain.

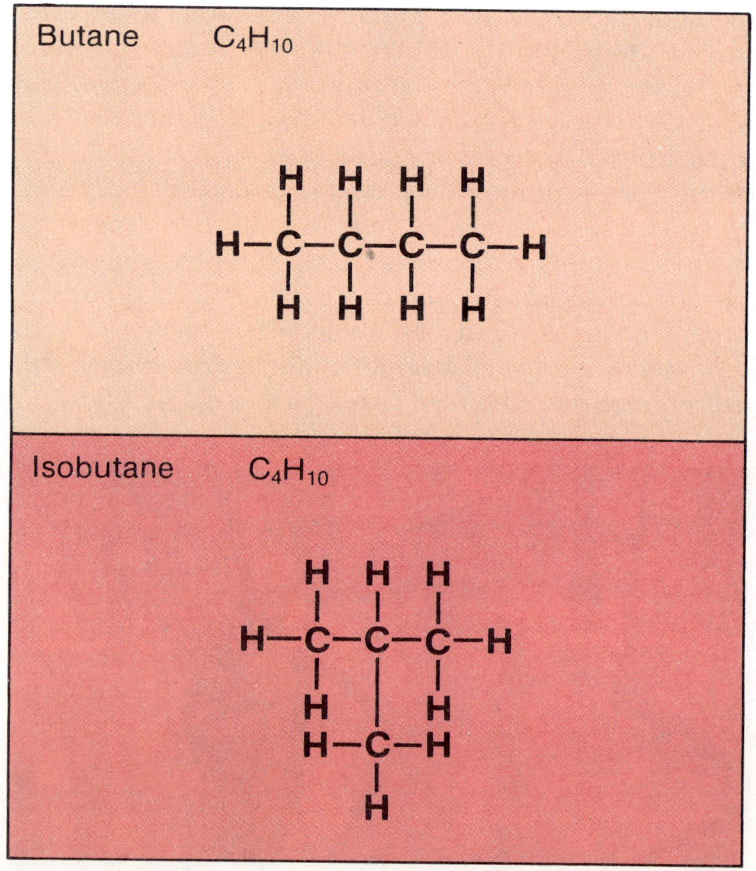

Do You Know?
It has been calculated that a hydrocarbon molecule with the formula $C_{30}H_{62}$ would have about 40 billion possible isomers.

Figure 9.9 A common ingredient in automotive antifreeze is ethylene glycol. Ethylene glycol contains carbon atoms joined by double bonds.

Sometimes the carbon atoms of a hydrocarbon are joined by double and triple bonds. Hydrocarbon compounds having double or triple bonds are grouped together. One of the more familiar members of this group is acetylene, C_2H_2. The two carbon atoms in the acetylene molecule are joined by a triple bond. The triple bond causes acetylene to burn with a very hot flame. For this reason, acetylene is used in welding torches.

Figure 9.10 Using an acetylene torch

SCIENCE BACKGROUND

The carbon atoms in other compounds may also link together in branched chains or rings. The way the carbon atoms join, in chains or in loops, and the numbers of hydrogen and carbon atoms involved determine what the properties and uses of the compound will be.

8. Explain that hydrocarbons often occur mixed with other hydrocarbons. Ask the students how mixtures of hydrocarbons can be separated (through the process of distillation). When is this method applied (when each part of the mixture has a different boiling point)? How is the process carried out (the mixture of liquids is heated to or just above the boiling point of the hydrocarbon with the lowest boiling point, and the gas is collected and condensed)? Explain that when your mixture contains many liquids as crude oil does and you wish to separate each, the process used is often called fractional distillation.

Figure 9.11 Cooking gas is a short chain hydrocarbon.

Hydrocarbon chains of different lengths have different properties and uses. In general, the shorter the chain is, the lighter the molecule is, and the lower the boiling point is. Because short chain hydrocarbons evaporate and burn easily, they are used as fuels. Cooking gas and gasoline are short chain hydrocarbons. The heavier long chain hydrocarbons are used as lubricating oils, as waxes, and as asphalt. They do not evaporate easily and are liquids or solids at room temperature.

Figure 9.12 Candles contain long chain hydrocarbons.

Because they have different boiling points, hydrocarbons in crude oil from wells can be separated in a process called **distillation.** In distillation a liquid is changed to the gas state, and then the gas is collected and cooled, returning it to the liquid state. You start with a liquid and end up with a liquid. How does this process separate the hydrocarbons in crude oil? The different hydrocarbons boil and change to a gas at different temperatures. As the temperature rises, each hydrocarbon evaporates to the gas state and separates from the liquid crude oil. The hydrocarbon gas is then cooled and returned to the liquid state. After the hydrocarbons are separated, they are put through other chemical processes. Many useful products such as gasoline, kerosene, diesel fuel, asphalt, lubricating oil, wax, and petroleum jelly are produced in this way. Because the distillation of crude oil separates the oil into many parts, or fractions, it is often called **fractional distillation.**

Figure 9.13 This refinery separates the hydrocarbons in crude oil by fractional distillation.

Study Questions for 9.2

1. What are hydrocarbons?
2. Draw the structural formula for ethane.
3. What are isomers?
4. How is crude oil distilled?

1. compounds containing only hydrogen and carbon
2. 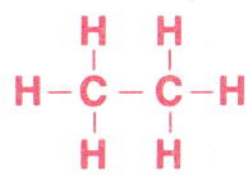
3. molecules with the same molecular formulas but different structural formulas
4. The crude oil is changed to the gas state. The different hydrocarbons change to a gas at different temperatures. As each hydrocarbon becomes a gas, the hydrocarbon is collected.

9. Conclude the lesson by using one of your models to review some basic information with the students. Hold up a model and ask the students to name the elements that are present (hydrogen and carbon). What type of compound contains these two elements (hydrocarbon)? In what process is crude oil broken into the hydrocarbons that compose it (fractional distillation)? On what principle does the process work (the parts have different boiling points)?

● **Word Skill: Structure**

FOLLOW-UP

Reinforcement: Hydrocarbons
Give students a list of different hydrocarbons. Ask each student to select one and do a research project on the compound. What is its structural and molecular formulas? What are its properties? What are its uses? Ask the students to create a museum display to present their research.

Enrichment: Structural Formulas
Have interested students report on the work of Friedrich Kekulé (1829–1886) in pioneering a system of notation for organic chemistry.

Tie-in: Engine Knock
Ask the students to find out what mechanics mean by engine knock. What causes it? How can it be prevented? Students should work in pairs. One should be a reporter and the other a mechanic. Together they should write a news report or make a tape to explain engine knock and its prevention.

SCIENCE BACKGROUND

A fractionating tower is used to separate the different hydrocarbons in crude oil. The crude oil is changed to a mixture of gases when it is heated to a very high temperature. The gases are pumped into the bottom of the fractionating column. As the mixture of gases rises, its parts cool and condense. Each condenses at a different temperature. As they condense, the liquids are caught on bubble trays and collected.

9.3

OBJECTIVES: The student should be able to

- **explain** how alcohols are formed from hydrocarbons;
- **list** some uses of methanol;
- **explain** how an organic acid is formed;
- **describe** the importance of fats, carbohydrates, and proteins to living things.

SCIENCE TERMS

alcohol	amino acids
organic acids	proteins

ADVANCED PLANNING

1. You will need the models of methane and ethane that were constructed earlier as well as the remaining Styrofoam balls.

2. You will need the transparency of the periodic table and an overhead projector or a wall chart of the periodic table.

LESSON PLAN

1. Introduce the lesson by showing the students the model of methane. Remind them that the black ball represents a carbon atom and the white balls represent hydrogen atoms. Explain that the hydrogen atoms can be replaced by other atoms or groups of atoms to form different molecules. Ask the students what atoms methane is composed of (1C and 4H). As you remove a hydrogen atom, tell the students that you are going to replace it with an atom of oxygen (red) attached to a hydrogen atom, or an OH group. Your model should now look like this.

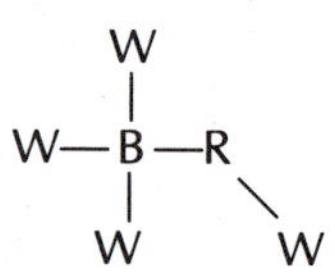

B = black (carbon)
W = white (hydrogen)
R = red (oxygen)

9.3 Adding Other Atoms to Hydrocarbons

You will find out

- how alcohols are formed;
- how organic acids are formed;
- how some organic compounds are important to living things.

It is possible to replace one or more of the hydrogen atoms in a hydrocarbon with other atoms or groups of atoms. In this way new types of molecules are formed. These new molecules form new organic compounds. Many of these compounds are important in your life.

Three chlorine atoms can replace three hydrogen atoms in a methane molecule. This replacement changes methane, CH_4, into chloroform, $CHCl_3$. Before safer and more effective drugs were found, chloroform was commonly used to put surgical patients to sleep.

Figure 9.14 Replacing atoms in methane

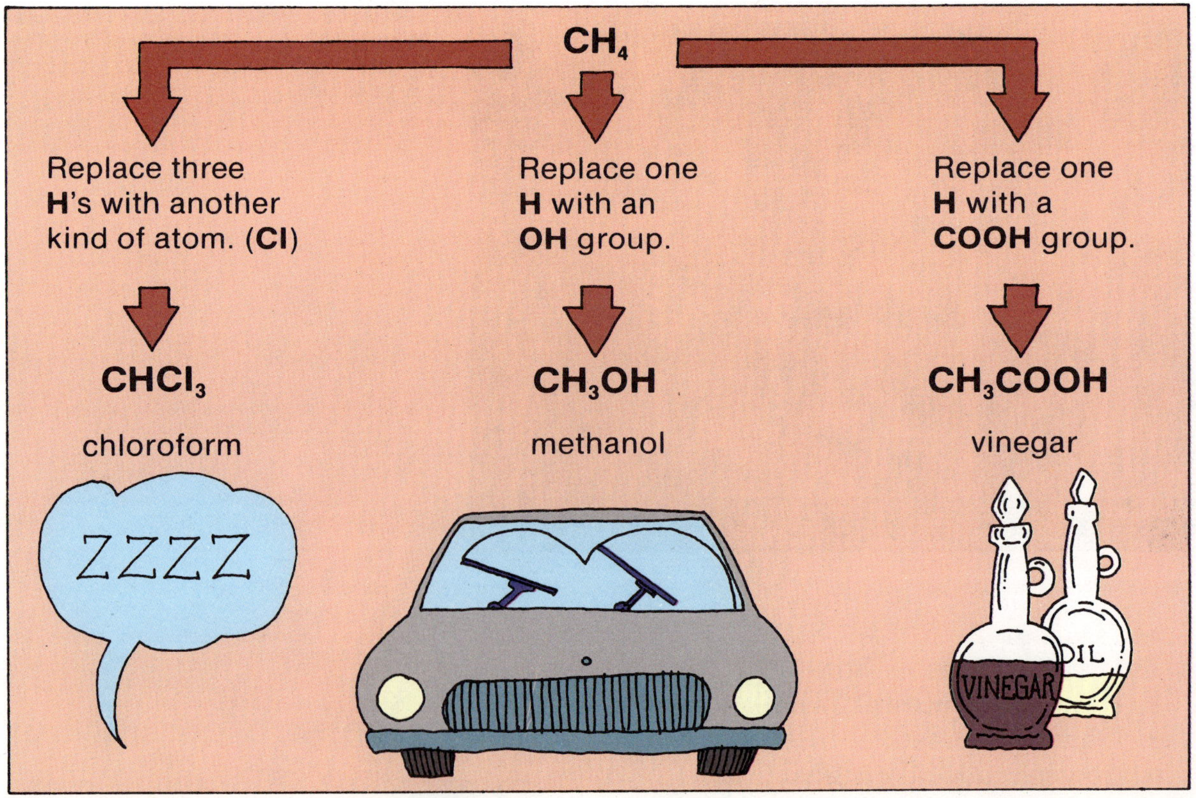

When an oxygen atom and a hydrogen atom, called the OH group, replace a single hydrogen atom in a hydrocarbon, an **alcohol** is formed. For example, if one of the hydrogen atoms in methane, CH_4, is replaced with an OH group, you get methanol, CH_3OH. Methanol, also known as methyl alcohol or wood alcohol, can be made from wood. It is a clear, poisonous liquid. Methanol freezes at −93.9°C. Because it freezes at a very low temperature, it is used in windshield washer fluid. Another alcohol, ethanol, can be used as a fuel. There are cars today running on ethanol instead of gasoline. Another gasoline substitute in current use is gasohol. Gasohol is a mixture of gasoline and alcohol. Most cars produced today have engines designed to burn gasoline. Those engines have to be modified to run on pure ethanol. Gasohol, however, can be used in gasoline-burning engines.

A carbon atom, two oxygen atoms, and a hydrogen atom, called the COOH group, form another group of atoms that can be added to hydrocarbons. Hydrocarbon chains with a COOH group are called **organic acids.** For example, when one of the hydrogen atoms in methane is replaced with a COOH group, the organic acid formed is acetic acid, CH_3COOH. Acetic acid is what gives vinegar its sour taste.

Figure 9.15 These citrus fruits contain the organic acid, citric acid.

2. Point out that the methanol molecule you have made is different from methane. Instead of one of the hydrogen atoms, an OH group is attached to the carbon atom. Tell the students that the suffix *-ol* means that the compound is an alcohol.

3. Point out that methanol, or wood alcohol, is a deadly poison. If you drink just 15 mL of it, you will become blind. If you drink 30 mL of it, you will die. (You may want to use a graduated cylinder to show what small amounts these are.)

4. Explain that another way to change straight chain hydrocarbons is to add a COOH group. Use your Styrofoam models to show that when this is done the oxygen joins to the carbon atom by forming a double bond (two toothpicks). Two pairs of electrons are shared. The OH group joins with a single bond as it did in the alcohols.

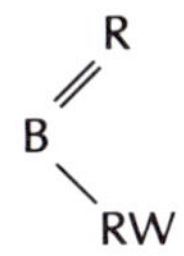

Ask what the compound is called whose chain ends with a COOH group (an organic acid).

SCIENCE BACKGROUND

The two atoms of the OH group are joined in such a way that they are very hard to separate. They act as if they were a single atom.

Methanol is used as a solvent for paint and shellac, in duplicating machines, and as a fuel.

When one of the hydrogen atoms in ethane is replaced with an OH group, ethane becomes ethanol. Ethanol is the alcohol that is found in alcoholic beverages. The replacement of a hydrogen atom with an OH group greatly changes the properties of the compound. Ethane is a gas that is insoluble in water. Ethanol is a liquid that mixes with water.

ACTIVITY: How Can You Prepare an Organic Molecule?

Go over the proper procedure for smelling an unknown substance. The substance should not be placed directly under the nose. It should be held out in front of the face. Any vapors can then be fanned toward the nose with an open hand. Tell the students to be sure the test tube is not pointing toward anyone while it is being heated.

◇ **Caution:** Students should wear safety glasses and aprons and use care in handling the acid.

● Laboratory Skills: Experimentation/Demonstration Manipulation of Materials

● ACTIVITY

How Can You Prepare an Organic Molecule?

Materials

test tube
250-mL beaker
hot plate
graduated cylinder
eyedropper
salicylic acid (an organic acid)
methyl alcohol (methanol)
concentrated sulfuric acid
test-tube holder
test-tube stand
safety glasses
apron

Procedure

1. Put on your safety glasses and apron.
2. Prepare a hot-water bath by placing the beaker, half filled with water on the hot plate. Turn the hot plate on medium.
3. Measure 1 g of salicylic acid and put it in the test tube.
4. Add 5 mL of methyl alcohol to the test tube.
5. Holding the test tube with the test-tube holder, carefully add 3 drops of sulfuric acid to the test tube.
6. Carefully place the test tube in the hot-water bath for 15 minutes. Be sure that the test tube is not pointed toward anyone.
7. After 15 minutes, use the test-tube holder to remove the test tube from the hot-water bath.
8. Hold the test tube about 12 cm in front of your nose and use your hand to fan some of the vapors toward your nose.

Questions

1. What odor did you detect from the test tube?
2. The material you made is an ester, methyl salicylate. Do library research to determine what an ester is. How is the ester methyl salicylate (oil of wintergreen) used?

1. wintergreen
2. An ester is a compound formed when the hydrogen of an organic acid is replaced by another group from an alcohol. Oil of wintergreen is commonly used as a flavoring.

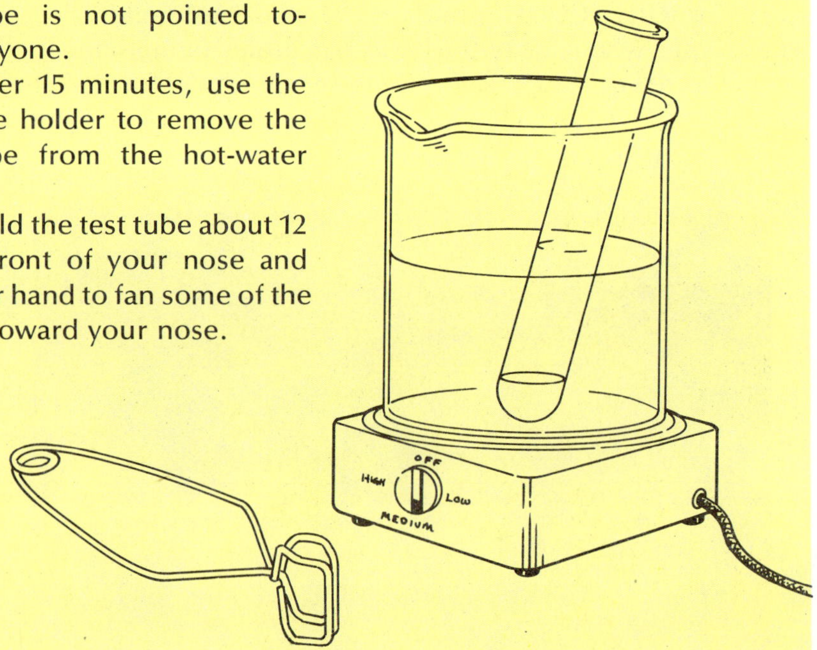

Figure 9.16 The human body contains proteins, fats, and carbohydrates.

An important group of organic acids are the **amino acids.** Amino acids are organic acids that contain nitrogen. The nitrogen atoms are present in NH_2 groups. Amino acids can join together to form long amino acid chains. These chains are called **proteins.** Proteins are a necessary part of living things. Thousands of different kinds of proteins are found in the human body. They are important parts of muscles, hair, bones, skin, blood, and other body organs.

Many organic compounds are made from different combinations of carbon, oxygen, and hydrogen atoms. Two categories of such compounds are fats and carbohydrates. Both fats and carbohydrates are found in living things. Plant carbohydrates provide animals with food energy and long chains of carbon atoms that are used to make other large molecules.

Study Questions for 9.3

1. How is methanol formed?
2. How is acetic acid formed?
3. How does your body use carbohydrates and proteins?

1. One of the hydrogen atoms of methane is replaced with an OH group.
2. One of the hydrogen atoms of methane is replaced with a COOH group.
3. Carbohydrates provide food energy. Proteins are important parts of body organs, bones, etc.

5. Ask the students how many bonds nitrogen will form with other atoms (3). Stick two toothpicks into a nitrogen atom (orange). Add white hydrogen atoms. Tell the students that this NH_2 group is called the amine group. When a carbon chain has the amine group on one end and the carboxyl group on the other end, it is an amino acid.

6. Ask the students to name the three important groups of organic compounds (fats, carbohydrates, and proteins). Ask why each of these groups is important (fats and carbohydrates contain carbon, hydrogen, and oxygen and are energy-rich compounds produced and used by plants and animals; proteins are very important in the body because they build body structures such as hair, nails, bones, skin, and blood).

FOLLOW-UP

Reinforcement: New Compounds
Ask the students to create collages showing hydrocarbons, fats, organic acids, alcohols, carbohydrates, and proteins. Have the students explain the structure of each on the back of the collage.

Enrichment: Preparing Ethyl Alcohol
Have each student prepare a test tube of a mixture of molasses and yeast. Let the mixture stand for several days at room temperature. What smell do the students notice coming from the test tube (ethyl alcohol)? How could they identify the gas that is bubbling off (bubble the gas into limewater; it is carbon dioxide)?

Tie-in: Food Chemists
Food chemists study the compounds in foods. Have the students research this career. What training is needed? What are the main responsibilities? Where do food chemists work? What are their working conditions? Does the profession pay well? Is there opportunity for advancement? Have the students prepare career pamphlets on food chemists.

SCIENCE BACKGROUND

Amino acids are the building blocks of proteins.

9.4

OBJECTIVES: The student should be able to

- **define** the term *polymer;*
- **explain** how a polymer is made;
- **name** two polymers;
- **describe** some useful properties of polymers.

SCIENCE TERM

polymers

LESSON PLAN

1. Introduce the lesson by explaining that scientists' progress in understanding the structure of matter has been helpful in many ways. It has helped scientists to cure diseases, find new and better ways to do things, and create new compounds to do specific jobs. The compounds that scientists create are synthetic compounds.

SCIENCE BACKGROUND

Synthetic polymers are usually plastics, synthetic rubber, and synthetic textiles. The properties of polymers allow scientists to make many useful products such as clothes, bottles, footwear, and recreational equipment.

9.4 Synthetic Materials

You will find out
- what a polymer is;
- how synthetic materials are made.

Scientists have made progress in understanding the structure of matter. They can change and rearrange molecules in new and useful ways. One of the best examples of this progress is the manufacture of synthetic materials from hydrocarbons.

Organic molecules containing a large number of atoms are called **polymers** [PAHL-uh-muhrs]. Some polymers are natural, like some of the proteins in your body. Many other polymers are synthetic. In order to make a polymer in the laboratory, smaller molecules must be linked together to form the long chains of polymers.

Figure 9.17 The clothing these people are wearing and the raft they are riding on are made of synthetic materials. Over 33 percent of the fiber and over 70 percent of the rubber used in the world today are synthetic.

Figure 9.18 The grass in this stadium is synthetic turf.

Consider, for example, the making of polyethylene, a tough, waxy solid used in plastics. Polyethylene is a synthetic material made by linking together smaller ethylene molecules. Ethylene, C_2H_4, is a very small molecule of six atoms compared to the polymer polyethylene. A polyethylene molecule can have thousands of atoms.

Polymers can be formed into almost any shape. These shapes become familiar objects such as combs, bowls, pipes, and bathtubs Some polymers can be made into coatings for packages. Other polymers are formed into thin threads that are woven into cloth. In this way synthetic fabrics such as nylon and polyester are made.

In the hydrocarbon benzene, C_6H_6, six carbon atoms are bonded in a ring structure. Molecules made with benzene are used in making rubber, plastics, acrylic fibers, varnish, and detergents.

Do You Know?
Many compounds made with the benzene ring structure have a noticeable smell. They are called aromatic compounds.

2. Ask the students what very large molecules can be made from smaller ones (polymers). Mention that a good example of polymer is polyethylene. Ask the students to describe polyethylene and tell how it is used (tough, waxy solid used to make plastic). How is polyethylene made (by heating many smaller ethylene molecules; when the molecules are heated, chemical bonds break and new bonds form, creating a long chain of ethylene molecules called polyethylene)?

3. Ask the students what properties polymers have that make them so useful (polymers can form fine threads and can be molded into many shapes).

● Comprehension Skill: Applied

ACTIVITY: How Do Three Substances Compare?

Before the students begin the experiment, ask them to do research on saccharin, sugar, and salt. They should explore the composition, uses, and potential dangers of each.

If students arrange the three substances in a triangular pattern on the can lid, they can burn all three at one time.

For saccharin samples, use any powdered commercial low-calorie sweetener.

◇ **Caution:** Students should wear safety glasses and be careful when heating the substances.

1. organic—burns leaving a black mass (carbon); inorganic—does not burn
2. sugar—cubic crystals, doesn't dissolve in alcohol, burns more slowly; saccharin—needlelike crystals, dissolves in alcohol, burns more quickly

● **Laboratory Skills: Experimentation/Demonstration Manipulation of Materials**

● ACTIVITY

How Do Three Substances Compare?

Materials

3 test tubes
salt
(natural, inorganic)
sugar
(natural, organic)
saccharin
(synthetic, organic)
isopropyl alcohol
fingernail polish remover
3 can lids
ring stand and ring
Bunsen burner
matches
tongs
safety glasses
apron
marking pencil
test-tube rack
magnifying glass

Procedure

1. Put on your apron and safety glasses.
2. Copy the chart and use it to record your observations.
3. Obtain a sample of salt, sugar, and saccharin.
4. Examine each substance with a magnifying glass.
5. Describe and diagram the crystals of each.
6. Label 3 clean test tubes *A*, *B*, and *C*.
7. Fill each test tube halfway with water.
8. Put 20 grains of salt in A; put 20 grains of sugar in B; put 20 grains of saccharin in C. Gently shake each test tube and observe.
9. Repeat steps 7 and 8 once with isopropyl alcohol and once with fingernail polish remover. Be sure the test tubes are clean between trials.
10. Set up the ring stand, ring, Bunsen burner, and can lid as shown in the diagram.
11. Place a small sample of sugar on the can lid.
12. Light the Bunsen burner and observe the sugar.
13. Turn off the burner. Let the can lid cool and remove it with tongs.
14. Put a clean can lid on the ring.
15. Repeat steps 11–14 once with saccharin and once with salt.

Questions

1. What comparison can you make between organic and inorganic substances?
2. How are sugar and saccharin alike? How are they different?

Material	Describe Crystals	Dissolves in Water?	Dissolves in Alcohol?	Dissolves in Remover?	Reaction When Heated
salt (natural, inorganic)	white cubes	yes	slightly	no	doesn't burn no change
sugar (natural, organic)	white cubes	yes	no	no	melts, burns turns black smells sweet
saccharin (synthetic, organic)	white needle-like	yes	yes	no	melts, burns turns brown smells slightly sweet

You use synthetic materials every day. They are in your home, in everything from furniture to kitchen utensils to sports equipment. In some ways synthetic materials may have advantages over comparable natural products because synthetics are produced under scientifically-controlled conditions. Synthetic materials can be made to have properties not found in natural products. Plastics, for example, are lightweight, can easily be shaped into different forms, and can be very strong.

Do You Know?
Americans use an average of 62 kg of plastic per person per year. Of this, 20 kg goes for packaging.

Figure 9.19 Can you find all the plastic items in this photograph?

Study Questions for 9.4

1. What is a polymer?
2. How are synthetic polymers made?

1. an organic molecule containing many atoms
2. Smaller molecules are linked together, making polymers that can be formed in different shapes.

4. Conclude the lesson by asking the students to name some synthetics they use or encounter daily (plastic, rubber, textiles, paints). You could also discuss with students how their lives would be different without synthetics.

FOLLOW-UP

Reinforcement: Polymers
Have the students create a series of diagrams with captions that explain what a polymer is, how it is made, and some of its useful properties. The last diagram should also include pictures of five items the student has used that are made of polymers.

Enrichment: Nylon
Ask the students to do research on nylon. They should create a list of as many uses of nylon as they can find.

Tie-in: Rubber Industry
Rubber tires are made in different ways to fulfill different purposes. Ask the students to work in groups to investigate the different types of tires and how each type is made.

SCIENCE BACKGROUND

Polymers are made up of smaller organic molecules called monomers.

What Did You Learn?

9.1 • Carbon atoms can form single, double, and triple covalent bonds.

9.1 • Compounds that contain carbon are called organic compounds.

9.2 • Hydrocarbons are compounds made of only carbon and hydrogen.

9.2 • Structural formulas are used to illustrate organic compounds.

9.2 • In straight chain hydrocarbons, the carbon atoms are bonded together in a single line.

9.2 • Isomers are compounds with the same molecular formulas but different structural formulas.

9.2 • Hydrocarbon chains of different lengths have different properties and uses.

9.2 • The hydrocarbons in crude oil can be separated by distillation.

9.3 • An alcohol is formed when an OH group replaces a hydrogen atom in a hydrocarbon.

9.3 • An organic acid is formed when a COOH group replaces a hydrogen atom in a hydrocarbon.

9.3 • Amino acids are organic acids with nitrogen atoms present in NH_2 groups.

9.3 • Proteins, fats, and carbohydrates are important organic compounds that are found in the human body.

9.4 • Polymers are organic molecules containing a large number of atoms.

9.4 • Synthetic polymers are made by linking together smaller molecules.

Key Terms

double bonds
triple bonds
organic compounds
hydrocarbons
structural formula
straight chain hydrocarbons
isomers
distillation
fractional distillation
alcohol
organic acids
amino acids
proteins
polymers

Biography

Wallace Hume Carothers (1896–1937)

Wallace Carothers began his higher education wanting to be an accountant. After only a year at business college, he changed schools to study science. He must have been a very good science student because he was soon given the job of teaching chemistry to his fellow students. The only chemistry teacher at the school had left during World War I, and it was impossible to get another.

After graduate school at the University of Illinois, Carothers taught there and at Harvard University. He later went to work at DuPont, a chemical company. At DuPont he developed synthetic materials, such as neoprene, a synthetic rubber. In the 1930's, he began joining together molecules in long chains similar to the chains found in silk. Eventually he found one which was stronger than silk. He had discovered nylon.

The discovery of nylon began the new industry of synthetic fibers. Today the nylon industry is one of the most important industries in the world. Unfortunately, Carothers did not live to see what became of his discovery. He died two days after his forty-first birthday.

Vocabulary

1. f
2. i
3. h
4. c
5. b
6. d
7. a
8. g
9. j
10. e

● Word Skill: Vocabulary

What Do You Remember?

1. carbon
2. carbon, hydrogen
3.
```
     H
     |
  H—C—H
     |
     H
```
4. single and multiple covalent bonds
5. by replacing a hydrogen atom with an OH group
6. It has a COOH group in place of a single hydrogen atom.
7. provide food energy
8. Amino acids contain nitrogen.
9. polyethylene, which is formed by linking many ethylene molecules together
10. different boiling points

TO THINK ABOUT AND DO

Vocabulary ●

Number a piece of paper from 1 through 10. Beside the appropriate number write the letter for the term that is described.

1. groups of these form proteins
2. compounds containing only carbon and hydrogen
3. example of a straight chain hydrocarbon
4. compounds containing carbon
5. process for separating the hydrocarbons in crude oil
6. large organic molecules made of many atoms
7. forms when OH replaces a hydrogen atom in a hydrocarbon
8. compounds with the same molecular formulas but different structural formulas
9. example of an alcohol
10. represents a hydrocarbon molecule

a. alcohol
b. distillation
c. organic compounds
d. polymers
e. structural formula
f. amino acids
g. isomers
h. ethane
i. hydrocarbons
j. methanol

What Do You Remember?

1. What element is in all organic compounds?
2. What two elements are in all hydrocarbons?
3. Write the structural formula for methane.
4. What type of bond is used to join carbon atoms?
5. How can a hydrocarbon be converted to an alcohol?
6. How does an organic acid differ from a hydrocarbon?
7. Why are carbohydrates important?
8. How does an amino acid differ from an organic acid?
9. Name a synthetic polymer and tell how it is formed.
10. What property of hydrocarbons is used in the distillation of crude oil?

Applying What You Have Learned ●

1. Name four types of compounds an organic chemist might study.
2. Why do the hydrocarbons butane and isobutane have different properties?
3. The molecular formula for octane is C_8H_{18}. Draw its structural formula.
4. For each of the following formulas tell whether it represents a hydrocarbon, an alcohol, an organic acid, or an amino acid.

(a) NH_2CH_2COOH (d) C_5H_{12}
(b) C_6H_6 (e) CH_3COOH
(c) C_2H_5OH (f) CH_3OH

Research and Investigation ●

Soap is a product that people use in their daily lives. Find out how soap is made. Is soap made today using the same method that the pioneers used? How does soap work to remove dirt from clothes?

Applying What You Have Learned

1. hydrocarbons, carbohydrates, fats, proteins, alcohols, organic acids
2. Their molecules are shaped differently, and they have different structural formulas.
3. 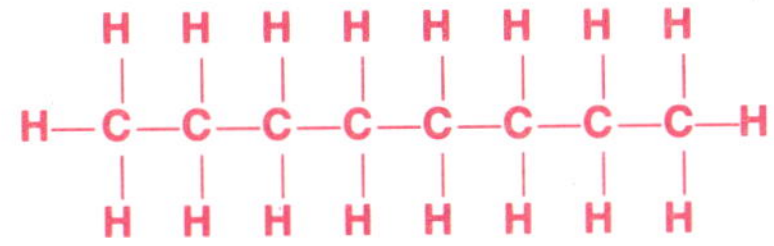
4. (a) amino acid
(b) hydrocarbon
(c) alcohol
(d) hydrocarbon
(e) organic acid
(f) alcohol

● Comprehension Skill: Applied

Research and Investigation

Soaps are long chain organic compounds. Today soap is made by the alkaline hydrolysis of a fat or an oil. The older method of producing soap consisted of treating animal fat with lye, a base, in a large open kettle. The mixture was heated and steam was bubbled through it. Salt was added to help the solid particles clump and fall to the bottom of the kettle. The solid was removed by filtration, washed, and formed into bars. While the mixture was in the kettle, sand was added if the soap was to be used for scouring. Perfume was added if it was to be a toilet soap. Antiseptic was added if it was to be a deodorizing soap.

Soaps are good cleaning agents because of the structure of their molecules. One end contains hydrogen and carbon. This end, called hydrophobic, dissolves in the grease or oil in clothes forming tiny droplets. The other end contains a negatively charged carboxylic (organic) acid group. This end, called hydrophilic, dissolves in water. Through the dissolving of both ends of the soap molecule, dirt or grease becomes broken into small pieces. Oil or grease droplets surrounded by soap molecules are called micelles. Because of the way the soap molecules arrange themselves about the oil or grease droplet, each micelle has a negative charge. Because like charges repel each other, the micelles remain apart from each other and are washed away by a stream of water.

● Reference/Research Skills

Long-Term Projects

1. Temperature, pH, and concentration all affect enzyme activity. Have the students design and perform controlled experiments to show how these factors affect chemical reactions. Some enzymes you may wish to use are rennilase, pectinase, protease, and amylase. Check the students' experiment plans before allowing them to proceed.
2. Have each student create a crossword puzzle using the vocabulary words that appear in the unit.
3. Have the students work in small groups to prepare oral presentations on the uses of radioisotopes in such areas as medicine and radioactive dating.
4. Have the students prepare scrapbooks of articles they find in newspapers and magazines that deal with home heating. Each article should be mounted and accompanied by a summary.
5. Ask each student to write a children's storybook that identifies and describes the different types of engines. The book should also show several uses of each type of engine.
6. Antacids work by helping to neutralize stomach acid. Have the students write letters to the makers of the major antacids asking for information on their product. Students should design and perform controlled experiments to compare the antacids. Check their experiment plans before allowing them to proceed. They can present their results by writing articles for their own consumer magazine.
7. Concern about the effects of salt on the body is continuously being expressed. Have the students write reports to explain how salt is harmful to the body and prepare a low-salt menu for a week's meals.

Unit Three

Matter, Energy, and Change

Bulletin Boards

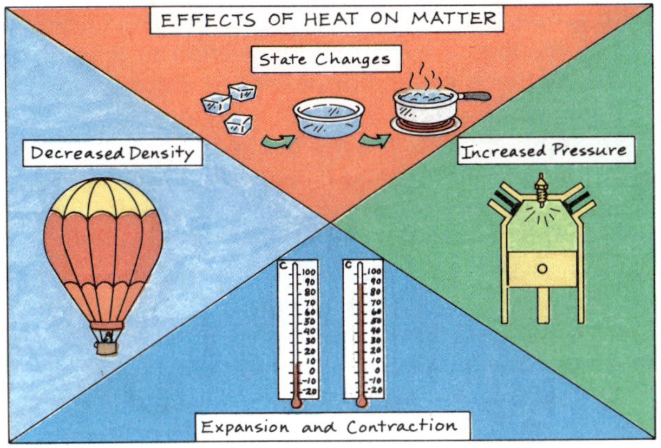

1. Write the title across the top of the board. Draw an X across the board to divide it into four sections. Draw the indicated pictures in each section. Print the title of each section on paper but do not add them to the board. Discuss with the students how heat can affect matter. As you discuss each section add the title and other examples.

2. To help students realize that some materials are better insulators than others, title your board *What Is a Good Insulator?* Cut from a magazine or draw pictures of objects that help to retain heat such as thermos bottles and down jackets. Mount them in a group on one side of the board. On the other side of the board mount pictures of objects through which heat travels easily such as plain glass and metal. Between the two groups print the question *How Do These Materials Differ?*

3. To reinforce the students' ability to recognize chemical reactions, title your board *Signs of Chemical Reactions*. Divide the board into four sections. In the first, write *Color Change* and show a rusty object. In the second, write *Gas Released* and show Alka-Seltzer in water. In the third, write *A Solid Forms* and show two clear liquids being mixed and a solid forming. In the last, write *Energy Released* and show a fire. Have the students add other examples to each section.

CHAPTER BACKGROUND

One way in which matter can change is by undergoing chemical change. When matter changes in this way, it loses its old properties, and new substances with properties all their own are formed. All chemical reactions obey the law of conservation of mass, that is, no matter is created or destroyed during a chemical reaction.

There are four basic types of chemical reactions. In synthesis, or combination, reactions, two elements combine to form a compound. For example, hydrogen gas and oxygen gas combine to form water.

In analysis, or decomposition, reactions, compounds are broken down into their elements. The decomposition of the orange powder mercuric oxide into mercury and oxygen is an example.

During single replacement, or substitution, reactions, an element or ion replaces and frees one of the elements in a compound. This occurs when a metal combines with an acid to form a salt as hydrogen gas is released. For example, zinc and hydrochloric acid combine to form zinc chloride and hydrogen gas.

In double replacement, or metathesis, reactions, two metals in two different compounds change places to form two new compounds. This occurs in the test for a salt. If silver nitrate is mixed with sodium chloride, sodium nitrate and a white precipitate, silver chloride, are formed.

Chemical reactions are divided into two groups based on whether energy is released or absorbed when the reactions occur. An exothermic reaction such as combustion releases energy into the environment. An endothermic reaction such as photosynthesis absorbs energy from the environment.

Chemical reactions occur at different rates. The study of the rates at which reactions occur is called kinetics. Factors such as temperature, concentration, movement, surface area, and the presence of catalysts all influence rates of reactions.

10 Chemical Reactions

Introduce the chapter by showing the students a piece of old yellowed newspaper and a burning piece of newspaper. Tell them that both are chemical reactions. One is occurring more quickly than the other. Show the students a piece of blueprint paper that was left in the sun. Compare it to an unexposed piece of blueprint paper. Tell them that a chemical reaction has occurred here too. Place a beaker on the front desk. Put on safety glasses and fill the beaker to a depth of about 1 cm with nitric acid. Carefully drop a penny into the beaker. Ask the students to observe the penny. Cut a shape from a piece of copper or copper mesh. Dip the piece into a beaker of silver nitrate for a few minutes. Remove the piece and ask the students what has happened to the copper.

Tell the students that each of the things you have shown them is a chemical reaction. In this chapter they will study chemical reactions and the factors that can influence them.

The world you live in is constantly changing. Nothing is ever quite the same. Changes are a good thing. Without them you could not live.

Some changes take place very rapidly. Think of the burning of gasoline in a car engine or the explosion of gunpowder. Other changes like the making and baking of bread take minutes or hours to occur. Still other changes such as the decay of wood or the yellowing of paper take days or even years.

Think about the many different kinds of changes in matter that are going on around you. Some changes are useful to people. Others are not. Some changes are even harmful. By understanding how changes happen and why, scientists can learn how to make useful changes work even better. Scientists can also learn how to prevent or slow down harmful changes.

Figure 10.1 Colorful changes occur in autumn leaves.

Lesson Organizer

10.1 Chemical Changes

Pages 206–209
Advanced Planning: Page 206
Follow-up: Page 209
- **R** Chemical versus Physical Change
- **E** Blueprints
- **T** Types of Chemists

10.2 Balancing Chemical Equations

Pages 210–213
Advanced Planning: Page 210
Activity: Is Mass Conserved in a Chemical Reaction? Page 213
Follow-up: Page 213
- **R** More Practice with Equations
- **E** Law of Conservation of Mass
- **T** Preventing Rust

10.3 The Rate of Chemical Reactions

Pages 214–217
Follow-up: Page 217
- **R** Increasing the Rate of Chemical Reactions
- **E** The Haber Process
- **T** Catalysts at Work

10.4 Energy and Chemical Reactions

Pages 218–219
Advanced Planning: Page 218
Follow-up: Page 219
- **R** How Do Exothermic and Endothermic Reactions Differ?
- **E** Energy from Oxidation
- **T** A Calorimeter

10.5 Acids, Bases, and Salts

Pages 220–225
Advanced Planning: Page 220
Activity: How Can You Form a Salt? Page 224
Follow-up: Page 225
- **R** Acid-Base Comparison
- **E** Antacids
- **T** Acid Rain

See ACTIVITIES book pages 55–60 for laboratory, skill, and review exercises for Chapter 10.

10.1

OBJECTIVES: The student should be able to

- **define** the term *physical changes;*
- **list** examples of physical changes;
- **define** the term *chemical changes;*
- **list** examples of chemical changes;
- **point out** the reactants and the products in an equation;
- **name** and **describe** the four types of chemical reactions.

SCIENCE TERMS

physical changes	chemical equation
chemical changes	reactants
chemical reaction	products

ADVANCED PLANNING

For demonstrations you will need three pieces of iron. Each piece should be a different shape, and one piece should be rusty. You will also need a beaker of water.

LESSON PLAN

1. Show the students a piece of iron and ask them to name some of its properties. Now ask the students to imagine that you have melted the iron and allowed it to cool into a new shape. Show the students another piece of iron with a different shape. Explain that both pieces of iron are the same material even though their shapes are different. This type of change to a new shape, size, or state is a physical change.

● Comprehension Skill: Applied

SCIENCE BACKGROUND

In a physical change the identifying properties of a substance remain unchanged. Melting ice, ripping paper, and dissolving sugar in water all result in physical changes. In each instance you still have the original substance—water, paper, or sugar. Only its form has been changed.

10.1 Chemical Changes

You will find out
- how physical and chemical changes are different;
- how to read chemical equations;
- what the four main groups of chemical reactions are.

● When an artist takes a block of marble and chips and polishes it to make a beautiful statue, the marble has changed shape. Some of the marble remains in the statue. The rest of it is left on the ground. But all of it is still marble. Only the size and shape of the marble have changed. An ice cube changes in size and shape when it melts. It also changes from the solid to the liquid state. But it is still the same chemical substance H_2O. Changes which affect the size, shape, or physical state of a substance but not its chemical properties and formula are called **physical changes.**

Figure 10.2 The faces on Mount Rushmore are the result of physical changes.

Figure 10.3 A chemical change occurs when iron rusts.

In changes such as burning, rusting, baking, and explosions chemical properties are changed. New substances are formed. Changes which affect the chemical properties of one or more substances are called **chemical changes.** For example, when wood burns, carbon in the wood combines with oxygen from the air to produce carbon dioxide.

The carbon dioxide is a new chemical substance. The properties of carbon dioxide are different from those of carbon and oxygen. When an iron nail is exposed to oxygen in the air, the oxygen combines with the iron to form rust. Note that chemical changes, like physical changes, can affect the size, shape, and state of substances. However, only chemical changes affect the chemical properties of substances.

People have been making use of chemical changes for thousands of years. They have burned wood and used the heat energy released to keep warm. The heat energy released has also been used to bring about other chemical changes such as those in cooking.

A chemical change is the result of a process called **chemical reaction.** Three different things can happen in chemical reactions. In some reactions, molecules break apart into atoms. In other reactions, atoms come together to form molecules. In a third type of reaction, atoms change places with other atoms to form new molecules.

Figure 10.4 A chemical reaction

2. Show the students a piece of rusty iron and ask them how, aside from its shape, it differs from the first piece of iron (color, flaky pieces). Explain that chemical properties refer to how matter behaves in changes where its identity is altered. When iron comes in contact with water and oxygen, it rusts. You no longer have iron in element form. You now have the compound iron oxide.
3. Ask the students which of the following are physical changes and which are chemical:

crushing a lump of sugar	(p)
burning a match	(c)
baking a cake	(c)
cutting wood into two-by-fours	(p)

4. Explain that scientists use the term *chemical reaction* to refer to the process by which one or more substances are changed into one or more new substances. A chemical equation is a statement in symbols that represents the reaction.

SCIENCE BACKGROUND

Burning wood, souring milk, and photosynthesis are all chemical changes. In each case you start with certain substances and end up with one or more new substances.

5. Explain that the word *synthesize* means "to make" or "to combine." In a synthesis reaction, elements or compounds combine to form a single compound.
6. Point out that in a decomposition reaction a compound is broken into simpler parts. Write the formula $CaCl_2$ on the board. Ask the students what elements will be formed when calcium chloride is decomposed (calcium and chlorine).
7. Explain that in a single replacement reaction one element is replaced in its compound by another element. On the board write the equation for the reaction between zinc and hydrochloric acid ($Zn + 2HCl \rightarrow H_2 + ZnCl_2$). Point out that zinc is the element that replaces the hydrogen in hydrochloric acid.

SCIENCE BACKGROUND

Written in its proper form, a chemical equation acts as a shorthand method of showing what changes take place during a reaction. It also indicates the relative amounts of the elements and compounds that take part in the reaction.

A **chemical equation** [ih-KWAY-shuhn] uses formulas and symbols to show what happens in a chemical reaction. For example, the chemical equation for the reaction that occurs when the carbon in coal burns to form carbon dioxide is the following:

$$C + O_2 \rightarrow CO_2$$

This equation tells you that one atom of carbon plus two atoms of oxygen produce one molecule of carbon dioxide. The arrow is read as *produces.*

In chemical equations, the substance or substances you start with are called the **reactants** [ree-AK-tuhntz]. They are put on the left side of the arrow. The results of the reaction, the **products,** are put on the right side. Thus, a chemical equation has this general form:

$$\text{reactants} \rightarrow \text{products}$$

There are four main groups of chemical reactions. In *synthesis reactions,* two or more elements or compounds unite to form one compound. The formation of salt from the elements sodium and chlorine is an example of a synthesis reaction.

$$Na + Cl \rightarrow NaCl$$

In *decomposition reactions,* one compound breaks down into two or more other substances. An example is the breakdown of zinc chloride into zinc and chlorine.

$$ZnCl_2 \rightarrow Zn + Cl_2$$

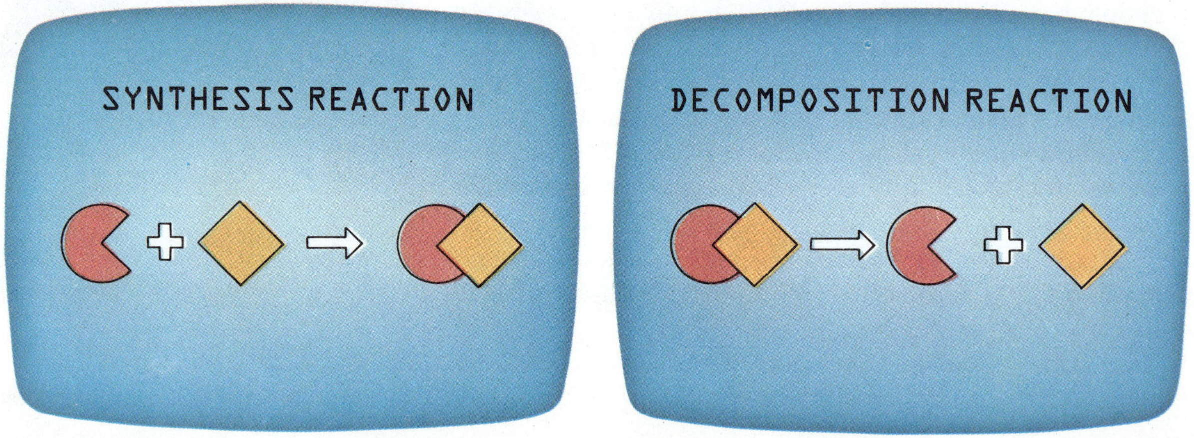

Figure 10.5 *left:* A compound is formed. *right:* A compound is broken down.

In a *single displacement reaction,* one element replaces another one in a compound. For example, magnesium can replace the beryllium in the compound beryllium fluoride. This reaction forms magnesium fluoride and beryllium.

$$BeF_2 + Mg \rightarrow MgF_2 + Be$$

In a *double displacement reaction,* two compounds react to produce two new compounds. For example, silver nitrate reacts with sodium chloride. Silver chloride and sodium nitrate are produced. As you can see, the sodium Na and the silver Ag change places. This results in two new compounds being formed.

$$AgNO_3 + (Na)Cl \rightarrow (Ag)Cl + NaNO_3$$

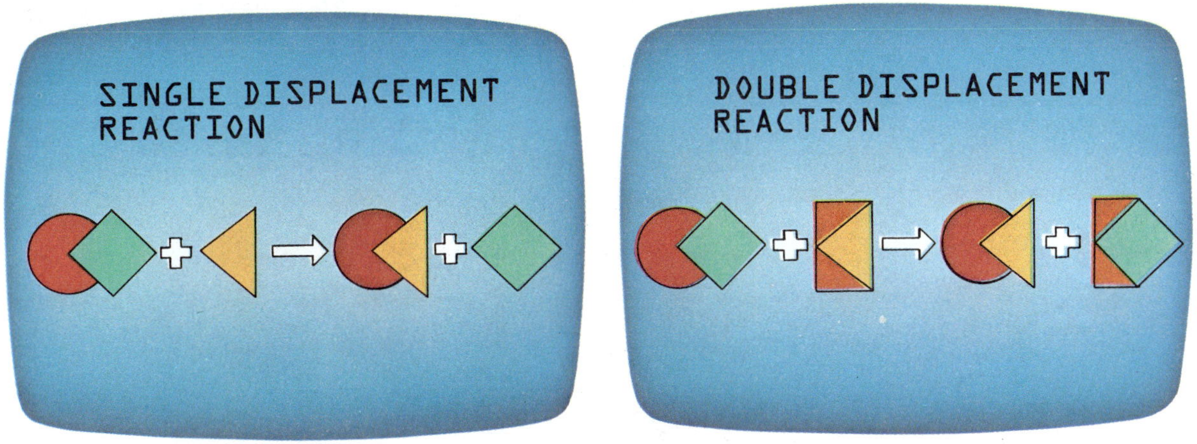

Figure 10.6 *left:* One compound is changed. *right:* Two compounds are changed.

Study Questions for 10.1

1. Give an example of a physical change and an example of a chemical change.
2. Identify the reactants and products in the following equation:

   $$BeF_2 + Mg \rightarrow MgF_2 + Be$$

3. Name the four main groups of chemical reactions.

1. Answers will vary. Some possibilities are the following: physical—evaporation of water; freezing of water; breaking of rock into pieces chemical—burning wood; rusting iron
2. reactants—BeF_2 and Mg; products—MgF_2 and Be
3. synthesis, decomposition, single displacement, double displacement

8. Explain that in a double replacement reaction two compounds react to form new compounds. On the board write the equation $NaCl + AgNO_3 \rightarrow AgCl + NaNO_3$. The two metals sodium and silver have exchanged places to form two new compounds.

9. Ask the students to identify the following reactions as synthesis, decomposition, single replacement, or double replacement reactions.

$Mg + H_2SO_4 \rightarrow MgSO_4 + H_2$	(single)
$4Al + 3O_2 \rightarrow 2Al_2O_3$	(synthesis)
$NaCl + AgNO_3 \rightarrow NaNO_3 + AgCl$	(double)
$CaCO_3 \rightarrow CaO + CO_2$	(decomposition)

FOLLOW-UP

Reinforcement: Chemical versus Physical Change

Have students make posters in which chemical and physical changes are illustrated. The poster should include two sections, chemical change and physical change, with pictures cut from magazines or hand drawn.

Enrichment: Blueprints

Blueprinting leaves demonstrates chemical change. Give students blueprint paper and a piece of glass whose edges are bound with masking tape. Have the students cut a piece of cardboard whose size matches that of the glass. Then have them tape the glass and cardboard together along one edge. The blueprint paper should then be inserted between the glass and cardboard so that the sensitive side of the paper is toward the glass. The leaves to be printed should be spread on top of this light-sensitive side and covered with the glass. Place in the sun until the paper darkens and turns blue. When this occurs, the paper should be removed. Wash in cold water so that the print can be seen.

Tie-in: Types of Chemists

Have the students select one type of chemist mentioned in the career item. Then ask the students to find out exactly what that type of chemist does and create a career poster. Use the posters for a careers-in-chemistry bulletin board.

10.2

OBJECTIVES: The student should be able to

- **explain** how the law of conservation of mass applies to a chemical reaction;
- **balance** a chemical equation.

SCIENCE TERMS

law of conservation of mass
coefficient

ADVANCED PLANNING

Prepare several work sheets of equations that the students can balance. Use a chemistry text for reference.

LESSON PLAN

1. Write the following equation on the board:

$$NaCl + AgNO_3 \rightarrow NaNO_3 + AgCl$$

Point out that in the reaction there are the same number of atoms of each element on both sides of the equation. Explain that the equation illustrates the law of conservation of mass, that is, the number of atoms of each element does not change in a chemical reaction. If the students need review in how to read a formula, refer them to the rules in the activity on page 166.

10.2 Balancing Chemical Equations

You will find out
- how to balance chemical equations.

In a chemical reaction, no atoms ever just disappear and no new atoms are formed. In other words, the number of atoms of each element does not change. This fact is called the **law of conservation of mass.** Mass is the amount of matter present in a substance. You can keep track of mass in chemical equations by counting the numbers of atoms on both sides of the equation.

Figure 10.7 The law of conservation of mass applies to all chemical reactions.

$H_2O \rightarrow H_2 + O_2$ $2H_2O \rightarrow 2H_2 + O_2$

Figure 10.8 In a balanced equation, the same number of atoms appears on both sides.

Because chemical reactions follow the law of conservation of mass, a chemical equation must show the same number of atoms of each element on both sides. In other words, the equation must be balanced. Consider, for example, the equation for the decomposition reaction in which water breaks up into hydrogen and oxygen molecules.

$$\mathbf{H_2O \rightarrow H_2 + O_2} \qquad \text{(not balanced)}$$

The equation is not balanced because there is only one oxygen atom on the reactant side and there are two on the product side. The balanced form of this equation is this:

$$\mathbf{2H_2O \rightarrow 2H_2 + O_2} \qquad \text{(balanced)}$$

The number 2 that you see in front of H_2O and H_2 is called a **coefficient** [koh-uh-FIHSH-uhnt]. A coefficient is a number placed in front of a formula or symbol that tells you how many atoms or molecules of that substance are present. If no coefficient appears in front of a symbol or formula, you know there is only one atom or molecule of that substance. To balance a chemical equation, you can change the coefficients but never the formulas or symbols.

2. Explain that if a coefficient is placed in front of a formula, the coefficient indicates the number of molecules that are present. For example, $6H_2SO_4$ means there are 6 molecules of H_2SO_4.

3. Explain that in order to determine the number of atoms of each element in a formula with a coefficient, you must multiply the coefficient by the number of atoms of each element. In $6H_2SO_4$, for example, there are 6 × 2 H atoms, or 12 H atoms; 6 × 1 (understood) S atoms, or 6 S atoms; and 6 × 4 O atoms, or 24 O atoms.

Mathematics Skill: Computation

4. Write this equation on the board:

$$2HCl + Zn \rightarrow ZnCl_2 + H_2$$

Tell the students that since the zinc and hydrochloric acid react when they are combined, they are called the reactants. Since zinc chloride and hydrogen gas are released in this reaction, they are the products.

5. Remind the students that according to the law of conservation of mass the number of each type of atom on the reactant side must be equal to the number of each type of atom on the product side. Write the following equation on the board:

$$Zn + HCl \rightarrow ZnCl_2 + H_2$$

As you ask each of the following questions, record the answer under the proper side of the equation. Ask how many Zn atoms are on the reactant side (1). How many H atoms (1)? How many Cl atoms (1)? Now move to the product side. How many Zn atoms are present (1)? Cl atoms (2)? H atoms (2)? Are there the same numbers of atoms on both the product and reactant sides of the equation (no)? Where is the difference (in the numbers of chlorine and hydrogen atoms)? Tell the students that to balance an equation the only mathematical operation they are allowed to perform is multiplication. Place the coefficient 2 in front of HCl. Point out that the addition of the coefficient affects the whole molecule. Ask how many H and how many Cl atoms are on the reactant side now (2 of each). Is the equation balanced now (yes)?

6. To conclude the lesson, assign homework examples in which the students must balance equations. Remind the students that the coefficients being inserted are usually small numbers.

EXAMPLE Balance the chemical equation for the single displacement reaction in which fluorine replaces the chlorine in sodium chloride to produce sodium fluoride.

$$\mathbf{F_2 + NaCl \rightarrow NaF + Cl_2} \qquad \text{(not balanced)}$$

1. Check the number of atoms of each element on both sides of the equation.

Reactants: 2 F atoms, 1 Na atom, 1 Cl atom
Products: 1 F atom, 1 Na atom, 2 Cl atoms

The equation is not balanced. The number of atoms of each element is not the same on both sides of the equation.

2. Balance the equation by changing the coefficients. First put 2 NaF molecules on the right. This balances F but puts Na out of balance. Balance Na by putting 2 NaCl molecules on the left. This balances both Na and Cl, and so you now have a balanced equation.

$$\mathbf{F_2 + 2NaCl \rightarrow 2NaF + Cl_2} \qquad \text{(balanced)}$$

3. Recheck the number of atoms of each element on both sides of the equation to be sure you have balanced it correctly.

Reactants: 2 F atoms, 2 Na atoms, 2 Cl atoms
Products: 2 F atoms, 2 Na atoms, 2 Cl atoms

In changing coefficients to balance equations, you should always use the smallest coefficient possible. As in this example, a change in a coefficient sometimes puts one element that was in balance out of balance. To account for this possibility, you should recheck the number of atoms of each element when you have finished balancing the equation.

ACTIVITY

Is Mass Conserved in a Chemical Reaction? ●

Materials

lead nitrate solution
sodium iodide solution
2 test tubes
test-tube rack
balance
safety glasses
apron

1. chemical
2. A color change occurred.
3. The mass of the test tube was included in the first measurement.
4. The mass did not change.

Procedure

1. Copy the chart and use it to record your observations.
2. Put on your safety glasses and apron.
3. Fill one test tube 1/3 full with lead nitrate solution.
4. Fill the other test tube 1/3 full with sodium iodide solution.
5. Put both test tubes in the test-tube rack.
6. Measure the total mass of the rack, both test tubes, and solutions.
7. Pour the lead nitrate into the sodium iodide. Be careful to avoid spilling.
8. Put the empty test tube back in the rack.
9. Measure the mass of the rack, of the test tubes, and of the solution.
10. Find the difference in the mass before and after mixing.
11. Share your results with those of friends and add those results to your chart.

Questions

1. Do you think the change you observed was a chemical or a physical change?
2. How do you know a new substance was formed?
3. Why was it necessary to put the empty test tube back in the rack?
4. How does this experiment illustrate the law of conservation of mass?

	Mass Before Mixing	Mass After Mixing	Difference In Mass
Results			
Results of Friend			

Study Questions for 10.2

1. What does the law of conservation of mass tell you about chemical reactions?
2. Balance the following equation:

$$KClO_3 \rightarrow KCl + O_2$$

1. The mass of the reactants equals the mass of the products.
2. $2KClO_3 \rightarrow 2KCl + 3O_2$

ACTIVITY: Is Mass Conserved in a Chemical Reaction?

Prepare the sodium iodide by dissolving 15 g of sodium iodide in 100 mL of water. Prepare the lead nitrate by dissolving 15 g of lead nitrate in 100 mL of water.

Before doing the experiment in class, mix a few milliliters of the solutions to be sure the precipitate forms. *Note:* The precipitate may be rinsed down the sink with running water.

Small cups or plastic containers that will fit on the pan of your balance can be substituted for test tubes.

● **Laboratory Skills: Manipulation of Materials Experimentation/Demonstration**

FOLLOW-UP

Reinforcement: More Practice with Equations

Create a learning station with work sheets of problems in balancing equations. Place the answers in a folder so that the students can have easy access to them. The posters that advertise the learning station should display several examples of the procedure by which equations are balanced.

Enrichment: Law of Conservation of Mass

Have some students look up and report on Lavoisier's experiments demonstrating the law of conservation of mass. A good reference is *A Short History of Chemistry* by Isaac Asimov.

Tie-in: Preventing Rust

A chemical reaction such as car rust can be detrimental. Ask the students to determine what the equation for car rust would be. They can think this problem through by answering the following questions. What metal are cars made from? What is rust? What compound is produced when rusting occurs?

10.3

OBJECTIVES: The student should be able to

- **explain** how molecules are broken apart;
- **explain** how temperature affects the rate of chemical reactions;
- **explain** how the concentration of reactants affects the rate of chemical reactions;
- **explain** how a catalyst affects the rate of chemical reactions;
- **describe** the role of enzymes in chemical reactions.

SCIENCE TERMS

catalysts
enzymes

ADVANCED PLANNING

For the demonstrations you will need fresh hydrogen peroxide (3%), test tubes, safety glasses, test-tube holders, test-tube racks, wooden splints, beakers, matches, ice, Bunsen burners, raw liver or other raw meat, and an ice-water bath.

LESSON PLAN

1. Open the lesson by asking the students why it is important to control the rate at which a reaction takes place (by controlling the rate of burning, for example, fuel can be made to last longer; by controlling the rate at which metals rust, metal products will last longer).
2. Explain that when you control the rate of a reaction you are controlling the speed at which the molecules of the reactants break apart and rejoin.

10.3 The Rate of Chemical Reactions

You will find out

- how the temperature affects the rate of chemical reactions;
- how the concentration of reactants affects the rate of chemical reactions;
- what is meant by the term *catalyst*.

It is important to know what the products will be in a chemical reaction. It is also important to know the rate of reaction. The rate of reaction is how fast the reaction occurs. For example, the rusting of iron releases heat energy. However, the reaction goes on so slowly at room temperature that no one would think of using a pile of rusting nails to keep warm. Other reactions proceed very quickly. Think of the explosion of dynamite. Knowing how fast a reaction will occur can be a life-or-death matter.

Figure 10.9 Some reactions proceed very quickly.

The rate at which a reaction occurs depends on many different things. Two very important considerations are the temperature and the concentration of the reactants. To understand why temperature and concentration are important, you must first understand what happens to the reactants in a chemical reaction.

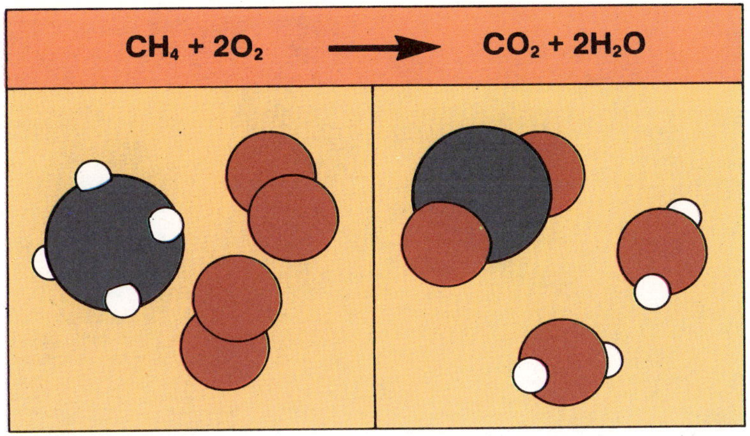

Figure 10.10 In the burning of methane, carbon dioxide and water are formed.

In chemical reactions, the reactants form new substances, the products. Reactant atoms and molecules must come in close contact with each other for this change to occur. When the reactants are molecules, the chemical bonds in those molecules must be broken so that the atoms involved can form new substances.

How are the bonds in reactant molecules broken? They are broken by collisions with other reactant molecules. If the reactant molecules are moving slowly, they will probably just bump into each other gently without breaking any bonds. Fast-moving molecules will collide with more force. These forceful collisions are more likely to break apart the molecules. So the faster the molecules are moving, the more likely it is that a collision will be strong enough to break one or both of them apart. Also, when the reactant molecules move faster, they collide more often. The rate of a reaction, then, depends on the force and frequency of collisions between reactants.

It is a general rule that reactions go faster when the temperature is raised. At higher temperatures the reactant molecules move faster. Thus, their chemical bonds are broken more quickly and easily than at lower temperatures. For example, most fires are started by raising the temperature of the material to be burned. You use a burning match to light a candle. The match raises the temperature of the reactants, the candle and the oxygen in the air around it. ●

Figure 10.11 Low temperatures slow down the chemical reactions that spoil food.

3. Tell the students that increasing the temperature is one way to increase the speed of the molecules, thereby speeding up a reaction. Demonstrate by putting hydrogen peroxide into three test tubes. Place one test tube in an ice-water bath, one in a test-tube rack, and one over a lighted Bunsen burner. What can be observed about the rates of the reactions (the heated test tube releases the most bubbles; the one in ice water has the least bubbles)? What differences exist among the three test tubes (temperatures)?

● **Comprehension Skill: Applied**

SCIENCE BACKGROUND

Molecules break apart when they collide with other molecules. Fast-moving molecules will collide with other molecules more often. Molecules that are moving fast are also more likely to break the bonds of molecules they collide with. Thus, if you can increase the speed of the molecules, you can increase the rate of molecular reaction.

4. Fill one test tube with hydrogen peroxide. In another test tube, pour 5 mL of hydrogen peroxide and add water to fill it to the same level as the first test tube. Ask the students which test tube has more hydrogen peroxide in it (the one without water). Tell the students that this test tube contains the more concentrated solution. Add a small piece of raw beef to each test tube. In which test tube is the reaction more violent (one without water)? Ask the students to explain why (there are more hydrogen peroxide molecules; the greater number of molecules results in more collisions).

SCIENCE BACKGROUND

A large surface area will increase the rate of reaction because there are more molecules available to be the targets of collisions. Stirring and shaking increase the rate of reaction, since they increase the total surface area.

Figure 10.12 The same principle that applies to the concentration of reactants in a chemical reaction applies to cars on a highway. The greater the concentration of cars, the greater the chance of collisions.

Figure 10.13 Coal miners must be careful because coal dust has a large surface area and will ignite very easily.

When many reactant atoms and molecules are crowded into a given space, the concentration of reactants is high. In general, the higher the concentration of reactants, the faster the reaction will proceed. It will proceed faster because contact and collisions among reactants are increased.

If one or more of the reactants is a solid, contact between reactants is increased when the surface area of the solid is increased. Reactant contact and collisions are increased and the reaction proceeds more quickly. Imagine you are making a fire in a fireplace. Would it be easier to start one large log burning or to start a pile of small kindling? The surface area is increased when you use kindling so the fire starts and burns more quickly.

Another way to increase the rate of reaction is to use **catalysts** [KAT-uh-lihstz]. A catalyst is a substance that affects the rate of a chemical reaction without itself being changed or used up in the reaction. Catalysts are like go-betweens. A catalyst can speed up a reaction or cause one to occur by increasing the contact between reactants. Some catalysts are widely used in the chemical industry to speed up reactions.

Catalysts that promote chemical reactions in living things are called **enzymes** [EHN-zymz]. The enzymes in your body are made of proteins. Enzymes make it possible for the complicated chemical reactions necessary for life to occur at body temperature. Thousands of enzymes are known. Each serves a specific purpose by promoting a specific reaction. Some enzymes help your body digest food. Others help in getting energy for your body. Research into the nature of enzymes can lead to a better understanding of the causes and cures of some diseases. This knowledge should lead to a better chance for good health for everyone.

Do You Know?
A recently synthesized form of sugar has a differently shaped molecule that your enzymes can't digest. It tastes sweet but releases no calories.

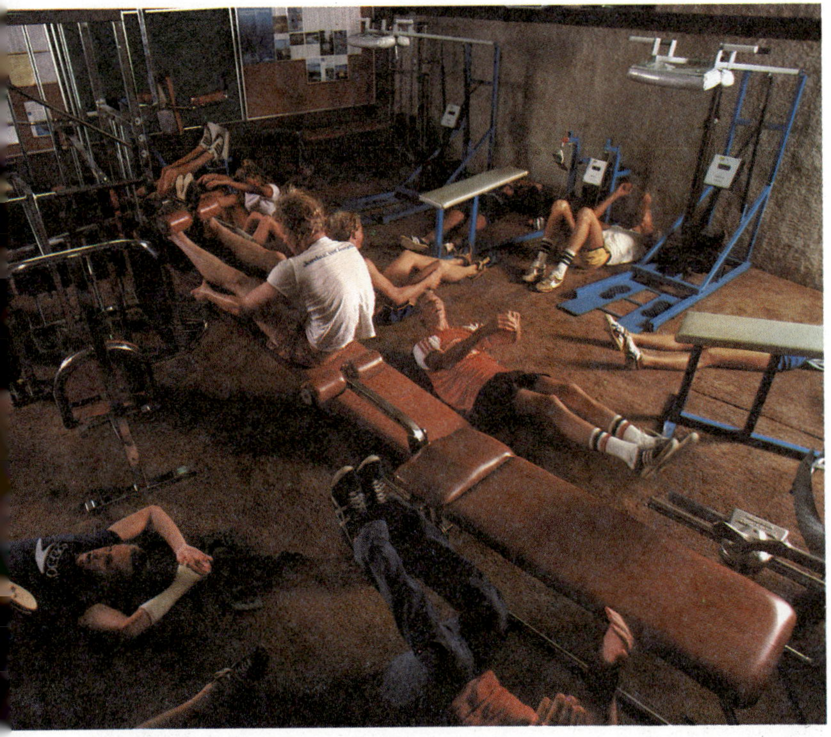

Figure 10.14 The functions of thousands of enzymes contribute to a healthy body.

Study Questions for 10.3

1. Name two factors that affect the rate of a chemical reaction.
2. What is a catalyst?
3. What is an enzyme?

1. temperature, concentration
2. substance that affects a reaction without being changed or used up
3. a catalyst in living things

5. Fill two test tubes half full of hydrogen peroxide. Have the students observe each tube (bubbles are slowly forming). Add a small piece of raw meat to one. What happens (faster bubbling)? Ask what is causing the rapid increase in bubbling (the meat). Explain that something in the meat is acting as a catalyst, thereby increasing the rate of the reaction.
6. Explain that living organisms contain catalysts called enzymes that help all body reactions to occur. Stress that the enzymes cause the reactions to occur at the proper temperature and rate for the body.
7. Conclude the lesson by reviewing the concept that chemical reactions occur when the bonds between molecules are broken and new bonds are formed. Review three methods of increasing the rate of the reaction (heat, increase in concentration, and catalytic action).

FOLLOW-UP

Reinforcement: Increasing the Rate of Chemical Reactions
Ask the students to create a mobile that shows at least three ways to increase the rate of a chemical reaction. Students should be prepared to give simple explanations and examples of each method.

Enrichment: The Haber Process
Ask students to do research on the Haber process, in which hydrogen and nitrogen combine to produce ammonia. How can the rate of this reaction be increased? Ask the students to use their findings to write articles for a new science encyclopedia in which the process is discussed.

Tie-in: Catalysts at Work
Ask the librarian to teach your students how to use the *Readers' Guide to Periodical Literature*. Have the students find articles that deal with ways in which catalysts are being used to help people.

10.4

OBJECTIVES: The student should be able to

- **define** the term *activation energy;*
- **describe** what energy change occurs during an exothermic reaction;
- **describe** what energy change occurs during an endothermic reaction.

SCIENCE TERMS

activation energy
exothermic reaction
endothermic reactions

ADVANCED PLANNING

For the demonstrations you will need matches, one test tube for each student, water, and ammonium chloride.

LESSON PLAN

1. To illustrate the rearrangement of atoms during a chemical reaction, write the following equation on the board:

$$2HCl + Zn \rightarrow ZnCl_2 + H_2$$

Point out that on the reactant side, hydrogen and chlorine are in chemical combination; on the product side, zinc and chlorine are chemically combined. In order for this rearrangement of atoms to occur the bonds between the reactant molecules of HCl must be broken. The energy used to break the bonds and start a reaction is called activation energy.

2. Light a match. Ask the students what they would feel if their hands were near the match (heat). Explain that reactions that give off heat are called exothermic reactions. *Exo* means "out of," and *thermic* refers to heat. Ask the students to name other exothermic reactions (dynamite explosion, burning oil at home, and burning gasoline in a car).

10.4 Energy and Chemical Reactions

You will find out

- what is meant by activation energy;
- how exothermic and endothermic reactions are different.

For many reactions to get started, energy must be added. This energy is necessary to begin breaking the bonds in the reactant molecules. The energy that must be supplied to start a chemical reaction is called **activation energy.** For example, to start a charcoal fire for a barbecue, you can supply the activation energy with a lighted match.

An important result of some chemical reactions is the release of heat energy. The heat energy comes from the formation of bonds in product molecules. In some reactions such as the burning of oil, the energy released in the formation of the product molecules is greater than the activation energy required. Such reactions produce a net release of energy.

A chemical reaction that releases energy is called an **exothermic** [ehk-soh-THUHR-mihk] **reaction.** All exothermic reactions do not produce large, noticeable amounts of energy in a short time as the burning of oil does. The rusting of iron, for example, is also an exothermic reaction. A rusting iron railing does not feel hot or even warm when you touch it. In this case, the energy is being released very slowly.

Do You Know?
Your body burns carbohydrates for fuel so why can't cars burn carbohydrates for fuel? Scientists are now working on methods to use beer and soybean oil as sources of fuel for cars.

Figure 10.15 Burning is an exothermic reaction.

In some chemical reactions, energy is absorbed instead of released. These reactions are called **endothermic** [ehn-doh-THUHR-mihk] **reactions.** The baking of a cake is an endothermic reaction. Heat energy is absorbed. The breakdown of water into hydrogen and oxygen is also an endothermic reaction. Electric energy is absorbed in this reaction.

While it is true that energy can be released or absorbed in a chemical reaction, energy is not being created or destroyed. It is just being changed from one form to another. The energy stored in the chemical bonds of reactant molecules may be released when the molecules break up and form product molecules. When energy is absorbed in a reaction, that energy is stored in the chemical bonds of the product molecules. Neither mass (matter) nor energy is created or destroyed in a chemical reaction.

Figure 10.16 Baking is an endothermic reaction.

Study Questions for 10.4

1. What is activation energy?
2. Describe and give an example of an exothermic reaction.
3. Describe and give an example of an endothermic reaction.

1. energy that must be supplied to start a chemical reaction
2. a reaction in which energy is released; example—burning wood
3. a reaction in which energy is absorbed; example—baking a cake

3. Explain that some chemical reactions absorb energy. These are called endothermic reactions. To occur, they must take in energy from the environment. Thus the environment becomes cooler.

FOLLOW-UP

Reinforcement: How Do Exothermic and Endothermic Reactions Differ?

Ask the students to pretend that they have been selected by the chief chemist in a laboratory to give a paper at an upcoming scientific conference. The topic of the paper is "How Do Exothermic and Endothermic Reactions Differ?" They should write a talk in which they explain the difference between the reactions and give examples of each.

Enrichment: Energy from Oxidation

The body maintains a relatively constant temperature because of the exothermic reactions that occur in it. Ask the students to make a chart that lists and explains some exothermic reactions that occur in the body.

Tie-in: A Calorimeter

Have the students construct a calorimeter. Tell the students that heat energy is measured in calories. The number of calories is found by multiplying the number of Celsius degrees by which the temperature of the water rises by number of grams of water present. The calorimeter should be saved for use in measuring the heat produced when the students use a base to make a salt in Lesson 10.5.

SCIENCE BACKGROUND

Photosynthesis is one example of an endothermic reaction. Energy from the sun must be taken in to allow photosynthesis to occur.

10.5

OBJECTIVES: The student should be able to

- **describe** what happens when neutralization occurs;
- **explain** how to determine the strength of an acid or a base;
- **explain** how an indicator is used;
- **name** two common indicators.

SCIENCE TERMS

acids	salt
base	pH scale
neutralization reaction	indicator

ADVANCED PLANNING

For demonstrations you will need safety glasses, test tube and rack, one-hole stopper with a rubber hose through it to fit a test tube, beaker, pan, water, zinc, hydrochloric acid, matches, acid such as vinegar, base such as ammonia, red litmus paper, blue litmus paper, and phenolphthalein.

LESSON PLAN

1. Introduce the lesson by writing the following formulas on the board:

   HCl (hydrochloric acid)
   HNO_3 (nitric acid)
   H_2SO_4 (sulfuric acid)

   Ask the students what they notice about each formula (each begins with H). As you circle the *H* in each formula, point out that the *H* is a characteristic of the formulas of acids. Name each of the acids. Tell the students that when an acid is in solution it releases H^+, or hydrogen ions.

2. Ask the students to name some characteristics of acids (contain H^+, contain a nonmetal or polyatomic ion, and have a sour taste). Stress to the students that they should never touch or taste an acid or any other substance whose identity is not known. If an acid should come in contact with the skin, it should be washed off immediately. Point out that when acids react with many metals the acids release hydrogen gas, which is explosive.

10.5 Acids, Bases, and Salts

You will find out
- what acids and bases are;
- what happens when an acid and a base react;
- how the strength of acids and bases is measured.

An interesting and important kind of chemical reaction involves two familiar and useful groups of compounds. Some of these compounds can be found in your home. To find them at home, you would look in the refrigerator, the kitchen cabinets, the cleaning closet, and the medicine cabinet. What are these groups of compounds? They are **acids** and **bases.** An acid is a substance that produces positive hydrogen ions, H^+, in liquid solution. A base is a substance that produces a negative ion in liquid solution, usually the OH^- ion. An ion is an atom or a group of atoms that has different numbers of positive protons and negative electrons. Acids and bases can be weak or strong depending on the concentration of H^+ and OH^- ions. The higher the concentration of H^+ and OH^- ions, the stronger the acid or base.

Almost all acids are combinations of hydrogen and one or more nonmetals. Acids react strongly with metals and are widely used in industry to process metals. Some acids are dangerous because they are poisonous and damage human skin.

Do You Know?
The acids in your stomach are strong enough to dissolve zinc and blister the skin.

Figure 10.17 Hydrangeas growing in basic soil produce pink flowers. The same plants produce blue flowers when growing in acidic soil.

Figure 10.18 Some familiar acids

You are personally familiar with many acids. You even eat and drink some of them! Obviously, the acids you eat and drink are not poisonous and do not damage human skin. Many foods contain some form of relatively weak acid. In general, the more acidic a substance, the more sour it tastes. For example, which is more sour-tasting: limes, tomatoes, or milk? Limes are very acidic, tomatoes less, and milk less still. Other foods and drinks that are acidic are oranges, lemons, vinegar, pickles, carbonated drinks, and tea.

Almost all bases are made of a metal, of hydrogen, and of oxygen. The primary exception is ammonia, NH_3. Substances that have a high concentration of a base in them have a bitter taste. Bases are slippery to touch. However, because many bases can damage human skin, you should always be careful when handling them. Relatively weak bases are widely used in cleaners and soaps. Weak bases are also used in medicines for upset stomachs. Stronger bases are used in products that unclog household drains.

Figure 10.19 Some familiar bases

3. On the board write the formulas NaOH, HN_4OH, and $Ca(OH)_2$. Ask the students what these formulas have in common (OH). Tell the students that most bases contain the OH ion. Underline *Na* and Ca. Ask the students if these elements are metals or nonmetals (metals). Point out that bases consist of a metal or a polyatomic ion joined to the OH^- ion. Ask the students to name some characteristics of bases (contain OH^-, have a bitter taste, and feel slippery). Stress with the students that they should never touch or taste a base. Bases can burn the skin. If a base happens to touch the skin, the base should be washed off immediately. Point out that bases are good cleaning agents because they react with grease and oil. Bases such as milk of magnesia can also be used as antacids, which settle the stomach.

SCIENCE BACKGROUND

An acid contains one or more hydrogen ions joined to a nonmetal or a polyatomic ion. A polyatomic ion consists of two or more atoms that act as one atom.

4. Explain that an acid and a base combine in a process called neutralization. Summarize this process on the board.

acid + base = salt + water

EXAMPLE HCl + NaOH = NaCl + HOH

hydrochloric acid + sodium hydroxide = sodium chloride + water

Point out that the hydrogen ion of the acid and the OH ion of the base have combined to form water. The chlorine from the acid and the sodium from the base have combined to form a salt.

When an acid and a base combine, they react in a **neutralization** [nu-truh-luh-ZAY-shuhn] **reaction.** The positive H^+ ion from the acid is neutralized by the negative OH^- ion from the base. An example of a neutralization reaction is the reaction of sodium hydroxide, NaOH, and hydrochloric acid, HCl.

$$\underset{\text{acid}}{\mathbf{HCl}} + \underset{\text{base}}{\mathbf{NaOH}} \rightarrow \underset{\text{salt}}{\mathbf{NaCl}} + \underset{\text{water}}{\mathbf{H_2O}}$$

The products of a neutralization reaction are always water and a salt. A **salt** is the compound produced in a neutralization reaction when the hydrogen in the acid is replaced with another element from the base. Usually a salt is a compound made of a metal and a nonmetal. You may have thought that there was only one kind of salt, table salt (sodium chloride). Although sodium chloride is the most common and familiar salt, it is not the only salt. Other familiar and useful salts are sodium bicarbonate, or baking soda, $NaHCO_3$ and silver nitrate, $AgNO_3$, which is used in photography.

Figure 10.20 Silver nitrate, a salt, is used in photography.

The strength of an acid or a base in a liquid solution is measured on the **pH scale.** The pH scale is a series of whole numbers from 0 to 14. The lower the pH number, the greater the concentration of H^+ ions. A neutral substance, such as pure water, has a pH of 7. A neutral substance is considered neither an acid nor a base. Any pH number over 7 indicates a base. The higher the number, the stronger the base. Similarly, any pH value under 7 indicates an acid. The lower the pH number, the stronger the acid.

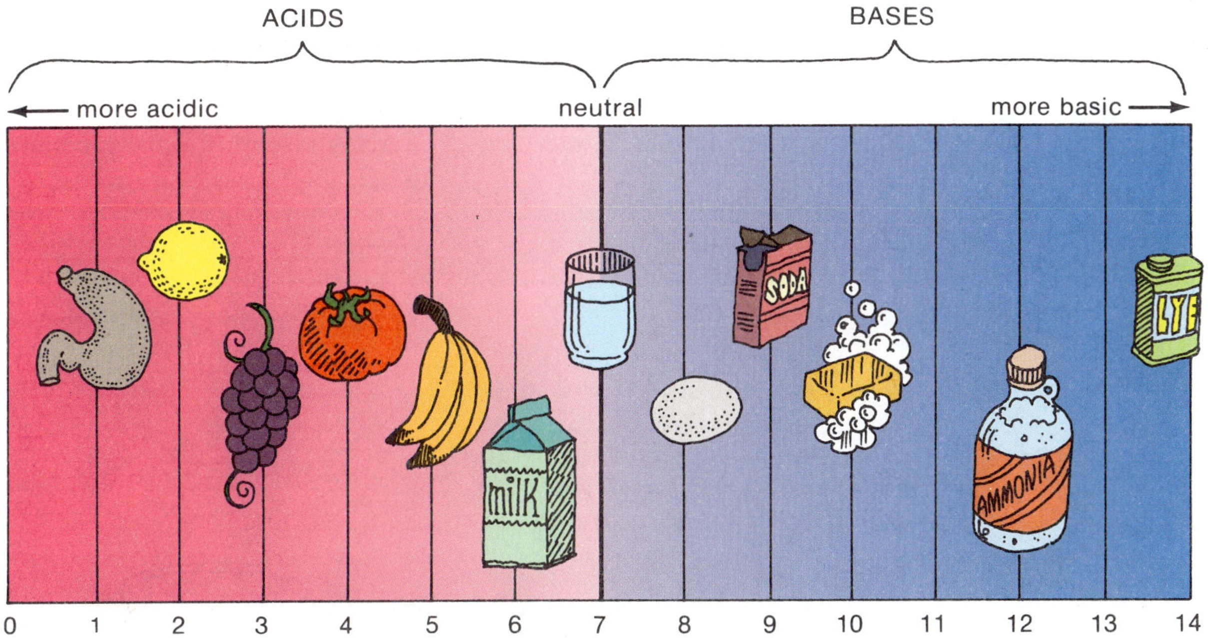

Figure 10.21 The pH scale

One way to tell whether a solution is an acid or a base is to use an **indicator.** An indicator is a special type of organic compound that changes color as the pH of a solution changes. A commonly used indicator is litmus paper. Litmus paper comes in both red and blue. Red litmus paper stays red in an acid or in a neutral solution but turns blue in a base. Blue litmus paper stays blue in a base or in a neutral solution but turns red in an acid. As you can see, litmus paper only tells you if a solution is acidic, basic, or neutral. It does not indicate a specific pH value.

5. Place a beaker of dilute HCl (2%) and a beaker of dilute NaOH (2%) on the desk. Clearly label each beaker so that the students can see what each contains. Stress that since you cannot taste or touch these substances to see whether they are acids or bases you must find another way to identify them. One way to identify them is to use an indicator. Dip a piece of red litmus paper into each solution. What do the students observe (in the base the litmus paper turns blue)? Now dip a piece of blue litmus into each solution. What do the students observe (in an acid the blue litmus turns red)? Now add a few drops of phenolphthalein to a sample of each. Note that the phenolphthalein turns bright pink in a base.

ACTIVITY: How Can You Form a Salt?

You should plan to complete this activity in two days, since the liquid referred to in step 8 will take overnight to evaporate. Caution the students that acids and bases can be harmful. If the students spill either on their skin, they should flood the area with water and immediately tell you. Give the students a dilute solution (1%) of HCl and a dilute solution (1%) of NaOH. Caution the students not to taste the salt they make.

Two beakers are needed for each student station. One beaker will be used for the sodium hydroxide solution, and the other beaker will be used for water in which to clean the eyedropper after each use.

Laboratory Skill: Manipulation of Materials

ACTIVITY

How Can You Form a Salt?

Materials

dilute hydrochloric acid
dilute sodium hydroxide
phenolphthalein
tin pie plate
graduated cylinder
2 beakers
eyedropper
magnifying glass
apron
safety glasses

Procedure

1. Put on your safety glasses and apron.

2. CAUTION: Do not let any of the liquids touch your skin. If they do, run cold water over the area immediately and tell your teacher. Immediately wipe up any spills. Never put your eyedropper down on the desk; always put it in a beaker of water. Rinse the eyedropper thoroughly in the water after each use.

3. Pour 5 mL of sodium hydroxide into a beaker.

4. Add a drop of phenolphthalein.

5. Add hydrochloric acid drop by drop until the liquid in the beaker turns clear.

6. After the liquid clears, add one drop of sodium hydroxide and the liquid should turn pink again.

7. Add one drop of hydrochloric acid to reclear the liquid.

8. Pour the liquid into the pie plate and allow the liquid to evaporate.

9. After the liquid has evaporated, use a magnifying glass to compare the crystals that remain with those of table salt. CAUTION: Do not taste any salt.

Questions

1. What happens when phenolphthalein is added to sodium hydroxide?

2. What forms in the pie plate when the liquid evaporates?

3. Where did the salt come from?

4. What is the role of the phenolphthalein in the experiment?

1. turns pink
2. salt
3. It was made during the reaction. When the water evaporated it was left behind.
4. The phenolphthalein was an indicator. It told you when you had added enough acid to neutralize the base.

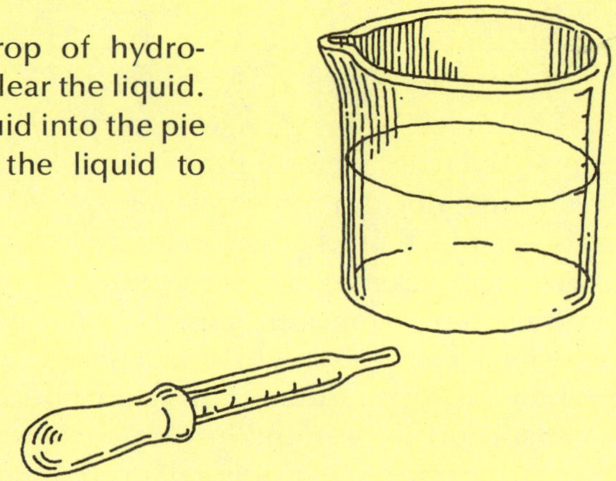

Figure 10.22 Limes contain citric acid. What would happen if you dipped a strip of red litmus paper and a strip of blue litmus paper into some lime juice?

Red would stay red; blue would turn red.

Another indicator is a liquid solution of phenolphthalein [fee-nahl-THAY-leen]. Phenolphthalein is colorless in an acidic solution. However, phenolphthalein turns bright pink in a basic solution. Like litmus paper, phenolphthalein does not indicate pH values, only whether a solution is acidic or basic.

Other indicators go through a range of colors as the pH of a solution changes. By comparing the color of the indicator with a special color chart, you can tell what the pH value of the solution is.

Figure 10.23 Phenolphthalein has just been added to this base. After a few moments the entire liquid will be pink.

Study Questions for 10.5

1. Give a definition and an example of an acid.
2. Give a definition and an example of a base.
3. What happens in a neutralization reaction?
4. What does the pH scale measure?
5. What does an indicator tell you?

1. produces positive hydrogen ions; example—vinegar, limes
2. produces negative ions, usually OH^-; example—ammonia, soap
3. An acid and a base combine to produce a salt and water.
4. strengths of acids and bases
5. whether a substance is acidic, basic, or neutral

6. Review with the students how each of the indicators acts in an acid and in a base (red litmus = blue in a base; blue litmus = red in an acid; phenolphthalein = pink in a base). Then summarize the pH scale (pH = 1–6 for acids, 7 for neutral, and 8–14 for bases).
7. Conclude the lesson by asking the students to explain or give examples of each property of the acids and of the bases that you discussed.

FOLLOW-UP

Reinforcement: Acid-Base Comparison
Ask the students to create a chart comparing the properties of acids and bases. Select several of the best charts to post in your room for use as reference in distinguishing between the two types of compounds.

Enrichment: Antacids
Ask the students to design and perform experiments to determine which antacid can absorb the greatest amount of acid before it is neutralized.

Tie-in: Acid Rain
Ask the students to pretend that they are newspaper reporters trying to find out what causes acid rain, why and how it is harmful, and what steps can be taken to prevent its harmful effects. The facts they discover should be written in the form of a newspaper article.

What Did You Learn?

10.1 • Physical changes affect the physical size, shape, or state of a substance.
10.1 • Chemical changes affect the chemical properties of a substance.
10.1 • The four types of chemical reactions are synthesis, decomposition, single displacement, and double displacement.
10.2 • The law of conservation of mass states that no atoms disappear and no new atoms are formed in a chemical reaction.
10.2 • The general form for a chemical equation is the following: reactants → products.
10.2 • To balance an equation, you can change the coefficients but not the formulas or symbols.
10.3 • The rate of reaction increases as the temperature and concentration of the reactants increase.
10.3 • A catalyst is a substance that affects the rate of reaction without being changed or used up in the reaction.
10.3 • Enzymes are catalysts in living things.
10.4 • Activation energy is the energy needed to start a chemical reaction.
10.4 • In exothermic reactions heat energy is released.
10.4 • In endothermic reactions heat energy is absorbed.
10.5 • An acid produces H^+ ions in liquid solution.
10.5 • A base produces negative ions, usually the OH^- ion, in liquid solution.
10.5 • In neutralization reaction, an acid and a base combine to produce water and a salt.
10.5 • The pH scale measures the strength of acids and bases.
10.5 • Indicators tell whether a solution is acidic or basic.

Key Terms

physical changes
chemical changes
chemical reaction
chemical equation
reactants
products
law of conservation of mass
coefficient
catalysts
enzymes
activation energy
exothermic reaction
endothermic reactions
acids
bases
neutralization reaction
salt
pH scale
indicator

Career

Chemist

Chemists search for new knowledge about substances and put this knowledge to practical use. About one half of today's chemists are engaged in research and development. Most of these chemists work to improve and create new products. Wonder drugs, synthetic materials such as nylon and polyester fabrics, and fuels for space travel are only a few examples of the many useful products that chemists have helped to develop.

Research chemists also work on research projects that extend scientific knowledge. Basic research often has practical uses. For example, synthetic rubber and plastics have resulted from research on polymers. Making polymers involves putting small molecules together to form larger ones.

Many other chemists work in analyzing and testing materials. In industry, tests must be made at practically every stage in the manufacture of a product. Industrial chemists also work in inspection and sales.

A chemist should be able to talk to and get along with others. This skill is important because chemists often work as teams on problems that require close cooperation. They must also be able to communicate clearly the results of their research. A chemist must have patience and the ability to stick to a job because many delays and problems come up during experiments.

Some of the specialties within chemistry include agricultural chemistry, paint chemistry, nutritional chemistry, petroleum chemistry, physical chemistry, and biochemistry.

Vocabulary

1. c
2. h
3. i
4. e
5. b
6. j
7. g
8. k
9. l
10. f

● Word Skill: Vocabulary

What Do You Remember?

1. A chemical reaction is a process that causes a chemical change to occur. Examples will vary. They may include the burning of wood or coal and the reaction of an acid and a base.
2. In physical changes, only the size, shape, or state of substances changes. Chemical properties remain the same. In chemical changes, chemical properties are changed.
3. $2Na + 2H_2 \rightarrow 2NaOH + H_2$
   $Mg + 2HCl \rightarrow H_2 + MgCl_2$
4. The same number of atoms of each element appear on both sides of a balanced equation.
5. synthesis—elements or compounds join to form a product: $2H_2 + O_2 \rightarrow 2H_2O$; decomposition—reactant material is broken down: $2HgO \rightarrow 2Hg + O_2$
6. energy that must be supplied to start a chemical reaction
7. exothermic—heat energy is released; endothermic—heat energy is absorbed
8. 1. Use a catalyst.
   2. Increase temperature of reactants.
   3. Increase concentration of reactants.
9. a substance that affects chemical reactions without being changed or used up in the reaction
10. the reaction that results when an acid and a base combine

TO THINK ABOUT AND DO

Vocabulary ●

Number a separate piece of paper from 1 through 10. Find the term in the second column that best describes the phrase in the first column. Write the letter for that term beside the appropriate number.

1. reaction that releases energy
2. substance to distinguish between acids and bases
3. change that affects the chemical properties of a substance
4. a substance that produces H^+ ions in solution
5. energy needed to start a reaction
6. reaction that absorbs energy
7. product of a neutralization reaction
8. number used to indicate how many molecules of a particular compound are present in a chemical equation
9. substance that produces OH^- ions in solution
10. substances formed in a chemical reaction

a. reactants
b. activation energy
c. exothermic reaction
d. physical change
e. acid
f. products
g. salt
h. indicator
i. chemical change
j. endothermic reaction
k. coefficient
l. base

What Do You Remember?

1. What is a chemical reaction? Give an example.
2. How do chemical changes differ from physical changes?
3. Balance these equations: $Na + H_2O \rightarrow NaOH + H_2$
   $Mg + HCl \rightarrow H_2 + MgCl_2$
4. How does a balanced equation illustrate the law of conservation of mass?
5. How do synthesis and decomposition reactions differ? Give an example of each.
6. What is activation energy?
7. How do exothermic and endothermic reactions differ? Give an example of each.
8. List three ways to increase the rate of a reaction.
9. What is a catalyst?
10. What is a neutralization reaction?

Applying What You Have Learned ●

1. How could you demonstrate the law of conservation of mass?
2. How could you determine if a substance is an acid or a base?
3. How could you determine the relative strength of an acid or a base?
4. Name two physical changes that happen in your home.
5. Name two chemical changes that happen in your home.

Research and Investigation ●

1. Throughout the ages, table salt has been a very valuable product. The term *salary* is derived from the word *salt* because at one time salt was so valuable it was used to pay debts. Look up the history of salt and find out some of the ways salt has been used.
2. Find out how acids are used in glass etching.

Applying What You Have Learned

1. Answers will vary, but all should include measurement of the mass of reactants and products.
2. Use an indicator such as litmus paper or phenolphthalein.
3. Use an indicator that goes through a range of colors as the pH of a solution changes.
4. Answers will vary. Some possibilities are boiling water, freezing water, melting ice, breaking objects, and dissolving sugar or salt in water.
5. Answers will vary. Some possibilities are baking, burning wood in a fireplace, burning oil or gas for home heating, burning candles, and using antacid medicines for upset stomachs.

● Comprehension Skill: Applied

Research and Investigation

1. Salt at one time had religious significance and was a symbol of purity. Among the Hebrews, it was a religious custom to rub newborn babies with salt to ensure their good health. Among many peoples, salt is still used as a sign of honor, friendship, and hospitality.

Salt was a chief economic product of the ancient world. It was important in the development of the earliest highways of trade. Some of the earliest roads were built from saltworks to major cities.

Salt was once so scarce and precious that it was used as money. Caesar's soldiers received part of their pay in salt. This part of their pay was called salarium. From it we have the word *salary*. It is from this term that we also have the expression "not worth his salt" meaning a man did not earn his wages.

2. The glass is coated with wax. A stylus is used to draw a picture in the wax. The glass is dipped in an acid bath. The acid reacts with the glass but not the wax. The wax is removed and the glass is cleaned. The picture is etched in the glass.

● Reference/Research Skills

CHAPTER BACKGROUND

Physical changes are changes in which only the state or form of matter is altered. Its chemical properties remain the same. Chemical changes are those in which the atomic structure of matter is rearranged, thereby creating an entirely new substance with properties unique to that substance. In nuclear reactions, the atomic nucleus itself is altered. Some nuclear reactions release tremendous amounts of energy that can be used in a number of ways.

Nuclear energy can be derived from fission—in which the nuclei of heavy atoms such as uranium atoms are split in two by neutrons—and fusion—in which the nuclei of two light atoms such as atoms of deuterium (an isotope of hydrogen called heavy hydrogen) are fused together by making them collide at high speeds.

Current research into the use of nuclear energy as a source of electric power for cities, industries, and agriculture is aimed at (1) improving the efficiency of uranium and thorium fuels used in the fission process by developing converter reactors and breeder reactors and (2) finding a method for controlling the nuclear fusion process.

Other applications of nuclear energy fall under the general headings of radiation biology and radiation medicine. In these areas, radioactive substances are used to study biological phenomena, to diagnose and treat diseases, and to improve agriculture. The effects of different kinds of radiation on biological materials are being studied, for example, to find out how radiation produces genetic changes. Isotopic tracers, or tagged atoms, are used to study the metabolism of plants and animals. For example, fertilizers containing radioactive isotopes can be used to determine at what point in the growth of a particular plant the fertilizer is necessary.

In medical research the use of tracers can help to determine the rate of blood flow through the heart and to the limbs, thereby aiding in the diagnosis of diseases. Although intense radiation can cause serious damage to living cells, it can also be used in controlled amounts to treat some diseases such as cancer.

Nuclear Reactions

11

Introduce the lesson by reviewing with the students physical and chemical changes. Ask the students how the two differ (P = properties retained, no new substance; C = properties change, new substances formed). List some examples of each (C = rust, burning, neutralization; P = melting ice, boiling water, torn paper).

Tell the students that in this chapter you are going to deal with a third type of change that can occur—nuclear change. Explain that nuclear changes affect the nucleus of an atom. Review the characteristics of the nucleus by drawing an atom on the board and having the students identify its particles and their charges (protons–positive; electrons–negative; neutrons–no charge).

Energy from the sun is essential to life on the earth. Without it, green plants could not produce the foods you and other animals must eat to live. Without the sun's energy, temperatures and weather patterns would change drastically. The earth would be cold, dark, and barren.

The sun produces enormous amounts of energy day after day, year after year, century after century. How is all this energy produced? Is there an exothermic chemical reaction going on in the sun? Scientists know of no chemical reaction that could produce enough energy to equal the amount produced by the sun. The sun's energy must come from changes other than chemical changes. The sun's energy comes from changes in the nucleus of the atom.

Since life on the earth could not exist without energy from the sun, nuclear changes are clearly very important. Nuclear changes occur on the earth as well as on the sun. Here they have been put to use in science, medicine, and industry.

Lesson Organizer

11.1 Nuclear Changes

Pages 232–233
Follow-up: Page 233
- **R** An Atomic Model
- **E** Health Physics Careers
- **T** Antimatter

11.2 Radioactivity and Radioactive Decay

Pages 234–237
Advanced Planning: Page 234
Activity: Can You See Nuclear Radiation? Page 36
Follow-up: Page 237
- **R** The Discovery of Radioactivity
- **E** Gamma Radiation
- **T** Measuring Radiation

11.3 Radioactive Isotopes and Half-Life

Pages 238–243
Activity: How Can You Demonstrate Half-Life? Page 241
Follow-up: Page 243
- **R** Half-Life
- **E** Nuclides
- **T** Radiation in Milk

11.4 Nuclear Fission

Pages 244–245
Advanced Planning: Page 244
Follow-up: Page 245
- **R** Splitting Atoms
- **E** Critical Mass
- **T** The Manhattan Project

11.5 Nuclear Fusion

Pages 246–247
Follow-up: Page 247
- **R** Fusion versus Fission
- **E** Positrons
- **T** Obtaining Further Information

See ACTIVITIES book pages 61–66 for laboratory, skill, and review exercises for Chapter 11.

11.1

OBJECTIVES: The student should be able to

- **describe** what happens in a nuclear reaction;
- **identify** the force that holds the atomic nucleus together;
- **describe** three kinds of nuclear reactions.

SCIENCE TERM

nuclear reaction

LESSON PLAN

1. Open the lesson by asking students to describe what happens at the atomic level during a chemical change (new substances are formed when electrons of different kinds of atoms interact). Point out that during a nuclear change the nucleus of the atom is involved. In this way one kind of atom can be transformed into a new kind of atom as the number of protons in its nucleus changes.

SCIENCE BACKGROUND

During a chemical change valence electrons are usually either transferred from the outer shell of one atom to the outer shell of another atom or shared among the outer shells of the combining atoms. This transfer or sharing of electrons produces chemical bonds.

11.1 Nuclear Changes

You will find out
- what happens in a nuclear reaction;
- what holds the nucleus of an atom together.

Chemical reactions involve the breaking and forming of chemical bonds. Chemical bonds between atoms involve electrons. Chemical reactions, therefore, affect the electrons of the reactant and product atoms. These reactions do not affect the nuclei of the atoms. However, it is possible for the nucleus of an atom to change. The nucleus of one kind of atom is changed into the nucleus of a different kind of atom in a **nuclear reaction.** Nuclear reactions can release enormous amounts of energy.

To understand how the nucleus of an atom can change, you must keep in mind some basic facts about the nucleus. Positive protons and neutral neutrons make up the nucleus of an atom. The number of protons, the atomic number, determines the kind of atom. If the number of protons changes, the kind of atom changes. The number of neutrons can change without changing the kind of atom. Atoms with the same number of protons but different numbers of neutrons are called isotopes.

Figure 11.2 Energy from nuclear reactions was used to power the *Voyager* spacecraft.

The nuclear force holds the protons and neutrons tightly in the tiny area of the nucleus. Since like electric charges repel, the positive protons tend to move apart from each other. The nuclear force must be very strong to counteract the electric forces that act to split the nucleus apart. The neutrons contribute to the binding strength of the nuclear force. Since the nuclear force is very strong, nuclear reactions can be a powerful energy source.

Figure 11.3 *left:* Electric forces act to split the nucleus apart. *right:* The nuclear force counteracts the electric forces and holds the nucleus together.

There are different kinds of nuclear reactions. In some, a single nucleus spontaneously releases particles of matter and energy. Other nuclear reactions involve the splitting apart of atomic nuclei in response to an outside influence. In a third kind of nuclear reaction, nuclei join together. No matter what kind of nuclear change occurs, one result is always the same. One kind of atom is changed into another kind of atom.

Study Questions for 11.1

1. What happens in a nuclear reaction?
2. Why don't the positive protons in the nucleus separate, splitting the nucleus apart?

1. The nucleus of one kind of atom changes into the nucleus of a different kind of atom.
2. The nuclear force counteracts the tendency of the positive protons to repel each other.

2. On the board draw a circle representing the nucleus of the atom. Ask what kinds of electric charges you should indicate within this nucleus (positive). Then ask why, if like charges repel each other, the nucleus does not fly apart (nuclear forces overcome the repulsion forces from the protons).
3. Have the students briefly differentiate among the three kinds of nuclear reactions described in this lesson. Some students may have questions about nuclear fission and fusion at this point. Aside from tying fission to the splitting of atomic nuclei and fusion to the joining of atomic nuclei, defer further discussion until Lessons 11.4 and 11.5.

FOLLOW-UP

Reinforcement: An Atomic Model

Have the students devise models of an atom and its isotopes. Students should create their own key to show nuclear forces and electric charges within the nucleus.

Enrichment: Health Physics Careers

Health physicists are specialists who establish standards for nuclear facilities. Ask the students to research this career and find out how the standards are determined and enforced. What type of background should a health physicist have? Students may write to the Health Physics Society, 4720 Montgomery Lane, Bethesda, Maryland 20014, for information.

Tie-in: Antimatter

Scientists agree that antimatter exists. Ask the students to look up the term *antimatter* and describe how antimatter differs from the matter that they are familiar with. Then have them create a model of a carbon atom and a model of an anticarbon atom. They should label the parts of each.

● **Comprehension: Applied**

SCIENCE BACKGROUND

Although the nuclear force holds the protons and neutrons together, the strength of this binding force varies from atom to atom. If the force is weak enough, it can be overcome, and the nucleus of that particular kind of atom can then be split.

11.2

OBJECTIVES: The student should be able to

- **describe** the discovery of radioactivity;
- **define** the process of radioactive decay;
- **list** the three forms of nuclear radiation.

SCIENCE TERMS

radioactivity	alpha particles
radioactive decay	beta particles
nuclear radiation	gamma radiation

ADVANCED PLANNING

You may want to ask someone from a local college or industry to discuss radiation.

LESSON PLAN

1. Review the story of Becquerel's discovery of radioactivity. Explain that the only things in Becquerel's drawer, other than the film, were a key and a rock. Ask the students what experiment they could do to determine whether it was the key or the rock that had caused the film to be exposed (in one drawer, place a piece of film and a key; in another drawer place a piece of film and a rock; wait a certain number of days; then examine both pieces of film and see which is exposed).
2. Point out that after Becquerel had determined that the rock had caused the film to be exposed he asked his friends Pierre and Marie Curie to examine the rock. Mention that the Curies hammered the rock, heated it, froze it, and even subjected it to other chemicals. Nothing that they did changed the fact that the rock was radioactive. The Curies therefore concluded that radioactivity is a property of the substance itself and is not affected by the environment.

SCIENCE BACKGROUND

Natural radioactivity was discovered by Antoine Henri Becquerel, who in 1903 received the Nobel prize in physics for his discovery.

Using a mineral containing uranium called pitchblende, the Curies then investigated the properties of radioactivity.

11.2 Radioactivity and Radioactive Decay

You will find out

- how radioactivity was discovered;
- what happens in radioactive decay.

The energy given off when a nuclear change occurs is called **radioactivity.** Radioactivity was discovered because it rained one day in Paris in 1896. Because of the clouds and rain, Antoine Henri Becquerel could not do his experiment on the effects of sunlight on uranium. So he wrapped the uranium in black paper and put it inside a cardboard box. He stored the box in his desk drawer on top of some photographic plates. When Becquerel returned to his desk sometime later, he found that the photographic plates had been exposed. Since the plates had not been outside his desk drawer, the only thing that could have affected the plates was the uranium. But how could a rock give off energy that could go through a cardboard box and a sheet of black paper?

Becquerel called in two other scientists to investigate. They were Pierre Curie and his wife, Marie. The three scientists did many tests on uranium. The three found that the amount of energy given off by the uranium depended only on how much uranium was present. It didn't matter if the uranium was in a solid, liquid, or gas state. Nor did it matter if it was combined with

Figure 11.4 *left:* Antoine Henri Becquerel wearing his uniform as a member of the French Academy of Sciences; *right:* The exposed photographic plates that Becquerel discovered

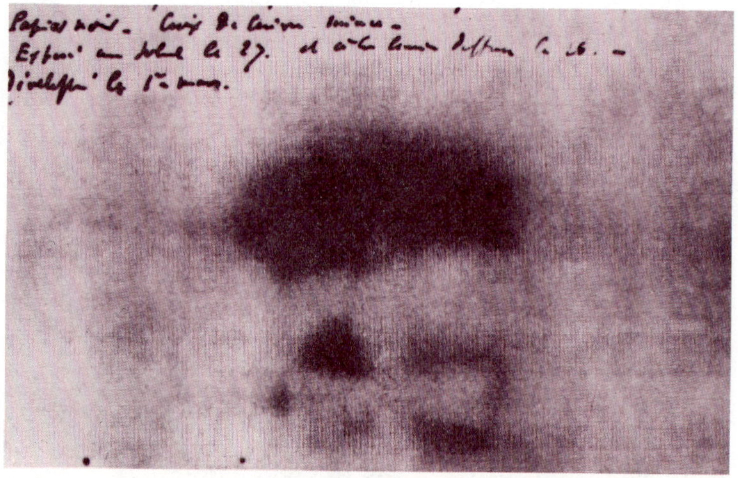

other elements. The mysterious release of energy had to be a property of the innermost part of the element. Scientists now call this innermost part the nucleus of the atom. Marie Curie gave this property the name *radioactivity.* In the years that followed, the Curies discovered two other radioactive elements, radium and polonium.

The radioactive elements the Curies discovered were undergoing a kind of nuclear reaction called **radioactive decay.** In radioactive decay, the nucleus of one kind of atom spontaneously changes into the nucleus of another kind of atom. For example, uranium nuclei can undergo radioactive decay and change into lead nuclei.

Radioactive decay produces **nuclear radiation.** Nuclear radiation can be in one of three forms: **alpha particles, beta particles,** or **gamma radiation.** An alpha particle, like a helium nucleus, has two protons and two neutrons. Beta particles are high-speed electrons. Gamma radiation is not in the form of particles. It is a form of energy without mass.

Do You Know?
Less than a thimbleful of radium gives off as much energy as the burning of 9 metric tons of gasoline.

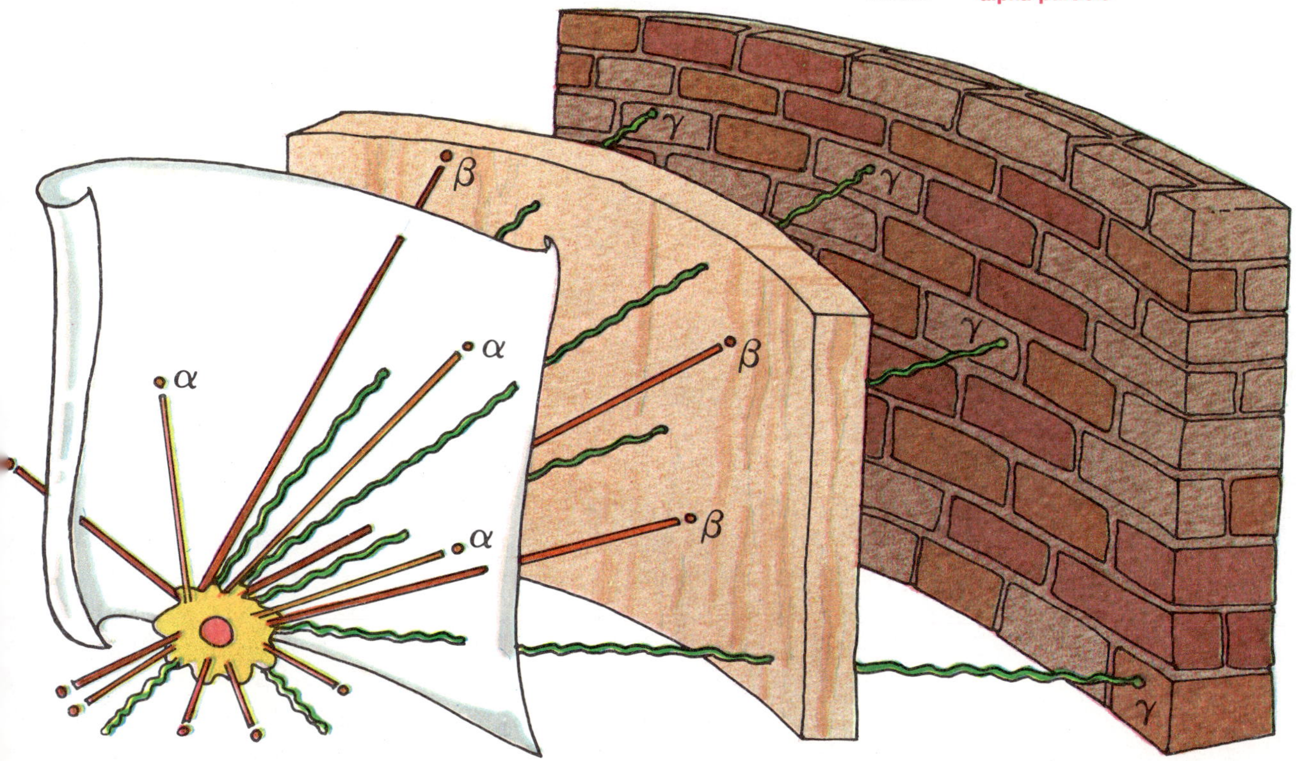

Figure 11.5 Alpha particles (α), beta particles (β), and gamma radiation (γ) have different penetrating abilities. Which form of radiation could be stopped by a page from this book? **alpha particle**

3. Ask the students what happens when radioactive elements undergo the process of radioactive decay (nuclei of one type of atom spontaneously change into the nuclei of another type of atom; during this process, particles and energy are released).

4. Remind the students that the particles and energy released during radioactive decay are classified into three main types. What are they (alpha particles, beta particles, and gamma radiation)?

SCIENCE BACKGROUND

Alpha particles are positively charged helium nuclei made of two protons and two neutrons. They lack penetrating power and can be stopped by a sheet of paper or aluminum foil or by a few centimeters of air. Beta particles are electrons that travel at almost the speed of light. They can pass through wood that is about 3 cm thick. Gamma radiation is in the form of rays similar to X rays. Gamma radiation is the most penetrating form of radiation. Even thick lead or concrete will not stop all gamma radiation. Gamma radiation is frequently given off at the same time that an alpha or a beta particle is emitted.

ACTIVITY: Can You See Nuclear Radiation?

The source of radioactivity may be a luminous dial from an old watch, or you can purchase a radioactive source from a supply house. The dry ice, which should be slightly larger than the jar lid being used, may be obtained from an ice-cream company.

◇ **Caution:** The students are not to touch the dry ice with their bare hands. Have tongs available for transferring it.

Explain that dry ice is solid carbon dioxide. At the end of the experiment observe that although there is less dry ice in the pan, there is no liquid in the pan. The dry ice has undergone sublimation; that is, it has changed from a solid directly into a gas.

The students should observe three types of vapor trails. Alpha particles leave the heaviest trails. They appear to run out of power as they fall to the bottom of the cloud chamber. Beta particles leave thinner trails that do not follow straight paths and may pass through the glass. The track of a particle that goes through the jar from one side to the other is probably caused by a cosmic ray from outer space. Cosmic rays are highly penetrating rays that originate outside the earth and consist of protons, neutrons, electrons, photons, and other particles.

Explain that the small amounts of nuclear radiation used in this activity pose no danger. Many people wear luminous watches!

● **Laboratory Skills: Manipulation of Materials**
Experimentation/Demonstration

● ACTIVITY

Can You See Nuclear Radiation?

Materials

glass jar with lid
black cloth
blotting paper
wire
dry ice
small baking pan
alcohol
luminous watch dial
flashlight
apron
tongs

Procedure

1. Put on your apron.

2. Cut the cloth so it fits inside the jar lid.

3. Cut the blotting paper so it fits in the bottom of the jar. Wet the blotting paper thoroughly with alcohol. Place the paper in the bottom of the jar and turn the jar upside down. If the paper does not stay on the bottom of the jar, make a wire ring that will fit inside the bottom of the jar and use it to hold the paper in place.

4. Using tongs, place the dry ice in the small pan. CAUTION: Do not touch the dry ice with your bare hands.

5. Remove the luminous watch dial from its case and place it on top of the black cloth.

6. Rewet the blotting paper with alcohol. Place the black cloth inside the lid. Holding the jar upside down, screw on the lid.

7. Place the jar upside down on the dry ice. Wait five minutes.

8. Your teacher will turn off the lights and make the room as dark as possible. Shine the flashlight through the side of the jar as you look down at the cloth. You should see vapor

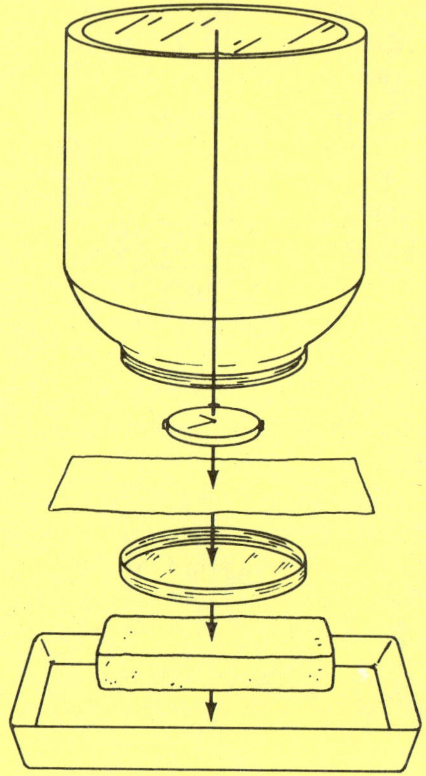

trails through the alcohol mist. The heavy trails are those of alpha particles. The weaker, wavy trails are made by beta particles.

Questions

1. Describe or draw the vapor trails that you see and identify them as those of alpha particles or beta particles.

2. Did all students see the same number of vapor trails? How can you account for this?

1. Heavy, straight paths that stop at the jar are alpha particles. Light, wavy paths are beta particles.
2. They probably did not. Watches could be different ages, have different types of luminous material, or have been exposed to different amounts of wear.

Many methods are used to detect nuclear radiation. One of the most commonly used devices is a Geiger counter. This device operates using a tube filled with a noble gas such as argon. Radiation passing through a thin window on the end of the tube knocks electrons off the argon atoms in its path. Electric forces are used to pull the electrons to a wire in the center of the tube. This creates a momentary electric charge that can be used to make a popping noise.

Too much nuclear radiation can be very harmful to living things. The radiation can cause sickness, suffering, and even death. For this reason, radioactive materials must always be handled with great care. Heavy shielding and protective material must be used. Unfortunately, the Curies did not know this. Marie and her daughter and son-in-law, who both worked with her, all died from overexposure to nuclear radiation.

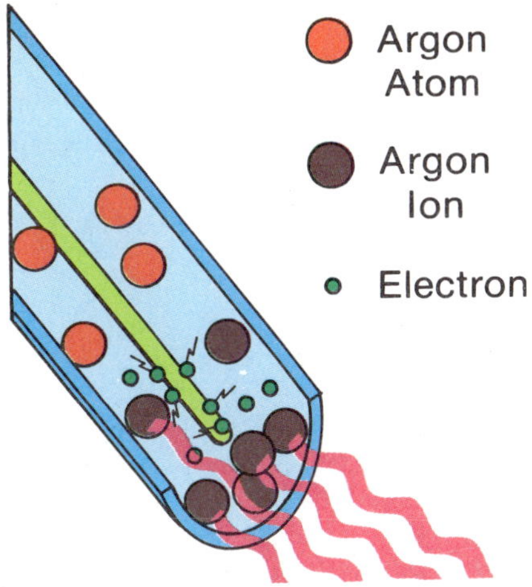

Figure 11.6 The tube of a Geiger counter

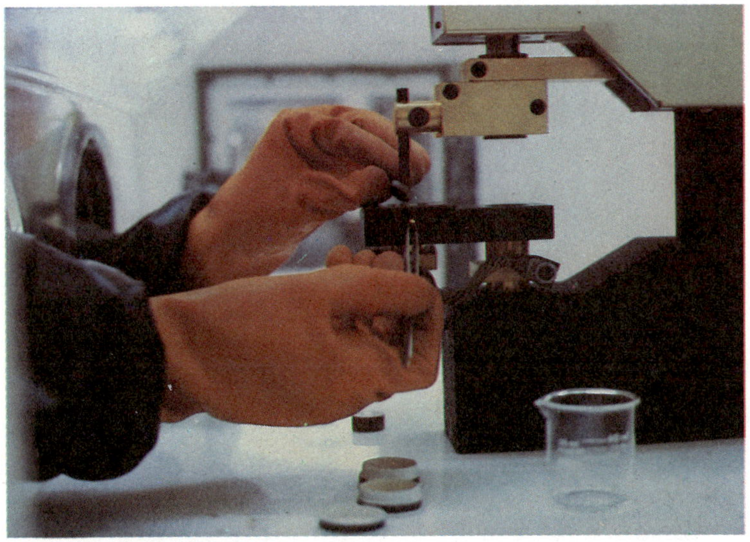

Figure 11.7 Using protective shielding to handle radioactive material

Study Questions for 11.2

1. Who were the discoverers of radioactivity?
2. What happens when a nucleus undergoes radioactive decay?
3. What are the three forms of nuclear radiation?

1. Antoine Henri Becquerel and Pierre and Marie Curie
2. It changes spontaneously from one kind of nucleus to another.
3. alpha particles, beta particles, and gamma radiation

5. Point out that there is one isotope of uranium that is radioactive. Ask the students to imagine that they are prospecting for uranium. What instrument would they use (Since Geiger counters are used to detect radiation, they can therefore be used to find uranium)? Ask the students how a Geiger counter indicates the presence of radiation (by clicking).

FOLLOW-UP

Reinforcement: The Discovery of Radioactivity
Ask the students to pretend that they have been hired to write a book about the history of science. Ask them to write a chapter in which they explain what radioactivity is and how it was discovered.

Enrichment: Gamma Radiation
Ask the students to do research on gamma radiation. How are they like X rays? How are they different? How can people who work with radiation protect themselves from gamma radiation? What are some of the uses of gamma radiation?

Tie-in: Measuring Radiation
The Geiger counter is an instrument used to detect and measure radiation. Ask the students to do research on the Geiger counter or on another instrument that measures radiation. Students should make a simple drawing of the instrument, explain how it operates, and describe its uses.

SCIENCE BACKGROUND

Marie Curie, her daughter, and her son-in-law died because of their exposure to nuclear radiation. This radiation can produce wounds that heal with great difficulty. It can cause disease, characterized by nausea, diarrhea, vomiting, internal bleeding, and a feeling of general weakness. Nuclear radiation can also cause cancer. If the amount of this radiation is controlled, however, it can be used to destroy diseased tissue in cancer and some skin disorders.

People who work with radioactive materials often wear badges that contain dosimeters. A dosimeter is a device that can be used to determine how much radiation the worker has been exposed to.

11.3

OBJECTIVES: The student should be able to

- **explain** why some isotopes are radioactive;
- **explain** the meaning of the term *half-life*;
- **list** some uses of radioactive isotopes.

SCIENCE TERM

half-life

LESSON PLAN

1. Remind the students that atoms sometimes contain a different number of neutrons than the number that would result if the atomic number (number of protons) were subtracted from the mass number (sum of protons and neutrons). These atoms are called isotopes. Although the isotopic forms of an element have the same number of protons and electrons and therefore the same electrical nature, the number of neutrons is different. This results in a different atomic mass.

11.3 Radioactive Isotopes and Half-Life

You will find out
- what makes some isotopes radioactive;
- what is meant by the term *half-life*;
- how radioactive isotopes are useful.

In the years since the Curies, scientists have learned that all isotopes of elements heavier than bismuth (element 83 on the periodic table) are radioactive. Many of the lighter elements also have isotopes that are radioactive.

Why are some isotopes radioactive while others are not? Radioactivity is the result of an unbalanced, or unstable, nucleus. Nuclear balance depends upon the number of neutrons compared to the number of protons. For atoms with less than 20 protons, a balanced nucleus means that the number of neutrons is equal or almost equal to the number of protons. For example, the most common carbon isotope, carbon 12, has six protons and six neutrons. This form of carbon has a balanced nucleus and is not radioactive. Another isotope of carbon, carbon 14, has an unbalanced nucleus. It has six protons and eight neutrons. Carbon 14 is radioactive.

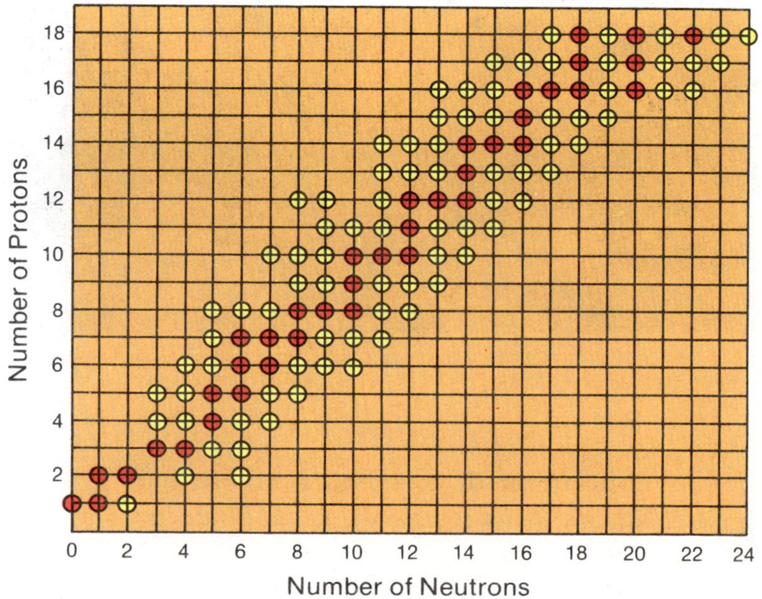

Figure 11.8 The number of neutrons associated with stable and radioactive isotopes of elements 1 through 18

Radioactive Isotope
Stable Isotope

The nuclei of atoms with 20 or more protons can have many more neutrons than protons and still be balanced. For example, the most common lead isotope has 82 protons and 125 neutrons. This isotope is a stable form of lead. It is not radioactive.

The nucleus of a radioactive isotope will undergo one or more nuclear changes in the form of radioactive decay. This radioactive decay will continue until the unbalanced nucleus has become balanced. As a result, a radioactive isotope of one element becomes a stable isotope of another element. In the process, energy is released. For example, consider the radioactive decay of an isotope of uranium. The uranium 238 nucleus goes through a series of 14 nuclear changes, or decays. As a result, the unstable uranium 238 becomes lead 206, a stable isotope.

It is impossible to say exactly when the radioactive decay of any one unstable nucleus will take place. That is like trying to say when a specific kernel of popcorn will pop. You can't be completely sure exactly when it will happen.

Figure 11.9 The radioactive decay of uranium 238

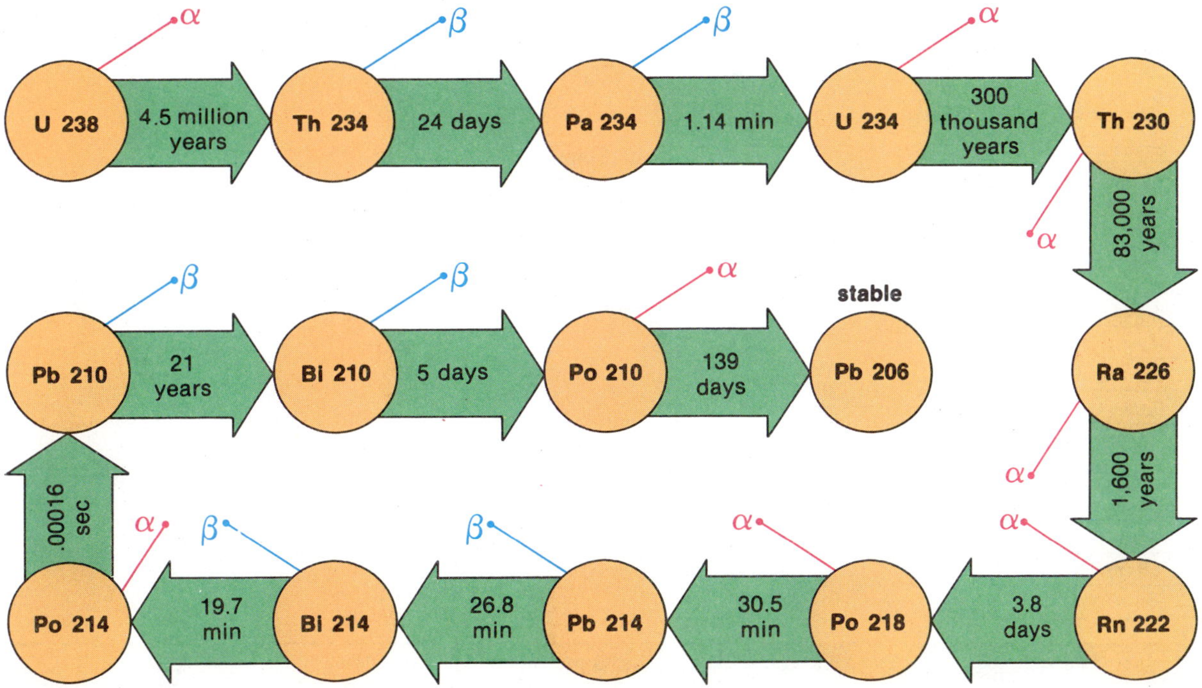

2. Explain that some isotopes are balanced, or stable; others are unstable. Whether an isotope is stable or unstable depends on the number of neutrons compared to the number of protons.

3. Point out that in radioactive decay radioactive isotopes give off alpha particles, beta particles, and gamma radiation until they reach stability, or become nonradioactive.

SCIENCE BACKGROUND

Scientists can now produce isotopes in the laboratory by using an accelerator to shoot neutrons into nuclei. The extra neutrons make the nuclei unstable. They give off alpha particles, beta particles, and gamma radiation to achieve stability.

4. Have a student define the term *half-life* as you write it on the board (the time it takes for half of a quantity of a radioactive material to undergo radioactive decay). Point out that this rate of decay differs from element to element. The half-life of Ra 226, for example, is 1,620 years. This means that if you have 1,000 g of Ra 226, only 500 g will still be Ra 226 in 1,620 years; in 3,240 years 250 g will still be Ra 226; and in 4,860 years, only 125 g will still be Ra 226. Every 1,620 years half of the remaining Ra 226 will decay and half will remain Ra 226.

5. Tell the students that Co 60 has a half-life of 5.3 years. If they have 200 g, how much will still be Co 60 in 5.3 years (100 g)? How much will have decayed (100 g)? Of the 100 g that are still Co 60, how much will be Co 60 after 5.3 more years (50 g)? How much will have decayed (50 g)? What is the total amount that has now decayed (100 g + 50 g = 150 g)? Point out that the sum of the remaining Co 60 and the decayed portion is always 200 g.

SCIENCE BACKGROUND

The half-life of an isotope is a property of the isotope itself. It does not depend on the amount of the isotope or on its physical state. Nor does it depend on whether the isotope is alone, part of a compound, or part of a mixture. Since the half-life is a characteristic property, it can be used to distinguish among isotopes.

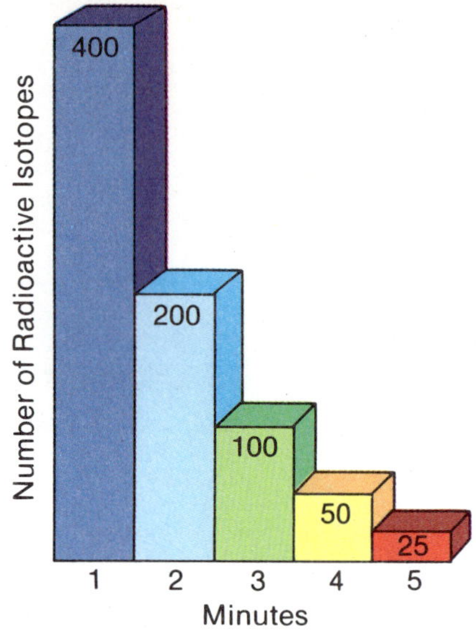

Figure 11.10 The half-life in this example is one minute.

You can, however, make a generalization about all the nuclei in a sample of a radioactive isotope. For example, suppose you have a large number of unstable carbon 14 nuclei. You can be sure that in about 5,700 years, half of those carbon 14 nuclei will have changed into stable nitrogen 14 nuclei. After another 5,700 years, half of the remaining carbon 14 nuclei will have changed into nitrogen 14 nuclei, and so on.

The time it takes for half the nuclei in a sample of a radioactive isotope to undergo radioactive decay is called the **half-life** of that isotope. The half-life of a particular isotope is a property of that isotope. This property does not depend on whether the isotope is in the solid, liquid, or gas state. An isotope's half-life is the same whether the isotope is by itself or in a compound.

The half-life of carbon 14 is 5,700 years. Some radioactive isotopes have half-lives of only a few seconds. Others have half-lives of billions of years. These isotopes were produced long ago when the material that later formed the earth was bombarded by neutrons in an explosion inside a star. Other radioactive isotopes, such as carbon 14, are produced continually by the bombardment of the top of the atmosphere with high-energy particles from space.

Figure 11.11 The half-life in this tennis tournament is one round. After each round, half of the players are eliminated.

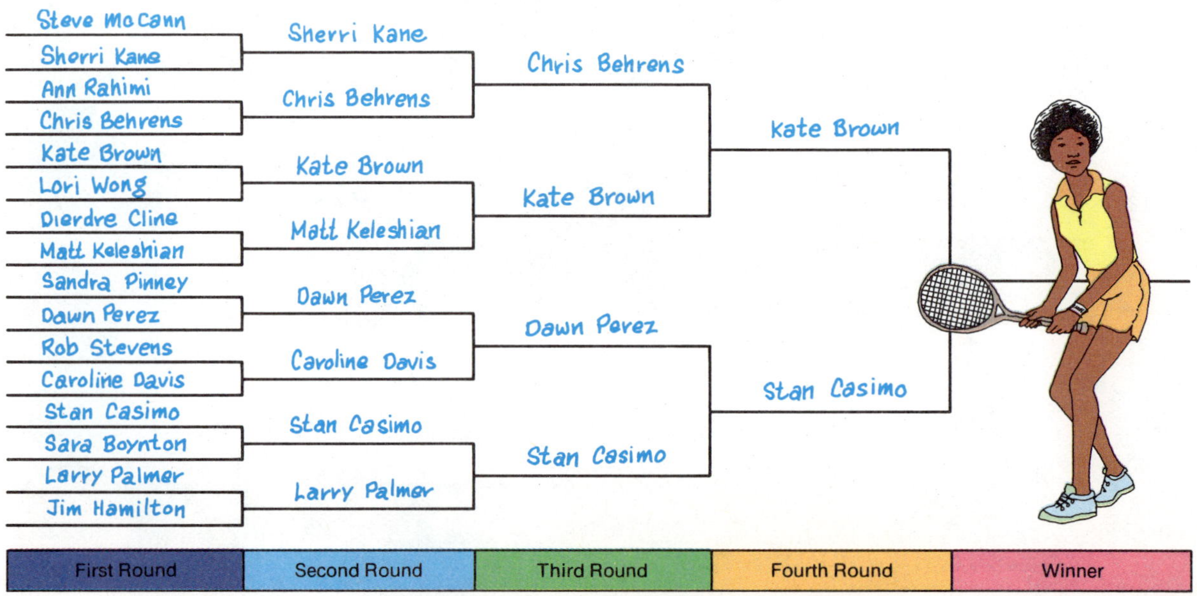

ACTIVITY

How Can You Demonstrate Half-Life?

Materials

100 blocks, each having one side painted
graph paper
pencil
large paper bag

Procedure

1. Copy the chart twice. Use one copy to record your group's results and the other copy for the total class's results.

2. Place the blocks in the bag. They represent a sample of a radioactive isotope.

3. Shake the bag. Dump the blocks out of the bag.

4. The blocks that fall with the painted side up represent nuclei that have undergone radioactive decay and are now stable. Count them and complete the information for trial 1 on your chart.

5. Do not put the stable blocks back into the bag. Put them aside. Put all the other blocks back in the bag. They represent nuclei that are still radioactive.

6. Repeat steps 3–5 until you have completed ten trials. Each time there should be fewer blocks to return to the bag.

7. Under your teacher's direction, share your group's results with those of the class. Record the total class's results on your class copy of the chart.

8. Make a graph of the class's results. Label the vertical axis *number of radioactive nuclei*. Label the horizontal axis *trials*.

Questions

1. Why wasn't the line on your graph straight?

2. What does your graph tell you about the number of radioactive nuclei left after each trial?

3. To what can you compare the number of trials?

1. Fewer atoms were dumped from the bag with each trial.
2. There are fewer left.
3. the passing of time

Trial	Started With	Decayed	Still Radioactive
1			
2			
3			
4			
5			

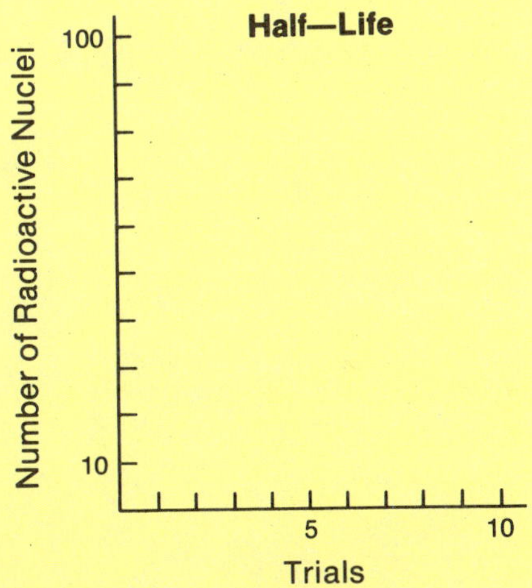

ACTIVITY: How Can You Demonstrate Half-life?

In this activity the students will demonstrate the influence of chance on half-life. It cannot be predicted which atoms in a group will decay during a given time. The atoms that decay, becoming stable, will be the blocks with the painted side up. Be sure that as the students do the experiment they do not put the stable atoms back into the sample. Point out that the number of trials that it takes for half of the blocks (50) to fall with the painted side up represents the half-life of the sample.

Have the students make two charts. One chart is to record the results of their group. The second is to record the results of the class as a whole. Put a chart on the board and have someone from each group write the group's results in the proper place. The students can then add up the numbers before plotting the graph. The graph should be completed by having the students draw a smooth line through the dots rather than by connecting each dot with a straight line.

● **Laboratory Skills: Manipulation of Materials
Graphing**

Do You Know?
Some scientists have suggested that the nucleus of a hydrogen 1 isotope is unstable and has a radioactive half-life of 1,000,000,000,000,000,000,000,000 years.

Many uses have been found for radioactive isotopes. In medicine, they are used to diagnose and treat illnesses. Radiation from radioactive isotopes of cobalt and radium is used to kill cancer cells. Radioactive iodine is used to study and treat problems with the thyroid gland. A number of different radioactive isotopes are used as tracers of chemical activity in the body. When a small amount of radioactive material is injected into the bloodstream, the flow of blood to various organs can be followed by observing the movement of the radioactive material.

Radioactivity is also used to study the process by which plants absorb carbon from the atmosphere. A fraction of the carbon in the carbon dioxide molecules in the atmosphere is in the form of radioactive carbon 14. By observing the radioactivity from carbon 14, scientists have been able to trace the absorption of carbon by plants.

Figure 11.12 The absorption and decay of carbon 14

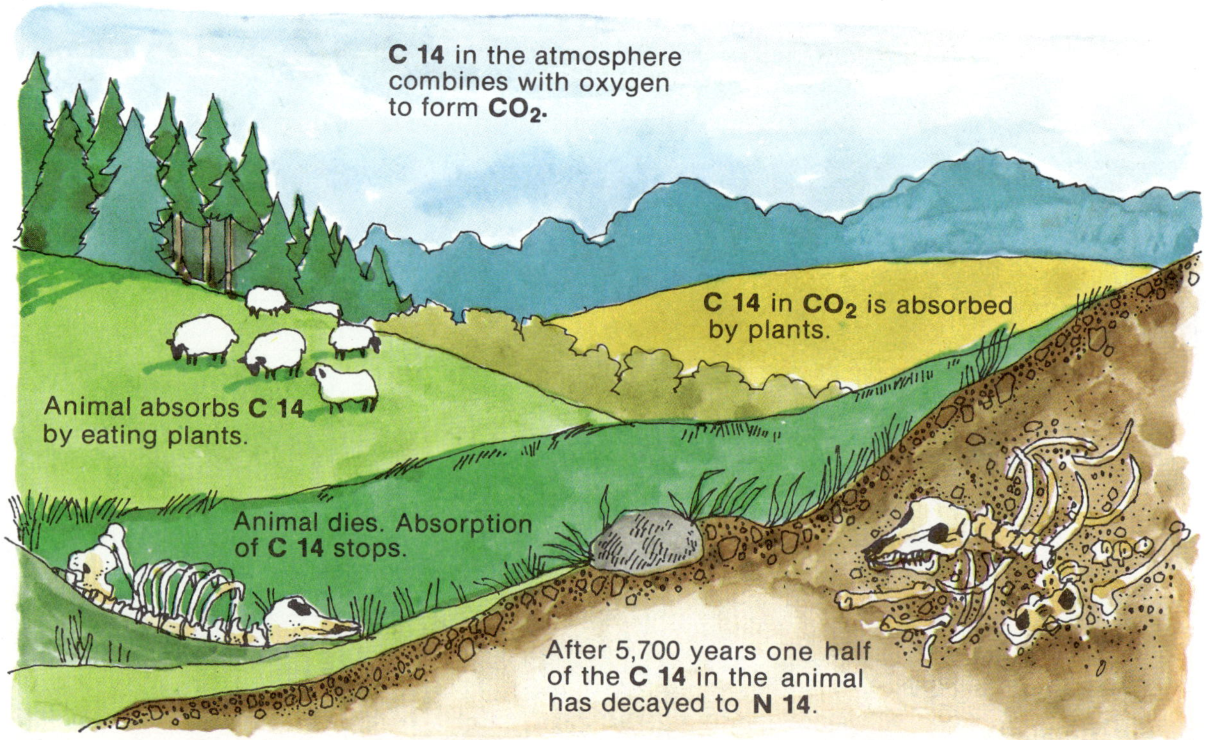

SCIENCE BACKGROUND

Fe 59 can be used to tag red blood cells. The cells can then be traced through the body. Na 24 can also be introduced into the blood and used to study diseases of the circulatory system.

Figure 11.13 Scientists study the level of carbon 14 radioactivity in ancient artifacts to determine their ages. This Egyptian sarcophagus dates back to 600 BC.

Radioactive carbon 14 is also used to estimate the age of old material that was once living. When a plant dies, it stops absorbing carbon dioxide from the atmosphere. When an animal dies, it stops absorbing carbon from its food. The carbon 14 in the dead organic matter slowly changes into nitrogen 14 through radioactive decay. The decreasing number of carbon 14 nuclei will show up as decreased radioactivity. By studying the level of carbon 14 radioactivity in dead organic matter, scientists can tell how much time has passed since the plant or animal died. This same method has been used to date bones, wood from coffins in the Egyptian pyramids, and paper in old manuscripts. Radioactive dating using carbon 14 and other isotopes is helping to answer many questions about the history of life on the earth and the history of the earth itself.

Do You Know?
Radioactive rubidium isotopes have been used to show that some rocks in Greenland are 3.9 billion years old.

Study Questions for 11.3

1. How is a radioactive isotope different from a stable isotope?
2. What is the half-life of a radioactive isotope?
3. Give three ways that radioactive isotopes are useful.

1. A radioactive isotope has an unbalanced nucleus. A stable isotope has a balanced, stable nucleus.
2. the time it takes for half the nuclei in a sample of a radioactive isotope to undergo radioactive decay
3. to kill cancer cells, to trace the path of materials in the body, to estimate the age of old materials

6. On the board construct a chart whose headings are *Radioactive Isotopes* and *Uses*. Ask several students to enter the name and symbol of a radioactive isotope and its use in the appropriate cc lumn. As an assignment, students could investigate the names and uses of other radioactive isotopes to be added to the chart.

FOLLOW-UP

Reinforcement: Half-Life

Make a poster of the following chart.

Number of Half-lives	Time (years)	Amount Stable N 14 (g)	Amount Radioactive C 14 (g)
0	0	0	5,000
1	5,700	2,500	2,500
2	11,400	3,750	1,250

Ask the students to complete the chart through eight half-lives.

Enrichment: Nuclides

Nuclides are all the nuclear species. Ask the students to construct a chart of the nuclides with their atomic numbers and masses.

Tie-in: Radiation in Milk

The atom bomb put strontium 90 into the atmosphere. A major concern is that Sr 90 falls to the ground and becomes incorporated in the food chain. Ask the students to find answers to the following questions: What foods are they most likely to eat that contain Sr 90? What damage might it cause? What is its half-life? How can people protect against it?

11.4

OBJECTIVES: The student should be able to

- **describe** what happens during nuclear fission;
- **describe** a chain reaction.

SCIENCE TERMS

nuclear fission
nuclear chain reaction

ADVANCED PLANNING

For a demonstration you will need 200 marbles and a cardboard border that is 2.5 cm (1 in.) high. The border should be long enough to enclose an area that is 1.2 m (4 ft) by 1.2 m (4 ft).

LESSON PLAN

1. Explain to the students that the nucleus of an atom can be made to split apart when neutrons are allowed to wander slowly (diffuse) around the uranium-containing material. The U 235 nuclei then catch (absorb) the wandering neutrons. The neutrons are ideal atomic bullets because they have no charge and will not be repelled.

● Comprehension: Applied

SCIENCE BACKGROUND

When an atomic nucleus is split, nuclei of lighter atoms, neutrons, and tremendous amounts of energy are released. For example, when U 235 is split, it may form Kr 92 and Ba 141. When you add the atomic masses of krypton and barium (92 + 141), the total is less than the mass of the uranium (92 + 141 = 233 versus 235). The missing mass is released as neutrons and as energy. The neutrons are shot off in all directions. They hit other nuclei and cause them to split. This releases more neutrons and more energy in a chain reaction.

11.4 Nuclear Fission

You will find out
- what is meant by the term *nuclear fission;*
- how a nuclear chain reaction proceeds.

An important kind of nuclear reaction is **nuclear fission** [FIHSH-uhn]. Nuclear fission is a reaction in which a heavy nucleus is split apart into nuclei of lighter elements and neutrons. Fission reactions are usually caused by the shooting of a beam of neutrons into a material containing a heavy radioactive isotope. Because they have no charge, neutrons are not repelled by the positive electric charge of the protons in the nucleus. So neutrons can enter the nucleus fairly easily. ● When a neutron enters a heavy nucleus, it upsets the nuclear balance, producing a fission reaction. For example, when a uranium 235 nucleus gains a neutron, the nucleus can split, or fission, into two lighter nuclei such as krypton 93 and barium 140 and three neutrons. The same nucleus could also split differently, producing different lighter nuclei and different numbers of neutrons.

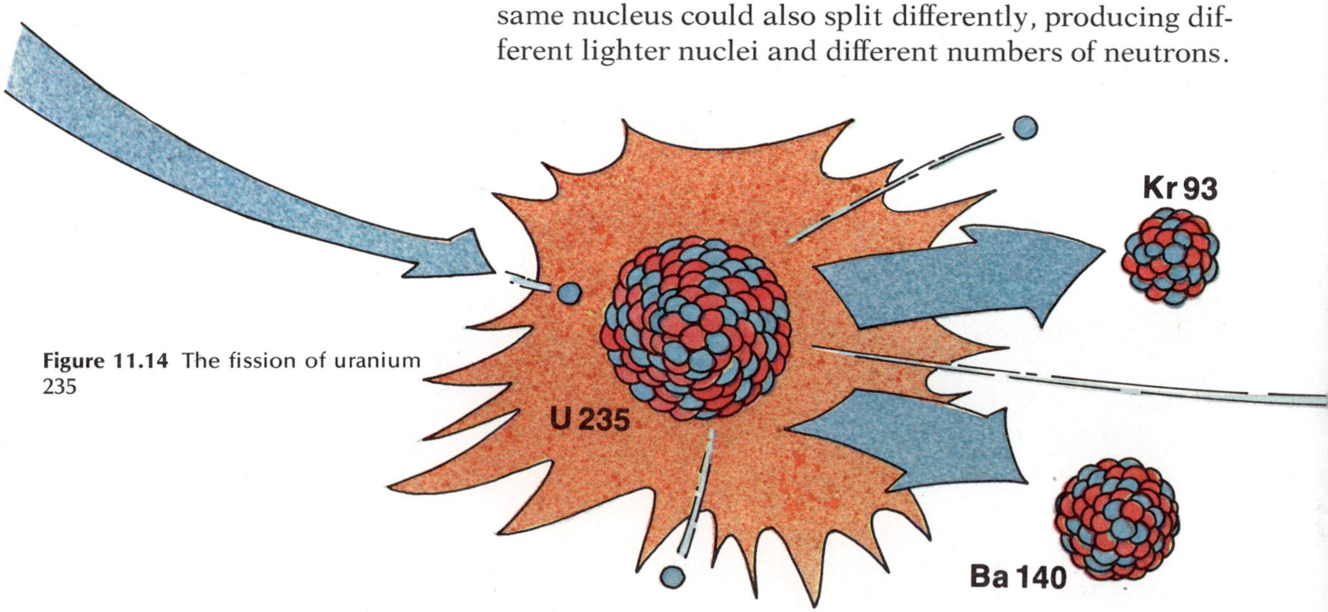

Figure 11.14 The fission of uranium 235

The energy released in fission is about 100 times greater than the energy released in radioactive decay. Therefore, nuclear fission can be a very powerful source of energy.

The energy of nuclear fission can be multiplied in a **nuclear chain reaction.** In a nuclear chain reaction, each fission reaction causes another fission reaction. For example, the neutrons released when a uranium 235 nucleus fissions can enter other uranium 235 nuclei, causing them to fission. Think about pushing over a long row of dominoes. As each domino falls, it knocks over the next domino, which falls and knocks over the next domino, and so on.

Once a chain reaction is started, it will keep going. The chain reaction can get out of hand very quickly. The first reaction may cause two more reactions. Each of these will cause another two reactions, making four. Each of these four will cause two more reactions, making eight, and so on. In a very short time most of the original nuclei will have fissioned. A very large amount of energy is released. This is what happens in an atomic bomb.

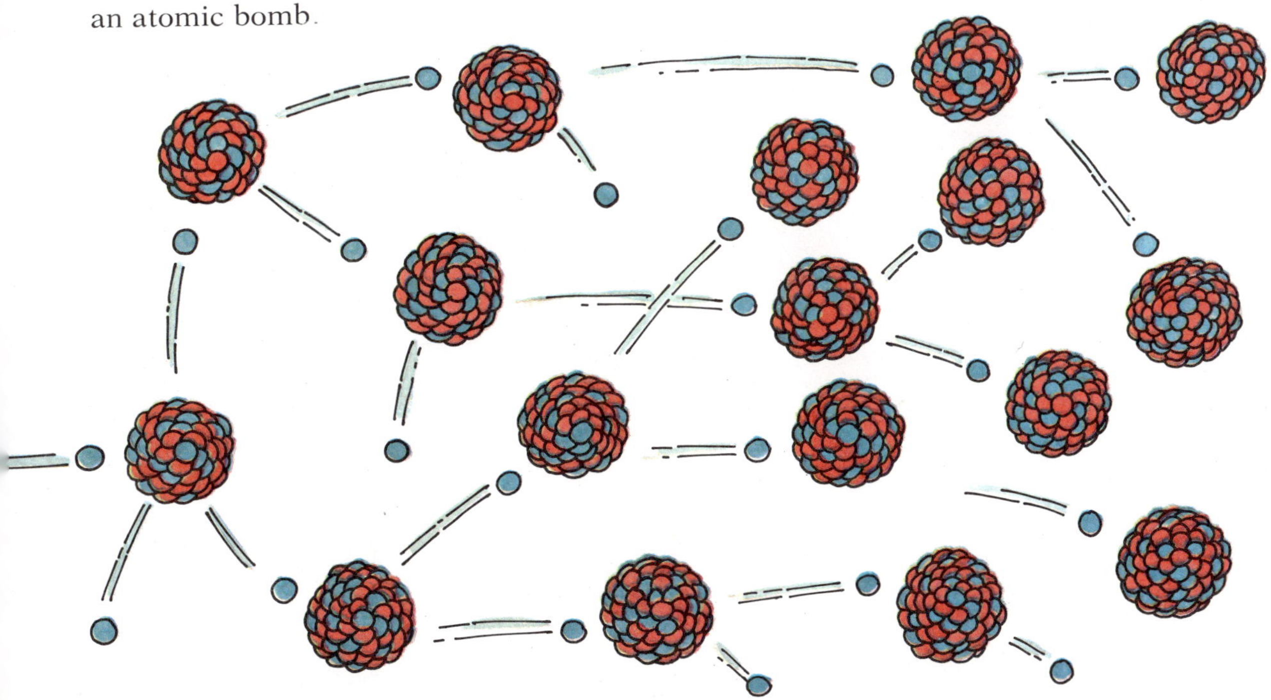

Figure 11.15 A nuclear chain reaction

Study Questions for 11.4

1. What is nuclear fission?
2. Describe a nuclear chain reaction.

1. the splitting of a heavy nucleus into lighter nuclei and neutrons
2. Energy and particles released in one fission reaction trigger more fission reactions.

2. Demonstrate a chain reaction by clearing an area in the center of the classroom. Put the cardboard border on the floor to block off a square that is 1.2 m (4 ft) by 1.2 m (4 ft). In the square place groups of ten marbles in a random pattern. Let no group be more than 15 cm (6 in.) from its neighbor. Forcefully roll a marble into the group closest to the front. The marble will split the group and cause other marbles to move and split into other groups. These, in turn, will split and release more marbles to split other groups. Ask the students how the demonstration illustrates a chain reaction (only one marble was moved, but when it hit a group, that group broke and released more marbles).

FOLLOW-UP

Reinforcement: Splitting Atoms

Ask the students to pretend that they are a well-known scientist. Have them write a speech which explains nuclear reactions, fission, and chain reactions to other interested students in the school.

Enrichment: Critical Mass

Critical mass is an important term for anyone who wants to create a chain reaction. Ask the students to look up the term and then write a definition of it for a new edition of a dictionary.

Tie-in: The Manhattan Project

One scientist, called to work on the development of the atom bomb during World War II, left a note for his students. It said "Gone fission." He had gone to work on the Manhattan Project, the project that developed the atom bomb. Several books that are understandable by students of this age have been written about the project. Ask the students to read one of the books and write a book report about it.

11.5

OBJECTIVES: The student should be able to

- **define** the term *nuclear fusion;*
- **explain** why it is more difficult to start a fusion reaction than it is to start a fission reaction.

SCIENCE TERM

nuclear fusion

LESSON PLAN

1. Introduce the lesson by reviewing what fission is (splitting of atomic nuclei, releasing energy and neutrons). Tell the students that just as the nuclei of atoms are split in nuclear fission the fusion process involves the joining of two light nuclei.

● Word Skill: Vocabulary

11.5 Nuclear Fusion

You will find out
- what is meant by the term *nuclear fusion.*

A third kind of nuclear reaction is **nuclear fusion** [FYOO-zhuhn]. In fusion reactions, nuclei of very light atoms such as hydrogen join together, forming a heavier nucleus. The energy released in nuclear fusion can be enormous. The words *fusion* and *fission* sound similar, but they are really opposites. In fusion, light nuclei join, or fuse, together. In fission, one heavy nucleus splits apart.

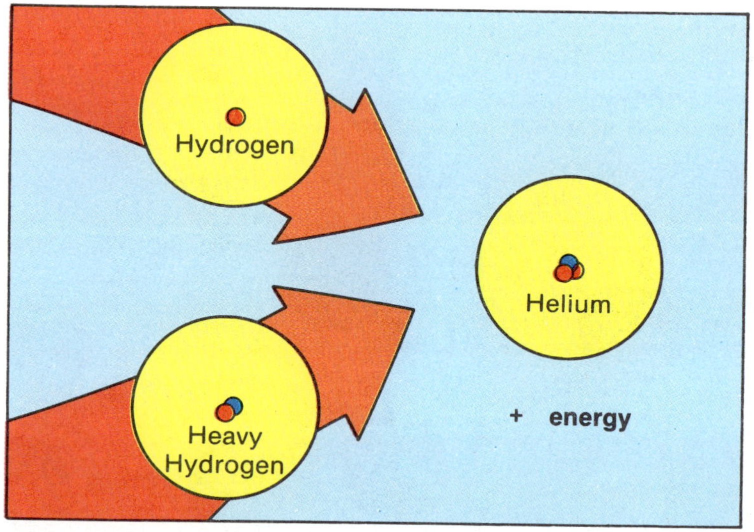

Figure 11.16 The fusion of hydrogen to form helium

It is much more difficult to start a fusion reaction than it is to start a fission reaction. The light nuclei must be moving very fast for the fusion reaction to proceed. Temperatures of 10,000,000°C or higher are needed. Such high temperatures exist deep in the centers of stars. The sun, a star, can be thought of as a giant nuclear furnace that is producing energy in nuclear fusion reactions.

In the sun, hydrogen nuclei join together, forming helium nuclei. Some of the hydrogen nuclei involved in this reaction are formed in another fusion reaction. In this other fusion reaction two light hydrogen nuclei combine to form a heavier hydrogen nucleus. Scientists

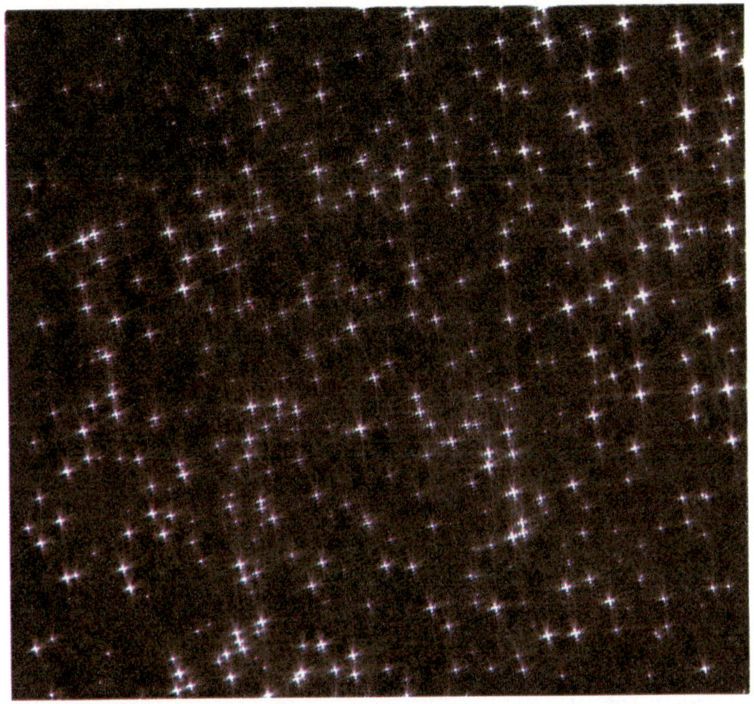

Figure 11.17 Fusion reactions in stars cause them to shine.

estimate that in the sun hundreds of millions of metric tons of hydrogen undergo fusion every second! As you might imagine, the amount of energy released every second is tremendous. The sun and all the other stars you see in the sky are shining due to the energy produced in nuclear fusion reactions.

On the earth the large amounts of energy released in fusion reactions could become an important source of energy. But first, scientists have to learn how to deal with the high temperatures required.

Fusion reactions have been used to provide the energy for hydrogen bombs. These weapons are triggered by a smaller fission bomb that heats the fusion material to very high temperatures. These temperatures start the fusion reaction.

Study Questions for 11.5

1. What is nuclear fusion?
2. How does the sun produce energy?

1. the joining together of light nuclei, forming a heavier nucleus
2. fusion of hydrogen nuclei into helium nuclei

2. Ask the students why it is more difficult to start a fusion reaction than it is to start a fission reaction (very high temperatures are needed to push nuclei close enough together to fuse).
3. Explain to the students that fusion occurs on stars. The stars, with their tremendous heat and strong gravitational forces, are natural fusion reactors that fuse the nuclei of hydrogen atoms to radiate heat and light.

FOLLOW-UP

Reinforcement: Fusion versus Fission
Ask the students to write a poem using either free verse or rhyming verse in which they distinguish between fission and fusion. They should mount their poems and create a border around them, illustrating the two terms.

Enrichment: Positrons
During fusion reactions in the sun, positrons join to form helium nuclei and other positrons. Ask students to do research on positrons. What are they (positive electrons whose mass is the same as that of the electron and whose charge has the same magnitude but opposite sign to that of the electron)? When and by whom was the positron discovered (in 1932 by the American physicist C. D. Anderson while studying cosmic rays)?

Tie-in: Obtaining Further Information
Later in the year the class will be studying nuclear energy as a source of power. Ask the English teacher to help the students write letters to government agencies and nuclear power plants requesting information about the future prospects of nuclear energy as a source of power.

SCIENCE BACKGROUND

When the hydrogen isotopes tritium and deuterium are subjected to the high temperatures needed for fusion, they become ionized, separating into negatively charged electrons and positively charged ions. This random mix of free positively and negatively charged atomic material is called a plasma. Plasma is considered by scientists to be a fourth state of matter. The plasma state is mentioned in Chapter 2.

What Did You Learn?

11.1 • In a nuclear reaction, the nucleus of one kind of atom changes into the nucleus of another kind of atom.
11.1 • The nuclear force counteracts the repelling electric forces and holds the nucleus together.
11.2 • Antoine Henri Becquerel, Pierre Curie, and Marie Curie discovered radioactivity while working with uranium.
11.2 • In radioactive decay, the nucleus of one kind of atom spontaneously changes into the nucleus of another kind of atom.
11.2 • Nuclear radiation is produced in radioactive decay.
11.2 • The three forms of nuclear radiation are alpha particles, beta particles, and gamma radiation.
11.3 • Radioactive isotopes are the result of an unbalanced nucleus.
11.3 • The half-life of an isotope is the time it takes for half the nuclei in a sample of that isotope to undergo radioactive decay.
11.3 • Radioactive isotopes are used in medicine to study and treat illnesses.
11.3 • The age of an organic object can be estimated by measuring the level of radioactive carbon 14 in that object.
11.4 • In nuclear fission, a heavy nucleus is split apart into lighter nuclei and neutrons.
11.4 • In a nuclear chain reaction, one fission reaction triggers other fission reactions.
11.5 • In nuclear fusion, light nuclei join together, forming heavier nuclei.
11.5 • The sun and other stars produce energy from nuclear fusion.

Key Terms

nuclear reaction
radioactivity
radioactive decay
nuclear radiation
alpha particles
beta particles
gamma radiation
half-life
nuclear fission
nuclear chain reaction
nuclear fusion

Biography

Marie Curie (1867–1934)

Marie Curie was the first person ever to be awarded the Nobel prize twice. In 1903 she shared the prize with her husband and Antoine Henri Becquerel. The prize was awarded in recognition of their work with radioactivity. In 1911 Marie Curie received her second Nobel prize for work in chemistry. Her studies led the way to the modern theory of the atom and to the many practical uses of nuclear physics and chemistry.

Marie Sklodowska was born in Warsaw, Poland. Because of political troubles in Poland, she moved to Paris in 1891 to attend the university there. Two years later she graduated at the top of her class in physics. In 1895 she married Pierre Curie, a French chemist. Shortly afterward, she and Pierre began working with Becquerel on radioactivity. It was Marie who gave the name *radioactivity* to the process by which uranium gave off radiation.

Working with what their daughter called terrible patience, in a small wooden shed with a dirt floor and a leaky roof, Pierre and Marie Curie discovered radium and polonium. They had to go through several thousand kilograms of uranium ore to get a tenth of a gram of pure radium!

In 1906, Pierre was run over by a horse-drawn wagon and killed. Marie was heartbroken but continued their work alone. She was given Pierre's professorship at the University of Sorbonne in Paris. She was the first woman to teach there.

During World War I, Marie Curie pioneered the use of X rays to find foreign objects such as bullets. She took her X-ray devices to the battle zone. When she had trouble getting someone to drive her, she got a driver's license and drove herself.

After the war, she set up institutes for the treatment of illnesses, using radium, in France and Poland. In 1934, she died of overexposure to nuclear radiation.

Vocabulary

1. c
2. a
3. b
4. b
5. a
6. a
7. a
8. c

● Word Skill: Vocabulary

What Do You Remember?

1. Antoine Henri Becquerel placed some uranium in his desk drawer near some film. When he later examined the film, it was exposed. He wondered how it had become exposed. The uranium must have affected the film. Becquerel and the Curies tested the uranium carefully. They learned that radioactivity was unaffected by outside influences.
2. a reaction that involves a change in the nucleus from one kind of atom to another
3. alpha particles—two protons and two neutrons
   beta particles—high-speed electrons
   gamma radiation—form of high energy without mass
4. any element after bismuth on the periodic table
5. different numbers of neutrons

TO THINK ABOUT AND DO

Vocabulary ●

Number your paper from 1 to 8. Select the one best term that completes each of the following sentences. Write the letter of the term next to the appropriate number.

1. The change of the nucleus of one element into the nucleus of another is called (a) physical change, (b) chemical reaction, (c) nuclear reaction.
2. Atoms of the same element with different numbers of neutrons are called (a) isotopes, (b) ions, (c) neutrons.
3. An isotope that can be used to determine the age of the organic matter in which that isotope is found is (a) uranium 235, (b) carbon 14, (c) lead 207.
4. The amount of time it takes for half the total amount of a radioactive isotope to decay is its (a) decay time, (b) half-life, (c) stability time.
5. The splitting of heavy nucleus into lighter nuclei and neutrons is (a) fission, (b) fusion, (c) decay.
6. A form of nuclear radiation with two protons and two neutrons is the ________ particle.
   (a) alpha (b) beta (c) gamma
7. Half-life is associated with (a) radioactive decay, (b) fusion, (c) fission.
8. The joining together of light nuclei with the release of energy is a reaction called (a) fission, (b) meltdown, (c) fusion.

What Do You Remember?

1. Describe how Becquerel and the Curies discovered radioactivity.
2. What is a nuclear reaction?
3. Distinguish among alpha particles, beta particles, and gamma radiation.
4. Name three radioactive elements.
5. How are isotopes of the same element different?

6. Why are some isotopes radioactive?
7. What is meant by the term *half-life?*
8. Describe carbon 14 dating.
9. What is nuclear fission?
10. What is a nuclear chain reaction?
11. What is nuclear fusion?
12. How do stars produce energy?

Applying What You Have Learned ●

1. Suppose that you must transport radioactive materials to a location 100 kilometers away from where you are. What safety measures would you take?
2. Pretend that you have found an ancient bone. What procedure would you use to determine its age?
3. How could radioactive phosphorus be used to determine the role of phosphorus in plant growth?
4. How is the sun a nuclear power plant?
5. You have 150 g of Fe 59, which has a half-life of 46 days. How many grams of Fe 59 will you have left at the end of 184 days?

Research and Investigation ●

1. If you want to use a radioactive isotope to trace the path of Ca in the human body, what characteristics should it have?
2. Robert W. Houghton and Richard G. Fairbanks of Lamont Doherty Geological Observatory have found that they can identify the source of water samples taken from the North American continental shelf by using isotopes. Each of the rivers that empty into the North Atlantic possesses a unique ratio of the stable oxygen isotopes oxygen 16 and oxygen 18. Why is it important for scientists to be able to identify the specific source of waters in the oceans? How could they use isotopes?

6. Nucleus is unbalanced.
7. time it takes for half of the nuclei in a sample of a radioactive isotope to undergo radioactive decay
8. By measuring the amount of radioactive carbon 14 in organic matter, scientists can determine how much time has passed since the material was formed.
9. the splitting of a heavy nucleus into lighter nuclei and neutrons
10. a nuclear reaction in which energy and particles released trigger more reactions of the same type
11. the joining together of light nuclei
12. through nuclear fusion

Applying What You Have Learned

1. Answers will vary. They should include using shielded leak-proof containers, using protective gloves and clothes for people handling the material, using a safe vehicle, and following a safe route.
2. a. Obtain a sample of the same type of bone from a living organism.
   b. Determine the percentage of C 14 in the bone.
   c. Determine the percentage of C 14 in the ancient bone.
   d. Compare the percentages of C 14 and determine how much less is in the ancient bone.
   e. Knowing that the half-life of carbon 14 is 5,700 years, determine the age of the bone.
3. Add radioactive phosphorus to the fertilizer and follow its path through the plant. The area of the plant where the most radioactivity is found is where the phosphorus is used.
4. Fusion reactions go on in the sun.
5. 9.375 = 9.4 g (rounded off)

● Comprehension Skill: Applied

Research and Investigation

1. You should be able to create radioactive Ca with a short half-life so the person won't be exposed to too much radiation.
2. Answers will vary depending on the resources used. Possible reasons for identifying the sources of waters are to track down sources of pollutants, to see where sediments originate, and to see how much water each river contributes to the ocean. There is a reference to the studies of Houghton and Fairbanks in *Science News*; Volume 121, No. 24; June 12, 1982; p. 392.

● Reference/Research Skills

CHAPTER BACKGROUND

Although the terms *heat* and *temperature* are sometimes used interchangeably, their meanings are quite different. Heat is a form of energy whose source is the motion of molecules. The amount of heat in a substance depends on the amount of matter in the substance and the speed at which its molecules are moving. Temperature is a measure of the average kinetic energy of the molecules. Two objects that have the same temperature may contain different amounts of heat. This is the case when the two objects at the same temperature have different masses or are made of different substances.

The centigrade, or Celsius, scale is the temperature scale used in all countries that follow the metric system. It is gradually being adopted in the United States, where the Fahrenheit scale is more common. On the Celsius scale, water boils at 100° and freezes at 0°. The Kelvin scale is used in all scientific work that deals with heat energy. Zero Kelvin is equal to −273°C. This is the lowest temperature possible, or absolute zero. Scientists have found that when the temperature of matter is near absolute zero its properties change drastically. For instance, normally good conductors become superconductors, and some materials that are normally insulators become better conductors than the metals normally used to conduct electricity. Rubber becomes so brittle that it shatters when you touch it. Helium, an inert gas, becomes a superfluid that climbs out the sides of any open container it is put in.

Heat energy is transferred from one substance to another in one of three ways. If the substances are in contact with one another, heat is transferred by conduction. In this process the fast-moving mlecules of the warmer substance bump into the slower molecules of the cooler substance. The resulting transfer of motion causes a transfer of heat energy.

Heat travels through fluids in convection currents. The heated fluid becomes less dense and rises away from the source of heat. Cooler, more dense fluid replaces the heated fluid until it too is heated and rises.

Heat from the sun travels through space as electromagnetic waves that are not dependent on molecular motion for transfer. These waves are referred to as radiation.

Heat 12

Ask the students what material is present in steam, liquid water, and ice (water). Point out that in all three states the very same compound is present, H_2O. Ask how the same compound can form substances that have such different physical properties (students will probably suggest that different temperatures are involved). Point out that since the molecules in all three states are the same, the difference in states must involve how those molecules are arranged. This arrangement is related to heat. In this chapter the students will learn what heat is and how it affects the molecules and atoms of matter on the earth.

Suppose your hands are cold and you want to warm them. What do you do? Rub them together? Hold them in the warm sunlight? Hold them over a fire? Hold them near an electric heater? Touch something warmer than your hands such as a mug of soup? Heat can be produced in many different ways. It has many different uses. Heat keeps you warm, cooks your food, and runs the cars, buses, planes, and trains you ride in.

You see and feel the effects of changes in heat all around you. Steam rises from a whistling teakettle. A windshield on a car or bus fogs up on a cold morning. A sea breeze begins to blow on a warm afternoon. Frost covers the landscape during a cold night and disappears in the sun.

Heat is absolutely necessary for you to survive. Without it, there would be no motion, no chemical reactions, no life.

Figure 12.1 Pouring molten metal

Lesson Organizer

12.1 Heat Energy
Pages 254–255
Advanced Planning: Page 254
Follow-up: Page 255
- **R** Count Rumford
- **E** Kinetic Energy and Heat
- **T** Calories in Food

12.2 Temperature and Specific Heat
Pages 256–257
Advanced Planning: Page 256
Follow-up: Page 257
- **R** Heat Versus Temperature
- **E** Absolute Zero
- **T** The Gulf Stream

12.3 The Effects of Heat Energy on Matter
Pages 258–263
Advanced Planning: Page 258
Activity: What Happens When Water Freezes? Page 262
Follow-up: Page 263
- **R** Changes of State
- **E** Cooking Racks
- **T** Sidewalks

12.4 The Flow of Heat
Pages 264–267
Advanced Planning: Page 264
Activity: Which Material Is the Best Insulator? Page 265
Follow-up: Page 267
- **R** Methods of Heat Transfer
- **E** Heat Transfer at Home
- **T** The Davy Lamp

See ACTIVITIES book pages 67–72 for laboratory, skill, and review exercises for Chapter 12.

12.1

OBJECTIVES: The student should be able to

- **describe** Rumford's discovery of the relationship between heat energy and the motion of particles;
- **explain** the relationship between heat energy and the motion of particles in the kinetic theory.

SCIENCE TERMS

kinetic theory
heat energy

ADVANCED PLANNING

For the demonstration you will need a hot plate, a heat-resistant glass baking dish, some cooking oil, and a small amount of uncooked popcorn.

LESSON PLAN

1. Ask the students to rub their hands together vigorously. What do they feel (heat)? Then ask what is causing the heat (motion, friction). Finally, ask the students what they think heat is.
2. Help the students to understand the concept of heat as a form of energy through the following series of questions. What was Rumford doing (directing the making of a brass cannon)? What did he observe (the brass became so hot from the drilling that it had to be cooled with water; when the drilling stopped, the metal cooled)? Why did Rumford decide that the heat was not in the drill or the brass (heat was being continuously produced; if it were confined to either the drill or the brass, it would have been used up and no longer given off)? Why did Rumford conclude that heat is a form of energy (heat appeared to be connected to the energy of motion of the drill rather than to the drill itself)?

● Word Skill: Vocabulary

12.1 Heat Energy

You will find out
- what the kinetic theory states;
- how heat energy and kinetic energy are related.

For a long time it was thought that heat was an invisible, weightless substance called caloric. This mysterious caloric passed from hot to cold objects. Then in 1798 an inventive scientist named Benjamin Thompson got a clue to what heat really is. Thompson, known as Count Rumford, was directing the drilling of a brass cannon in Germany. The brass became very hot as the drill ground through it. When the drilling stopped, the brass cooled. Rumford figured that the heat must be connected somehow to the energy of the drill in motion. Heat must be a form of energy.

To understand how heat can be a form of energy, you must keep in mind some facts about the basic particles of matter. Remember that matter is made of atoms and molecules. The **kinetic theory** states that these particles are always moving and that they often bump into each other. This random motion is present in varying degrees in solids, liquids, and gases. The particles in a solid are very close together. They do not move from one place to another. Instead they vibrate back and forth. In liquids, the particles move very small distances but still remain close together. In gases, the particles are quite far apart and move freely about. No matter what the state, all atoms and molecules are in motion. Energy of motion is called kinetic energy.

Figure 12.2 The heat in a desert, like all heat, is a form of energy.

Figure 12.3 The people in these photographs represent particles in matter. *left:* Particles in a solid are very close together and move very little. *center:* Particles in a liquid are farther apart and move more. *right:* Particles in a gas are very far apart and move freely.

The **heat energy** of a substance is the total kinetic energy of all the molecules or atoms of that substance. When a substance is heated, the atoms or molecules move faster. Faster-moving particles have more kinetic energy. The more kinetic energy the particles have, the greater the heat energy of that substance.

In the case of Rumford's cannon drill, the grinding motion of the drill against the brass cannon added energy to the particles of the brass. The particles moved faster. Their kinetic energy increased. This increase showed up as an increase in heat energy. Thus, the brass became hotter.

It is important to remember that the law of conservation of energy applies to heat energy. When energy changes forms, that energy may appear to have been created or destroyed. This is not the case. For example, the heat energy produced when Rumford's cannon drill was working was not newly created energy. The kinetic energy of the drill was changed into the heat energy of the brass.

Do You Know?
The water at the bottom of Niagara Falls is about 0.12°C warmer than the water at the top of the falls because the kinetic energy of the falling water is changed to heat energy when the water hits the bottom of the falls.

Study Questions for 12.1

1. State the kinetic theory.
2. What is heat energy?

1. All matter is made of atoms and molecules that are moving.
2. the total kinetic energy of all the molecules or atoms of a substance

3. To illustrate the relationship between heat energy and motion, perform the following demonstration. Place a small amount of cooking oil in the bottom of a heat-resistant glass dish. Add a few kernels of popcorn. Cover the dish, place it on a hot plate, and turn the hot plate on. What happens to the kernels as they get hotter (they begin to move; they pop)? What causes the kernels to move (heat from the burner)? Point out that the kernels represent particles of matter, either molecules or atoms. When molecules or atoms are heated, they too begin to move. As the temperature rises, they acquire more energy of motion.

FOLLOW-UP

Reinforcement: Count Rumford
Ask the students to pretend that they work for the Smithsonian Institution in Washington, D.C. They have been directed to prepare a display illustrating Count Rumford's investigation of the concept that heat is caused by motion. Have the students create a model of their display in a shoe box and write a brief explanation of the model.

Enrichment: Kinetic Energy and Heat
Have the students measure the surface temperature of a piece of sheet lead. Then ask them to pound on the lead with a hammer for a few moments. They should again measure the temperature of the lead. How can they explain the difference in the two temperatures? Note: If sheet lead is not avilable, the activity can be done with a can of lead shot. Students should measure the temperature of the shot, shake the shot for several minutes, and measure its temperature again.

◇ **Caution:** Students should wear safety glasses.

Tie-in: Calories in Food
Ask the students what a food calorie is (kilocalorie). Have them create charts that compare foods with a high kilocalorie content and foods with a low kilocalorie content. How can such a chart be used to help select foods for a proper diet?

12.2

OBJECTIVES: The student should be able to

- **describe** the relationship between molecular speed and temperature;
- **distinguish** between heat energy and temperature;
- **name** two units used to measure heat energy;
- **define** the term *specific heat*.

SCIENCE TERMS

temperature	calorie
absolute zero	specific heat

ADVANCED PLANNING

1. For the demonstration you will need three beakers, one each filled with ice water, hot water, and room-temperature water; and three thermometers.
2. Be sure the students know how to read a thermometer.

LESSON PLAN

1. Place the three beakers, each with water of a different temperature, on your desk. Place a thermometer in each. On the board draw a chart with a narrow column entitled Beaker Number and a wide column entitled Temperature °C. Subdivide the temperature column into four smaller columns: Reading 1, Reading 2, Reading 3, and Average. Ask three students to read the temperature of each beaker three times at two-minute intervals and record the temperatures on the chart. Have three other students find the average for each beaker and record the average temperatures.
2. Ask the students why, with respect to molecular movement, the temperature in each beaker is different (molecules are moving with different amounts of energy). In which beaker is there the greatest amount of molecular movement (beaker with highest temperature)? In which is there the least amount of movement (beaker with lowest temperature)?

12.2 Temperature and Specific Heat

You will find out

- how temperature and heat energy are different;
- what is meant by specific heat.

Figure 12.4 Measuring temperature in degrees Celsius and degrees Fahrenheit

The **temperature** of a substance is a measure of the average kinetic energy of the atoms or molecules that make up that substance. Temperature is measured in degrees Celsius. As the atoms or molecules of a substance move faster, the temperature of the substance increases. Similarly, when the atoms or molecules of a substance slow down, the temperature decreases.

On a hot day (30°C) the average speed of the molecules in air is about 440 m/s. On a very cold day (−20°C) the molecules average about 400 m/s. At a low enough temperature, (−273°C) the average speed would be almost 0 m/s. There would be only slight vibrational motion. Since −273°C is the lowest possible temperature, it is called **absolute zero.** Scientists have been able to create temperatures close to absolute zero.

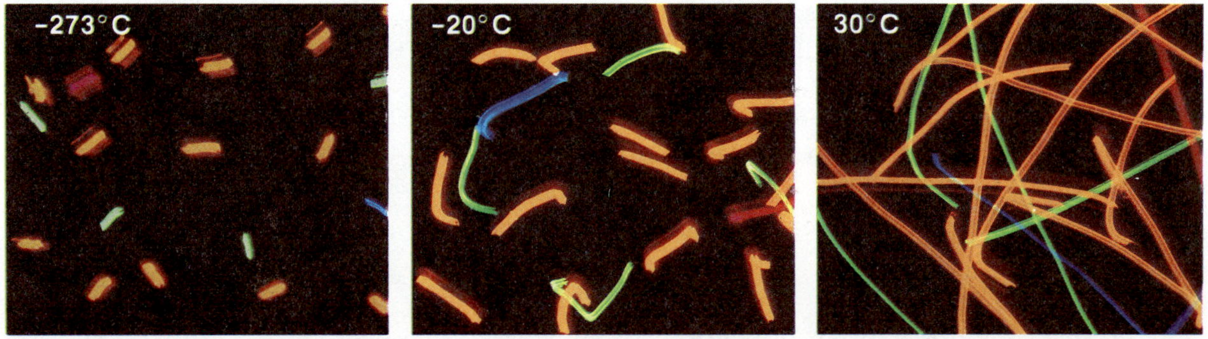

Figure 12.5 The lights in these photographs represent the movement of molecules in air. The higher the temperature, the faster the molecules are moving.

Suppose a large rock and a small pebble have the same temperature. Do they have the same amount of heat energy? No. The larger rock has more heat energy. Remember that heat energy is the total kinetic energy of all the particles in an object. The rock has more mass than the pebble. So the rock has more moving particles and therefore more kinetic energy than the pebble. Thus, the rock has more heat energy. It is important to remember that substances can have the same temperature but different amounts of heat energy. Temperature

Do You Know?
Scientists have produced temperatures as low as one twenty-millionth of a degree C above absolute zero.

Figure 12.6 All the soup is the same temperature. However, the soup in the pot has more heat energy because it has more mass.

refers to average kinetic energy. Heat energy refers to total kinetic energy.

Since heat is a form of energy, the unit for measuring heat is the same unit used for measuring any other type of energy or for measuring work. This unit is the joule. Another unit often used to measure heat is the **calorie.** A calorie is the amount of heat required to raise the temperature of one gram of water one degree Celsius. One calorie equals 4.186 joules. A calorie, then, is a larger unit of heat energy than a joule. Calories are measured using a device called a calorimeter.

You may have noticed that it is easier to raise the temperature of some substances than it is to raise the temperature of others. The amount of heat needed to raise the temperature of one gram of a substance one degree Celsius is called the **specific heat** of the substance. Specific heat is a property of matter. Iron has a specific heat only one tenth as large as that of water. This means that it takes about one tenth as much heat to raise the temperature of 1 kg of iron as it does to raise the temperature of 1 kg of water by one degree Celsius. The specific heat of nearly all gases, liquids, and solids is less than that of water.

Table 12-1 Specific Heat of Several Substances

Substance	Specific Heat (J/g)
water	4.18
alcohol	2.43
wood	1.75
sugar	1.15
marble	0.88
copper	0.39
silver	0.23

Study Questions for 12.2

1. What is temperature?
2. How can two substances have the same temperature but different amounts of heat energy?
3. What is specific heat?

1. the average kinetic energy of the atoms or molecules of a substance
2. Since heat energy is total kinetic energy and temperature is average kinetic energy, a larger object has more heat energy because it has more mass than a smaller object of the same temperature.
3. the amount of heat needed to raise the temperature of 1 g of a substance 1°C

3. As an assignment have the students draw two diagrams with captions. One should illustrate what heat is (total kinetic energy) and the other should illustrate what temperature is (average kinetic energy). Students should include the units that each is measured in (heat can be measured in joules or, more commonly, in calories; temperature is measured in degrees Celsius and in degrees Kelvin).
4. Explain to the students that the amount of heat needed to increase the temperature of an object depends on both its mass and the material of which it is composed. Some materials absorb heat better than others. This ability of a substance to absorb heat is the specific heat of that substance.

● Mathematics Skill: Computation

FOLLOW-UP

Reinforcement: Heat Versus Temperature

Ask the students to pretend that they are James Joule, the British physicist for whom the joule, a unit of energy, is named. Each student should write a letter to Count Rumford explaining the difference between heat and temperature.

Enrichment: Absolute Zero

Have the students find out more about the concept of absolute zero. How do the properties of matter change at this temperature? Under what conditions could absolute zero be reached? Have scientists been able to reach absolute zero?

Tie-in: The Gulf Stream

Ask the students to use geography books to draw maps that show the path of the Gulf Stream. They should write a one-page explanation of why this warm ocean current is so important.

SCIENCE BACKGROUND

Students have probably heard of calories in reference to food. Food calories are really kilocalories. One thousand calories make a kilocalorie.

12.3

OBJECTIVES: The student should be able to

- **explain** why most objects expand when heated and contract when cooled;
- **describe** how melting occurs;
- **define** the term *heat of fusion*;
- **explain** how evaporation occurs;
- **define** the term *heat of vaporization*;
- **explain** how condensation occurs;
- **explain** how freezing occurs.

SCIENCE TERMS

melting	boiling point
melting point	heat of vaporization
heat of fusion	condensation
evaporation	freezing

ADVANCED PLANNING

For demonstrations you will need a Bunsen burner, some matches, a tripod stand with wire gauze, a heat-resistant glass flask fit with a one-hole stopper that has a glass tube extending 2 inches on the end that enters the bottle and 1 foot on the end that is exposed, some water, some food coloring, a rubber band, a pair of tongs, a hot plate, a pan of boiling water, a pan of ice water, a bottle, and a balloon that has been blown up several times and that will fit on the neck of the bottle.

LESSON PLAN

1. For the first demonstration fill a heat-resistant glass flask about halfway with a solution of water and food coloring. In the neck of the flask place a one-hole stopper that has a piece of glass tubing inserted through it. (The tubing should extend about 5 cm (2 in.) below the stopper into the colored water and about 30 cm (1 ft) above the stopper.) Set up a tripod with a piece of wire gauze on top. Heat the contents of the flask by placing the flask on top of the wire gauze and lighting the Bunsen burner under it. The colored water should rise in the glass tubing to a level above the stopper. As this occurs, mark the level of the colored water in the glass tubing by putting a rubber band around the tubing at that level.

12.3 The Effects of Heat Energy on Matter

You will find out

- how substances are affected by temperature changes;
- what happens when substances change state.

You have seen some of the effects of heat on matter. Puddles evaporate and disappear. Snow melts. Water boils. Lakes freeze and later melt. Both the physical state and the volume of substances are affected by the amount of heat energy present. If you keep the kinetic theory in mind, it will be easy to see how heat energy affects matter. Since all matter is made of particles that are moving, the effects of heat on matter can be understood in terms of how fast and how far those particles move.

Suppose you heat the air in a balloon. The kinetic energy of the molecules in that air increases. The air in the balloon expands. The increase in kinetic energy causes the molecules to bump into the walls of the balloon harder and more often. The result is an increased pressure on the walls of the balloon. This increased pressure causes the balloon to expand.

Most solids, liquids, and gases expand when heated. The atoms or molecules gain kinetic energy. They move farther apart. As you might guess, when most substances are cooled, the opposite changes occur. The particles lose kinetic energy. They move closer together. The volume of the substance gets smaller, or contracts.

Figure 12.7 The air inside a balloon expands when heated and contracts when cooled.

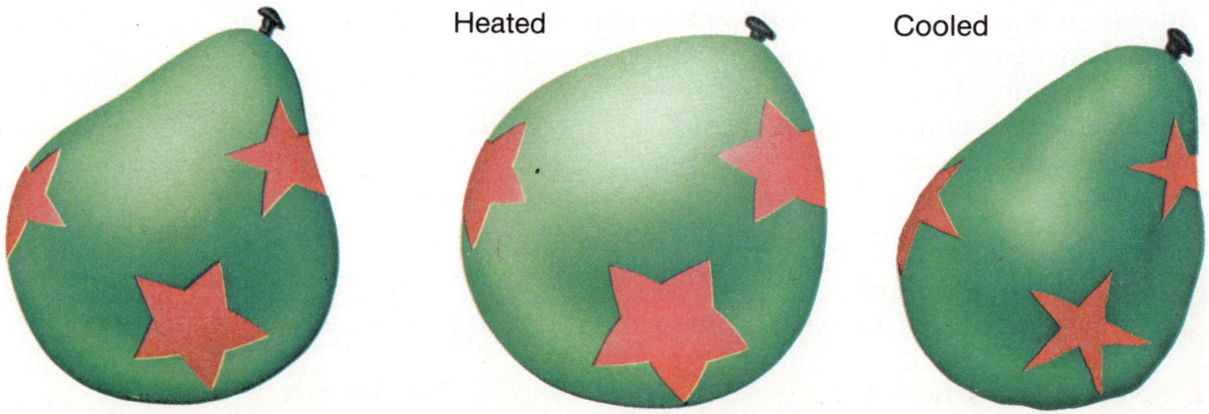

The mercury thermometer makes use of the expansion of liquid mercury to measure temperature. As the temperature around the thermometer increases, the mercury atoms gain kinetic energy and move apart. The mercury expands in a thin column up the thermometer. As the temperature decreases, the mercury atoms lose kinetic energy and move closer together. The liquid volume contracts. The column of mercury becomes shorter.

Different substances expand by different amounts when they are heated. An iron bar one meter long increases in length by about one millimeter when heated from 0°C to 100°C. Under the same conditions an aluminum bar increases in length by about two millimeters. A Pyrex glass rod expands by less than half a millimeter. The amount that a substance expands or contracts in response to changes in temperature is a property of that substance.

Construction engineers must consider the change in the volume of solids with temperature. For example, a 300-meter steel bridge will increase in length by about 6 centimeters when the temperature increases 20°C. Long steel bridges have special structures at the ends to allow for the expansion. Concrete also expands when heated. Concrete bridges are built with expansion joints between the concrete slabs. The expansion joints are needed to keep the bridge from buckling when the concrete expands.

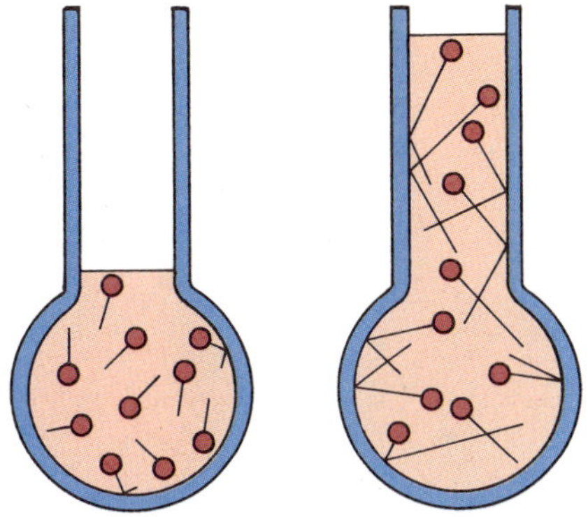

Figure 12.8 Atoms of mercury move more when they gain heat energy. As a result, the mercury expands.

Figure 12.9 Expansion joints

2. Ask the students what happened to the level of the water in the glass tube (it rose). What caused the level of the water to rise (the expansion of the air in the upper part of the flask)? Turn off the burner and use the tongs to remove the flask from the tripod. Allow the flask to cool for a few minutes. Ask the students what they observe happening to the water in the tube (level of water in tube is getting lower). Why is this happening (water molecules are moving closer together)?

3. Heat a pan of water to a boil over a hot plate. In the pan of boiling water place a bottle whose neck is covered with a balloon. The balloon should inflate. Why does the balloon inflate (the pressure increases)? Then place the bottle-and-balloon apparatus in an ice bath. What happens (the balloon deflates)? Why (the air molecules slow down and move closer together)?

4. Ask the students what they can conclude about the effect of heat on the different states of matter (heat causes matter to expand). How does heat cause matter to expand (the molecules of matter become more active and move farther apart)?
5. Ask the students what happens when butter is heated (it melts). Explain that for melting to occur, the butter must absorb heat. The heat is used to overcome the attraction between the molecules in the butter. This heat is called the heat of fusion.

Figure 12.10 Unlike most substances, water does not contract when cooled. It expands.

The different rates of expansion for different metals can be used to make a thermostat. Strips of iron and brass are welded together making a single strip. This strip will bend when heated. It bends because the brass side expands more than the iron side. This bending can be used to turn a pointer, to regulate a valve, or to turn a switch. In this way a thermostat can start and stop a heating system at whatever temperatures are desired.

Increases in heat energy account for the changes from solids to liquids and liquids to gases. Similarly, decreases in heat energy account for the opposite changes. As heat energy is added to a solid, the particles move farther apart. If enough heat is applied to the solid, the particles will move so much that they are no longer closely held in any particular arrangement. When this happens, the solid changes into a liquid. This process is called **melting.** The temperature at which a solid changes to a liquid is the **melting point** of that substance.

Once a solid reaches its melting point, its temperature does not change even as more heat is added. The temperature remains at the melting point as long as there is material to melt. The addition of heat only causes the substance to melt faster. The amount of heat energy used to melt a substance without changing its temperature is called the **heat of fusion** of that substance.

Figure 12.11 This strip of two metals bends when heated because one metal expands more than the other.

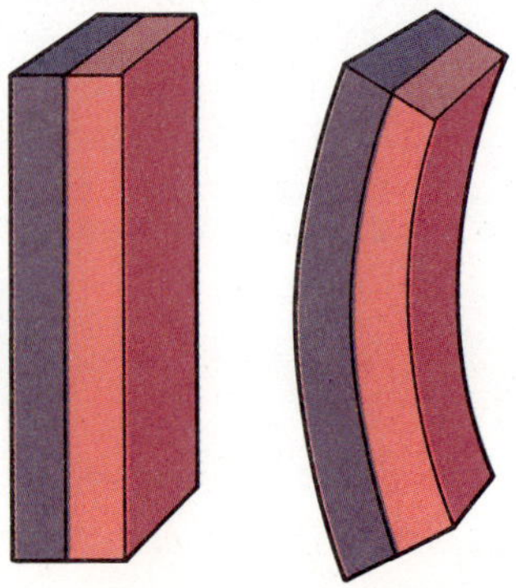

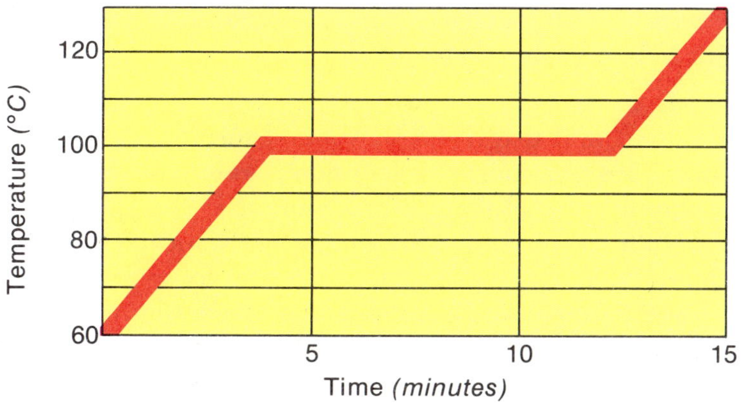

Figure 12.12 This graph shows the change in temperature of water as it boils and evaporates. The line remains level at the boiling point until all the water has evaporated.

You have observed the effects of the heat of fusion many times. When? You see the effects when you put ice cubes in a glass of water. The heat energy used to melt the ice cubes (heat of fusion) is taken from the water. The water cools as the ice cubes melt.

Once a solid has completely melted, the addition of more heat can change the liquid to a gas. As the temperature of a liquid increases, the particles move faster. Some of the particles will reach speeds so high that they can escape from the liquid into the air. The change from the liquid to the gas state is called **evaporation.**

Sometimes liquids boil when they evaporate. The temperature at which a liquid begins to boil is called the **boiling point** of that substance. Adding heat to a boiling liquid does not increase its temperature. The temperature remains at the boiling point until all the liquid has evaporated. Both the boiling and melting points of a substance are properties of that substance.

It is important to remember that liquids can evaporate without boiling. The puddles that remain after a rain shower eventually disappear. Have you ever seen a boiling puddle? Water does evaporate at temperatures below its boiling point. It does so more slowly, however.

Evaporation, like melting, is a cooling process. The fastest-moving liquid particles escape into the gas state. As these particles evaporate, they take energy with them. The heat energy used to evaporate a liquid is called the **heat of vaporization** [vay-puhr-ih-ZAY-shuhn] of that substance.

Figure 12.13 Puddles evaporate at temperatures below the boiling point of water.

6. Ask the students what change of state occurs during the process of evaporation (liquid to gas). Explain that evaporation occurs because some of the molecules are moving rapidly enough to break all attractive forces that exist among them and escape into the atmosphere. Explain that the heat used to change liquids to gases is called the heat of vaporization.
7. Point out the importance of evaporation. The water that evaporates from oceans, rivers, and lakes later returns to the earth as rain. The human body is constantly perspiring. As the fluid evaporates, it cools the body, thereby allowing it to maintain a temperature at which the body's chemicals can work most effectively.

SCIENCE BACKGROUND

The boiling point of a liquid changes with pressure. For example, at sea level, water boils at 100°C; on a mountain, where air pressure is less, water boils at less than 100°C.

ACTIVITY: What Happens When Water Freezes?

In this activity any timepiece that clearly indicates minutes can be used.

Stress that students must be very careful to avoid breaking the thermometers. At the end of the experiment some thermometers may be locked in ice, which should be allowed to melt before attempting to remove the thermometers.

Help the students to prepare graphs to illustrate the temperature changes that occur. Then compare the graphs. There should be a plateau on each. Ask students what was happening to the temperature during the time of the plateau (it was not changing). Explain that when the state of matter is changing the temperature does not change.

● **Laboratory Skills: Manipulation of Materials Experimentation/Demonstration**

● ACTIVITY

What Happens When Water Freezes?

Materials

watch with a second hand
600-mL beaker
test tube
ice
salt
2 thermometers

1. The water temperature line should show a period where the temperature did not change.
2. It froze.
3. Energy was being released during freezing. This energy is equal to the heat of fusion for the water.

Procedure

1. Copy the chart and use it to record your observations. Be sure to handle the thermometers carefully to avoid breaking them.

2. Fill the test tube with water to a depth of about 2.5 cm.

3. Place one thermometer in the test tube. Record the water temperature on your chart beside *0 minutes*.

4. Set the test tube and your second thermometer in the beaker of ice as shown in the diagram. Immediately record the temperature of the ice on your chart beside *0 minutes*.

5. Pour salt over the ice. Record the temperature of both thermometers every minute until both show the same temperature.

6. When the temperatures are equal, remove the test tube from the beaker. If a thermometer is locked in ice, allow that ice to melt before removing the thermometer for cleanup.

Questions

1. Make a graph to show your results. Label the horizontal axis *time* and the vertical axis *temperature*. Use different colors to plot the water temperatures and the ice temperatures.

2. What happened to the water in the test tube?

3. Why was there no change in temperature while the water was freezing?

Time *(minutes)*	Temperature of Water *(°C)*	Temperature of Ice *(°C)*
0		
1		
2		
3		
4		

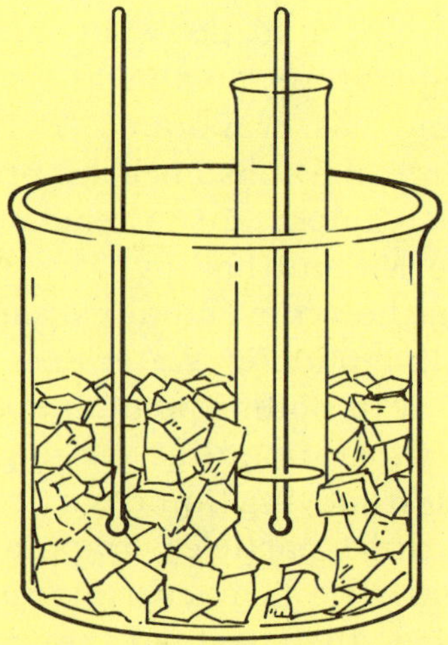

Figure 12.14 In cold weather, water vapor in the breath of people and animals often condenses forming a cloud you can see.

You have noticed the effects of the heat of vaporization many times. How often have you felt chilled when you stepped out from a shower or a swim dripping wet? Did you feel warmer after you dried off? The water on your skin was evaporating. The heat energy used to evaporate the water (heat of vaporization) was taken from you! When you perspire, your body is cooling itself following the same principles. As the perspiration evaporates from your skin, you feel cooler.

As you might guess, when temperatures decrease, the changes occur in reverse. A gas changes to a liquid. A liquid changes to a solid.

The change of a gas to a liquid is called **condensation** [kahn-duhn-SAY-shuhn]. Condensation is the opposite of evaporation. As the temperature drops, the particles in a gas move more slowly and cover shorter distances. They eventually come close together condensing into a liquid. Drops that form on the outside of a glass of cold water, fog, clouds, and dew are all examples of the condensation of water vapor.

The change of a liquid to a solid is called **freezing.** As the temperature of a liquid drops, the molecules in the liquid move more and more slowly. Finally, they come very close together, barely moving at all.

Do You Know?
On a brisk 5-km walk your body burns fuel, producing enough heat energy to raise the temperature of 10 kg of water by 21°C.

Study Questions for 12.3

1. How can heating cause a substance to expand?
2. What happens when a solid melts?
3. What happens when a liquid evaporates?

1. The particles in the substance gain kinetic energy and move farther apart, causing the substance to take up more space.
2. The particles gain kinetic energy and move more.
3. The particles gain enough kinetic energy to escape into the atmosphere.

8. Draw a glass of ice water on the board. Ask the students what they would notice on the outside of the glass if they came back in about five minutes (drops of water). Ask where the water comes from (air). What is the process in which a gas becomes a liquid (condensation)? Explain that as the water molecules from the air were cooled by the cold glass they came closer together and a liquid was formed.
9. Ask the students the name of the process by which a liquid becomes a solid (freezing). Does the temperature change when a liquid freezes (no)?

FOLLOW-UP

Reinforcement: Changes of State
Have the students create a collage that shows at least two examples of each of the following processes: evaporation, condensation, freezing, and melting. On the back of the collage the students should write a paragraph to explain the role of heat energy in each process.

Enrichment: Cooking Racks
Many cooking product catalogs show racks that will help cook food more quickly. Have students construct their own rack by sticking an aluminum nail through a potato. Have them bake one potato with an aluminum nail stuck through it and another potato without a nail. Which potato bakes more quickly?

Tie-in: Sidewalks
When masons are laying cement sidewalks, they leave a gap every meter or so. Have students interview a mason and ask the reason for this. What would happen if the gap were not left?

SCIENCE BACKGROUND

Farmers often wet down their plants on a cold night. As the water freezes, it gives off heat. This heat prevents the fruit itself from freezing, since the fruit freezes at a lower temperature than does the water. The heat released from the freezing water keeps the fruit from freezing.

12.4

OBJECTIVES: The student should be able to

- **describe** the process of heat conduction;
- **distinguish** between conductors of heat and insulators of heat;
- **describe** the process of convection;
- **explain** how radiation differs from conduction and convection;
- **list** examples of heat transfer by convection, conduction, and radiation.

SCIENCE TERMS

conduction convection insulators radiation

ADVANCED PLANNING

1. For demonstrations you will need several different metals cut in 15-cm (6-in.) pieces, a tripod, some wax, several tacks, a Bunsen burner, some matches, some wire gauze, a 1,000-mL beaker, some sawdust or dots of colored paper, a pair of scissors, some colored construction paper, some thread, and a high-intensity lamp.

2. Begin gathering cans and insulating materials needed for the activity.

3. Before the class begins, assemble the equipment for the first demonstation. To one end of each metal piece fasten a tack with wax. Place a thin wire gauze on top of the tripod and place the metal pieces on top of the gauze so that one end of each of the pieces is directed toward the center. The end with the tack should be pointed away from the center and positioned so that it hangs slightly over the edge of the wire gauze. Set up the tripod over a Bunsen burner so that the flame will be in the center, equidistant from the metal pieces.

LESSON PLAN

1. As the class begins, light the burner and ask the students to watch. What do they notice (the tacks fall from the metals)? Why do they fall (the bars get hot and the wax melts)? Do the tacks all fall at the same time (no)? Why not (some bars conduct heat faster)? Point out that you heated only one end of each metal piece yet the other end of the metal piece became so hot that the wax melted and the tack fell. Explain that the heat traveled through the metals by conduction.

12.4 The Flow of Heat

You will find out

- how heat energy is transferred in conduction, convection, and radiation.

When hot cocoa is poured into a mug, the mug itself gets warm. Heat is transferred from the cocoa to the mug by **conduction** [kuhn-DUHK-shuhn]. In heat conduction heat energy is transferred from a hot object to a cooler one by collisions between particles. The two objects must be touching for this transfer to occur. The faster-moving particles on the boundary of the hot object collide with the slower-moving particles on the boundary of the cooler object. The slower-moving particles gain energy. These particles move faster. They strike other particles inside the cooler object. In this way the heat energy is spread throughout the object. You feel the effects of heat conduction when you touch something hot. The heat is transferred from the surface of the hot object to your hand.

Figure 12.15 Only the bottom of this pan is touching the burner. However, you would still need a pot holder to grip the handle because heat is transferred from the burner to the bottom of the pan to the handle by conduction.

Some materials are better conductors of heat than others. Metals are very good conductors. Most gases are poor conductors of heat because the particles in a gas are far apart. They don't bump into each other very often, and so the heat is transferred slowly. Poor conductors of heat are called **insulators.** Gases such as air are good insulators. A knit sweater keeps you warm because the air trapped between the fibers slows down the flow of heat from your body to the cold air outside.

Figure 12.16 The air trapped between the fibers of a knitted hat acts as an insulator.

ACTIVITY Which Material Is the Best Insulator? ●

Materials

2 beakers
tin can
hot plate
asbestos gloves
thermometer
insulating material, such as Styrofoam chips, cloth, newspaper
safety glasses

Procedure

1. Put on your safety glasses. Fill the can with insulating material. Place an empty beaker in the can and mold the insulation around it so the beaker fits snugly. Be sure that there is insulation under the beaker and all around it.

2. Use the hot plate to heat 100 mL of water in a beaker to 75°C.

3. Carefully pour the water into your insulated beaker. Use asbestos gloves to lift the beaker of hot water.

4. Immediately measure and record the temperature of the water and the time.

5. Watch the temperature of the water and determine how long it takes to cool to 40°C.

6. Copy the chart. Write your name, the kind of insulating material, and the time it took for the water to cool to 40°C on the chart. Share your information with your classmates and add their information to your chart.

Name	Insulating Material	Time Until Cooled to 40° C

Questions

1. Make a bar graph to compare the time it took for each group's water to cool to 40°C.

2. The best insulating material is the one that required the most time for the water to cool to 40°C. List the materials in order from the best insulator to the worst insulator.

2. Ask the students if all metals conduct heat at the same rate (no). How did the demonstration show this (the metals were all the same distance from the burner but the tacks did not all fall off at the same time)?

3. Explain that just as some metals are not as good conductors as other metals, many other materials are not good conductors of heat at all. These materials are called insulators. Conductors allow heat to pass quickly along them; insulators do not.

4. Ask how good insulators can be useful (the handles of cooking pans are often made of wood, a good insulator, so that the pan can be picked up without burning the hands; in the same way, potholders are also helpful because the cloth is an insulator).

5. Stress that conductors transfer heat through molecular collisions or the collision of electrons and that insulators do not allow heat to pass through them easily. Ask the students to make a list of conductors of heat and insulators of heat (some conductors are iron, copper, silver, steel; some insulators are wood, cloth, paper, air).

ACTIVITY: Which Material Is the Best Insulator?

All beakers and tin cans must be identical. Otherwise, additional variables such as the different thicknesses of the insulation must be considered when the students discuss their results.

◇ **Caution:** Students should wear safety glasses during this experiment and should be extremely careful with the hot water. Do not let them heat the water to boiling. (You may wish to tell the students to heat the water to about 75°C and let it cool 65°C from the starting point).

1. Bar graphs should have the materials on the horizontal axis and time on the vertical axis.
2. The material with the longest bar in the graph should be listed as the best insulator. Others should follow in order from next longest bar to shortest bar.

● **Laboratory Skills: Manipulation of Materials**
Experimentation/Demonstration

7. Fill the 1,000-mL beaker with 800 mL of water and add sawdust or the colored paper dots. Allow the dots to settle to the bottom of the beaker. Place the beaker on the tripod and light the Bunsen burner. As the water heats up, ask the students to watch the path that the sawdust follows. What do they observe (particles rise in the center and fall at the edges)? Cut a spiral from a piece of colored paper. Tie a piece of thread to the center and suspend it above the Bunsen burner flame. Ask the students to describe the motion of the spiral (it rotates).
8. Ask the students what path the sawdust followed (up from the bottom to the surface and back down, a circular motion). Why did the sawdust follow a circular path (the sawdust was carried by the water; as the water was heated it became less dense; the cooler, denser water sank beneath the heated water and pushed it to the top; when this water was heated it became the less dense water, while the top water cooled and became more dense)?
9. What happened to the spiral (it rotated)? How was the action of the spiral like the action of the sawdust (the air above the flame was heated; as the air was heated it became less dense, rose, and was replaced by cooler, more dense air; as the air rose it carried the spiral up with it)?
10. Explain to the students that they have been observing heat transfer by convection. Point out that convection is important in heating homes. The air near a radiator is heated. It becomes less dense and rises as the cooler air near the ceiling falls to replace it. As the two portions of air continuously change places, they carry air that heats the room.

● **Comprehension Skill: Applied**

Figure 12.17 In this example of convection, hot air rises and is replaced by cool air.

Conduction is a good way to spread heat energy in solids. In liquids and gases, **convection** [kuhn-VEHK-shuhn] is usually more efficient. Convection is the spreading of heat energy by large movements of a liquid or a gas.

To understand how convection works, picture a fire in a fireplace. The chimney opening is much smaller than the opening of the fireplace into the room. Why does the smoke so conveniently go up the chimney instead of pouring out into the room? The fire heats the air around it, causing the air to expand. As the hot air takes up more space, it becomes less dense. The less dense hot air floats on, or rises above, the more dense cool air surrounding it. This hot air rises up the chimney, taking the smoke and ashes with it. Cool air from the room flows into the fire from all sides to take the place of the rising hot air. This cool air is soon heated and rises to be ● replaced by more cool air. A continuous circulation pattern of hot and cool air is set up in the room. Heat energy is moved around. This circulation of air is an example of convection.

For heat to flow by conduction or convection, matter must be present to carry the heat. Can heat flow in empty space, that is, in the absence of matter? The answer must be yes. When you stand in the sunshine, you feel warm. Heat from the sun has reached you. Most of the space between the earth and the sun is empty. The sun's energy travels through empty space on its way to you on the earth. This energy is often called radiant energy because it flows by means of **radiation** [ray-dee-AY-shuhn]. Radiation is a form of energy transfer that occurs through matter and through empty space.

The sun produces radiant energy. So does a fire, a radiator, or a pan of hot water. So do you! You may have noticed that a room full of people is much warmer than an empty room. Radiant energy from the people heats the room. Radiant energy travels in all directions. It can be transferred from one object to another without contact between the two objects. The sun's energy warms the earth even though the earth and the sun never touch.

Figure 12.18 This collared lizard is warmed by the radiant energy of the sun.

Study Questions for 12.4

1. What is heat conduction?
2. What is heat convection?
3. How is radiation different from conduction and convection?

1. the transfer of heat energy from a hot object to a cooler one by collisions between particles
2. the spreading of heat energy by large movements of a liquid or a gas
3. In radiation, energy can be transferred through empty space. To transfer energy by conduction or convection, matter must be present.

11. Mention the third method of heat transfer—radiation. Ask for two students to volunteer to help you in a demonstration. Arrange a high-intensity lamp so that the bulb is facing down onto the table top. Ask one student to place a thermometer on the table so that the bulb of the thermometer is directly under the light bulb. Ask the other student to construct a chart on the board that can be used to record the temperature at 15-second intervals. Tell the students that when you say "Start," you want the light turned on. Then every 15 seconds you want the student at the board to say "Time" and record the temperature the other student reads out. The thermometer should not be moved or touched. Record the temperature for two minutes.
12. Analyze the chart with the students. How long did it take for the temperature to rise (it rose immediately)? Explain that the heat from the light was transferred through the air to the thermometer by radiation. Why couldn't the heat transfer have occurred by conduction (because gases are poor conductors of heat)?

FOLLOW-UP

Reinforcement: Methods of Heat Transfer

Have students work in groups of three. Ask each group to prepare a three-chapter report on the methods of heat transfer. Each student should be responsible for one chapter. Each chapter must define and explain the method of heat transfer being discussed, give three examples of the method, and have two diagrams to illustrate the ideas presented in the chapter.

Enrichment: Heat Transfer at Home

Ask each student to make a model of the living room in his or her home. Use the model to explain how the three methods of heat transfer work in the room. How could the student use insulation to improve the heating conditions in the room?

Tie-in: The Davy Lamp

Show the students that when a piece of wire gauze is held above a flame, the flame will not go above the wire. Ask the students to find out about the Davy lamp. When is it used? By whom? Why does it work? How is it like the wire gauze you just showed them?

What Did You Learn?

- 12.1• Count Rumford discovered that heat is a form of energy.
- 12.1• The kinetic theory states that the atoms and molecules that make up matter are always moving.
- 12.1• The heat energy of a substance is the total kinetic energy of all the particles of that substance.
- 12.2• Temperature is a measure of the average kinetic energy of the particles that make up a substance.
- 12.2• At absolute zero, the lowest possible temperature, almost all motion of particles stops.
- 12.2• Heat energy can be measured in joules or calories.
- 12.2• The specific heat of a substance is the amount of heat needed to raise the temperature of one gram of that substance one degree Celsius.
- 12.3• Most substances expand when heated and contract when cooled.
- 12.3• When solids gain enough heat energy, they change to a liquid, or melt.
- 12.3• The heat energy required to melt a substance without changing the temperature is the heat of fusion of that substance.
- 12.3• When liquids gain enough heat energy, they change to a gas, or evaporate.
- 12.3• The heat energy used to evaporate a liquid is the heat of vaporization of that substance.
- 12.3• The change of a gas to a liquid is called condensation.
- 12.3• The change of a liquid to a solid is called freezing.
- 12.4• In conduction, heat energy is transferred from a hot object to a cooler one by collisions of particles.
- 12.4• Poor conductors of heat are called insulators.
- 12.4• In convection, heat energy is spread by large movements of a liquid or gas.
- 12.4• Radiation is a form of energy transfer that occurs through matter and space.

Key Terms

kinetic theory
heat energy
temperature
absolute zero
calorie
specific heat
melting
melting point
heat of fusion
evaporation
boiling point
heat of vaporization
condensation
freezing
conduction
insulators
convection
radiation

Career

Mechanical Engineer

Mechanical engineers work on methods to produce and use power. They design and develop machines that produce power. Some examples of these machines are internal combustion engines, jet and rocket engines, steam turbines, and nuclear reactors. Mechanical engineers also design and develop machines that use power, such as automobiles, heaters, air conditioners, printing presses, and steel-rolling mills.

A mechanical engineer may work in any one of many specialized branches. He or she could be working on the testing and design of cars, locomotives, or jet engines. Mechanical engineers also build precision instruments and machines for highly specialized work. This equipment is used in many fields such as underwater prospecting and rocket engines for the space shuttle.

Some mechanical engineers figure out how to build machines. Others test the machines. Still others oversee the actual manufacture of the machines. Highly complicated or technical machines are also sold by mechanical engineers, since they are the ones who best understand how the machines work.

In the years to come, industrial machines and processes will get more complicated. Mechanical engineers will be needed to build new and better machines. Mechanical engineers will also be needed to develop new energy systems and to help solve environmental pollution problems.

To prepare for a mechanical engineering career, you should take as much mathematics as possible in high school. Physics and chemistry also provide good background. English should not be overlooked, since a mechanical engineer must be able to communicate skillfully with others about his or her work. Most jobs in mechanical engineering require a college degree.

Vocabulary

1. kinetic
2. temperature
3. specific heat
4. calorie
5. fusion
6. vaporization
7. evaporation
8. condensation
9. convection
10. radiation

● Word Skill: Vocabulary

What Do You Remember?

1. Count Rumford. The students should describe Rumford's observations while drilling the cannon.
2. the total kinetic energy of the particles in a substance
3. a measure of the average kinetic energy of the particles in a substance
4. joule and calorie
5. the amount of heat needed to raise the temperature of 1 g of water 1°C
6. degrees Celsius
7. All matter is made of atoms and molecules. These atoms and molecules are in constant motion.
8. The molecules gain kinetic energy and move farther apart.
9. The molecules slow down and position themselves together.
10. the heat energy used to melt a substance without changing the temperature

TO THINK ABOUT AND DO

Vocabulary ●

Number your paper from 1 to 10. Beside each number write the word or words from the list that best complete each sentence.

1. Energy of motion of particles is ________ energy.
2. ________ is the measure of the average kinetic energy of particles.
3. The amount of heat required to raise the temperature of 1 g of a substance 1°C is its ________.
4. A unit used to measure heat energy is the ________.
5. The energy used to melt a substance is the heat of ________.
6. The energy used to change a liquid to a gas is the heat of ________.
7. A liquid changes to a gas in the process of ________.
8. A gas changes to a liquid in the process of ________.
9. The method of heat transfer that involves the movement of a liquid or a gas is ________.
10. Heat energy travels through empty space in ________.

calorie
radiation
fusion
kinetic
evaporation
temperature
specific heat
condensation
boiling point
convection
vaporization
absolute zero

What Do You Remember?

1. Name the scientist who proposed that heat results from the motion of molecules. Why did he believe this?
2. What is heat energy?
3. What is temperature?
4. What units are used to measure heat?
5. What is a calorie?
6. What units are used to measure temperature?
7. Explain the kinetic theory of matter.
8. Why does ice melt when it is heated?
9. Why does water freeze when it is cooled?
10. What is heat of fusion?

11. What is heat of vaporization?
12. Explain how evaporation occurs.
13. Define conduction, convection, and radiation as methods of heat transfer.
14. Name three good conductors.
15. Name three good insulators.

Applying What You Have Learned ●

1. Study the drawing. If the ball is heated, will you be able to pull it through the ring? Explain your answer. How can you get the ball through the ring?
2. How can the way a thermos is constructed be used to illustrate convection, conduction, and radiation?
3. Why do railroad tracks have spaces between the rails at certain intervals?
4. If you pour hot water into a glass, it may break. But if there is a metal knife in the glass, it won't. Why?
5. Why do people put jars under hot water to help open them?
6. How does the motion of the molecules of a solid at 30°C compare to the motion of molecules in the same substance at 50°C?

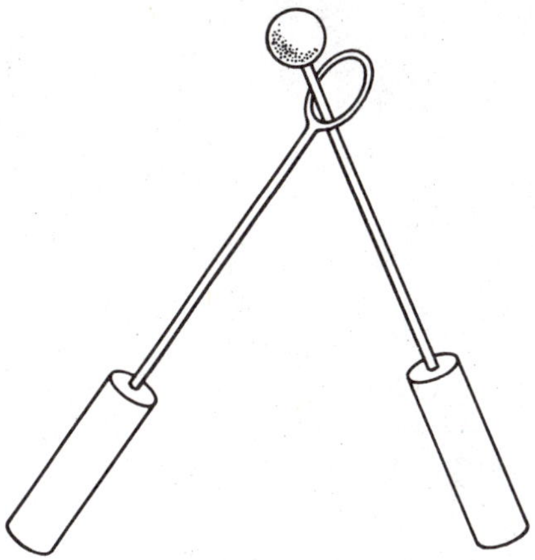

Research and Investigation ●

1. Three temperature scales are used—Fahrenheit, Celsius, and Kelvin. Compare the boiling points and freezing points of water on each scale. Where or when is each scale used?
2. Cooling substances to near absolute zero often changes their properties. Give some examples of these changes.
3. What is sublimation? Give two examples.

11. the heat energy used to evaporate a liquid without changing the temperature
12. The particles gain so much energy that they escape into the air.
13. conduction—transfer of heat when particles collide with each other, passing energy along
convection—spreading of heat energy by large movements of a liquid or a gas
radiation—transfer of energy through empty space
14. Students will probably name three metals.
15. Answers will vary. They may include air and other gases, liquids, and solids that are nonmetals.

Applying What You Have Learned

1. No, the ball will expand. Heat the ring or cool the ball to separate the two.
2. It is double walled with a partial vacuum between the walls to prevent heat loss by convection or conduction. The cork prevents heat loss by conduction. The inside is silver in color to reflect radiant heat back into the contents.
3. The space allows for expansion from heat without bending. A bend could cause a train to derail.
4. The glass breaks because the inside expands faster than the outside. It won't break with a knife in it because the knife absorbs some of the heat.
5. to expand the metal in the cover
6. The molecules will have more kinetic energy at 50°C and will be moving farther and faster at that temperature.

● Comprehension Skill: Applied

Research and Investigation

1. The Fahrenheit scale is used in the United States. Most other countries use the Celsius scale. On the Fahrenheit scale water freezes at 32°F. This is the same as 0°C. Water boils at 212°F or 100°C. The Kelvin scale is used to measure very low temperatures. On the Kelvin scale 0°K is equal to −273°C. On the Kelvin scale water freezes at 273°K and boils at 373°K.
2. Helium, normally an inert gas, becomes a liquid at temperatures near absolute zero. It will move up the inside of its container. Lead at very low temperatures will ring like a bell when struck. Normally lead absorbs sound. Rubber becomes brittle and breaks when touched. Some materials that are conductors under normal conditions become excellent conductors at very low temperatures.
3. In the process of sublimation a solid turns directly into a gas without becoming a liquid first. Two examples are dry ice and mothballs. Similarly, a gas can turn into a solid without ever becoming a liquid. This change is also called sublimation. Two examples are frost and snow.

● Reference/Research Skills

CHAPTER BACKGROUND

An engine is any device that can change energy into mechanical work. Engines use the energy stored in fuels to create motion.

Steam, diesel, and gasoline engines are all heat engines since they convert heat energy into mechanical energy. In gasoline and diesel engines, fuel is burned in the cylinder, thereby creating a high pressure that moves a piston to turn a shaft. The burning of the fuel within the engine further classifies the diesel and gasoline engines as internal combustion engines. This distinguishes them from steam engines, where the fuel is burned away from the cylinders of the engine. This kind of engine is known as an external combustion engine.

Heat pumps are machines that reverse the natural flow of heat energy from a warmer region to a cooler region. A refrigerator is a kind of heat pump that pumps heat energy from the cool air inside it and ejects the heat energy into the warmer air of the kitchen. Air conditioners act as heat pumps when they pump heat from the relatively cool indoors to the warm outdoors.

A heat pump can also be used to warm a house by pumping heat from the outside into the house. Such a pump is essentially the same device as a refrigerator. It contains a set of outdoor coils that absorb heat from the air outside. This heat is then carried indoors and released in the indoor coils. Even though the air may be very cold in the winter, the coils can still absorb heat from it. Once the heat is pumped from the outside into the house, fans are used to circulate the warm air throughout the house.

13

Heat Energy in Your Life

Introduce the lesson by setting up a test tube of water over a Bunsen burner. Place a marble in the test tube. Light the burner and ask the students to observe what happens to the marble (rises and falls in the test tube). Ask what is providing the energy that causes the marble to move (heat from the burner). Tell the students that in this chapter they will learn how heat energy is converted to work and how it is able to move the parts in the engines of cars and trains.

Imagine for a moment what it would be like if all the machines and engines that depend on heat energy for their power suddenly stopped working. The first thing you might notice would be that the electricity would go off. Heat energy is a necessary part of most electrical power plants. You would have no lights, no television or radio, and none of the appliances that depend on electricity for their power.

Going outside, you would notice that all traffic had stopped. Automobiles, trucks, buses, and trains depend on heat energy for their power too. There would be no airplanes flying overhead and no subways running underground. The work in most factories would grind to a halt.

In short, the world as you know it would no longer exist. The modern industrial world of today has come about in large part because of the work heat energy has done and continues to do.

Figure 13.1 An early steam-powered tractor

Lesson Organizer

13.1 Heat Engines

Pages 274–275
Advanced Planning: Page 274
Follow-up: Page 275
- **R** A Heat Engine Model
- **E** The Hero Engine
- **T** The Miner's Friend

13.2 External Combustion Engines

Pages 276–279
Advanced Planning: Page 276
Activity: How Does a Steam Turbine Work? Page 278
Follow-up: Page 279
- **R** How Does a Steam Engine Operate?
- **E** Model Engines
- **T** Settling the West

13.3 Internal Combustion Engines

Pages 280–283
Advanced Planning: Page 280
Follow-up: Page 283
- **R** Types of Engines
- **E** Engine Efficiency
- **T** Wankel Engines

13.4 Cooling and Heating Systems

Pages 284–287
Advanced Planning: Page 284
Activity: How Does Evaporation Affect Temperature? Page 285
Follow-up: Page 287
- **R** Refrigerator Manual
- **E** Wet- and Dry-Bulb Hygrometers
- **T** Bacterial Spores

See ACTIVITIES book pages 73–78 for laboratory, skill, and review exercises for Chapter 13.

13.1

OBJECTIVES: The student should be able to

- **describe** how a heat engine does work;
- **identify** the parts of a heat engine;
- **distinguish** between external and internal combustion engines;
- **list** some sources of heat for heat engines.

SCIENCE TERMS

heat engine	crankshaft
cylinder	external combustion engine
piston	internal combustion engine

ADVANCED PLANNING

With your students' social studies teacher plan a classroom discussion of how steam engines spurred the rise of industry. The discussion should be set up to coincide with your study of steam engines in the first and second lessons of this chapter.

LESSON PLAN

1. Introduce the lesson by reviewing the concept of heat as a form of energy caused by the motion of particles. Point out that the higher the temperature of a gas the more quickly the particles move and the more frequently they collide.
2. On the board draw a simple diagram of a heat engine that includes the cylinder containing the heated gas, the piston, and the crankshaft.

● **Comprehension Skill: Applied**

SCIENCE BACKGROUND

Many heat engines used today are steam engines. The first person to demonstrate the power of steam was Hero, a Greek scientist in Alexandria, more than 2,000 years ago. He created a device called the Ball of Winds.

13.1 Heat Engines

You will find out
- how heat engines do work;
- where the heat for heat engines comes from.

Figure 13.2 Trucks use heat engines.

Imagine that you are in charge of moving a piano to a place 400 kilometers away. You have to do work. To do work, a force must move an object a distance. You must supply a force to move the piano 400 kilometers. How will you do it? You can't carry it on your back. The distance is too far, and the piano is too heavy. You will probably use a truck. How is a truck able to do work? How does any truck or car operate? Many people know how to drive. They know how to turn the engine on and off and how to add fuel. But what makes a car or a truck move? If you look under the hood of a car or a truck, you will see a heat engine. Heat engines move cars and trucks.

A **heat engine** is a device that uses heat energy to do work. The engine does work by changing heat energy into energy of motion. How does this change occur? ● When a gas is heated, its pressure increases. The pressure increases because the gas molecules gain kinetic energy when heated. This increase in kinetic energy causes the molecules to move faster. The result is an increase in pressure.

If a container is elastic such as a balloon, that container will expand as the pressure increases. But what if the container is rigid and cannot expand? In that case, the pressure inside increases when heat is added.

Figure 13.3 Molecules are heated in an open, an expanding, and a sealed container. Inside which container is the pressure greatest? sealed container

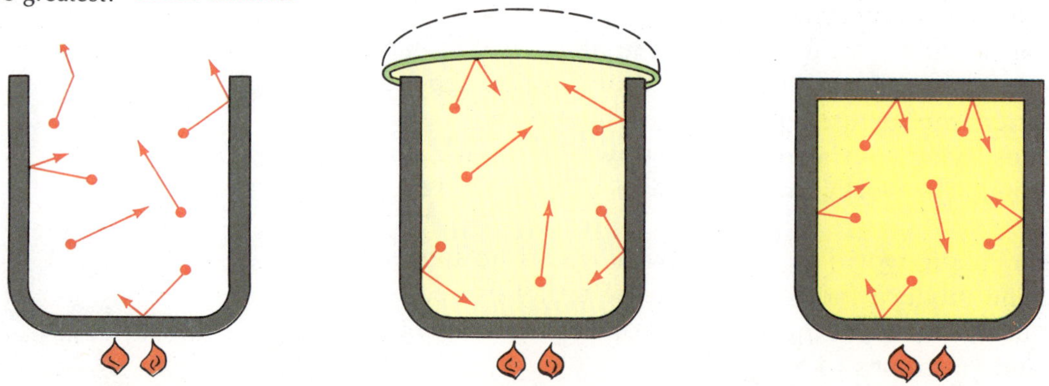

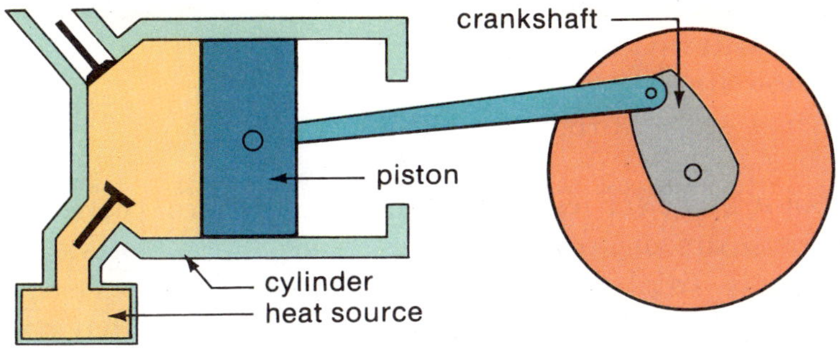

Figure 13.4 Parts of a heat engine

How does an increase in pressure produce energy of motion? You have probably observed the answer to this question. Suppose you boil water in a covered pan. The boiling water changes to steam, a gas. The pressure increases. Since the cover is not held tightly, the increased pressure forces it up. You can see the cover on a pan of boiling water move up.

Many heat engines work in a similar way. In a heat engine the container that confines the heated gas is a **cylinder.** The movable top of the cylinder, called a **piston,** corresponds to the cover on the pan. In these heat engines, the outward movement of the piston is used to turn a rod that turns a **crankshaft.** The turning crankshaft can be used to turn one or more wheels.

Where does the heat needed to increase the pressure inside the cylinder come from? It is released in the burning, or combustion [kuhm-BUHS-chuhn], of a fuel. If the burning takes place outside the cylinder, the engine is an **external combustion engine.** If the burning takes place inside the cylinder, it is an **internal combustion engine.** The fuel that burns can be wood, coal, gas, or oil.

In recent years the heat energy released in nuclear reactions has been used to supply the heat for some heat engines. Nuclear-powered submarines and aircraft carriers are in use today.

Do You Know?
A liter of water will boil into about 1,700 liters of steam if the pressure is constant.

Study Questions for 13.1

1. What causes the piston in a heat engine to move?
2. How are internal and external combustion engines different?

1. A gas in the cylinder is heated. Its pressure increases.
2. Fuel is burned in the cylinder in internal combustion and outside the cylinder in external combustion.

3. Ask the students what happens when the gas molecules in the cylinder are heated (the molecules move faster).
4. Point out the piston and explain that a piston is a solid cylinder that fits snugly inside the larger cylinder and moves as a result of pressure from the heated gas. What will happen to the piston as the gas in the cylinder expands (it will move out)?
5. Remind the students that an engine whose fuel is burned outside the cylinder is called an external combustion engine. Where is the fuel burned in an internal combustion engine (inside the cylinder)?

● **Word Skill: Vocabulary**

FOLLOW-UP

Reinforcement: A Heat Engine Model
Have students construct a cutaway model of a heat engine. Toothpicks with paper flags can be used for labeling the basic parts: cylinder, piston, and crankshaft. An index card that explains how the heat engine operates should also be included.

Enrichment: The Hero Engine
Interested students can make a Hero engine from a small tea or spice tin can. Have the students punch holes in two sides of the can that are diagonally opposite each other. The students should then hang the can from a ring stand by a fish line swivel or pull-chain. A small amount of water should be placed at the bottom of the can and a Bunsen burner lighted under it. When the water boils and steam forms, the steam should be directed to the side as it escapes from the holes. Students should be able to observe that this causes the can to rotate in a direction away from that of the steam jet.

Tie-in: The Miner's Friend
Practical engineering problems have always stimulated the growth of scientific knowledge. Have students research the work of Thomas Savery of England. What was the problem that Savery was initially trying to solve (with the growing demand for coal in the 1700's, there was great need for pumping groundwater out of mines)? How did the "Miner's Friend," Savery's steam engine, operate?

13.2

OBJECTIVES: The student should be able to

- **describe** the development of the steam engine;
- **explain** how a steam turbine operates.

SCIENCE TERM

steam turbine

ADVANCED PLANNING

1. To demonstrate the principle upon which the steam turbine operates, you will need a small spouted teakettle into which you place a cup of water. Make a turbine wheel from the round top of a tin can. Every 0.6 cm (¼ in.) along the circumference draw a line to the center of the circle. Into each line cut a 1 cm (⅜ in.) line with metal cutting snips. Use a nail to punch a hole in the center of the tin circle. Mount the turbine wheel on the end of a stick with a nail that fits loosely into the center hole. When the water in the kettle is boiling vigorously, hold the turbine in the jet of steam escaping from the spout.

2. See activity notes on student home preparation of turbine disks.

LESSON PLAN

1. Explain to the students that the early steam engines were hard to use because they had to be heated and then cooled to allow the piston to move. Point out that the English inventor James Watt improved the early steam engines by discovering that the piston could be pushed back and forth by changing the flow of steam in and out of the engine.

13.2 External Combustion Engines

You will find out

- how steam engines developed;
- how steam turbines work.

Some of the earliest heat engines were external combustion steam engines. The first external combustion steam engines were not very successful. The engines were large and inefficient. They worked so poorly that one observer complained that it took an iron mine to build the engine and a coal mine to run it. Finally in 1769, James Watt, a Scot, made a steam engine that was much more efficient. Soon Watt's steam engines were in use in factories all over Great Britain.

Figure 13.5 This painting illustrates a 21-kilometer race that took place in 1830 between a horse-drawn cart and the *Tom Thumb,* a locomotive powered by Watt's steam engine. The *Tom Thumb* was in the lead but broke down before the finish line.

The first locomotives were steam engines. These steam locomotives, nicknamed iron horses, quickly replaced boats and horses in moving goods and people. As you might guess, the early steam locomotives were not without problems. One of the first passenger trains in the United States was the *DeWitt Clinton.* It ran between Albany and Schenectady, New York. The *DeWitt Clinton* had a woodburning steam locomotive and open-air passenger cars. The train's jerk as it started knocked most of the passengers out of their seats. Sparks from the smokestack set fire to the umbrellas that passengers raised as shields from the smoke and wind. Through

Figure 13.6 The *De Witt Clinton*

trial and error the trains were improved. The trains in the United States were recognized as some of the best in the world.

The success of steam trains soon made other forms of land transportation in the United States out of date. The pony express, stagecoaches, and horse-drawn wagons all became part of history as the iron horses took over. The development of railroads was an important factor in the growth of industry and the westward expansion of the United States in the nineteenth century.

Figure 13.7 This modern plant uses steam in an external combustion engine to produce electricity.

2. Explain that the change in the flow of steam was achieved by the sliding valve. Once this improvement was made, the steam engine became practical. In fact, it became the basis for the rise of industry in England. At that time it was also found that the steam engine could be attached to a movable platform and the locomotive, or iron horse, became a reality.

Reference/Research Skills

SCIENCE BACKGROUND

When the steam engine was developed, manufacturers of the machine decided to talk about their new product in terms of the horses it would replace. The unit *horsepower* was standardized by James Watt to mean the amount of work a horse could deliver each second. Today the horsepower unit is used to rate the performance capability of electric motors and automobile engines. One horsepower is equal to 746 newton•meters per sec.

ACTIVITY: How Does a Steam Turbine Work?

◇ **Caution:** Students should wear safety glasses at all times during this experiment. Students should be reminded not to direct the steam toward anyone, since steam can burn the skin.

Glass tubing should always be held in a thick paper towel or glove when inserting it into a stopper. You may wish to insert the glass tubing into the stopper for the students. The tubing should be twisted (not pushed) into the stopper.

Have the students prepare their turbine disks at home. To do this, they should cut three circles—15 cm, 10 cm, and 5 cm in diameter—from a piece of tagboard. In the center of each disk they should draw two circles, one with a diameter of 2.5 cm and one with the same diameter as the thimbles they will use.

Have the students use a ruler to divide each disk into eight sections. One section should be colored red. The students should cut along the eight section lines until they reach the circle whose diameter is 2.5 cm. In this way the edge of the disk is in sections but the center remains intact. Each disk section should be folded in half to form a blade.

To avoid having each student use each turbine, you may wish to have different students do the experiment with turbines of different sizes and then share their results.

Another variation of the experiment would be to use turbines of the same size that are made from materials of different densities such as cardboard, balsa wood, or tin pie plates.

● **Laboratory Skills: Manipulation of Materials**
Experimentation/Demonstration

● ACTIVITY

How Does a Steam Turbine Work?

Materials

turbine disks
eyedropper
ring stand and clamp
Bunsen burner
one-hole rubber stopper
test tube
long pencil with eraser
thimble
straight pin
safety glasses
apron

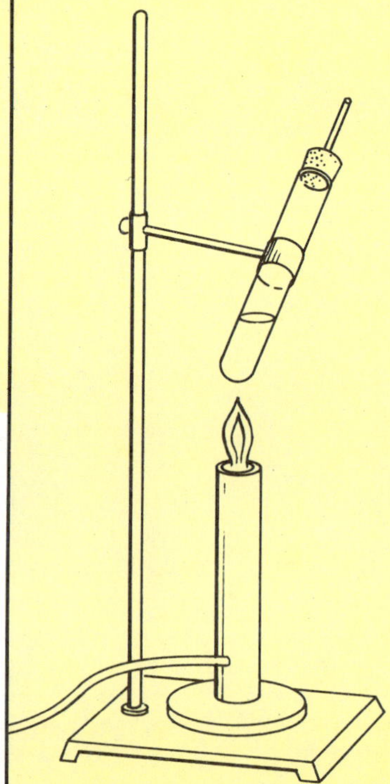

Procedure

1. Put on your apron and safety glasses. Remove the rubber cap from the eyedropper, leaving only the glass tube. Wet the outside of the glass tube and carefully insert the wider end into the hole of the rubber stopper.

2. Fill the test tube 1/4 full with water. Carefully place the stopper in the test tube. The glass tube should extend out. CAUTION: The stopper should not be pushed into the test tube too tightly.

3. Set up the test tube, Bunsen burner, ring stand, and clamp as shown in the diagram.

4. Use the pin and thimble to mount the turbine disk on the pencil as shown in the diagram.

5. Light the Bunsen burner and wait for the water to boil. CAUTION: Do not stand or place your hand in front of the steam.

6. Holding the pencil, place the blades of the turbine disk in the direct path of the steam. Count how many times the red blade passes over your hand in one minute.

7. Repeat step 6 twice and find the average of the three trials.

8. Replace your disk with a different disk. Repeat steps 6 and 7. If you need to add water to the test tube, turn off the burner and let the test tube cool before touching it. Never fill the test tube more than 1/4 full.

1. Heat energy from the burner caused the water to evaporate. The moving steam struck the blades of the disk and made them turn.

Questions

1. How was heat energy used to produce a rotational motion?

2. Why was the steam directed through the glass tube instead of just out the top of the test tube?

3. How did the size of the turbine disk affect the number of rotations per minute?

2. The glass tube had a narrower opening. It made the steam flow faster. It also made it easier to direct the path of the steam.
3. The smallest disk rotated the most.

Today the steam engine is not used so much as another kind of external combustion engine, the **steam turbine** [TUHR-byn]. In a steam turbine, steam under pressure is used to produce a rotational motion. The steam is heated in a cylinder, creating a very high pressure. Then the hot steam is allowed to escape from the cylinder through a small opening. It rushes out in a forceful stream which is directed at blades mounted on a wheel. The steam hits the blades, causing the wheel to turn, or rotate.

You have seen the creation of rotational motion like the kind that occurs in a steam turbine. Falling water turns a water wheel in much the same way. Similarly, wind moves the blades of a windmill. Young children create a similar rotational motion when they blow on a pinwheel.

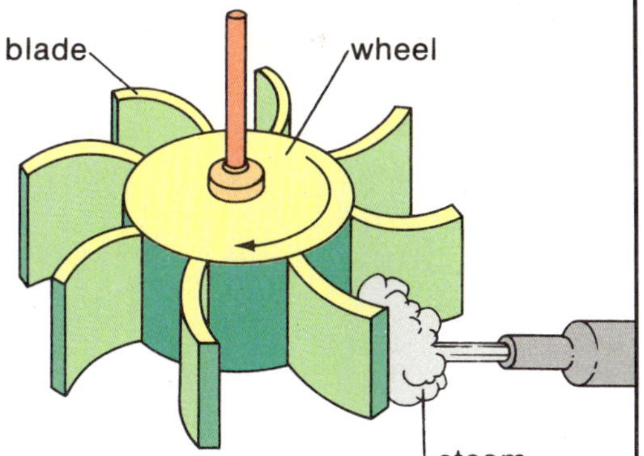

Figure 13.8 A steam turbine

Figure 13.9 When wind strikes the blades of a windmill, a rotating motion similar to the motion of a steam turbine is created.

Do You Know?
A large steam turbine can produce the power of a million horses.

Study Questions for 13.2

1. Describe the early trains.
2. How does a steam turbine work?

1. steam engines, open-air cars, rough ride
2. Steam under pressure is directed at blades mounted on a wheel, causing the wheel to turn.

3. Point out that the steam engine has been largely replaced by the steam turbine. Set up the steam turbine that you made during the Advanced Planning. Ask the students to describe what they observe (steam hits the turbine and causes it to spin).
4. Explain that the turbine is a wheel with blades. The wheel moves when pressure from steam hits the blades. Ask the students how the steam is produced (by boiling water). What heat sources can be used for making the steam that turns the blades (variety of fuels such as coal, gas, and oil; heat released in nuclear fission)?

FOLLOW-UP

Reinforcement: How Does a Steam Engine Operate?
Ask the students to pretend that they are James Watt, the scientist who perfected the steam engine. Ask the students to write letters to Robert Fulton, the inventor of the steamboat. In their letters they should describe how the engine works and how they believe it will change life in England.

Enrichment: Model Engines
Have interested students construct models of steam engines or steam turbines. Ask the students to use their models to demonstrate how the engine or turbine operates.

Tie-in: Settling the West
Ask students to investigate the way that steam engines helped Americans to settle the West. The students should express their research in the form of diary entries that describe how they used the steam engine to help start their new lives.

13.3

OBJECTIVES: The student should be able to

- **list** the advantages of internal combustion engines over most external combustion engines;
- **identify** the four basic parts of most gasoline engines;
- **explain** how the gasoline engine operates;
- **name** the four strokes of a gasoline engine;
- **compare** diesel engines to other internal combustion engines;
- **explain** why no machine is 100 percent efficient.

SCIENCE TERMS

carburetor spark plug strokes

ADVANCED PLANNING

1. You will need the steam turbine from 13.2.

2. Prepare a transparency of an internal combustion engine. Obtain an overhead projector.

3. Make a set of transparencies to illustrate the four strokes in the internal combustion engine.

4. Ask your students if anyone has a model of a gasoline engine. Encourage these students to bring in their models to demonstrate how an internal combustion engine works.

LESSON PLAN

1. Explain that there is another way to make hot gases for producing mechanical energy. Instead of burning fuel under a boiler, the fuel could be burned in the cylinder itself. On the overhead projector, show your transparency of an internal combustion engine. Point out that the fuel is burned inside the cylinder, thereby creating hot combustion products that move the piston. Stress that since the burning takes place inside the engine, the engine is called an internal combustion engine.
2. Emphasize that the distinction between an internal and an external combustion engine is based on where the fuel is burned. Mention that internal combustion engines are smaller and cheaper to use than steam engines. Ask how both kinds of engines do work (by converting heat energy to mechanical energy).

13.3 Internal Combustion Engines

You will find out
- how gasoline engines operate;
- how diesel engines operate;
- why heat engines cannot be perfectly efficient.

The idea of using a steam engine to power a carriage was first suggested many years before Watt perfected his steam engine. However, these early suggestions were not taken very seriously. In fact, one of the first men to propose a steam-driven carriage was considered so extreme he was certified as insane. Still, as years went by, the idea of a steam-powered carriage, or car, gained interest and support.

It is generally agreed that the first true car was a steam-driven vehicle invented in 1769. The inventor was a Frenchman, Nicolas-Joseph Cugnot. Cugnot's car had three wheels. A huge, heavy boiler placed out in front supplied the steam. Cugnot's car could run at a speed of 3.6 km/h. Later steam-driven cars were much better designed. In 1906, two Americans, the Stanley brothers, set the world speed record. Their Stanley steamer was clocked at 205.4 km/h.

Figure 13.10 This early car was powered by an internal combustion engine that used gasoline.

Do You Know?
Before the term *automobile* was in general use, many other terms were considered such as *motorig, motor fly, oleo locomotive,* and *electrobat.*

In the late 1800's cars using internal combustion engines had begun to draw attention. These cars soon became more popular than the external combustion steam cars. The internal combustion engines could be made smaller. They were also cheaper to operate than most external combustion engines.

Most gasoline engines have four basic parts: a carburetor, pistons, cylinders, and spark plugs. Gasoline evaporates and mixes with air in the **carburetor** [KAHR-buh-rayt-uhr]. The gasoline and air mixture is injected into a cylinder. The mixture is ignited by a spark from the **spark plug.** The burning releases heat. This heat increases the pressure inside the cylinder, causing the piston to move.

Most gasoline engines operate on a series of four **strokes,** or movements of the piston. The first stroke is the *intake stroke*. The piston moves out, allowing the air and gasoline mixture to enter the cylinder. Next comes the *compression stroke*. The piston moves in. This movement shrinks, or compresses, the space available to the mixture of air and gasoline. The *power stroke* follows. A spark ignites the compressed gasoline and air mixture. The explosion and heat released as the mixture burns forces the piston out. This outward movement turns the crankshaft. The fourth stroke is the *exhaust stroke*. The piston moves in. This movement pushes the waste products of the burning out of the cylinder. After the exhaust stroke, the cycle is complete and can be repeated. A typical gasoline engine has several cylinders. Each one goes through the four-stroke cycle thousands of times a minute.

Figure 13.11 The four basic parts of a gasoline engine

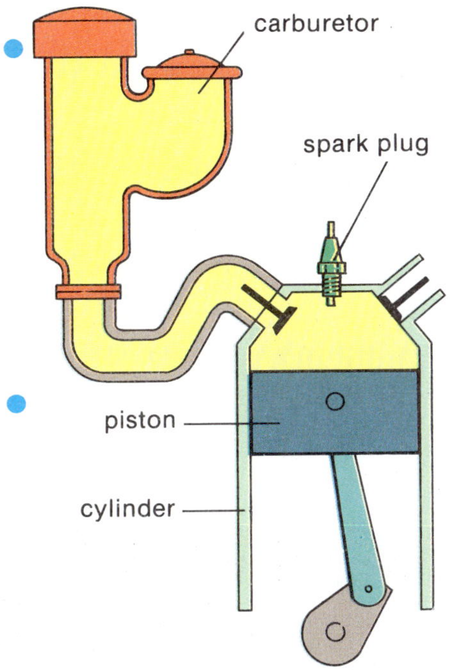

Figure 13.12 The four strokes of a gasoline engine

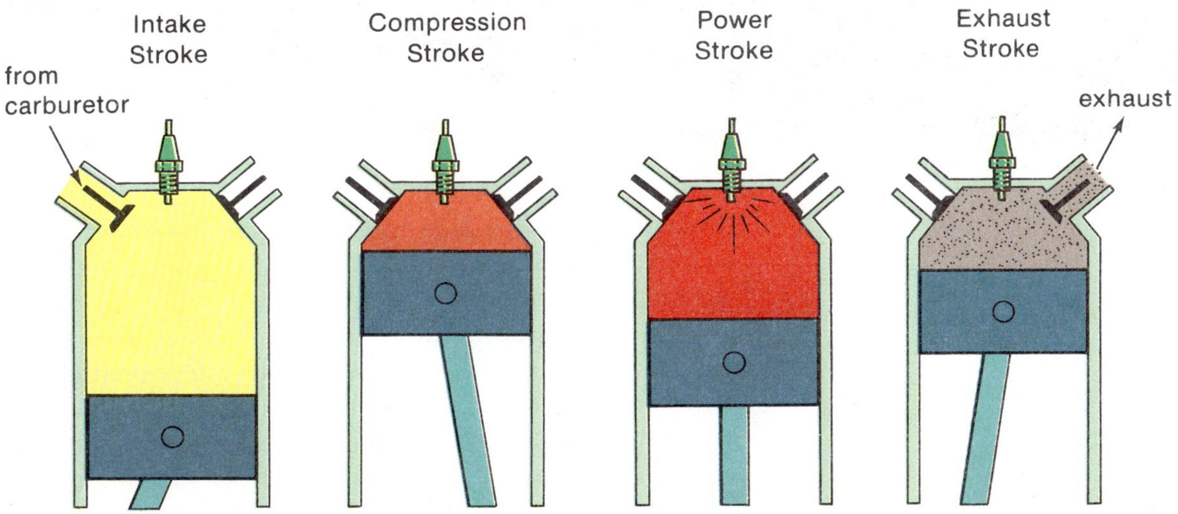

Comprehension Skill: Interpretative

3. On the board list the four basic parts of most gasoline engines (carburetor, pistons, cylinders, spark plugs). If any students have brought in a model of a gasoline engine, ask them to point out the four parts.

4. Have the students explain the four strokes of the engine by constructing a four-column chart on the board. Across the top of the board write the label *gasoline engine–four strokes*. In each of the columns write the name of one of the strokes—intake, compression, power, and exhaust. Ask one student to describe what happens to the piston in the cylinder during each of the strokes. Have another student record this information in the appropriate column.

5. As an assignment have the students create a series of diagrams with captions explaining how internal and external combustion engines differ and how internal combustion engines operate.

Word Skill: Vocabulary

SCIENCE BACKGROUND

The gasoline engine was developed over a number of years in the nineteenth century. Finally, in 1876, Dr. Nikolas A. Otto obtained a patent for an improved four-stroke engine. Most internal combustion engines today use the four-stroke sequence.

The efficiency of gasoline engines has been improved by squeezing the gasoline-air mixture into a smaller cylinder space. One problem with this technique is the premature burning of gas caused by the high pressure. This burning causes a "knocking" noise from the engine and a loss of engine power. To overcome engine knock, alcohol or lead tetraethyl is added to gasoline. The use of lead compounds, however, adds unwanted pollutants to the air.

6. Explain that diesel engines operate like gasoline internal combustion engines. How are they different (they burn diesel fuel rather than gasoline, require no spark plugs to ignite fuel, and have no carburetor to mix fuel and air). What are the advantages of diesel engines (they are more powerful and use less fuel)?
7. On the board draw a single cylinder of a steam engine with the piston out. Ask what has moved the piston out (steam pressure). Then ask what must happen for the piston to move into the cylinder (steam must decrease in volume due to cooling). Explain that some heat energy is lost when the steam cools enough to cause the piston to go back down. This heat loss means that the engine cannot produce as much energy as was put into it.
8. Explain that energy is lost whenever any part of a machine becomes hot and must be cooled. Energy is also lost in friction between rubbing machine parts and in electricity needed to run the machine.

SCIENCE BACKGROUND

In railway locomotives, diesel engines are often used to drive a dynamo. The dynamo generates electricity to drive the train.

You may know someone who has a diesel car. Many trucks and some cars use diesel fuel instead of gasoline. Diesel engines operate without the carburetor and spark plugs necessary in gasoline engines. In diesel engines, air is drawn into the cylinder and compressed. This compression is greater than the compression that occurs in gasoline engines. It causes the temperature of the air to rise to about 800°C. Diesel fuel is then injected into the cylinder. Since the temperature is so high, no spark is needed to ignite the fuel. It burns almost instantly (0.01 second).

Because the compression and temperature are greater than in gasoline engines, diesel engines are more powerful. They use fuel more efficiently. They must also be built with very strong materials to withstand the high temperature and pressure.

Figure 13.13 A diesel truck

All the heat produced in a heat engine cannot be used to do work. The engine must be cooled so the heating and expansion cycle can start again. Also, the engine would become too hot if it were not cooled. Some heat is lost to the air around the engine. But a cooling system is also needed to carry away more heat. These heat losses, although necessary and unavoidable, limit the efficiency of heat engines. Thus, no engine is perfectly efficient. The most efficient heat engines are those that have the smallest amount of heat loss.

Figure 13.14 The ears of an elephant act like a radiator. Excess heat from the elephant's body is transferred to the air through the ears.

The cooling system in a car uses water or some other fluid to absorb heat. The fluid carries this heat to the radiator where it is quickly transferred to the air outside. If the cooling system fails, the four-stroke cycle will stop and the engine can become damaged. Most cars have a temperature gauge to alert drivers to a cooling system failure.

A car's exhaust system carries the waste products of burning to the outside. These waste products carry with them some of the heat released when gasoline is burned. The temperature of a car's exhaust can be 1,000°C or higher. If a gasoline engine were perfectly efficient, 100 percent of the heat energy produced would be used to do work. However, some of this heat must be lost to the cooling system. Additional heat is lost in exhaust. In the end, only about 25 percent of the heat energy produced is used to move the car. Cars are only about 25 percent efficient.

Study Questions for 13.3

1. What are the four basic parts of a gasoline engine?
2. What are the four strokes in the cycle of a gasoline engine?
3. How are gasoline and diesel engines different?
4. Why is no heat engine perfectly efficient?

1. carburetor, pistons, cylinders, spark plugs
2. intake, compression, power, exhaust
3. Diesel engines do not use carburetors or spark plugs.
4. Heat losses prevent heat engines from being perfectly efficient.

9. Explain that when you measure the efficiency of a machine you actually measure the amount of heat that is lost. On the board write the following formula:

   % efficiency = 100 % − % of energy lost as heat

10. Stress to the students that no machine is 100 percent efficient. All machines become hot and thereby lose heat energy to their surroundings. Ask the students how much energy a car is losing if its efficiency is rated at 40 percent (100 percent, or the ideal efficiency minus 40 percent, the actual efficiency, is equal to 60 percent, the energy that is lost).

FOLLOW-UP

Reinforcement: Types of Engines
Ask each student to construct a chart that lists the advantages and disadvantages of gasoline engines and diesel engines. Some students may wish to draw simple models to illustrate their charts.

Enrichment: Engine Efficiency
Have students visit a local garage and interview the mechanic about the efficiency of different car engines. Ask the students to find out which type of engine the mechanic thinks is the most efficient and why.

Tie-in: Wankel Engines
Some scientists believe that the Wankel engine will help to end the air pollution caused by gasoline and diesel engines. Have students do research to determine how a Wankel engine differs from other internal combustion engines. How will it help to end air pollution? Are these engines being used in any cars today? Which ones?

SCIENCE BACKGROUND

Drivers should check the fluid level in their radiators from time to time. If the fluid level is allowed to drop, the cooling system cannot do its job.

Like gasoline engines, other heat engines are not 100 percent efficient. Diesel engines are about 37 percent efficient. Steam turbines are 47 percent efficient. One challenge that scientists and engineers face as fuel supplies dwindle is to develop more efficient heat engines.

13.4

OBJECTIVES: The student should be able to

- **describe** the function of Freon;
- **explain** how a heat pump can be used both to heat and cool a home;
- **explain** how heat pumps use less energy than other heating systems.

SCIENCE TERM

heat pump

ADVANCED PLANNING

In the activity you may substitute cotton balls for the gauze.

LESSON PLAN

1. Introduce the lesson by explaining that heat flows from warmer objects to cooler objects, causing them to warm up. Point out that refrigerators and air conditioning systems use energy to remove heat from objects and to pump the heat into the surrounding air.

13.4 Cooling and Heating Systems

You will find out

- how heat is transferred and moved in some cooling and heating systems.

You might think that an appliance that provides heat must be very different from one that cools. Surprisingly, these kinds of appliances can be very similar. In both cases the same process is occurring. Heat is being absorbed from one place, carried, and then released in another place. The heat energy is simply being moved from where it isn't wanted to where it is wanted.

Refrigerators and freezers produce low temperatures by moving heat from their food compartments to the air outside. To do this, these appliances make use of the special properties of certain compounds such as Freon. Freon both evaporates and condenses easily. When it evaporates, it absorbs heat. When it condenses, it releases heat. Freon is moved through a series of pipes that run both inside and outside the refrigerator. Freon evaporates in the pipes near the food compartment. As Freon evaporates, it absorbs heat from the compartment, causing it and the foods in it to cool. When Freon

Figure 13.15 In a refrigerator, Freon carries heat from one place to another.

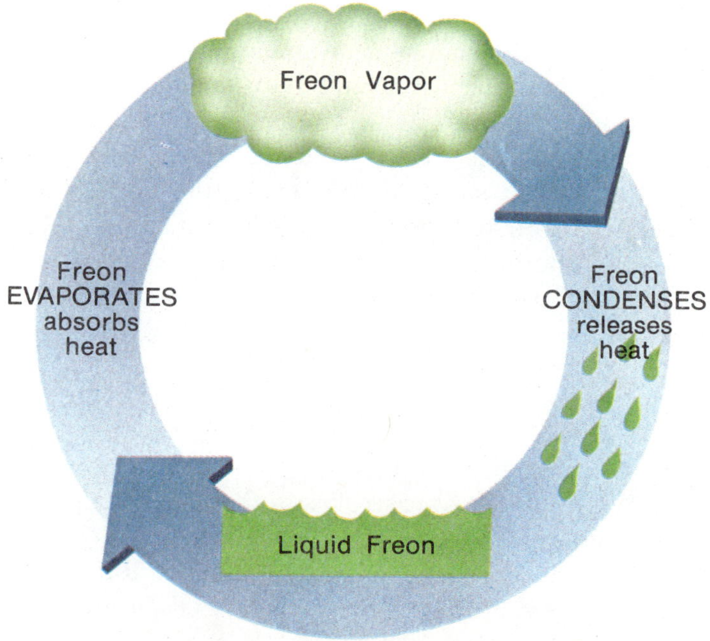

flows in the pipes away from the compartment, it condenses, releasing that heat. In short, Freon acts as a heat carrier. It absorbs heat through evaporation from inside the refrigerator. Then it carries the heat to pipes outside the refrigerator and releases it through condensation. If you touch the back of a refrigerator that is running, you will feel the heat that is being released.

ACTIVITY

How Does Evaporation Affect Temperature? ●

Materials

thermometer
beaker
gauze
rubber band
alcohol
eyedropper
safety glasses

Procedure

1. Copy the chart and use it to record your observations. Put on your safety glasses.
2. Carefully wrap the bulb of the thermometer with gauze. Secure the gauze with a rubber band.
3. Place the thermometer in the beaker. Drop 10 drops of alcohol on the gauze. Record the temperature on your chart under *0 Minutes*. Continue to record the temperature every minute for 5 minutes.
4. Repeat steps 2 and 3 with water instead of alcohol. Use new gauze.

Questions

1. Make a graph of your results. The horizontal axis should show time. The vertical axis should show temperature. Use different colors to plot the results for alcohol and for water.
2. Which liquid required more heat to evaporate?
3. What do you call the amount of heat required to evaporate a liquid?

1. The water line should appear above the alcohol line on the graph.
2. alcohol
3. heat of vaporization

Time *(minutes)*	Temperature of Alcohol *(°C)*	Temperature of Water *(°C)*
0		
1		
2		
3		
4		
5		

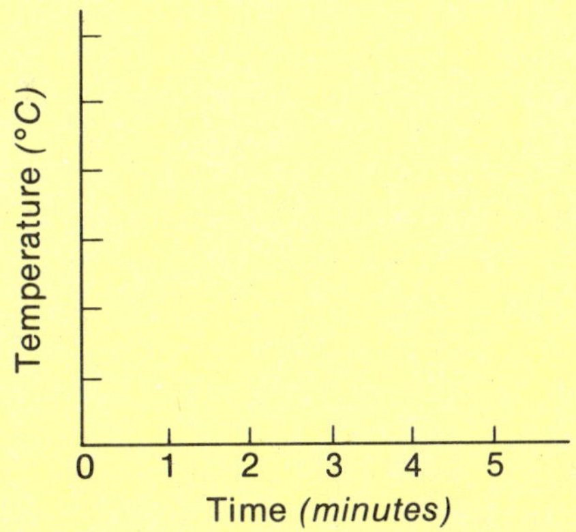

2. Ask the students what happens when the coolant Freon is moved through a coil of pipes that run inside the food compartment of a refrigerator (it evaporates, absorbing heat from the compartment). What change of state occurs when Freon evaporates (liquid to gas). Ask why heat is needed to convert Freon to a gas (energy is needed to overcome the attraction between the molecules, thereby freeing them to become a gas). What is the term that refers to the amount of heat needed to change a liquid to a gas (heat of vaporization)?
3. Tell the students that the coiled pipes they may have noticed on the back or bottom of a refrigerator are the condenser coils. Ask the students what they noticed about the temperature of the coils (warm). Explain that the coils are warm because they are expelling the heat of vaporization. The gaseous coolant, Freon, was pumped into the coils by the compressor. In the coils it liquifies and then moves into the storage tank for the cycle to begin again.

ACTIVITY: How Does Evaporation Affect Temperature?

From the activity the students should observe that the water lowers the temperature more than the alcohol does. In other words, more heat is required to evaporate water than to evaporate alcohol.

Some students may need help in making the graph.

● **Laboratory Skills: Manipulation of Materials Experimentation/Demonstration**

SCIENCE BACKGROUND

A refrigeration system has four main parts—compressor, condenser coils, storage tank, and cooling unit. The four parts are connected so that the refrigerant, or coolant, is pumped in a continuous cycle, changing states as it cools or warms the surrounding air.

Liquid Freon is a nonpoisonous, nonexplosive fluorocarbon.

4. Ask the students to use their knowledge of a heat pump to explain how an air conditioner operates (the fluid evaporates inside the house, absorbing heat; it condenses outside the house, releasing the stored heat).

An air conditioner works on exactly the same principle as a refrigerator. In fact, an air-conditioned house could be thought of as a big refrigerator. The heat inside the house is absorbed through evaporation and released through condensation outside the house. Like a refrigerator, if you touch the part of a working air conditioner that is outside a house, the air conditioner will feel warm. You are feeling the heat that is being released from condensation.

It is possible to use the same principles that apply to refrigerators and air conditioners to take heat from outdoors and use it to heat a house. A machine that does this is called a **heat pump.** Although it is hard to imagine, even on very cold days the air outside does contain some heat. A heat pump absorbs that heat through evaporation and releases it through condensation inside the house. Most heat pumps can work in reverse in hot weather. Thus, the same machine can work both to heat and to cool a home.

Figure 13.16 Air conditioners carry heat from inside a building to the air outside.

Figure 13.17 Even on very cold days a heat pump absorbs heat from the outside air.

Heat pumps use less energy than other heating systems. A heat pump moves heat energy from one place to another. In other heating systems, energy is changed from one form to another to produce heat. For example, in many systems, heat is supplied by burning a fuel, such as natural gas, coal, or oil. Chemical energy in the bonds of the fuel molecules is being changed to heat energy. This heat is used to heat water, to boil water, which makes steam, or to heat air. The hot gas or water then moves to other parts of the building by convection or with the help of a fan. As the gas or water cools, it returns to the furnace where it is reheated.

Study Questions for 13.4

1. How are an air conditioner and a heat pump similar?
2. How is Freon used to cool the inside of a refrigerator?
3. How is a heat pump different from other home-heating systems?

1. Both work by carrying heat from one place to another.
2. Freon absorbs heat in the refrigerator by evaporation, carries it outside the refrigerator, and releases it by condensation.
3. A heat pump moves heat energy from one place to another.

5. Ask why heat pumps are more energy efficient than other heating systems (because they only move heat energy from one place to another rather than converting energy to different forms. Such conversions not only require energy, they also waste energy in the process).

FOLLOW-UP

Reinforcement: Refrigerator Manual

Ask the students to pretend that they have been hired by a refrigerator company to write a manual that will be put inside their refrigerators to explain how the machine works. Tell the students to remember that they are writing the manual for people who may have little scientific background.

Enrichment: Wet- and Dry-Bulb Hygrometers

Have students construct wet- and dry-bulb hygrometers by carefully attaching two thermometers to a piece of board and fastening a wick made from linen or muslin to one of the thermometers. The free end of the wick should be in a glass of rainwater. The device can be used to measure relative humidity.

Students should fan the wet-bulb thermometer until the temperature will go no lower. They should then make a reading of both the wet and the dry bulbs. To determine the relative humidity, students should subtract the wet-bulb reading from the dry-bulb reading and look up this result on a standard chart that gives the corresponding relative humidity.

Have the students explain the meaning of the term *relative humidity* and keep a record of the relative humidity for a specific period of time.

Tie-in: Bacterial Spores

Although many forms of bacteria are killed by the low temperatures in refrigerators, others lie dormant until the temperature rises again. Have students do research to find out how bacteria protect themselves during adverse conditions such as low temperature.

What Did You Learn?

13.1 • Heat engines use heat energy to do work.
13.1 • Heat engines operate on the principle that when a gas is heated, its pressure increases.
13.1 • The pressure of a heated gas in a cylinder pushes out a piston.
13.1 • In external combustion engines fuel is burned outside the cylinder.
13.1 • In internal combustion engines fuel is burned inside the cylinder.
13.2 • James Watt made the steam engine more efficient.
13.2 • A steam turbine uses steam under pressure to produce a rotational motion.
13.3 • Nicolas-Joseph Cugnot invented the first true car in 1769.
13.3 • Cars and trucks use internal combustion engines.
13.3 • The four basic parts of a gasoline engine are a carburetor, pistons, cylinders, and spark plugs.
13.3 • The four strokes in most gasoline engines are intake, compression, power, and exhaust.
13.3 • Heat losses prevent heat engines from being perfectly efficient.
13.3 • Gasoline cars have a cooling system to prevent damage to the engine due to high temperatures.
13.4 • Refrigerators, freezers, air conditioners, and heat pumps all transport heat from one place to another.
13.4 • Heat can be absorbed, moved, and released by fluids such as Freon.
13.4 • Freon absorbs heat through evaporation.
13.4 • Freon releases heat through condensation.
13.4 • Most heat pumps can operate as both a heater and an air conditioner.

Key Terms

heat engine
cylinder
piston
crankshaft
external combustion engine
internal combustion engine
steam turbine
carburetor
spark plug
strokes
heat pump

Biography

Elijah McCoy (1843–1929)

Elijah McCoy was born in Ontario, Canada. He was the son of runaway slaves. His parents were very poor. They worked hard and sacrificed in order to send Elijah to Scotland. There he trained to become a mechanical engineer.

When McCoy finished his studies, he came to the United States. Because of his race, he could not find work as a mechanical engineer. Eventually he found a job with Michigan Central Railroad, oiling the moving parts on locomotives.

McCoy soon got bored with this work on the railroad. He began to think about ways to improve the oiling of locomotives. Working at night in his tiny workshop at home, he invented a lubricator cup. This device allowed small amounts of oil to drip continuously onto the moving parts of the engine. This invention would make locomotives run more efficiently. It would also eliminate the need to stop the train for oiling.

The buildup of heat caused by the friction of moving parts is a serious problem for all machines. It wasn't long before industries all over the world were using McCoy's lubricating cups. They were called "the real McCoy." The invention was a big money saver because the machines didn't have to be stopped to be oiled. Today, versions of McCoy's lubricating cup are widely used in locomotives, cars, buses, rockets, ships, and almost every other kind of modern machine.

Vocabulary

1. h
2. e
3. f
4. l
5. g
6. c
7. k
8. b
9. a
10. d

● Word Skill: Vocabulary

What Do You Remember?

1. The pressure increases.
2. The temperature in the cylinder is so hot that the fuel ignites spontaneously.
3. intake—gasoline and air mixture enters cylinder
   compression—mixture is compressed
   power—spark ignites the mixture pushing piston
   exhaust—waste products of combustion are expelled
4. Heat losses are unavoidable.
5. Freon
6. Heat is absorbed inside the house, carried, and released outside the house. The inside of the house corresponds to the compartment of a refrigerator.
7. It absorbs heat through evaporation and releases it through condensation.
8. to prevent engine damage due to excessive heat
9. piston—moves out when pressure inside cylinder increases
   cylinder—contains the gas that is heated
   crankshaft—turns one or more wheels
10. Watt—efficient steam engine
    Cugnot—first true car

TO THINK ABOUT AND DO

Vocabulary ●

Number your paper from 1 to 10. Match the phrase on the right with the description on the left. Write the letter of the phrase beside the appropriate number.

1. a device that changes heat energy into work
2. the burning of a fuel inside a cylinder
3. a device perfected by James Watt
4. the stroke in which the piston moves in to shrink the space available to the gasoline and air mixture
5. the stroke in which the piston moves out, allowing the gasoline and air mixture to enter the cylinder
6. that part of a gasoline engine where gasoline evaporates and mixes with air
7. fluid that absorbs, carries, and releases heat in a refrigerator
8. home-heating device that moves heat from one place to another
9. a device that ignites the gasoline and air mixture in the cylinder
10. that which prevents heat engines from being perfectly efficient

a. spark plug
b. heat pump
c. carburetor
d. heat loss
e. internal combustion
f. steam engine
g. intake
h. heat engine
i. external combustion
j. efficiency
k. Freon
l. compression

What Do You Remember?

1. What happens to the pressure of a gas when it is heated in a confined space?
2. Why don't diesel engines need spark plugs?
3. List and explain the four strokes of a gasoline engine.
4. Why aren't heat engines perfectly efficient?
5. What is an example of a substance that carries heat away from the food compartment of a refrigerator?
6. How is an air-conditioned house like a refrigerator?
7. How does Freon absorb and release heat?
8. Why do gasoline engines need a cooling system?
9. What are the functions of the piston, cylinder, and crankshaft in a heat engine?
10. Name the inventions credited to Watt and Cugnot.

Applying What You Have Learned

1. If a heat engine loses 38 percent of the heat it produces, how efficient is it?
2. If you were selecting a fluid to carry heat in a refrigerator, would you want one that required a lot of heat energy or a little heat energy to evaporate? Why?
3. What problems do you think the developers of gasoline cars faced?
4. Why do you think tune-ups are simpler for diesel engines than for gasoline engines?
5. What area in the food compartment of a refrigerator would you expect to be the coldest? Why?

Research and Investigation

1. Before they had refrigerators, people used iceboxes to keep food cold. Find out how iceboxes chilled food.
2. Visit one or more car dealerships. Ask for information on diesel cars and how they compare with gasoline cars. A salesperson may discuss this with you. There should also be pamphlets available detailing diesel and gasoline engines. Present the information you gather in a written report.

Applying What You Have Learned

1. 100 % − 38 % = 62 %
2. A fluid that required a lot of heat energy to evaporate would remove heat more effectively from the compartment of the refrigerator.
3. Answers will vary. Answers may include finding materials that tolerate high temperature, determining the best proportions of gasoline and air to burn, keeping exhaust away from passengers, finding materials that were both tough and light, and determining how to carry the fuel.
4. Diesel engines do not have carburetors or spark plugs to be cleaned, checked, or replaced.
5. Students may suggest the bottom of the compartment because warm air is less dense and rises. Or they may suggest the area of the compartment nearest the pipes where Freon is evaporating. The evaporation is absorbing heat.

● Comprehension Skill: Applied

Research and Investigation

1. Most iceboxes had two compartments. Block ice was put in the top compartment. The ice cooled the air by absorbing heat from it during melting. A drain collected the liquid water and carried it to a pan under the icebox. This pan had to be emptied from time to time. Food was kept in the compartment under the ice. The warmer, less dense air rose to the ice compartment. The denser, cool air sank and surrounded the food.
2. Depending on the dealership visited, reports will vary. They will probably include comparisons of estimated mileages, differences in maintenance schedules, differences in engine structure, and differences in price.

● Reference/Research Skills

Long-Term Projects

1. Divide the class into small groups. Have each group design a trail to a destination on the school grounds. The directions must be described by using a compass. The distances should be measured in metric units. Have the teams exchange directions and try to find each others' destinations.
2. Have the students collect different types of batteries and fuses. They should label the main parts of each and explain their uses in a museum display.
3. Transformers are used by electric companies to increase and decrease current. Ask the students to draw diagrams and write reports to explain how transformers work. If students have old train transformers at home they may wish to use them in an oral presentation to the class.
4. Computer language is based on the binary system. Ask the students to investigate the binary system. They should prepare lessons that explain how to use the system to add and subtract. They should present the lessons to the class.
5. Ask the students to build telegraphs to show the relationship between electricity and magnetism. Check their plans and procedures before allowing them to proceed. The students should write reports to explain how the telegraph works. They should use the telegraphs to send messages using Morse code.
6. Have the students prepare a flowchart to show how steel is made.
7. Have a group of students prepare a game of *Password* using the vocabulary words in the unit. Allow the class to play the game to reinforce the vocabulary words.
8. Ask the students to do research on the life and inventions of Thomas Edison. Each should write a biography of the scientist.

Unit Four

Magnetism and Electricity

Bulletin Boards

1. Set up the board as in the drawing. Attach an envelope containing 17 index cards. The cards should contain the directions for baking a cake. The students must arrange them in logical order. The 17 steps are: 1. Choose a recipe; 2. Buy ingredients; 3. Preheat oven; 4. Sift and measure flour; 5. Measure shortening; 6. Measure milk; 7. Beat eggs; 8. Combine milk, shortening, and eggs; 9. Combine all ingredients; 10. Grease baking pans; 11. Pour batter into pans; 12. Bake for 45 minutes; 13. Remove cake from oven; 14. Remove cake from pans; 15. Make frosting; 16. Frost cake; 17. Eat cake.

2. To emphasize the difference the use of electricity has made in our lives, title one half of your board *Then* and the other half *Now*. Have the students mount illustrations of activities not requiring electricity such as hand stirring and using a candle on the Then side. They should mount illustrations of activities involving electricity such as using an electric mixer and a flashlight on the Now side.

3. Title your board *Using Electricity Safely*. Have the students illustrate steps that should be taken to avoid accidents with electricity. They will find safety tips in their text and in booklets obtainable from most local electric companies.

CHAPTER BACKGROUND

Magnetism is the force produced by the motion of electric charges in the atoms of matter. Scientists believe that the atoms of matter are arranged in groups called magnetic domains. If the domains are lined up, the material is magnetized. If the arrangement of the domains is random, the substance will not exhibit magnetic properties. A magnet can be used to induce magnetism in another material by lining up its magnetic domains.

Magnetism does not depend only on the arrangement of the magnetic domains. It depends also on the structure of individual atoms within the domains. Each atom has electrons that move in energy shells around the nucleus as they spin on an axis. In order for a material to be magnetic, there must be an imbalance in the way the electrons are spinning on their axes. That is, more of the electrons in the outer shell must spin in one direction than are spinning in the other direction. Only in nonmagnetic materials do equal numbers of outer-shell electrons spin to the right and to the left.

Although the phenomenon referred to as static electricity was recognized for many years, it was not explained until Sir William Crookes observed that gases under low pressure conduct an electric current. In his experiment in 1870, Crookes used a glass tube containing a small quantity of gas. Metal electrodes that could be attached to a high-voltage battery were sealed in each end of the tube. When the electrodes were connected to the battery, Crookes observed a green glow on the walls of the tube. He reasoned that something from the electrodes must have been hitting the sides of the tube. Hundreds of experiments were needed before it could be established that the electrodes were transmitting negatively charged particles called electrons and that positively charged particles called protons also exist. The identification of these particles began the understanding of the electrical nature of matter.

14 Magnetic and Electric Forces

Two days before you introduce this chapter, prepare a demonstration using a small bar magnet, clear corn syrup, a test tube that the magnet will fit in, a jar, iron filings, and string. Sift the iron filings to remove any dust. Fill the jar with syrup and stir in the filings until they are evenly distributed. Put the magnet in the test tube. Use the string to suspend the tube in the jar. Be sure the magnet is neither touching the bottom of the jar nor extending above the level of the syrup. The filings will slowly arrange themselves in a pattern around the magnet.

Show the demonstration to the students. Ask them what has caused the filings to be arranged in a pattern. If they do not recognize that the bar is a magnet, place a paper clip against the jar. Ask what type of material would attract the paper clip (magnet). Tell the students that in this chapter they will learn about magnetism and how it is related to electricity. Do not throw away the demonstration.

How important are magnets to you? Do you use them every day? Could you live without them? You could live without magnets. But your life would probably be changed much more than you realize. Magnets are an essential part of most appliances in your home. For example, hair dryers, tape recorders, televisions, and record players involve magnets. In many homes, even the electricity to run these appliances comes from power plants that use magnets.

Magnetic and electric forces are very closely related. In many ways the behavior of magnetic and electric forces is similar. Also the action of one force can affect the action of the other. For this reason, the study of magnetism is usually linked to the study of electricity. In this chapter you will learn about some of the basic properties of magnetized and electrified matter. In later chapters you will see how these properties affect your daily life.

Figure 14.1 A compass needle is a magnet. Rapid expansion of sea trade and exploration resulted from the development of the compass.

Lesson Organizer

14.1 Magnets
Pages 296–299
Advanced Planning: Page 296
Follow-up: Page 299
- **R** Testing Materials for Magnetism
- **E** Loss of Magnetism Through Heat
- **T** Magnetic Bacteria

14.2 The Magnetism of the Earth
Pages 300–301
Advanced Planning: Page 300
Follow-up: Page 301
- **R** The Earth as a Magnet
- **E** Navigational Correction
- **T** Aurora Borealis

14.3 Static Electricity
Pages 302–305
Advanced Planning: Page 302
Activity: How Do Electric Charges Act? Page 303
Follow-up: Page 305
- **R** Static Charges
- **E** Electrostatic Precipitators
- **T** Photostatic Copiers

14.4 Conductors and Electric Discharges
Pages 306–309
Activity: Which Materials Are Conductors? Page 309
Follow-up: Page 309
- **R** Grounding Buildings
- **E** Capacitors
- **T** What Are the Best Insulators?

See ACTIVITIES book pages 79–84 for laboratory, skill, and review exercises for Chapter 14.

14.1

OBJECTIVES: The student should be able to

- **define** the term *magnetism;*
- **describe** the poles of a magnet;
- **explain** what is meant by the term *magnetic field;*
- **identify** the basic difference between electricity and magnetism;
- **tell** what is meant by magnetic domains;
- **explain** how magnetism can be induced;
- **explain** how magnets can lose their magnetism;
- **describe** the iron ore lodestone.

SCIENCE TERMS

magnetism	magnetic field
magnet	magnetic domains
magnetic poles	induced magnetism

ADVANCED PLANNING

For demonstrations you will need a dish of iron filings, an iron bar, two bar magnets, a ring stand with a ring, some string, an overhead projector, a piece of glass or acetate the size of the projector glass, a test tube loosely corked and filled with iron filings, and a needle.

LESSON PLAN

1. Introduce the lesson by reviewing the term *force* (push or pull that affects matter). Explain to the students that they are going to study another force—magnetic force, or magnetism.

2. Place a dish of iron filings on the table. Hold up two iron bars and ask the students how they can tell which bar is the magnet (will attract the iron). Place both bars near the dish to test the students' answer. Stress that a magnet is an object that has the ability to attract certain materials such as iron or steel.

SCIENCE BACKGROUND

It is commonly believed that the name *magnetism* is derived from Magnesia, a province in northern Greece where deposits of magnetic iron ore were found.

The theoretical subatomic particle called a magnetic monopole is predicted to behave like a single pole, either north or south, cut from a magnet. Such a particle would carry magnetic charge in much the same way that electrons carry electric charge.

14.1 Magnets

You will find out
- what magnetism is;
- how magnetism and electricity are alike;
- why some substances are magnetic.

Physical scientists have made hundreds of observations about matter. Many of these observations can be explained through the action of forces. You have already learned about some of these forces, such as nuclear force, gravitational force, and friction. Another very important force is magnetic force, or **magnetism.** Magnetism is a force produced by the motion of electric charges in a material. It shows up, for example, as the ability of a magnetized object, or **magnet,** to attract small pieces of iron.

Suppose you dip a magnet into a pile of iron filings. When you pull the magnet out, you see that the filings form a mossy covering on both ends. The rest of the magnet remains bare. The magnetic force is greatest at the ends of the magnet. The filings cling to the area of greatest magnetic force. The areas of greatest magnetic force are called **magnetic poles.** Every magnet has two poles. One is the north pole (N). The other is the south pole (S). It is easy to identify the two poles in a bar magnet. When the magnet is held by a thin wire, one end turns toward the north. This end is the north pole. The other end is the south pole.

In some ways magnetic force is similar to electric force. For example, if you bring the north poles of two magnets together, they repel each other. On the other hand, if the north pole of one magnet is brought near the south pole of another magnet, the magnets attract each other. It is a general rule that like poles of a magnet repel each other and unlike poles attract each other. This rule is very similar to one you learned earlier for electric charges. Opposite charges attract and like charges repel. Another similarity exists between magnetic and electric forces. Like the force between electric charges, the force between two magnets decreases as the magnets are moved farther apart.

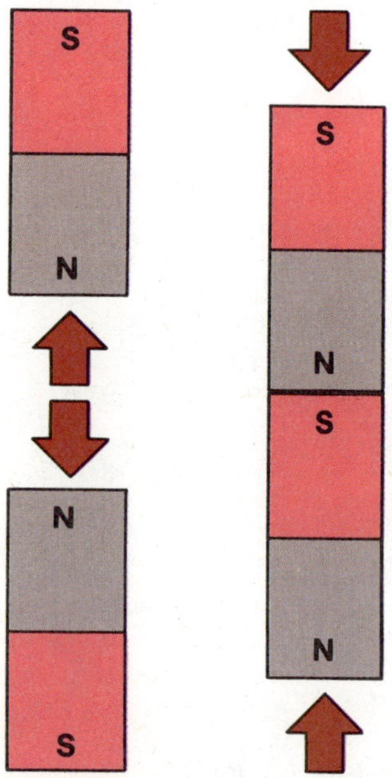

Figure 14.2 Like poles repel each other. Opposite poles attract.

A magnetic force acts between two magnets even though they may not be touching. Objects that experience the force of a magnet are said to be in the magnetic force field of that magnet. A magnetic force field, or **magnetic field,** is the area around a magnet where its magnetic force can be experienced. The magnetic field is strongest near the poles and weak farther away from the poles.

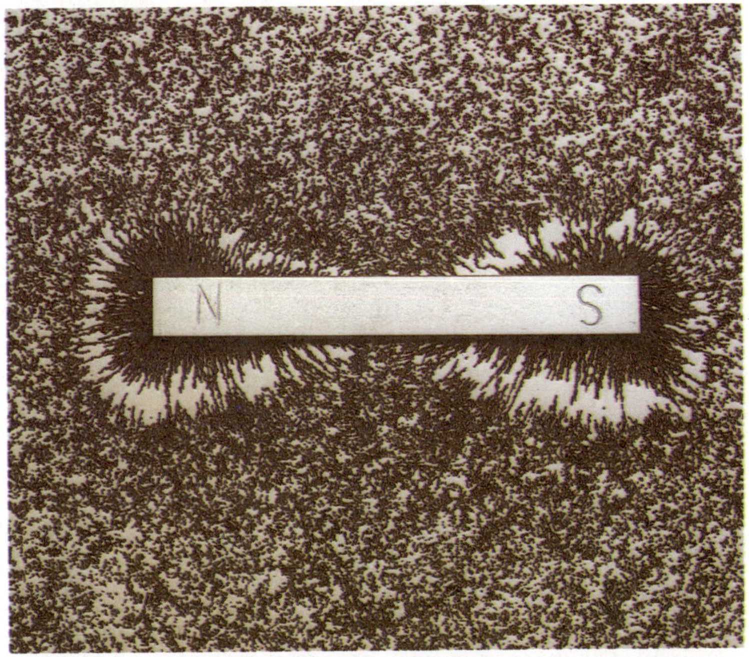

Figure 14.3 Iron filings are concentrated near the poles where the magnetic force is strongest.

An important difference between electricity and magnetism is that a single kind of electric charge can be put on an object. An object can be either positive or negative. This cannot happen with magnets. A single pole is never found on one magnet. The north and south poles cannot be separated. If you cut a magnet in half, each half instantly becomes a whole magnet. Both new magnets will have a north and a south pole. The poles cannot be separated because the basic units of matter—atoms and the particles they contain—have both a north and a south pole. No particles with a single magnetic pole are known to exist. Therefore, no form of matter with a single pole is known to exist.

Figure 14.4 No matter how you cut a magnet, the remaining pieces will always have both a north and a south pole.

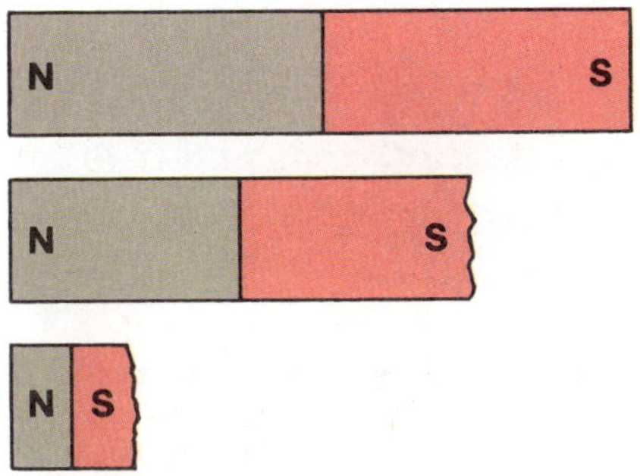

3. Tie a single piece of string around each end of a bar magnet so that a loop is formed above the magnet. Tie another piece of string to the midpoint of the loop and attach this string to the ring on the ring stand. Allow the magnet to turn freely. In what position does it come to rest (facing the north pole)? Use a compass to demonstrate that the magnet has lined itself up to face the north pole. If the poles on the magnet are not clearly marked, use a crayon to mark each of the two ends so that they can be easily identified.

4. Bring the north pole of another magnet close to the north pole of the magnet attached to the ring stand. What happens (the two poles move away from each other, are repelled)? Ask the students to predict what will happen if you bring the two south poles close together (they will be repelled). Try bringing the two south poles together to test the students' prediction. Explain that just as two electrons repel each other, the like poles of a magnet repel each other.

5. Ask the students what they think will happen when a north and a south pole come in contact. Then suspend one magnet from the ring on the ring stand as before and point out the north end of the magnet. Bring the south end of the other magnet near the north end of the suspended magnet. Ask the students what they observe (magnets join).

6. Have the students observe as you place a bar magnet on an overhead projector and cover it with a piece of glass or acetate. Sprinkle iron filings on the glass. Tell the students that the area marked by the iron filings is called a magnetic field. Objects that fall within this field will be attracted to the magnet.

7. Explain that although magnetism and electricity have similarities—both depend on electrons and both obey the same laws of attraction and repulsion—there is a major difference between them. While particles can have either a positive or a negative charge, no particles have been shown to have a single pole.

8. On the board copy the square for iron from the periodic table. Ask the students what an atom of iron looks like. As they describe it to you, draw it on the board. Tell the students that scientists believe that there is an unbalanced rotation of electrons in the M shell of iron. The M shell has 14 electrons. Nine are rotating in one direction, and only five rotate in the other direction. This imbalance accounts for the fact that iron is magnetic.

9. Explain that scientists believe that the atoms of magnetic materials are organized into groups of atoms called magnetic domains. When a substance is unmagnetized, the magnetic domains are scattered and without any order. When a substance is magnetized, the magnetic domains are lined up.

10. Tell the students that they can magnetize a needle by stroking it in the same direction 30 or 40 times with a magnet. Demonstrate this. The needle becomes magnetized because the magnetic domains are lined up. Prove that the needle is magnetized by placing it in the bowl of iron filings. When you remove the needle, it should be covered with iron filings.

● Word Skill: Vocabulary

SCIENCE BACKGROUND

A steel magnet will lift 500 times its own weight. An alnico magnet will support 4,450 times its own weight.

Do You Know?
In ancient times many people thought magnets were magic. Magnets were thought to cure the sick and cause people to fall in love.

Most materials cannot be magnetized. Iron and a few other materials such as steel, nickel, and cobalt can be magnetized. These substances have regions called **magnetic domains** [doh-MAYNZ]. Magnetic domains, which are clusters of many atoms, can be thought of as tiny magnets. When the domains are randomly arranged, the material is not magnetic. However, if those domains are lined up, the material becomes a magnet. For example, when a steel bar is stroked in one direction with a magnet, the steel bar becomes magnetized. The magnetic domains in the steel are lined up by the action of the magnet. This magnetizing of a material by a magnet is called **induced** [ihn-DOOST] **magnetism.**

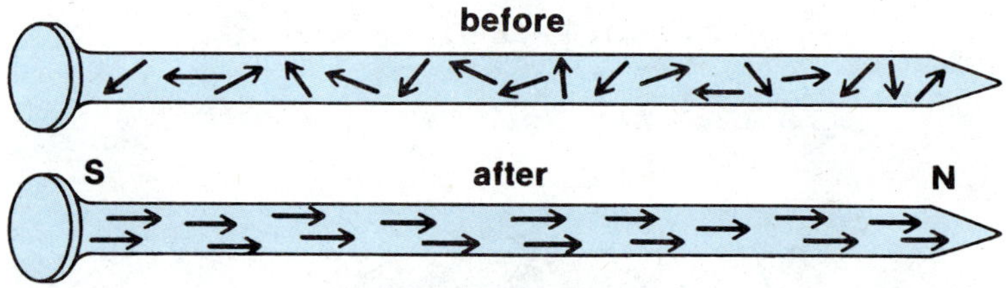

Figure 14.5 The magnetic domains in a magnetized object are lined up in the same direction.

Nails, paper clips, and other objects made of iron can be temporarily magnetized. However, they keep their magnetism for only a short time. In a few materials, such as steel, and in some special mixtures of metals, the magnetism remains long after the inducing magnet is removed. These materials make permanent magnets. Some of the best permanent magnets are made of a mixture of iron, aluminum, nickel, and copper. These magnets are called alnico magnets. Can you guess ● where the name *alnico* comes from? Look at the first letters of the elements that form it.

If permanent magnets are heated to temperatures of from several hundred to a thousand degrees Celsius, they will lose their magnetism. Their magnetism is lost because the increased atomic motion produced by heating causes the magnetic domains to get out of line. Permanent magnets can also lose their magnetism if they are dropped too many times. The magnetic domains are bumped out of line when the dropped magnet strikes the floor.

Figure 14.6 A permanent magnet holds these paper clips at the top where they can be picked up easily.

Figure 14.7 In the outdoor sport of orienteering, compasses are used to determine location and direction.

The Greeks and the Chinese discovered more than 2,000 years ago that some rocks are natural magnets. These rocks are a form of iron ore called lodestone, or magnetite. The magnetic domains in lodestones are naturally lined up.

Lodestones were very important to sailors during the Middle Ages. A lodestone was used to magnetize the sailing ship's compass, keeping it in working order. Without a working compass, the sailors could become lost at sea. Because the ship's lodestone was so important, any sailor who tampered with it was tortured or put to death.

Study Questions for 14.1

1. What is magnetism?
2. How are magnetism and electricity alike?
3. How can a steel rod be magnetized?

1. a force produced by electric charges in motion
2. Both are forces. In both, opposites attract and like poles or charges repel. Both forces decrease as the poles or charges are moved farther apart.
3. stroke it in one direction with a magnet

FOLLOW-UP

Reinforcement: Testing Materials for Magnetism

Set up a strong horseshoe magnet on the ring of a ring stand. Tie a string to the base of the ring stand and tie a paper clip to the free end of the string. Hold the paper clip so that it is just inside the magnetic field of the horseshoe magnet. Adjust the length of the string so that the paper clip appears to be dancing in air. Have the students hold pieces of zinc, copper, glass, paper, iron, and rubber between the magnet and the paper clip and observe when the paper clip falls and when it doesn't (falls when iron or any other substance that can be magnetized is between the magnet and the paper clip).

Enrichment: Loss of Magnetism Through Heat

Have the students magnetize a needle and then use the needle to determine how heat affects magnetism. As they do this activity, they should form a hypothesis, design a controlled experiment, record their observations, and form a conclusion.

Tie-in: Magnetic Bacteria

It has been found that some bacteria are magnetic and orient themselves on a north-south axis. Ask the students to do research on the bacteria and determine why they are magnetic and how this adaptation helps them. Do any other organisms respond to magnetism?

SCIENCE BACKGROUND

In 1600 William Gilbert, an English scientist, published an essay in which he claimed that the earth behaved like a giant magnet. To support his view, he shaped a piece of lodestone into a sphere. Regardless of where Gilbert placed a compass on the lodestone sphere, the compass always pointed north.

The word *lode* means "course" in Middle English. Lodestones were given their name because they helped keep sailors on course.

14.2

OBJECTIVES: The student should be able to

- **state** how the earth is like a magnet;
- **differentiate** between the north geographic pole and the north magnetic pole.

SCIENCE TERMS

north magnetic pole
north geographic pole

ADVANCED PLANNING

Obtain a world globe.

LESSON PLAN

1. Point to the equator on the globe and ask what would happen if you held a compass there (compass needle would point north; the other end of the needle would point south). Point to a place in China and ask the students what would happen if they stood there with their compasses (compass needle again lines up in the north-south position). What would cause the compass needle to point north in each case (the earth's magnetic field)?

2. On the board draw a cross section of the earth. Label the earth's molten iron core. Explain that scientists are not sure why the earth acts like a magnet. They believe that this magnetic property is somehow related to the rotation of the molten iron core.

14.2 The Magnetism of the Earth

You will find out
- how the earth acts like a magnet;
- how the north geographic pole and the north magnetic pole differ.

Have you ever wondered why a compass needle always lines up in the north–south direction? It does so because the earth has a magnetic field. The needle of a compass is magnetized. The earth's magnetic field acts to line up the compass needle. How can the earth have a magnetic field? The answer to this question is unclear. Scientists believe the earth's magnetism is somehow related to the rotation of the liquid iron center of the earth.

Because the earth has a magnetic field, the earth is acting as if there is a huge bar magnet stuck in it. This imaginary magnet does not lie perfectly straight up and down inside the earth. Instead the magnet is tilted. The **north magnetic pole** is the north pole of this imaginary magnet. A compass needle points to the north magnetic pole.

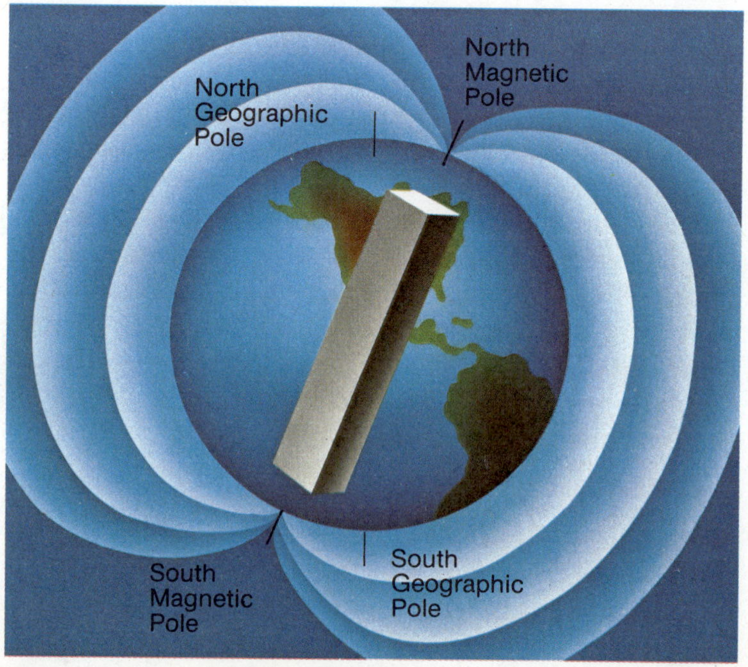

Figure 14.8 The earth acts as if there is an imaginary bar magnet stuck in it.

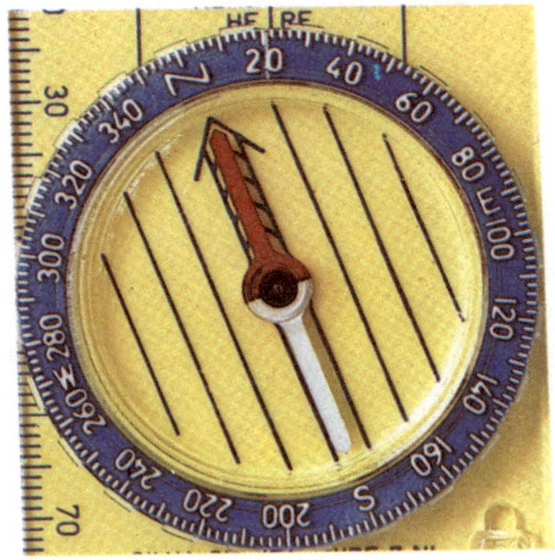

Figure 14.9 The magnetic needle of a compass always points to the north and south magnetic poles.

The north pole you are probably familiar with is the **north geographic pole.** The north geographic pole is on maps and globes at the very top, or northernmost point, of the earth. The north magnetic pole is near but not at the north geographic pole. The north magnetic pole is about 1,700 km south of the north geographic pole. On a map, this location is in northern Canada almost directly above the middle of the United States.

When compasses at different places point toward the north magnetic pole, they are not necessarily pointing in the same geographic direction. For example, a compass in Boston points between geographic northwest and north. But one in Los Angeles points between geographic northeast and north. Navigators must use special charts to allow for the difference between geographic and magnetic north. To make matters worse, the north magnetic pole of the earth moves. It slowly wanders around. For this reason navigation charts must be changed every ten years or so.

Do You Know?
Pigeons and dolphins have tiny particles of magnetite in their heads. Scientists think that these particles may act as compasses that help these animals navigate.

Study Questions for 14.2

1. How is the earth like a magnet?
2. Which north pole does a compass needle point to and why?

1. The earth has a magnetic field and acts as if a huge bar magnet were stuck in it. 2. It points to the north magnetic pole because this pole, the north pole of the imaginary earth magnet, exerts a magnetic force.

3. Ask the students what the north pole of the earth's imaginary magnet is called (north magnetic pole). Point to the north pole on the globe. Where is the north magnetic pole with respect to the north geographic pole on the globe (about 1,700 km south of the north geographic pole)?

● **Comprehension Skill: Applied**

FOLLOW-UP

Reinforcement: The Earth as a Magnet
Have students create posters to illustrate the magnetic properties of the earth. The students should show the north geographic pole, the north magnetic pole, and the earth's magnetic field.

Enrichment: Navigational Correction
Navigators must correct the courses of their vessels as they distinguish between the north geographic pole and the north magnetic pole. Ask the students to contact the local Navy recruiter to find out how this is done. The recruiter may be able to supply some charts that will help explain how vessels are kept on course.

Tie-in: Aurora Borealis
The aurora borealis, or northern lights, is a visible extension of the earth's magnetic field. Ask the students to investigate the aurora and prepare a diorama to demonstrate how it might appear to an observer.

SCIENCE BACKGROUND

From time to time the north and south magnetic poles change places. At least nine of these reversals have occurred in the last 3.5 million years.

14.3

OBJECTIVES: The student should be able to

- **define** the term *static electricity;*
- **explain** how an object can build up static electric charge;
- **describe** how an electroscope detects electric charge;
- **describe** how electrostatic induction is used to produce a charge on an object.

SCIENCE TERMS

static electricity
electroscope
electrostatic induction

ADVANCED PLANNING

For the demonstration you will need an electroscope, a glass rod, a piece of silk, a rubber rod, and a piece of wool. You can find directions for building an electroscope in the *World Book Encyclopedia.*

LESSON PLAN

1. Introduce the lesson by pointing to the square for lithium on the periodic table and asking a student to draw an electron shell diagram of a lithium atom on the board. Is the atom charged (no)? Why not (the number of positively charged protons is equal to the number of negatively charged electrons)? Erase the L shell from the lithium atom. Point out that lithium still has three protons but only two electrons. Is lithium still neutral (no)? Stress that when atoms gain or lose electrons they become charged atoms, or ions.

SCIENCE BACKGROUND

Thales is said to be the first person to have experimented with electricity. About 600 B.C. he rubbed a piece of amber with a woolen cloth and saw that it would pick up bits of paper or straw.

Van de Graaff generators produce large static electric charges.

14.3 Static Electricity

You will find out

- how an object can build up a static electric charge;
- how static electricity can be detected;
- how an electric charge can be induced in an object.

Electric force, like magnetic force, can be traced to the atoms that make up matter. An atom has an equal number of positive protons and negative electrons. It has no overall electric charge. However, an atom can sometimes lose or gain electrons. If enough of the atoms in an object lose electrons, that object will have a noticeable positive electric charge. On the other hand, if the atoms in an object gain electrons, the object will become negatively charged.

The electric charge that builds up on an object is called **static electricity.** Static electricity is an electric charge that does not flow. You can produce static electricity with a plastic comb. When you comb your hair with a plastic comb, your hair loses electrons to the plastic. When this happens, your hair becomes positively charged. The comb becomes negatively charged. You can see the effects of these charges by holding the comb over your hair. Since opposite electric charges attract, the negative comb attracts the positive hair. Strands of your hair separate and move up toward the comb.

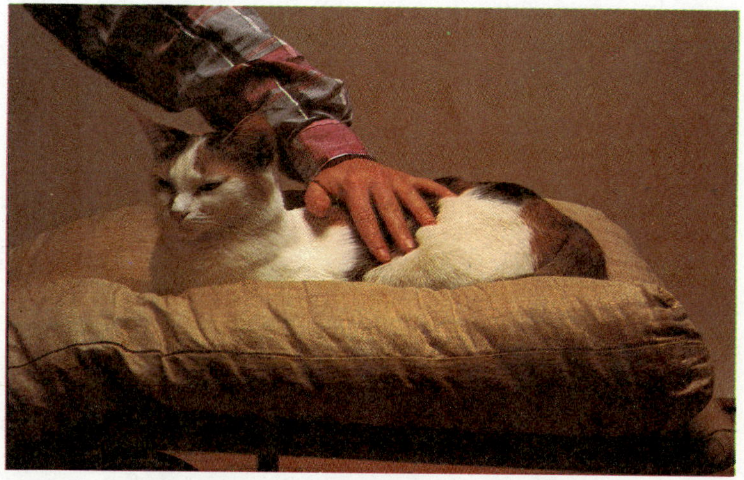

Figure 14.10 When you stroke a cat's fur, electrons move from the fur to your hand.

Another example of static electricity occurs in the laundry room. Clothes taken from a dryer often cling together. Why does this happen? The clothes rub against each other in the dryer. Electrons move from one piece of clothing to another. A positive charge builds up on the clothes that lose electrons. A negative charge builds up on the clothes that gain electrons. The oppositely charged clothing clings together.

ACTIVITY

How Do Electric Charges Act? ●

Materials

ring stand and ring
string
3 test tubes
silk
2 plastic combs
fur or flannel

Procedure

1. Use string to suspend a test tube from the ring of the ring stand.
2. Rub the test tube with silk for two minutes. Then let the test tube hang freely from the string.
3. Rub a second test tube with silk for two minutes. Now bring this test tube near the one hanging from the string. What happens? Put the test tube you are holding aside.
4. Again rub the hanging test tube with silk for two minutes.
5. Rub a comb for two minutes with fur or flannel. Bring it near the hanging test tube. What happens? Remove the test tube from the string.
6. Hang the plastic comb from the string. Rub it for two minutes with fur or flannel.
7. Rub a second comb for two minutes with fur or flannel. Bring this comb near the comb hanging from the string. What happens?

Questions

1. What happened when the two test tubes were brought together? Why?
2. What happened when the test tube and comb were brought together? Why?
3. What happened when the two combs were brought together? Why?
4. If the silk had a negative charge after it was rubbed on the test tube, what charge did the test tube have?
5. If the fur or flannel had a positive charge after it was rubbed on the comb, what charge did the comb have?

1. They repelled each other. Both had the same charge.
2. They attracted each other. They had opposite charges.
3. They repelled each other. Both had the same charge.
4. positive
5. negative

2. Explain that when ions build up on an object, the object becomes charged with static electricity. Mention as an example the shock that you sometimes feel when you get out of a car and touch its metal door handle. Ask the students to name some other examples of static electricity (after walking across a rug, you feel a shock when you touch a light switch; you hear crackling sounds as you comb your hair on a cold dry day).

ACTIVITY: How Do Electric Charges Act?

If you do not have the students do this activity, you may want to do it for them as a demonstration.

This activity or demonstration will reinforce the concept that like charges repel and unlike charges attract. When the glass is rubbed with silk, the glass becomes positively charged and the silk becomes negatively charged. When the comb is rubbed with the fur or flannel, the comb becomes negatively charged and the fur or flannel becomes positively charged.

Although the procedure suggests that objects should be rubbed for two minutes, temperature and humidity may cause variation in the time necessary to charge an object. Before class on the day of the activity or demonstration you may want to try rubbing the objects for less than two minutes and observe whether or not they become charged.

● **Laboratory Skills: Experimentation/Demonstration Manipulation of Materials**

3. Tell the students that static electricity can be detected by using a device called an electroscope. Show your electroscope to the students and identify its parts.
4. Rub the glass rod vigorously with the silk for a few minutes. Ask the students to observe the foil leaves of the electroscope as you touch the knob with the rod. What happens (leaves move apart)? Explain that the rod became positively charged as you rubbed it with the silk. When you touched the rod to the knob of the electroscope, the rod attracted electrons from the leaves. This left the leaves positively charged. They moved apart because they had the same charge.
5. Rub the rubber rod with the wool. Ask the students to predict what will happen when you touch the rubber rod to the electroscope (leaves will move apart; electrons move from rubber rod, which was negatively charged). What will happen when you touch the knob of the electroscope with your hand (leaves will move together)?

The presence of an electric charge on an object can be detected with an **electroscope** [ih-LEHK-truh-skohp]. An electroscope is a jar with a metal rod fitted into it. The top of the rod is in the form of a knob. Two metal leaves are attached to the bottom of the rod inside the jar. If you touch the knob with a negatively charged object, electrons will move from the object to the knob and down the rod into the metal leaves. The leaves will become negatively charged. Since like charges repel, the leaves will spread apart. If you touch a positively charged object to the knob, then electrons will move from the rod onto the object, leaving the metal leaves positively charged. Again the leaves move apart, since they have the same charge. If an uncharged object touches the knob, no electrons will move. The leaves will not become charged and will not separate. In this way an electroscope can tell you if the object you are testing has an electric charge. It cannot, however, tell you if the charge is negative or positive.

Figure 14.11 *left:* The leaves of the electroscope hang straight down. *right:* When a charged object touches the knob, the leaves separate.

Electric forces can be described with electric fields in much the same way that magnetic fields describe magnetic forces. The electric field is strongest near the object carrying an electric charge. The field is weaker farther away from the object.

Figure 14.12 Balloons with a static charge produce an opposite charge on the wall through electrostatic induction.

Just as magnetism can be induced, an electric charge can be produced in an object if that object is brought near a charged object. This is called **electrostatic induction** [ih-lehk-truh-STAT-ihk in-DUHK-shuhn]. The electric field of the charged object will either attract or repel electrons in the neutral object. If the charged object is negative, it will repel electrons. If it is positive, it will attract electrons.

For example, suppose you bring a negatively charged plastic comb near an electrically neutral scrap of aluminum foil. The electrons in the foil will be repelled by the negatively charged comb. Some of them will move to the side of the foil away from the comb. This leaves the side of the foil nearer the comb with a positive charge. The electric forces will attract the foil toward the comb.

Study Questions for 14.3

1. Describe and explain an example of static electricity.
2. How does an electroscope detect an electric charge?
3. Give an example of electrostatic induction.

1. When you comb your hair with a plastic comb, electrons go from your hair to the comb. The resulting negatively charged comb attracts the positively charged hair.
2. Both leaves become either positively or negatively charged when a charged object touches the knob. This causes the leaves to spread apart.
3. A negatively charged plastic comb attracts scraps of aluminum foil. The charged comb induces an opposite electric charge on the side of the foil nearest the comb.

6. Explain that just as you can induce magnetism by rubbing a steel bar with a magnet, you can induce an electric charge by rubbing a neutral object with a charged object. In electrostatic induction, electric fields attract opposite charges on neutral objects, thereby causing the objects themselves to become charged.

FOLLOW-UP

Reinforcement: Static Charges

Ask the students to pretend that they have been asked to teach a third grade class about static electricity. Have your students write what they would tell the third graders to explain the phenomenon. Call a nearby elementary school and offer your students' services.

Enrichment: Electrostatic Precipitators

The burning of fossil fuels produces a great deal of pollution. Electrostatic precipitators help to remove some of the pollutants from the smoke released by power plants. Ask the students to find out how electrostatic precipitators operate. Then have the students report their findings to the class.

Tie-in: Photostatic Copiers

Photostatic copiers operate because of static electricity. Ask the person who services the copying machine in your school to come in and explain the process to the students.

14.4

OBJECTIVES: The student should be able to

- **explain** what causes an electric charge to flow;
- **distinguish** between conductors and insulators;
- **describe** what happens in an electric discharge;
- **explain** how lightning is produced;
- **describe** how people can be protected from lightning;
- **explain** how grounded conductors and electrical grounds are used together as a safety device.

SCIENCE TERMS

electric discharge
grounded conductors
electrical ground

LESSON PLAN

1. Introduce the lesson by writing the words *conductors* and *insulators* on the board as column headings. Ask the students to distinguish between conductors and insulators with respect to the movement of electrons (electrons move easily through conductors and poorly through insulators). Write the definition of each word under the appropriate heading. Ask the students to name some examples of conductors and insulators (conductors—copper, iron bar, quarter; insulators—rubber, glass, paper). Write each example under the appropriate heading.

SCIENCE BACKGROUND

Whether or not a material is a conductor depends on its electron shell configuration. Elements that have outer electron shells more than half full are insulators. Elements with few electrons in their outer shells are conductors. Elements like silicon and germanium, whose outer shells are exactly half full, are called semiconductors.

14.4 Conductors and Electric Discharges

You will find out

- what moves an electric charge around;
- how electric discharges occur;
- how you can be protected from lightning.

Many objects, including your body, can have an electric charge. Sometimes this charge stays put as static electricity. What happens if the charge on an object does not stay put? The charge may quickly spread out over the entire object. Or the charge may flow from one object to another. Whether a charge will spread out, flow, or stay put depends on how easily the electrons in that object can move. It is the movement of electrons that moves a charge.

The same terms that apply to conduction of heat are used to describe the movement of electrons, or electric charges. Materials in which the electrons can move easily are called conductors. Ones in which electrons cannot move easily are called insulators. The best conductors are metals, such as silver, copper, and gold. Helium, argon, sulfur, rubber, and glass are examples of materials that are commonly used as insulators.

In conduction, a charge can flow from one part of an object to another part. A charge can also flow from one object to another when those objects are touching. It is also possible for a charge to flow from one object to another object even though the objects are not touching. For example, if you walk across a carpet, your body may

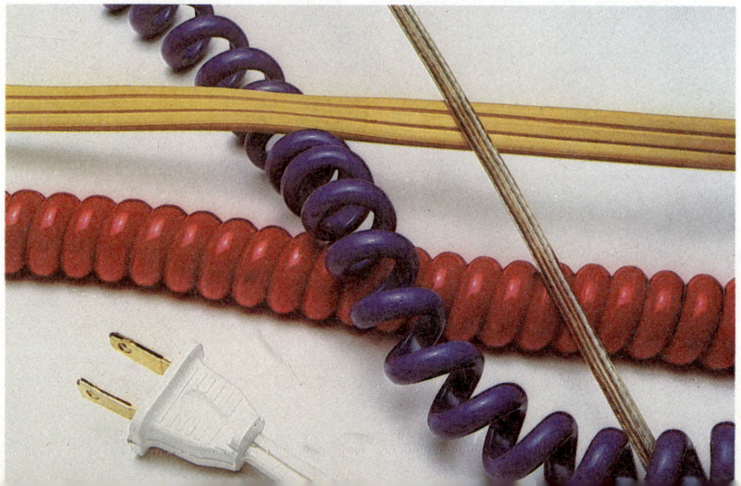

Figure 14.13 Colorful insulation covers these electric cords.

pick up electrons from the carpet. The air around you is an insulator. Therefore, the electrons you gained from the carpet do not move from your body to the air. You become a charged object. What happens if you now bring your hand near a metal object such as a doorknob? You see and feel a spark. The electric field between your charged fingers and the doorknob acts to produce a charge on the atoms in the air. The air now becomes a conductor and carries the electric charge from your body to the doorknob. The flow of an electric charge through a gas from a charged object to a conductor is called an **electric discharge.**

Figure 14.14 Particles of blowing sand or snow can cause a buildup of electric charges on a wire fence. Discharges from these fences can be powerful enough to knock people and cattle to the ground.

The electric discharges you experience from walking on a rug are harmless. However, not all electric discharges are harmless. For example, when the inside of a ship is washed, electric charges can build up on water drops as they rub against the inner sides of the ship. The buildup of charges on these drops can cause electric discharges. Sparks from these discharges have ignited fumes and caused explosions in oil tankers.

The movement of particles in the air by wind and rain can cause a buildup of charges in a cloud. If this charge gets too great, it will discharge to the ground or to another cloud. This discharge, or movement of electrons, appears as a bolt of lightning.

2. Draw a circle and a teardrop shape on the board. At the top of the circle draw several plus signs. Explain that the plus signs represent positive electric charges. Ask how these like charges will act toward each other (they will repel). Explain that since like charges repel, they will spread themselves out evenly over the surface of the circle. Spread out the charges in your drawing. Explain that in the teardrop the charges cannot spread out evenly because of the object's shape. They tend to concentrate at the point of the teardrop. From there they jump off and the object loses its charge, or becomes discharged.
3. Ask the students what happens when they rub their feet on a rug and then touch a metal doorknob (they receive a small shock). Explain that they are experiencing an electric discharge. They are losing the electric charge that was built up on them. The charge that was concentrated on their fingers jumps off, or discharges, onto the metal object.
4. Point out that lightning is simply a large spark, or the passage of electric charge between two oppositely charged objects. Ask the students what conditions must exist for lightning to occur (movement of particles in the air by the wind and rain causes a buildup of charge in a cloud; if the charge becomes too great as the cloud passes near an object with the opposite charge, the cloud will discharge).

SCIENCE BACKGROUND

In 1752 Benjamin Franklin flew a kite into a thunderstorm to demonstrate that clouds contain an electric charge. Franklin held the kite by a thin wire, which was attached to a string around a metal key. When a spark jumped from the key to his hand, Franklin had evidence that the clouds contained an electric charge.

5. As an assignment have the students create posters illustrating the various hazards of lightning and ways people can protect themselves. They might show, for example, what to do if you find yourself riding in a boat or walking in a field when lightning begins. They might also include a brief explanation of why it is safer to be in a car during a lightning storm (metal body of car keeps the charges outside the car). Finally, the students might include a diagram showing the use of lightning rods as grounded conductors.

SCIENCE BACKGROUND

Lightning can flow from the ground to clouds and from clouds to the ground. If the upper region of a cloud is negatively charged, the lower region will be positively charged. If the earth is negatively charged, electrons will move from the earth up to a cloud.

The thunder that accompanies lightning is caused when air is heated to the point that it expands suddenly. As cooler air rapidly moves into the space left by the sudden expansion of the heated air, thunder is produced.

Lightning takes the easiest path to the ground. Thus it usually strikes the tallest object around. In flat regions, a tree may be the tallest object. You should, therefore, stay away from trees during a thunderstorm. Lightning may strike the tree. The electrons can move down the tree and discharge to your body. You should avoid standing near good conductors, such as metal fences, for the same reason.

Do You Know?
Temperatures near 30,000°C can be produced by bolts of lightning.

Tall buildings also make good targets for lightning. The Empire State Building is usually struck dozens of times during a thunderstorm. Most tall buildings are protected by lightning rods. A lightning rod is a metal rod connected to conductors that lead to the earth. Electrons can move from a lightning bolt, to the metal lightning rod, down the conductor attached to it, and into the earth. A lightning rod does not prevent lightning strikes. It attracts them. The lightning rod makes a safe path for the electric charge to flow to the earth.

Lightning rods are examples of **grounded conductors.** A grounded conductor is one that is connected to another very large conductor such as the metal frame of a very large building or the earth. The very large conductor is called the **electrical ground.**

Figure 14.15 Lightning rods provide a safe path for electric charges to flow to the earth.

ACTIVITY

Which Materials Are Conductors? ●

Materials

light bulb and socket
3 pieces of insulated copper wire
battery

Procedure

1. Arrange your apparatus as shown in the diagram. Test the bulb to be sure it is working by bringing the free wires together. The bulb should light.
2. Use your apparatus to locate ten conductors and ten insulators in the room.

Questions

1. What similarities do you see among the materials that were conductors? **shiny, metallic, solid**
2. What similarities do you see among the materials that were insulators? **dull, nonmetallic, solid**

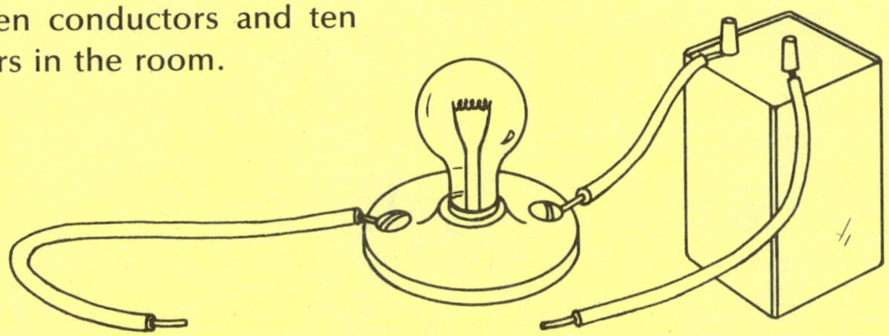

A grounded conductor cannot hold a charge. If the conductor is negatively charged, the extra electrons move from the conductor into the electrical ground. If the conductor is positively charged, then electrons move up from the electrical ground to cancel out the positive charge. The electrical ground acts as a very large storehouse of electric charge. Since it is so large, some electrons gained or lost do not have any overall effect. By grounding a conductor, you can prevent dangerous electric discharges from occurring. Electrical grounds are an important safety feature of many machines and appliances as well as buildings.

Do You Know?
The separation of charges in sandstorms can cause lightning.

Study Questions for 14.4

1. What makes some substances good conductors of electricity?
2. Describe and explain an example of an electric discharge.
3. Explain how a lightning rod works.

1. Substances with electrons that can move easily are good conductors.
2. An electric discharge may occur when you walk across a carpet and then bring your hand near a metal object. You pick up electrons from the carpet. When your hand approaches a metal object, the air conducts the charge from your body to the object.
3. It attracts lightning. When lightning strikes the rod, the electrons flow down the rod, down the conductor attached to the rod, and into the earth, an electrical ground.

ACTIVITY: Which Materials Are Conductors?

Suggest that the students keep a list of conductors and insulators they find in the room. Conductors they may find are metal chair legs, paper clips, and metal bands on pencils. Insulators they may find are paper, chalk, wood, and glass.

● **Laboratory Skills: Experimentation/Demonstration Manipulation of Materials**

FOLLOW-UP

Reinforcement: Grounding Buildings
Ask the students to determine whether or not their building is grounded. In a class discussion have the students express their ideas on whether or not it should be grounded. If the building should be grounded, how would they go about grounding it?

Enrichment: Capacitors
Capacitors, or condensers, are used to store electric charge. Where are they used? Why are they used? Ask the students to do research to determine the answers to these questions.

Tie-in: What Are the Best Insulators?
When wires in homes become frayed, people often mend them with tape. Is this a good idea? Have students write a booklet of tips for electrical safety in the home. They should include information about improving home safety by using the best insulators.

What Did You Learn?

- 14.1 • Magnetism is a force produced by electric charges in motion.
- 14.1 • Every magnet has a north and a south pole.
- 14.1 • The magnetic field is the area around a magnet where its magnetic force can be experienced.
- 14.1 • In magnets the magnetic domains are lined up.
- 14.1 • There are temporary magnets, permanent magnets, and natural magnets.
- 14.2 • Compass needles point to the north magnetic pole.
- 14.2 • The earth has a magnetic field and behaves as if a huge bar magnet is stuck through it.
- 14.3 • Static electricity is an electric charge that builds up on an object and does not flow.
- 14.3 • An electroscope detects static electricity.
- 14.3 • In electrostatic induction an electric charge is produced on an object that is brought near a charged object.
- 14.4 • The electrons in conductors can move around.
- 14.4 • The flow of electricity through a gas from a charged object to a conductor is an electric discharge.
- 14.4 • Lightning is an electric discharge.
- 14.4 • A grounded conductor cannot hold a charge because any charge that passes through it flows into the electrical ground.

Key Terms

magnetism
magnet
magnetic poles
magnetic field
magnetic domains
induced magnetism
north magnetic pole
north geographic pole
static electricity
electroscope
electrostatic induction
electric discharge
grounded conductors
electrical ground

Biography

Mildred Dresselhaus (1930-)

Mildred Dresselhaus is an electrical engineer and a physicist. Her work involves the study of how different types of materials conduct electricity under different conditions. The results of her studies could prove very useful in communications, computers, or other devices that involve magnetic and electric forces.

Dresselhaus was born in New York City. Her parents were immigrants who had come to the United States to find a new life. When Mildred was born, her parents were poor. She had to work even while in grade school at several dirty, dull jobs to help her family. Her parents taught her to have faith in herself, and they encouraged her to do well in school.

She did very well. She graduated near the top of her class from Hunter College High School. After that she graduated with highest honors from Hunter College and received a doctorate in physics from the University of Chicago.

As a result of her excellent original research, Mildred Dresselhaus is known and honored around the world. She is now the director of the Center for Materials Science and Engineering at the Massachusetts Institute of Technology. She is also a member of the National Academy of Engineering. Dr. Dresselhaus advises the government on research and serves on several governmental committees.

Vocabulary

1. b
2. c
3. a
4. a
5. c
6. b
7. b
8. c

● **Word Skill: Vocabulary**

What Do You Remember?

1. a force produced by electric charges in motion
2. Its magnetic domains are lined up.
3. Two north poles repel. Two south poles repel. A north pole and a south pole attract.
4. The earth has a magnetic field and acts as if there is a huge bar magnet stuck through it.

TO THINK ABOUT AND DO

Vocabulary ●

Number your paper from 1 to 8. Beside each number write the letter of the word or words that best complete the sentence.

1. In a magnet the regions of greatest magnetic force are the (a) magnetic domains, (b) magnetic poles, (c) lodestones.
2. The attraction of scraps of aluminum foil to a negatively charged plastic comb is an example of (a) magnetism, (b) conduction, (c) electrostatic induction.
3. A lightning rod (a) attracts lightning, (b) repels lightning, (c) is an insulator.
4. A compass needle points to the ________ pole. (a) magnetic (b) geographic (c) closer
5. A device that detects the presence of an electric charge is (a) a conductor, (b) an electric field, (c) an electroscope.
6. Clusters of many atoms which can be thought of as tiny magnets are (a) insulators, (b) magnetic domains, (c) magnetic poles.
7. The electric charge that builds up on an object and does not flow is (a) an electroscope, (b) static electricity, (c) electrostatic induction.
8. An example of a grounded conductor is (a) a copper penny, (b) a negatively charged plastic comb, (c) a lightning rod.

What Do You Remember?

1. What is magnetism?
2. Why is some matter magnetic?
3. What happens when you bring the north poles of two magnets together? The south poles? The north pole of one magnet and the south pole of another magnet?
4. How is the earth like a magnet?

5. What are two ways that magnetism and electricity are alike?
6. How are magnetism and electricity different?
7. What is magnetic induction?
8. How does an electroscope detect an electric charge?
9. What is a grounded conductor?
10. Why do clothes sometimes cling together when they come out of the dryer?

Applying What You Have Learned ●

1. How can you magnetize an iron nail? Will the nail be a permanent or a temporary magnet?
2. If you rub a glass rod with a piece of silk, the silk will cling to the rod. Why does this happen?
3. Why are metals good conductors of both heat and electricity?
4. Why is it a good idea to keep strong magnets away from watches?

Research and Investigation ●

1. Benjamin Franklin used something called a Leyden (Leiden) jar in some of his experiments. What is a Leyden jar?
2. What is a Van de Graaff generator?
3. The earth has a magnetic field. Do any other planets have a magnetic field?

5. Like poles or charges repel and opposites attract. The magnetic or electric force between two objects decreases as the objects are moved apart. Both forces are experienced in a force field.
6. An electrified object can hold only one kind of charge, positive or negative. A magnetized object has two poles, north and south.
7. the magnetizing of a material by a magnet
8. A charged object will cause the leaves of the electroscope to become charged with the same charge. The leaves will repel each other, spreading apart. Any charged object that touches the knob of the electroscope will cause the leaves to spread apart.
9. A grounded conductor is a conductor connected to a very large conductor such as the earth or the metal frame of a very large building.
10. The clothes rub together in the dryer. Some clothing loses electrons, becoming positively charged. Other clothing gains electrons, becoming negatively charged. The positively charged clothing clings to the negatively charged clothing.

Applying What You Have Learned

1. Rub a magnet in one direction against the nail to make it a temporary magnet.
2. One material loses electrons to the other. They become oppositely charged and are attracted to each other. The glass rod loses electrons to the silk.
3. The metallic bond in metals involves wandering electrons. These wandering electrons can move easily, carrying both heat and electric charge.
4. A strong magnet may cause some of the moving parts of the watch to become temporary magnets. The parts would then stick together and the watch would stop.

● Comprehension Skill: Applied

Research and Investigation

1. A Leyden jar is a glass jar lined inside and out with tinfoil. Since the jar contains a conducting rod connected to the inner foil lining, it can be used to store electric charges.
2. A Van de Graaff generator produces static electricity. A motor moves a belt. The belt picks up electric charges and moves them to a metal ball at the top of the generator. The metal ball picks up the charges from the moving belt. As the belt continues to move, charges build up on the metal ball. A person can pick up the static charge by touching the metal ball. Can de Graaff generators used in research can produce an electric potential of several million volts.
3. Magnetic fields exist around Mercury, Jupiter, Saturn, Venus, and Mars.

● Reference/Research Skills

CHAPTER BACKGROUND

The first important experiments with electric currents were conducted in the late eighteenth century by Luigi Galvani and Alessandro Volta. Galvani had seen a dissected frog's leg muscle twitch without any connection to a known source of current. After a number of experiments to investigate the phenomenon, Galvani decided that the current came from the muscle itself and called it "animal electricity."

As Volta did experiments of his own, he began to doubt Galvani's animal electricity theory and to suspect that the electricity came from something else. Volta produced the first steady flow of electric current when he converted chemical energy into electric energy in a device known as a voltaic cell or an electrochemical cell.

In a voltaic cell, two different metals are suspended in a conducting fluid called an electrolyte. For example, the metals zinc and copper may be suspended in saltwater. As a result of a chemical reaction, zinc accumulates an excess of electrons and becomes negatively charged. Copper loses electrons to the saltwater electrolyte and becomes positively charged. When a wire is connected between the two metals, the excess electrons flow from the negatively charged zinc strip to the positively charged copper strip. The result is a flow of electrons.

Using voltaic cells, the German physicist George S. Ohm studied voltage, current, and resistance during the early nineteenth century. He discovered that in closed circuits the current and voltage are directly proportional to each other and the current and resistance are inversely proportional to each other. Thus, if the voltage, or potential difference, is increased, there will be a corresponding increase in the current produced. However, as it becomes more difficult for the electrons or ions to move because of increased resistance, less current will be produced.

Ohm's discovery makes it possible to predict accurately how large the current is in the circuit. With this knowledge we can manipulate the parts of a circuit to produce the electric power we need.

15 Electricity

Have on your desk two sets of decorating lights, one a parallel circuit and one a series circuit, and a simple circuit that contains a switch. Ask the students to observe as you open and close the switch. Can they explain why the light turns off and on? Now ask them to observe while you remove one light from both the parallel and the series decorating lights. Can the students explain their observations? In which set of decorating lights was the removal of the bulb like the opening of the switch?

Tell the students that in this chapter they will learn about the circuits they have observed and more about how electricity affects their lives.

Every now and then, because of a storm or an accident, the supply of electric power to your home may be cut off. Can you remember one of these power failures, or blackouts? What did you do? Did you use candles for light? For a while the blackout may have been fun because it was so different from your everyday life. However, you probably soon tired of it and were happy when things went back to normal. Certainly if you were in an elevator when the power went out, you were glad when the blackout was over!

The invisible flow of electric power is a basic part of your everyday life. You flip a switch, and the power is there to light your home and to power your radio and television, as well as practically every other appliance in your home. Electricity is important. Benjamin Franklin had no idea how important when he spoke about it over 200 years ago. He said, "Electrical fluid . . . may . . . be of use to mankind."

Esto Photographics Inc.

Lesson Organizer

15.1 Electric Currents
Pages 316–317
Advanced Planning: Page 316
Follow-up: Page 317
- **R** Electric Current and Its Production
- **E** Transformers
- **T** A Century of Light

15.2 Electric Circuits
Pages 318–323
Advanced Planning: Page 318
Activity: How Can You Construct Electric Circuits? Page 322
Follow-up: Page 323
- **R** Ohm's Law
- **E** Measuring Current and Voltage in Circuits
- **T** Resistors

15.3 Electric Power
Pages 324–325
Advanced Planning: Page 324
Follow-up: Page 325
- **R** Determining Electric Costs
- **E** An Energy Audit of Your Room
- **T** Power for Tomorrow

15.4 Electric Safety
Pages 326–329
Activity: How Can You Compare the Resistance of Different Wires? Page 327
Follow-up: Page 329
- **R** Electric Safety
- **E** An Electrician's Tour
- **T** Safety At Home

See ACTIVITIES book pages 85-90 for laboratory, skill, and review exercises for Chapter 15.

15.1

OBJECTIVES: The student should be able to

- **define** the term *electric current;*
- **explain** what causes electric current to flow;
- **define** the term *voltage;*
- **define** the term *volt;*
- **describe** the structure of a dry cell;
- **explain** how a dry cell produces a voltage;
- **define** the term *amperes.*

SCIENCE TERMS

electric current	positive terminal
voltage	negative terminal
volts	amperes
dry cell	

ADVANCED PLANNING

For demonstrations you will need a miniature light-bulb socket and light bulb and a dry cell.

LESSON PLAN

1. Attach the wires to your light-bulb socket and dry cell so that the light bulb lights. Point out that there are two types of electricity—static electricity and electric current, illustrated by the dry cell and light bulb.

● **Word Skill: Vocabulary**

SCIENCE BACKGROUND

Batteries can be classified as wet cells or as dry cells. Wet cells contain a liquid; dry cells contain a paste. You can have students create a simple wet cell with a lemon or a potato. Have the students insert a copper strip and a zinc strip about 1 cm apart into the lemon or potato.

15.1 Electric Currents

You will find out
- what is meant by the term *electric current;*
- what causes electric current to flow;
- how a dry cell produces voltage.

It's very easy to use electricity. Almost anyone can flip a switch or turn a dial. You have probably had experience plugging in cords, replacing light bulbs, and changing batteries in a flashlight. But what are you really doing when you do these things? To understand how electricity acts to do all that it does in your life, you must first understand how electric charges move.

Electric charges can flow from one end of an object throughout that object. Charges can also flow from one object to another. The flow of electric charges from one point to another point is called **electric current.**

The flow of electric charges, or current, is similar to ● the flow of water in a hose. Water flows inside a hose because of differences in water pressure. When you turn on the faucet, the water pressure is higher at the faucet end of the hose than at the open end. Water flows from the area of high pressure to the area of low pressure. Similarly, current flows from a point of high electric potential, or pressure, to a point of low electric potential. This difference in electric potential that causes current to flow is called **voltage.** Voltage is measured in units called **volts.**

The currents in your home are created by voltages of 120 or 240 volts. By comparison, a flashlight battery has a voltage of about 1.5 volts. How does a flashlight battery produce a voltage? A flashlight battery is a device known as a **dry cell.** A dry cell produces voltage as a result of chemical reactions between a moist paste and a metal.

Figure 15.2 Differences in water pressure cause water to flow in a hose. Voltage causes current to flow in a conductor.

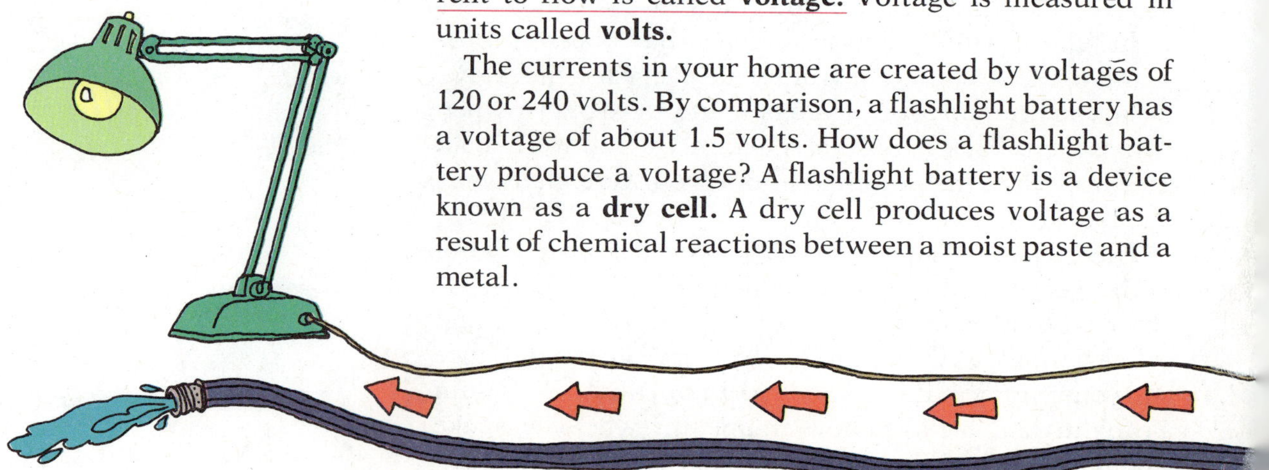

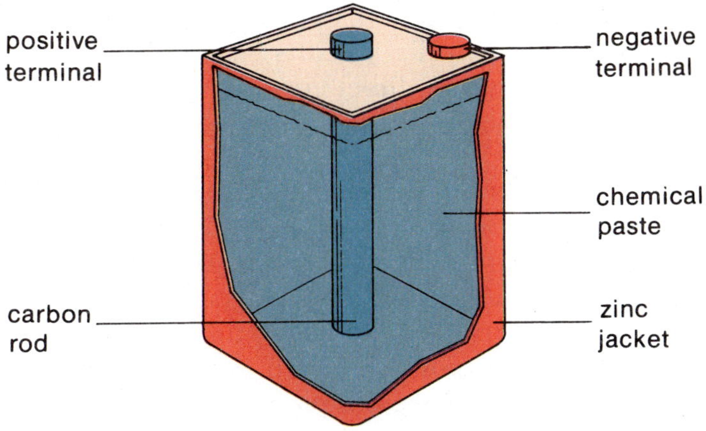

Figure 15.3 The separation of charges in a dry cell produces a voltage.

A dry cell is made of a carbon rod surrounded by a mixture of chemicals that forms a paste. The paste is contained in a zinc jacket. Chemical reactions between the zinc, the carbon, and the paste cause a buildup of positive ions on the carbon rod. These positive ions give the carbon rod a positive charge. At the same time the zinc jacket is left with a negative charge. This separation of charges produces a voltage.

Since the carbon rod is positive, the place on it where contact is made with a conductor is called the **positive terminal.** Similarly, the place on the negative zinc jacket where contact is made is called the **negative terminal.** If the positive and negative terminals are connected with a conductor such as a metal wire, a current flows. Electrons move through the wire from the negative to the positive terminal.

Currents vary in strength. A current gets stronger as more charges flow through a conductor in the same amount of time. The amount of electric current, or number of charges, that passes a point in a conductor in one second is measured in **amperes** [AM-pihrz].

Do You Know?
A single bolt of lightning can carry 160,000 amperes of current through a voltage of 100 million volts.

Study Questions for 15.1

1. What is electric current?
2. What is voltage?
3. How does a dry cell produce a voltage?

1. Electric current is the flow of charges from one point to another.
2. Voltage is the difference in electric potential that causes current to flow.
3. through chemical reactions between a moist paste and a metal

2. Remind the students that scientists often compare the flow of electric current through a wire to the flow of water through a hose. Ask a student to explain this analogy (water flows through a hose from the area of high pressure at the faucet to the area of low pressure at the nozzle; electrons flow from a point of high electric potential, or pressure, to a point of low electric potential).
3. Have a student define the terms *voltage* (difference in electric potential that causes current to flow) and *volt* (unit used to measure voltage).
4. Mention that one way to produce a voltage is by a chemical reaction. Use flashlight and radio batteries as examples of devices in which a voltage is produced by a chemical reaction.

● **Comprehension Skill: Applied**

FOLLOW-UP

Reinforcement: Electric Current and Its Production
Ask students to construct posters to illustrate what electric current is and how it is similar to water moving through a pipe. Their posters should include a diagram that shows how electric current is produced in a dry cell.

Enrichment: Transformers
Electric current generated by utility companies passes through both step-up and step-down transformers before it reaches a home. Ask students to determine the purpose of these transformers and report their findings to the class.

Tie-in: A Century of Light
Ask the students to write letters to the Thomas Alva Edison Foundation (2000 Second Avenue, Detroit, Michigan 48226) requesting information about Edison and his work. They should use the information they receive to create a bulletin board about Edison.

SCIENCE BACKGROUND

The positive pole of a battery is called the anode. The negative pole is the cathode.

The ampere is named after André Ampère. One ampere is the movement of 6.24×10^{17} electrons past a point in one second. The volt is named after Alessandro Volta.

15.2

OBJECTIVES: The student should be able to

- **define** the term *electric circuit;*
- **explain** how to construct a simple electric circuit;
- **compare** and contrast series and parallel circuits;
- **distinguish** between alternating and direct currents;
- **explain** what determines how much current flows through a circuit;
- **describe** two ways the strength of a current depends on the type of wire;
- **define** the term *resistance;*
- **state** Ohm's law;
- **solve** problems using Ohm's law.

SCIENCE TERMS

electric circuit	alternating current
series circuit	resistance
parallel circuits	ohms
direct current	Ohm's law

ADVANCED PLANNING

1. For demonstrations you will need the dry cell, the wires, the light-bulb socket, and the light bulb from Lesson 15.1. You will also need a 1.5-volt dry cell and a 9-volt dry cell.

2. Prepare two work sheets of problems related to Ohm's law. One work sheet should contain numbers the students can easily manipulate in their heads so that it can be used in a class exercise. The other work sheet can be used as an assignment.

LESSON PLAN

1. Attach the wires from the dry cell to the light-bulb socket and light bulb. Explain that electrons move from the negative terminal, along the wire, through the light bulb, and on to the positive terminal. The path that the electrons follow, or the circuit, is a circle that begins and ends with the dry cell.
2. Stress that as long as the circuit is closed, the electrons will move and light the bulb. Ask why the bulb will not light in an open circuit (current cannot move; electrons have no place to go and therefore pile up).

15.2 Electric Circuits

You will find out
- how series and parallel circuits differ;
- how alternating and direct currents differ;
- what is meant by resistance.

When a voltage is created, charges do not automatically begin to flow. For example, a spare flashlight battery that is not being used has a voltage. But no current is flowing between the two terminals. No charges are flowing. For charges to flow there must be a path from the negative to the positive terminal. A conductor such as a metal wire can make this path. A path that connects the negative and positive terminals without any breaks is called a closed path. The path of an electric current is called an **electric circuit** [SUR-kiht].

You can build a simple electric circuit with a dry-cell battery, two wires, and a light bulb. You connect one of the wires to the light bulb and the positive terminal of the battery. You connect the other wire to the light bulb and the negative terminal of the battery. When these connections are made, the path for the current is closed. The circuit is closed, or complete. Current flows and the light bulb goes on. If a connection is broken anywhere along the current's path, that path is no longer closed. The circuit is open, or broken. Current cannot flow through an open circuit.

Figure 15.4 Current cannot flow through an open circuit.

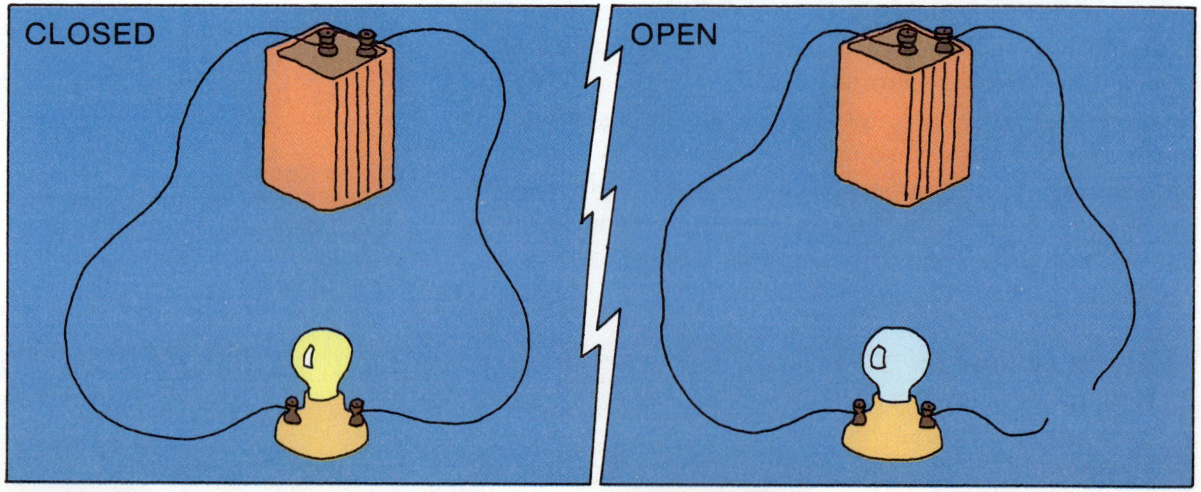

A circuit made of a battery, two wires, and a light bulb is a very simple circuit. It has only a few parts. Most circuits are much more complicated. These circuits have many parts and may have several different paths for the current to flow through. Circuits can be grouped according to the kind of path the current takes.

If a circuit has only one path for the current, it is a **series circuit.** In a series circuit, all the current must go through each part of the circuit. A series circuit has the advantage of being easy to hook up and study. However, it has a major problem. If one part of the circuit—for example, a light bulb—goes out, the circuit is open and the current stops. The whole circuit goes dead. Some strings of decorating lights use series circuits. If one light goes out, all the lights go out.

Parallel circuits are more complicated than series circuits. In parallel circuits, the current divides and flows in two or more separate paths, or branches. If the current in one branch stops, the current still flows in the other branches. For example, many light fixtures have several light bulbs on a parallel circuit. When one bulb goes out, all the others stay lighted.

Figure 15.5 *left:* A parallel circuit offers more than one path. *right:* A series circuit offers only one path.

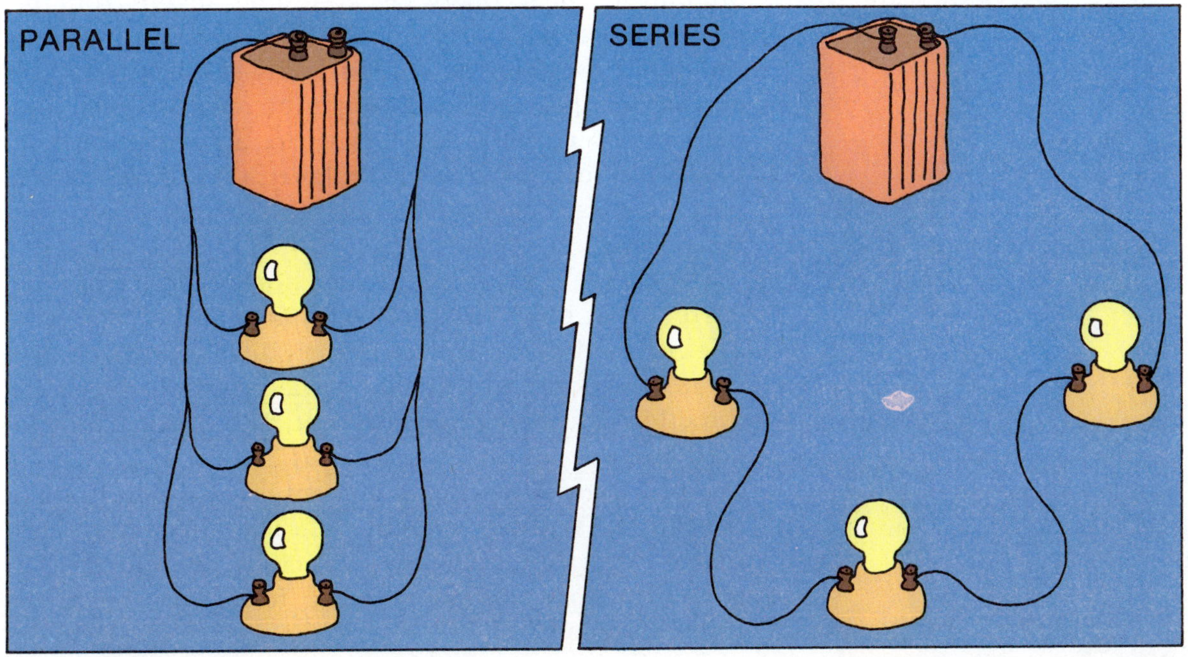

3. Ask the students to state the difference between a series circuit and a parallel circuit (series—current flows through each cell and the voltage is cumulative; parallel—current flows in two or more separate branches and the total voltage of all the cells is the same as the voltage of one cell).

4. As an assignment have the students draw diagrams of a series circuit and a parallel circuit. They should label the positive and negative terminals on the dry cells and use arrows to show the path of the electric current.

5. As the students correct their diagrams, explain that the current should be shown to flow in only one direction. Ask what this current is called (direct current).
6. Point out that another type of current, alternating current, is produced by generators in which a coil of wire spins between two magnets. The current flows first in one direction and then in the other. Ask the students which kind of current, alternating or direct, is found in their homes (alternating current).

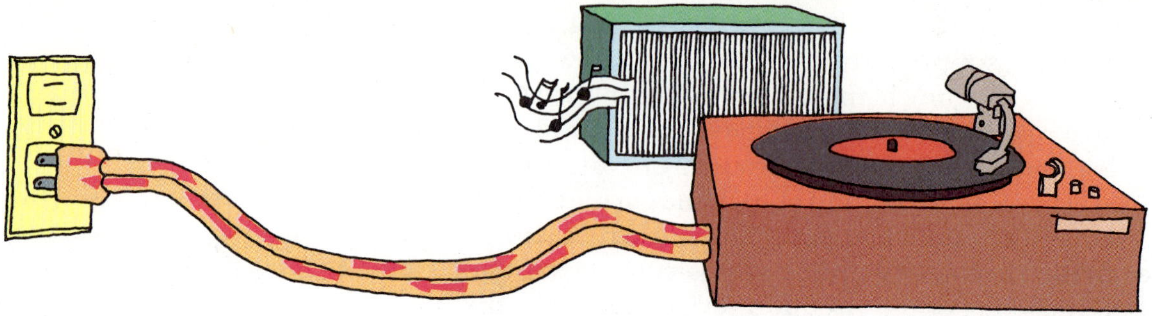

Figure 15.6 One branch of a parallel circuit

The electric wiring in your home is in the form of many parallel circuits. These circuits are designed so that new circuit parts such as lamps and appliances can be added by plugging them into wall outlets. The wall outlets work like the terminals of a battery.

Suppose you want to plug in and use a new record player. The record player has a cord made of two wires. The cord ends in a plug that has two prongs, one for each wire. You plug the record player into a wall outlet. When you turn on the record player, the current flows from the outlet into one prong of the plug. The wire connected to that prong carries the current into the record player. The second wire carries the current from the record player back to the outlet. The current flows into the outlet through the second prong. The path for the current from the outlet, through the record player, and back to the outlet is one branch of a parallel circuit. When you turn on the record player, that branch of the parallel circuit is complete, closed. Current flows. You hear music.

The current that flows in your home differs from the current that flows from batteries in an important way. The current flowing from a battery always moves in the same direction. Current that always flows in the same direction is called **direct current,** DC. The current in your home is not direct current. It is constantly switching directions. This kind of current is called **alternating** [awl-tur-NAY-tihng] **current,** AC. The alternating current in your home changes direction very quickly. It goes from forward to backward to forward again 60 times a second!

All circuits do not carry the same amount of current. What determines how much current flows in a circuit? This question was answered by the German scientist George S. Ohm in the early 1800's. Ohm found that voltage determines the amount of current that flows in a circuit. More current is produced by higher voltages. This is what you might expect. A higher voltage produces more current in much the same way that greater pressure differences produce more water flow in a hose.

You can make another useful comparison between a wire as a conductor of electricity and a hose as a conductor of water. The diameter of the hose affects the amount of water that can flow through it. A large-diameter hose carries more water than one with a small diameter. Ohm found that the amount of current that flows through a wire depends on the wire. A wire with a large diameter carries more current than a wire with a small diameter. Also, certain kinds of wire such as those made of copper, silver, or gold carry current more easily than wires made of other materials such as iron or lead.

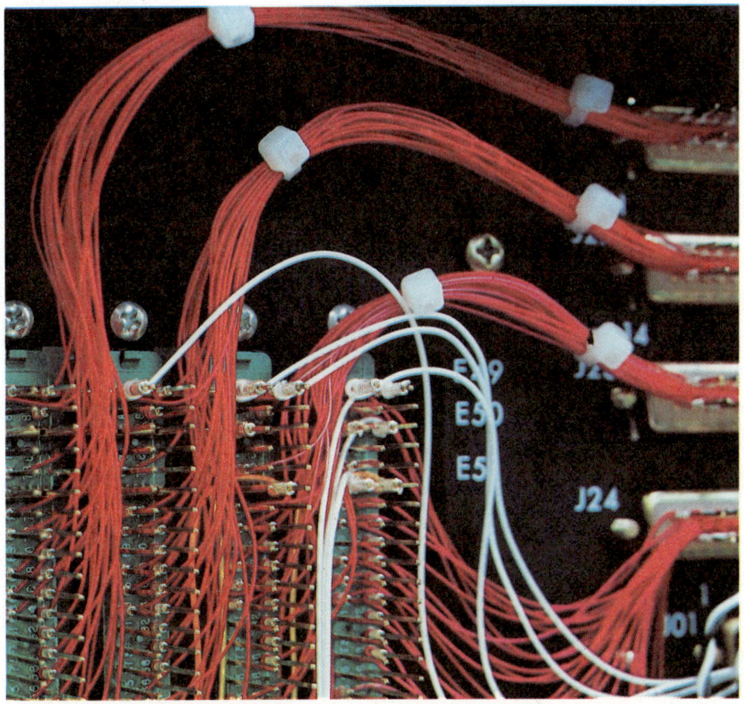

Figure 15.7 Wires of different sizes and materials have different resistances.

Do You Know?
An electron in a circuit moves a distance of 1 mm per second.

7. Show the students the 1.5-volt dry cell and the 9-volt dry cell. Ask which cell will produce a stronger electric current (9-volt cell).

8. Remind the students that voltage is not the only factor that determines how much current flows through a circuit. What is another factor that determines the amount of current (the type of wire)?

9. Point out that wire made of certain materials allows electric current to flow through it more easily than wire made of other materials. Ask the students to name some materials that allow current to flow easily (silver, copper, and gold). What do these materials have in common (all are transition metals)? Why does the current flow through them more easily than through other metals (they offer the least resistance or opposition to the flow)? If silver offers less resistance than copper, why isn't it used instead of copper (too expensive)?

SCIENCE BACKGROUND

Currents in metals are carried by electrons. In some other materials, especially liquids and gases, currents can be carried by both positive and negative ions.

ACTIVITY: How Can You Construct Electric Circuits?

Review the way a series circuit and a parallel circuit are hooked up (series—wire from positive terminal of one cell to negative terminal of next; parallel—wire from positive terminal of one cell to positive terminal of next and negative terminal of one cell to negative terminal of next).

If the students work in groups, you will need less equipment. Check each circuit the students make. Have them compare the series and the parallel circuits. Stress the advantages of the parallel circuits (lights remain bright as additional ones are added; when one lamp goes out the others remain on).

After students have completed the activity, draw each circuit on the board and ask the following questions:

(a) Why are the lights in the series circuit dimmer than the lights in the parallel circuit (more resistance as the current flows through each bulb)?

(b) In which circuit will all the lights burn out if one burns out (series; the circuit is broken)?

(c) Which kind of circuit is in homes (parallel)? Why (lights do not get dimmer as more are turned on; not necessary to have all lights on at one time in order to have one light on)?

● **Laboratory Skills: Manipulation of Materials Experimentation/Demonstration**

SCIENCE BACKGROUND

The resistance of a conductor depends on its thickness, composition, length, and temperature.

The total resistance of a series unit is the sum of the individual resistances of all the parts in that circuit. As parts are added to the circuit, the total resistance increases. According to Ohm's Law, when the resistance increases, the current decreases.

The current in each branch of a parallel circuit has a different strength depending on the resistance of that branch. Most of the current flows along the path of least resistance. The voltage across each branch is the same.

The light bulbs in a series circuit do not light as brightly as those in a parallel circuit. The series lights are dimmer because the voltage must push against the sum of all the resistances in the series circuit.

● ACTIVITY — How Can You Construct Electric Circuits?

Materials

3 batteries
7 sockets and light bulbs
small screwdriver
12 pieces of insulated copper wire (30 cm each)

1. The series lights are dimmer.
2. They are the same.
3. They all go out.
4. The others remain lighted.

Procedure

Follow the diagrams to construct three different circuits. You will need to refer to your circuits as you answer the following questions.

5. Each light increases the resistance in the circuit. According to Ohm's law, when the resistance increases and the voltage remains the same, the current decreases.
6. The series circuit offers only one path for the current to follow. If that one path is broken, no current can flow.

Questions

1. How does the brightness of the lights in the series circuit compare to the brightness of the light in the simple circuit?
2. How does the brightness of the lights in the parallel circuit compare to the brightness of the light in the simple circuit?
3. If you unscrew one light bulb in the series circuit, what happens to the others?
4. If you unscrew one light bulb in the parallel circuit, what happens to the others?
5. How can you explain the dimness in the series circuit?
6. How can you explain the fact that when one light bulb goes out, all the light bulbs in a series circuit go out?

Simple Circuit

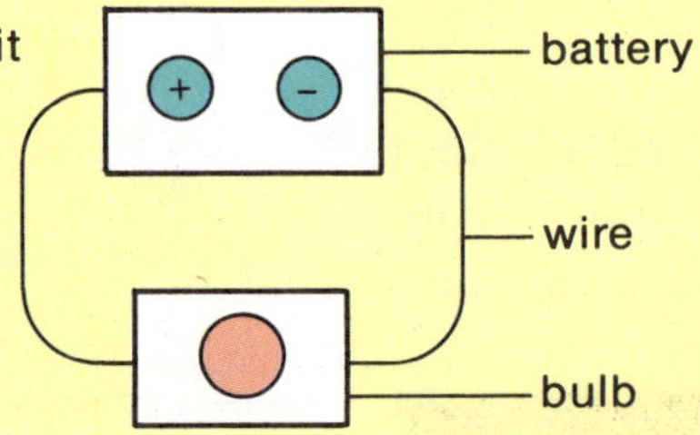

Parallel Circuit

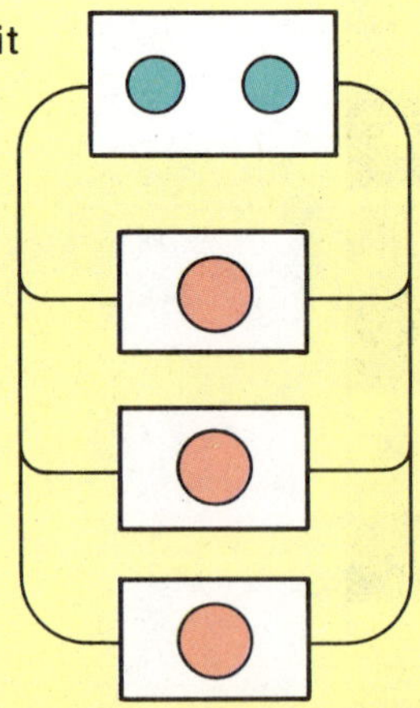

Series Circuit

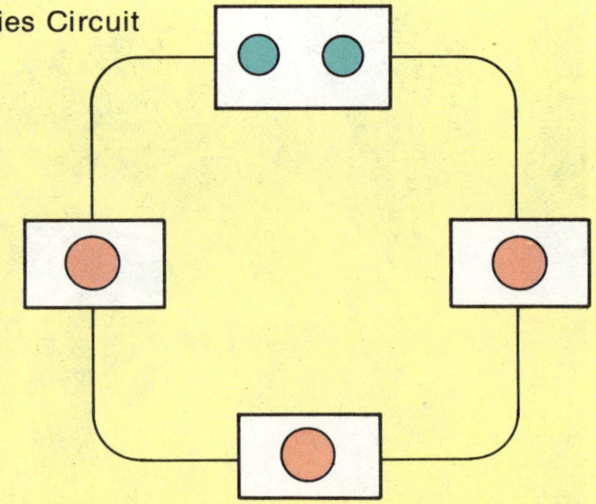

The current-carrying properties of a material are summed up in the **resistance** [rih-ZIHS-tuhns] of that material. The resistance is the opposition a material offers to the flow of electric charges. All materials, even good conductors, offer some resistance. Therefore, all the parts of a circuit have some resistance. The wires used to carry current usually have a low resistance. Other parts of a circuit such as lamps and appliances have a high resistance. Resistance is measured in **ohms.**

Material	Resistance	Material	Resistance
silver	0.032 ohm	tin	0.22 ohm
copper	0.033 ohm	steel piano wire	0.23 ohm
gold	0.047 ohm	lead	0.42 ohm
aluminum	0.055 ohm	carbon	66 ohms

Table 15-1 Resistance of Several 20-Gauge Wires

The resistance in metals is related to voltage and current in a formula called **Ohm's law.** Ohm's law tells you that the greater the voltage, the higher the resistance. It also tells you that the more current, the lower the resistance. The formula for Ohm's law is the following:

$$\textbf{Resistance} = \frac{\textbf{voltage}}{\textbf{current}}$$

EXAMPLE Find the resistance of an automobile headlight that uses 4 amperes of current in a 12-volt circuit.

SOLUTION $\text{Resistance} = \frac{\text{voltage}}{\text{current}} = \frac{12 \text{ volts}}{4 \text{ amperes}} = 3 \text{ ohms}$

Study Questions for 15.2

1. How are series and parallel circuits different?
2. How are alternating and direct currents different?
3. What is resistance?
4. What does Ohm's law tell you?

1. A series circuit offers only one path for current. A parallel circuit offers two or more paths. When one part of a series circuit goes dead, the whole circuit goes dead. When one part of a parallel circuit goes dead, current still flows through the other parts.
2. Alternating current constantly switches directions. Direct current always goes in one direction.
3. Resistance is the opposition a material offers to the flow of electric charges.
4. Ohm's law shows the relationship of resistance to current and voltage:

$$\text{Resistance} = \frac{\text{voltage}}{\text{current}}$$

10. Point out that the resistance, or the opposition a material offers to the flow of electric charges, is measured in units called ohms. Explain that in the early 1800's George Ohm described mathematically the relationship between the resistance, voltage, and current in a circuit. He stated that the resistance of a conductor can be found by dividing the voltage by the current.
11. Distribute the work sheet of Ohm's law problems to be done orally in class. You may want to state the problems in the following format: What is the resistance of a doorbell that uses 6 amps in a 9-volt circuit (1.5 ohms)? Distribute the second work sheet as an assignment.
12. Conclude the lesson by stressing that all parts of a circuit offer some resistance to the flow of electric charges. Mention that in some circuits, devices called resistors are used to control this flow.

● Mathematics Skills: Computation
Problem Solving

FOLLOW-UP

Reinforcement: Ohm's Law

In one corner of your classroom set up the Ohm's Law Corner. In it place batteries, wires, and bulbs that offer different amounts of resistance. Allow the students to work with the materials to strengthen the concept of resistance. Also place in this corner some work sheets with answer keys of Ohm's law problems.

Enrichment: Measuring Current and Voltage in Circuits

Provide students with ammeters and voltmeters as well as batteries, bulbs, and wire. Have the students attach the meters to series and parallel circuits to determine the voltage and amperes of different combinations of bulbs and batteries in each type of circuit. Students should keep pictorial records of their experiments.

Tie-in: Resistors

Resistors are marked with bands of different colors to show their sizes. Have the students visit an electric supply store and learn the size of resistors and the markings and use for each size. Then have the students construct charts to report the information they get.

15.3

OBJECTIVES: The student should be able to

- **define** the term *electric power;*
- **state** the relationship between electric power, voltage, and current;
- **solve** problems to determine the amount of electric power used by appliances;
- **define** the term *kilowatt;*
- **solve** problems to determine the amount of energy used when power and time are known;
- **calculate** the cost of running an appliance for a certain amount of time.

SCIENCE TERMS

electric power
kilowatt
kilowatt-hour

ADVANCED PLANNING

1. Prepare two work sheets to be used as practice in determining the amount of power used by appliances.

2. Prepare two work sheets as practice in determining the amount of energy used by appliances and the cost of using an appliance a given amount of time.

LESSON PLAN

1. Introduce the lesson by reviewing the term *power* (the rate at which work is done). Ask the students what is meant by electric power (the amount of electric energy, or work, that flows through a circuit in a given time).
2. On the board write the equation for electric power (Electric power = voltage × current). Ask what units are used to measure any kind of power (watts, kilowatts). What is the meaning of the term *voltage* (difference in electric potential that causes current to flow)? What unit is used to measure the amount of current (amperes)?
3. Distribute the first work sheet of power problems. You may want to state the problems in the following format: How much power is used by a dryer in a 120-volt circuit that is drawing 30 amperes of current (3,600 watts)? Distribute the second power work sheet for homework.

15.3 Electric Power

You will find out
- what is meant by electric power;
- how electric companies measure and charge for electric energy.

Electric currents deliver electric energy to the parts of a circuit. This energy is then used in many different ways. For example, it can light lamps, run appliances, or produce heat. Different circuit parts operate on, or draw, different amounts of current when they are turned on. For example, an air conditioner draws more current than a blender. **Electric power** refers to the amount of electric energy that flows through a circuit at a given time. The electric power in an electric circuit depends on the voltage and the current.

Electric power = voltage × current

Electric power, like other kinds of power, is measured in watts. To measure large amounts of power, a larger unit, the **kilowatt,** is used. A kilowatt is 1,000 watts.

EXAMPLE A toaster in a 120-volt circuit draws 5 amperes of current. How much power does it use?

SOLUTION voltage × current = Power
120 volts × 5 amperes = 600 watts

Table 15-2 Annual Energy Use of Several Appliances

Appliance	Average Wattage	Kilowatt-hour Use
dishwasher	1200	363
microwave oven	1450	190
toaster	1150	39
washing machine	500	103
clothes dryer	5000	933
vacuum cleaner	1000	46
blow dryer	1000	14
color television	200	440

Figure 15.8 How many of the electric appliances in this photograph have you used?

Electric companies bill according to how much energy is used. The basic unit they use to measure electric energy is the **kilowatt-hour.** A kilowatt-hour is the amount of energy produced by one kilowatt of power in one hour. Electric energy can be expressed in the following formula:

Electric energy = power × time

EXAMPLE Suppose a utility company charges 10¢ per kilowatt-hour. How much does it cost to leave a 60-watt light bulb on for 100 hours?

SOLUTION

$$\text{power} \times \text{time} = \text{Energy}$$
$$60 \text{ watts} \times 100 \text{ hours} = 6{,}000 \text{ watt-hours}$$
$$= 6 \text{ kilowatt-hours}$$

Since 1 kilowatt-hour costs 10¢, 6 kilowatt-hours cost 60¢. It costs 60¢ to leave a 60-watt light bulb on for 100 hours.

Do You Know?
Scientists at Columbia University have reported that electric currents as small as 20 millionths of an ampere can help to heal broken bones.

Study Questions for 15.3

1. What is electric power?
2. What unit do electric companies use to measure energy use?

1. Electric power refers to the amount of electric energy that flows through a circuit at a given time.
2. kilowatt-hour

4. On the board write the equation for electric energy (Electric energy = power × time) and ask a student to solve a sample problem like the following: How much energy is used by a 3,600-watt electric dryer that is run for 2 hours (3,600 watts × 2 hours = 7,200 watt-hours = 7.2 kilowatt-hours)?
5. Explain that power companies bill you by measuring the amount of energy you use in kilowatt-hours and multiplying that amount by the cost per kilowatt-hour.
6. Ask the students to imagine that the electric company charges five cents per kilowatt-hour. How much will it cost to leave on a 40-watt light bulb for 100 hours (to determine the number of kilowatt-hours, or energy used, multiply power by time, or 40 × 100 = 4,000 watt-hours = 4 kilowatt-hours; to find the cost, multiply the kilowatt-hours by the price per kilowatt-hour, or 4 kilowatt-hours × $0.05/kilowatt-hour = $0.20).
7. Distribute the electric energy and cost work sheets and use them as you did the power work sheets.

● Mathematics Skill: Computation
Problem Solving

FOLLOW-UP

Reinforcement: Determining Electric Costs
Have each student construct a flow chart to illustrate how to determine how much it costs to run an electric appliance for one year.

Enrichment: An Energy Audit of Your Room
Have students conduct an energy audit of one of the rooms at home. They should report their findings by drawing a plan of the room and attaching a cost tag to each electric device.

Tie-in: Power for Tomorrow
To help keep energy costs low, utility companies are experimenting with new sources of power. Two of the more promising are fuel cells and biomass. Ask the students to select one of these new sources of energy for research. They should report the results of their research by writing or taping an interview with a utility company official.

15.4

OBJECTIVES: The student should be able to

- **identify** the greatest danger with electricity;
- **describe** the purpose of a fuse and a circuit breaker;
- **explain** how a short circuit occurs;
- **describe** what should be done if a live wire is encountered;
- **explain** why birds can sit on high-voltage wires;
- **explain** why it is dangerous to use an electric current near water.

SCIENCE TERMS

fuse	short circuit
circuit breaker	live wire

LESSON PLAN

1. Introduce the lesson by having the students play a game of charades to illustrate electric safety. One student should act out the method of dealing with a hazard in the home. The other students should try to guess the hazard and the method of dealing with it.

SCIENCE BACKGROUND

The expression *to blow a fuse* refers to the action of a fuse or circuit breaker when it opens a circuit, thereby stopping the flow of current.

Fuses help to protect people from fires caused by defective wiring. All fuses contain a short piece of wire that has a lower current capacity than the circuit it protects. The melting temperature of the fuse wire is also low. If the current becomes too great, the fuse will get hot and melt, thereby breaking the circuit.

Circuit breakers protect a circuit in the following way: As the current in the heating element increases, the coil gets hotter and hotter until a piece of metal expands and releases the catch that holds the switch in the on position. The circuit is then broken.

15.4 Electric Safety

You will find out
- why fuses and circuit breakers are important;
- what happens in a short circuit;
- how you can avoid accidents with electricity.

You use electricity every day in many ways. Even though electric power is familiar to you, you should always treat it with care and respect. When electric power is not used carefully, it can be very dangerous.

One danger involves the heat produced by electric currents. All electric wires have some resistance. This resistance causes some electric energy to be lost when current is flowing. Remember, energy never disappears. It only changes form. Therefore, the electric energy that is lost has really only changed form. It changes to heat energy. If too much current flows through a wire, too much electric energy may be changed to heat energy. The wire may become so hot that a fire starts.

A circuit can overheat if too many appliances are plugged in and used on it at the same time. This practice can cause too much current to flow and may start a fire. To prevent this from happening, all circuits should have a **fuse** or a **circuit breaker.** Fuses and circuit breakers are devices that break, or open, the circuit to stop the current. These devices do this automatically when there is a danger of overheating.

Figure 15.9 *left:* Circuit breakers; *right:* Too many appliances are plugged into this outlet.

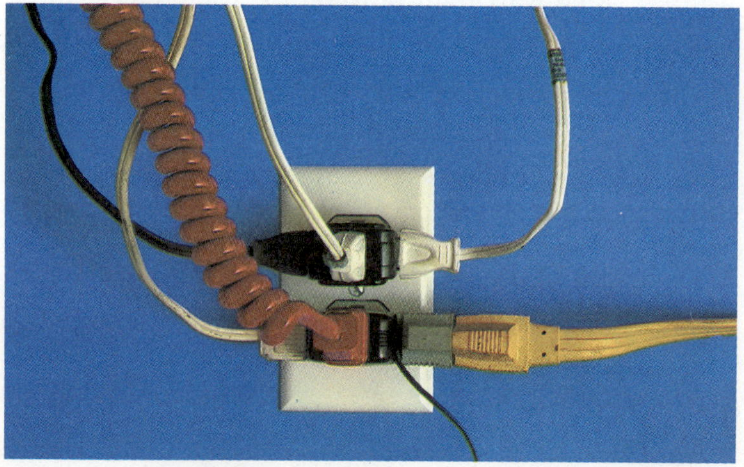

Most circuits in today's homes have circuit breakers that open the circuit when the current is greater than 15 or 20 amperes. Normally the currents drawn by lights or appliances are only a few amperes. However, some appliances such as electric heaters and hair dryers can draw 10 or more amperes. Care should be taken not to use more than one of these appliances on the same circuit at the same time.

ACTIVITY How Can You Compare the Resistance of Different Wires? ●

Materials

ammeter
voltmeter
rheostat
switch
2 alligator clamps
7 pieces of insulated copper wire (30 cm each)
iron, aluminum, and copper wires of the same length and thickness
battery

Procedure

1. Set up the apparatus as shown in the diagram.
2. Set the rheostat so that the maximum resistance is in the circuit.
3. Place the aluminum wire between the alligator clamps.
4. Close the switch.
5. Adjust the rheostat so that the ammeter indicates a current is flowing. Record the ammeter and voltmeter readings.
6. Divide the voltage by the current to determine the resistance of the aluminum wire.
7. Repeat steps 3-7 with the copper wire and with the iron wire.

Questions

1. Create a bar graph to compare the resistance of the three wires. ●
2. Which wire had the highest resistance? Which wire had the lowest resistance?

1. The bar graphs should show the kinds of wires across the bottom and resistance in ohms along the side.
2. The copper wire has the lowest resistance. The iron wire has the highest resistance.

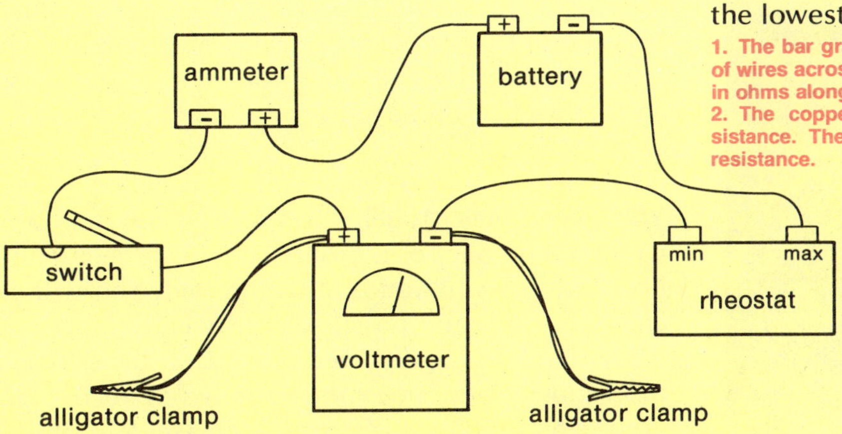

ACTIVITY: How Can You Compare the Resistance of Different Wires?

Before the students do the activity, review the terms *ampere, volt,* and *ohms.* Mention that an ammeter measures current in amperes and a voltmeter measures the amount of energy each unit of charge can deliver.

The rheostat is used to change the resistance in a circuit. Review Ohm's law. The students should be able to calculate the resistance from their measurements of current and voltage.

The gauge of a wire refers to its cross-section area. The higher the gauge the smaller the cross-section area. Since the only variable desired in this activity is the kind of wires, the gauge of the wires must be constant.

Check each student's apparatus to be sure it is set up correctly.

● **Laboratory Skills: Manipulation of Materials; Experimentation/Demonstration**

● **Mathematics Skills: Computation; Problem Solving; Graphing**

SCIENCE BACKGROUND

A third prong, called the ground, is found in many plugs today. It is a wire that goes from the case of the appliance through the circuitry of a building to the ground, where an electric current can be discharged.

To prevent small children from inserting objects in electric outlets, plastic covers are now available to cover wall outlets.

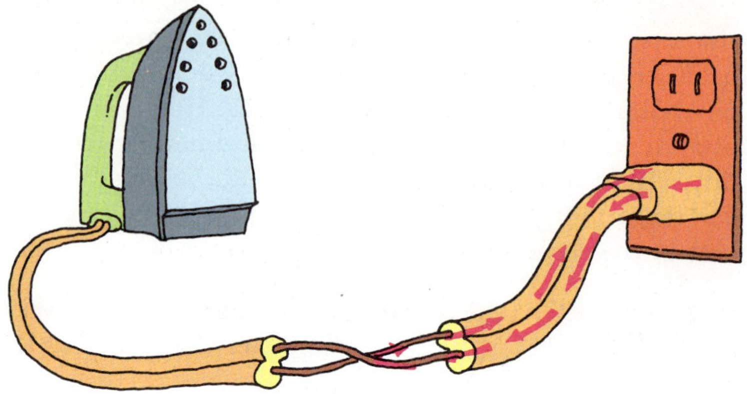

Figure 15.10 A short circuit

A circuit can also overheat if a **short circuit** occurs. In a short circuit, the current flows over an unintended path that has very low resistance. This could happen if the two wires in a cord accidentally touched. Then the current would not flow through whatever the cord was connected to at all. It would flow from the outlet to the point where the wires touched and back to the outlet. This circuit would have very low resistance. A large current would go through it. The result would be rapid overheating and possibly a fire.

Ordinarily, cords have a covering of insulation. The insulation keeps the wires from touching and creating a short circuit. Sometimes, however, the insulation becomes damaged and a short circuit occurs. When this happens, a fuse or circuit breaker should act to shut off the current. To avoid any problems or risk, any cord with frayed or damaged insulation should be repaired or replaced.

Figure 15.11 Power company employees must be very careful when working with live wires.

Any wire connected to a voltage is called a **live wire.** You should never touch a live wire. If you do, your body will become part of a circuit, and a large current will flow through you and into the ground. You will experience a painful and possibly deadly electric shock.

Live high-voltage power lines often have little or no insulation. Sometimes these lines fall to or near the ground as a result of a storm or an accident. If you ever see one of these lines down, do not touch it. Call the power company and your local authorities.

Figure 15.12 Birds can perch safely on power lines because no part of them offers the current a path to the ground.

Have you ever wondered why the same high-voltage power lines that are so dangerous to you make good, safe perches for birds? A bird on a wire is safe because no part of it offers the current a path to the ground. If, however, any part of the bird touches a pole or anything else that reaches the ground, then the bird will die of electric shock. It dies because its body now is providing the current with a path to the ground. The current takes that path going through the bird, through whatever it is touching, and into the ground. Sometimes birds and other animals, such as squirrels, are killed in this way.

There is a similar danger to you if you use an electric appliance while you are touching or standing in water. The water conducts electricity. Electric current from an appliance may take a path through you to the water you are touching. You could be seriously hurt or killed. It is for this reason that you are warned never to use electric appliances or touch sources of electricity such as light-bulb sockets and outlets while in contact with water.

Study Questions for 15.4

1. Why are fuses and circuit breakers necessary?
2. What is a short circuit?
3. Why is a bird safe sitting on a high-voltage line?
4. What can happen if you operate an electric appliance in water?

1. They open the circuit and stop the flow of current when there is a danger of overheating.
2. an unintended path of low resistance over which current may flow
3. As long as no part of the bird is touching anything that leads to the ground, the current will not flow through the bird.
4. Current may flow through you to the water.

2. Have each student create a collage that uses words and pictures to illustrate the greatest danger of electric currents and the importance of fuses. The collage should also show how short circuits occur, why birds can sit on high-voltage wires, why electric current is so dangerous when it is near water, and how electric disasters can be prevented in a home.
3. Conclude the lesson by asking the students to sponsor an electric safety poster contest for grades 1, 2, and 3 in neighboring schools. Allow the students to draw up the rules and judging criteria. Contact the town paper and the parents' organizations of participating schools to advertise your contest.

FOLLOW-UP

Reinforcement: Electric Safety

Have the students plan and produce a play about electric safety for a neighboring elementary school.

Enrichment: An Electrician's Tour

Arrange for an electrician to take small groups of students to see the master control panel, fuse boxes, electric meter, etc., in the school. A discussion of the function of each of the devices and the rules of electric safety should follow.

Tie-in: Safety at Home

Have the students prepare a list of five things they propose to do to make their homes safer from the potential hazards of electric current. They should ask their parents to sign the list as each thing is accomplished.

What Did You Learn?

15.1 • The flow of electric charges from one point to another is called electric current.
15.1 • Voltage causes charges to flow.
15.1 • A dry cell produces voltage as a result of chemical reactions between a moist paste and a metal.
15.1 • Current is measured in amperes.
15.2 • The path of an electric current is called an electric circuit.
15.2 • In a series circuit all the current follows one path.
15.2 • In a parallel circuit the current divides and flows in two or more separate paths.
15.2 • Direct current always flows in the same direction.
15.2 • Alternating current constantly switches direction.
15.2 • Resistance is the opposition a material offers to the flow of electric charges.
15.2 • Resistance is measured in ohms.
15.3 • Electric power refers to the amount of electric energy that flows through a circuit at a given time.
15.3 • Electric power is measured in watts and kilowatts.
15.3 • A kilowatt-hour is the amount of energy produced by one kilowatt of power in one hour.
15.4 • Resistance causes some electric energy to be changed into heat energy.
15.4 • Fuses and circuit breakers open a circuit when too much current is flowing and there is a danger of overheating.
15.4 • A short circuit is an unintended path of low resistance over which current may flow.
15.4 • Live wires that are not well insulated are dangerous and should never be touched.
15.4 • You should never use electricity while in contact with water.

Key Terms

electric current
voltage
volts
dry cell
positive terminal
negative terminal
amperes
electric circuit
series circuit
parallel circuits
direct current
alternating current
resistance
ohms
Ohm's law
electric power
kilowatt
kilowatt-hour
fuse
circuit breaker
short circuit
live wire

Biography

Thomas Alva Edison (1847–1931)

Thomas Alva Edison is a good example of someone with a good mind who became rich and famous through hard work. Edison used his mind to educate himself at home.

After reading a few science books, Edison wanted to do experiments of his own. He got a job as a newsboy on a train between Port Huron and Detroit, Michigan. Using the money he made, he soon set up a chemical laboratory in the baggage car of the train. Unfortunately, the laboratory caught fire and the railroad banned further experiments.

Accidents played a large role in young Edison's life. In one train accident he was pulled to safety by his ears. As a result Edison became permanently deaf. In another accident Edison rescued a small boy. The boy's father was so grateful that he offered to teach Thomas telegraphy. Edison's knowledge of the telegraph led him to several inventions. These inventions allowed him to set up his own business by the time he was 23 years old!

Before he was through, Edison had patented 1,100 inventions. In one four-year stretch he received 300 patents. This accomplishment is an average of one patent every five days! No other inventor has ever come close to Edison's patent record. He improved the telephone, invented the phonograph and the first workable light bulb, and helped to create motion pictures.

How did Edison do so much? He worked almost continually. During his peak, he was working 20 hours a day and sleeping in catnaps. As Edison once said, "Genius is 1 percent inspiration and 99 percent perspiration."

Vocabulary

1. an electric current
2. true
3. positive terminal
4. true
5. short circuit
6. series circuit
7. Voltage
8. Alternating current
9. true
10. Fuses or circuit breakers

● Word Skill: Vocabulary

What Do You Remember?

1. the flow of electric charges
2. voltage
3. ampere—the number of charges that pass a point in one second; volt—the difference in electric potential at two points; ohm—resistance; watt—power
4. the path over which current can flow
5. In a short circuit a very large current flows over a path of very low resistance. The wires may overheat, causing a fire.
6. In series circuits all the current flows through all the parts of the circuit. If one circuit part goes dead, the whole circuit goes dead. In parallel circuits the current divides and goes through different branches. If one branch goes dead, the other branches still carry current.
7. The opening for one prong acts like the negative terminal of a battery. The opening for the other prong acts like the positive terminal.
8. Ohm's law is a formula that relates resistance to voltage and current. The laws says the following:

$$\text{Resistance} = \frac{\text{voltage}}{\text{current}}$$

9. Chemical reactions between the zinc, the carbon, and the paste cause a buildup of positive ions on the carbon rod. The carbon rod becomes positively charged. The zinc jacket becomes negatively charged. This separation of charges produces voltage.
10. Direct current always flows in the same direction. Alternating current constantly switches direction.

TO THINK ABOUT AND DO

Vocabulary ●

Number your page from 1 to 10. Read each of the following sentences. If the sentence is true, write *true*. If it is false, make it true by changing the underlined term.

1. Moving electric charges produce static electricity.
2. The unit used to measure resistance is the ohm.
3. The carbon rod in a dry cell is the negative terminal.
4. The path of an electric current is called an electric circuit.
5. A series circuit occurs when current flows over an unintended path with low resistance.
6. When one part of a parallel circuit goes dead, the whole circuit goes dead.
7. Energy causes electric charges to flow.
8. Direct current constantly switches directions.
9. Electric power can be measured in kilowatts.
10. Terminals break, or open, a circuit, stopping the current.

What Do You Remember?

1. What is electric current?
2. What moves electric current?
3. What do the following units measure: ampere, volt, ohm, watt?
4. What is a circuit?
5. Why are short circuits dangerous?
6. How are series and parallel circuits different?
7. In what way is an electric outlet like the terminals of a battery?
8. What is Ohm's law?
9. How is voltage created in a flashlight battery?
10. How do direct current and alternating current differ?

Applying What You Have Learned

1. What is the resistance of a lamp that uses 2 amperes of current in a 120-volt circuit?
2. How many kilowatts of power are used by a heater that draws 10 amperes of current in a 120-volt circuit?
3. How much energy would you use if you ran the heater in question 2 for 100 hours?
4. Give at least three safety hints for using electricity in the home.

Research and Investigation

1. Contact the power company that provides electricity to your school. Request information about electric safety and what to do in electrical emergencies. Present this information in a report.
2. Wall switches control the flow of current through the rooms of many homes. Do research to find out how these switches work.

Applying What You Have Learned

1. Resistance $= \frac{\text{voltage}}{\text{current}} = \frac{120 \text{ volts}}{2 \text{ amperes}} = 60$ ohms

2. Power = voltage × current = 120 volts × 10 amperes = 1,200 watts = 12 kilowatts
3. Energy = power × time = 12 kilowatts × 100 hours = 1,200 kilowatt-hours
4. Answers will vary. Some possibilities are the following: repair damaged electric cords, do not plug in and use too many appliances at the same time on the same circuit, do not use a heater and a hair dryer at the same time on the same circuit, never use electricity while in contact with water, and never touch uninsulated live wires.

● Comprehension Skill: Applied

Research and Investigation

1. Most power companies have a free pamphlet written for young adults, detailing safety precautions and steps to be taken in emergencies.
2. The most common wall switch is a toggle switch. Wires are attached to screws which lead to metal strips inside the switch itself. When the switch is off, the strips do not touch so there is an incomplete, or open, circuit. However, when the switch is on, a metal bar presses against the strips completing the circuit. Current flows.

● Reference/Research Skills

CHAPTER BACKGROUND

Electromagnetism is based on the fact that an electric current will produce a magnetic field and a changing magnetic field will produce an electric current. Since Oersted's discovery of the link between electricity and magnetism in 1820, we have learned that there is a magnetic field around any straight wire that carries current. When the wire is bent into a series of coils, the magnetic field is strengthened because each coil acts as a separate magnet. Adding a metallic core further strengthens the magnetic field because the core itself becomes magnetized through induction.

The discovery in the 1830's of the phenomenon of electromagnetic induction led to the development of the first successful electric generator. In 1832, Michael Faraday found that a moving magnet can cause an electric current in a coil of wire. He applied this principle to build the first generator.

With very little formal education, Faraday was able to produce electricity from magnetism and build a generator. He explained that when any conducting material moves across a magnet, an electromotive force, the "push" (measured in volts) produced by a generator, is generated in the conductor. This happens when the magnet is thrust into a coil of wire, resulting in an electromotive force that causes current to flow.

When the current through the coil of an electromagnet is reversed, the poles of the electromagnet are reversed as well. If the poles of a bar magnet that is suspended between two other magnets were suddenly changed, the magnet would swing on its pivot. If this pole reversal were continued, the magnet would spin. This is the principle of an electric motor, a device that changes electric energy to kinetic energy.

The transformer was another Faraday discovery, although Joseph Henry is credited with building the first transformer in 1838. Faraday's transformer was an iron ring with two coils of wire wound around it, one on either side. When a current was induced in one coil, the current in the other coil changed and vice versa creating an alternating current.

Since it is neither safe nor convenient to use very high voltages in most applications of electric current, especially in homes, a transformer is needed. With a transformer the voltage can be "stepped up" for transmission and "stepped down" again for use at the consumer end.

16 Electromagnetism

Build a simple electromagnet by winding insulated copper wire around an iron nail and then attaching the free ends of the wire to the terminals of a battery. Show the students that the nail will attract paper clips. Disconnect the wire from one of the terminals to open the circuit. Now show the students that the nail no longer attracts paper clips. Ask the students to explain their observations. Tell them that in this chapter they will learn about how electricity and magnetism can be used.

What did you do today? Did you listen to a tape recorder? Did you use the telephone? Did you see or hear a motor at work in a mixer, blender, refrigerator, washing machine, hair dryer, fan, vacuum cleaner, or elevator? Did you hear a doorbell or a buzzer sound?

If you saw or heard any of these devices or machines, you observed the combined effects of electricity and magnetism at work. In fact, if you used electric current in any way, you were using the combined effects of electricity and magnetism.

A little over a hundred years ago, scientists and inventors were discovering the relationship between electricity and magnetism. They learned how to use this relationship to produce electric currents and to make machines that would run on electricity. As you can see from the hundreds of electric devices and machines in use today, what those scientists and inventors did has made amazing changes in the way you live.

Lesson Organizer

16.1 Magnetism from Electricity

Pages 336–339
Advanced Planning: Page 336
Activity: Can Electricity Produce Magnetism? Page 338
Follow-up: Page 339
- **R** Oersted's Discovery
- **E** How Does a Telephone Work?
- **T** Floppy Disks

16.2 Electricity from Magnetism

Pages 340–343
Advanced Planning: Page 340
Activity: Can Magnetism Produce Electricity? Page 341
Follow-up: Page 343
- **R** Electric Currents and Magnetic Fields
- **E** Production of Electricity
- **T** Technology Advances Discoveries

16.3 Electric Motors and Transformers

Pages 344–345
Advanced Planning: Page 344
Follow-up: Page 345
- **R** Generators and Motors
- **E** Transformers
- **T** Motors in Our Lives

See ACTIVITIES book pages 91–96 for laboratory, skill, and review exercises for Chapter 16.

16.1

OBJECTIVES: The student should be able to

- **define** the term *electromagnetism*;
- **explain** how Oersted discovered the connection between electricity and magnetism;
- **list** some characteristics of the magnetic field produced by an electric current;
- **describe** some ways to increase the strength of the magnetic field produced by an electric current;
- **explain** how an electromagnet is made;
- **list** some uses of electromagnets.

SCIENCE TERMS

electromagnetism electromagnets

ADVANCED PLANNING

For demonstrations you will need an overhead projector, a liquid compass, three pieces of insulated wire each 60 cm (23 in.) long, three dry cells, some paper clips, and an iron nail.

LESSON PLAN

1. Introduce the lesson by asking the students to list some ways that electric energy is useful (for lighting, for operating household appliances). Point out that some of the uses they name, e.g., television tube, copying machine, depend on both electric forces and magnetic forces produced by electric currents. Explain that the physics of electricity and magnetism is called electromagnetism.
2. Set up the overhead projector. Put the compass on the stage and ask the students to note the position of the compass needle. Hold the wire above the compass and ask the students to watch the needle (nothing happens). Ask whether the wire has magnetic properties (no, since the needle does not move).
3. Attach the wire to a dry cell. Now hold the wire over the compass and ask the students to watch the needle (it moves). Point out that the moving needle indicates the presence of a magnetic field. If the electric current is the only thing that has been added to the wire, what must be producing the magnetic field (the electric current)?

● **Word Skill: Vocabulary**

16.1 Magnetism from Electricity

You will find out
- how electric current and magnetism are related;
- how electromagnets are made;
- how electromagnets are used.

You know from experience that electric energy carried by electric currents does much useful work. But did you know that most electric devices involve not only electricity but magnetism as well? The branch of physical science that involves the combined effects of electricity ● and magnetism is called **electromagnetism** [ih-lehk-troh-MAG-nuh-tihz-uhm].

The relationship between electricity and magnetism was discovered in the early 1800's by Hans Christian Oersted, a Danish scientist. As part of a classroom experiment, Oersted placed a compass just above a wire carrying an electric current. To his surprise, the compass needle—a small magnet—moved from a position pointing north to a position at a right angle to the wire. When he reversed the direction of the current, the needle reversed direction. But it was still at a right angle to the wire. When the current was turned off, the needle pointed north again. Oersted had discovered that electric currents produce magnetic fields.

Figure 16.2 When current is flowing, the compass needle moves to a position at a right angle to the wire.

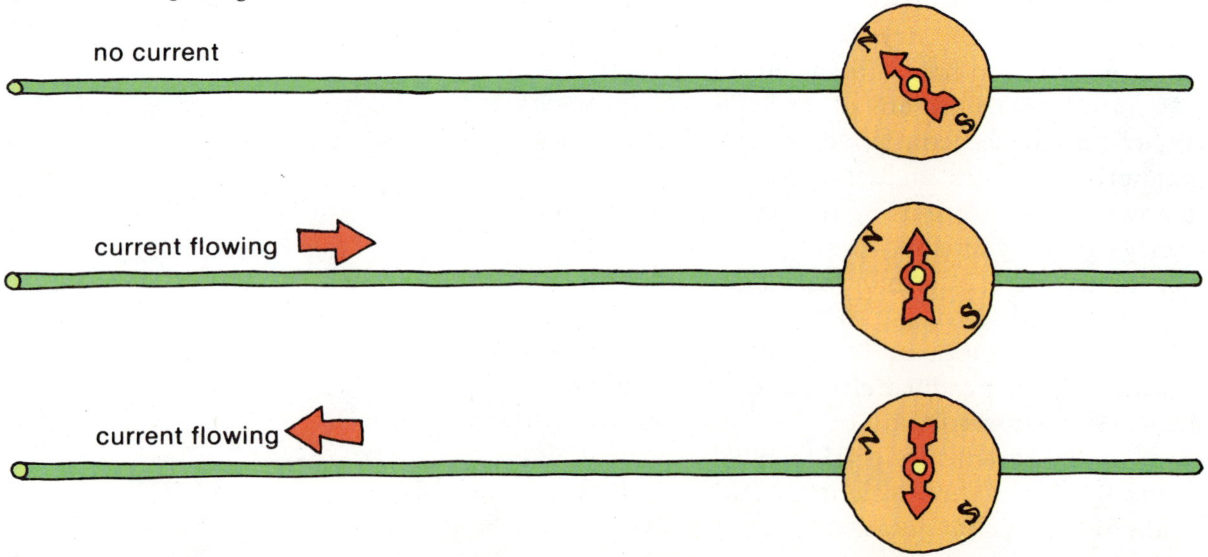

Oersted also discovered things that affect the magnetic field produced by an electric current. The direction of the magnetic field depends on the direction of the current. The magnetic field always acts in a direction at right angles to the direction of the current. When the current reverses direction, the direction of the magnetic field also reverses. The north magnetic pole becomes the south magnetic pole and vice versa. Oersted also found that the magnetic field is strongest near the wire. The strength of the field depends on the amount of current flowing through the wire.

A very strong magnetic field can be produced by a current in a wire that is wound into a tight coil. The magnetic field produced is directed along the center of the area inside the coil. What will happen if you put a bar made of a metal that can be permanently magnetized inside the coil? The bar will be magnetized by this magnetic field. When the current is turned off, the bar will remain magnetized. Most permanent magnets are made in this way.

Figure 16.3 The Japanese maglev train uses the magnetic forces of electromagnets on the train and on the track to attain speeds of up to 520 km/h.

4. Set up the overhead projector. Place the wire, attached to a dry cell, across the compass. Ask the students how they would describe the position of the compass needle in relation to the wire (perpendicular).
5. Hold the dry cell so that the negative terminal is on the students' left. Ask the students to watch the compass needle as you bring the wire towards the compass. Then turn the dry cell so that the negative terminal is on the students' right and bring the wire toward the compass. What happens to the way the compass needle moves when you reverse the dry-cell terminal positions (it moves in different directions). Conclude that the direction of the magnetic field depends on the direction of the current.
6. Ask the students to observe as you move the wire close to the compass and then away from it. When does the needle move (only when the wire is close to the compass)? What can the students conclude about the strength of the magnetic field generated by the current (the magnetic field is strongest close to the wire through which the current is flowing)? Ask what other factor the strength of the field depends on (the strength of the current flowing through the wire).
7. Ask the students how they could increase the strength of the magnetic field (wind wire into a tight coil; coil wire around an iron bar).

SCIENCE BACKGROUND

The "right-hand" rule can be used to show the relationship between the direction of the current and the position of the magnetic poles. Hold a coil of wire in your right hand with the fingers pointing with the current. (Current runs from positive to negative pole.) Your thumb will point toward the north pole of the coil.

8. Point out that in many applications—from doorbells to atom smashers—magnetic fields are produced by coils of wire wound around iron cores. Ask the students what happens to the iron core when a current is switched on (it becomes magnetized and increases the strength of the field). What is the wire/iron core arrangement called (an electromagnet)?

ACTIVITY: Can Electricity Produce Magnetism?

Try to use new dry cells for this experiment. Remind the students to disconnect the wires when they are not making observations. In this way, they will not wear down the dry cells.

◇ **Caution:** The wires will become hot during the experiment. Students should be careful when they touch them.

Ask the students to sprinkle only a few iron filings on the plate. Explain that they will be able to see the concentric circles more easily with only a few filings on the plate.

1. The filings moved and became arranged in concentric circles around the wire. The needle moved so it was at a right angle to the wire.
2. The filings stayed in the same arrangement. The needle again moved to a position at a right angle to the wire, but the ends of the needle had changed places.
3. The changes were caused by changes in the magnetic field created by the current in the wire.

● **Laboratory Skills: Manipulation of Materials Experimentation/Demonstration**

SCIENCE BACKGROUND

The first electromagnet, invented in England in 1824 by William Sturgeon, could lift a weight of nine pounds. A modern electromagnet can lift 22,500 kilograms (50,000 pounds).

Suppose the bar you put inside the coil is made of a metal like iron that loses its magnetism easily. When current flows through the coil, the bar will be magnetized. But the bar will lose its magnetism as soon as the current is turned off. If the direction of the current is reversed, the bar will be magnetized in the opposite direction. This behavior can be used to make a very useful magnet. It can make a magnet whose magnetism can be turned on and off by turning the current on and off. Also, the direction of the magnetic field can be reversed by reversing the direction of the current. Magnets in which magnetism is produced by an electric current are called **electromagnets.**

● ACTIVITY

Can Electricity Produce Magnetism?

Materials

ring stand
ring clamp
fine iron filings
insulated copper wire (75 cm)
compass
6-volt battery
paper plate about 16 cm in diameter

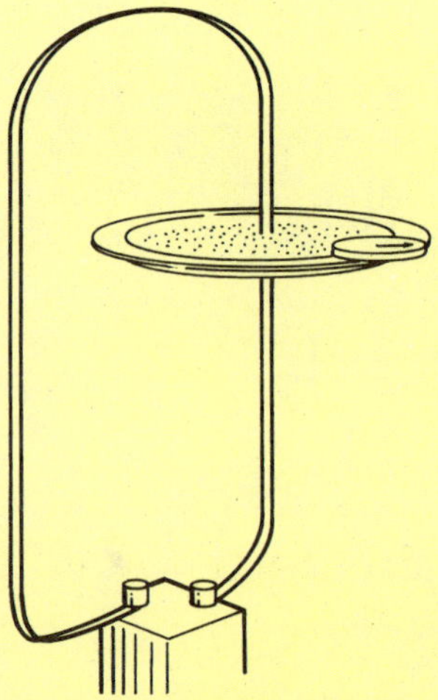

Procedure

1. Attach the ring clamp to the ring stand.

2. Use the point of a pencil to poke a tiny hole in the middle of the paper plate. Place the plate on the ring stand. Thread the wire through the hole in the plate.

3. Sprinkle some filings on the plate. Place a compass on the plate.

4. Attach the ends of the wire to the battery. The end coming from under the plate should be attached to the negative (−) terminal.

5. Observe the compass needle. Remove the compass and tap the plate gently. Observe the filings.

6. Reverse the attachment of the wire ends to the battery. The end coming from under the plate should now be attached to the positive (+) terminal.

7. Again place the compass on the plate.

8. Repeat step 5.

Questions

1. How did the filings and the compass needle change when the wire ends were first attached to the battery?

2. Did you notice any change in the filings or the compass needle when the battery attachments were reversed?

3. What caused the changes you observed?

Figure 16.4 Using an electromagnet

The magnetic force of electromagnets can be very strong. This force is often used to move very large masses. You can see electromagnets at work in junkyards, lifting tons of scrap metal onto railroad cars or trucks. A strong electromagnet the size of this book can lift a mass of 4,000 kg!

Because their magnetic force can be switched on and off, electromagnets are used as switches in telephones, doorbells, and hundreds of other devices. You are using the combined effects of electricity and magnetism when you use these devices.

Do You Know?
When you look at an object, the nerve cells in your brain generate electric currents that produce voltage changes of about a millionth of a volt.

Study Questions for 16.1

1. Describe Oersted's experiment and discoveries.
2. How are electromagnets made?
3. What are some uses of electromagnets?

1. Oersted placed a compass near a wire carrying an electric current. He discovered that the current produced a magnetic field. The strength and direction of the magnetic field produced depends on the strength and direction of the current.
2. Electromagnets are made by putting a metal that can be magnetized inside a coil of wire that carries an electric current.
3. They are used to lift large masses and as switches in telephones and doorbells.

9. Ask three students to try to pick up paper clips with one of the following versions of an electromagnet: a wire, a coil of wire, and a piece of wire wound around an iron nail. Each of the electromagnets should be connected to a dry cell. Have the other students predict which electromagnet will attract the most paper clips (piece of wire wound around the iron nail).
10. Have each of the three students disconnect one end of their wire from the dry cells. What do the other students observe (paper clips fall off)? Stress that one of the main advantages of electromagnets is that the magnetic properties of objects used as the core of electromagnets are lost as soon as the current is turned off.
11. Conclude the lesson by asking the students to list some practical uses of electromagnets (in salvage yards for lifting scrap metal, in telegraphs, and in telephones).

FOLLOW-UP

Reinforcement: Oersted's Discovery
Ask the students to pretend that they are Hans Christian Oersted. In a letter to the Academy of Sciences, they should describe their discovery that an electric current can produce a magnetic field.

Enrichment: How Does a Telephone Work?
Ask a repairperson from your local telephone company to meet with a group of interested students to explain the role of electromagnets in telephones. Have the students who participate build models to illustrate how the telephone works.

Tie-in: Floppy Disks
Computer disks work on the principles of electricity and magnetism. Have students determine how the disks are produced and how the magnetic signals are converted to electric impulses.

16.2

OBJECTIVES: The student should be able to

- **define** the term *electromagnetic induction;*
- **explain** why electromagnetic induction is important;
- **describe** how generators produce electricity;
- **define** the term *alternator.*

SCIENCE TERMS

electromagnetic induction
electric generators
alternator

ADVANCED PLANNING

In preparation for the activity, collect cardboard cylinders from rolls of paper towels.

LESSON PLAN

1. Introduce the lesson by relating the story of the almost-simultaneous discovery—the production of electricity from magnetism—by Faraday and Henry (although Henry actually produced electricity from magnetism in 1831 shortly before Faraday, Faraday is generally credited with the discovery, because he had established priority by first publication and through exhaustive investigations into many aspects of the subject for seven years before the discovery; Henry made the discovery during a one-month vacation from the Albany Academy in New York, where he taught mathematics and philosophy).
2. Ask the students how they think a magnet and a coil of wire can be used to produce an electric current (the ideas should include the movement of the magnet in the coil). How does the movement of the magnet cause a current to flow (the magnetic field is changed in relation to the coil).
3. Ask a student to define the term *electromagnetic induction* (the creation of electric current by a moving magnet or any other type of changing magnetic field). Point out that the discovery of Henry and Faraday led to the invention of electric generators and transformers.

16.2 Electricity from Magnetism

You will find out
- what is meant by electromagnetic induction;
- how electric generators produce electricity.

Electric currents can produce magnetic forces. Can the reverse happen? Can magnetic forces produce electric currents? Yes, they can.

Not long after Oersted's discovery, two men on opposite sides of the Atlantic Ocean made another important discovery at about the same time. Joseph Henry, an American, and Michael Faraday, an Englishman, found that a change in a magnetic field produces an electric current. When a magnet is moved into a coil of wire, a current starts to flow in the wire. The current stops flowing when the magnet stops moving. When the magnet is pulled out of the coil, a brief current flows again but in the opposite direction.

A current is produced in a coil of wire when a magnet is moved into or out of the coil. A current is also produced when the magnet is still and the coil is moved over it. As long as the magnetic field changes in relation to the coil, a current will flow. The creation of a current by a changing magnetic field is called **electromagnetic induction.**

Figure 16.5 *left:* The galvanometer indicates no current is flowing. *right:* The movement of the magnet produces a current in the coil.

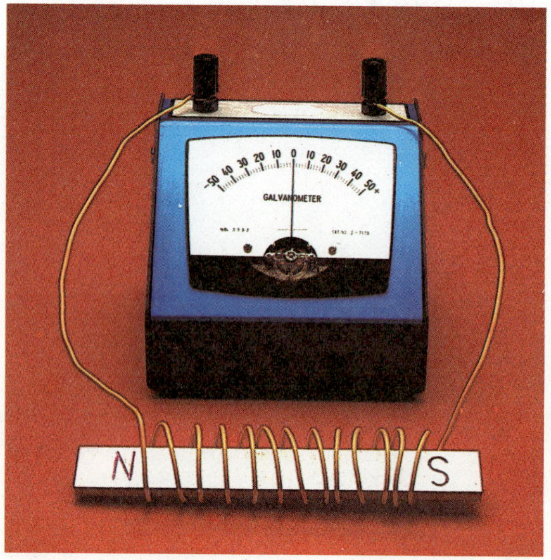

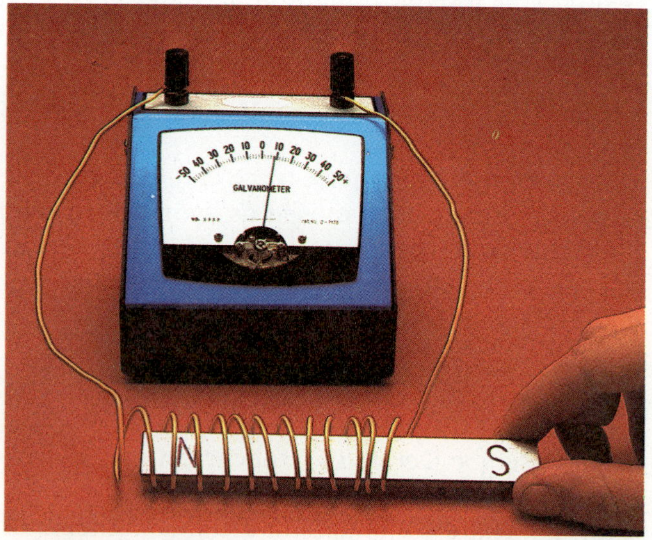

ACTIVITY

Can Magnetism Produce Electricity? ●

Materials

4 m of insulated copper wire
bar magnet
galvanometer
cardboard cylinder from a roll of paper towels
cellophane tape

Procedure

1. Wrap the wire around the cylinder 20 times. Wrap the coils close together.
2. Remove the insulation from the ends of the wire. Attach both ends to a galvanometer.
3. Observe the galvanometer as you insert the magnet and move it rapidly back and forth in the cylinder.
4. Detach the wires from the galvanometer. Wrap the wire around the cylinder 20 more times. Reattach the wires to the galvanometer.
5. Again insert the magnet in the cylinder and move it rapidly back and forth.
6. If a stronger magnet is available, repeat the experiment with it.

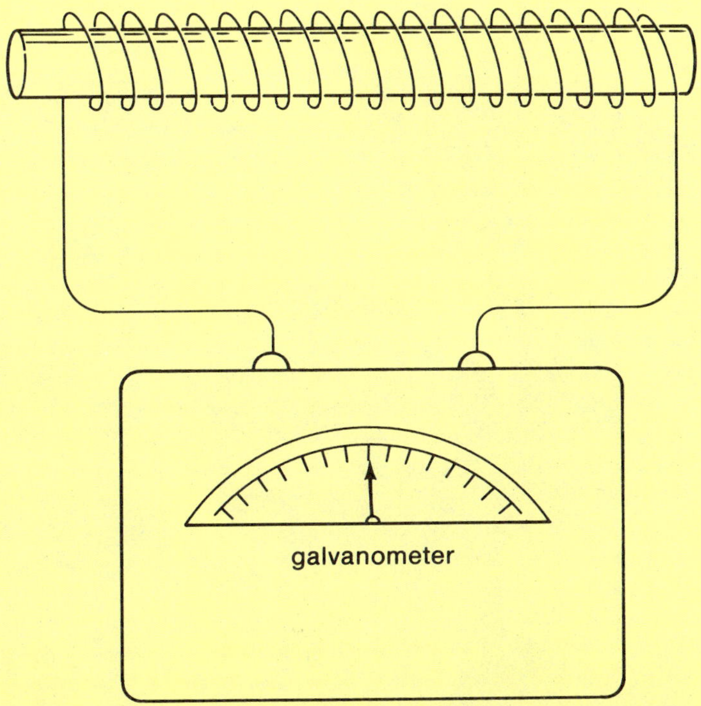

Questions

1. Why did you scrape the insulation off the wire?
2. What happened to the galvanometer needle when you moved the magnet back and forth in the cylinder?
3. Did the movement of the needle change when you used more coils of wire? Why?

1. The insulation was removed so current could flow into the galvanometer.
2. It moved to show that an electric current was flowing in the coil.
3. Yes. The more coils the magnet moves through, the more current it produces.

ACTIVITY: Can Magnetism Produce Electricity?

Review the purpose of a galvanometer (measurement of weak electric currents). Explain that the back-and-forth movement of the galvanometer needle indicates the presence of an alternating current.

Students should notice that more coils produce more current. Students can tape the ends of the coil to prevent them from unraveling.

If magnets of different strengths are available, have the students try them. The stronger magnets will produce more current.

You may want to relate this experiment to electric generators (machines that spin coils of many turns of wire in strong magnetic fields; coils with large numbers of turns are used in order to generate high voltage).

● **Laboratory Skills: Manipulation of Materials Experimentation/Demonstration**

4. Tell the students that electric generators convert kinetic energy into electric energy. Then have them read the part of Section 16.2 that deals with simple generators. Ask the students to draw a series of diagrams with captions to explain the operation of a generator.

● **Comprehension Skill: Applied**

SCIENCE BACKGROUND

Electric current itself is not a source of power as are coal, gas, and oil. Electric current must be generated.

A typical power-plant generator may have a capacity of 1 million kilowatts.

A 100,000-kilowatt generator can supply enough electricity to light a million 100-watt light bulbs.

The discovery of electromagnetic induction was very important to science. It showed that a separation of charges, such as in a battery, is not the only way to ● produce electric current. In fact, most of the electric current you use comes from electromagnetic induction rather than batteries. The discovery of electromagnetic induction made it possible for practically everyone in the United States to have the benefit of electric power.

Electromagnetic induction is used in **electric generators.** An electric generator is a machine that changes kinetic energy into electric energy. To do this, a generator uses a permanent magnet or an electromagnet and metal wire to carry the current.

In a simple generator the magnet is fixed and the coil of wire moves. The coil revolves in the magnetic field between the north and south poles of the magnet. The result is the same as if the magnet were moving instead of the coil. The effect of the magnetic field on the coil changes as the coil moves. This change in the magnetic field causes a current to flow in the coil.

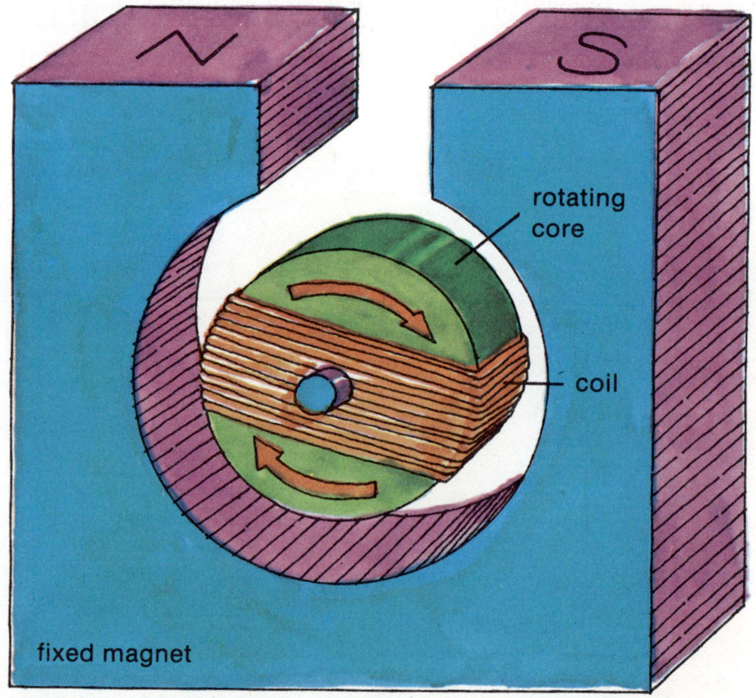

Figure 16.6 A simple generator

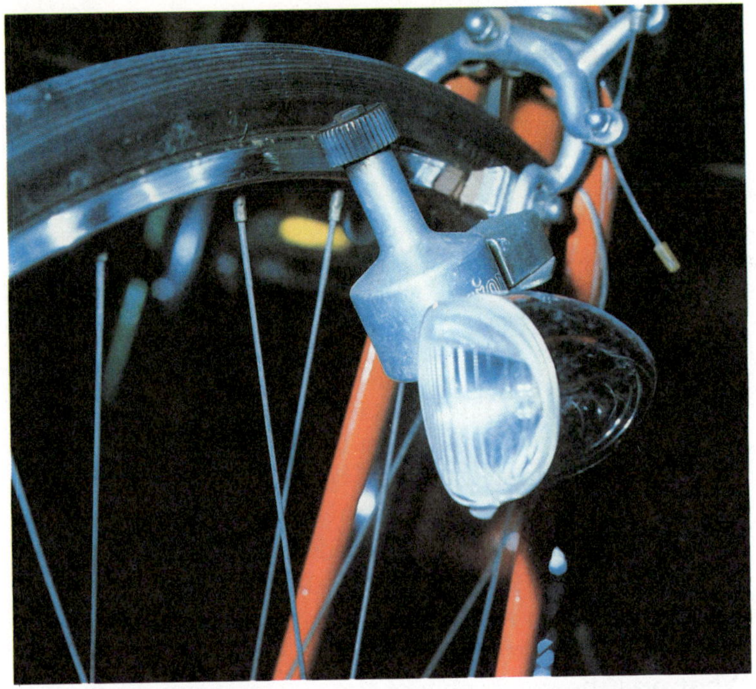

Figure 16.7 Some bicycles have a small generator. When you pedal, the generator produces a current large enough to turn on a light.

Because the coil rotates, the direction of the magnetic field changes in relation to the coil. The changing direction of the magnetic field causes the current that flows in the coil to change direction as well. Thus, the current produced is alternating current. A generator that produces alternating current is often called an **alternator.**

In the simple generator the moving coil has kinetic energy. This kinetic energy is changed to electric energy because an electric current flows in the coil. The large generators that produce most of the electricity you use have more complicated parts and steps than the generator just described. But the same basic principle still applies. Electric energy is produced, or generated, from kinetic energy.

Do You Know?
Sharks have sensors that can detect induced voltages as weak as 5 billionths of a volt. They use these sensors to hunt for prey.

Study Questions for 16.2

1. How is electric current produced by electromagnetic induction?
2. How is electricity produced by an electric generator?

1. A current is produced when a metal wire is placed in a changing magnetic field.
2. A generator produces electricity by electromagnetic induction.

5. Conclude the lesson by asking what is indicated by the swinging of the galvanometer needle from right to left (electric current moves in one direction as you push the magnet into the coil and in the opposite direction as you pull it from the coil; the current alternates or changes its direction). Tell the students that a generator that produces an alternating current is called an alternator.

FOLLOW-UP

Reinforcement: Electric Currents and Magnetic Fields
Have the students create posters that illustrate how magnetism can produce an electric current and how an electric current can produce a magnetic field.

Enrichment: Production of Electricity
Have students design and perform experiments to determine how they can produce the most electric current from a magnet and a coil of wire. Variables the students could test include size of coil (length) versus length of magnet, number of loops, and types of wire. Each experiment should include a hypothesis, experimental design, observations, and a conclusion.

Tie-in: Technology Advances Discoveries
For Henry's and Faraday's discovery to be beneficial, technological advances had to be made. Ask the students to do research to determine what advances were necessary to make Henry's and Faraday's discovery useful.

16.3

OBJECTIVES: The student should be able to

- **identify** an electric motor;
- **explain** how an electric motor changes electric energy to kinetic energy;
- **list** some uses of electric motors;
- **explain** how electric current is moved from generators to motors for use;
- **identify** the role of a transformer in a circuit;
- **explain** why transformers can be dangerous.

SCIENCE TERMS

electric motor
transformer

ADVANCED PLANNING

1. You may want to arrange for a field trip to a local power plant in order for students to see how electricity is produced.

2. Electric utility companies often provide a wide range of free materials and speakers for schools. Contact your local utility for supplementary materials.

LESSON PLAN

1. Ask the students to recall the law of magnetism (like poles repel; unlike poles attract). Point out that the electromagnet in a motor is simply a coil of wire between the poles of a stationary magnet. The coil spins as a result of being attracted and repelled by the fixed magnet. Explain that the coil moves back and forth in half-turn steps so rapidly that it appears to move in a continuous circle.

SCIENCE BACKGROUND

The three main parts of a motor are the field magnet, a stationary magnet; the armature, an electromagnet that is free to rotate between the poles of the stationary magnet; and the commutator, a device that reverses the direction of the current in the armature as the armature rotates.

16.3 Electric Motors and Transformers

You will find out
- how an electric motor changes electric energy into kinetic energy;
- why transformers are important.

An **electric motor** is the reverse of a generator. A generator changes kinetic energy into electric energy. An electric motor changes electric energy into kinetic energy.

The heart of an electric motor is an electromagnet that rotates on an axle. The electromagnet rotates continuously between the two poles of a fixed permanent magnet. Like magnetic poles repel. So when the north pole of the electromagnet nears the north pole of the fixed magnet, the electromagnet is pushed away. It moves toward the south pole of the fixed magnet. Then, just as the north pole of the electromagnet reaches the south pole of the fixed magnet, the current is reversed. What was the north pole of the electromagnet becomes its south pole. So it is pushed away again. The continuous switching of the current causes the electromagnet to be continuously repelled by the fixed magnet. It spins around and around. The moving electromagnet has kinetic energy. This kinetic energy comes from the electric current flowing in the coil. The kinetic energy of the electromagnet can be used to do work.

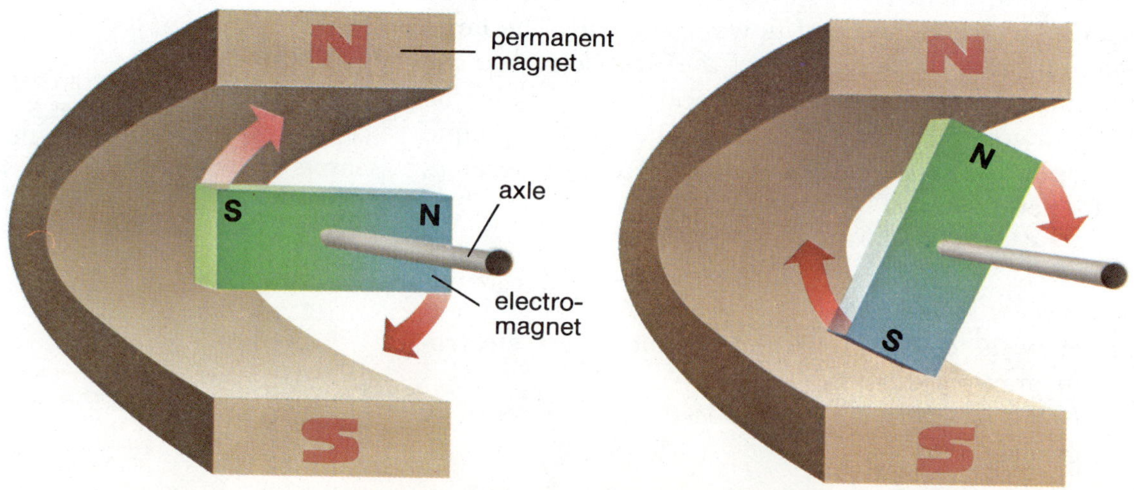

Figure 16.8 The continuous repulsion of the electromagnet causes it to spin.

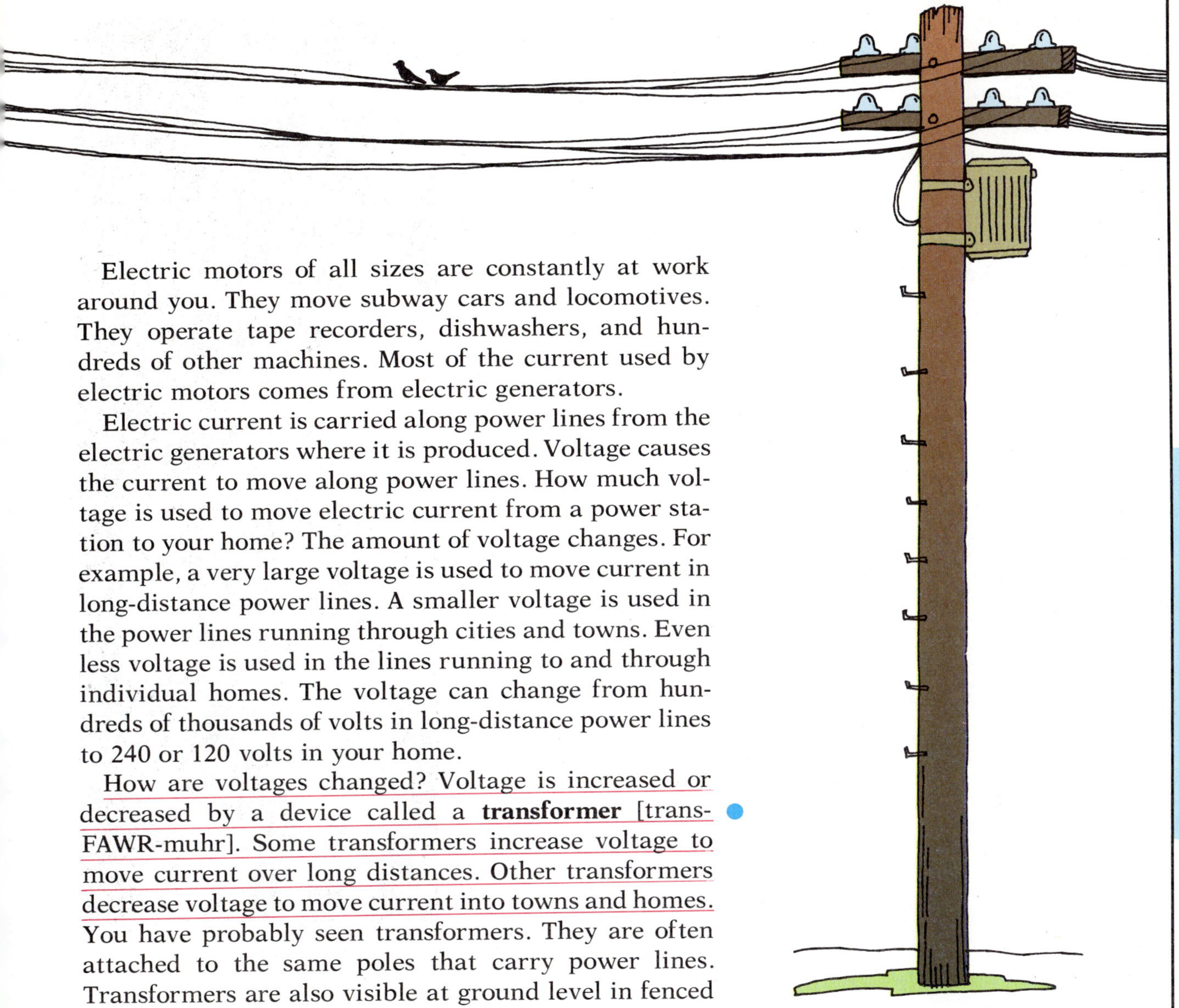

Electric motors of all sizes are constantly at work around you. They move subway cars and locomotives. They operate tape recorders, dishwashers, and hundreds of other machines. Most of the current used by electric motors comes from electric generators.

Electric current is carried along power lines from the electric generators where it is produced. Voltage causes the current to move along power lines. How much voltage is used to move electric current from a power station to your home? The amount of voltage changes. For example, a very large voltage is used to move current in long-distance power lines. A smaller voltage is used in the power lines running through cities and towns. Even less voltage is used in the lines running to and through individual homes. The voltage can change from hundreds of thousands of volts in long-distance power lines to 240 or 120 volts in your home.

How are voltages changed? Voltage is increased or decreased by a device called a **transformer** [trans-FAWR-muhr]. Some transformers increase voltage to move current over long distances. Other transformers decrease voltage to move current into towns and homes. You have probably seen transformers. They are often attached to the same poles that carry power lines. Transformers are also visible at ground level in fenced electrical yards. Since very high voltages are involved in transformers, they are very dangerous devices. You should never go near them.

Figure 16.9 The transformer attached to the pole changes voltage.

1. An electromagnet rotates between the poles of a fixed magnet. The current in the electromagnet continuously changes direction, causing its poles to be continuously repelled by the fixed magnet. The electromagnet spins. This kinetic energy is used to do work.
2. They are used to increase or decrease the voltages in power lines.

Study Questions for 16.3

1. How does an electric motor do work?
2. How are transformers used?

2. Ask the students how current is moved from the power plant to their homes (along power lines). What moves current along power lines (voltage)?
3. Point out that the voltage produced by generators at power plants is very high. How is the voltage lowered before it enters a home (with transformers)? Where are transformers located (on the same poles as the power lines; outside power stations on the ground)?
4. Stress that transformers carry very high voltages of electric current and therefore can be very dangerous. Students should never go near them.

● Word Skill: Vocabulary

FOLLOW-UP

Reinforcement: Generators and Motors
Ask the students to construct a chart to compare and contrast motors and generators.

Enrichment: Transformers
Ask the students to find out the differences between step-up and step-down transformers. Provide an old train transformer for the students to take apart. Is it a step-up or a step-down transformer? What is its purpose?

Tie-in: Motors in Our Lives
Ask the students to construct collages to illustrate all the uses of motors in their lives.

SCIENCE BACKGROUND

The transmission systems of many electric power companies are often linked together to form a grid. If demand for electricity in a particular region of the grid is high, part of the supply can be provided by other parts of the grid. In this sense, the grid provides pools, or reserves, of electric power.

Transformers are of two types: step-up transformers increase voltage and decrease current; step-down transformers decrease voltage and increase current for use in homes.

What Did You Learn?

16.1 • Electromagnetism is the branch of physical science that involves the combined effects of electricity and magnetism.

16.1 • Oersted discovered that electric currents produce magnetic fields.

16.1 • Electromagnets are magnets in which magnetism is produced by an electric current.

16.2 • Faraday and Henry discovered that a change in the magnetic field of a magnet produces an electric current.

16.2 • In electromagnetic induction, a current is created by a changing magnetic field.

16.2 • An electric generator uses electromagnetic induction to change kinetic energy into electric energy.

16.2 • An alternator is a generator that produces alternating current.

16.3 • An electric motor changes electric energy into kinetic energy.

16.3 • Voltage moves electric current along power lines.

16.3 • A transformer increases or decreases voltage.

Key Terms

electromagnetism
electromagnets
electromagnetic induction
electric generators
alternator
electric motor
transformer

Career

Electrical Engineer

An engineer uses scientific principles to solve practical problems. Electrical engineering is a large and rapidly growing branch of engineering. Electrical engineers help to design, develop, and produce electric equipment. They are also concerned with the generation and transmission of electric power.

Some electrical engineers help design electric appliances for people and businesses. Others work in the design and production of electric equipment for computers, aircraft, guided missiles, and spacecraft. Still other electrical engineers work in the planning and construction of power plants and power stations.

Good electrical engineers like to solve problems. They are able to carry out difficult calculations and do very careful work. They are also able to work well with others as part of a team.

To prepare for a career in electrical engineering, you should take as much mathematics as possible. Physics and chemistry are also important. English is important since engineers must be able to communicate well with others, both in conversation and in writing. Most electrical engineering positions require at least a college degree in engineering.

Vocabulary

1. d
2. f
3. a
4. g
5. b
6. e
7. i

● Word Skill: Vocabulary

What Do You Remember?

1. Oersted placed a compass above a wire carrying an electric current. The compass needle moved, indicating the presence of a magnetic field around the wire.
2. Electromagnets are used to lift large masses and as switches in telephones and doorbells. Electromagnets are also used in generators and motors.
3. Each man, working independently, showed that a change in a magnetic field produces an electric current.
4. The electric motor changes electric energy into kinetic energy. The generator changes kinetic energy into electric energy.
5. A coil of wire rotates around a magnet. The changing magnetic field in relation to the coil causes a current to flow in the wire.
6. An electromagnet rotates between the poles of a fixed magnet. The current in the electromagnet continuously changes direction. This change continuously reverses the north and south poles of the electromagnet. The poles are pushed away by the fixed magnet again and again causing the electromagnet to spin. This spinning motion is used to do work.
7. voltage
8. They increase or decrease voltage.
9. the creation of an electric current by a changing magnetic field.
10. alternating current

TO THINK ABOUT AND DO

Vocabulary ●

Number a piece of paper from 1 through 7. Beside the appropriate number write the letter for the term that is described.

1. The branch of physical science that involves the combined effects of electricity and magnetism
2. A magnet in which magnetism is produced by an electric current
3. The creation of a current by a changing magnetic field
4. A machine that changes kinetic energy into electric energy
5. A generator that produces alternating current
6. A machine that changes electric energy into kinetic energy
7. A device that increases or decreases voltage

a. electromagnetic induction
b. alternator
c. battery
d. electromagnetism
e. electric motor
f. electromagnet
g. electric generator
h. voltage
i. transformer

What Do You Remember?

1. How did Oersted discover that electricity could produce magnetism?
2. How are electromagnets used?
3. What did Henry and Faraday contribute to the study of electromagnetism?
4. How does an electric motor differ from an electric generator?
5. How does a generator produce electricity?
6. How does a motor use electric energy to do work?
7. What causes electric current to move along power lines?
8. How are transformers used?
9. What is electromagnetic induction?
10. What kind of current is produced by an alternator?

Applying What You Have Learned ●

1. How is an electromagnet different from a permanent magnet?
2. Why would alnico, which makes a strong permanent magnet, not be useful in making an electromagnet?
3. How could you vary the strength of an electric current produced by a magnet?
4. How could you prove an electromagnet is a magnet only while current is passing through it?

Research and Investigation ●

1. What are solenoid coils and how are they used?
2. How are electromagnets used in relays?

Applying What You Have Learned

1. The magnetism of an electromagnet can be turned on and off. Also, the strength of its magnetic force can be changed by changing the strength of the current. A permanent magnet cannot be turned on and off. The strength of its magnetic field cannot be changed.
2. The alnico would become a permanent magnet in the presence of an electric current. When the current was turned off it would remain magnetized. Its magnetism could not be turned on and off.
3. You could vary the number and tightness of the coils wound around the magnet, vary the strength of the magnet, or vary the speed of the magnet's motion inside the coil.
4. Allow an iron object to cling to the electromagnet. Then turn off the current going to the electromagnet. If the iron object drops when the current stops, the electromagnet is no longer a magnet.

● Comprehension Skill: Applied

Research and Investigation

1. A solenoid coil is a coil of wire with a loosely sliding rod of magnetized metal inside. Solenoids can be used to open valves or trip switches.
2. Relays are used to strengthen telegraph signals so they can travel long distances. A relay consists of an electromagnet which receives electric current from batteries. The electromagnet has many coils of wire which strengthen telegraph signals. Relays are also used in lighting systems for large rooms or halls.

● Reference/Research Skills

CHAPTER BACKGROUND

The beginnings of the electronic age can be traced to the discovery of the Edison effect by Thomas Alva Edison in 1883. Edison's discovery occurred as he was perfecting the light bulb. He suspended a small electric plate within a bulb. A wire passing through the glass bulb connected the plate to a dry cell. The other terminal of the dry cell was connected to a carbon filament that was also within the bulb. Only when the apparatus was connected so that the plate was joined to the positive terminal and the filament to the negative terminal was an electric current observed to flow through the bulb's vacuum from the filament to the metal plate. The circuit would only pass current in one direction.

Edison could not explain his discovery. In fact, it remained a scientific curiosity until Sir Ambrose Fleming tried to put it to practical use in 1904. Fleming used the Edison effect in his attempt to build a wireless station at Poldhu, Wales. The station was to be capable of sending messages across the Atlantic. As he attempted to build a sensitive detector of wireless signals, Fleming built several improved Edison lamps that did respond to wireless waves. He referred to his device as a valve because it passed current in only one direction. Unfortunately, the valve was only able to pass signals; it could not amplify them for increased transmission.

In 1906 Lee de Forest, an American inventor, succeeded where Fleming had failed. He added a perforated metal grid between the filament and the plate. Voltages applied to the grid were able to modify and control the current.

De Forest's audion, or the triode as it is called today, became most useful after 1914. At that time E. H. Armstrong published the first correct explanation of its operation. He used his own improved tubes to build circuits that revolutionized radio communications and gave rise to electronics as we know it.

Electronics 17

Obtain a darning needle, a square of paper small enough to fit through the eye of the needle, a pair of tweezers, and an overhead projector. While the needle is on the stage of the overhead projector use the tweezers to pass the paper through the eye of the needle. Tell the students that some of the electronic chips used in computers today are so small that they can fit through the eye of a needle. These tiny electronic chips can do the same things that it took a roomful of equipment to do years ago. In this chapter the students will be introduced to electronics and some of the ways that it affects their lives.

Would you believe that thin squares smaller than postage stamps could change your life? Believe it or not, they already have. Pocket calculators, tape recorders, televisions, radios, telephones, home computers, and video games all work by using tiny squares, or chips of silicon. These chips can do things that a room full of equipment could not do 25 years ago.

What do silicon chips do? How do computers work? How does your television make a picture on the screen? How can complicated machines such as calculators, radios, and televisions be so small? You will be able to answer these questions when you have finished this chapter.

Figure 17.1 Electronic equipment in a jumbo jet

Lesson Organizer

17.1 Electronic Devices

Pages 352–355
Advanced Planning: Page 352
Follow-up: Page 355
- **R** Electronic Devices
- **E** The Edison Effect
- **T** Careers in Electronics

17.2 Functions of Electronic Devices

Pages 356–359
Activity: How Do Shortwave Radio Operators Communicate? Page 359
Follow-up: Page 359
- **R** Functions of Electronic Devices
- **E** High Fidelity
- **T** Electronic Devices in the Home

17.3 Integrated Circuits and Microelectronics

Pages 360–361
Advanced Planning: Page 360
Follow-up: Page 361
- **R** Integrated Circuits
- **E** The Printed Circuit
- **T** Photoelectric Devices in Industry

17.4 Computers

Pages 362–367
Advanced Planning: Page 362
Activity: How Can You Create a Flowchart? Page 366
Follow-up: Page 367
- **R** Computer Vocabulary Bee
- **E** Computer Programs
- **T** Computer Constellations

See ACTIVITIES book pages 97–102 for laboratory, skill, and review exercises for Chapter 17.

17.1

OBJECTIVES: The student should be able to

- **describe** functions of *electronic devices;*
- **list** some electronic devices;
- **describe** a vacuum tube;
- **list** some disadvantages of vacuum tubes;
- **describe** a cathode-ray tube;
- **describe** how a picture is formed on a television screen;
- **define** the term *semiconductor;*
- **name** some common semiconductors;
- **discuss** the role of impurities in semiconductors;
- **list** some advantages of electronic devices that contain semiconductors;
- **identify** a transistor.

SCIENCE TERMS

electronic devices	semiconductors
signals	impurities
vacuum tubes	transistor
cathode-ray tube	solid-state
fluorescent	

ADVANCED PLANNING

1. For demonstrations you will need a transistor radio, a tape recorder, a phonograph, a vacuum tube, and an old tube radio.

2. Cut some circles from colored paper and label each with a negative sign. Glue some magnets on the backs of the circles so that they will stick to the chalkboard.

LESSON PLAN

1. As the lesson begins, have on your desk a transistor radio, a tape recorder, and a phonograph. Ask the students to classify them (electronic devices). How do they operate (by rapidly and accurately controlling the movement of electrons)? Ask the students to name some other electronic devices (television, computers, digital watches).

17.1 Electronic Devices

You will find out
- what electronic devices do;
- what vacuum tubes are;
- how a picture is made on a television screen;
- what semiconductors are.

Electronic [ih-lehk-TRAHN-ihk] **devices** have rapid and accurate control of the flow of electrons. Televisions, radios, record players, tape recorders, computers, cameras, and guidance systems for spacecraft are just a few of many different instruments, machines, and appliances that use one or more electronic devices.

Most electronic devices respond to short bursts of electric energy called **signals.** The signals may come from any one of many different sources such as television stations, the keyboard of a computer, or a microphone. An electronic device directs or strengthens these signals to get useful information from them. For example, your television has electronic devices that pick up very weak signals from many miles away. Those signals are strengthened so they can make an accurate copy of a picture and sound.

Figure 17.2 When you use a computer, you are using many electronic devices.

Machines that use electronic devices need a source of electric power in order to run. Often this power comes from the current in an electric outlet. Some electronic machines do not have a cord and a plug. Instead they operate on current produced by batteries. You have probably seen and used small calculators that operate on batteries.

The first electronic devices were glass tubes from which nearly all the air had been removed. These tubes are called **vacuum tubes.** A vacuum tube contains at least two metal parts. One part is a negatively charged metal wire. Another part is a positively charged metal plate. When a vacuum tube is put in an electric circuit, electrons are released from the wire and flow to the plate. The current in that circuit can be controlled by controlling the flow of electrons from the wire to the plate.

Until the 1950's vacuum tubes were the heart of the electronics industry. They were used in radios, televisions, and all other kinds of electronic equipment. However, vacuum tubes have limitations. Have you ever seen a very old radio or television? If you have, you were probably surprised at how big it is. It has to be big because the vacuum tubes inside it take up lots of space. Vacuum tubes also require a lot of electric power and give off lots of heat. Finally, they wear out after only a few thousand hours of use.

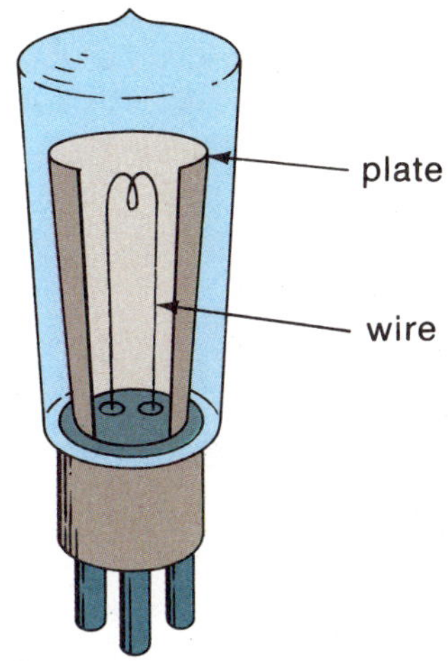

Figure 17.3 In a vacuum tube, electrons flow from the wire to the plate.

Figure 17.4 The radio in this photograph is big because the vacuum tubes inside it take up a lot of space.

2. Ask the students where the signals for programs on their televisions and radios come from (from radio and television stations). How does a television or a radio use these signals to get useful information (directs or strengthens the signals so that an accurate copy of a picture and sound is made)?

3. Show the students your vacuum tube and explain that the inside of the tube is a vacuum, or contains no air. Pass the tube around the room so that the students can see its structure for themselves.

4. Show the students your old tube radio and the new transistor radio. Ask the students to compare the two devices. Why do they think vacuum tubes were replaced by other electronic devices (their size meant that radios and televisions had to be much larger than they are today; they also required more electricity, became hotter, and wor out sooner than today's devices)?

5. On the board draw a simple cutaway view of a vacuum tube. Label the cathode "C" and the anode "A." Explain that the cathode is the negatively charged metal wire and the anode is the positively charged metal plate. Ask the students on which part—anode or cathode—there is an excess of electrons (cathode).

6. On the cathode of your vacuum tube drawing place several of the colored circles that represent electrons. Draw arrows from the cathode to the anode and move some of the "electrons" to the anode. As you do this, explain that the vacuum tube controls the current by controlling the rate at which the electrons move from cathode to anode.

SCIENCE BACKGROUND

The negatively charged wire in a vacuum tube is called the cathode. The positively charged plate is called the anode.

7. Remind the students that one type of vacuum tube still in use today is found in television sets. What is it (a cathode-ray tube)? What is the principle upon which it operates (an electron beam from the cathode strikes a fluorescent screen, thereby making it glow; the beam is then moved by a signal to produce patterns of dots of light that are seen as pictures)?

Word Skill: Vocabulary

8. Explain that most electronic devices today rely on semiconductors. Write the term *semiconductor* on the board. Underline the root word, *conductor*. Ask what a conductor is (material through which an electric current can flow). What are the names of two common conductors (copper, iron)? Circle the prefix *semi*. Ask what it means (partly). What, then, is a semiconductor (a material that is a better conductor than an insulator but not as good as a conductor)?

9. Ask the students to name the most commonly used semiconductor (silicon; germanium is used less today because it is relatively expensive). Point out that in order to increase the conductivity of these semiconductors, phosphorus is often added. Since phosphorus is a foreign material whose atoms are not found in pure crystals of silicon or germanium, it is said to be an impurity.

One kind of vacuum tube is still in use and very familiar to you. It is the **cathode** [KATH-ohd] **ray tube.** The picture tube in your television is a cathode-ray tube. It is a vacuum tube in which a stream of electrons is directed onto a screen coated with a **fluorescent** [floo-REHS-ehnt] material. The word *fluorescent* describes a material that glows when struck by electrons. When the stream of electrons hits the fluorescent screen, a dot of light forms on the screen. A signal fed into the cathode-ray tube moves the stream of electrons around producing a pattern of dots. You see this pattern as a picture. Cathode-ray tubes are used to make pictures not only in televisions but also on computer display screens and other machines.

In a color television, the inside surface of the fluorescent screen is covered with tiny dots. These dots are arranged in three sets. One set of dots glows red when struck by electrons. Another set glows green. The third set glows blue. The cathode-ray tube sends out three streams of electrons, one for each set of dots. Together, the three sets of glowing dots form the color television picture you see.

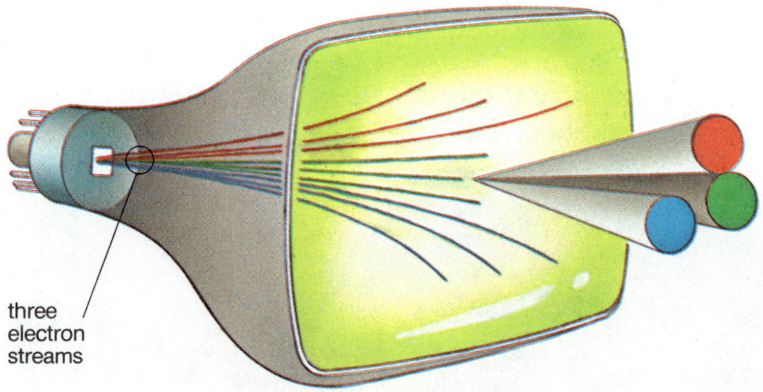

Figure 17.5 The cathode-ray tube of a color television

A television may be the only appliance or machine you are familiar with that still uses a kind of vacuum tube. Vacuum tubes have limitations, so scientists looked for and developed new and better ways to control the flow of electrons. Most electronic devices today use **semiconductors.** A semiconductor is a solid material

Figure 17.6 Assembling television sets

that has a conductivity in the range between that of an insulator and that of a good conductor. The most commonly used semiconductors involve crystals of germanium and silicon. When tiny amounts of other elements are added to a pure crystal of germanium or silicon, the conductivity of the crystal can be increased. The elements that are added are called **impurities.** The conductivity of a crystal is regulated by the kinds, amounts, and placement of impurities on it. Electronic devices that use semiconductors are smaller, require less power, produce less heat, and last much longer than vacuum tubes. Such electronic devices quickly replaced vacuum tubes in most electronic equipment.

The **transistor** [tran-ZIHS-tuhr] is the electronic device that replaced many kinds of vacuum tubes. In vacuum tubes electrons flow through a vacuum. A transistor uses semiconductors to control the flow of electrons. Semiconductors are solids. Because the electrons are flowing through a solid, transistors became known as **solid-state** devices. You may have seen or heard the term *solid-state* in reference to radios, televisions, or other electronic equipment.

Study Questions for 17.1

1. What is an electronic device?
2. What is a vacuum tube?
3. How is a picture made on a television screen?
4. What are semiconductors?

1. a device that has rapid and accurate control of the flow of electrons
2. an electronic device made of a glass tube from which nearly all the air has been removed and having at least two metal parts
3. A stream of electrons in a cathode-ray tube forms a pattern of dots on a fluorescent screen.
4. materials that have a conductivity in the range between that of an insulator and that of a good conductor

10. Show the students your transistor radio and explain that the transistor is really a semiconductor that controls the flow of electrons just as the vacuum tube did. What is an essential difference between the two (the electrons are moving through a solid in a transistor rather than through a vacuum in the vacuum tube)?
11. Conclude the lesson by placing the tube and transistor radios side by side. Point out that the transistor radio contains semiconductors rather than vacuum tubes. Ask the students what advantages the transistor radio has (smaller, produces less heat, requires less power, lasts longer).

FOLLOW-UP

Reinforcement: Electronic Devices
Have students pretend that they are reporters. Each sudent should write an interview with an electronics engineer. As reporters, the students should ask the engineer to explain what an electronic device is and how electronic devices are used in the home.

Enrichment: The Edison Effect
Ask the students to do research on the Edison effect. What is it? How was it discovered? Why is it important?

Tie-in: Careers in Electronics
Ask some representatives of the high-technology industries in your area to speak to students concerning possible careers in the electronics industry.

SCIENCE BACKGROUND

Atoms of silicon and germanium have four valence electrons. Phosphorus atoms have five valence electrons. When phosphorus atoms are added to a pure silicon crystal, the extra valence electron wanders freely through the crystal, greatly increasing the conductivity.

In solid-state devices, a signal flows through a solid material rather than a gas. Most solid-state devices are made of semiconductor materials.

17.2

OBJECTIVES: The student should be able to

- **describe** what a rectifier does;
- **describe** what an oscillator does;
- **describe** what an amplifier does.

SCIENCE TERMS

rectifier
oscillator
amplifier

LESSON PLAN

1. On the board construct a three-column chart entitled *Functions of Electronic Devices*.
2. In the first column write the word *rectifier*. Ask the students to state the difference between the electric current from a dry cell and the current from a wall socket (battery is DC; wall socket is AC). Explain that since most electronic devices require a direct current and wall sockets provide an alternating current, rectifiers are used to convert alternating currents to direct currents. Under *rectifier* write *AC to DC*.

SCIENCE BACKGROUND

Electronic devices are being used to solve many human medical problems. Research in this area is called bionics. Today research is being done to help people control artificial limbs with microcomputers. Your students may be interested in seeing this in the film *Adaptation*, which is part of the *Search for Solutions* series.

17.2 Functions of Electronic Devices

You will find out
- what rectifiers do;
- what oscillators do;
- what amplifiers do.

What would you do if you wanted to find out the latest weather report or who was leading in an election or a football game? You would probably turn on a radio or television. Someone in a broadcast studio far away from your home would answer your question. How did that answer reach you? Many different electronic devices were involved in bringing the answer to you.

Electronic devices control the flow of electrons. However, this does not mean that all electronic devices are the same. Different electronic devices control the flow of electrons to perform different functions. One function is to change the alternating current supplied to an appliance to direct current. The current produced by electric generators for homes and businesses is alternating current. However, many of the appliances and machines in homes and businesses require direct current. An electronic device called a **rectifier** [REHK-tuh-fy-uhr] changes alternating current to direct current. For example, when you turn on a radio, alternating current flows into the machine. A rectifier quickly changes that current to direct current. You hear the results. Most rectifiers today are solid-state devices.

Figure 17.7 Electric signals from distant broadcast studios are picked up by antennas.

The electronic signals that are picked up by machines such as radios and televisions are produced by an electronic device called an **oscillator** [AHS-uh-layt-uhr]. An oscillator produces electronic signals from direct current. Oscillators are used in radio and television broadcasting. They are also used to produce timing signals in watches and heart pacemakers.

As electronic signals travel, they become weaker. When a weak signal is picked up by a machine such as a radio, an electronic device is used to strengthen the signal. Such a device is an **amplifier** [AM-pluh-fy-uhr].

Did you watch television or listen to a radio or a tape recorder today? Did you hear someone talk over a microphone? If so, you made use of amplifiers. Most electronic equipment could not operate without fast and efficient amplification of weak signals. For example, without amplifiers the signals that reach your television are so weak that the cathode-ray tube could not make an accurate copy of the original picture.

Figure 17.8 The functions of a rectifier, an oscillator, and an amplifier

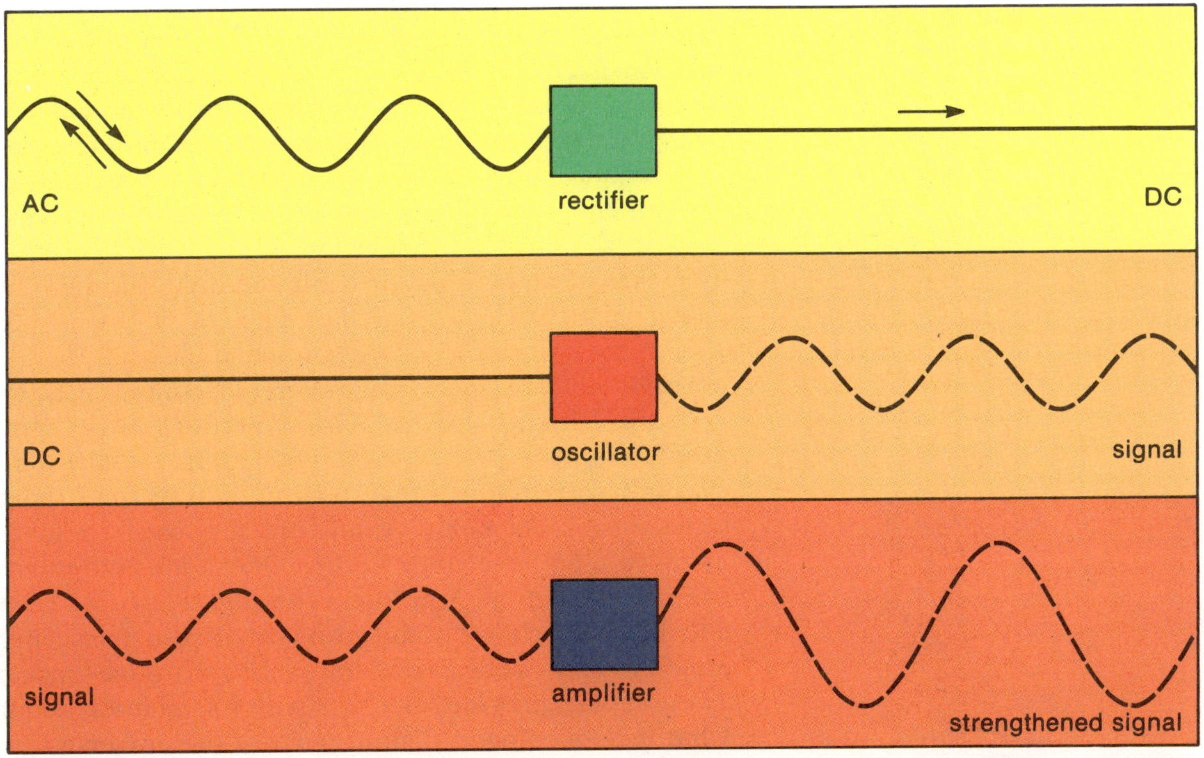

3. At the top of the second column write *oscillator*. What is its function (produces electronic signals or short bursts of energy from direct current)? Under *oscillator* write *DC to signals*.

4. At the top of the third column write the word *amplifier*. Ask the students what the word *amplify* means (to make larger, to strengthen). Explain that an amplifier adds energy to the signal, thereby strengthening it.

5. Conclude the lesson by stressing that electronic devices fulfill three main functions. Use your chart to review these functions.

● **Comprehension Skill: Applied**

FOLLOW-UP

Reinforcement: Functions of Electronic Devices

Have the students construct a poster to illustrate the three functions of electronic devices, the specific instrument that fulfills each function, and examples of its use.

Enrichment: High Fidelity

High fidelity started as an engineer's hobby. However, the sound of music from the early records and phonographs was very different from the music heard in concert halls. Engineers wondered why concert hall sounds could not be duplicated in the home. Have students find out how engineers solved the problem. Students should draw a series of diagrams with captions to illustrate their research.

Tie-in: Electronic Devices in the Home

Ask students to list the electronic devices in their homes. Have each student write a diary for one week in which life without the devices is described.

Figure 17.9 Many amplifiers are used to televise a baseball game.

Several amplifiers are used to make a picture on a television screen. Suppose you are watching a live broadcast of the World Series. When a television camera is pointed at the scene on the playing field, light from the scene enters the camera. The light is directed onto a fluorescent screen where it makes a pattern of dots. An electronic device changes these dots into a pattern of electronic signals. An amplifier in the camera strengthens these signals. The amplified signals are sent to a broadcasting station. An amplifier in the station strengthens the signals again. They are broadcast through the air. Your television picks up these signals by means of an antenna or cable. An amplifier in your television increases the signal as much as a billion times ● so that the cathode-ray tube can make the picture.

Amplifiers are used everywhere in communication networks. Radio and television stations use them. Amplifiers are used in communications satellites to strengthen the signals they pick up and to send those signals back to earth. Amplifiers are used in long-distance telephone lines and in the radar equipment used in navigation. Amplifiers in radio telescopes keep people in touch with the universe. Radio telescopes amplify very weak signals traveling through space. These signals provide scientists with information about what is happening in distant parts of the universe.

ACTIVITY How Do Shortwave-Radio Operators Communicate?

Materials

paper
pencil

A	B	C	D	E	F
•—	—•••	—•—•	—••	•	••—•
G	H	I	J	K	L
——•	••••	••	•———	—•—	•—••
M	N	O	P	Q	R
——	—•	———	•——•	——•—	•—•
S	T	U	V	W	X
•••	—	••—	•••—	•——	—••—
Y	Z	1	2	3	4
—•——	——••	•————	••———	•••——	••••—
5	6	7	8	9	0
•••••	—••••	——•••	———••	————•	—————
.	,	?	S.O.S.		Start
•—•—•—	——••——	••——••	•••———•••		—•—
End of Message		Understand		Error	
•—•—•		•—•		••••••••	

Procedure

Morse code signals are broadcast and amplified by electronic devices. Study the code and use it to write two messages. One message must be about electronics.

Questions

1. Trade messages with a friend. What do your friend's messages say?

2. What does the following message say?

•—• • —•—• — •• ••—• •• • •—• •••

—•—• •••• •— —• ——• •   •— —•—•

— ———   —•• —•—•

1. Answers will vary. One answer should relate to electronics.
2. The message translates to "Rectifiers change AC to DC."

Study Questions for 17.2

1. What is the function of a rectifier?
2. What is the function of an oscillator?
3. What is the function of an amplifier?

1. changes alternating current to direct current
2. produces electronic signals from direct current
3. strengthens electronic signals

ACTIVITY: How Do Shortwave Radio Operators Communicate?

Some of your students may be interested in receiving their ham radio operator's license. To do so they must learn the international Morse code. The system is used all over the world.

You may wish to extend the activity by preparing messages for the students to convert to code.

17.3

OBJECTIVES: The student should be able to

- **describe** an integrated circuit;
- **identify** microelectronics;
- **discuss** the advantages of integrated circuits.

SCIENCE TERMS

integrated circuits
microelectronics

ADVANCED PLANNING

1. For a demonstration prepare a parallel circuit with at least five parts. Make the wiring loose enough so that the circuit will fall apart as you handle it.

2. For a demonstration you will need a computer, transistor radio, or calculator with a circuit that the students can see.

LESSON PLAN

1. Introduce the lesson by showing the students your parallel circuit. As you show it to them, point out that older televisions and radios also had circuits in which each part was separately wired. Hold up the circuit and be sure that part of it falls apart as you handle it. Ask why these circuits were inconvenient (all parts had to be wired separately, which took a long time; there were many wires in very complicated circuits).

● **Comprehension Skill: Applied**

2. Show the students one of the integrated circuits from a computer, radio, or calculator. On the board write the term *integrated circuit*. Point out that the circuit does not contain recognizable discrete elements such as resistors, transistors, or capacitors.

17.3 Integrated Circuits and Microelectronics

You will find out
- how an integrated circuit is made;
- what is meant by the term *microelectronics*.

The electrons that flow through electronic equipment follow a carefully laid-out path, or circuit. Each amplifier, rectifier, or other electronic device in the equipment must be connected to that circuit. In older electronic equipment such as early radios and televisions, all the parts of the circuit had to be wired together separately. This wiring took a long time to set up. It also resulted in a jumbled mess of wires in very complicated circuits. These problems disappeared with the use of **integrated** [ihn-tuh-GRAYT-ehd] **circuits.** An integrated circuit is a circuit that is put on a semiconductor chip by placing impurities in a very precise arrangement. The impurities are arranged in such a way that each different area of the chip conducts current differently. Each area on the chip becomes a different and specific electronic device that is part of the integrated circuit. Each circuit part has a specific function. Some parts are rectifiers. Other parts are amplifiers, oscillators, or switches.

Figure 17.10 This person is watching television! Advances in microelectronics have made it possible to produce a television so tiny that you must hold it up to your eye to see the picture.

Figure 17.11 This photograph illustrates the tiny size of a semiconductor chip.

When all the impurities have been added to a chip, all the electronic devices in that particular integrated circuit are in place. However, the circuit is still incomplete. The electrons that flow through each electronic device in the circuit must have a path to follow from one device to the next. A metal, usually aluminum, makes the desired connections between the different devices. A thin coating of aluminum is put over the entire chip. The aluminum is then carefully removed from some areas and left in place in others to complete the circuit. In this way it is possible to put an entire circuit containing over a thousand different electronic devices on a chip the size of a baby's thumbnail.

Integrated circuits form the basis of a new branch of electronics called **microelectronics.** Although integrated circuits are very tiny and complicated, they are easy to produce. They cost much less to make than other earlier kinds of circuits. Integrated circuits work faster, use less power, and are more reliable. Integrated circuits have made possible the amazing pocket calculators, electronic watches, computers, and other electronic machines that are in general use today.

Do You Know?
All the words in this book can be stored on a microchip this size:

Study Questions for 17.3

1. How is an integrated circuit made?
2. What is microelectronics?

1. Impurities are arranged on a tiny semiconductor chip to form different electronic devices. A metal such as aluminum makes the connections between the devices.
2. a branch of electronics involving integrated circuits

3. Ask the students to describe the process that is used to miniaturize integrated circuits (after the impurities are added, the entire chip is covered with aluminum; then the aluminum is removed from all areas except those where it will be used to complete the circuit on the chip).
4. Point out that integrated circuits have become the basis for a new branch of electronics. What is it called (microelectronics)? What does this term imply about the size of the integrated circuits that are used (extremely small; some are small enough to fit on a baby's thumbnail)?
5. Conclude the lesson by asking several students to list the advantages of integrated circuits (can be made easily and relatively cheaply; work faster and use less power than those containing vacuum tubes).

FOLLOW-UP

Reinforcement: Integrated Circuits
Ask the students to write an illustrated report in which they explain what integrated circuits are and why they are important.

Enrichment: The Printed Circuit
Before integrated circuits were invented, printed circuits were regarded as the major advance in the electronics industry. Ask the students to do research on printed circuits. Where are they used? What are their advantages? Disadvantages? How have the disadvantages been largely overcome? Ask the students to pretend that they work for a television repair shop. Have them write a pamphlet in which printed circuits are discussed.

Tie-in: Photoelectric Devices in Industry
Industries use photoelectric devices to count, sort, and test products on conveyor belts. Ask the students to research this process. What devices allow it to work? Why is it more efficient than manual methods?

17.4

OBJECTIVES: The student should be able to

- **describe** some ways in which computers are changing people's lives;
- **list** the three basic functions of a computer;
- **list** the five basic units of a computer.

SCIENCE TERMS

computer	memory unit
computer program	control unit
input unit	processing unit
processing	output unit

ADVANCED PLANNING

1. Have available materials that the students can use to make a bulletin board collage. As an assignment, ask students to bring in pictures that illustrate ways in which computers affect their lives, e.g., video games, code on items at store, direct dial system.
2. Try to have a computer available for the students to use.
3. Try to have available computer programs, printouts, and disks that you can show the students.

LESSON PLAN

1. To introduce the lesson, have the students group their computer-related pictures in categories according to how they help or affect their lives. Then have the students arrange the pictures in a collage on the bulletin board.

17.4 Computers

You will find out
- what the three basic functions of computers are;
- what units are used to perform the three basic functions of computers;
- how computers are used.

Imagine that in 1960 someone chose one hundred people at random. Those one hundred people were asked one question, "Have you ever used a computer?" How many would have said yes? There is no way of knowing exactly how many. But you can be sure that if the same survey were made today, many more people would say yes.

The computers in use today are very different from the huge, expensive earlier computers that could be used only by highly trained experts. Today's computers are everywhere. They are in offices, government agencies, stores, banks, libraries, schools, and homes. They are operated by people like you. If you haven't seen or used a computer yet, the chances are very good that you will have that opportunity soon.

Figure 17.12 This early computer took up a whole room and could be used only by specially trained operators.

A **computer** is a machine that can carry out a detailed set of instructions to perform a task. The set of instructions is called a **computer program.** The computer program is written in a special language the computer understands. Computer language makes it possible to store and use large amounts of information in very small amounts of time. A computer programmer makes up the programs.

A computer is just a tool. It does not have a mind of its own. It can only follow instructions. Any computer, no matter how complicated, operates by performing three basic functions. It takes in information, does something with that information, and gives out new information. Different parts, or units, of a computer perform these three functions.

A computer uses an **input unit** to perform the first basic function. The input unit takes in information. This unit usually is a keyboard. Computer programmers use the keyboard to type programs into the computer. A computer operator also uses the keyboard to type in the information needed for the computer to perform a task. In the input unit, the information that is typed in is changed into the special computer language.

Do You Know?
Computers can process information so quickly that the speed of their operation is often measured in nanoseconds. One nanosecond is one billionth of a second.

Figure 17.13 Using a computer

2. Ask the students what computers do. As students suggest their ideas, write them on the board. Do not reject any ideas at this point. Point out that in order to do all these things a computer must be able to follow instructions. A computer, then, is simply a machine that can follow complicated instructions.

3. Explain that the computer follows instructions called a program. Ask how the instructions, or program, come about (a computer programmer uses electrical impulses to write instructions on a magnetic tape or disk; the impulses act like a series of words, giving directions to the computer).

SCIENCE BACKGROUND

Just as letters make up the words we use in our language, computer language is coded in binary terms—all zeros and ones. Each zero or one is called a bit.

4. On the board draw a flowchart entitled "How a Computer Solves Problems." As you construct five boxes for the chart, explain that the computer has five basic parts involved in solving problems.
5. Number your boxes 1 through 5. In box 1 write *input unit*, in box 2 *memory unit*, in box 3 *control unit*, in box 4 *processing unit*, and in box 5 write *output unit*. Ask a student to group the five boxes in the flowchart to reflect the three basic functions of a computer (input: box 1, processing: boxes 2–4, output: box 5).
6. Explain that in the input step you tell the computer what the problem is. This is usually done by typing the problem on a keyboard. The information is then changed into computer language.
7. Once the computer has the information in its input unit, it must use, or process, it. Ask the students to tell you where the processing is done (memory unit, control unit, processing unit).
8. As the students describe each of the three units, write the information next to the appropriate box on the flowchart (memory unit—computer looks through all of its programs to see whether it has a program or set of directions that can be used to solve the problem; control unit—computer brings to the foreground the information it will need to solve the problem; processing unit—program (directions) is used to solve the problem).

SCIENCE BACKGROUND

Often computer information is stored on magnetic tape. The magnetic tape is sectioned into tiny segments, sometimes as many as 12,000 per square inch. Each of these tiny segments, or bits, can be magnetized in one of two possible directions, corresponding to 0 or 1. These 0's and 1's are coded programs or data used by the computer.

The second basic function of a computer is to do something with the information that has been received by the input unit. This function is called **processing.** Three units are involved in the processing of information in a computer. The **memory unit** stores all the programs and information that have been put into the computer. The **control unit** regulates the flow of information in the computer. When information to be processed is received by the input unit, the control unit brings that information together with the appropriate program in the **processing unit.** The processing unit carries out the program by following its instructions.

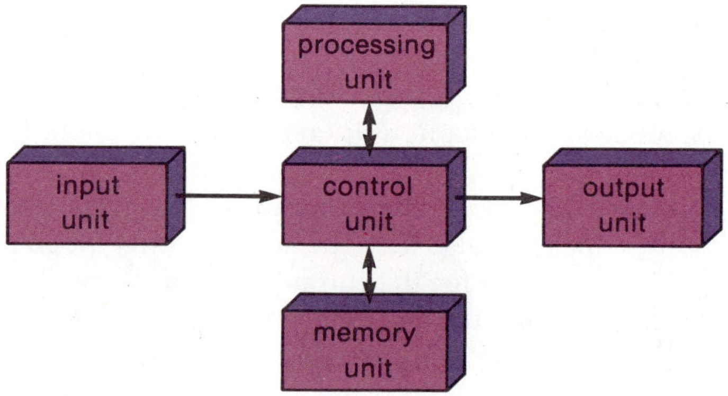

Figure 17.14 The flow of information in a computer

The third basic function of a computer is to give out new information. This function is carried out by the **output unit.** The output unit changes the results of the processing unit from computer language back into words, numbers, and symbols and prints this new information on a screen or a paper.

It is important to remember the three basic computer functions: input, processing, and output. Computers use electronic devices to perform these functions. Using electronic devices allows computers to solve problems millions of times faster than you could with a pencil and paper. Simple problems, such as finding the average of a list of numbers, can be solved so quickly that the output unit shows the answer almost as soon as the problem is typed in. Problems that require complicated programs, such as those used to interpret weather patterns, can take many minutes, even hours, to solve.

Figure 17.15 Computers are being used in sports and the arts. Ballet dancers use computers to design their dances.

Computers are used to solve scientific and engineering problems, to keep records, and to organize files. Even if you have never used a computer yourself, computers are involved in your life. For example, when you call the telephone operator for a number, a computer is used to search rapidly through the listings for the number. All new telephone numbers are entered into the computer system, keeping the lists up-to-date. At the end of the year the computer is hooked up to a machine that sets the type for the new telephone books. Computers are also used in the direct-dial system that almost instantly connects your telephone to others across the country.

The airlines use computers both in the air and on the ground. Suppose you live in Los Angeles and want to reserve a seat on a flight to New York City. You call a reservation agent in Los Angeles and explain what you want. The agent checks with a computer for the ticket price, the flight times, and whether or not seats are available. In a matter of seconds the computer displays the information. If you reserve a seat, the computer prints out a ticket, stores the information that another seat has been taken, and is ready for the next request.

Do You Know?
It would take half the people in the United States between the ages of 18 and 45 to run the nation's telephone system if it were not computerized.

9. Point out that the solution is given in the output stage. Ask whether the output unit itself has a problem-solving function (yes, in the sense that the output unit must change the results of the processing unit from the language of the computer to words, numbers, and symbols).

SCIENCE BACKGROUND

In CAT scans, "Computerized Axial Tomography," computers can reassemble thousands of individual X-ray measurements to reproduce a two-dimensional cross-sectional view of any part of the body. By putting a number of these views together, doctors can get a three-dimensional view of any part of the body.

ACTIVITY: How Can You Create a Flowchart?

For this activity, you may want to provide dry cells, wire, and bulbs. Students can then exchange flowcharts and try to build a parallel circuit by following the directions they receive.

You may want to substitute something else of interest to your students for the Bunsen burner in the second question.

Stress that computers do not think; they only follow directions. Therefore, the flowcharts should include all necessary steps no matter how obvious they may seem.

Laboratory Skill: Experimentation/Demonstration

ACTIVITY

How Can You Create a Flowchart?

Materials

paper
pencil
ruler

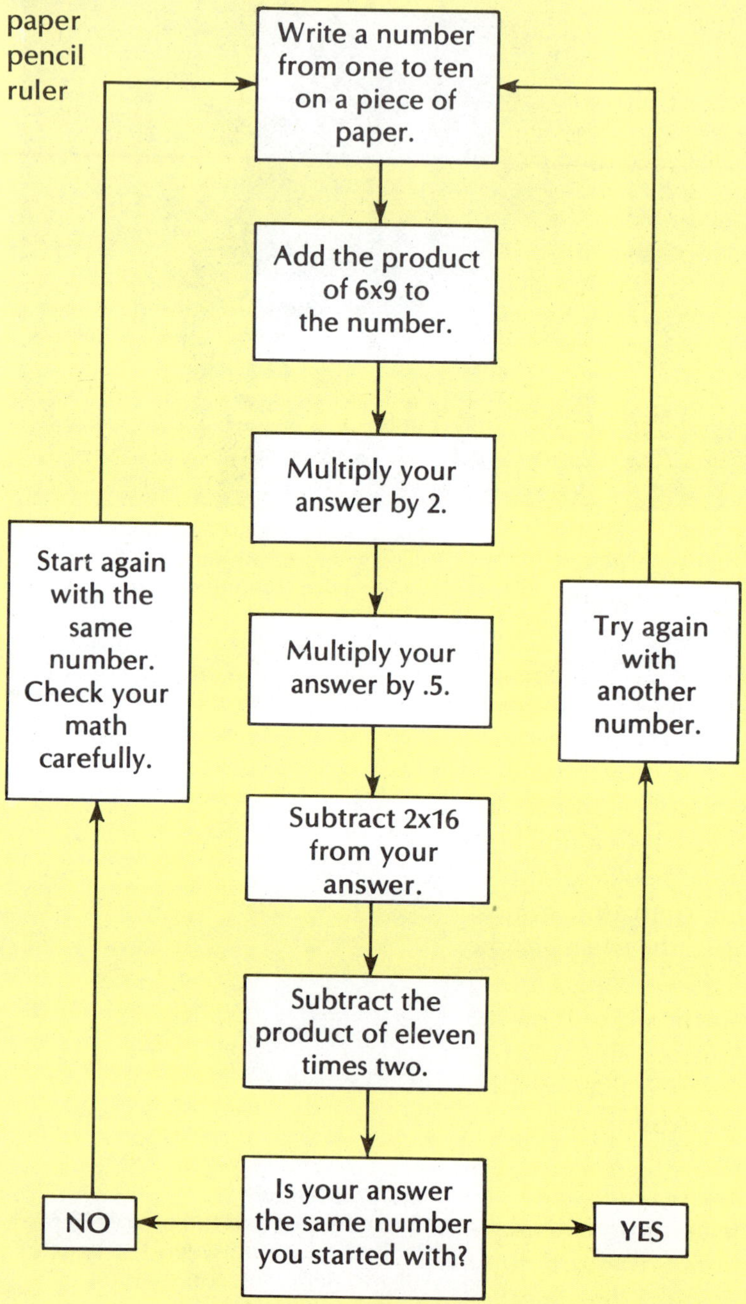

Procedure

Computer programmers write the directions they wish a computer to follow on a flowchart. Study the flowchart given. Notice that each box contains only one direction. Follow the flowchart's directions.

Questions

1. Make a flowchart telling how to construct a simple circuit with a battery, two wires, and a light bulb and socket.
2. Make a flowchart telling how to go from your home to school.

Flowcharts will vary. They should have all the necessary details in the proper sequence. To test the circuit flowcharts you may wish to have students actually follow them step by step with the materials involved.

0 41280 03252 7

Figure 17.16 *left:* A computerized supermarket checkout; *above:* The supermarket product code

Airline pilots use computers, too. Just before take-off, the latest weather information is put into a computer on the ground. The computer uses this information to determine the best route, speed, and altitude for greatest safety and fuel economy. Computers on board the airplane help to guide the plane in flight.

Computers are used in supermarkets, too. Have you ever noticed the little rectangle filled with black lines on the items you buy at supermarkets? The lines are a coded message. The message tells the name, manufacturer, and size of the item. When the cashier slides the item over a small window set into the checkout counter, an electronic device picks up the code and relays the information to a computer. The computer reads the code and prints the information on a paper tape. The computer also notes in its memory unit that there is one less of that item on the shelf. The store manager can use that information to decide when to reorder.

Study Questions for 17.4

1. What are the three basic functions of a computer?
2. Name and define the units of a computer.
3. Give an example of a way that computers affect your life.

1. take in information, process that information, give out new information
2. control unit—regulates the flow of information; input unit—receives information; memory unit—stores programs and information; processing unit—carries out programs; output unit—gives out new information
3. Answers will vary. Some possibilities are computer games, to store and locate telephone numbers, to reserve airline tickets, and to analyze weather data.

10. Give the students a quick succession of simple math problems to solve such as 5 × 4; 2 × 3; and 3 × 3. Stress that although the students' minds had to search for information, recall it, and use it before any output or answer was given, they required only a fraction of a second to solve each problem. Point out that a computer can solve a very complicated problem even more quickly than this.

FOLLOW-UP

Reinforcement: Computer Vocabulary Bee
Have a computer vocabulary bee by forming two student teams. Then define a computer term for a student and have the student tell you the term and spell it. Conversely, give a student a computer term to be spelled as well as defined.

Enrichment: Computer Programs
Ask interested students to create computer programs that will quiz the users on physical science information. Students should write out their programs and have them approved before putting them on the disks.

Tie-in: Computer Constellations
"Name the Constellations" is a computer program that students should enjoy using. With it students can learn about 12 major constellations. It can be obtained from James Palmer, 46 E. Main Street, Stafford Springs, CT 06076.

What Did You Learn?

17.1 • Electronic devices have rapid and accurate control of the flow of electrons.
17.1 • Most electronic devices respond to short bursts of electric energy called signals.
17.1 • The first electronic devices were vacuum tubes.
17.1 • The cathode-ray tube is a vacuum tube that directs a stream of electrons onto a fluorescent screen.
17.1 • Semiconductors have a conductivity in the range between that of an insulator and that of a good conductor.
17.1 • Transistors use semiconductors to control the flow of electrons.
17.2 • Rectifiers change alternating current to direct current.
17.2 Oscillators produce electronic signals from direct current.
17.2 • Amplifiers strengthen electronic signals.
17.3 • An integrated circuit is a circuit that is put on a semiconductor chip by placing impurities in a very precise arrangement.
17.3 • Microelectronics is a branch of electronics which involves electronic devices connected in integrated circuits.
17.4 • A computer is a machine that can carry out a detailed set of instructions to perform a task.
17.4 • The three basic functions of a computer are to take in information, to process that information, and to give out new information.
17.4 • Computers have five units: the input unit, the memory unit, the control unit, the processing unit, and the output unit.
17.4 • Computers are used to do many things such as store and locate telephone numbers, make airline reservations, and analyze weather information.

Key Terms

electronic devices
signals
vacuum tubes
cathode-ray tube
fluorescent
semiconductors
impurities
transistor
solid-state
rectifier
oscillator
amplifier
integrated circuits
microelectronics
computer
computer program
input unit
processing
memory unit
control unit
processing unit
output unit

Career

Broadcast Technician

Broadcast technicians operate and maintain the electronic equipment used to record and transmit radio and television programs. They work with microphones, sound recorders, television cameras, videotape recorders, and other equipment.

Broadcast technicians operate equipment in the control room that regulates the quality of sounds and pictures being recorded or broadcast. They also operate controls that switch broadcasts from one camera to another, from film to live programming, or from network to local programs. On occasion broadcast technicians may be required to go out and set up a remote broadcast in a football stadium, an airport, or wherever news is breaking.

In small studios, broadcast technicians perform a variety of duties that may range from giving technical direction to people in the studio to operating the record turntable as a disk jockey. In large studios, the work is more specialized. For example, a broadcast technician might be an audio-control specialist who operates the controls that regulate sound pickup, transmission, and switching. Audio-control technicians also find work at sound recording studios.

Jobs for broadcast technicians can be found wherever there are broadcasting stations or recording studios. This means most jobs are in large cities. About a third of the jobs are in towns having fewer than 10,000 people.

Students interested in becoming broadcast technicians should take physical science and mathematics courses in high school. After high school they should take electronics and communications courses in a technical school, college, or university.

Vocabulary

1. b
2. a
3. b
4. c
5. b
6. c
7. b

● Word Skill: Vocabulary

What Do You Remember?

1. a device that accurately controls the flow of electrons; record players, cameras, radios, televisions, computers
2. a glass tube from which most of the air has been removed and having at least two metal parts, a wire and a plate
3. require a lot of current, too big, give off too much heat, wear out after a few thousand hours of use
4. A stream of electrons in a cathode-ray tube forms a pattern of dots on a fluorescent screen.
5. a material that glows when struck by electrons
6. rectifier—changes alternating current to direct current; amplifier—strengthens weak electronic signals; oscillator—produces electronic signals from direct current

TO THINK ABOUT AND DO

Vocabulary ●

Number your paper from 1 to 7. Select the one best term that completes each of the following sentences. Write the letter of the term next to the appropriate number.

1. Devices that have rapid and accurate control over the flow of electrons are (a) circuit parts, (b) electronic devices, (c) machines.
2. Short bursts of electric energy used by most electronic devices are called (a) signals, (b) beeps, (c) bits.
3. The first electronic devices were (a) radio tubes, (b) vacuum tubes, (c) cathode-ray tubes.
4. A material that has a conductivity in the range between that of an insulator and that of a good conductor is called (a) an amplifier, (b) an oscillator, (c) a semiconductor.
5. An electronic device that strengthens weak electronic signals is (a) a rectifier, (b) an amplifier, (c) an oscillator.
6. The part of a computer that receives information is the (a) output unit, (b) memory unit, (c) input unit.
7. The part of the computer that carries out a program is a (a) memory unit, (b) processing unit, (c) output unit.

What Do You Remember?

1. What is an electronic device? Give three examples of machines that use electronic devices.
2. Describe a vacuum tube.
3. List three limitations of vacuum tubes.
4. How is the picture on a television screen produced?
5. What is a fluorescent material?
6. Identify the function of each of the following: rectifier, amplifier, oscillator.

7. What is an integrated circuit?
8. What is a computer program?
9. What are the three basic functions of a computer?
10. Name the five basic parts of a computer and identify the function of each.

Applying What You Have Learned ●

1. Name at least two appliances or machines in your home that use amplifiers.
2. Why are rectifiers not found in battery-powered radios?
3. In what ways might a bank use a computer?
4. How do computers help people to deal with large amounts of information?

Research and Investigation ●

1. Electric eyes are often used to open doors, ring buzzers, or set off burglar alarms. How do they work?
2. By using remote control you can open a garage door while in a car or change channels on a television without touching the set. How does remote control work?

7. a circuit that is put on a semiconductor chip by placing impurities in a very precise arrangement
8. the set of of instructions a computer follows to perform a task
9. take in information, process information, give out new information
10. input unit—takes in information; memory unit—stores programs and information; control unit—regulates the flow of information; processing unit—carries out the program; output—gives out new information

Applying What You Have Learned

1. radio, television
2. A battery produces direct current.
3. Answers will vary. They might include the following: to keep an updated list of bank customers, to keep a record of checking and savings account activity, to monitor cash flow, and to make predictions for future cash requirements.
4. Computers can store large amounts of information in a small space. Information can be retrieved quickly. Computers can find trends, patterns, and sequences in data.

● Comprehension Skill: Applied

Research and Investigation

1. An electric eye is controlled by a phototube. This is a vacuum tube that conducts electricity when light strikes it. It has a cathode surrounding an anode. When light strikes the treated surface of the cathode, electrons leave its surface and are attracted to the anode. The flow of current is amplified and operates a relay that works a switch.
2. Remote control is actually a series of electric eyes. For televisions, one controls the on-off switch, one the volume, and one the channel. A flashlight or even sunlight can operate simple systems. For this reason more complex systems work on sound with sensitive microphones replacing the electric eye. A tuning fork produces a sound that is well above the hearing ability of people or dogs. The microphone picks up the sound and converts it to a weak electric signal that is amplified so it can operate a relay that controls a switch.

● Reference/Research Skills

Long-Term Projects

1. Have the students make a collage showing examples of how waves act. On the back of the collage, students should explain each picture.
2. Ask the students to research the uses of ultrasonic waves. They should prepare posters or models and a tape recording to express the results of their research.
3. Have the students make a display of optical illusions. Encourage them to use the display to investigate how people see the illusions. Do boys see them differently than girls do? Do older people see them differently than younger people do?
4. Have the students create wave tanks to investigate the properties of waves. The industrial arts teacher may be able to help them. Suggest reference books such as *PSSC Physics* (D. C. Heath and Company) to help the students.
5. Inquire to see if any company in the area can lend you a decibel meter. Have the students use the meter to measure the amounts of noise in different areas around your school and town. Take the readings at different times of the day. Mark the readings on maps of the school and the town. Ask the students to suggest ways to reduce noise in areas where the readings are high.
6. Some students may be interested in photography. Ask these students to prepare posters and a demonstration to explain how light makes an image on film and how film is developed.
7. Have students build periscopes and use them to demonstrate how light is reflected.
8. Have the students work in groups to select objects from their homes and use them to make musical instruments. They should play a song for the class.

Unit Five

The Action of Waves

Bulletin Boards

1. Title the board *Wave Action*. Divide the board into four parts and label each as shown in the diagram. Under Interference, draw two record players with their speakers facing. In the center, where the notes would meet, draw the notes that appear to be distorted. In the second section, Refraction, draw a rainbow and some water drops. Under Diffraction, draw a boy calling around the corner of a building to two girls. Under Reflection, mount a mirror or a very flat piece of foil. At the bottom of each section, leave room for the students to add other examples.

2. Title your board *Lenses in Our Lives*. On the board, mount pictures of eyeglasses, a magnifying glass, a telescope, and a microscope. Have each directed at an object. Use arrows to show the path of light rays from the objects through the lenses. Encourage the students to bring in pictures of other objects that contain lenses.

3. Hearing losses often go undetected. To help students understand the importance of detecting and remedying hearing loss, title your board *How You Hear*. In the center, draw a diagram of the ear. Have the students label the parts. Help the students make a list of sounds they should be able to hear. Have them illustrate the sounds on the board.

CHAPTER BACKGROUND

Waves are motions that carry energy from one place to another. They are caused when something disturbs the medium through which they travel. The medium may be matter, such as the air through which sound waves travel, or it may be the electromagnetic field through which radio and television signals travel.

Waves are generally divided into two main groups—transverse and compressional waves. Light and rope waves are transverse waves. Any particle on these waves travels perpendicularly to the direction of the wave. Sound waves are compressional. Any particle on these waves travels in the same direction as the wave.

All waves have certain common characteristics. The following terms are used to describe these characteristics: amplitude, the vertical distance from the equilibrium point (point halfway between the crest and the trough) to the crest; wavelength, the horizontal distance between the crests; frequency, the number of crests that pass a particular point in a given period of time.

There is a direct relationship between frequency and wavelength. For example, in waves traveling from the ocean toward the shore, the number of crests reaching the shore in a given period of time (frequency of wave) is determined by the speed with which the waves are moving and by the distance between the crests (wavelength). If the waves are coming in at a fast rate, the number of crests reaching the shore in a given amount of time will be large. If, on the other hand, each crest is coming in fairly slowly with a long distance between crests, the number of crests reaching the shore in a given period will be small.

The shortening or lengthening of sound waves caused by the relative motion of the source of waves past an observer is the Doppler effect. Most students have experienced this effect when a car with the horn blowing approached and then passed them. As the sound of the horn approaches, the apparent pitch of the sound increases. As the sound moves away, the pitch decreases.

The "radar guns" used by police officers to catch speeding drivers utilize the Doppler effect. The moving car bounces back gun-emitted waves. The change in the frequency of these waves as the car approaches the gun is used to calculate the car's speed.

Waves

18

Ask the students what a drummer, a surfer, and a person using a flashlight have in common. Point out that all three persons are involved with waves. Ask in what way each is involved with waves. The students will understand easily how the surfer is involved but may have difficulty with the others. Tell them that sound (from the drums) and light (from the flashlight) are waves. In this chapter they will learn how waves can be measured and what things all waves have in common.

You may have thought that all waves occur in water. In fact, when you picture waves you probably think of an ocean. But you do not have to be near an ocean or any water at all to encounter waves. Every morning when you open your eyes, you encounter waves. Every sound you hear, from a whisper to a siren to a song, involves waves. You can make waves in a rope by tying one end to a doorknob and moving the other end up and down.

You can see waves in water and in a rope. Many waves, however, are invisible. These waves are in the air around you, in the earth under your feet, or in the empty space among the stars. In this chapter, you will learn how to describe waves and how the characteristics of waves can change.

Figure 18.1 Detail from a Japanese print called *The Wave* by Hokusai

Lesson Organizer

18.1 The Nature of Waves

Pages 376–381
Advanced Planning: Page 376
Activity: How Can You Make a Model of a Wave? Page 380
Follow-up: Page 381
- **R** Illustrating Waves
- **E** Types of Waves
- **T** Light and Sound

18.2 Changes in Waves

Pages 382–385
Advanced Planning: Page 382
Activity: How Can You Observe Wave Behavior? Page 383
Follow-up: Page 385
- **R** Properties of Waves
- **E** Surfaces and Wave Activity
- **T** Creating Rainbows of Light

18.3 The Doppler Effect

Pages 386–387
Advanced Planning: Page 386
Follow-up: Page 387
- **R** The Doppler Effect
- **E** Ambulance Warning
- **T** Star Movements

See ACTIVITIES book pages 103–108 for laboratory, skill, and review exercises for Chapter 18.

18.1

OBJECTIVES: The student should be able to

- **identify** a wave;
- **define** the term *medium* in relation to waves;
- **distinguish** between transverse and compressional waves;
- **identify** the crest, trough, and amplitude of a wave;
- **determine** the speed of a wave when given the wavelength and frequency.

SCIENCE TERMS

wave	trough
medium	amplitude
transverse waves	wavelength
compressional wave	frequency
crest	hertz

ADVANCED PLANNING

1. For demonstrations you will need an overhead projector, a glass pie plate filled with water, a pebble, two 2 m (6 ft) long ropes, and a Slinky.

2. You will need a work sheet of problems for student practice in determining speed when given wavelength and frequency.

LESSON PLAN

1. Place the pie plate filled with water on the overhead projector. Ask the students to watch the water as you drop the pebble into it. What happens to it (ripples appear on the surface)? Ask what the pattern of ripples is called (a wave). What caused the ripples (pebble disturbed the water)? Explain that a wave is a disturbance that travels through matter or space.
2. Tell the students that the substance through which a wave travels is its medium. What was the medium of the wave formed by the pebble (water)? Ask what medium sound waves are traveling through as you speak (air, gas).

18.1 The Nature of Waves

You will find out

- what a wave is;
- how waves are grouped;
- what some characteristics of waves are.

Have you ever tossed a stone into a pond and watched what happened? A pattern of ripples spreads away in circles from the point where the stone enters the water. These ripples are an example of a **wave.** A wave is a disturbance that travels through matter or space. A disturbance refers to anything that changes the situation of matter or space. Waves can travel through solids, liquids, or gases. Some waves travel through empty space. The matter or space through which a wave travels is called a **medium.** Often waves come in a series of disturbances, as in the ripples of a pond. This series is also called a wave.

Figure 18.2 An example of a wave

Waves can be divided into two groups. The groups are based on the direction of the wave motion and the direction of the wave disturbance. Imagine a bottle floating in the ocean. As waves move by the bottle, it bobs up and down. The up-and-down motion of the bottle indicates the direction of the disturbance. This bobbing up and down is at right angles to the direction of the waves which are moving in the water. Ocean waves are an example of **transverse** [trans-VURS] **waves.** In transverse waves, the disturbance occurs at right angles to the direction of the wave motion.

Now imagine a long, closely packed line of people waiting to get into a movie. Suppose someone in the back of the line gives the next person a shove. That person bumps into the next person and so on down the line. No one person moves very far, but the disturbance moves all along the line. As the wave goes by, the people move forward in the direction of the wave and then back again as they get their balance. This is an example of a **compressional** [kuhm-PRESH-uhn-uhl] **wave.** In compressional waves, the disturbance occurs back and forth along the direction of the wave motion. The medium—people in the example—is pushed together, or compressed, as the wave goes by.

Figure 18.3 The two kinds of waves

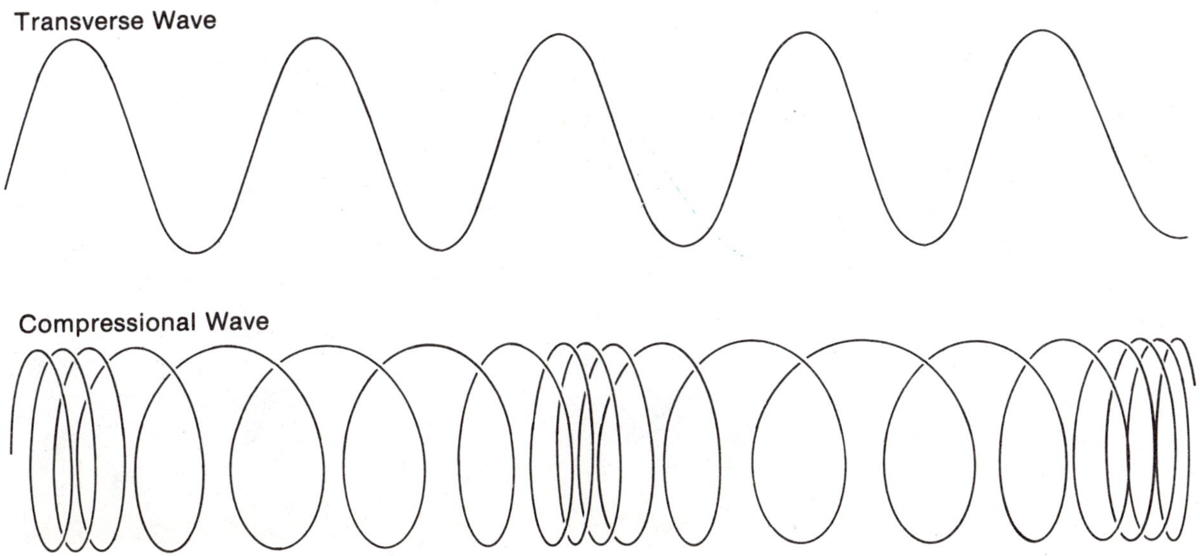

3. Give two students a rope. Ask one of the students to pretend to be a pole and merely hold the end of the rope at waist height. Have the other student stand at such a distance that the rope is stretched parallel to the ground but is not taut. Then ask this student to make waves in the rope by flicking the wrist up and down.
4. Give two students a Slinky. Ask one of the students to be a pole as the other student makes waves by sharply tapping the end of the Slinky.
5. Ask the students to imagine a red ball sitting on each of the two waves. When the rope makes waves, how will the ball move (up and down)? How will the wave move (across, right to left)? When the Slinky makes waves, how will the ball travel (across, right to left)? How will the wave travel (across, right to left)?
6. On the board construct a chart whose two columns are labeled *Rope* and *Slinky*. Ask the students to describe and compare the rope wave and the Slinky wave. Record their observations on the chart.

SCIENCE BACKGROUND

A compressional wave may be called a longitudinal wave.

The compression on a compressional wave is equivalent to the trough on a transverse wave.

7. Above the word *Rope* write *Transverse Wave*. Above the word *Slinky* write *Compressional Wave*. Point out that light travels in transverse waves and sound travels in compressional waves.
8. On the board make a diagram of a transverse wave. As you explain that the transverse wave has certain identifiable parts, ask the students what the "hills" are called (crests). Label one of the crests. What are the "valleys" called (troughs)? Label a trough.
9. Draw a horizontal line through all the crests and another horizontal line through all the troughs. Draw a third horizontal line halfway between these two lines. (This third line is the equilibrium or zero point.) Ask what the distance from this line to the crest or trough line is called (amplitude). Point out that the amplitude is the height of the wave. Ask how amplitude and energy are related (the greater the amplitude, the more energy the wave has).

SCIENCE BACKGROUND

Crests are the areas of greatest disturbance or energy; troughs are the areas of least disturbance or energy.

Two kinds of waves that are of special interest to physical scientists are sound waves and light waves. Light waves are transverse waves. Sound waves are compressional waves. Because light and sound are so important, each will be discussed in a separate chapter.

Waves have several characteristics that can be measured. It is easy to see how this can be done if you think of a wave as a series of hills and valleys. The highest point of a wave is called the **crest.** The crest corresponds to the top of a hill. The lowest point of a wave is called the **trough** [trawf]. The trough corresponds to the bottom of a valley. The height of a wave is called its **amplitude** [AM-pluh-tood]. The amplitude is the distance from a point halfway between the crest and the trough straight up to the crest. Amplitude is often measured in meters.

Figure 18.4 A wave with a large amplitude

Do You Know?
Sightings of unusually high ocean waves—up to 30 meters high—have been reported by sea captains. Such waves are thought to be the result of interference involving many ocean waves.

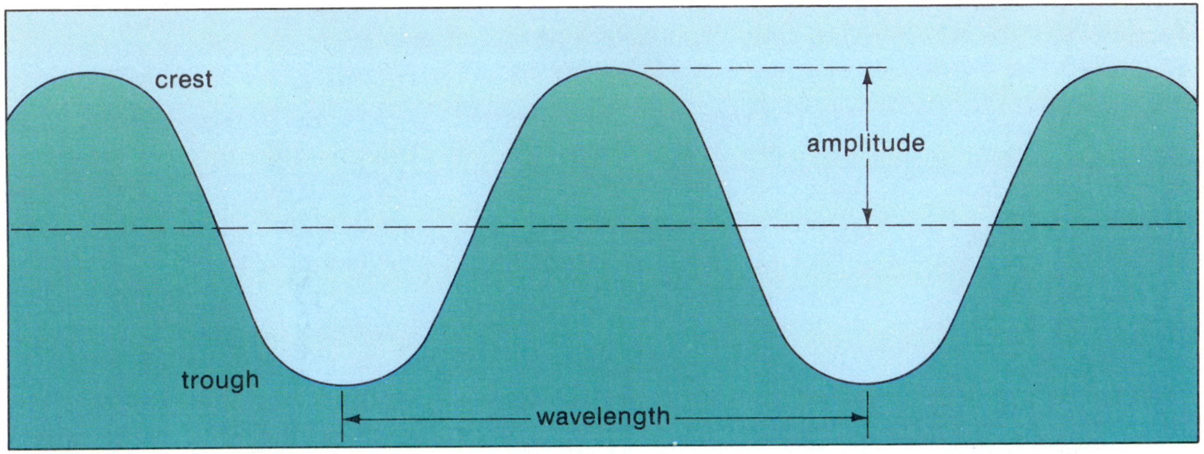

Figure 18.5 Characteristics of a transverse wave

Waves carry energy from one place to another. The amplitude of a wave can indicate how much energy it is carrying. The greater the amplitude, the more energy the wave has. Expert surfers enjoy riding on waves that have a large amplitude. The energy of such waves offers surfers long and exciting rides.

The distance from one crest of a wave to the next can be measured. This distance is called the **wavelength.** Wavelength varies greatly depending on the waves. For example, the wavelength of waves on the open ocean might be several meters. However, the ripples made when you toss a stone into a pond could have wavelengths of only a few centimeters.

Wavelength is directly related to another wave measurement called **frequency** [FREE-kwuhn-see]. The frequency of a wave is the number of crests that pass a point in a second. Frequency is measured in **hertz** [hurts]. One hertz (Hz) is one crest or trough past a point in a second.

Figure 18.6 These waves have equal amplitudes. What measurement is unequal? wavelength

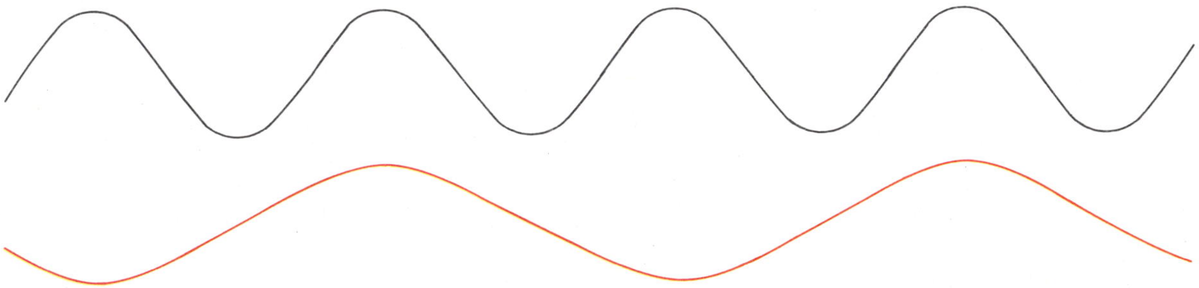

10. Draw a line from the center of one crest to the center of the next. Ask a student to label this distance (wavelength, which is usually measured in meters).

11. Ask two pairs of students to help you. Give each pair a rope. Have one student in each pair act as a "pole." Have the other student in each pair flick his or her wrist to make waves. Pair 1 should make waves every fifteen seconds. Pair 2 should make waves every 5 seconds. Ask all the students to imagine a pole halfway between each of the pairs. In which pair will more waves pass the pole in 30 seconds (Pair 2)?

12. Explain that the number of crests that pass a certain point in a second is called the frequency of the wave. It can be labeled as the number of waves per second or as hertz (Hz).

SCIENCE BACKGROUND

Although the units for measuring amplitude and wavelength are given in meters, other metric units of length may be more appropriate, depending on the size of the wave.

Frequency can be measured in hertz or in cycles per second (cps). When speed is calculated from wavelength and frequency, speed is expressed in meters per second (m/s).

ACTIVITY: How Can You Make a Model of a Wave?

◇ **Caution:** Students should extinguish the candles as soon as they have finished with them.

Stress that the pin should touch the paper very lightly or the tuning fork will stop vibrating. Students may need several trials before they produce a usable wave model.

● **Laboratory Skills: Manipulation of Materials**
Experimentation/Demonstration

● ACTIVITY

How Can You Make a Model of a Wave?

Materials

ring stand
clamp
straight pin
candle
matches
tuning fork
tuning fork hammer
carbon paper
safety glasses
ruler
tape

Procedure

1. Put on your safety glasses.
2. Light the candle. Use a drop of wax from the candle to fasten a straight pin to one prong of the tuning fork as shown in the diagram. Put out the candle.
3. Set up the clamp, ring stand, and tuning fork as shown in the diagram.
4. Place the carbon paper under the pin. Adjust the height of the tuning fork so the pin lightly touches the paper.
5. Use the hammer to strike the prong of the tuning fork that is not holding the pin. The tuning fork will vibrate.
6. With a slow, steady motion, pull the carbon paper under the pin to make a model of a wave. You may have to practice several times to make a satisfactory model.
7. Cut out your best wave model and tape it to your paper.

Questions

1. Label the crests and troughs on your wave model.
2. Use an arrow to indicate wavelength. How long is the wavelength in mm?
3. Draw a horizontal line through the center of your wave model. Use an arrow to indicate the amplitude. How large is the amplitude in mm?
4. Study your wave model. Does the wavelength change? Explain your answer.
5. Does the amplitude of the wave change? Explain your answer.

1. The highest points should be the crests. The lowest points should be the troughs.
2. The wavelength should be the distance from one crest to the next or one trough to the next.
3. The amplitude should be the distance from the horizontal line straight up to a crest or straight down to a trough.
4. As the vibrations of the tuning fork diminish, the wavelength will change.
5. As the vibrations of the tuning fork diminish, the amplitude will change.

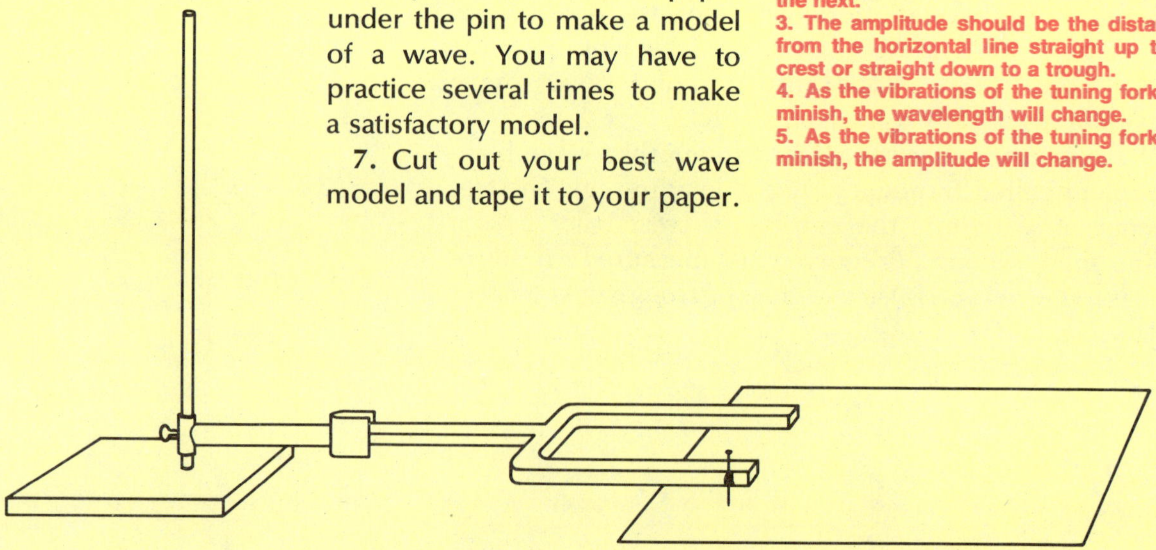

Figure 18.7 The waves are traveling. The water is not.

Wavelength and frequency are related to another characteristic of waves, speed. Wave speed is measured in meters per second. The speed of a wave can be calculated using this formula:

Speed = wavelength × frequency

EXAMPLE Find the speed of a wave with a wavelength of 0.5 m and a frequency of 20 Hz.

SOLUTION Speed = wavelength × frequency
= 0.5 m × 20 Hz = 10 m/s

The speed at which a wave travels depends on the properties of the medium it travels through. Remember, a wave may move through matter, but a wave itself is not matter. A wave is a disturbance in a medium. The medium is not traveling. The disturbance is traveling. Thus, the speed of a wave is the speed of a disturbance.

Do You Know?
Tidal waves caused by earthquakes can have speeds of up to 800 km/h.

Study Questions for 18.1

1. What is a wave?
2. How do transverse and compressional waves differ?
3. Name four characteristics of waves that can be measured.

1. a disturbance that travels through matter or space
2. In transverse waves, the disturbance occurs at right angles to the direction of the wave motion. In compressional waves, the disturbance occurs back and forth along the direction of the waves.
3. amplitude, wavelength, frequency, speed

13. Review the meaning of the term *wavelength* (distance from center of one crest to center of next crest). To what two wave characteristics is wavelength related (frequency, speed)? How can the speed of a wave be determined (by multiplying its wavelength by its frequency)?
14. Give the students some sample problems in which they are asked to find the speed of a wave. For example, what is the speed of a wave whose wavelength is 2 m and frequency is 100 Hz?

Speed = wavelength × frequency
= 2 m × 100 Hz
= 200 m/s

15. Distribute your work sheet of speed problems. You may want to assign these problems for homework, or you may choose to work through them orally during class.

Mathematics Skills: Problem Solving Computation

FOLLOW-UP

Reinforcement: Illustrating Waves
Ask the students to create a poster that illustrates the two groups of waves (transverse, compressional) and the parts of each (crest, trough, wavelength, amplitude).

Enrichment: Types of Waves
Have students make a list of different types of waves—for example, light waves, sound waves, and water waves. Then ask students to classify the waves as transverse or compressional.

Tie-in: Light and Sound
Have students investigate the relationship between amplitude and loudness of sound or amplitude and brightness of light. How can knowledge of the relationship be used to benefit people?

18.2

OBJECTIVES: The student should be able to

- **describe** the reflection of a wave;
- **distinguish** between the angle of incidence and the angle of reflection;
- **discuss** the similarity between the angle of incidence and the angle of reflection;
- **describe** the diffraction of a wave;
- **describe** the refraction of a wave;
- **explain** the interference of waves.

SCIENCE TERMS

reflection	diffraction
angle of incidence	refraction
angle of reflection	interference

ADVANCED PLANNING

For demonstrations, you will need a mirror, an overhead projector, a pie plate filled with water, a block, a pencil, a glass of water, and a 2 m (6 ft) long rope.

LESSON PLAN

1. Introduce the lesson by asking the students to list some ways that waves can be changed (by hitting barriers, by traveling through a medium that has changed).
2. On the board construct a four-column chart entitled *Changes in Waves*.
3. Hold up the mirror toward your students. Ask what happens when light hits it (bounces off). Tell the students that this bouncing off of waves is called reflection.

18.2 Changes in Waves

You will find out
- how waves are changed by boundaries and barriers;
- how waves are changed when the medium changes;
- how waves are changed by other waves.

Waves can change. You have observed some of these changes. You observe them every time you use a mirror, hear an echo, see a rainbow, or watch water waves come together. Since a wave is a moving disturbance, anything that affects the wave motion changes the wave. Sometimes the changes are caused by a boundary or barrier that the waves encounter. Changes can also result from changes in the medium. A third way waves can change is by encountering other waves.

If you watch the waves in a swimming pool carefully, you will see that they bounce back when they hit the sides of the pool. The bouncing back of a wave when it hits a boundary is called **reflection** [rih-FLEHK-shuhn]. Waves bounce off the surface of a boundary in much the same way as a rubber ball bounces off a hard surface. If a wave hits a surface head on, it is reflected straight back. Similarly, a ball that is thrown straight at a wall bounces straight back. When a wave hits a surface at an angle, it is reflected back at the same angle to the other side. This behavior is similar to the behavior of a ball thrown against a surface at an angle. Basketball players know that a ball thrown against a backboard at an angle will rebound at the same angle to the other side.

Do You Know?
The structure of a cat's eyes allows them to reflect light waves very effectively. This is why cats' eyes seem to shine in the dark.

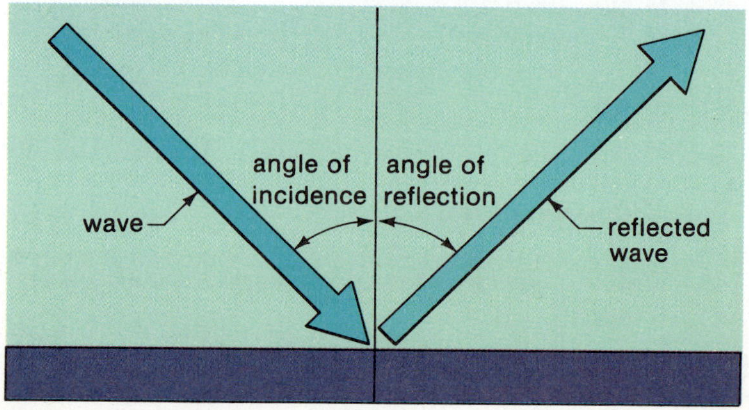

Figure 18.8 Reflection

ACTIVITY

How Can You Observe Wave Behavior? ●

Materials
Slinky
book

Procedure

1. Stretch out a Slinky between you and a partner so that there is a space of about 1 cm between each coil. As you perform the following steps, carefully observe the results.
2. Have your partner hold one end steady while you give your end a series of sharp, steady taps.
3. Have your partner hold one end of the Slinky against a book while you give your end one sharp tap.
4. Have your partner hold one end of the Slinky against his or her stomach while you give your end one sharp tap.
5. On the count of three you and your partner should give each end of the Slinky one sharp tap.

Questions

1. Describe the Slinky as a wave passed through it.
2. Was the wave transverse or compressional?
3. What happened when the wave hit the book?
4. What happened when the wave hit your partner's stomach?
5. Why do you think the wave behaved differently when it struck the book than it did when it struck your partner's stomach?
6. What happened when two waves met each other?

1. The coils moved together (compressed), then apart.
2. compressional
3. It was reflected.
4. It seemed to stop. (It was absorbed.)
5. The body is not a hard, smooth surface so it did not reflect the wave.
6. They appeared to pass through each other. Where the waves met, a larger compression took place.

The angle at which a wave hits a surface is called the **angle of incidence** [IHN-sih-duhns]. The angle at which it is reflected is called the **angle of reflection.** The general rule for waves is that the angle of incidence equals the angle of reflection.

You are able to see yourself in a mirror because the mirror reflects light waves. Reflection of light waves also allows you to see yourself in a shiny plate, a smooth, shiny piece of metal, and the glassy surface of a calm pond. The reflection of sound waves causes the echoes that you hear when you call out in a large, empty room.

ACTIVITY: How Can You Observe Wave Behavior?

For this activity students should work in pairs.

You may want to have the students place their Slinky coils on a table top or the floor. Students should be reminded to be careful not to tangle the coils.

● **Laboratory Skill: Manipulation of Materials**

4. Label the first column of the chart *Reflection*. Explain that a wave striking a surface is referred to as an incident wave. What is the wave called when it bounces off the surface (a reflected wave)? Add the information to the first column.
5. Explain that scientists who work with waves have found that a reflected wave bounces off a surface at the same angle that an incident wave hits the surface. In other words, the angle of reflection and the angle of incidence are equal.

6. Label the second column of the chart *Diffraction*. Place the block in the pie plate of water that is on the overhead projector. Ask the students to watch as you create some waves by sharply tapping the side of the dish. Are the waves stopped by the block (no, they go around it)? Explain that diffraction is the bending of a wave as it moves around an object. Add this explanation to your chart.
7. Place a pencil in a glass of water on your desk. Ask the students to draw what they see. Have several students put their diagrams on the board. Stress that the pencil appears to be bent in the glass on your desk and in the diagrams on the board.
8. Ask the students what happens when waves travel through different media (their speed changes). What media are the light waves traveling through in the glass on your desk (air, glass, water)? Ask the students how a change in light wave speed affects what they see (the changing speed causes light waves to bend, thereby causing an object to appear bent).
9. Ask what characteristic of light waves is being illustrated by the pencil that appears to be bent in the glass of water (refraction). Label the third column of the chart *Refraction* and write its definition (bending of waves as they pass from one medium to another).

Figure 18.9 Diffraction

Waves sometimes bend around objects. For example, imagine water waves approaching a pier. When the waves arrive against the posts of the pier, the waves bend and move around them. The posts act as a barrier. The bending of waves around the sides of a barrier is called **diffraction** [dih-FRAK-shuhn].

The speed of a wave changes when the wave medium changes. This change in speed can cause the wave to bend. The bending of waves due to a change in the medium is called **refraction** [rih-FRAK-shuhn]. A rainbow is the result of the refraction of light waves.

A simple way to understand refraction is to compare the motion of a wave to the motion of a car traveling from one surface to another. Suppose a car is driven off a paved parking lot onto a soft, sandy beach. The sand will slow down the motion of the car. If the car approaches the beach at an angle, one front wheel will hit the sand sooner than the other front wheel. For a short time the wheel in the sand will move slower than the wheel still on the pavement. This difference in speed causes the car to make a slight turn. In the same way, a wave moving from one medium to another changes speed. The wave can bend, or refract. The amount of refraction depends on two things. First, it depends on

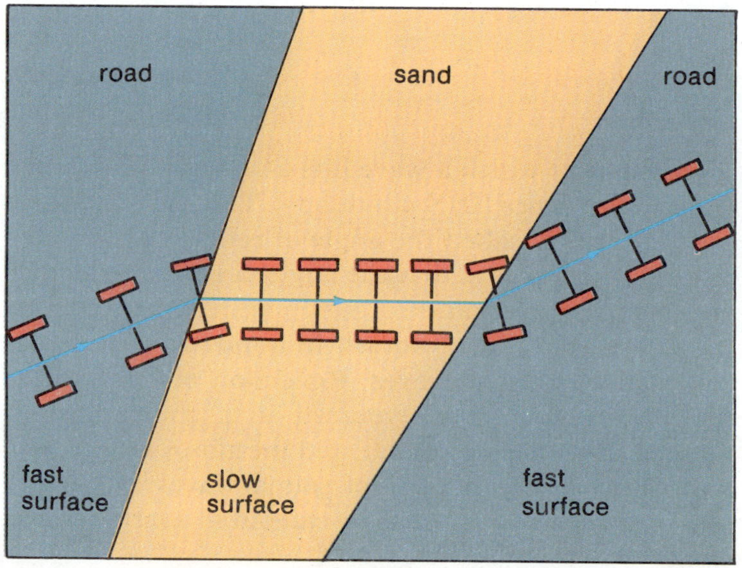

Figure 18.10 Waves bend, or refract, when their medium changes. A car may turn when the surface it is traveling on changes.

how much the speed of the wave changes when it enters the new medium. Second, it depends on the angle of incidence. Waves that hit a new medium almost head-on are refracted less than waves that hit at a large angle. The sharper the angle of incidence, the greater the refraction.

Waves can affect each other. What happens when two or more waves meet? Different things happen depending on the characteristics of the waves. The combined effects of two or more waves coming together is called **interference** [ihn-tuhr-FIHR-uhns]. The word *interference* may suggest to you that waves that meet must cancel or oppose each other. This sometimes happens. However, the energy of waves that meet is in some cases added together. For example, interference occurs when the crests of two waves meet. When this happens, the energy of the two waves is added together.

Figure 18.11 Interference

Study Questions for 18.2

1. What is reflection?
2. What happens when waves encounter a barrier?
3. What is refraction?
4. Give an example of interference.

1. the bouncing back of a wave when it hits a boundary
2. The waves bend around the sides of the barrier.
3. the bending of waves due to a change in the medium
4. When the crests of two waves come together, their energies are added together.

10. Remove the block from the pie plate of water. Then ask the students to observe the water as you sharply tap opposite sides of the pie plate. Do the waves block each other (no, they pass through each other)? Does the new wave look like the original waves (no)?

11. Ask one student to hold one end of a rope and another student to hold the other end of the rope. When you say "Go," each student should start a new wave. Ask the other students to describe what happens when the waves meet (a new pattern appears).

Word Skill: Vocabulary

12. Label the fourth column *Interference*. Explain that the result of two or more waves coming together is called interference. Write this definition under the word *Interference*.

FOLLOW-UP

Reinforcement: Properties of Waves

Have the students pretend they are writing some entries for a science dictionary. Each student should write a definition and draw an illustration for each of the following terms: *reflection*, *refraction*, *diffraction*, and *interference*.

Enrichment: Surfaces and Wave Activity

Ask the students to find out which types of surfaces or materials can cause reflection, refraction, diffraction, and interference. Students should construct a chart in which they list at least two surfaces that will cause each wave activity.

Tie-in: Creating Rainbows of Light

Rainbows are created by separating white light. Many stores now sell crystal teardrops that break light into its spectrum of colors. Ask the students to investigate how the crystals create their effect and have them demonstrate their findings to the class.

18.3

OBJECTIVES: The student should be able to

- **describe** how the frequency of waves can appear to change;
- **explain** what is meant by the Doppler effect;
- **describe** two applications of the Doppler effect.

SCIENCE TERM

Doppler effect

ADVANCED PLANNING

For the demonstration you will need a whistle.

LESSON PLAN

1. Walking across the room as you blow a steady sound on the whistle. Ask the students what happens as you get closer to them (sound becomes apparently higher in pitch). What happens as you get farther away (sound becomes apparently lower in pitch)? Tell the students that they have just observed the Doppler effect.

18.3 The Doppler Effect

You will find out

- how the frequency of waves can appear to change;
- what the Doppler effect is;
- how the Doppler effect can be used.

Imagine you are sitting in a boat tied to a pier. Waves are moving directly toward you. You measure the frequency of those waves by counting the number of crests that pass the front of your boat. Now imagine that your boat starts moving directly into the oncoming waves. You again measure the frequency of the waves just as you did before. Nothing about the waves has changed. But this time you measure a higher frequency. More crests pass the front of your boat in the same amount of time because you are moving toward them. Remember, the actual frequency of the waves never changed. Your measurement of that frequency changed because you were moving. If you had not been moving, the frequency you measured would not have changed.

Figure 18.12 The observed frequency of sound waves from a moving car increases as the car approaches and decreases as it moves away.

What would happen if your boat were moving in the same direction as the waves? If you again measure the frequency, you will find it to be lower than the frequency you measured when your boat was tied to the pier. Fewer crests pass the front of your boat now because you are moving with the waves.

The same changes in measured frequency occur when the observer is stationary and the source of the waves is moving. What do you hear when a fast moving truck approaches you, passes you, and speeds on away from you? The sound of the truck changes in loudness and in quality. The frequency of sound waves in relation to you changes because the source of those waves, the truck, is moving. You can't measure those changes in frequency by counting the crests of the sound waves because you can't see them. But you do observe the changes with your ears. You hear a higher sound as the truck approaches and a lower sound as it moves away. The frequency of the sound waves is greater as the truck approaches than it is as the truck moves away.

The change in the frequency of a wave as measured by an observer due to the motion of the wave source or of the observer is called the **Doppler effect.** The principle behind the Doppler effect can be used to determine whether a wave source is moving away from or toward an observer. If you know the actual frequency of a wave and its measured frequency is different, you can figure out the speed of the wave source.

Many drivers have been caught for speeding by radar equipment that uses the Doppler effect. The radar equipment sends out waves of a certain frequency in the direction of a moving car. When those waves reach the car, they are reflected back by the car. The car becomes a moving source of reflected waves. The change in measured frequency of those reflected waves as the car moves is used to calculate the car's speed.

Figure 18.13 Using the Doppler effect to detect speeding

Study Questions for 18.3

1. What is the Doppler effect?
2. Name a way the Doppler effect can be used.

1. the change in the frequency of a wave as measured by an observer due to the motion of the wave source or of the observer
2. to measure the speed of cars

2. Explain that the Doppler effect is based on the fact that as a noise-making object moves toward you more sound waves will reach your ear, thereby making the sound seem higher. Explain that the Doppler effect can also be observed with light waves.
3. Ask if any of the students can explain why the sound becomes apparently lower as a noise-emitting object moves away from them (fewer sound waves reach the ear in a given period of time because they have farther to travel as the object moves farther away).
4. Conclude the lesson by asking how the Doppler effect can be used (in astronomy to study the motion of stars, which emit light waves; on highways when the speed of cars is determined by using radar).

● Comprehension Skill: Applied

FOLLOW-UP

Reinforcement: The Doppler Effect
Ask students to write an interview with a train engineer in which they ask the engineer questions about the Doppler effect.

Enrichment: Ambulance Warning
Ask students to explain how a moving ambulance siren exhibits the Doppler effect. What other examples of the Doppler effect can they name? Have them prepare collages to illustrate their examples.

Tie-in: Star Movements
Invite an astronomer from a nearby observatory to come to your school and explain how astronomers use the Doppler effect to trace the movement of stars.

What Did You Learn?

18.1 • A wave is a disturbance that travels through matter or space.
18.1 • The matter or space through which a wave travels is called a medium.
18.1 • In transverse waves, the disturbance occurs at right angles to the direction of the wave motion.
18.1 • In compressional waves, the disturbance occurs back and forth along the direction of the waves.
18.1 • The crest is the highest point of a wave.
18.1 • The trough is the lowest point of a wave.
18.1 • Amplitude is the distance from a point halfway between the crest and trough straight up to the crest.
18.1 • The larger the amplitude of a wave, the more energy it has.
18.1 • The wavelength is the distance from one crest to the next.
18.1 • Frequency, measured in hertz, is the number of crests that pass a point in a second.
18.1 • The speed of a wave equals wavelength times frequency.
18.2 • Reflection is the bouncing back of a wave when it hits a boundary.
18.2 • Diffraction is the bending of waves around the sides of a barrier.
18.2 • Refraction is the bending of waves due to a change in the medium.
18.2 • Interference is the combined effects of two or more waves coming together.
18.3 • The Doppler effect is the change in the frequency of a wave as measured by an observer due to the motion of the wave source or of the observer.
18.3 • The Doppler effect is used to measure the speed of cars.

Key Terms

wave
medium
transverse waves
compressional wave
crest
trough
amplitude
wavelength
frequency
hertz
reflection
angle of incidence
angle of reflection
diffraction
refraction
interference
Doppler effect

Career

Physical Oceanographer

Oceanography is the scientific study of the ocean. Different kinds of oceanographers study different aspects of oceans. Physical oceanographers study the motions of ocean waters and the exchange of energy between the land, the atmosphere, and the ocean. Some physical oceanographers specialize in the study of waves. For example, they may forecast the frequency and amplitude of the waves produced by storms and earthquakes. Or they may study the impact and effects of wave energy on shorelines.

Physical oceanographers can study the ocean in different ways. They often go to sea on research ships. In the laboratory, physical oceanographers use wave tanks to study the motion and behavior of waves. Physical oceanographers also use computers to carry out complicated calculations relating to wave motion.

A good educational background in physics is necessary to be a physical oceanographer. Generally, a college degree is required. If you are interested in becoming a physical oceanographer, you should take math and science courses in high school.

Although not all physical oceanographers spend time at sea, many do. For those who do, the adventurous life at sea is one of the main attractions of their career. These physical oceanographers are able to combine an exciting outdoor life with important and challenging work.

Vocabulary

1. transverse
2. Frequency
3. true
4. Doppler effect
5. Interference
6. true
7. true
8. wave
9. refraction
10. compressional

● Word Skill: Vocabulary

What Do You Remember?

1. a disturbance traveling through matter or space
2. hertz
3. the matter or space through which a wave travels
4. transverse and compressional
5. wavelength and frequency
6. Waves are bent due to a change in the medium.
7. The frequency of a wave as measured by an observer changes if the observer or the wave source is moving.
8. It is used to study the motion of stars and to measure the speed of cars and trucks.
9. wavelength
10. the combined effects of two or more waves coming together

TO THINK ABOUT AND DO

Vocabulary ●

Number your paper from 1 to 10. Read each of the following sentences. If the sentence is true, write *true*. If the sentence is false, make it true by changing the underlined expression.

1. An ocean wave is an example of a <u>compressional</u> wave.
2. <u>Wavelength</u> is measured in hertz.
3. Waves with a large <u>amplitude</u> have a large amount of energy.
4. The change in the frequency of a wave as measured by an observer due to the motion of the wave source or the observer is called the <u>frequency shift</u>.
5. <u>Refraction</u> is the combined effects of two or more waves coming together.
6. The bouncing back of a wave when it hits a boundary is called <u>reflection</u>.
7. The highest point of a wave is the <u>crest</u>.
8. A <u>trough</u> is a disturbance that travels through matter or space.
9. The bending of waves due to a change in the medium is called <u>the Doppler effect</u>.
10. In <u>transverse</u> waves the disturbance occurs back and forth along the direction of the waves.

What Do You Remember?

1. What is a wave?
2. What units are used to measure frequency?
3. What is a medium?
4. What are the two groups of waves?
5. What two wave characteristics are related to speed?
6. What causes refraction?
7. What is the Doppler effect?
8. How is the Doppler effect used?
9. What is the distance from crest to crest called?
10. What is interference?

Applying What You Have Learned

1. What wave change is involved in each of the following situations?
   (a) You hear voices of people around the corner.
   (b) A dentist uses a small, round mirror to observe your teeth.
   (c) A person standing on a cliff shouts out a word and hears an echo.
2. Tell whether each of the following produces transverse or compressional waves.
   (a) a stone thrown in a pond
   (b) a ringing telephone
   (c) a glowing light
3. What is the frequency of a wave with a speed of 30 m/s and a wavelength of 0.5 m?

Research and Investigation

1. Earthquakes produce invisible waves that travel through the earth. Do research on earthquake waves.
2. What does absorption of waves mean? How does absorption of ocean waves affect shorelines?

Applying What You Have Learned

1. (a) diffraction
   (b) reflection
   (c) reflection

2. (a) transverse
   (b) compressional
   (c) transverse

3. $\frac{30 \text{ m/s}}{0.5 \text{ m}} = 60 \text{ Hz}$

● Comprehension Skill: Applied

Research and Investigation

1. Earthquake waves are called seismic waves. There are three kinds of seismic waves: P waves, S waves, and L waves.
2. Absorption is the loss of energy from a wave. The wave's energy is changed to another form. Along shorelines, the absorption of ocean waves has caused beach erosion and deposited sediments.

● Reference/Research Skills

CHAPTER BACKGROUND

Sound waves are disturbances that are transmitted by producing pressure changes in the medium through which they travel. The prongs of a tuning fork, for example, move rapidly back and forth when struck. As each prong moves in one direction, the air molecules near it are compressed. When the prong moves back, the air molecules spread apart.

Sound must travel through matter in waves in order to be transmitted. The matter may be in a solid, liquid, or gaseous state. Generally, sound travels faster through solids because their molecules are closer together. The speed of sound waves also depends on the temperature and elasticity of the medium.

If sound waves strike the ear, they can produce the sensation of hearing. The pitch of a sound we hear gets higher as the frequency of the sound wave increases. Loudness of a sound is related to the intensity of sound, measured in decibels. A whisper at 20 dB is barely audible, while a cannon shot at 225 dB is far beyond the limit of endurance of the human ear.

Like all waves, sound waves can be reflected and absorbed. Dolphins, whales, porpoises, and bats use a system called ultrasonics to navigate. Ships use reflected sound called sonar to locate schools of fish, map the ocean floor, and find objects hidden underwater.

When a sound wave strikes some objects, it causes them to vibrate at the same frequency. The two sounds are said to be in resonance. If, for example, a tuning fork is struck and its stem is touched to a piano, the same note can be heard from the piano once the tuning fork has stopped.

Although much has been written to characterize musical sound and noise, the distinction is largely psychological. Pleasant sounds are called music; unpleasant sounds are called noise. Here the term *music* is used to mean a series of regular vibrations, and the term *noise* denotes a series of irregular vibrations. A fine symphony orchestra may, then, be capable of noise as they tune up for a performance and music as they play a concert.

19 Sound

Ask the students the familiar question, "If a tree falls in a forest and there is no one to hear it, will its crashing to the ground produce a sound?" Ask the students to justify their answers. Then strike a tuning fork to produce a sound. Stretch an elastic between your fingers and pluck it to produce a sound. Ask the students why the tuning fork and the elastic produced sound (vibration, movement). Was there anything in the forest to vibrate and produce sound when the tree fell (air, ground, branches)? Tell the students that for sound waves to be produced there must be vibrations and matter for the sound to travel through. In this chapter they will learn about the vibrations that produce sound and the characteristics of sound.

Things that move usually make sounds. Many of those sounds can be heard. Think of things that move. What words are used to describe the sounds they make? Water trickles. Streams gurgle. Brooks babble. Waterfalls roar. You use moving vocal cords to make sounds. People whistle, sneeze, cry, yell, and whisper. Animals make many sounds. Dogs bark and wolves howl. Cats purr. Snakes rattle and hiss. Elephants trumpet and bellow.

Moving parts of machines make sounds as well. A dentist's drill whines. Jet engines rumble and roar. Wheels squeak. Typewriters clack.

This chapter will tell you how these and all sounds are produced. You'll learn about sounds you hear and sounds you don't hear. You will also find out what makes the sounds you hear loud and soft and high and low.

Lesson Organizer

19.1 Sound Waves

Pages 394–395
Advanced Planning: Page 394
Follow-up: Page 395
- **R** Sound Waves
- **E** Does Sound Travel Through a Vacuum?
- **T** Hearing Loss

19.2 The Sounds You Hear

Pages 396–401
Advanced Planning: Page 396
Activity: What Do You Hear? Page 400
Follow-up: Page 401
- **R** Sound Characteristics
- **E** Sterilizing with Ultrasonics
- **T** Locating Schools of Fish

19.3 The Sound of Music

Pages 402–405
Advanced Planning: Page 402
Activity: How Can You Change Pitch? Page 403
Follow-up: Page 405
- **R** Resonance
- **E** A Musical Demonstration
- **T** Electronic Music

19.4 Noise

Pages 406–407
Advanced Planning: Page 406
Follow-up: Page 407
- **R** Silence Please
- **E** How Noisy?
- **T** Designing a Playroom

See ACTIVITIES book pages 109–114 for laboratory, skill, and review exercises for Chapter 19.

19.1

OBJECTIVES: The student should be able to

- **define** the term *vibration;*
- **describe** how sound waves are produced;
- **identify** a sound wave as a compressional wave;
- **explain** why sound cannot travel through a vacuum;
- **explain** why sound travels at different rates through different states of matter;
- **identify** some factors that affect the speed of sound.

SCIENCE TERMS

sound waves
vibration

ADVANCED PLANNING

For the demonstrations you will need a tuning fork, an overhead projector, a glass pie plate filled with water, a ring stand, a glass of ice water, and a glass of room-temperature water.

LESSON PLAN

1. As you strike a tuning fork, ask what causes the sound (vibrations, rapid back-and-forth motion). Since the students will only be able to see a faint blur of the moving prongs, demonstrate that they are actually moving rapidly back and forth by dipping a corner of one prong into the plate of water that has been placed on the overhead projector (the vibrations of the prong should cause the water to spray). Alternatively, demonstrate the motion of the prongs by touching them to a sheet of paper held in your hand (the paper will rattle).

● Comprehension Skill: Applied

SCIENCE BACKGROUND

A compressional wave has two parts, the compression and the rarefaction. The compression occurs where the molecules are more closely packed. The rarefaction occurs where the molecules are more loosely packed.

19.1 Sound Waves

You will find out
- how sound waves are produced;
- why sound waves must travel in matter;
- what affects the speed of sound waves.

Listen. What do you hear? Is a clock ticking? Can you hear traffic? Is someone talking? All the sounds you hear involve **sound waves.** A sound wave is a compressional wave produced by a back-and-forth motion. For example, when a bell rings, the sides of the bell move back and forth rapidly, making sound waves. The rapid back-and-forth motion is called a **vibration.** As the bell vibrates, it pushes together, or compresses, the air around it, making a compressional wave.

In Chapter 18, a compressional wave was compared to a disturbance that travels along a closely packed line of people. Someone in the back of the line gives the next person a push. People are then pushed together as the disturbance moves down the line. Just as the people bumped into each other as a result of a push, the molecules in matter bump into each other as the result of a vibration. In this way a sound wave is produced.

Sound waves move outward through matter from the source of the vibration. They cannot travel in empty

Figure 19.2 A vibrating bell produces sound waves.

space. For sound waves to exist, molecules or atoms must be present to bump into each other.

All sound waves are compressional waves that travel through matter. However, all sound waves do not travel at the same speed. Two important factors affecting the speed of sound waves are the medium through which the sound wave is traveling and the temperature.

It is said that when hunters tracked buffalo, they put their ears to the ground and listened. In this way they could hear a herd of buffalo approaching. Why could a hunter with an ear to the ground hear the buffalo better and sooner than a hunter on a horse? The ground is a solid. The molecules in the ground are closer together than the molecules in air. Sound waves are able to move faster when molecules are closer together and can bump into each other more easily. The hunter on the horse would eventually hear the buffalo. But the hunter with an ear to the ground would hear them first. Sound waves travel faster through solid ground than they do through air. In general, sound waves travel fastest in solids, less fast in liquids, and slowest in gases.

The kind of medium present also affects the speed of sound waves. For example, iron and glass are both solids. But sound waves travel 5,103 m/s in iron and only 4,540 m/s in glass. Both the kind of matter present and the state of that matter affect the speed of sound waves.

Temperature also affects the speed of sound waves. For example, when the air temperature is higher, the molecules in the air move faster. They bump into one another more often. Thus, the speed of sound in air at 20°C is 343 m/s. But at 0°C, the speed is only about 331 m/s. In general, the higher the temperature, the faster sound waves travel.

Table 19-1 The Speed of Sound in Different Mediums

Medium	Speed (m/s)
granite	6,000
steel	5,200
iron	5,103
aluminum	5,000
glass	4,540
oak	3,850
seawater at 25°C	1,531
water at 25°C	1,498
cork	500

Study Questions for 19.1

1. What are sound waves?
2. Why can't sound travel in a vacuum?
3. What two factors affect the speed of sound?

1. compressional waves produced by a back-and-forth motion
2. No molecules or atoms are present in a vacuum to bump into each other and create a sound wave.
3. the matter it travels through and the temperature

2. Explain that sound consists of repeated pushes of molecules against their neighbors. In this way energy is transferred. Can sound travel through a vacuum (no; there would be no molecules to collide)?
3. On the board draw diagrams of a rock, a bowl of water, and a balloon filled with air. How are the molecules arranged in each (rock—orderly, tightly packed; water—less orderly, more loosely packed; air—randomly, far apart)? Through which does sound travel the fastest (solid)? Why (molecules most tightly packed)?
4. Show the students your glass of ice water and your glass of room-temperature water. Remove the ice from the ice water. How do the molecules in the warmer water differ from those in the colder water (they are moving more rapidly)? Which kinds of molecules will transfer sound energy faster (warm water molecules)?

FOLLOW-UP

Reinforcement: Sound Waves
Ask the students to create a series of captioned diagrams that illustrate how sound waves travel through matter.

Enrichment: Does Sound Travel Through a Vacuum?
Have students do the following demonstration to determine whether sound can travel through a vacuum. An electric bell should be placed inside a bell jar and connected to a dry cell. The air from the jar should be evacuated, and students should listen for their "answer" (bell will not be heard in this nearly complete vacuum).

Tie-in: Hearing Loss
Contact a hearing aid company in your area. Ask a representative from the company to speak to the students about different types of hearing losses and why they occur.

SCIENCE BACKGROUND

Sound travels at a speed of 331 m/s (1,090 ft/s) at 0°C in air. For every 1°C increase in air temperature, sound travels 60 cm (2 ft) faster per second.

19.2

OBJECTIVES: The student should be able to

- **explain** how the intensity of a sound is determined;
- **identify** the term *decibels*;
- **explain** the relationship between frequency and vibration;
- **explain** what is meant by pitch;
- **state** one factor that affects pitch;
- **identify** ultrasonic waves;
- **describe** how sound waves can be used to locate objects.

SCIENCE TERMS

intensity	pitch
decibels	ultrasonic

ADVANCED PLANNING

For the demonstrations you will need a glass of water, a spoon, a decibel meter, and tuning forks of at least three different frequencies.

LESSON PLAN

1. Ask the students to listen to the difference between the sounds you produce when you gently strike a glass of water with a spoon and when you strike it more sharply (a sharp tap produces a louder sound). Explain that the increased loudness is due to the increased intensity of the sound.

● **Word Skill: Vocabulary**

2. Show the students your decibel meter. Tell them that it is used to measure the intensity of sounds in units called decibels. Allow the students to measure and compare different classroom sounds.
3. On the board, draw two waves with different amplitudes. Point out to the students that these waves differ in height, or amplitude. Explain that the wave with greater amplitude has more energy. What effect will greater energy have on the intensity of the sound (greater intensity)?

19.2 The Sounds You Hear

You will find out

- what determines the loudness of a sound;
- what factors affect pitch;
- what ultrasonic sounds are;
- how sound waves can be used to locate objects.

You can make a loud noise by shouting. Or you can make a soft noise by whispering. Both sounds are produced by the same kind of vibration. Why is one sound louder than the other?

When you shout, your vocal cords vibrate with more energy than they do when you whisper. A vibration with more energy produces sound waves with more energy. The more energy a sound wave has, the greater its amplitude. The loudness, or **intensity** [ihn-TEHN-suh-tee], of a sound depends on the amplitude of the sound waves involved. The sound waves of a shout have greater amplitude than the sound waves of a whisper.

Intensity is measured in **decibels** [DEHS-uh-behlz]. ● The decibel (dB) is named after Alexander Graham Bell, the inventor of the telephone. You can barely hear sounds of less than 1 dB. Sounds of over 120 dB are so loud they can be painful.

Do You Know?
Mosquitoes, grasshoppers, crickets, butterflies, and moths are among the few kinds of insects that have the ability to hear.

Table 19-2 The Intensity of Some Familiar Sounds

Sound	Intensity (dB)
jet plane taking off	150
chain saw	115
power mower	100
noisy restaurant	80
vacuum cleaner	75
average home	50
purring cat	25
rustling leaves	20
whisper	15
faintest sound that can be heard	0

Figure 19.3 In which of the situations pictured would you be likely to speak in a voice of low intensity? library

The intensity of any sound decreases as you move away from the source of that sound. For example, a thunderclap right over your head will sound louder to you than it will to a person 2 km away.

An important characteristic of sound waves is frequency. The frequency of a sound wave depends on how rapidly the source of the sound is vibrating. The faster the vibration, the higher the frequency.

Frequency is closely related to **pitch.** Pitch describes the highness or lowness of a sound that the ear hears. High-frequency sound waves produce high-pitched sounds. Low-frequency sound waves produce low-pitched sounds. Frequency is measured in hertz. Pitch cannot be measured. It can only be heard.

4. Review the meaning of frequency that the students learned in Chapter 18. Ask the students to listen carefully while you strike each of your tuning forks. Are the sounds the same (no)? Explain that the sounds aren't the same because each tuning fork is designed to vibrate at a different frequency.
5. Ask the students to listen again while you tap first the tuning fork with the highest frequency and then the tuning fork with the lowest frequency. What difference do the students hear (high frequency = high sound; low frequency = low sound)?
6. Tell the students that the highness or lowness of a sound is called the pitch. Stress that the pitch of a sound depends on the frequency, or number of vibrations per second. The greater the number of vibrations per second the higher the pitch.

SCIENCE BACKGROUND

When sound waves of differing frequencies alternately cancel and reinforce each other, an effect called "beats" is produced.

7. Ask the students to imagine a guitar. They have just plucked a string very gently. How can they increase the intensity of the sound (pluck harder)? Why (more energy present)? Tell the students that when the intensity of a sound increases, it may seem to have a higher pitch. However, the frequency has not changed.

8. Ask the students to name some high-pitched sounds (chalk on a board, hinge that needs oiling, soprano voice) and some low-pitched sounds (kettle drums, bass).

Figure 19.4 A bird's song is high-pitched.

Frequency is not the only characteristic of sound waves that affects pitch. Intensity can also affect pitch. A sound of high intensity may seem to have a slightly higher pitch. The frequency stays the same whether the intensity changes or not.

You use pitch to tell the difference between one sound and another. For example, one way to recognize the difference between a siren and a foghorn is by the difference in pitch. The siren has a high pitch while the foghorn has a low pitch.

Figure 19.5 A bullfrog's croak is low-pitched.

If you or someone you know has a dog, you may have noticed that dogs can hear sounds that you cannot. A pet dog may show signs that it hears a person or car approaching long before any person hears a thing. The human ear can hear sound waves of frequencies from about 20 to 20,000 Hz. Dogs and cats can hear frequencies as low as 15 Hz. They can also hear frequencies of over 50,000 Hz. Porpoises and bats can hear even higher frequencies, over 100,000 Hz.

People and other animals not only hear sounds of different frequencies, but they also produce sounds of different frequencies. It might surprise you to learn that you and other animals can hear more frequencies than you can produce. For example, you can hear frequencies of 20 to 20,000 Hz. But your vocal cords can only produce frequencies of 85 to 1,100 Hz.

Do You Know?
Sound waves with frequencies as high as several billion hertz have been produced.

Figure 19.6 The frequencies of sounds heard and produced by people and some other animals

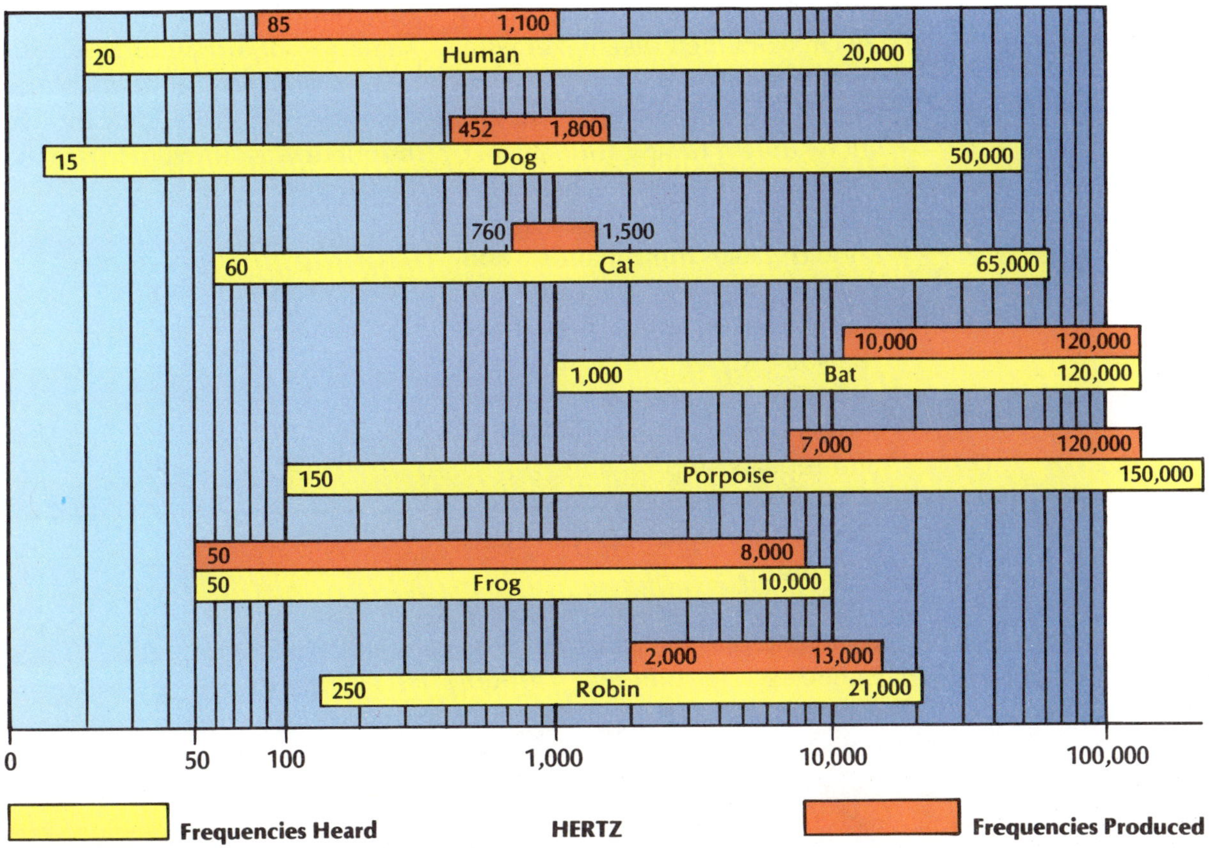

9. Ask the students why a dog sometimes pricks its ears even though there appears to be no disturbance (dog hears a sound people do not hear). Explain that different animals hear sounds of different frequencies. Dogs and cats hear a wider range of frequencies than people do.

10. Ask each student to use the information in Figure 19.6 to prepare two charts. One chart should compare the frequency range of sounds produced by each animal. The other chart should compare the frequency range of sounds heard by each animal. Do the animals hear and produce sounds in the same range (no)?

SCIENCE BACKGROUND

Motion that travels faster than sound is referred to as *supersonic*, a word that means above (super) sound (sonic). Objects that travel at supersonic speeds produce such strong pressure waves that the result is a loud, explosive sound called a sonic boom.

11. Write the word *ultrasonics* on the board. Ask the students what *ultra* means (above, highest, best). Point out that ultrasonic waves have frequencies of more than 20,000 Hz.

ACTIVITY: What Do You Hear?

Tell the students that a tuning fork produces a single tone of a specific frequency. Try to have available sets of tuning forks of at least three different frequencies.

Stress that this activity is not a hearing test. Students should not be concerned if they find that others can hear the tuning forks at greater distances than they can. You may wish to have the school nurse discuss hearing tests with the class and demonstrate the school's hearing test equipment.

The groups should be widely separated or take turns so that the sounds of one group do not confuse the listeners in other groups.

● **Laboratory Skills: Manipulation of Equipment Experimentation/Demonstration**

SCIENCE BACKGROUND

Scientists have found medical applications for ultrasonic waves. They can be used to explore and produce pictures of the body. They can also be used to heal injuries by massaging inner body tissues.

Sound waves above 20,000 Hz are called **ultrasonic.** Ultrasonic waves have a very short wavelength. Because of their short wavelength, ultrasonic waves are reflected by small objects. Bats are able to produce and hear ultrasonic sounds. While flying, bats produce ultrasonic squeaks with wavelengths of only a few millimeters. These squeaks are reflected by tree limbs, insects, and other things. Bats use the reflected sounds to find their way in total darkness.

● ACTIVITY

What Do You Hear?

Materials

tuning forks of different frequencies
meterstick

Procedure

1. Work in groups of five. One person will be the noisemaker and recorder. The other four people will be listeners.

2. The listeners should stand 30 cm from the noisemaker with their eyes closed.

3. The noisemaker should then strike a tuning fork against the palm of his or her hand. All listeners who hear the sound should raise their hands. The noisemaker then records the number of listeners who hear the sound.

4. The noisemaker then moves back 1 m from the group and repeats step 3.

5. Keep repeating step 4 until none of the listeners can hear the sound.

6. Repeat the activity using tuning forks of different frequencies. A different group member should act as the noisemaker for each tuning fork used.

Questions

1. Why were the listeners told to close their eyes?

2. How far away was the noisemaker when no one heard the sound? Was the distance the same for all tuning forks used?

3. Were the listeners able to hear the high-frequency or low-frequency sounds at greater distances?

1. so they could not see the tuning fork being struck
2. and 3. Answers will vary.

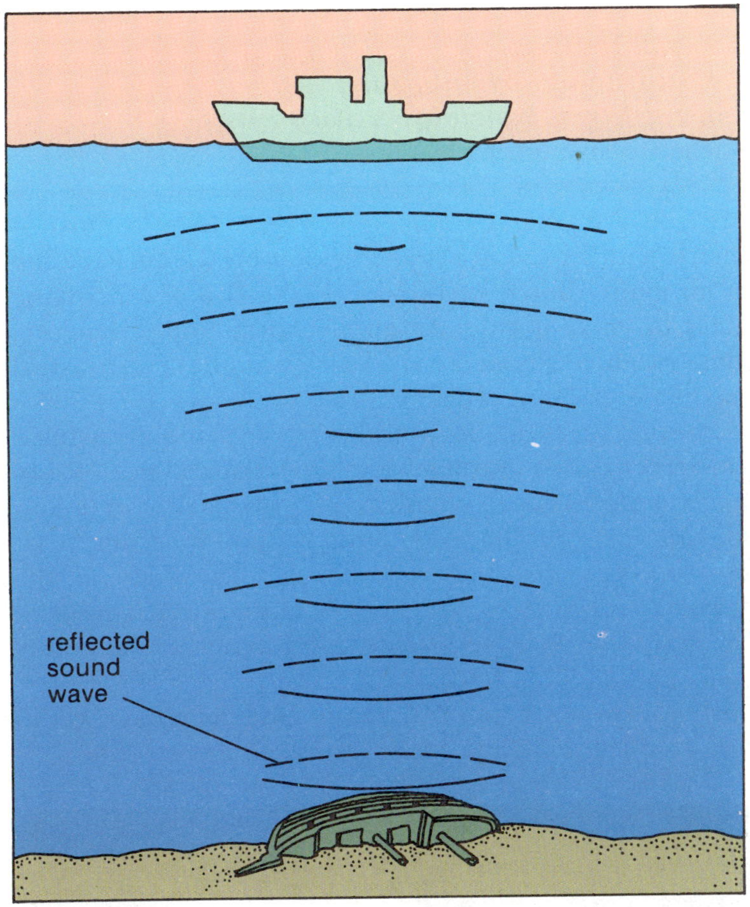

Figure 19.7 Sound waves sent down from a ship are reflected back to the ship by an object underwater. The use of reflected sound waves to gather information is called **sonar**—***so***und ***n***avigation ***a***nd ***r***anging.

All sound waves, not only ultrasonic ones, can be reflected. You produce sound waves that are reflected when you shout into a canyon and hear an echo. Research ships send sound waves deep underwater and study the waves that are reflected back up. Such studies are done to locate objects underwater and to measure the depth of the ocean.

Do You Know?
Experiments with bats have been conducted in totally dark rooms having many piano strings stretched across them. Using reflected sound waves, the bats were able to detect the piano strings and fly around them.

Study Questions for 19.2

1. What affects the loudness of a sound?
2. What factors affect pitch?
3. What are ultrasonic waves?
4. How do bats find their way in darkness?

1. the amplitude of the sound waves involved
2. frequency and intensity
3. sound waves with frequencies above 20,000 hertz
4. They produce ultrasonic squeaks. The reflections of these ultrasonic sounds indicate the location of objects.

12. Explain that animals use ultrasonic waves to survive. By studying the animals' use of ultrasonic waves, scientists have been able to use them to benefit people. Ask the students to list some ways that ultrasonic waves can be used (to kill bacteria, to locate objects, to detect flaws inside metals).

FOLLOW-UP

Reinforcement: Sound Characteristics
Ask the students to write an interview with Alexander Graham Bell. During the course of the interview, students should ask Bell questions that lead him to explain the meaning of the terms *intensity*, *frequency*, and *pitch*.

Enrichment: Sterilizing with Ultrasonics
Ultrasonics has been used to clean dishes and to sterilize medical instruments. Have students do library research to find out how ultrasonics is used to do these things. Are there other practical uses of ultrasonics (homogenize milk, for example)? Ask the students to report their findings orally.

Tie-in: Locating Schools of Fish
Ask the students to write letters to large coastal fishing companies asking them how they use ultrasonics to locate schools of fish, what equipment is needed and how it is used, and what other benefits are derived from this technology.

19.3

OBJECTIVES: The student should be able to

- **define** *resonance*;
- **explain** the relationship between the length of a vibrating string and its pitch;
- **explain** how different kinds of musical instruments produce musical notes;
- **explain** what is meant by overtones.

SCIENCE TERMS

resonance
overtones

ADVANCED PLANNING

For the demonstration, obtain a glass cylinder filled with water, a hollow glass tube, a tuning fork, and a string, wind, percussion, or electronic instrument.

LESSON PLAN

1. To demonstrate resonance to the students, hold an open glass tube in a glass cylinder three-quarters full of water. Hold a vibrating tuning fork over the tube and slowly raise and lower the tube until the intensity of sound is at a maximum. Tell the students that at this point the air column is resonating. Why (the frequency of the vibrating air column exactly matches the frequency of the vibrating tuning fork)?
2. Explain to the students that the air in their mouth, larynx, and nose resonates, thereby increasing the intensity of the sounds the vocal cords produce. Their voices change when they have a cold because the resonating air cavities change.

19.3 The Sound of Music

You will find out

- what is meant by resonance;
- how different instruments make music;
- what is meant by overtones.

Are you a musician? Have you ever played a musical instrument? You have if you've ever blown over the top of an empty bottle or tapped on the rim of a drinking glass to make it ring. An empty bottle and a drinking glass produce sounds in the same way that the musical instruments of an orchestra do.

Musical instruments produce musical sounds using a property called **resonance** [REHZ-uh-nuhns]. Resonance amplifies the sound that a musical instrument makes. For example, if you pluck a guitar string, it vibrates. The vibrating string causes the body of the guitar to vibrate, or resonate. This vibration amplifies the sound. If you were to stuff the body of the guitar with cloth, it would not vibrate. The guitar string would still vibrate when you plucked it. But the guitar would not produce the same sound. It would not resonate.

Figure 19.8 Piano strings of different lengths produce notes of different frequencies.

Musical instruments can produce notes of different pitch. For example, you can play different notes on a guitar by pressing your finger down at different places on a string and plucking it. When you press your finger on the string, you are shortening the part that will vibrate. A change in a string's vibrating length changes the frequency of the sound wave produced. The change in frequency produces a change in pitch. The shorter the vibrating part of the string, the higher the pitch of the note produced.

ACTIVITY

How Can You Change Pitch? ●

Materials

shoe box
scissors
three rubber bands of different thicknesses

Procedure

1. Cut three pairs of slits in each end of the shoe box. Each slit should be 1 cm long. The two slits in each pair should be 1 cm apart. Stretch the rubber bands over the length of the open side of the shoe box and fasten each to a pair of slits.

2. Pluck each rubber band and listen to the pitch.

3. Press the rubber band of medium thickness in the center to make a shorter vibrating part. Now pluck it again. Listen for a change in pitch.

4. Cut a pair of slits opposite each other across the width of the shoe box. Place your medium-thickness rubber band in these slits. Again, pluck the rubber band and listen to the pitch.

1. thinnest rubber band
2. thickest rubber band
3. The pitch gets higher.
4. The pitch gets lower.

Questions

1. Which of the three rubber bands has the highest pitch?

2. Which of the three rubber bands has the lowest pitch?

3. How does the pitch change when the vibrating part of the rubber band is shortened?

4. How does the pitch change when the rubber band is not as tightly stretched?

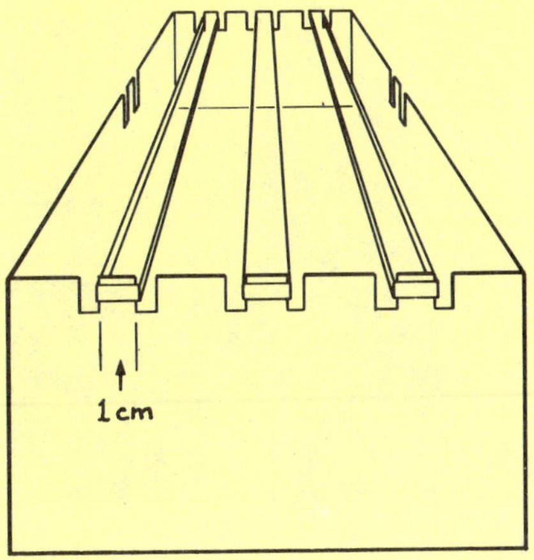

3. Ask the students how different sounds are produced in a string instrument (fingers are used to shorten strings, thereby increasing pitch). Explain that the length of the vibrating string determines the frequency and therefore the pitch of the sound.

ACTIVITY: How Can You Change Pitch?

Try to use rubber bands with very different thicknesses. For accurate results, keep have students try to keep all the rubber bands under the same tension.

Remind the students that since they must listen carefully in this activity, they should be as quiet as possible.

If you have students who play string instruments, ask them to demonstrate at the end of the experiment so that other students can check their results.

● **Laboratory Skills: Manipulation of Equipment
Experimentation/Demonstration**

SCIENCE BACKGROUND

Musical notes are often called tones.

4. Ask the students to name some wind instruments (clarinet, trumpet, saxophone). Point out that in a wind instrument an air column vibrates. How does a trumpet player change the length of the air column to produce different sounds (opening and closing valves)?
5. Use the drum as an example of a percussion instrument. How does a drum produce sound (the taut skin is hit, which causes it to vibrate)?
6. Ask the students to name some examples of electronic instruments (electric guitars and electric organs). How is their sound amplified (electronic devices)?

Figure 19.9 Different notes are produced on a clarinet by opening and closing holes in the instrument.

Figure 19.10 The valves in a French horn are used to change the frequency of notes produced.

In wind instruments, the body of the instrument contains a column of air which resonates. The simplest kind of wind instrument is an empty bottle. If you blow across the top of a bottle, you can cause the air in the bottle to vibrate. The length of the column of air in the bottle determines the frequency of the note you make. Short air columns produce high notes. Tall ones produce low notes. In instruments such as flutes or clarinets, musicians change the length of the column of air by opening and closing holes in the sides of the instruments. In this way the musicians produce different notes.

Brass wind instruments, such as trumpets and tubas, have tubes of different lengths through which the sound resonates. The musicians shift from one tube to another by opening and closing valves. In this way notes of different frequency are produced. With a trombone, it is possible to shift gradually from one frequency to another by sliding the tube in and out. Notes are made on drums, cymbals, and xylophones by hitting a surface and causing vibrations.

Most musical instruments produce more than simple notes of only one frequency. These instruments also produce notes of frequencies that are multiple frequencies of simple notes. Such notes are called **overtones.** Different musical instruments vibrate and resonate differently. They also produce different overtones. This is why the same note played on a piano and a flute sound different.

Figure 19.11 Notes are made on drums and cymbals by striking them.

Study Questions for 19.3

1. What is resonance?
2. How are sounds of different frequencies produced in a guitar, a clarinet, and a trumpet?
3. What are overtones?

1. the property of a musical instrument that amplifies the sound it makes
2. guitar—by changing the length of the vibrating part of the string; clarinet—by changing the length of a column of air; trumpet—by shifting to tubes of different lengths
3. notes with frequencies that are multiple frequencies of simple notes

7. Explain that most instruments produce sounds of more than one frequency. The frequencies are multiples of the simple frequencies. Ask what they are called (overtones).

8. Conclude the lesson by asking each student to create a musical instrument from milk cartons and rubber bands; glasses, each with a different amount of water inside; reeds to be blown; or various kinds of metal pieces to be struck. Students should be prepared to demonstrate the ability of their instruments to produce a tone(s), identify the group to which the instrument belongs, and explain how the instrument produces sound.

FOLLOW-UP

Reinforcement: Resonance

Have the students demonstrate the concept of resonance. Ask one student to strike a tuning fork and another student to stand about 30 cm (1 ft) away with a tuning fork of the same frequency. Ask the students to note what happens (tuning fork that has not been struck begins to vibrate and sound is produced). Do the same demonstration with tuning forks of different frequencies and have students observe (second fork does not begin to vibrate).

Enrichment: A Musical Demonstration

Have the members of the school band provide students with a demonstration showing how a wide variety of sounds can be produced with musical instruments.

Tie-in: Electronic Music

Have the students review Chapter 17. Ask them to find out how the principles of electronics are used to produce music. Then have an interested group of students present a panel discussion of their findings.

19.4

OBJECTIVES: The student should be able to

- **explain** what is meant by noise pollution;
- **describe** what happens when sound waves are reflected;
- **define** the term *reverberations;*
- **explain** how echoes can be prevented.

SCIENCE TERMS

noise pollution
reverberations
echo

ADVANCED PLANNING

For the demonstration obtain a tape or record of some popular music and several metal pie plates.

LESSON PLAN

1. Play a tape or record of pleasant sounding music for 30 seconds; then strike several tin pie plates for the same amount of time. Ask the students how the sounds varied (tape—pleasant, pie plates—unpleasant). Tell the students that unpleasant sounds are referred to as noise. An excess of unpleasant sounds is called noise pollution.
2. Ask a student to come to the board and draw a diagram of an empty gymnasium. The diagram should explain why sounds echo in the room. To do this, the student should show in the diagram the reflection of sound waves from many surfaces. Tell the students that the mixture of reflected sound waves causes reverberations. Have a second student come to the board and add things that would reduce the reverberations to the diagram. Discuss the characteristics of these additions with the students.

19.4 Noise

You will find out
- what is meant by noise pollution;
- what can happen when sound waves are reflected;
- how echoes can be prevented.

Figure 19.12 Using ear protection

People hear and appreciate sounds differently. The same sounds that are beautiful music to one person may be just unpleasant noise to someone else. There are, however, many sounds that are unpleasant to most people. High-pitched sounds, loud sounds, and disturbing echoes in a room are generally considered unpleasant. An excess of unpleasant sounds is called **noise pollution.**

Recent studies have found that noise pollution can do more than just keep people awake or bother them. It can actually make them sick. Constant exposure to sound levels over 100 dB can cause permanent hearing losses. These sound levels can also cause headaches, an upset stomach, and high blood pressure. To avoid these problems, workers in factories wear ear protection. So do workers who use noisy machines, such as chain saws and jack-hammers, and airport employees who work near loud jet engines.

Have you ever heard a speaker in a large ball park or stadium? If you listened and watched carefully, you probably noticed that the speaker's voice seemed to continue even after the speaker stopped. You were hearing **reverberations** [rih-vur-buh-RAY-shuhnz]. Reverberations are mixtures of reflected sounds. They can occur in any stadium, auditorium, or large hall.

A reflected sound that is heard more than 1/20 of a second after the original sound is called an **echo.** Echoes can be fun. When you yell into a canyon, you can hear yourself two, three, or maybe four times as the sound is reflected back to you. However, echoes in an auditorium can cause a speech or concert to sound jumbled and unpleasant.

Most large halls and auditoriums are built and decorated to prevent echoes. Materials are used that absorb sounds instead of reflecting them. Fabrics and special wall and ceiling tiles absorb sounds. Cloth-covered

Figure 19.13 Which of these settings would reduce echoes and reverberations?

the hall with carpeting and fabric-covered chairs

chairs, drapes, and curtains are used to reduce echoes as well as to provide comfort and decoration.

A good concert hall is designed not only to reduce echoes but to allow the music to reach everyone in the audience. This is done by a design that allows the musical sounds to reverberate in a certain way. Concert halls are designed for a specific use. For example, opera houses are designed differently than halls for orchestra concerts.

Proper design in factories and other workplaces can reduce noise pollution. The same principles that apply to the design of a quiet workplace also apply to the design of concert halls. However, in the workplace, the main goal is to absorb sound. No echoes or reverberations of any kind are desired.

Study Questions for 19.4

1. What is noise pollution?
2. Name two kinds of reflected sound waves you can hear.
3. How can echoes and reverberations in a hall be prevented?

1. an excess of unpleasant sounds
2. reverberations and echoes
3. by using wall and ceiling tiles that absorb sound, by using cloth-covered chairs, by hanging drapes and curtains

3. Take the students to the auditorium. Ask them to imagine an auditorium with flat cement walls, a wooden floor, and straight-backed wooden chairs. How has their auditorium been designed to prevent the excess noise and echoing that would occur in the imaginary auditorium (soft padded chairs, curtains, carpets, chairs attached to floor, layered ceiling)?
4. Explain that concert halls are designed a bit differently from auditoriums. They are designed so that the reverberations move in certain paths. This allows the music to reach everyone in the concert hall.

● **Comprehension Skill: Literal**

FOLLOW-UP

Reinforcement: Silence Please
Ask each student to prepare a list of 10 sources of noise pollution and suggestions of how the noise pollution from each source can be decreased.

Enrichment: How Noisy?
Have students use the decibel meter to measure the sounds in different areas of the school and in their neighborhoods. What areas are too noisy? How could the sound in these places be decreased?

Tie-in: Designing a Playroom
Ask the students to create shoe box models of a playroom that is acoustically designed to absorb excess noise.

SCIENCE BACKGROUND

Acoustics is the scientific study of sound. Its major areas of concern are the generation, propagation, and perception of sound.

Acoustical materials are generally porous and inelastic. They absorb sounds but do not transmit them readily.

What Did You Learn?

19.1 • A sound wave is a compressional wave produced by a back-and-forth motion.
19.1 • Matter must be present for sound waves to be produced and travel.
19.1 • The matter present and the temperature affect the speed of sound waves.
19.2 • The loudness, or intensity, of a sound depends on the amplitude of the sound waves involved.
19.2 • Intensity is measured in decibels.
19.2 Pitch describes the highness or lowness of a sound that the ear hears.
19.2 • Frequency and intensity affect pitch.
19.2 • Sound waves above 20,000 hertz are called ultrasonic.
19.2 • Reflected sound waves can be used to locate objects.
19.3 • Musical instruments amplify sounds by resonating.
19.3 • Most musical instruments produce overtones.
19.4 • An excess of unpleasant sounds is called noise pollution.
19.4 • Reverberations are mixtures of reflected sounds.
19.4 • An echo is a reflected sound that is heard more than 1/20 of a second after the original sound was made.
19.4 • Concert halls are designed and furnished to prevent echoes and unwanted reverberations.

Key Terms

sound waves
vibration
intensity
decibels
pitch
ultrasonic
resonance
overtones
noise pollution
reverberations
echo

Career

Industrial Hygienist

Industrial hygienists are involved with health and safety concerns in the workplace. They do research on work-related health problems and inspect the workplace to see that conditions there promote good health and safety.

Some workers have become ill as a result of the jobs they do. Some jobs expose workers to high noise levels and other factors affecting health, such as dust, chemicals, and high heat. These jobs must be modified to protect the workers involved. Industrial hygienists study the ill effects of such jobs and help to develop ways to eliminate those effects.

In 1970 the federal government set health and safety standards for the workplace in the Occupational Safety and Health Act (OSHA). Studies and reports from industrial hygienists helped to determine those standards. One of the standards requires that the noise level of a workplace be no higher than 85 decibels. Noise pollution is not the only concern of OSHA. The quality of the air that workers must breathe and the dangers of equipment that workers must use are also concerns.

Some industrial hygienists work for private companies. Others are employed by governmental agencies. In many cases an industrial hygienist is required to have a college degree in engineering or in physical or biological science. As workers and employers become more aware of health and safety factors on the job, more industrial hygienists will be needed to inspect and evaluate conditions in the workplace.

Vocabulary

1. d
2. c
3. h
4. f
5. l
6. g
7. k
8. j
9. a
10. b

● Word Skill: Vocabulary

What Do You Remember?

1. the matter the sound waves travel through, the temperature
2. Molecules or atoms must be present to bump into each other.
3. The guitar can't resonate.
4. You change the length of the vibrating part of the string and produce a note of a different frequency and pitch.
5. They can suffer headaches, upset stomachs, high blood pressure, and hearing losses.
6. Cloth-covered chairs, drapes, curtains, and special wall and ceiling tiles are used.
7. intensity
8. They are guided by reflected ultrasonic waves which allow them to locate objects.
9. a high-pitched sound
10. frequency and intensity

TO THINK ABOUT AND DO

Vocabulary ●

Number a piece of paper from 1 through 10. Beside the appropriate number write the letter for the term that is described.

1. unit used to measure intensity
2. quality that describes the highness or lowness of a sound
3. property of a musical instrument that amplifies its sound
4. a reflected sound that is heard more than 1/20 of a second after the original sound was made
5. notes produced by most musical instruments
6. a mixture of reflected sounds
7. a compressional wave produced by a back-and-forth motion
8. an excess of unpleasant sounds
9. a rapid back-and-forth motion
10. loudness

a. vibration
b. intensity
c. pitch
d. decibel
e. frequency
f. echo
g. reverberation
h. resonance
i. ultrasonic
j. noise pollution
k. sound wave
l. overtones

What Do You Remember?

1. What two factors affect the speed of sound waves?
2. Why can't sound waves travel in empty space?
3. Why does a guitar sound different if its body is stuffed with cloth?
4. What happens when you press a finger on a guitar string and pluck it?
5. What can happen to people who are exposed to noise pollution?
6. How are auditoriums built to reduce echoes?
7. What quality of sound is affected by the amplitude of the sound waves involved?
8. How do bats find their way in darkness?
9. What kind of a sound is produced by a high-frequency sound wave?
10. What two factors affect pitch?

Applying What You Have Learned

1. Name two sounds you might encounter daily that have a low decibel rating. Name two that have a high decibel rating.
2. How could sound waves be used to locate a sunken ship?
3. How would you furnish a large room to prevent echoes?
4. How is a musical instrument such as a clarinet different from a bottle you blow on?
5. The word *ultrasonic* means "above sound." Why do you think the frequency of 20,000 Hz was chosen to mark the beginning of ultrasonic frequencies?

Research and Investigation

1. How can sound be recorded?
2. In 1975, scientists used a special sonar camera unit to try to photograph Nessie, the Loch Ness Monster. In 1981, the Goodyear blimp *Europa* flew over the waters of Loch Ness in search of Nessie. The blimp located an object moving underwater at a depth of 14 m to 18 m. *Europa* used sonar to track the object. Do research on the Loch Ness Monster and how sonar has been used to try to find Nessie.

Applying What You Have Learned

1. Check answers against the chart of decibel ratings in the chapter.
2. A ship would send sound waves underwater. Studies of the reflection of those sound waves would indicate the position of a sunken ship.
3. Use carpeting, cloth drapes, cloth-covered furniture, and sound absorbent ceiling tiles.
4. The clarinet can produce overtones and different pitched notes. It also resonates more.
5. Humans cannot hear sounds of over 20,000 Hz.

● Comprehension Skill: Applied

Research and Investigation

1. Sound can be recorded on discs or cylinders made of plastic, on film, or on magnetic tape.
2. The sonar camera unit used a strobe light and sound waves. Sound waves were sent underwater. When the waves encountered a large object, the strobe light was triggered to give a quick, bright flash. The flash exposed the film in the camera unit. Using reflected sound waves, the blimp located and tracked an object for about half an hour. However, a barge working with the blimp was unable to get closer than 66 m to the object.

● Reference/Research Skills

CHAPTER BACKGROUND

Light is a form of energy called radiant energy that travels in straight lines. As the rays get farther from the source, they spread farther apart and their intensity decreases.

Although the exact nature of light remains a mystery, several theories have been proposed to explain it. Sir Isaac Newton proposed that light consists of particles, or corpuscles, that move out from their source in all directions. When they hit a surface, they bounce off it. Newton also suggested that light travels at different speeds through different materials.

Christian Huygens, a contemporary of Newton's, proposed a different theory about light. Huygens believed that light is a series of waves. Like Newton, he believed that light spreads out in all directions from its source. Huygens also incorrectly believed that light waves are not affected by gravity, and he correctly stated that light moves through air faster than it moves through water.

Thomas Young, an English linguist and physician as well as an expert in many fields of science, helped to prove Huygens' idea by finding that light shows the phenomenon of interference. Interference patterns, or the overlapping of two sets of waves at one instant, could not easily be explained by the particle theory of light.

Huygens' theory was weakened by his belief that light, like sound, needs a medium of transmission. Near the end of the nineteenth century, James Maxwell disproved Huygens' theory. He showed that light has electrical and magnetic properties that allow it to travel through a vacuum. Later experiments showed that visible light is only one type of electromagnetic wave. Today it is known that there is a spectrum of electromagnetic waves that includes gamma rays, X rays, ultraviolet rays, infrared rays, microwaves, and radio waves.

In the twentieth century Max Planck combined the ideas of Newton and Huygens in a further attempt to explain the nature of light. Later confirmed by Albert Einstein, Planck's explanation suggested that electromagnetic radiation is quantized as discrete packets of energy called photons. Planck proposed that in some situations light acts like waves and in others like particles. His quantum theory describes light as electromagnetic waves made of bundles of energy.

Light 20

Introduce the lesson by darkening the room and flashing the Morse code SOS signal (••• - - - •••) with a flashlight. Use long flashes for dashes and short flashes for dots. Tell the students that you are using light to communicate. Until World War II, naval ships often used light and the Morse code to communicate. Point out that light has many important properties that make it useful to people. In this chapter the students will learn about light and its uses. (The activity—How Do Shortwave Radio Operators Communicate?—in Chapter 17 includes the Morse code alphabet. You might want to have students make up and send messages to each other using flashlights and the Morse code.)

Lights! Camera! Action! In order to film a scene, you must first see it. To see it, you must have light. There are many sources of light. Of course, there is sunlight and starlight. Fireflies give off light. Volcanic eruptions can light the sky.

People have devised many methods to produce light. They have burned wood, whale oil, candle wax, kerosene, and natural gas. They have used electricity to produce light in flashlights, table lamps, spotlights, automobile headlights, traffic lights, decorating lights, neon lights, and lighthouses.

All the light you see, no matter what the source, is alike in many ways. In this chapter you will find out about the light waves you can see and about other, invisible waves that also affect your life.

Lesson Organizer

20.1 The Electromagnetic Spectrum
Pages 414–419
Follow-up: Page 419
- **R** Light Tic-Tac-Toe
- **E** Discovery of X Rays
- **T** Microwaves

20.2 Visible Light
Pages 420–423
Advanced Planning: Page 420
Activity: How Does Color Affect Heat Absorption? Page 422
Follow-up: Page 423
- **R** Color and Wavelength
- **E** How the Eye Sees Color
- **T** Color Printing

20.3 Changing Visible Light Waves
Pages 424–427
Advanced Planning: Page 424
Activity: How Can You Produce a Mirage? Page 426
Follow-up: Page 427
- **R** Bending Light
- **E** The Disappearing Stamp
- **T** The Best Way to Fish

20.4 Diffraction and Polarization
Pages 428–429
Advanced Planning: Page 428
Follow-up: Page 429
- **R** Polarization and Diffraction
- **E** Polarizing Light
- **T** Seeing Out/But Never In

See ACTIVITIES book pages 115–120 for laboratory, skill, and review exercises for Chapter 20.

20.1

OBJECTIVES: The student should be able to

- **explain** what is meant by the term *electromagnetic waves;*
- **compare** electromagnetic waves and sound waves;
- **describe** how electromagnetic waves behave like waves and like particles;
- **describe** photons;
- **list** some ways that electromagnetic waves are used.

SCIENCE TERMS

visible light	electromagnetic spectrum
electromagnetic waves	photons

LESSON PLAN

1. Introduce the lesson by asking the students to sit quietly and listen to the sounds around them. Explain that just as there are sound waves all around them there are also electromagnetic waves. Just as they can hear some sound waves and not others, they can see some electromagnetic waves and not others.

● **Word Skill: Word Structure**

SCIENCE BACKGROUND

The study of light is called optics.

20.1 The Electromagnetic Spectrum

You will find out

- what electromagnetic waves are;
- how sound waves and electromagnetic waves differ;
- how electromagnetic waves behave;
- how some kinds of electromagnetic waves are used.

Right now, as you sit reading this book, waves are traveling all around you. You have already learned about the sound waves around you. You can hear some sound waves. Others you cannot. Light waves are also all around you. As with sound waves, you can see only some light waves. The light waves you can see make up what is called **visible light.** What you cannot see are many other kinds of waves that, along with visible light waves, are called **electromagnetic** [ih-lehk-troh-mag-NEHT-ihk] **waves.** Electromagnetic waves are transverse waves that result from the action of electric and ● magnetic forces. The word *electromagnetic* comes from the names of these two forces. All the kinds of electromagnetic waves together make up the **electromagnetic spectrum** [SPEHK-truhm].

Figure 20.2 Visible light from the sun

Electromagnetic waves differ from sound waves in some important ways. Sound waves are compressional. Electromagnetic waves are transverse. Sound waves can only travel through matter. Electromagnetic waves can travel through empty space as well as through matter. This ability allows you to see stars on a clear night. Electromagnetic waves of light from distant stars travel through empty space, through the atmosphere, and to your eyes.

Electromagnetic waves travel much faster than sound waves. The speed of sound waves in air at 0°C is 331 meters per second. Electromagnetic waves can travel through that same air at 300,000,000 meters per second! This speed is often referred to as the speed of light. You have probably noticed the difference in the speeds of sound waves and electromagnetic waves during a thunderstorm. You see the lightning before you hear the thunder. The light from the lightning travels faster than the sound of the thunder.

The behavior of electromagnetic waves is much more difficult to describe than the behavior of sound waves. Sound waves always behave like waves. But electromagnetic waves behave in some ways like particles and in other ways like waves.

Do You Know?
Because electromagnetic waves behave in some ways like waves and in other ways like particles, some people have suggested calling them *wavicles*.

Figure 20.3 It is obvious to people watching a rocket launch from a distance that light waves travel faster than sound waves. They see the launch begin before they hear the roar of the engines.

2. On the board draw a transverse wave and a compressional wave and review the difference between them (transverse—particles travel at right angles to direction of wave; compressional—particles travel with wave). Ask the students which of these types of waves describes electromagnetic waves (transverse waves). What causes electromagnetic waves (the action of electric and magnetic forces)?
3. On the board construct a two-column chart entitled *Electromagnetic Waves Versus Sound Waves*. Label one column *Electromagnetic (Light) Waves* and the other column *Sound Waves*. Tell the students that they will use the chart to compare sound waves and electromagnetic waves.
4. Review the wave categories that sound waves and electromagnetic waves belong to (compressional and transverse, respectively). Enter this information on the chart.
5. Ask if sound can travel through a vacuum (no). On the chart note that sound must travel through matter. Can light travel through a vacuum (yes)? Note this on the chart. Ask the students to give some evidence that light can travel in a vacuum (the sun and stars can be seen).
6. Remind the students that during a thunderstorm they see lightning and then hear thunder. Although both occur at almost the same time, the sound of thunder is always received later. On the chart note that sound travels more slowly than light. You may also want to indicate the approximate speed of sound (344 m/s, or 1,125 ft/s) and light (300 million m/s, or 982 million ft/s).

Comprehension Skill: Interpretative

7. Explain that unlike sound waves, which always behave as waves, electromagnetic waves have been found to act both as particles and as waves. (You may want to provide the students with some of the historical background on these concepts that is provided in the Chapter Background. Question 2 in Research and Investigation may also be assigned at this time.)
8. Have the students construct a two-column chart entitled *Behavior of Light*. Since they will be summarizing the behavior of light as waves and as particles, have the students label one column *Waves* and the other column *Particles*.
9. Ask a group of students to list the characteristics that apply to the wave behavior of light (reflection, refraction, diffraction, interference, measurable frequency and wavelength). After everyone has completed the wave column, ask the students to summarize the particle nature of light in the particle column (particle nature based on behavior of photons, which have the following properties: tiny bundles of energy, not matter, released as electrons move from higher to lower energy level).

SCIENCE BACKGROUND

The aurora borealis, or northern lights, is produced when charged particles from the sun are injected into the earth's magnetic field.

The color on soap bubbles is caused by interference of light waves reflected from the inner and outer surfaces of the soap film.

Figure 20.4 A photon is released when an excited electron returns to its original energy level.

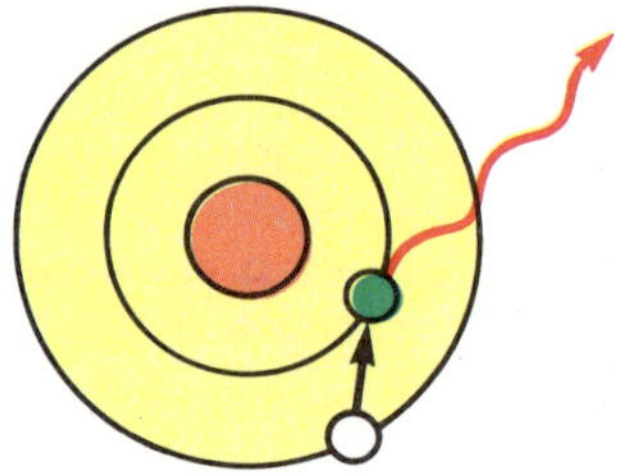

The particle nature of electromagnetic waves involves **photons.** Photons are tiny bundles of energy. Photons are not matter. But they come from matter. They are produced after atoms gain energy, usually in the form of heat or electricity. A particle that has gained energy is described as excited. When an atom is excited, the electrons in that atom are affected. In an unexcited atom, all the electrons are moving around the nucleus in certain specific energy levels. When the atom gains energy, some of the electrons become excited and move temporarily to higher energy levels. As those electrons return to their original energy levels, they release photons. Electromagnetic waves involve photons. Photons behave like particles.

Electromagnetic waves behave like waves by exhibiting reflection, refraction, diffraction, and interference. Also, like other waves, their frequency and wavelength can be measured. Frequency and wavelength are related. Electromagnetic waves with high frequencies have short wavelengths. Similarly, those with low frequencies have long wavelengths.

Figure 20.5 The electromagnetic spectrum

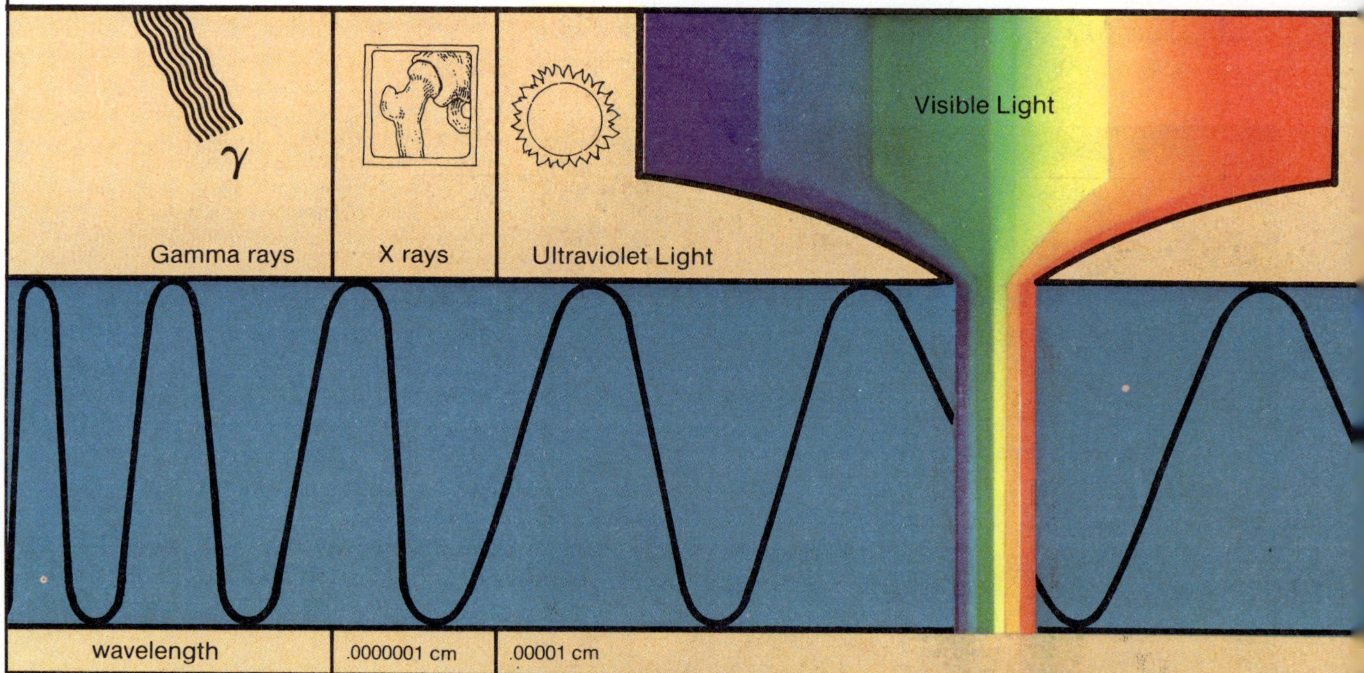

Different kinds of electromagnetic waves are identified by their wavelengths. Figure 20.5 shows the different kinds of waves in the electromagnetic spectrum in order according to their wavelengths. They range from gamma rays, which have the shortest wavelengths, to radio waves, which have the longest.

Different kinds of electromagnetic waves affect you differently. Gamma rays are able to penetrate many substances. They can cause damage to the cells of your body. Doctors use gamma rays to kill cancer cells.

X rays can penetrate all but the very dense parts of your body. Doctors and dentists use X rays to examine your bones and teeth. X rays are also used to examine the contents of containers. Airport security guards use X rays to inspect luggage and other cargo.

Ultraviolet light can be used to sterilize objects. Hospitals use ultraviolet light to sterilize equipment. Some school science departments use it to sterilize safety glasses. Light from the sun and from sunlamps contains ultraviolet light. These waves can penetrate your skin, causing suntans and sunburns. Recently, scientists have

Do You Know?
Scientists have used X rays to determine the diseases present in Egyptian mummies at the time of their death.

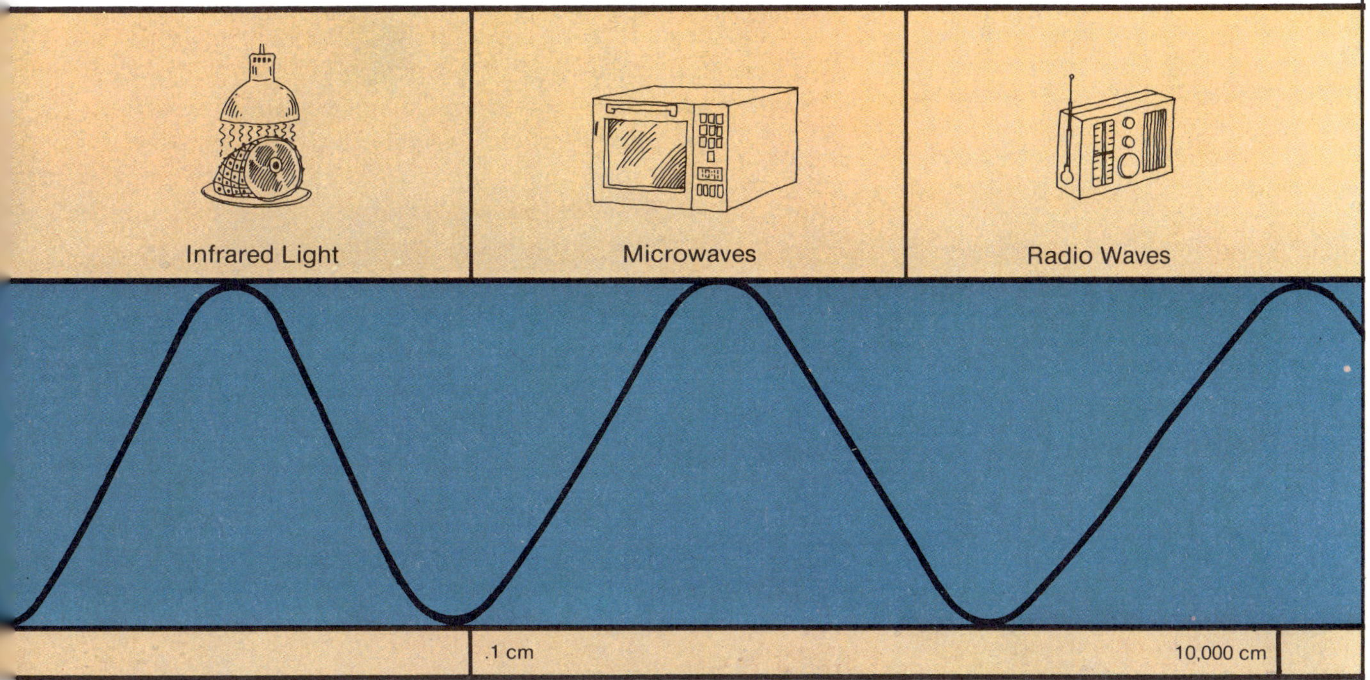

10. As the students look at the diagram of the electromagnetic spectrum in their texts, have them name the different types of waves in the electromagnetic spectrum. Stress that each type of wave has a different wavelength and frequency.

11. Point out that some electromagnetic waves have wavelengths so short that they cannot be seen. Ask the students to use the diagram in their texts to name some (gamma rays, X rays, and ultraviolet rays).

12. Mention that even though X rays and ultraviolet waves cannot be detected with the senses, they are still useful. Ask the students to name some uses of X rays and ultraviolet rays (X—examining bones and inspecting contents of containers; UV—killing germs).

SCIENCE BACKGROUND

The wavelengths of light are so short that they are measured in units called angstroms. There are 100 million angstroms in one centimeter.

13. Explain that other electromagnetic waves have a frequency so low and a wavelength so long that people cannot detect them with their eyes. Have the students find the infrared rays and radio waves on the diagram. Ask for uses of the infrared rays (heating, cooking) and radio waves (radio and television signals, microwave cooking).

Figure 20.6 To preserve valuable paintings, museums are lighted by sources that do not produce significant amounts of ultraviolet light.

discovered a link between the ultraviolet light in sunlight and skin cancer. To minimize this danger, people are advised to limit their exposure to sunlight.

Have you ever noticed that the colors of clothing, curtains, and upholstered furniture often fade after long exposure to sunlight? Ultraviolet light is absorbed by the substances that give color to some fabrics, paints, and other materials. This absorption causes the colors to fade. To keep the colors in priceless paintings from fading, museum workers hang them away from direct sunlight. The paintings are also kept away from the light of fluorescent lamps. Like sunlight, unfiltered fluorescent light contains a significant amount of ultraviolet light.

You cannot see infrared light, but you can feel it. You feel warmer as your body absorbs infrared light. Your body also gives off infrared light. Some animals that hunt at night can see the infrared light given off by warm bodies. This ability helps them to track their prey. Special cameras that can detect infrared light take heat, or infrared, photographs. Infrared photographs are in color. But the colors of the subject are not the same as the colors you see when you look at the subject. In infrared photographs, different colors represent different temperatures.

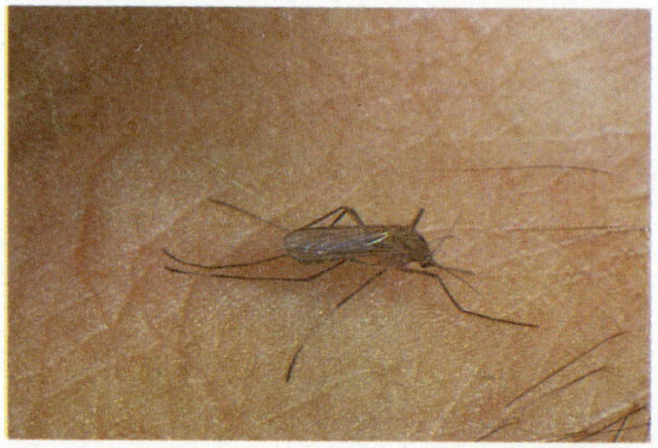

Figure 20.7 Female mosquitos have antennae that detect the infrared light given off by warm-blooded animals. These antennae can lead them to you!

Microwaves are very useful. In microwave ovens, microwaves are produced and are absorbed by food but not by the material that holds or contains that food. Foods cook quickly because they cook all at once instead of from the outside in, as happens with conventional ovens. Microwaves are also used in communications. Telephone and television signals can be carried by microwaves. Some telescopes also use microwaves.

Figure 20.8 An infrared photograph of a girl and her dog

Study Questions for 20.1

1. What are electromagnetic waves?
2. How do sound waves and electromagnetic waves differ?
3. What are photons?
4. Name three kinds of electromagnetic waves and tell how each kind affects people.

1. transverse waves that result from the action of electric and magnetic forces
2. sound waves—compressional, travel more slowly, can't travel in a vacuum, always behave like waves; electromagnetic waves—transverse, travel very quickly, can travel in a vacuum, behave in some ways like waves and in other ways like particles
3. tiny bundles of energy
4. gamma rays—can cause cell damage; X rays—used to examine bones and teeth; ultraviolet light—causes suntans and sunburns and is linked to skin cancer

FOLLOW-UP

Reinforcement: Light Tic-Tac-Toe

Cut four strips of paper that are each 1.25 cm (0.50 in.) wide and 20.3 cm (8 in.) long. Cut out a series of about 20 circles and 20 triangles. Arrange the strips to form a tic-tac-toe board on your overhead projector. To play the game, divide the class into two teams—one team is represented by circles, the other team by triangles. For each move on the tic-tac-toe board, ask questions that will help to reinforce the concepts and vocabulary of this lesson.

Enrichment: Discovery of X Rays

Have students research the life of Roentgen and his discovery of the X ray. Why did he name them X rays?

Tie-in: Microwaves

Ask the students to pretend that they are security experts. A foreign nation has been shooting microwaves at one of the other embassies. What damage can the microwaves do? How can they be stopped? What materials are impossible for microwaves to penetrate?

20.2

OBJECTIVES: The student should be able to

- **describe** how the colors that make up white light can be separated;
- **explain** why an object appears to be a certain color;
- **explain** why the sky appears to be different colors at different times.

SCIENCE TERMS

prism
white light

ADVANCED PLANNING

1. For the demonstration you will need a piece of cardboard, some tape, a filmstrip projector, a piece of white cardboard, and a prism.

2. Prepare the projector cover described in the first step of the lesson plan and practice this demonstration so that you will know the best angle at which to hold the prism.

LESSON PLAN

1. From a piece of cardboard cut a circle the size of the projector lens. Cut a slit in the center of the circle. Tape the circle to the projector lens so that the slit will be parallel to the desk top. Place the projector on a desk in the center of the classroom. Set up a piece of white cardboard about 60 cm (2 ft) from the projector to act as a screen.
2. Have the students gather around the setup. Turn on the projector and place the prism in front of the slit so that a spectrum appears on your screen. Ask the students what they observe (rainbow colors of the spectrum). What colors are present in order from bottom to top (red, orange, yellow, green, blue, indigo, and violet)?

SCIENCE BACKGROUND

The progression of colors in the visible spectrum corresponds to the mixing of colors the students may have learned in art.

R-O-Y-G-B-I-V   R + Y = O
Y + B = G
B + V = I

20.2 Visible Light

You will find out
- what makes a rainbow;
- why objects have the colors they do;
- why the sky changes color.

The electromagnetic waves of visible light fall between ultraviolet light and infrared light in the electromagnetic spectrum. When visible light waves encounter a new or changed medium, they either pass through it or are refracted, reflected, or absorbed. This behavior of visible light waves accounts for everything you see.

An example of the refraction of visible light that you have probably seen many times is a rainbow. Did you know that you do not need rain to see a rainbow? You can make one by directing a fine spray of water from a hose into the sunlight. In fact, you do not even need water to make a rainbow. If you direct light through a triangular piece of glass called a **prism,** you will see a rainbow of colors.

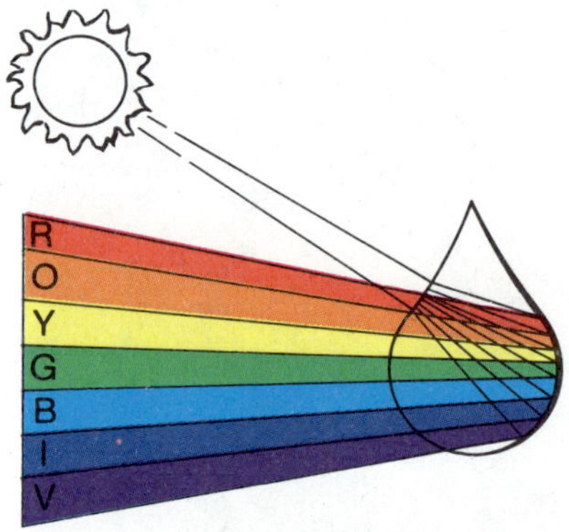

Figure 20.9 You can remember the order of the seven distinct colors that appear when white light is separated by remembering the name *Roy G. Biv.*

Where do the colors of the rainbow come from? Visible light includes light waves of seven distinct colors: red, orange, yellow, green, blue, indigo, and violet. Each color is made up of electromagnetic waves of a different wavelength. Red has the longest wavelength. Violet has the shortest. When all the colors of visible light are mixed together, you see colorless or **white light.** When the colors are separated, as in a rainbow, you see each individual color.

Prisms and drops of water separate white light into its colors by refraction. When white light passes from the air to a prism or drop of water, the light waves slow down. This change in speed causes the waves to bend, or refract. Waves of different wavelengths are bent to different degrees. Since each color involves waves of a different wavelength, each color is bent to a different degree. Violet waves are bent the most. Red waves are bent the least. In this way the colors that make up white light are separated.

White light can pass through some materials, such as air. But many materials reflect the waves of all or some

Figure 20.10 What color of light is reflected by these tulip blossoms?

mostly red

of the colors in white light. The waves of colors that are not reflected are absorbed. The color of an object you see depends on which colors that object reflects and which colors it absorbs. A green leaf is green because it reflects mostly green light. It absorbs most of the waves of the other colors. A banana is yellow because it reflects mostly yellow light.

What do you see when an object absorbs all colors? You see black. A black cat's fur absorbs almost all the white light that strikes it. What if an object reflects all colors? You see white. A polar bear's fur reflects almost all the white light that strikes it.

Only a few colors have been mentioned. But scientists estimate that the human eye can see about 17,000 different colors. Where do all these colors come from? What makes mud brown, fish silver, and flowers pink? All the colors you see are mixtures of the colors in white light. When objects reflect a mixture of more than one of these colors, you see a new color. The 17,000 different colors you can see come from 17,000 different mixtures. All the colors you see in this book come from different mixtures of magenta (pinkish red), cyan (a shade of blue), yellow, and black.

Do You Know?
In an experiment, special lights shone on the food at a dinner party. Celery looked pink; steak, gray; and peas, black. Most of the guests could not eat at all. Those who did felt ill.

3. Point out that the light from the projector is white light. Ask the students where the colors came from (prism separates the white light into its colors).

4. Explain that many materials do not allow light to pass through them. They reflect some wavelengths and absorb others. Ask the students to explain why people see an object as a particular color (object reflects the color that is seen and absorbs the other colors of the spectrum).

SCIENCE BACKGROUND

The colors in visible light make up what is called the visible spectrum. The visible spectrum is often described as having seven distinct colors. It actually has a continuum of colors from red to violet.

Color TV sets make all the colors you see with combinations of dots that produce red, blue, or green light.

5. Have the students explain why the following objects appear to be a particular color: red apples (reflect red light); orange oranges (reflect orange); black objects (absorb all the colors of the spectrum); white objects (reflect all the colors of the spectrum).

ACTIVITY: How Does Color Affect Heat Absorption?

Try to have available at least four different colors of cellophane. If possible, use small cocktail ice cubes. They will melt quickly, allowing the students to complete three trials within the class period.

After the activity, have the students share their results with each other by having each student write his or her color and average melting time on the board. Then discuss how the result can be applied to energy conservation ideas (darker colors of clothing and homes will absorb more solar heat).

● **Laboratory Skills: Manipulation of Materials**
Experimentation/Demonstration

● **Mathematics Skill: Graphing**

In white light, a red apple is always red. A black and orange tiger is always black and orange. But the same sky overhead is not always the same color. Around noontime, it is blue. At sunrise and sunset, it may show shades of pink, orange, and red. What causes these changes in color? The sky is lighted by white light from the sun. The sun's white light must travel through the atmosphere to reach your eyes. Two things about the atmosphere affect the color of the sky you see. First, the color of the sky depends on the amount of atmosphere the sun's light travels through to get to your eyes. Second, it depends on the groups of molecules and tiny particles that are present in the atmosphere.

The sky is blue when the sun is high in the sky. With the sun in this position, light waves from it pass through

● ACTIVITY

How Does Color Affect Heat Absorption?

Materials

3 squares (12 cm by 12 cm) of one color of cellophane
3 rubber bands
3 ice cubes of the same size
3 clear plastic cups
clock or watch

Procedure

1. Place an ice cube in the center of a square of cellophane. Pull up the four corners and secure them with a rubber band so that the ice cube is in a sack.

2. Put the sack in a cup and place the cup in direct sunlight or in direct lamplight.

3. Measure the time it takes for the cube to melt completely.

4. Repeat steps 1–3 two more times.

5. Find the average time of your three trials. Share your results with your classmates. Collect the results of classmates who used colors of cellophane that were different from yours.

Questions

1. Make a bar graph to compare the melting times of the different colors. ●

2. If you wanted a house to warm up quickly in the sunshine, which color would you select for the roof?

3. Why do people who live in hot, desert climates wear light colors?

1. The graphs should have the different colors across the bottom and the time up the side.
2. A dark color will absorb more heat.
3. Light colors absorb less heat.

Figure 20.11 In colorful sunsets, you can see only the longer wavelengths of visible light.

fewer kilometers of atmosphere to reach your eyes than they do when the sun is low in the sky. Small groups of molecules and very tiny particles in the atmosphere are able to reflect in all directions, or scatter, only the shorter wavelengths contained in white light. The shorter wavelengths involve blue shades. The scattered blue light waves make the sky look blue.

A different situation exists at sunrise and at sunset. Where do you see the sun at sunset? It is very low in the sky. When you see the sun low in the sky, light waves from it travel through many more kilometers of atmosphere to reach your eyes. The shorter-wavelength blue waves are scattered many kilometers farther away from your eyes. Thus, you do not see them. You see only the longer wavelengths of reds and oranges. Forest fires, dust storms, and volcanoes release a lot of particles into the atmosphere, producing unusually brilliant red sunsets. The added particles scatter more of the longer wavelengths, allowing only the longest, red wavelengths to reach your eyes.

Study Questions for 20.2

1. How do raindrops make a rainbow?
2. Why does a green leaf look green?
3. What two things about the atmosphere affect the color of the sky you see?

1. Raindrops refract white light from the sun. Since each color in white light has a different wavelength, each is bent to a different degree.
2. It reflects mostly green light and absorbs most of the waves of the other colors in white light.
3. the amount of atmosphere the sun's light travels through to reach your eyes and the groups of molecules and tiny particles present in the atmosphere

6. Explain that at different times of the day the sky appears to be different colors. Ask what color the sky often appears to be at sunrise, noon, and sunset (red, blue, and red, respectively).
7. Point out that two factors affect the color of the sky. Ask what they are (the amount of atmosphere the sun's rays must travel through and the kinds of molecules and particles that are present in the sky). Ask the students which color light rays are reflected to their eyes at noon when the sky appears to be blue (blue).

FOLLOW-UP

Reinforcement: Color and Wavelength

Have the students create posters to show white light being dispersed by a prism and to show the relationship that exists between color and wavelength.

Enrichment: How the Eye Sees Color

Ask the students to prepare written reports to explain how the eye sees color. Students should try to find answers to the following questions: What is the function of the cones? Why are some animals colorblind? Why are come people colorblind?

Tie-in: Color Printing

Ask the printing teacher to show the students how four-color printing is done. The students can do library research on the process and then see it in action in the shop.

SCIENCE BACKGROUND

When you see a red sky or a blue sky, not everyone on earth is seeing the same colors. Depending on your position on the surface of the earth, you see different color skies at different times.

20.3

OBJECTIVES: The student should be able to

- **describe** what happens when light is reflected off rough surfaces and smooth surfaces;
- **explain** how a mirror image differs from the object it represents;
- **explain** why the image in a fun-house mirror is distorted;
- **explain** what happens when light waves are refracted;
- **explain** how a mirage is formed.

SCIENCE TERM

mirage

ADVANCED PLANNING

1. For the first demonstration get a projector, a mirror, a box of aluminum foil, some cardboard, and a piece of manila paper.

2. For the second demonstration, you will need a large beaker of water and a plastic ruler.

LESSON PLAN

1. Introduce the lesson by showing the students the cardboard, the mirror, a piece of smooth aluminum foil, a piece of wrinkled aluminum foil, and a piece of manila paper. Ask the students which of these objects are good reflectors (mirror, smooth aluminum foil) and review with them the meaning of the word *reflection*.
2. Ask what properties make the mirror and smooth aluminum foil good reflectors (smooth, shiny). What happens to light rays when they hit a smooth surface such as a mirror (they are reflected at the same angle)? What happens to light rays when the surface they hit is not smooth (light rays are reflected in many different directions)?

20.3 Changing Visible Light Waves

You will find out

- how mirrors reflect light waves;
- how refraction of light waves can fool your eyes.

Why do some mirrors show you just as you are, while others make you look distorted? Why do you sometimes see puddles ahead on the highway when no water is present? The visible light waves that reach your eyes travel in straight lines until they encounter changes in their medium. These changes sometimes cause you to see things differently from what they really are.

Many surfaces reflect light. A rough surface may appear not to reflect light. It looks dull because it is reflecting light in many different directions. A smooth surface reflects most light in the same direction. A mirror has a smooth surface. Light waves that hit anywhere on the surface of a mirror are reflected in the same way. If you look at yourself in a mirror, your reflection looks just like you. The only difference is that, in the reflection, the right and left sides are reversed. You may not notice this difference when you see your own reflection. But

Figure 20.12 The water's glassy surface acts like a mirror.

Figure 20.13 The word *ambulance* is spelled backward on the hood of this emergency vehicle. When drivers see it in their rear-view mirrors, the word appears normally written and is easy to read.

the difference is obvious when you look at a word in a mirror. The word is spelled backwards. Also, the letters appear backwards. Dental students have to learn to allow for the backward reflection they see when using a mirror to examine a patient's teeth.

The reflection you see of yourself in a fun-house mirror looks nothing like the real you. Why? Although the mirror may look smooth, its surface is uneven. When you look into such a mirror, light waves from different parts of your body are reflected at different angles. The net result is a very distorted and funny view of your body.

Figure 20.14 The uneven surface of this mirror causes the distortion you see.

3. Ask the students how the image of the person they see in a mirror is different from the real person (reversed right to left).
4. On the board draw a side view of a fun-house mirror. Ask a student to indicate with arrows the incident and reflected light rays on the mirror. Have another student explain why an image will be distorted in this mirror (light rays from different parts of the body are reflected at different angles).

5. Put the beaker of water on your desk and place the ruler in the beaker. Have the students observe the equipment at eye level. Then ask them to draw what they saw (the ruler appears to be broken because the light waves are bent).
6. Explain that the light waves from the part of the ruler underwater are bent as they pass from water to air. This is what refraction is—the bending of light waves as they pass from one medium to a new medium.

ACTIVITY: How Can You Produce a Mirage?

◇ **Caution:** Students should be very careful with the lighted candles. Have students wear safety glasses throughout the activity.

The glasses used should be clear with no decoration. The candles should be slightly shorter than the glasses. If the candles are too thin or uneven on the bottom to stand alone, let some wax drip onto a cardboard square and stand the candle in it. Make sure the students exercise caution with the lighted candles.

Before the students begin, ask them to predict when they will see a mirage. Review with them the fact that mirages are caused by refraction of light waves.

If the students have difficulty seeing the mirage when they look up through the glass, check to be sure that the candle is close to the glass and that the flame is below the surface of the water.

● **Laboratory Skills: Manipulation of Materials**
Experimentation/Demonstration

Figure 20.15 Light waves from the part of the pencil underwater are bent as they pass from the water to the air.

One way to get an inaccurate view of an object is to look at it in an uneven mirror. Another way is to look at an object underwater. If you look at an object that is in water, you may notice that its shape or location appears different from what is really the case. To understand how this happens, imagine a pencil in a glass half full of water. You see the pencil when light waves reflected from it reach your eyes. You see the part of the pencil out of the water exactly as it is. But the light waves from the part of the pencil underwater are bent, or refracted, as they pass from one medium—water—to a new medium—air. This refraction makes the part of the pencil that is underwater appear to be in the wrong place in relation to the part above water.

Light waves can be refracted when they travel from one medium to another. They can also be refracted

● ACTIVITY How Can You Produce a Mirage?

Materials

glass with straight sides
candle
matches
safety glasses

Procedure

1. Put on your safety glasses. Fill the glass with water until it is almost full. Place it near the edge of a table.

2. Place the candle behind the glass. Light the candle. The top of the flame should be just below the level of the water. CAUTION: Be sure the candle cannot tip over. Also be sure someone watches it all the time that it is lighted.

3. Look down on the surface of the water. Do you see the flame?

4. Kneel down so that you are lower than the glass of water. Look up through the water at the candle. Do you see the flame?

Questions

1. How many flames did you see when you looked down on the glass of water?

2. How many flames did you see when you looked up through the glass of water?

3. Where were the flames you saw when you looked up through the glass?

4. What caused you to see two flames?

5. What was the mirage that you saw?

1. one
2. two
3. One was behind the glass. The other appeared at the top of the glass.
4. Light waves from the flame were bent as they passed from the water to the glass to the air.
5. the second flame that appeared at the top of the glass

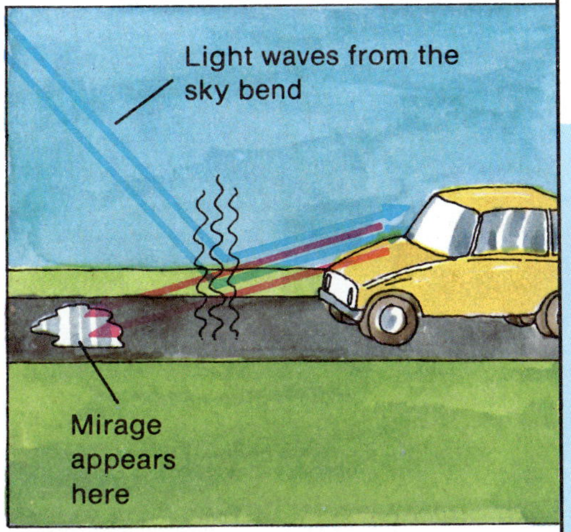

Figure 20.16 *left:* The water that appears on the road is a mirage. *right:* The mirage is produced when light waves are bent as they pass from denser, cooler air to less dense, warmer air.

within the same medium if the density of that medium changes. For example, the density of the air around you is affected by temperature. When the temperature increases, the density decreases. In hot weather, the air near the surface of a highway may become hotter and less dense than the air above. Imagine you are riding in a car on such a highway. Light waves traveling to your eyes from patches of blue sky will be refracted as they pass from the cooler, denser air above to the warmer, less dense air near the highway. As a result, you see what appear to be puddles of water on the road ahead. You are seeing an example of a **mirage.** A mirage is an illusion caused by the refraction of light waves, making objects that are far away appear to be nearby, floating in air, or upside down.

Study Questions for 20.3

1. How are fun-house mirrors different from ordinary mirrors?
2. Why does a pencil in a glass of water look broken at the water's surface?
3. What is a mirage?

1. The surfaces of ordinary mirrors are perfectly smooth. The surfaces of fun-house mirrors are uneven.
2. Light waves traveling to your eyes from the part of the pencil underwater are refracted as they pass from the water to the air. The part of the pencil underwater appears to be unconnected from and just beside the part of the pencil above the water.
3. an illusion caused by the refraction of light waves, making objects that are far away appear to be nearby, floating in air, or upside down

7. Ask the students what a mirage is and where it occurs most often (an illusion caused by the refraction of light waves; occurs primarily on roadways during the summer and in the desert).

FOLLOW-UP

Reinforcement: Bending Light
Ask the students to prepare a series of diagrams that show how light waves are refracted as they go from one medium to another. Students can show this by drawing the path of a dune buggy as it goes off the paved road into the sand and onto the harder packed sand at the edge of the ocean. Students may need to use small toy cars to plot the course of the dune buggy.

Enrichment: The Disappearing Stamp
Place a postage stamp on a table. Set a glass of water on the stamp. Cover the glass with a saucer. The stamp will appear to have disappeared. Ask the students to make up other tricks that are based on refraction.

Tie-in: The Best Way to Fish
Whether they are using a pole or a net, people who fish must make allowances for the refraction of light before they cast their lines or drop their nets. The refraction of light waves makes the fish appear to be where they are not. Have students interview some people who go fishing to learn how they must adjust their nets and lines in order to catch fish.

20.4

OBJECTIVES: The student should be able to

- **describe** a diffraction grating;
- **explain** how a diffraction grating separates the colors in white light;
- **describe** some uses of a diffraction grating;
- **explain** how a polarizing material affects light;
- **describe** how glare is reduced with polarizing sunglasses.

SCIENCE TERMS

diffraction grating
polarizing

ADVANCED PLANNING

Have available for student use diffraction gratings or spectroscopes and polarizing lenses. For the light source use a single bulb at a large distance.

LESSON PLAN

1. Pass out the diffraction gratings and ask the students to use them to look at a light source. What do they observe (rainbow of colors)? Explain that when light waves pass through the grating the waves spread out and interfere with each other. Where crests cross or where troughs cross, the bright rainbow occurs.
2. As the students observe a light through the diffraction grating, tell them to look for dark bands. If the meeting of crests or the meeting of troughs produces a bright band of color, what do the students think produces the dark bands (crests meet troughs)?

20.4 Diffraction and Polarization

You will find out
- how a diffraction grating separates the colors in white light;
- how polarizing sunglasses reduce glare.

You have learned that refraction of white light can separate it into colors. Refraction is not the only way to separate the colors in white light. You can also separate them by interference.

Imagine a thin piece of glass with thousands of parallel slits close together. Such a device is called a **diffraction grating.** When white light passes through a diffraction grating, different wavelengths are diffracted, or bent, differently, causing them to spread out. As the light waves are spread out, they run into each other. Interference results. Where crests meet troughs, the waves are cancelled out and dark bands appear. Where crests meet crests or troughs meet troughs, bright bands in a rainbow of colors appear.

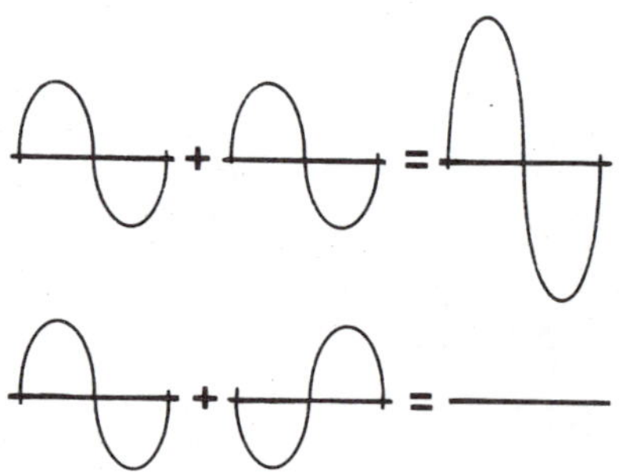

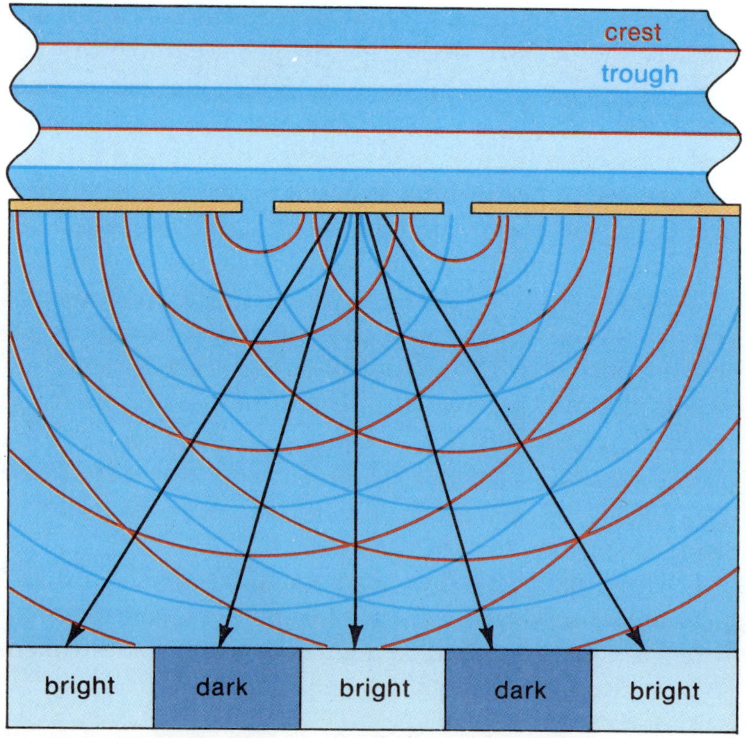

Figure 20.17 The path of light waves through a diffraction grating

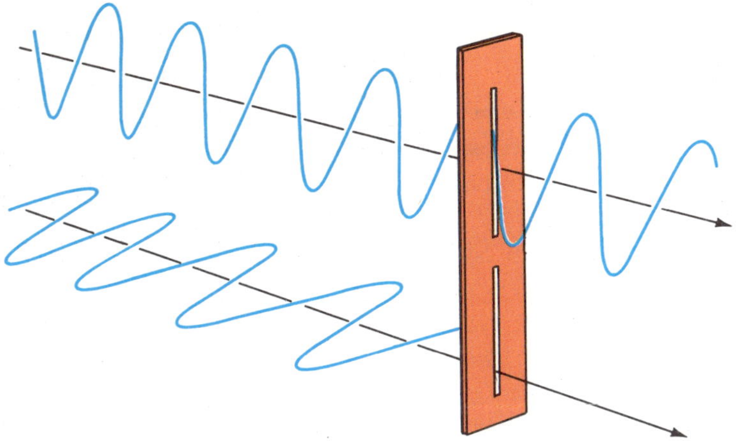

Figure 20.18 Polarizing light waves

A diffraction grating can be used to separate the different wavelengths of light given off by a substance. For example, neon gives off light when it is electrified. The light appears red. But when it is passed through a diffraction grating, other colors are apparent as well. A diffraction grating can be used to identify a substance by identifying the different wavelengths of light given off by that substance. The different wavelengths given off identify a substance as surely as fingerprints identify a person.

Although you probably enjoy bright sunny days, you may remember some days when the sun's light was so bright that your eyes were bothered. You may have used sunglasses or wished you had a pair. Have you ever heard of **polarizing** [POH-luh-ry-zing] sunglasses? A material that is polarizing allows only light waves with crests and troughs at a certain angle to pass through. All other light waves are reflected or absorbed. If very bright sunlight is shining in your eyes, polarizing sunglasses can be helpful. Since only some of the light waves will be able to pass through to your eyes, glare is reduced.

Figure 20.19 Some areas of the sky polarize sunlight. Honeybees are able to detect this polarization. This ability helps them to navigate.

Study Questions for 20.4

1. How does a diffraction grating separate the colors in white light?
2. How do polarizing sunglasses reduce glare?

1. by diffraction, which causes interference
2. They allow only light waves with crests and troughs at a certain angle to pass through.

3. Have the students observe a light through the polarizing lens. What change do the students observe in the light (light is less bright)?

4. To help explain how polarizing lenses screen light to reduce glare, draw a picture of a light on the board. Beside it draw a picture of a piece of glass. On the other side draw a picture of a piece of glass with several parallel lines. Label this glass *polarizing*. Draw light rays coming from the light in all directions. Explain that all the rays can go through the glass. However, only the rays that are vibrating in the same direction as the parallel lines in the polarizing piece will pass through it. Thus, some light rays are screened out and the light becomes dimmer.

● Word Skill: Vocabulary

FOLLOW-UP

Reinforcement: Polarization and Diffraction
Ask each student to write a report in which he or she explains what is meant by the polarization and diffraction of light and why each is important.

Enrichment: Polarizing Light
Have students obtain several different types of sunglasses. Ask them to design an experiment in which they determine which glasses reduce glare the most. They should include a statement of the problem, a hypothesis, an experimental design, some observations, and a conclusion.

Tie-in: Seeing Out/But Never In
Some types of surfaces allow only one-way viewing. The surfaces are used in some windows and in some eyeglasses. Ask the students to investigate how the material from which these surfaces are made works and to list as many practical uses as possible for the material.

What Did You Learn?

20.1 • The light waves you can see make up visible light.

20.1 • Electromagnetic waves are transverse waves that result from the action of electric and magnetic forces.

20.1 • The electromagnetic spectrum is made up of all the kinds of electromagnetic waves together.

20.1 • Photons are tiny bundles of energy that are given off when excited electrons return to their original energy levels.

20.1 • The photons of electromagnetic waves behave like particles.

20.1 • Electromagnetic waves behave like waves by exhibiting reflection, refraction, diffraction, and interference.

20.1 • Different kinds of electromagnetic waves are identified by their wavelengths.

20.2 • Raindrops and prisms separate the colors in white light by refraction.

20.2 • The color of an object depends on which colors that object reflects and which colors it absorbs.

20.2 • The color of the sky depends on the amount of atmosphere the sun's light travels through to get to your eyes and the groups of molecules and tiny particles that are present in the atmosphere.

20.3 • All light waves are reflected from the smooth surface of a mirror in the same way.

20.3 • An object underwater may appear to be in the wrong place because light waves from it are refracted as they pass from the water to the air.

20.3 • Light waves can be refracted if the density of their medium changes.

20.3 • A mirage is caused by the refraction of light waves making objects that are far away appear to be nearby, floating in air, or upside down.

20.4 • A diffraction grating separates the colors in white light by interference.

20.4 • Polarizing sunglasses reduce glare by allowing only light waves with crests and troughs at a certain angle to pass through.

Key Terms

visible light
electromagnetic waves
electromagnetic spectrum
photons
prism
white light
mirage
diffraction grating
polarizing

Career

X-ray Technician

X-ray technicians work in hospitals, clinics, and public-health agencies. They work with both doctors and patients. Following a doctor's order, an X-ray technician prepares and positions a patient for an X ray. The technician also operates the equipment used to take the X ray. It is the X-ray technician who is responsible for taking a picture at the correct exposure. Therefore, the technician must know about X-ray equipment as well as about the human body.

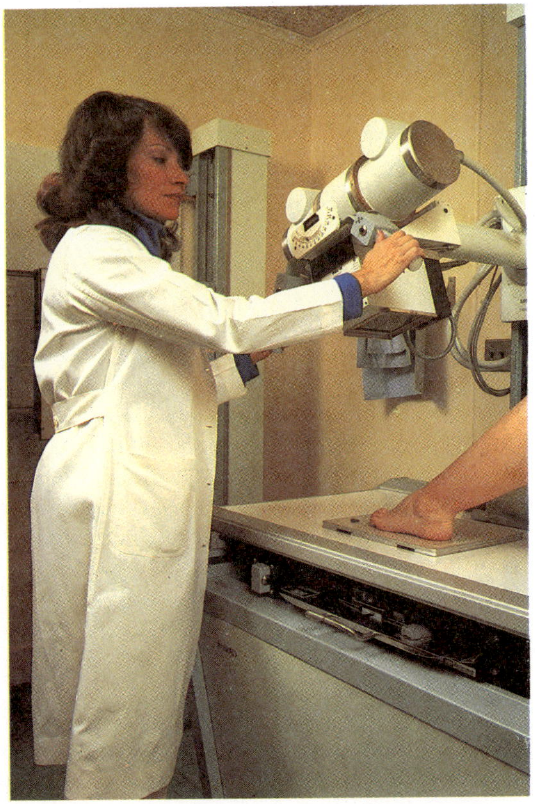

Many patients needing X rays are very ill or are experiencing pain. An X-ray technician must be understanding of a patient's condition and feelings. The technician must enjoy working closely with and helping people. He or she must also follow directions exactly and pay close attention to the operation of X-ray equipment. Doctors rely on proper X rays to help them diagnose conditions and prescribe treatments.

There is a great demand for qualified X-ray technicians. To become an X-ray technician, you must complete a two-year training program after high school. Such programs are offered by many hospitals, medical schools, colleges, junior colleges, and vocational schools and in the military services.

Vocabulary

1. photons
2. electromagnetic spectrum
3. infrared light
4. white light
5. prism
6. reflects
7. polarizing
8. mirage
9. diffraction grating
10. electromagnetic waves

● Word Skill: Vocabulary

What Do You Remember?

1. transverse waves produced by the action of electric and magnetic forces
2. When excited electrons return to their original energy levels, they produce bundles of energy called photons.
3. all colors
4. by refraction
5. by diffraction, then interference
6. Polarizing sunglasses let through only light waves with crests and troughs at a certain angle. This reduces the number of light waves that reach your eyes.
7. Electromagnetic waves travel much faster than sound waves.

TO THINK ABOUT AND DO

Vocabulary ●

Number your paper from 1 to 10. Unscramble the words. Beside each number write the unscrambled word or words that best complete the sentence.

1. When excited electrons return to their original energy levels, they release bundles of energy called ________.
2. All the kinds of electromagnetic waves together make up the ________.
3. Your body absorbs ________ and you feel warmer.
4. When all the colors of visible light are mixed together, you see colorless or ________.
5. A triangular piece of glass that refracts white light is called a ________.
6. A banana is yellow because it ________mostly yellow light.
7. ________ sunglasses allow only light waves with crests and troughs at a certain angle to pass through.
8. When you see puddles ahead on a highway even though no water is present, you are seeing a ________.
9. A thin piece of glass with thousands of parallel slits close together is called a ________.
10. ________ are transverse waves that result from the action of electric and magnetic forces.

gearim
zipinalrog
mortteaelcecign sawev
shootnp
dreafrin gilth
thiew thilg
fatciifdrno aggritn
misrp
tteeeccgarmonil mustrepc
cleerstf

What Do You Remember?

1. What are electromagnetic waves?
2. How are photons produced?
3. What colors of white light are reflected by snow?
4. How does a prism separate the colors in white light?
5. How does a diffraction grating separate the colors in white light?
6. How do polarizing sunglasses help to reduce glare?
7. Why, during a thunderstorm, do you usually see the lightning before you hear the thunder?

8. How can the sky be different colors at different times of the day?
9. Why do you look strange in a fun-house mirror?
10. Give an example of a mirage.

Applying What You Have Learned ●

1. What part of the electromagnetic spectrum would be used for each of the following?
   (a) taking pictures of a person's lungs to look for evidence of lung disease
   (b) making a surface free of germs
   (c) keeping food warm in a cafeteria
2. Why do polar bears have black skin?
3. When you look at people standing in a swimming pool, often their legs look strange? Why?

Research and Investigation ●

1. Write a report on why electromagnetic waves are described as behaving both like particles and like waves. What makes gamma rays seem more particle-like and radio waves more wave-like?
2. What contribution did each of the following people make to the study of light: Michelson, Roemer, Einstein, Newton, and Huygens?

8. The sky's color changes depending on how much atmosphere the sun's light travels through to reach your eyes, and the size and number of particles in the atmosphere.
9. The surface of a fun-house mirror is uneven, so light waves from different parts of your body are reflected at different angles.
10. You see puddles ahead on a highway when light waves from patches of blue sky are refracted as they pass from denser, cooler air to less dense, warmer air.

Applying What You Have Learned

1. (a) X rays
   (b) ultraviolet light
   (c) infrared light
2. Polar bears have black skin because black absorbs most of the sun's light waves.
3. Light waves traveling from a person's legs (which are under water) are refracted as those waves pass from the water to the air.

● Comprehension Skill: Applied

Research and Investigation

1. The photoelectric effect can only be explained by the particle theory of light. Polarization and other phenomena can only be explained by the wave theory of light. Gamma rays are described as photons of high energy. Diffraction and interference are observed in gamma rays only with difficulty. Radio waves very clearly exhibit properties of waves.
2. In 1926 Michelson measured the speed of light to be 299,796 kilometers per second (186,285 miles per second). In 1676 Roemer devised a method to determine the speed of light. He calculated the speed of light to be 300,000 kilometers per second (186,000 miles per second)—a remarkably accurate figure considering the lack of technology available. Albert Einstein helped to formulate and confirm the quantum theory. Newton proposed the corpuscular theory. This idea stated that the light emitted by luminous bodies consists of tiny particles of matter called corpuscles. Huygens proposed the wave theory of light. He thought that luminous objects emitted a series of waves that spread out in all directions from the source.

● Reference/Research Skill

CHAPTER BACKGROUND

Optics is a branch of physics that describes the production, transmission, detection, and measurement of light. Devices such as binoculars, cameras, magnifying glasses, and projectors depend on the principles of optics. They all have lenses that transmit and control light. This chapter is concerned with geometrical optics—the study of optical instruments and what happens when light hits them. (The other main branch of optics, dealing with the nature and behavior of light is physical optics, which was discussed in Chapter 20.)

A lens is a disk of transparent material whose opposite sides are smooth, curved surfaces. There are two main types of lenses. A convex lens is thicker in the center than at the edges. Its function is to bring light waves together. Convex lenses are found in eyes as well as in cameras, microscopes, telescopes, and projectors. A concave lens has sides that curve inward. It is thinner in the middle than at the edges. It causes light waves to diverge, or spread apart. Concave lenses are used to correct nearsightedness. They are also found in binoculars.

Lenses form images. Convex lenses can be used to produce both real and virtual images. Virtual images are formed by the extension of light waves rather than by the actual waves. Since they are imaginary images, they cannot be projected on a screen. Real images are formed by actual waves of light and can be projected on a screen. Real images are always inverted. Concave lenses always produce virtual images that are smaller than the object being viewed.

A laser is a device that amplifies light and makes it brighter. Lasers can be used to produce intense light. A beam of light produced by a laser can weld a detached retina onto the inside back of an eye or burn a hole in a diamond.

Optical fibers can be used to transmit light along curved paths. The fibers have applications in medicine and in communication.

21

The Uses of Light

To introduce the lesson, have available a filmstrip projector and filmstrip, a pair of eyeglasses, a magnifying glass, a ring stand and clamp, a high-intensity lamp, and a piece of paper. Attach the magnifying glass to the ring stand and adjust it so that when the light from the high-intensity lamp passes through the magnifying glass, it focuses on the paper in a sharp point. Ask the students to observe the apparatus. What is the effect of the magnifying glass on the light (focuses the light waves to a point)? Turn on the filmstrip projector and ask the students to observe while you adjust the focus. What do they see happening (the picture clears and blurs as you move the lens)? Ask the students why some people wear glasses (to see more clearly, to focus light for their eyes). Explain that the magnifying glass, projector, and glasses all contain curved pieces of glass or plactic called lenses. In this chapter the students will study lenses and their uses.

Before the 1600's, no one knew that germs caused diseases or that pond water was full of tiny plants and animals. People did not know about germs and tiny plants and animals because they could not see them. In the early 1600's, the microscope was invented. A whole new world opened up to people.

At about the same time that the microscope was being invented, the telescope was being developed. With the telescope, people could see distant stars and planets. Microscopes and telescopes direct light waves so people can see objects that are not visible to their eyes alone.

Recently, there has been a development that uses light in a way that has nothing to do with seeing. This special kind of light can be used to carry television signals and telephone messages. It is powerful enough to cut diamonds and precise enough to perform surgery.

In this chapter you will find out how microscopes, telescopes, and other devices help you to see. You will also find out about light that is used to carry messages and perform surgery.

Figure 21.1 A laser-light show from the Empire State Building

Lesson Organizer

21.1 Lenses

Pages: 436–439
Advanced Planning: Page 436
Activity: How Can You See Images? Page 438
Follow-up: Page 439
- **R** Types of Lenses
- **E** A Camera Without a Lens
- **T** The Manufacture of Lenses

21.2 Using Lenses

Pages: 440–443
Advanced Planning: Page 440
Activity: Why Do You See Motion in Movies? Page 442
Follow-up: Page 443
- **R** How Do You See?
- **E** The Blind Spot
- **T** Microscopes for Specific Jobs

21.3 Lasers and Optical Fibers

Pages: 444–447
Advanced Planning: Page 444
Follow-up: Page 447
- **R** Laser Light
- **E** Holography
- **T** Lasers in Our Lives

See ACTIVITIES book pages 121–126 for laboratory, skill, and review exercises for Chapter 21.

21.1

OBJECTIVES: The student should be able to

- **describe** a lens;
- **identify** an image;
- **compare** concave and convex lenses;
- **locate** the focal point and focal length of a convex lens;
- **distinguish** between a real image and a virtual image.

SCIENCE TERMS

lenses	focal point
image	focal length
convex lens	virtual image
concave lens	real image

ADVANCED PLANNING

For the demonstrations you will need two pieces of glass no smaller than 10 cm (4 in.) by 8 cm (3.2 in.), a convex lens, a concave lens, and a filmstrip projector.

LESSON PLAN

1. Show the students the pieces of glass and the convex and concave lenses. Ask how the glass differs from the lenses (lenses are curved).
2. Darken the room and shine the projector light through the glass, then through the convex lens. How does the surface of each material affect the light (glass—no effect; lens—bends light waves so that they come together at a point)? Summarize for the students that lenses are curved pieces of glass or plastic that bend light waves.

SCIENCE BACKGROUND

Quartz is used in lenses because it transmits ultraviolet rays better than many other materials do.

21.1 Lenses

You will find out
- what a lens is;
- what the two basic kinds of lenses are;
- how to find the focal length of a convex lens;
- how virtual images and real images differ.

Do you or does someone you know collect stamps, or coins, or butterflies? A collector of small objects usually uses a magnifying glass to examine his or her collection. Only with the help of a magnifying glass can the interesting and important details be seen on a stamp, a coin, or the wing of a butterfly. A magnifying glass is an example of a lens.

Do You Know?
Most insects have compound eyes, each with hundreds of lenses.

Lenses are curved pieces of glass or plastic that refract light waves. Lenses can bring light waves together or spread them apart. A lens does this by refracting, or bending, light waves. Light waves that pass through a lens often form a picture you can see. The picture is called an **image.**

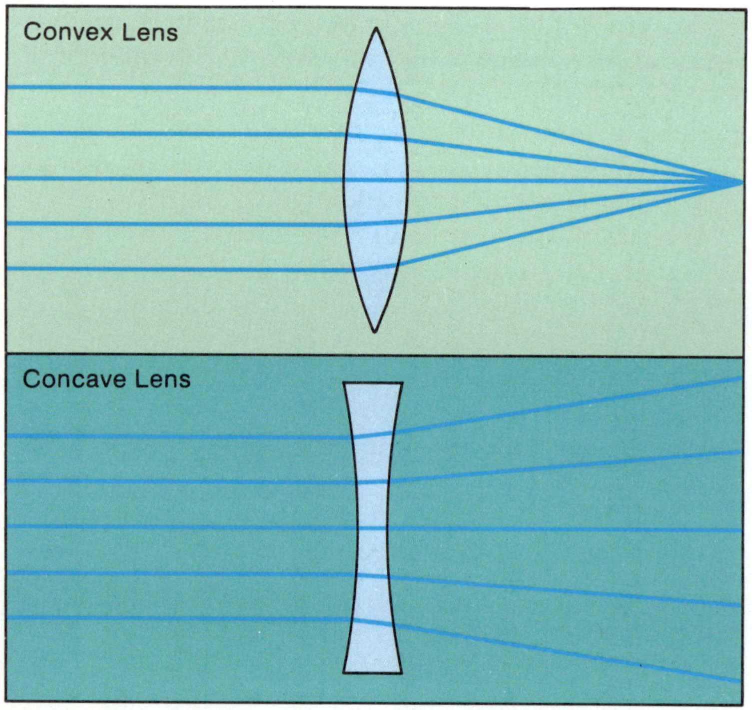

Figure 21.2 Light waves are brought together by convex lenses and spread apart by concave lenses.

There are two basic kinds of lenses. A **convex** [kahn-VEHKS] **lens** has curved surfaces that bulge out in the middle. The lens is thicker at the middle and thinner at the edges. Convex lenses bring light waves together. A **concave** [kahn-KAYV] **lens** has curved surfaces that dip in at the middle. The lens is thinner in the middle and thicker at the edges. A concave lens spreads light waves apart.

Light waves that pass through a convex lens come together at a point. This point is called the **focal** [FOH-kuhl] **point** of the lens. The distance from the center of the lens to the focal point is called the **focal length** of the lens. The focal length and the thickness of a convex lens are related. The thicker the lens, the shorter the focal length.

A magnifying glass uses a convex lens. You can find the focal length of a magnifying glass by holding it between the sun and a flat surface. If you move the magnifying glass slowly back and forth, you will eventually see a sharp point of light. This point is the focal point. The distance between your magnifying glass and the focal point is the focal length.

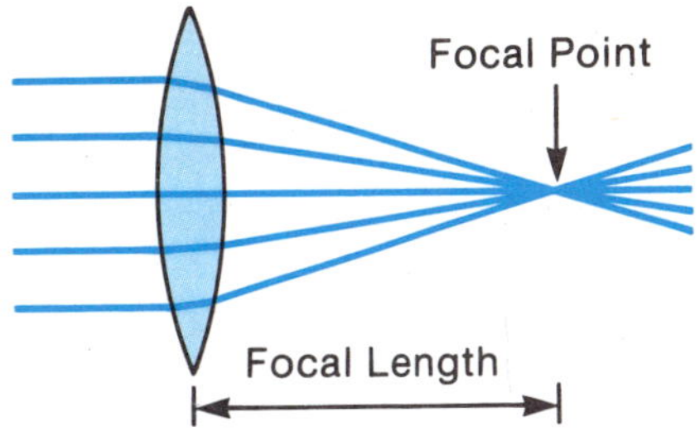

Figure 21.3 The focal point and focal length of a convex lens

Figure 21.4 *left:* To get the background in focus, the camera's lens was moved so that the film was at the focal point. *right:* To get the foreground in focus, the photographer increased the distance between the film and the lens.

3. Ask what happens when light waves pass through a lens (they often form a picture you can see). What is the picture called (an image)? What are some devices that use lenses to produce images (cameras, projectors)?

● **Word Skill: Structure**

4. On the board draw diagrams of a convex lens and a concave lens. Label each. Tell the students that they should remember a con*cave* lens by its *caved*-in surface in the middle.
5. Ask the students what the focal point is (where the parallel rays meet). On the diagram of the convex lens and its light waves label the focal point. Ask a student to draw a line on the diagram to show the focal length (distance from a lens to the focal point).
6. Point out the relationship between the focal length and the thickness of the lens. Thick lenses have shorter focal lengths than do thin lenses.

SCIENCE BACKGROUND

Convex lenses are also called converging, or positive, lenses. Concave lenses are also called diverging, or negative, lenses.

ACTIVITY: How Can You See Images?

Review the terms *real image* and *virtual image*. Encourage the students to experiment with their magnifying glasses, holding them at different distances from their eyes and the objects they are observing. The distance that students will have to raise the magnifying glass before they see a clear inverted image will vary. When students see the virtual image, they will see the number 5338 magnified. When they see the real image, the number will be upside down and will look like the word *bees*.

● **Laboratory Skills: Manipulation of Materials**
Experimentation/Demonstration

SCIENCE BACKGROUND

A drop of water in a loop of wire can be used as a magnifying glass.

Two kinds of images can be formed by lenses. You can see both kinds of images with a common magnifying glass. When you hold a magnifying glass over an object and move the lens up just a little, you see a **virtual** [VUR-choo-wuhl] **image** of the object. A virtual image is an image that appears right side up and cannot be focused on a screen. The virtual image appears magnified as long as you hold the magnifying glass a distance from the object that is close to but not equal to the focal length of the lens. When you move the magnifying glass a focal length away, you see a blur.

● ACTIVITY

How Can You See Images?

Material
magnifying glass

Procedure

1. Place the magnifying glass over the number below.

2. Pick the glass up and slowly move it toward your eye. Observe the number as you do so.

3. Stand up. Hold the magnifying glass at arm's length over the number. Slowly move the magnifying glass up. You will see the same things you saw in step 2. Continue to move the magnifying glass up until you see another clear image.

Questions

1. What happens to the number as you raise the magnifying glass? Does its size change? Does the image remain clear?

2. What happens to the number as the magnifying glass is moved farther from it?

3. When the image became clear again, how did it differ from the first image?

4. What kind of image did you see in step 2?

5. What kind of image did you see in step 3?

1. It becomes larger, then blurs and disappears.
2. It becomes larger, then blurs and disappears, then reappears as a clear image that is upside down.
3. It was upside down.
4. virtual image
5. real image

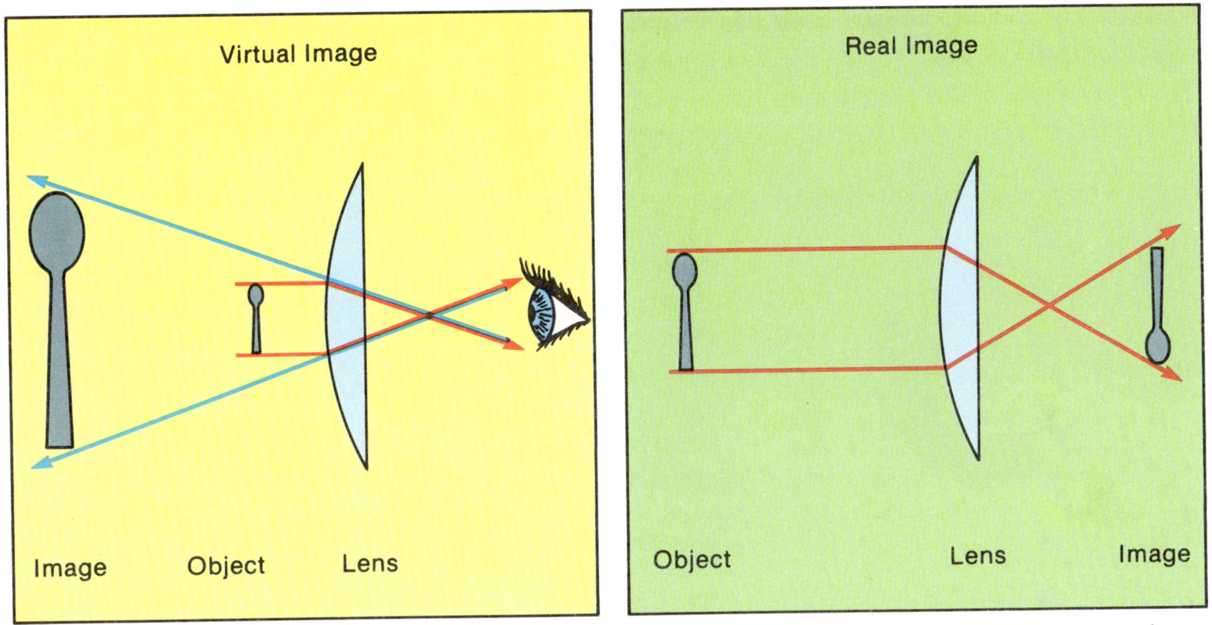

Figure 21.5 *left:* Forming a virtual image; *right:* Forming a real image

You see the second kind of image formed by lenses when you hold a magnifying glass at arm's length over an object and slowly move it up. First, you see the virtual image of the object. Next, you see a blur. Finally, you see an image of the object again. But this time it appears upside down. You are seeing a **real image.** A real image is an image that appears upside down and can be focused on a screen. A real image is upside down because light waves from the object are bent sharply when they pass through the lens. Light waves from the top of the object pass through the lens and are bent down. Light waves from the bottom of the object pass through the lens and are bent up.

Study Questions for 21.1

1. Describe a convex lens and tell how it affects light waves.
2. Describe a concave lens and tell how it affects light waves.
3. How is the focal length of a lens related to the focal point of that lens?
4. How are virtual images and real images different?

1. thicker in the middle and thinner at the edges; brings light waves together
2. thinner in the middle and thicker at the edges; spreads light waves apart
3. The focal length is the distance from the center of the lens to the focal point.
4. virtual image—is right side up, cannot be focused on a screen; real image—is upside down, can be focused on a screen

7. On the board form two columns with the terms *Real image* and *Virtual image* as headings. Ask the students to distinguish between real and virtual images (real—can be formed on a screen, appears upside down; virtual—cannot be formed on a screen, appears right side up). Write this information under the appropriate column on the board.

FOLLOW-UP

Reinforcement: Types of Lenses

Ask the students to create posters that illustrate the types of lenses and how each affects light rays.

Enrichment: A Camera Without a Lens

To make a simple pinhole camera, students should cover the open end of an oatmeal box with tissue paper and use a straight pin to punch a hole in the center of the closed end of the box. Have the students observe a candle flame or other light by holding the camera with the hole facing the light and observing through the tissue paper. Have them note what type of image forms on the tissue paper. Ask the students how this image forms.

Tie-in: The Manufacture of Lenses

Have students research the methods by which lenses are manufactured. Suggest that they consult an optician in addition to their library research. Students should present their findings in a written report.

21.2

OBJECTIVES: The student should be able to

- **explain** how the eyes see;
- **identify** two common vision problems;
- **describe** how lenses correct vision problems;
- **compare** a camera to the eye;
- **describe** how microscopes and telescopes use lenses;
- **distinguish** between refracting telescopes and reflecting telescopes.

SCIENCE TERMS

retina
refracting telescopes
reflecting telescopes

ADVANCED PLANNING

1. For the demonstrations you will need a model or a chart of the human eye, several microscopes, and one or more telescopes.

2. Ask students to bring in their cameras. They should not have film in them.

3. Prepare two transparencies. One transparency should show the parts of a microscope. The other should show the parts of a reflecting telescope and of a refracting telescope.

4. Obtain an overhead projector.

LESSON PLAN

1. Introduce the lesson by showing the students the model or chart of the eye. Point out the pupil, lens, retina, and muscles surrounding the lens. Explain that light waves enter the eye through the pupil and strike the lens. How does the lens affect the light waves that pass through it (bends them and brings them to a point)?

● Comprehension Skill: Interpretative

2. As you point out the retina on the model or chart, remind the students that a real image appears upside down on the retina. Ask why people don't see everything upside down (brain interprets the image right side up).

21.2 Using Lenses

You will find out
- how your eyes see;
- how a camera produces a photographic image;
- how microscopes and telescopes use lenses.

Figure 21.6 The eye

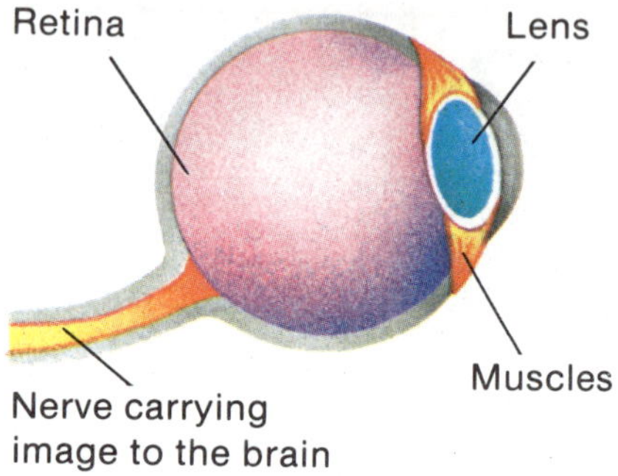

What do cameras, microscopes, telescopes, and your eyes all have in common? They all have lenses. The lenses in your eyes, however, are unique because their thickness can change depending on what you are looking at.

Light waves entering the eye are directed by a lens onto a surface called the **retina** [REHT-uhn-uh]. A real image is formed on the retina. Remember, real images are upside down. Does everything you see look upside down? The world you see is not upside down because your brain receives that real image and interprets it right side up.

Each of your eyes has a clear, changeable lens. The ● lens is not rigid. Muscles in your eye change the thickness of the lens. This allows you to clearly see objects that are close and objects that are far away. When you look at this page, the muscles in your eyes cause the lenses to thicken. You see the printed page clearly. If you look up from this book and across the room to a poster on the wall, your eye muscles stretch out the lenses and they become thinner. You then see the poster clearly.

Figure 21.7 *left:* When you look at something distant, the lenses in your eyes stretch out. *right:* When you look at something close up, the lenses thicken.

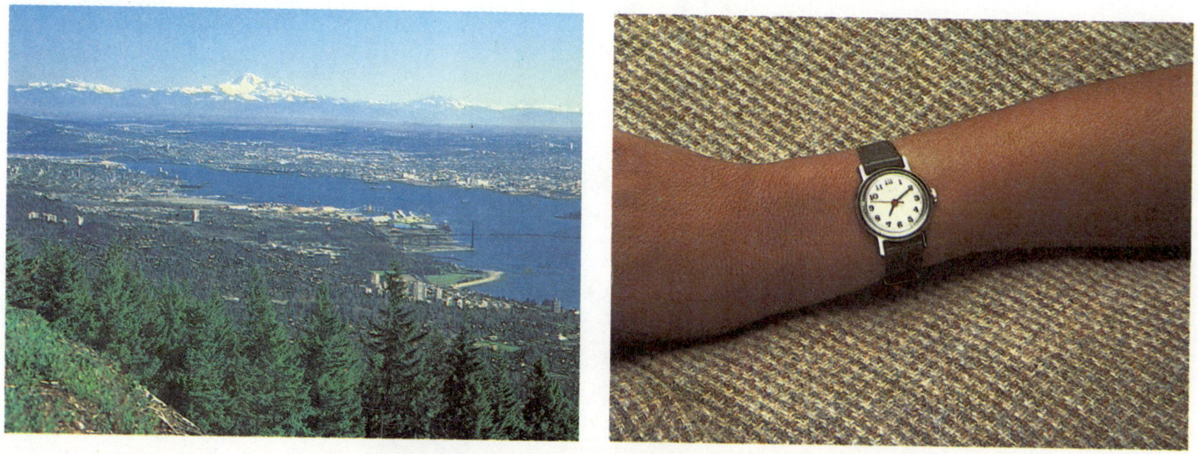

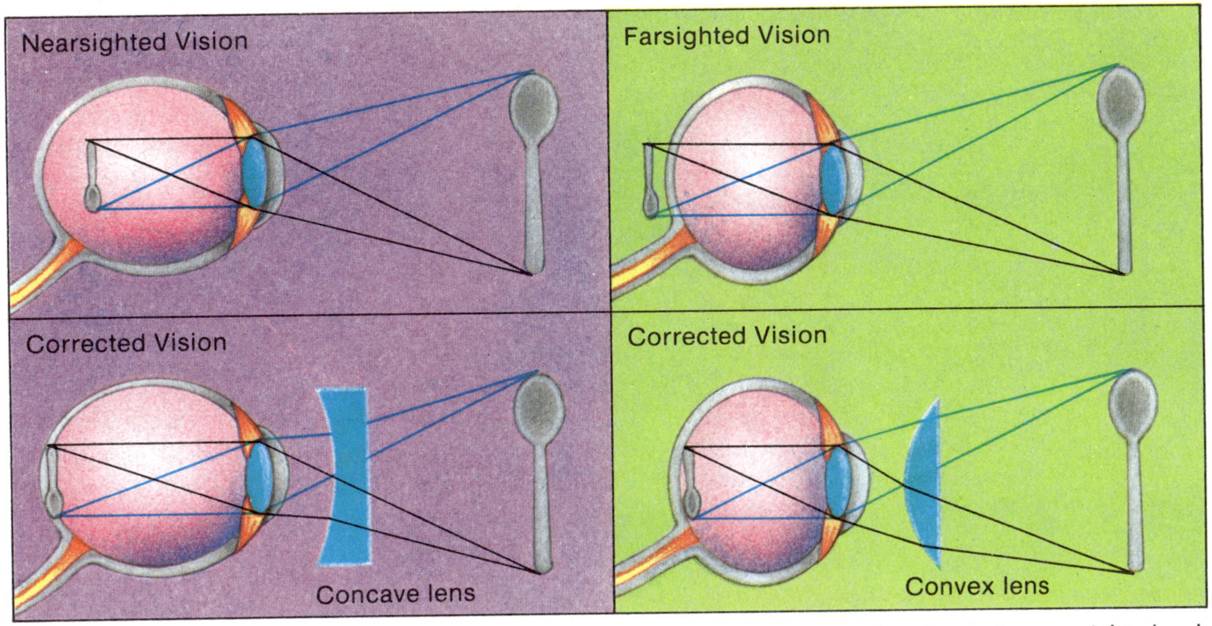

Figure 21.8 Correcting nearsighted and farsighted vision

The distance between the lens and the retina is very important to good eyesight. If this distance is too short, a person is farsighted. The real image is formed behind the retina instead of on it. Farsighted people can clearly see faraway objects. But nearby objects are unclear. Farsightedness can be corrected by convex lenses. Convex lenses change the path of light waves entering the eye, causing them to form an image on the retina instead of behind it.

Many older people develop farsightedness even though the distance between the lens and the retina in their eyes may be normal. The muscles in the eyes in older people may become weak and unable to thicken the lens for clear vision of close objects. Corrective lenses can solve this problem.

If the distance between the lens and the retina is too long, a person is nearsighted. The real image is formed in front of the retina instead of on it. Nearsighted people can clearly see objects nearby. But faraway objects are unclear. Nearsightedness can be corrected by concave lenses. Concave lenses change the path of light waves entering the eye, causing them to form an image on the retina instead of in front of it.

Do You Know?
Many experts believe that the Chinese were the first to invent and develop eyeglasses.

3. Ask the students to name the two main reasons why people wear glasses (farsightedness and nearsightedness). Have a student draw a diagram of a farsighted eye on the board. Explain that because the distance between the lens and the retina is too short, the image forms behind the retina, not on it. As a result nearby objects appear fuzzy.
4. Have a student draw a diagram of a nearsighted eye on the board. What has happened in this case (the distance between the lens and the retina is too long; the image forms in front of the retina; distant objects appear fuzzy)?
5. In front of the diagram of the farsighted eye draw a convex lens. Have students come to the board and diagram the light waves that come through each of the lenses (convex lens brings the light waves together sooner, causing the image to form on the retina; concave lens spreads light waves apart, thereby pushing the image back).

SCIENCE BACKGROUND

Accommodation is the property of the eye that enables it to adjust to see both near and far objects.

Astigmatism is a defect that occurs when the surface of the lens is curved more in one direction than in another.

6. Have the students form small groups. Each group should have a camera. Have the students examine the cameras. Each group should use the camera and text to prepare a poster that compares the parts of the eye with the parts of the camera. Alternatively, or as a home assignment, have each student prepare a report with diagrams to explain how the eye and the camera operate.

7. Use the overhead projector to show the students the transparency of the microscope. Ask what kinds of lenses are used (convex). Where are the lenses located (eyepiece and objectives)? How are they used to magnify objects (refract light waves from object being observed)?

ACTIVITY: Why Do You See Motion in Movies?

You can prepare index cards for the students to use by cutting unlined 7.5-cm × 12.5-cm (3 in. × 5 in.) index cards in half.

Some students may be interested in making "flipbook" movies based on the activity. To do this, each student will need 25 to 50 cards. The cards should be firmly stapled together on one side. The pictures should be drawn near the edge of the cards opposite the staples. Each picture should be only slightly different from the previous one.

● **Laboratory Skills: Manipulation of Materials**
Experimentation/Demonstration

SCIENCE BACKGROUND

The Dutch lens grinder Anton van Leeuwenhoek (1632–1723) made over 250 different simple microscopes. He used each microscope to examine a different material.

Oil immersion microscopes increase the magnification with a drop of oil between the lens and the slide.

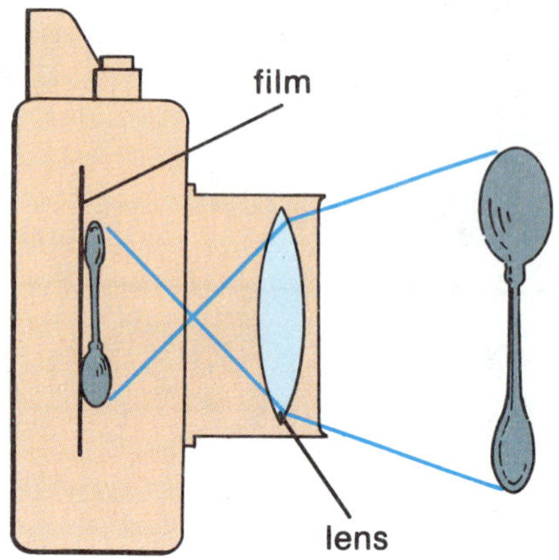

Figure 21.9 In a camera, light waves pass through a lens and form an image on photographic film.

A camera is similar to any eye. In an eye, a real image is formed on the retina. In a camera, a real image is formed on photographic film. The film contains chemicals that are affected by light. Light waves that strike the film cause chemical reactions to take place. These chemical reactions produce a photographic image.

A camera, like your eyes, has a lens. Your eyes focus on an object by changing the thickness of the lens. A photographer may focus a camera on an object by moving the lens back and forth.

Most microscopes use two or more convex lenses. Light waves from very tiny objects are refracted by the lenses so that a magnified image is formed. Many objects that were once mysterious and invisible to human eyes are now very familiar sights when seen through a microscope.

● ACTIVITY

Why Do You See Motion in Movies?

Materials
pencil
cellophane tape
white card

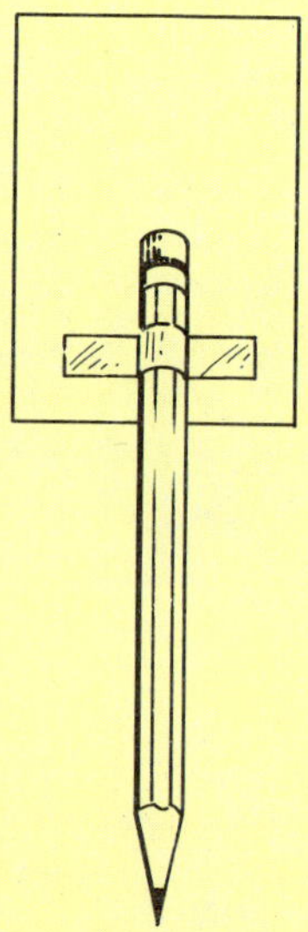

Procedure

1. Hold the card so the longer edge is parallel to the edge of your desk that is closest to you.
2. On one side of the card draw a tree. Make your lines heavy and dark.
3. On the other side of the card over the same area that has the tree, draw 10 small apples. Make your lines heavy and dark.
4. Tape the card to the top of your pencil as shown in the diagram.
5. Hold the pencil between the palms of your hands. Rub your hands together to spin the pencil. Observe the card.

1. The apples appeared to be on the tree and falling from it.

Questions

1. What happened to the apples when you spun the card?
2. Persistence of vision means that an image remains on the retina of your eye. Can you explain why the apples seemed to be on the tree and falling from it?
3. In a movie there are hundreds of individual pictures. Each differs only slightly from the previous picture. When you watch a movie, why do the objects in the still pictures appear to be moving?

2. Before one image disappears from the retina, the next arrives. The images overlap and you see them together.
3. Each image is retained by the retina for a fraction of a second. When the next image appears, they overlap. Since each is slightly different from the previous ones, the picture appears to move.

Figure 21.10 These binoculars use lenses to help the viewer see distant objects.

A microscope uses lenses to help you to see very tiny objects. The lenses in a telescope help you to see objects of enormous size. Your eyes need help in seeing these objects, not because they are small, but because they are so far away. Most **refracting telescopes** use two convex lenses to concentrate and bend light waves from distant objects. The light waves are bent so that they produce a magnified image of the object. **Reflecting telescopes** use a curved mirror and a convex lens. Thus, both reflection and refraction of light waves are involved in forming a magnified image.

Study Questions for 21.2

1. How is each of the following involved in seeing: lens, retina, muscles in the eye?
2. How do eyeglass lenses help farsighted and nearsighted people to see better?
3. How is a photographic image produced?
4. How do microscopes and telescopes help you to see tiny or distant objects?

1. lens—forms a real image on the retina; retina—is surface on which a real image is formed; muscles—change the thickness of the lens
2. farsighted—convex lenses cause light waves entering the eye to form a real image on the retina instead of behind it; nearsighted—concave lenses cause light waves entering the eye to form a real image on the retina instead of in front of it
3. The camera's lens directs light waves on a film. Chemical reactions occur on the film when the light waves strike it making a photographic image.
4. microscopes—use convex lenses to magnify tiny objects; telescopes—use convex lenses, and sometimes mirrors, to concentrate and bend light waves to magnify distant objects

8. If a telescope is available, allow the students to examine it. Ask them how its lenses are used (to bend light waves in such a way as to enable a person to see objects of enormous size from tremendous distances).
9. Show the students the transparency that compares the refracting and reflecting telescopes. What difference do they see between the two (refracting telescope has two convex lenses; reflecting telescope has a curved mirror and a convex lens)?

FOLLOW-UP

Reinforcement: How Do You See?

Ask the students to pretend that they are optometrists. Have them read the career in this chapter. Then have each student write a magazine article to explain how people see, what it means to be nearsighted and farsighted, and how these defects are corrected.

Enrichment: The Blind Spot

Refer students to a biology laboratory manual or a book of optics experiments that shows how to locate the blind spot. Have the students perform the experiment and explain what the blind spot is.

Tie-in: Microscopes for Specific Jobs

There are many types of microscopes. Some of these are the simple microscope, the compound microscope, the phase contrast microscope, the oil immersion microscope, and the electron microscope. Have the students work in groups of five. Each student should investigate one type of microscope and explain its importance.

SCIENCE BACKGROUND

Galileo invented the refracting telescope. Isaac Newton invented the reflecting telescope.

The largest telescopes are reflecting telescopes. Problems arise in trying to make their convex lenses any larger than one meter in diameter. At that size the lens becomes too heavy. It is also difficult to make a lens so large without imperfections.

21.3

OBJECTIVES: The student should be able to

- **describe** the properties of laser light;
- **list** the components necessary to produce laser light;
- **describe** the ruby laser and the helium-neon laser;
- **describe** some uses of lasers;
- **define** the term *hologram*;
- **describe** how a hologram is produced;
- **explain** how optical fibers carry signals and messages;
- **explain** how an endoscope is used.

SCIENCE TERMS

laser light
laser
holograms
optical fibers
endoscope

ADVANCED PLANNING

For the demonstrations you will need several photographs of a person; four 2-cm × 4-cm rectangles of paper labeled *laser*, *mirror*, *object*, and *film*; several paper arrows to represent light waves; an overhead projector; and a fiber optics lamp.

LESSON PLAN

1. Explain that by using a device called a laser it is possible to produce a narrow beam of light called laser light. Ask how laser light differs from white light (in laser light all light waves have the same wavelength and travel in the same direction; the waves are lined up).
2. Ask the students to list the three components that are necessary to produce laser light (presence of medium that can produce laser light, form of energy to excite atoms of medium, device to control the direction of laser light). List the three components on the board.

21.3 Lasers and Optical Fibers

You will find out
- what laser light is;
- how lasers produce laser light;
- how lasers are used;
- what optical fibers are.

Most light sources produce light waves of different wavelengths and those light waves travel in many directions. It is possible to have light waves all of the same wavelength and all traveling in the same direction. Every crest is lined up with every other crest. Every trough is lined up with every other trough. The result is an intense, narrow beam of light called **laser light.** Since all the light waves in laser light have the same wavelength, they are all of the same color. A device that produces laser light is called a **laser.**

A laser must have three basic components to produce laser light. First, a substance, or medium, that can produce laser light must be present. Second, a form of energy must be available to excite the atoms of the medium. The atoms must be excited in order to produce the light waves of laser light. Third, the direction of the laser light produced must be controlled. The controlling is usually done with mirrors.

Do You Know?
The word, ***laser,*** comes from the words ***l****ight* ***a****mplification by* ***s****timulated* ***e****mission of* ***r****adiation.*

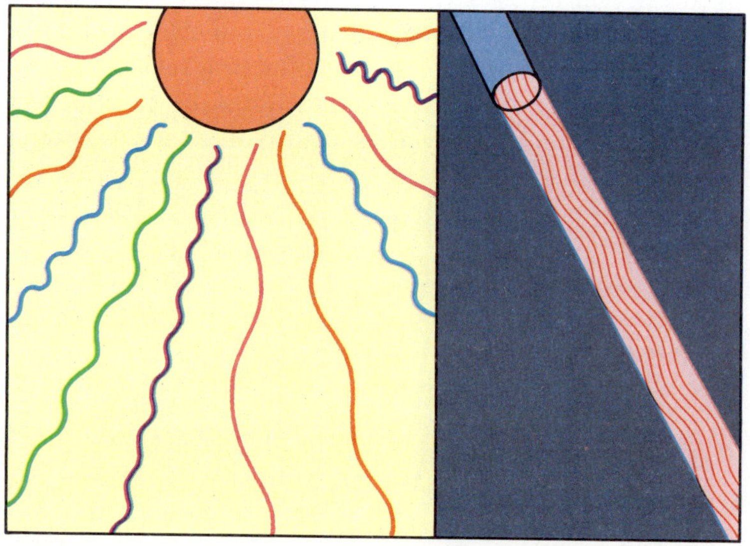

Figure 21.11 In laser light, all the waves are of the same wavelength and are traveling in the same direction.

The first laser developed was the ruby laser. In this laser, the substance that produces laser light is a solid ruby rod containing chromium atoms. Energy in the form of bursts of light is supplied to the ruby rod. A coating of silver on each end of the rod acts like a mirror to control the direction of the laser light produced.

One of the most commonly used lasers is the helium-neon laser. Helium and neon gases in a glass tube produce the laser light. An electric current is used to excite the atoms of helium and neon. Mirrors are used to control the direction of the laser light produced.

Lasers were first developed in 1960. Since then, their importance has grown greatly. For example, they are used to cut diamonds and weld metals. Eye doctors are using lasers to reattach torn retinas. Lasers are used in other kinds of surgery as well.

Figure 21.12 *left:* Using laser light to measure and align in the construction of a tunnel; *right:* Using laser light to cut a diamond

3. Beside the first item, *Medium*, write the words *solid ruby rod* and *helium and neon gases*. Explain that in the first laser a solid ruby rod containing chromium atoms was used to produce light. Today's lasers often use helium and neon gases in a glass tube.
4. Beside the second item, *Energy*, write the words *light* and *electric current*. As you draw a line to connect *ruby rod* and *light*, point out that the first laser used light energy to excite the atoms in the ruby rod. As you draw a line from *helium and neon gases* to *electric current*, point out that today an electric current is used to excite the atoms of helium and neon gas.
5. Beside the third item, *Devices to control laser light direction*, write *silver coating and mirrors*. Explain that in the ruby laser the energy produced is controlled or directed by coatings of silver on each end of the ruby rod.
6. Ask the students to name some uses for lasers (cut metal and gems, weld, perform surgery). Assign groups of students to research each topic in the tie-in activity and report to the class.

SCIENCE BACKGROUND

The laser was preceded in its development by a device called the maser, invented in the 1950's. The maser produced coherent microwave radiation.

7. Show the students a photograph of a person. Ask them how the photograph differs from the image of the person they see when they look at him or her (a person has three dimensions; a photograph has only two).

Comprehension Skill: Interpretative

8. Use the overhead projector to explain how holograms are produced. At the bottom of the overhead projector place a cutout of a laser with an arrow (light wave) coming from it. Have the laser light wave arrow split in two after it hits a mirror. One arrow should hit an object and then be reflected to a piece of film. The second arrow should go directly from the mirror to the film. Explain that when the two light wave arrows meet on the film, interference occurs and a hologram is produced.
9. Ask how the hologram can be seen (must be observed with laser light that is the same wavelength as the laser light used to produce the hologram).
10. Ask each student to create a series of diagrams with captions to show how a hologram is produced. Direct interested students to the enrichment activity.

SCIENCE BACKGROUND

Dennis Gabor was awarded the Nobel prize in 1971 in recognition of his invention of holography.

Do You Know?
A laser can burn a hole in a diamond in a thousandth of a second.

Lasers can produce three-dimensional photographs called **holograms** [HOH-luh-gramz]. A hologram shows all sides of a subject instead of just one. To make a hologram, a single beam of laser light is split into two beams by a mirror. One beam is directed at the subject of the photograph. When the laser light strikes the subject, it is reflected from the subject onto a photographic film. The second laser beam never strikes the subject. Instead, it is reflected by a mirror directly onto the same photographic film. The two beams of light meet on the photographic film. Interference occurs and creates a pattern on the film. The film is developed producing a hologram.

To see the hologram, you must direct laser light onto it. The laser light must be of the same wavelength that was used to produce the hologram. You see a three-dimensional image of the subject. In fact, if you change your viewing angle, you see another view of the subject.

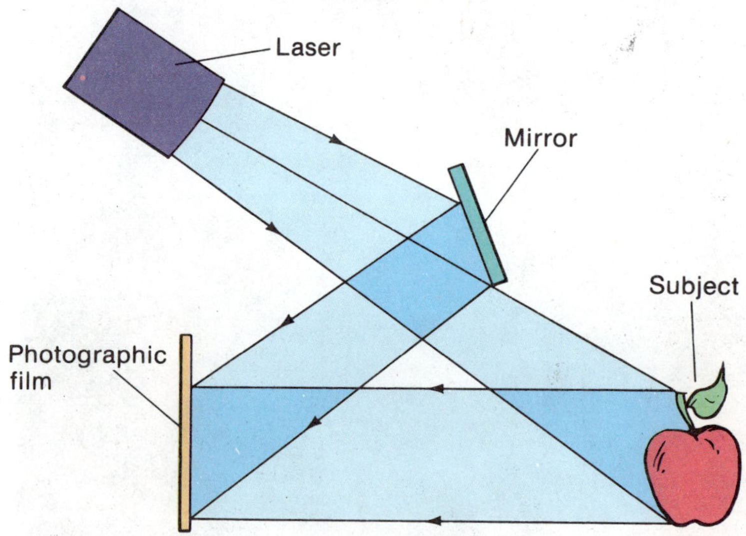

Figure 21.13 Producing a hologram

One of the biggest breakthroughs in the field of communications has been the use of laser light and ordinary light to carry messages. For example, laser light can carry 10,000,000 telephone messages at the same time! Conventional metal wires are able to carry far fewer messages.

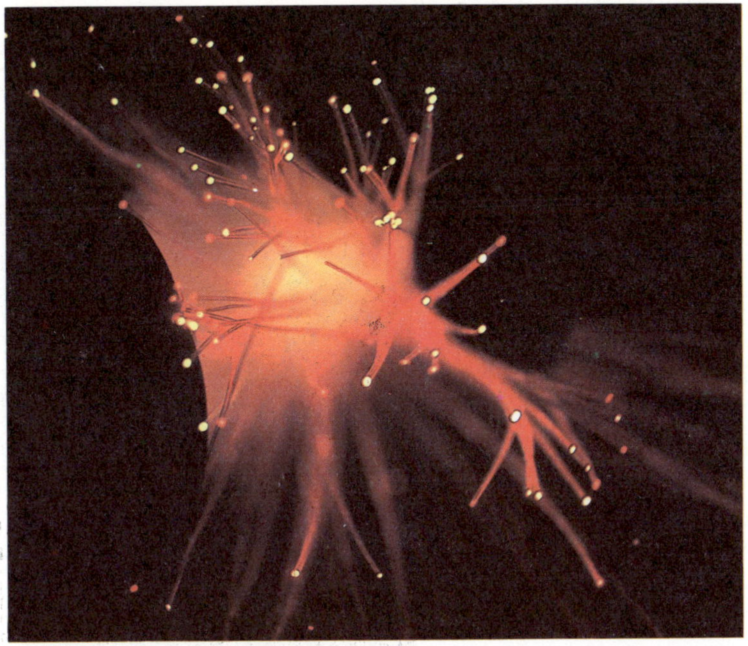

Figure 21.14 Optical fibers

Signals and messages are carried by light waves along thin strands of glass called **optical fibers.** Using the principle of refraction, scientists have designed optical fibers so that light waves become trapped inside and are able to move along them. The light waves travel very quickly. They can carry a large amount of information in a short amount of time. Optical fibers using laser light carry messages over long or short distances. Ordinary light can also be used for short distances.

Optical fibers carrying ordinary light are used by doctors in a device called an **endoscope** [EHN-duh-skohp]. An endoscope can be inserted into the body. It allows a doctor to examine internal organs without having to perform surgery. Endoscopes are frequently used to examine the lungs and the digestive system.

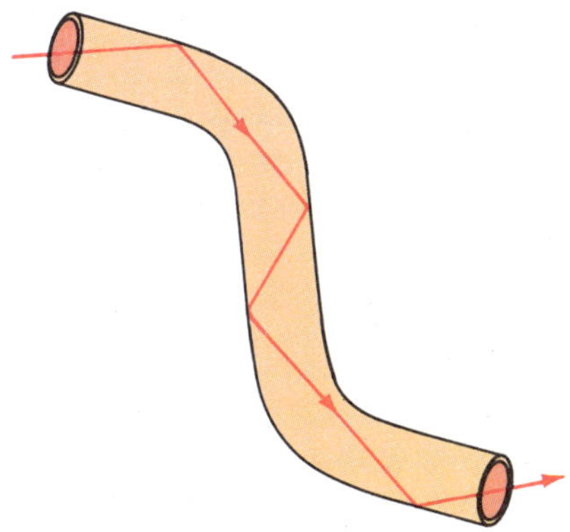

Figure 21.15 The path of light waves in optical fibers

Study Questions for 21.3

1. What is laser light?
2. What are the three basic components of a laser?
3. Name two uses of lasers.
4. What are optical fibers?

1. light waves all of the same wavelength traveling in the same direction
2. a substance that can produce laser light, a form of energy to excite the substance, a way to control the direction of the laser light
3. to carry messages in optical fibers; to reattach torn retinas
4. thin strands of glass that use light waves to carry signals and messages

11. Darken the room and turn on the fiber optics lamp. Point out that the light source is in the center of the lamp. Ask how the light travels from the center to the tips of the fibers (light is reflected along the inside of the fiber). What implications does the use of optical fibers have for communication (more messages can be carried by optical fibers than by wires; fibers are less expensive than wires)?

FOLLOW-UP

Reinforcement: Laser Light

Ask each student to prepare a chart entitled *The Production of Laser Light*. The chart should include the three components needed to produce laser light and an explanation and example of each.

Enrichment: Holography

The use of holograms is new but scientists are finding many uses for them. Ask students to research the use of holograms in microscopy, interferometry, bioholography, and acoustic holography. They should report their findings to the class.

Tie-in: Lasers in Our Lives

Today lasers are used in surgery, communication, meteorology, defense, and light shows. Ask an interested group of students to prepare a series of models or diagrams to describe the use of lasers in people's lives. Do they predict that laser use will increase? How do they justify their position?

What Did You Learn?

- 21.1 • Lenses are curved pieces of glass or plastic that refract light waves.
- 21.1 • A picture formed by a lens is called an image.
- 21.1 • A convex lens has curved surfaces that bulge out at the middle to bring light waves together.
- 21.1 • A concave lens has curved surfaces that dip in at the middle to spread light waves apart.
- 21.1 • The focal length of a lens is the distance from its center to its focal point.
- 21.1 • A virtual image appears right side up and cannot be focused on a screen.
- 21.1 • A real image appears upside down and can be focused on a screen.
- 21.2 • The lens in your eye forms a real image on the retina.
- 21.2 • Muscles in your eye change the thickness of the lens so you can see both close and distant objects clearly.
- 21.2 • In farsighted people, the real image is formed behind the retina instead of on it.
- 21.2 • In nearsighted people, the real image is formed in front of the retina instead of on it.
- 21.2 • A camera, like your eye, uses a lens to form a real image on a surface.
- 21.2 • Microscopes and telescopes use convex lenses to form magnified images.
- 21.3 • Laser light is made up of light waves all of the same wavelength traveling in the same direction.
- 21.3 • Lasers involve a substance that can produce laser light, an energy source, and a way to control the direction of the laser light.
- 21.3 • Lasers are used to make holograms.
- 21.3 • Optical fibers are thin strands of glass that use light waves to carry signals and messages.

Key Terms

lenses
image
convex lens
concave lens
focal point
focal length
virtual image
real image
retina
refracting telescopes
reflecting telescopes
laser light
laser
holograms
optical fibers
endoscope

Career

Optometrist

Optometrists are health professionals who specializc in the diagnosis of vision problems and the prescription of lenses to correct those problems. Optometrists are the major providers of vision care. They are concerned with how general health problems affect vision, color vision, and depth perception and how well a person sees. Some optometrists work in private practice. Other optometrists work in groups with other health professionals to provide complete health care.

An office visit to an optometrist generally involves a 45-minute to 1-hour examination of the eye and vision capabilities. Since many instruments are used in the examination, a person planning to be an optometrist should have a good mechanical aptitude as well as an interest in physics, math, and biological science.

To become an optometrist, you need 5 to 6 years of study at a school of optometry. There are 10 accredited schools of optometry in the United States. The course of study includes the basic health sciences, optics and visual science, behavioral and community science, and clinical science and services. After graduation, you must take clinical and written examinations in order to obtain a license.

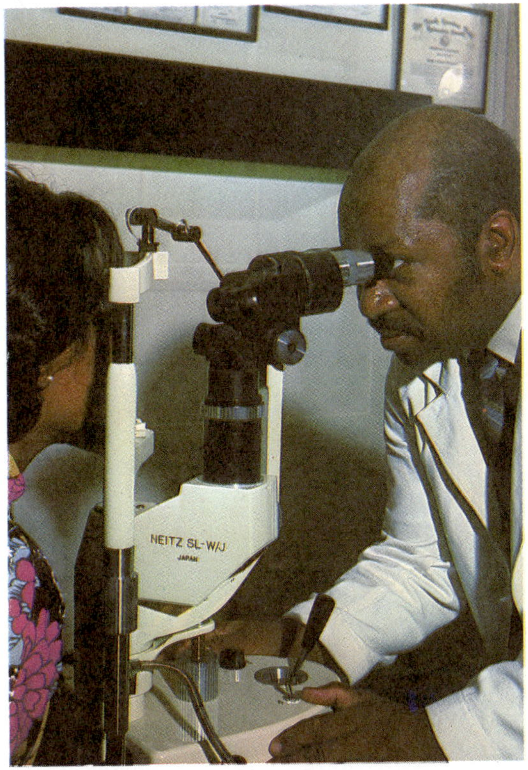

Vocabulary

1. reflecting telescope
2. true
3. retina
4. real image
5. true
6. true
7. convex lens
8. optical fibers
9. lasers
10. true

● Word Skill: Vocabulary

What Do You Remember?

1. a curved piece of glass or plastic that refracts light waves
2. convex—thicker in the middle and thinner at the edges; concave—thinner in the middle and thicker at the edges
3. brings them together
4. spreads them apart
5. real image
6. The muscles in the eye change the thickness of the lens so you can see close objects and distant objects.
7. In nearsighted people, the distance between the lens and the retina is too long. This condition causes the real image to form in front of the retina instead of on it.
8. Both use a lens to focus a real image on a surface. In the eye, that surface is the retina. In the camera, that surface is the film.
9. light waves all of the same wavelength, traveling in the same direction
10. thin strands of glass that carry messages using light waves

TO THINK ABOUT AND DO

Vocabulary ●

Number your paper from 1 to 10. Read each of the sentences. If the sentence is true, write *true*. If it is false, make it true by changing the underlined expression.

1. A refracting telescope uses a system of lenses and mirrors to view stars and planets.
2. The distance between a lens and the point where the light waves come together is called the focal length of the lens.
3. Photographic film in a camera is like the lens of the eye.
4. A virtual image can be focused on a screen.
5. A three-dimensional photograph produced by laser light is called a hologram.
6. A concave lens spreads light waves apart.
7. A magnifying glass has a concave lens.
8. Images are used to send telephone messages.
9. Lenses produce light waves of one wavelength that all travel in one direction.
10. A doctor can use an endoscope to see inside the lungs.

What Do You Remember?

1. What is a lens?
2. Name and diagram the two basic kinds of lenses.
3. How does a convex lens affect the light waves that pass through it?
4. How does a concave lens affect the light waves that pass through it?
5. What kind of image is found on the retina?
6. How do your eyes change to clearly see objects at different distances?
7. Why are some people nearsighted?
8. How is a camera like an eye?
9. What is laser light?
10. What are optical fibers?

Applying What You Have Learned

1. Lighthouses use lenses to concentrate light into a bright beam that is easily seen. What kind of lenses do they use?
2. You may have noticed some people holding a book or paper out at arm's length while reading. Who might do this and why?
3. How is a hologram different from an ordinary photograph?
4. Birdwatchers often use binoculars to enable them to examine birds from a distance. What kind of lenses are used in binoculars?

Research and Investigation

1. In 1982 Alan Scott, an ophthamologist, reported that a derivative of the poison that causes botulism can be used to correct eye defects. Investigate Scott's drug, Oculinum, and report on how it functions.
2. Some people wear glasses that are bifocal or trifocal. What are bifocal and trifocal glasses?

Applying What You Have Learned

1. convex
2. A farsighted person might hold a book or paper at arm's length to read it. Without glasses, that person sees nearby objects unclearly but more distant objects clearly.
3. hologram—allows one to see a three-dimensional image, is made using laser, must be viewed in laser-light; ordinary photograph—is two-dimensional, is made using ordinary light, can be viewed in ordinary light
4. convex

Comprehension Skill: Applied

Research and Investigation

1. Oculinum is used to treat crossed eyes. When injected into the eye muscle, it blocks nerve signals, allowing the muscle to relax and stretch. The muscle on the opposite side of the eye can then pull the eye into line. The effects wear off after several months. But because the muscles have been altered during this time, the eye is held in a balanced position.
2. Bifocals and trifocals aid in the vision of both nearby and distant objects. Bifocals have two lenses. Trifocals have three. The top lens is designed for distance vision. The bottom lens is designed for close vision. In trifocals, the middle lens is for vision between close and distant vision.

Reference/Research Skills

Long-Term Projects

1. Have the students identify and bring in recordings of popular songs that deal with energy.
2. Have the students work in groups to create dioramas to illustrate how the fossil fuels were formed.
3. Ask the students to decide whether they are for or against the use of nuclear energy. Select students to prepare and present a debate concerning the use of nuclear energy. Encourage the students to use visual aids when they present their arguments.
4. Have students do research on the relationship between our current energy sources and the pollution and health problems they cause. Have them write a series of articles for the local newspaper in which they discuss the problems and present possible solutions.
5. Have the students design and build kites of different sizes and shapes. They should use the kites to demonstrate wind energy. Sponsor a kite flying contest for your students.
6. Authors of science fiction stories use their imaginations to see a new world based on known scientific principles. Have each student write a science fiction story based on one of the principles you discussed this year.
7. Have each student create a set of cards for a game of *Concentration* using the vocabulary words from this unit. One card in each pair should contain a word and the other card should have an illustration or definition of the word. Allow the students to play the game to reinforce the vocabulary words.

Unit Six

Energy and the Future

Bulletin Boards

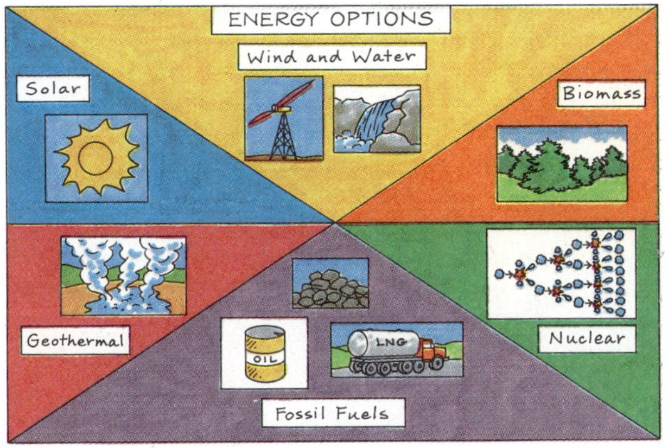

1. Divide the board into six parts as in the diagram. In the section marked *Solar*, draw the sun. Under *Wind and Water*, show a modern windmill and a waterfall. In the biomass section, mount a picture of a tree surrounded by several plants. Above *Geothermal*, place a picture of Old Faithful. In the fossil fuel section have a picture of a lump of coal, a can of oil, and a railroad car labeled LNG. In the nuclear section, show a diagram of a chain reaction or a picture of a nuclear power plant.

2. Obtain a map of your city from city hall. Mount it in the center of your board. Title the board *Acid Rain in* ______(city)______. Provide each student with some pH paper and a pH scale. Each time it rains have the students take the pH of the rainwater near their home. Instruct them to compare their pH paper to the scale while it is wet and to write the pH on the paper. Attach each paper to the location of the students' homes on the map.

3. To help students become familiar with the different objects in space, title your board *Objects in Space*. Mount and label pictures of the objects discussed in Chapter 24. Have students add their descriptions.

CHAPTER BACKGROUND

Over three thousand years ago the Chinese were puzzled by the discovery of "black rocks," oil, and gas seeping from the ground. They used the black rocks, or coal, to melt their copper and iron ore. They collected the oil and gas for use in heating and cooking.

Today we know that coal, oil, and natural gas are fossil fuels that formed when heat and pressure caused the remains of primitive plants and animals to acquire a new chemical structure. We know, too, that the formation of fossil fuels occurred over millions of years and our modern society uses the resources too quickly for nature to replace them. We use the fossil fuels for heating and cooking food as well as for producing plastics, synthetic fibers, gasoline, jet fuel, oils, waxes, fertilizers, cleaning fluids, and insecticides. Today we are concerned with meeting the need for an equivalent of 95 million barrels of oil a day in the year 2000.

One answer to this tremendous need may lie in nuclear energy. Today's nuclear reactors can produce as much electricity from 1 kg (about 2 lbs.) of uranium as a conventional power plant produces from 5,000 kg (about 10,000 lbs.) of fossil fuel. The major difference between the two plants is in how the steam is produced that drives the turbines. Nuclear power is not a trouble-free source of energy. The hazards of radioactivity cause some people to prefer another energy source.

While technology is applied to solve the problems of nuclear energy, conservation appears to be the best short-term answer to energy deficiencies. Government, industry, and consumers must work together to conserve the present energy supply. Each group must realize that its actions affect the world as we know it today and the world as future generations will know it.

22 Energy and Society

To introduce the chapter, put the information that follows on the board. You may want to express it in bar-graph form. Discuss with the students the meaning and significance of this information.

How is energy used in the United States (36 percent industry, 26 percent transportation, 21 percent residential, 17 percent business and commercial)? What energy sources are used in the United States (47 percent oil, 25 percent natural gas, 18 percent coal, 4 percent nuclear, 4 percent hydroelectric, 2 percent other)?

During the study of this chapter you may wish to refer to other chapters that present related information. Chapter 9 relates to the discussion of fossil fuels. It covers hydrocarbons and the distillation of crude oil. Chapter 11 relates to the discussion of nuclear energy. It covers fission and fusion in detail. Chapter 16 relates to the use of energy to generate electricity. It explains how a generator produces electricity.

Did an electric alarm clock wake you this morning? How was your breakfast cooked? Did you get to school by car or bus? Think for a moment about your life-style. You probably take for granted the use of cars, buses, heating systems, cooling systems, lights, and other things. Where does the energy come from to provide you with this life-style?

In this chapter you will learn about the two main sources of energy in the United States. You will find out how your life-style is maintained by these energy sources. You will also learn about what is being done to ensure that there will be enough energy available in the future. While these energy sources do answer the current need for energy, they also create problems. The solution to those problems should be an interest and a concern of everyone.

Figure 22.1 A supertanker entering Boston Harbor

Lesson Organizer

22.1 Fossil Fuels

Pages 456–461
Advanced Planning: Page 456
Follow-up: Page 461

- **R** The Fossil Fuels
- **E** Offshore Drilling
- **T** Using Genetics to Clean up Oil Spills

22.2 Nuclear Energy

Pages 462–465
Advanced Planning: Page 462
Follow-up: Page 465

- **R** Nuclear Power
- **E** Nuclear Power Plants
- **T** The Efficiency of Power Plants

22.3 Energy Conservation

Pages 466–469
Advanced Planning: Page 466
ACTIVITY: Do You Have Too Many Windows? Page 467
Follow-up: Page 469

- **R** Conservation Scrapbook
- **E** Comparing Insulating Materials
- **T** Energy Storage

See ACTIVITIES book pages 127–132 for laboratory, skill, and review exercises for Chapter 22.

22.1

OBJECTIVES: The student should be able to

- **explain** what is meant by the term *fossil fuels*;
- **explain** how fossil fuels were formed;
- **identify** the original source of energy in fossil fuels;
- **explain** how energy is stored in fossil fuels;
- **list** some uses of fossil fuels;
- **describe** how fossil fuels are obtained;
- **discuss** how oil is transported;
- **explain** why natural gas is liquified for transport;
- **discuss** some problems associated with fossil fuels.

SCIENCE TERMS

fossil fuels	tar sands
shaft mining	oil shale
strip mining	acid rain

ADVANCED PLANNING

1. Obtain a large map of North America on which to locate Prudhoe Bay and Port Valdez in Alaska, a large map of the United States, and a map of your city or town.

2. Have pH paper available.

3. You may want to obtain the activity master program, *The Energy Challenge*, from the Technical Information Center, P.O. Box 62, Oak Ridge, TN 37830.

LESSON PLAN

1. Introduce the lesson by asking the students what type of fuel they burn in their homes (oil, natural gas, coal, wood). Tell the students that oil, natural gas, and coal are called fossil fuels.

● **Comprehension Skill: Interpretative**

2. Ask why coal, oil, and natural gas are called fossil fuels (they were formed from the remains of plants and animals that lived millions of years ago).

SCIENCE BACKGROUND

Ninety-three percent of our energy comes from fossil fuels—23 percent from coal; 30.4 percent from oil; and 39.6 percent from natural gas.

22.1 Fossil Fuels

You will find out
- what fossil fuels are;
- how fossil fuels are obtained;
- what some of the problems associated with fossil fuels are.

Close to 90 percent of the energy used in the United States comes from coal, oil, and natural gas. These energy sources are called **fossil** [FAHS-uhl] **fuels.** Fossil fuels are fuels formed from the remains of plants and animals that lived millions of years ago. Energy from the sun was stored in the plants. When the animals ate the plants, they obtained and stored the energy. The energy was stored in the chemical bonds of molecules in the plants and animals.

After the plants and animals died, they were buried under heavy layers of mud and sand. They only partially decayed. The chemical bonds and the energy contained in them were preserved. Depending on the specific kind of material and the conditions, oil, coal, or natural gas was formed. These fossil fuels are used to do many things such as power cars, heat and cool buildings, and generate electricity. ●

Figure 22.2 Drilling for oil offshore

Do You Know?
The United States has 6 percent of the world's population and uses 35 percent of the total energy used in the world.

Figure 22.3 Strip mining

Coal is the most plentiful fossil fuel in the world. Some coal deposits are deep underground and must be taken out through tunnels, or shafts, cut into the earth. This method, called **shaft mining,** is expensive and can be dangerous. When the coal lies near the surface, tunnels are not needed. This surface coal can be exposed with explosives. It is then removed with giant shovels. This practice is called **strip mining.** Strip mining is the least costly and safest way to get coal. However, it completely destroys the environment in and around the area being mined. Mine operators are now required to repair the damage done to the environment.

Oil is found in many places around the world in underground pools. Sometimes the oil is mixed with other substances. Deposits of oil mixed with sand are called **tar sands.** Oil is also found in a rock called **oil shale** [SHAYL]. Scientists are working on ways to separate out and use the oil in tar sands and oil shale.

3. Ask what provides the energy that plants need to live (sun). Ask how animals obtained and stored this energy (they ate the plants). Explain that when people burn coal today they are releasing the sun's energy that was stored in plants and animals millions of years ago.
4. Have the students compare the process of strip mining and shaft mining (in strip, or surface, mining, bulldozers, and power shovels remove coal from shallow coal seams near the ground's surface; in shaft, or tunnel, mining, coal is removed from several hundred meters below the surface, then passed through a shaft to the surface).
5. Ask the students which type of mining, strip or shaft, would allow each miner to remove more coal per day (strip mining). In which type of coal mine can a greater percentage of the total be mined (strip, 80%–90% versus 50%–80%)? Which type can operate less expensively (strip)? Which type is less damaging to the environment (shaft)?
6. Stress that strip miners are now required to repair the land they damage. This is called land reclamation. Have the students prepare for a debate in which they argue the pros and cons of each type of mining. Can they think of a compromise between the two?
7. Write the terms *tar sands* and *oil shale* on the board. Ask the students to identify each (tar sands—oil mixed with sand; oil shale—oil locked in rocks).
8. Have the students find Colorado, Utah, and Wyoming on the map of the United States. Point out that these states contain deposits of oil shale.
9. Have the students find the province of Alberta on the map of North America. Explain that in Alberta, huge swamps contain tar sands that are only 1 m to 6 m (3 ft. to 18 ft.) below ground. The tar sands are dug and washed with steam and hot water to remove the oil.

SCIENCE BACKGROUND

Coal has been mined in the United States since 1673. Bituminous is the most plentiful coal. It can provide 24,000 million joules of energy per metric ton. Anthracite is the most carbon-rich coal. It burns with almost no smoke. Peat is the youngest coal. It is a lightweight, crumbly material that contains leaves and bark.

10. Ask the students how oil and natural gas are obtained (drilling). Ask what methods are used today to locate oil and natural gas (aerial and satellite photographs, sound waves, geological studies).
11. Ask the students to locate Prudhoe Bay and Port Valdez in Alaska on the map of North America. How is the oil drilled in Prudhoe Bay transported to Port Valdez (Alaskan pipeline)? Explain to the students that in addition to pipelines, supertankers are also used to carry oil and gas to refineries.

SCIENCE BACKGROUND

Tar sand is also called bituminous sandstone.

Tar sand deposits in Alberta, Canada, contain 820 billion barrels of oil. This is more oil than has been produced in the history of the United States and Canada together.

Oil that has just been pumped from under the ground is called crude oil, or crude.

Engineers considered not only the environment but also the structure of the Alaskan pipeline when they chose not to bury it. If the pipeline were buried, it could melt the permafrost. If the permafrost melted, the pipes would sink into the ground, rupture, and spill the oil.

A lot of time, effort, and expense are involved in finding oil. Oil prospectors study aerial and satellite photographs. They use sound waves to study the earth's outer layer. This and other information is used to determine the kinds of rocks present in an area and the layering of those rocks. Certain rock structures are known to be associated with oil deposits. If the indications suggest that oil is present, a well may be drilled. Some wells strike oil. Others come up dry.

The oil that comes out of a well is called crude oil. The crude oil must be transported to a refinery. At a refinery, it is separated into gasoline, kerosene, diesel fuel, heating oil, and other products.

Crude oil is often transported over land through pipelines. The recently completed Alaskan pipeline covers 1,287 km. It goes from the Arctic Ocean to the southern coast of Alaska. It took 20,000 people three years to build. More than half of the pipeline is on stilts. If the

Figure 22.4 The Alaskan pipeline

Figure 22.5 An LNG tanker

pipeline were buried, heat from the oil flowing through it could have a negative effect on Arctic plants and animals. If the pipeline were on the surface of the ground, it would block the path of migrating caribou and reindeer.

Giant ships called supertankers transport crude oil at sea. Some supertankers are over 350 m long. Crew members often use bicycles to get from one end of the ship to the other!

Like oil, natural gas is found deep underground. To make transportation easier, the natural gas is often cooled and compressed above ground to form a liquid. This liquid is called liquified natural gas or LNG. You may have heard of LNG tankers or seen trucks carrying LNG.

The chemical energy stored in the chemical bonds of fossil fuels is released when they are burned. Your lifestyle depends on this energy. Unfortunately, some unhealthy compounds are produced and released into the air when fossil fuels are burned. One such compound is carbon monoxide. Carbon monoxide is a deadly poison. It prevents your body from getting the oxygen it needs. People have died from breathing air having too high a concentration of carbon monoxide.

12. Write the letters *LNG* on the board. What do the letters stand for (liquified natural gas)? Remind the students that gas is liquified before it is transported. Can the students explain why this is done (takes up less space, safer)?

13. Ask how the burning of fossil fuels can damage the environment (ashes, sulfur oxides, nitrogen oxides, carbon monoxide, and acid rain can result).

14. Assign students to bring in samples of rainwater. Test the samples with pH paper. On a map of your city or town, mark the locations tested and their pH values. Could precipitation harm these areas (yes, if it is very acidic)? How (it can damage plant and animal life, and contaminate the water supply)?

SCIENCE BACKGROUND

A danger to coal miners is black lung disease. The miner's lungs become filled with coal dust, making breathing difficult.

Figure 22.6 The burning of fossil fuels contributes to air pollution.

Figure 22.7 Much of the damage to this statue was caused by air pollution.

Two other unhealthy compounds produced by the burning of fossil fuels are oxides of sulfur and nitrogen. The breathing of these oxides can be harmful. When sulfur and nitrogen oxides combine with water in the atmosphere, they form acids. These acids fall to the ground as **acid rain.** Acid rain can damage both plant and animal life. It can also contaminate water supplies.

Small particles are also released into the air when fossil fuels are burned. These particles, together with the unhealthy compounds produced, create air pollution. People who breathe polluted air can suffer from eye, nose, and throat irritations, breathing problems, and even cancer.

Steps are being taken to control air pollution. Devices that control or prevent the release of pollutants into the air are being used in cars, factories, and other places where fossil fuels are burned. Even so, the problem is far from solved. It remains a concern for everyone.

The transportation of fossil fuels can also cause damage to the environment. Oil pipelines are designed to minimize their effects on the environment. Even so, their construction and mere existence change the natural state. Supertankers carrying crude oil sometimes accidentally spill their cargo. These oil spills cause serious damage to the sea and coastal environments. Much plant and animal life is destroyed. On land, plant and animal life are also destroyed by the practice of strip mining.

Another problem related to the use of fossil fuels is their availability. Fossil fuels are being used up much more quickly than they are forming. New fossil fuels are forming today. But the formation process takes millions of years! Clearly, at the rate fossil fuels are being used, there will certainly come a time when they are no longer available.

One possible solution to the problem of dwindling fossil fuel resources is the development of synthetic fuels. Scientists can make alcohol from sugar cane and other plants. They are currently exploring ways to use alcohol to fuel cars. Both alcohol alone and in combination with gasoline (gasohol) are possible energy sources for the future.

Do You Know?
Some oil wells extend more than 8 km into the earth.

Figure 22.8 Oil spills can cause serious environmental damage.

Study Questions for 22.1

1. What are fossil fuels?
2. Where are coal, oil, and natural gas found?
3. Name two problems associated with the use of fossil fuels for energy.

1. fuels formed from the remains of plants and animals that lived millions of years ago
2. coal—deposits deep underground and surface deposits;
oil—underground pools, tar sands, oil shale;
natural gas—deep underground
3. air pollution, environmental damage, dwindling supplies

15. Stress that the environmental pollution caused by fossil fuels is only one of the problems associated with their use. Another problem is the fact that we are running out of fossil fuels. One answer to this problem is gasohol. Is gasohol sold where your students live? What is its main use (replaces or supplements gasoline in cars)? How is gasohol produced (from sugarcane and other plants)?

FOLLOW-UP

Reinforcement: The Fossil Fuels

Have the students form three groups. Assign each group one of the fossil fuels. Have each group prepare a series of posters or models to be used to explain how their fuel is obtained, produced, and used. What are the problems associated with its use?

Enrichment: Offshore Drilling

In an effort to find more domestic oil, oil companies are drilling for oil in offshore areas. Ask the students to research offshore drilling methods and compare them to drilling methods used on land. They should create charts to compare the two methods.

Tie-in: Using Genetics to Clean Up Oil Spills

Genetic engineering has been used to produce bacteria that can digest oil. It is hoped that these bacteria will be able to clean up oil spills. Ask the students to do research to learn more about these bacteria, how they are produced, and how they can be used.

22.2

OBJECTIVES: The student should be able to

- **explain** what is meant by the term *nuclear energy*;
- **distinguish** between fission and fusion;
- **explain** how a nuclear-fission power plant produces electricity;
- **describe** how nuclear chain reactions are controlled;
- **explain** the importance of a breeder reactor;
- **discuss** some problems associated with nuclear reactors;
- **describe** the difficulties of designing and operating fusion reactors.

SCIENCE TERMS

nuclear reactor
control rods
breeder reactor
thermal pollution
deuterium

ADVANCED PLANNING

1. You may want to have some models of atoms available as you review atomic structure.

2. Build a cardboard model of a nuclear reactor with movable control rod sticks.

LESSON PLAN

1. Introduce the lesson by reviewing the structure of the atom. If you have some models of atoms, have the students identify the nucleus, protons, neutrons, and electrons. If models are not available, have students refer to the illustrations of the atom in Chapter 6.
2. Ask the students what nuclear energy is (the energy released when an atom's nucleus undergoes a nuclear reaction). Then have the students identify two kinds of nuclear reactions (fission and fusion). How can they be distinguished (fission—a heavy nucleus is split into lighter nuclei; fusion—two light nuclei are joined together to form a heavier nucleus)?

22.2 Nuclear Energy

You will find out

- how fission is used to generate electricity;
- what some of the problems associated with nuclear reactors are;
- why fusion reactors are difficult to design and operate.

Most of the electricity in the United States is generated with fossil fuels. Another energy source that is also being used to generate electricity is nuclear energy. Nuclear energy is released when the nucleus of an atom undergoes a nuclear reaction.

There are two main kinds of nuclear reactions. In fission, energy is released when a heavy nucleus is split into lighter nuclei. In fusion, energy is released when two light nuclei join together forming a heavier nucleus. Society's demand for energy is high and continues to grow. Use of the large amounts of energy released in nuclear reactions is one possible way to meet this growing demand.

Scientists continue to explore ways to obtain and use fission and fusion energy. So far, efforts involving fission have been more successful than those involving fusion. Fission-powered submarines, aircraft carriers, and space probes have been launched. In the United States and in other countries, electricity is being generated using energy from fission.

Figure 22.9 The Indian Point nuclear power plant in New York

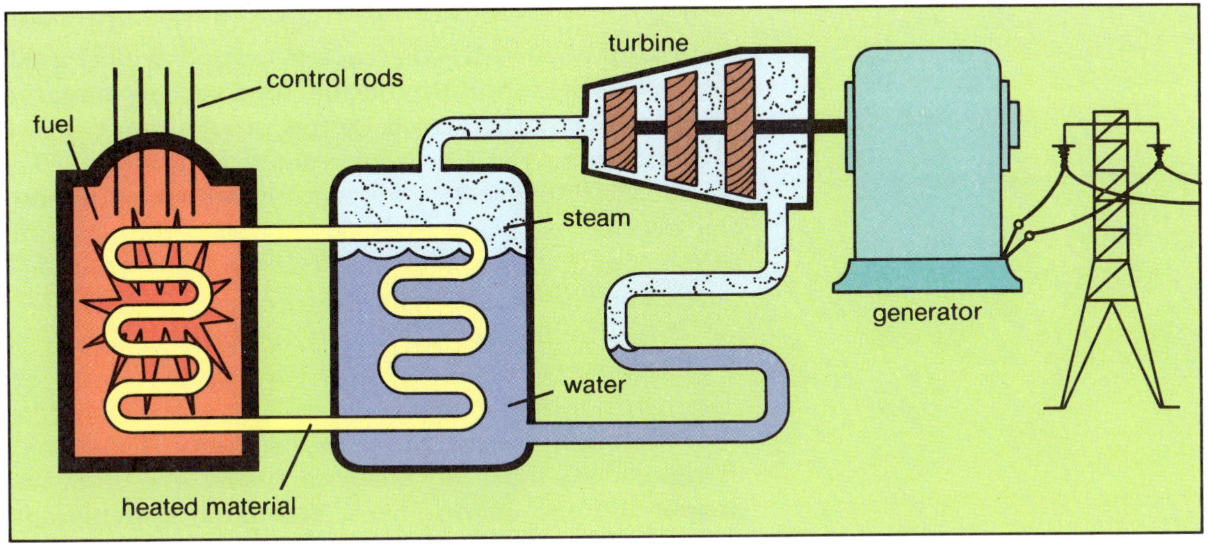

Figure 22.10 A nuclear power plant

A nuclear-fission power plant generates electricity. The power plant uses a fissionable material for fuel. A material that is fissionable is one that is able to undergo fission. The fuel most commonly used is uranium 235. The uranium fuel undergoes fission in a chain reaction. The energy released from the chain reaction is absorbed as heat by a material such as water. The heat is used to produce steam. The steam is directed at the blades of a turbine. The rotating turbine is used to generate electricity.

The chain reactions take place in a **nuclear reactor.** It is important to have effective control of chain reactions because they proceed very quickly and involve large amounts of energy. In the reactor, **control rods** made of steel mixed with another element, such as boron or cadmium, control the chain reaction. They control the reaction by absorbing neutrons.

To stop the chain reaction, control rods are lowered into the reactor. They absorb neutrons, preventing those neutrons from bombarding the uranium nuclei and continuing the chain reaction. As the rods are raised, fewer neutrons are absorbed and the chain reaction can proceed. By positioning the control rods at just the right level, the stopping, starting, and speed of a chain reaction can be controlled.

Do You Know?
A sample of uranium 235 the size of a golf ball can provide the same amount of energy as more than 500,000 liters of oil.

3. Ask how electricity is produced in a traditional power plant (steam spins a turbine, which turns a generator). Explain that the main difference between the traditional plant and a nuclear plant is in how the steam is produced. How is steam produced in traditional power plants (by burning fossil fuels)? Explain that in a nuclear power plant, uranium 235 is bombarded with neutrons, producing a chain reaction. The heat released by the reaction is used to produce steam.

4. Show the students the nuclear reactor model. Explain that by lowering and raising the reactor's control rods, the nuclear reaction can be controlled. What happens when the rods are lowered (they absorb neutrons, slowing down the reaction)? What happens when the rods are raised (fewer neutrons are absorbed and the reaction continues at a faster rate)?

5. What is a breeder reactor (a reactor that produces nuclear fuel; in such a reactor, the uranium produces plutonium 239, which can then be used as a fuel)? How will it help solve the problem of an insufficient supply of uranium (by producing plutonium 239, which can be used as a fuel, the uranium supply can be made to last longer)?

6. Review with the students the dangers of using fossil fuels. What are some problems associated with the use of nuclear reactors (radioactivity, radioactive waste, thermal pollution)?

How much uranium is there on Earth? Scientists predict that if nuclear-fission reactors become widely used, the uranium supply could be used up in about a hundred years. The answer to this problem may be the **breeder reactor.** In a breeder reactor, plutonium 239 is produced from the fission of uranium. The plutonium can also be used as a fuel. In this way, the uranium supply could be made to last longer. However, plutonium fuel is especially dangerous to work with. So far, this problem has prevented any large-scale use of breeder reactors.

As with other energy sources, there are concerns about nuclear energy. Since radioactivity is involved, extreme care must be taken to insure the safety of people and the environment. Protection systems and back-up systems must be designed, thoroughly tested, and used.

The waste products of nuclear reactors are radioactive. These wastes will remain radioactive for thousands of years. How and where can they be safely discarded? Many people are concerned about the safety of radioactive waste dumps. Scientists and engineers are working on a solution to this problem.

Figure 22.11 The control room of a nuclear power plant

Enormous amounts of heat are involved in the production of nuclear energy. Water from a lake or river near a nuclear reactor is often used to absorb and take away excess heat. This practice raises the temperature of the water and the land around it. This unnatural increase in temperature is called **thermal** [THUR-muhl] **pollution.** Thermal pollution can damage or destroy plant and animal life.

Scientists are working on ways to produce useful energy from nuclear fusion. One kind of fusion reaction involves an isotope of hydrogen having one proton and one neutron. This isotope is called **deuterium** [doo-TIHR-ee-uhm]. The oceans contain a large supply of deuterium. The fusion of deuterium produces no radioactive wastes. It releases a large amount of energy. With the supply of deuterium plentiful, why isn't fusion energy being used?

For deuterium to undergo fusion, it must be in the plasma state. This state requires temperatures of many millions of degrees Celsius. Such a condition exists naturally on the sun but is very difficult to produce and control here on Earth. Scientists at the Princeton Plasma Physics Laboratory have succeeded in producing the plasma state in their experimental fusion reactor. Researchers at Princeton and elsewhere are making progress in the development of fusion energy.

Another problem with fusion energy is that a nuclear-fusion reactor becomes highly radioactive. It would require very expensive shielding to protect the environment. As a result, even the most optimistic scientists agree that it will be many years before a practical fusion reactor is built and operating.

Study Questions for 22.2

1. How is fission used to generate electricity?
2. Why are control rods important?
3. Name two problems that are associated with nuclear reactors.
4. What difficult problems do scientists face in designing fusion reactors?

Figure 22.12 The fusion test reactor at the Princeton Plasma Physics Laboratory

Do You Know?
It has been estimated that there is enough deuterium in one cubic meter of seawater to produce an amount of energy equal to the energy produced from 1,500 barrels of oil.

1. Uranium fuel in a nuclear reactor undergoes fission in a chain reaction. The energy released is absorbed as heat and used to produce steam. The steam turns a turbine which generates electricity.
2. They control the starting, stopping, and speed of a chain reaction.
3. radioactive contamination of the environment; thermal pollution
4. managing the high temperatures involved, shielding the reactor

Word Skill: Vocabulary

7. Ask the students what deuterium is (isotope of hydrogen that contains one neutron and one proton in its nucleus). Explain that there is sufficient deuterium in the oceans to solve many of our energy problems. How could deuterium help bring about a solution (it can be used in nuclear fusion to create tremendous amounts of energy)? If this is true, why don't we rely on fusion (fusion requires higher temperatures than scientists are presently able to control; it also involves a highly radioactive reactor)?

8. Have the students form small groups based on whether or not they favor the use of nuclear power. Have the groups of students write letters to obtain information to support their viewpoint. Allow a reasonable amount of time for students to gather information and display their research efforts. Then invite other students in the school to visit the display.

FOLLOW-UP

Reinforcement: Nuclear Power
Have each student prepare an illustrated report explaining how nuclear power plants provide energy.

Enrichment: Nuclear Power Plants
Provide a map of the United States. Ask the students to locate on the map the nuclear power plants that are currently in operation. Have each student research one power plant by writing to the plant to learn how it produces energy, how it protects its workers from radiation, and how it disposes of nuclear waste. (You may want to obtain the names and locations of nuclear power plants by writing to the Nuclear Regulatory Commission.) If there is a power plant in your area, ask a representative from the plant to visit your classes and discuss nuclear power with the students.

Tie-in: The Efficiency of Power Plants
Ask students to find out about the efficiency of each of the following types of power plants: fossil fuel, water-cooled nuclear reactor, gas-cooled nuclear reactor, and breeder reactor. Students should prepare a graph to compare the efficiencies.

22.3

OBJECTIVES: The student should be able to

- **explain** what is meant by the term *nonrenewable*;
- **explain** why energy conservation is necessary;
- **discuss** some ways energy is being conserved.

SCIENCE TERM

nonrenewable

ADVANCED PLANNING

Have some magazines available for use in making collages.

LESSON PLAN

1. Introduce the lesson by writing the terms *renewable* and *nonrenewable* on the board. Ask the students to name some sources of energy. As they do so, write the name of the source under the proper term (renewable—solar, wood, wind; nonrenewable—coal, oil, gas, uranium).
2. Ask the students to differentiate between renewable and nonrenewable sources (a renewable source continues to be available for use; a nonrenewable source cannot be replaced).
3. Ask why energy conservation is necessary (some resources cannot be replaced; as the nonrenewable sources are used up, they become more and more expensive).

22.3 Energy Conservation

You will find out
- why energy conservation is necessary;
- what is being done to conserve energy.

Scientists are exploring the oceans and examining data from satellites in their search for sources of energy. As present supplies of fossil and nuclear fuels are used, new supplies are needed. Fossil and nuclear fuels are **nonrenewable.** Something that is nonrenewable cannot be replaced. If you cut down a tree, you can plant a seed to grow a new tree. You can replace, or renew the tree. But if you burn coal or use uranium as a nuclear fuel, you have no way of replacing that coal or uranium.

The energy from fossil and nuclear fuels is becoming more and more expensive. This fact, together with the understanding that these fuels are nonrenewable, has made people realize they can't afford to waste energy.

Only a little more than a third of the energy from the fuel in an electric power plant is used to produce electricity. The rest of the energy is lost in the form of heat. Scientists and engineers are working on ways to reduce these energy losses. They are also trying to develop ways to use the energy that is lost.

Figure 22.13 Coal, like all other fossil fuels, is a nonrenewable resource.

ACTIVITY

Materials
meterstick
pencil
paper

Do You Have Too Many Windows?

Procedure

1. Copy the chart and use it to record your measurements and calculations.

2. Windows can account for a significant amount of heat loss. To see how much heat may be lost from your classroom, calculate the area of the classroom floor and of each window in the room. To calculate area, measure the length and width of the floor or window. Then multiply the two numbers together.

3. Find the total window area by adding together the area of each window in the room.

1. The percentage will vary depending on the classroom.
2. weather stripping, storm windows, thermopane glass, keeping shades down, keeping drapes closed
3. Answers will vary.

Questions

1. Experts have devised a formula that tells whether a large amount of heat could be lost through the windows of a room. Apply this formula to your classroom data.

$$\frac{\text{total window area}}{\text{floor area}} \times 100 = ?\%$$

2. The higher the percentage calculated from the formula above, the greater the potential for heat loss from the room. What steps could be taken to compensate for a large potential for heat loss?

3. Repeat the procedure for a room in your home and apply the formula in question 1 to your data for that room.

	Length (m)	Width (m)	Area (m^2)
Floor			
Window A			
Window B			

ACTIVITY: Do You Have Too Many Windows?

Explain that plain glass has little insulation value. Question 3 must be answered at home.

- Laboratory Skill: Manipulation of Materials
- Mathematics Skill: Computation

SCIENCE BACKGROUND

The three keys to energy conservation are don't waste, use less, and develop efficiency.

WAYS TO SAVE ENERGY

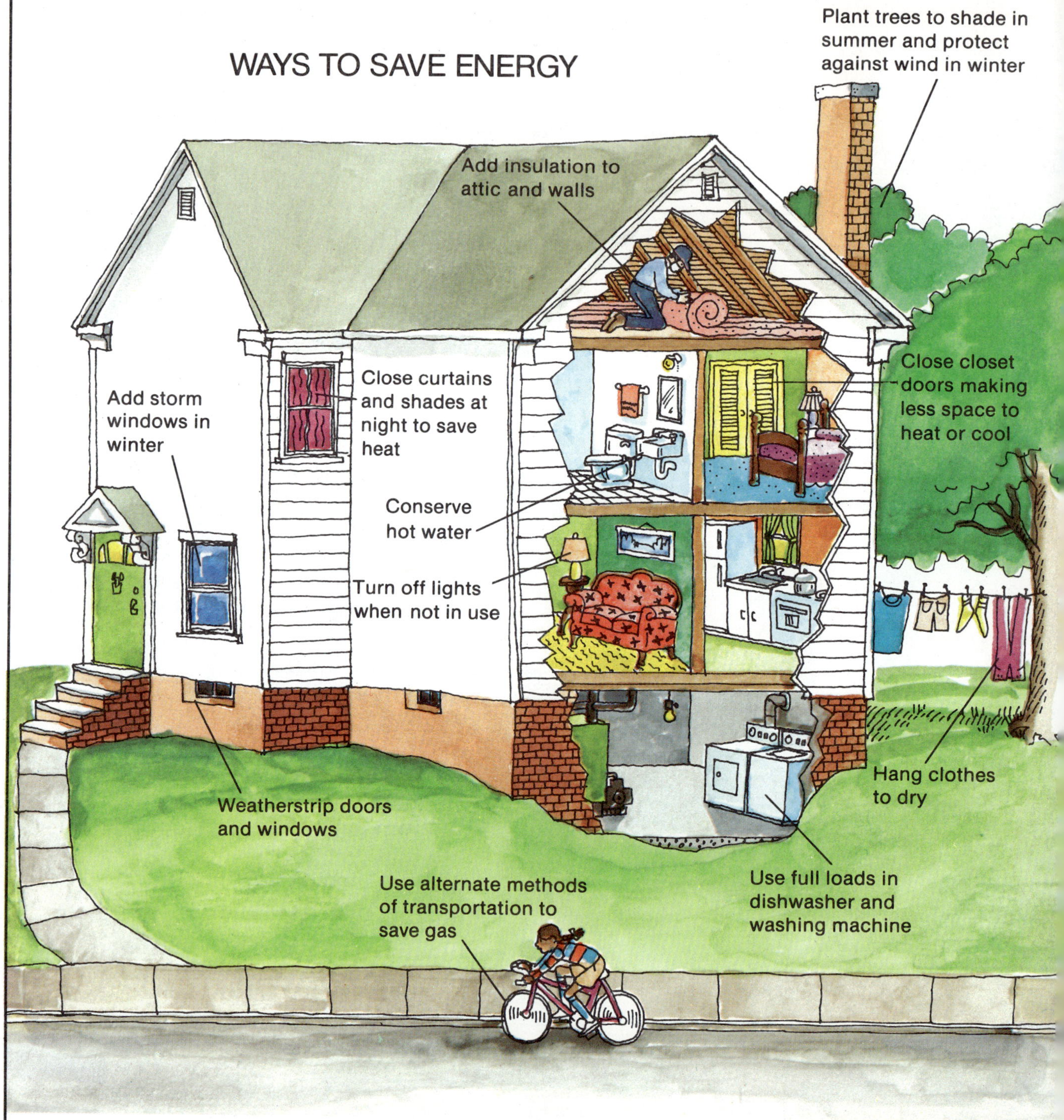

Figure 22.14 Some cities have constructed special bicycle lanes, promoting the use of bicycles to save gasoline.

Government laws and recommendations encourage energy conservation. The highway speed limit is set at 88 kilometers per hour (55 mph) because higher speeds cause more energy to be wasted. The temperatures in government and public buildings are restricted to no more than 20°C in winter and no less than 25.5°C in summer. These restrictions cut down on the energy used for heating and air conditioning. Tax cuts have been made available to people who spend money to conserve energy in their homes.

Homeowners have discovered many ways to conserve energy. This conservation saves them money. Insulating a home keeps it warmer in the winter and cooler in the summer. Energy and money are saved. If the attic floor of a house has good insulation, and weather stripping, storm windows, and storm doors are used, fuel bills can be cut by up to 50 percent. Electric lights and appliances that were once carelessly left on are now being turned off when not needed. Also, energy-efficient appliances are now available.

Do You Know?
Most cars get about 21 percent better mileage at 88 km/h (55 mph) than at 113 km/h (70 mph).

Study Questions for 22.3

1. Why is energy conservation necessary?
2. How is energy waste being reduced?

1. Fossil and nuclear fuels are nonrenewable and expensive.
2. reduction of waste heat from power plants, laws and government recommendations for energy efficiency and reduced consumption, home improvements to save energy

4. Explain that the production of electricity is one of the main uses of fossil fuels. Ask why only about a third of the energy from fuel in power plants is used to produce electricity (the rest of the energy is lost in the form of heat).
5. As you point out that scientists and engineers are trying to find ways to recapture and use the lost energy, ask what other things are being done to save energy (highway speed limit of 88 kilometers (55 miles) per hour, restrictions on temperatures in government buildings, home insulation, storm windows, energy-efficient appliances).
6. As an assignment have each student list at least ten things that can be done to save energy. The student should then prepare a collage illustrating the ten items on the list.

FOLLOW-UP

Reinforcement: Conservation Scrapbook
Ask each student to prepare a scrapbook of pictures and commentary to show ways energy can be conserved in the home.

Enrichment: Comparing Insulating Materials
Have students design experiments to compare the insulating capabilities of different materials. Students should include a statement of the problem or hypothesis, an experimental design, their observations, a chart and/or graph of comparisons, and a conclusion.

Tie-in: Energy Storage
People store money in the bank for use when they need it. In a similar way scientists believe that people can store energy. Four suggested methods for storing large amounts of energy are batteries, electrolysis, compressed air, and pumped storage. Ask interested students to research each method. Which would be the most economically feasible? Have the students report their findings to the rest of the class.

What Did You Learn?

- 22.1 • Fossil fuels are formed from the remains of plants and animals that lived millions of years ago.
- 22.1 • Coal is obtained through shaft mining and strip mining.
- 22.1 • Oil is found in underground pools, tar sands, and oil shale.
- 22.1 • Natural gas is often cooled and compressed to form a liquid, called liquified natural gas.
- 22.1 • The burning of fossil fuels releases particles and unhealthy compounds such as carbon monoxide and sulfur and nitrogen oxides.
- 22.1 • The transportation of fossil fuels can damage the environment.
- 22.1 • Fossil fuels are being used up much more quickly than they are forming.
- 22.2 • A nuclear power plant uses the energy released from fission to produce steam that is used to generate electricity.
- 22.2 • The position of the control rods in a nuclear reactor controls the starting, stopping, and speed of the chain reaction.
- 22.2 • In a breeder reactor, uranium produces plutonium fuel.
- 22.2 • The protection of people and the environment from the radioactivity and thermal pollution associated with nuclear power plants is a serious problem.
- 22.2 • Scientists are making progress in the development of fusion reactors.
- 22.3 • Fossil and nuclear fuels are nonrenewable.
- 22.3 • The energy from fossil and nuclear fuels is becoming more and more expensive.
- 22.3 • Energy conservation is being encouraged and practiced at the governmental and the individual level.

Key Terms

fossil fuels
shaft mining
strip mining
tar sands
oil shale
acid rain
nuclear reactor
control rods
breeder reactor
thermal pollution
deuterium
nonrenewable

Biography

Chien-Shiung Wu (1912-)

Chien-Shiung Wu is a well-known nuclear physicist. She has a keen understanding of the significance of theory in scientific research and a deep appreciation of the importance of experiments. Her methods and results are highly respected. One scientist said of her, "She has never made a mistake in her measurements."

Chien-Shiung Wu was born in China, near Shanghai. Her father, principal of the local school, encouraged Chien-Shiung to read about any subjects that interested her. After graduating from the National Central University in Nanking, China, she came to the United States to continue her studies. She earned her doctor's degree in physics from the University of California in 1940.

During World War II, Dr. Wu went to Columbia University in New York to work on the secret Manhattan Project for the development of the atom bomb. After the war she stayed at Columbia and continued her work in nuclear physics. Her reputation as a good experimenter grew.

In 1956, two young Chinese scientists, Dr. T. D. Lee and Dr. C. N. Yang, came to Dr. Wu with a new theory that they wanted to test. The theory questioned one of the basic laws of nuclear physics. Dr. Wu and her co-workers proved that Lee and Yang were right. The news shocked physicists all over the world. This kind of work shows that science is always growing and changing. As Dr. Wu told a group of young science students, "It is the courage to doubt what has long been established . . . that has pushed the wheel of science forward."

Vocabulary

1. f
2. b
3. g
4. c
5. e
6. h
7. j
8. a
9. k
10. d

● Word Skill: Vocabulary

What Do You Remember?

1. lowered
2. The area above surface coal deposits is totally removed. All plant and animal life in the area is destroyed.
3. The supply of uranium fuel for nuclear reactors will not last forever. Breeder reactors produce and use plutonium fuel from uranium, making the supply of uranium last longer.
4. The remains of plants and animals that died millions of years ago were buried. These remains partially decayed forming coal, oil, and natural gas.
5. The waste products are radioactive. Their safe disposal is a problem.
6. Fission involves the splitting of a heavy nucleus into lighter nuclei. Fusion involves the coming together of two lighter nuclei forming a heavier nucleus.
7. by installing insulation, weather stripping, storm doors and windows; by using energy-efficient appliances; by turning off lights and appliances when they are not in use
8. They locate oil by studying rock formations and layers. Certain rock structures are associated with oil deposits.
9. It releases unhealthy compounds and particles into the air, causing air pollution.
10. a source that cannot be replaced as it is used

TO THINK ABOUT AND DO

Vocabulary ●

Number a piece of paper from 1 to 10. Beside the appropriate number write the letter for the term that is described.

1. the exposure and removal of surface coal
2. a rock containing oil
3. an unnatural increase in temperature
4. an isotope of hydrogen
5. a reactor that produces plutonium
6. fuel formed from the remains of dead plants and animals
7. something that controls chain reactions
8. that which describes an energy source that cannot be replaced
9. that which is formed from sulfur and nitrogen oxides
10. a combination of oil and sand

a. nonrenewable
b. oil shale
c. deuterium
d. tar sands
e. breeder reactor
f. strip mining
g. thermal pollution
h. fossil fuel
i. shaft mining
j. control rods
k. acid rain
l. nuclear reactor

What Do You Remember?

1. In what position are the control rods of a nuclear reactor when the chain reaction is stopped?
2. How does strip mining affect the environment?
3. What problem may be solved by the use of breeder reactors?
4. How were fossil fuels formed?
5. What problem exists concerning the waste products of nuclear reactors?
6. How are fission and fusion different?
7. How can homeowners conserve energy?
8. How do prospectors locate oil?
9. How does the burning of fossil fuels affect the air you breathe?
10. What is a nonrenewable energy source?

Applying What You Have Learned ●

1. How would you describe a perfect energy source?
2. How is the energy in fossil fuels related to the energy of the sun?
3. People in the United States first used wood to heat their homes. Coal was used at the beginning of this century. Now people rely on oil and natural gas. Experts estimate that it takes approximately 50 years for the public to change from one energy source to another. Can you explain why this is so?

Research and Investigation ●

1. Throughout history, people have relied on many different energy sources. How many can you name? Give an example of how each energy source is or was used.
2. Do research and report on the dangers involved in shaft mining.

Applying What You Have Learned

1. Answers will vary but might include the following: is in plentiful supply, is safe and easy to obtain and transport, is inexpensive, and does not damage the environment.
2. Plants used the sun's energy to form compounds. These compounds in plants and in animals that ate the plants produced fossil fuels. The energy in the chemical bonds of the compounds in fossil fuels comes indirectly from the sun.
3. Answers will vary. They may include habits that are hard to break, expense that is involved in converting to different energy sources, and problems with a new energy source that must be worked out before it can be widely used.

● Comprehension Skill: Applied

Research and Investigation

1. Answers and examples will vary. Some energy sources mentioned might be people's muscles, animals such as horses and oxen, wood, sunlight, coal, hot springs, water power, wind power, whale oil, and petroleum.
2. Answers will vary. Some possibilities include a discussion of coal dust explosions, mine collapses, and black lung and other respiratory problems.

● Reference/Research Skills

CHAPTER BACKGROUND

As the people of the nation look to the future, they find current energy sources inadequate. This realization has led scientists to search for energy sources that would replace or supplement the nonrenewable fossil fuels we now depend on.

The use of solar energy is one alternative. Each day the sun provides 100 times more energy than we use. The problem is to find ways to collect, save, and convert the energy to usable forms. Today's technology is based on passive solar heating, or the heating of homes by the heating of water or air. In the future scientists expect to rely on more direct uses of solar energy, such as the use of solar cells, devices that convert sunlight to electricity.

In the United States and other well-developed nations, most of the rivers that can be used to provide electricity are being used to do so. This has caused scientists to look for other sources of constantly moving water. The answer was thought to be the tides. In 1966 engineers in France successfully built a tidal-electric power plant at the mouth of the Rance River. Although the plant produces only about 25 percent of the output of a fuel-operated plant, it is an important supplementary power source. Today engineers are searching for other coastline areas where the tides can be used to produce electricity without upsetting the ecology of the area.

Technologists are experimenting with windmills that can be used to pump water to great heights where it can be dropped through penstocks onto turbines. They are trying to find better ways to use the wind to compress air that can later be blown into gas turbines. Some scientists have estimated that if properly developed, the wind could provide us with 1.5 trillion kilowatt hours of electricity by the year 2000. This is the amount of electric power currently being used in the United States each year.

In the United States the methods of using geothermal energy are, for the most part, experimental. It is hoped that by the end of the 1980's geothermal energy will provide three percent of our total electrical needs. Scientists are looking for better ways to harness the energy as well as searching for hot rocks in areas in the western states. The rocks, constantly heated by the earth, may be able to replace fossil fuels in some steam-powered electrical plants.

23 Energy Alternatives

Tell the students that in this chapter they will learn about some sources of energy that are not being widely used at present. Today's world is clearly dependent on fossil fuels. Ask what things about fossil fuels might make people look for alternatives to them as energy sources (nonrenewable, expensive, pollution, environmental damage, uneven world distribution).

In the future, people probably will not depend on any one energy source the way they currently depend on fossil fuels. No one energy source will be as dominant as fossil fuels are now. Tell the students that this chapter will introduce some energy sources that may eventually meet some of the energy needs currently being met by fossil fuels.

The perfect energy source would be cheap, plentiful, efficient, easy to use, and would not cause pollution. Is there such an energy source? Maybe not, but there are things in the environment that do supply energy without showing signs of ever running out. What gives off heat and light every day? What causes a sailboat to move over the water? How does a raft get downstream?

People have used energy from the sun, wind, and water for centuries. Before the extensive use of fossil and nuclear fuels, these sources of energy were very important. Now the problems associated with fossil and nuclear fuels are becoming better known. As a result, people are beginning to think about using other energy sources.

In this chapter you will find out how the sun, the wind, and water are being used as sources of energy. You will also learn about other energy sources that are minor now but might become more important in the future.

Figure 23.1 An apartment building that has been modified to use solar energy

Lesson Organizer

23.1 Solar Energy

Pages 476–481
Advanced Planning: Page 476
Activity: How Can You Make a Solar Collector? Page 479
Follow-up: Page 481

- **R** How Can Solar Energy Be Used?
- **E** Solar Problems
- **T** A Solar Home

23.2 Energy from Water

Pages 482–485
Advanced Planning: Page 482
Follow-up: Page 485

- **R** Depending on the Sun
- **E** Hydroelectric Power in the United States
- **T** Water Power in United States History

23.3 Wind and Other Sources of Energy

Pages 486–489
Activity: How Can You Measure Wind Speed? Page 487
Follow-up: Page 489

- **R** Wind, Geothermal Energy, and Biomass
- **E** Prevailing Winds
- **T** Bioconversion

See ACTIVITIES book pages 133–136 for laboratory, skill, and review exercises for Chapter 23.

23.1

OBJECTIVES: The student should be able to

- **name** some early uses of solar energy;
- **explain** why people are interested in solar energy;
- **explain** what is meant by the term *renewable*;
- **describe** some ways solar energy is collected;
- **explain** how solar energy can be used;
- **discuss** the advantages and disadvantages of solar energy.

SCIENCE TERMS

solar energy	solar collectors
renewable	solar cells

ADVANCED PLANNING

1. To perform the small-group student demonstration, you will need the following materials for each group of students: a piece of corrugated cardboard; two pieces of screen, each approximately the same size as the cardboard; several grapes and other small fruits; and a triple beam balance. In each piece of cardboard cut a square so that the cardboard becomes a frame whose border is about 3 cm (1.2 in.) wide.

2. For the teacher demonstrations you will need a ring stand, a funnel lined with aluminum foil, a test tube clamp, a ring clamp that will hold the funnel, a test tube half filled with water, a thermometer, and a solar cell device such as a solar cell music box or any one of several solar cell toys.

LESSON PLAN

1. Introduce the lesson by writing the term *solar energy* on the board. Ask the students what the source of solar energy is (sun). Review with them the meaning of the term *energy* (ability to do work).
2. Ask the students to pretend that it is the year 1775. What are some things they would use the sun for (heating, drying, and baking foods)?

23.1 Solar Energy

You will find out
- why people are interested in solar energy;
- what devices are being used to collect solar energy;
- how solar energy can be used.

The idea of using energy from the sun, or **solar energy**, to answer people's energy needs is not a new one. Solar energy has been used to dry and bake foods for centuries. In the early 1600's, a French inventor reportedly used solar energy to heat air in an engine that pumped water. The Pueblo Indians built their homes out of adobe, a mixture of mud and sand. The adobe absorbed heat from the sun during the day and gave off that heat at night. Clearly, use of solar energy is not new. What is new, however, is an increasing and intensified interest in solar energy.

Why are people suddenly so interested in solar energy and other energy alternatives? They have come to recognize that while fossil and nuclear fuels can and do effectively answer their energy needs, these fuels are nonrenewable. They are also becoming more and more expensive. In addition, the use of fossil and nuclear fuels

Figure 23.2 The walls of these adobe homes absorb solar heat during the day and release that heat at night.

Figure 23.3 A solar home

creates some serious environmental problems. The question people are asking now is whether there is another energy source that could answer their energy needs without presenting the same serious problems. Many people believe that solar energy is the answer or, at least, part of the answer.

Solar energy is a **renewable** energy source. A renewable energy source is one that can be replaced. Every day the earth receives an enormous amount of solar energy. The use of solar energy does not appear to threaten the environment. Scientists and engineers are working on ways to collect, store, and change solar energy into other forms of usable energy. As ways to do these things become more practical, efficient, and affordable, more businesses and individual home owners are choosing to use solar energy.

The simplest way to use solar energy to heat a building is to allow sunlight in through windows or glass panels. If you have ever entered a car that had been closed up while sitting in the sun, you have surely noticed the results of this kind of solar heating. The temperature of a greenhouse can also be dramatically affected in the same way. This use of solar energy, while simple and relatively inexpensive, only affects areas that receive direct sunlight.

Do You Know?
If the sun's energy came from coal, it would shine at its present brightness for less than 5,000 years. By that time, all the coal would have been burned up.

3. To demonstrate the usefulness of the sun's energy, have each student or group of students build a fruit dryer. To built the dryer, tape a piece of screen to the cardboard prepared in the advanced planning. Place several pieces of fruit inside the square and tape a second piece of screen over the fruit. For one week the dryer should be hung in a sunny place during the day and brought in at night.

4. Tell the students that you would like them to determine how much water evaporates from the fruit during the week. What procedure should they follow (use a balance to find the mass of the fruit, dry it for one week, and find its mass again; then subtract the second mass from the original mass)?

5. Point out that people have always used the sun for heat and light. Ask why people are so interested in solar energy now (fossil fuels are nonrenewable). Stress that the sun, unlike fossil fuels, is a renewable source of energy.

Word Skill: Vocabulary

6. Set up the foil-lined funnel and test tube of water in the following way: Attach the ring clamp about 8 cm (about 3 in.) from the bottom of the ring stand. Insert the funnel. Attach the test tube clamp and test tube above the funnel so that the sun striking the mirror-lined funnel will be reflected onto the test tube. With the thermometer, measure the initial temperature of the water. Record this reading. Then ask the students what will happen to the water when the sun shines on the funnel (its temperature will increase). Check their prediction by measuring the water temperature after three minutes. Why did the temperature increase (sun's heat was reflected onto the test tube holding the water, thereby heating it)?
7. Discuss some of the problems associated with using the sun's energy (collecting it, storing it, and changing it into usable energy that can meet people's needs). You may want to assign the enrichment activity at this time.

SCIENCE BACKGROUND

There are many different kinds of solar collectors. Some use water to absorb solar energy. Others use sand, rocks, gravel, or air.

A device that produces electricity from light is called a photovoltaic cell. Solar cells are photovoltaic cells that use sunlight.

Scientists have devised ways to collect solar energy in one area of a building and carry it to other areas. **Solar collectors** collect and change solar energy into heat energy. These devices are positioned on a building in a place that allows for maximum exposure to sunlight. Some can even be moved or tilted to follow the sun. Many collectors are flat metal plates painted black. They are painted black because a dark surface absorbs solar energy better than a light one. Water or air passing in pipes behind the plates is heated. The warm water or air is then pumped to places where it is needed. The same water or air can be used over and over to absorb and transport heat. Other solar collectors have curved mirrors that direct sunlight onto tanks or pipes carrying water or air. Like the flat plates, the tanks and pipes are often painted black for better absorption of solar energy.

Solar cells are used to collect solar energy and change it into electric energy. These devices are used on spaceflights. Large panels covered with thousands of solar cells provide electric power for the spacecraft. On Earth, solar cells are in use in homes and businesses. However, they are not in widespread use. One reason for this is the current high cost of making and installing the cells.

Do You Know?
The average amount of solar energy that falls on a square meter in the United States in one year is equal to the amount of energy produced by the burning of one barrel of oil.

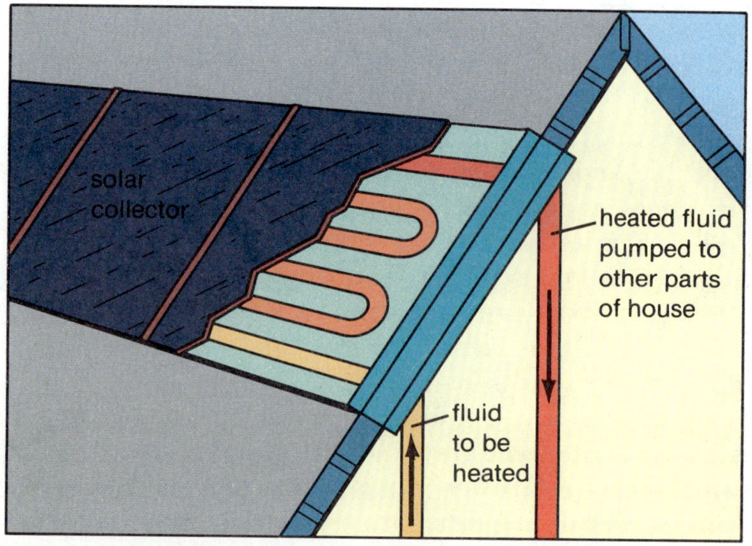

Figure 23.4 Using solar collectors in a solar home

Buildings using solar cells or solar collectors or both exist today. It is probable that more will be built in the future. However, the conversion of already existing buildings to solar energy is not realistic in most cases due to the expense involved. Also, most existing buildings are not located or positioned for maximum exposure to sunlight. Still, the use of solar energy to answer the energy needs of a single building is growing.

ACTIVITY How Can You Make a Solar Collector? ●

Materials

shoe box
newspaper
black spray paint
rubber or plastic tubing with a small opening (about 1 m)
funnel to fit into tubing
tape
plastic wrap
2 beakers
graduated cylinder
thermometer

1. One axis of the graph should show the temperature. The other should show the trial number.
2. increases
3. Answers will vary. Some possibilities include more and better insulation, longer tubing, metal tubing, more trials, and less time between trials.

Procedure

1. Half fill the shoe box with crumpled newspaper. Spray paint on the newspaper and the inside of the box.

2. Punch a hole wide enough for the tubing at each end of the box. Arrange the tubing as shown in the diagram. Tape it in place. Cover the box tightly with plastic wrap.

3. Put 100 mL of water in a beaker. Measure and record the temperature of the water.

4. Place the box in direct sunlight. Rest one end on a thin book to slightly tilt the box. Set an empty beaker under the lower end of the tubing. Insert the funnel into the other end.

5. Pour the water into the funnel. Allow it to flow through the tubing and collect in the empty beaker.

6. Repeat step 5 fifteen times. After every fifth trial, measure and record the temperature of the water and the trial number.

Questions

1. Make a line graph to show the temperatures you measured. ●

2. How does the water temperature change?

3. How could you make your solar collector produce hotter water?

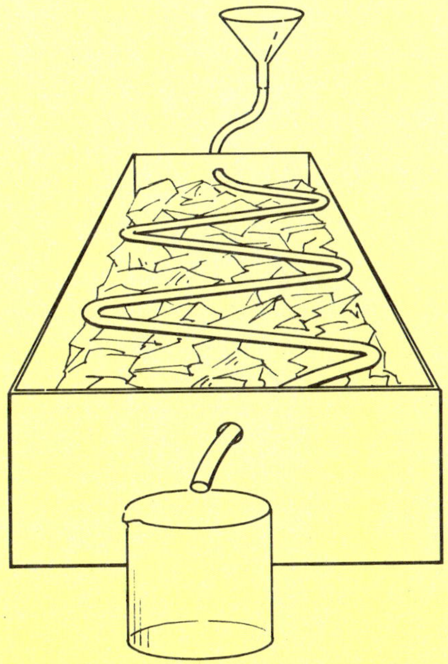

8. Stress that solar collectors, such as the ones in the activity, convert solar energy to heat energy. Explain that another solar device, the solar cell, converts sunlight directly into electricity. Ask students to list some applications of solar cells (to power spaceships and satellites and to provide electricity for military outposts far from other sources of electricity).

ACTIVITY: How Can You Make a Solar Collector?

You may want to have students make their collectors at home. If black spray paint is not available, it can be omitted.

When the students arrange the tubing in the box, they should be sure it is not bent so sharply that the flow of water would be obstructed. At the beginning of the experiment the water should be several degrees cooler than the air.

Students can do this activity inside or outside as long as they are in direct sunlight. If students go outside, be sure they are adequately supervised. Students should be careful not to break the thermometers.

The activity will also work using light from a light bulb. If the students use a light bulb instead of sunlight, point out that they are not collecting *solar* energy.

◇ **Caution:** Adequate ventilation must be used while using spray paint.

● **Laboratory Skills: Manipulation of Materials Experimentation/Demonstration**

● **Mathematics Skill: Graphing**

9. Demonstrate your solar cell device to the students. Allow them to experiment with it to determine how much of the solar cell must be in the sun for the device to work. Ask the students what they would do about cloudy days (a method of storing the electricity must be found; batteries may be used). Ask what they would do about using solar cell devices if there were shadows caused by trees or tall buildings (new laws would have to be made regulating the height of buildings and other objects that cause shadows).

Figure 23.5 Collecting solar energy in a New Mexico desert

What about using solar energy on a larger scale to provide electricity, not just for a single home, but for a whole community? The problem here is that the sun's energy, although plentiful, is spread out over an enormous area. Suppose solar cells were being used to produce as much electricity as the amount currently being produced in the United States by other means. Experts estimate that this would require solar cells covering from 25,000 to 70,000 square kilometers of land! Where is there an area of land this size that is available? Not any area of land will do. The land must receive a consistently large amount of solar energy.

The most appropriate areas for collecting solar energy in large amounts are in the desert areas of the world near the equator. Within the United States, the desert areas of Arizona and New Mexico are most appropriate. However, the use of large areas of these states to collect solar energy would probably conflict with other interests.

Experts have suggested two possible ways to collect solar energy without taking over large land areas. One way would be to construct solar collection stations in ocean areas. The second way is to use solar collection satellites. As a satellite orbited Earth, the satellite would receive a constant, reliable supply of solar energy. This energy could then be sent down to Earth.

As the use of solar energy grows and is further investigated, more practical and effective applications of it will be developed. Although it does not seem to have a negative effect on the environment, there is a possibility that the widespread use of solar energy might bring about some change in the overall distribution of heat from the sun. However, at the present time, solar energy does appear to be a promising energy alternative. People should also remember that once they have paid the price for the purchase and installation of solar energy devices, the sunlight is free.

Figure 23.6 A solar music box

Study Questions for 23.1

1. What makes solar energy a desirable energy source?
2. What is a solar collector?
3. What is a solar cell?
4. What problem has prevented the use of solar energy to produce electricity for whole communities rather than single buildings?

1. plentiful, renewable, does not threaten the environment
2. device that collects solar energy and changes it to heat energy
3. device that collects solar energy and changes it to electric energy
4. cost and the large amount of appropriate land area required

10. Encourage the students to do outside reading on solar power. Based on this reading have each student prepare a chart that lists at least five advantages and five disadvantages of solar energy (advantages: renewable, nonpolluting, no embargoes, inexpensive after the initial investment, available in most areas; disadvantages: cells are expensive to produce and install, sun doesn't always shine, new building codes would be needed, solar collectors would have to be very large).

FOLLOW-UP

Reinforcement: How Can Solar Energy Be Used?

Ask the students to pretend that they are newspaper reporters who must do research on how solar energy can be used. To report their findings, students should prepare the front page of the *Solar Times*. This page should include headlines, articles, and illustrations.

Enrichment: Solar Problems

Have groups of students do research and prepare written solutions to each of the following problems of solar energy: amount of solar energy varies with the season, the geography, and the cloud cover; there is no sun at night; energy from the sun is diffuse.

Tie-in: A Solar Home

If there is a solar development in your area, invite the contractor or a homeowner from the development to speak to students about solar homes. What are the advantages? What are the disadvantages? Consider arranging for your students to visit one of the homes.

23.2

OBJECTIVES: The student should be able to

- **describe** how hydroelectric power is produced;
- **explain** how hydroelectric power is related to solar energy;
- **explain** why hydroelectric power is not more widely used;
- **explain** how tidal energy is used;
- **explain** why tidal energy is not more widely used.

SCIENCE TERMS

hydroelectric power
tidal energy

ADVANCED PLANNING

1. For the small-group student demonstration, each group will need the following materials: a 4-liter (about 1-gal) jar, a 30 cm × 1 cm (12 in. × 0.4 in.) heavy aluminum foil strip, a plastic straw, some adhesive tape, a piece of heavy wire from a coat hanger, a piece of black construction paper, a pair of scissors, a box of heavy-duty aluminum foil, and some ice.

2. Obtain a map or transparency showing the major rivers of the United States. If you use a transparency, obtain an overhead projector and some marking pens.

LESSON PLAN

1. Introduce the lesson by writing the term *hydroelectric power* on the board. Divide the word *hydroelectric* between *hydro* and *electric*. Ask the students what the term *hydro* refers to (water). How is hydroelectric power produced (water spins a turbine that turns a generator)?

● **Word Skill: Structure**

2. Have the students form several small groups. Have each group prepare and use a water cycle jar by following the procedure in steps 3–10.

SCIENCE BACKGROUND

In 1960, hydroelectric power provided 20 percent of the energy of the United States. Today it produces only 4 percent.

23.2 Energy from Water

You will find out

- how hydroelectric power is produced;
- how hydroelectric power is related to solar energy;
- why hydroelectric power is not more widely used;
- how tidal energy is used.

For years people have used moving water to carry logs downriver to sawmills and paper mills. Rivers have also moved rafts of goods and people from place to place. The major application of moving water as an energy source, however, has been in the generation of electricity. When falling water strikes the blades of a turbine, a rotational motion is produced. This motion is used to generate electricity. The electricity generated by the use of moving water is called **hydroelectric power.**

Figure 23.7 Falling water can be used to generate electricity.

Do You Know?
One of the earliest hydroelectric power plants was built in Niagara Falls, New York, in 1896.

In order to produce hydroelectric power at a steady rate, the water in rivers is dammed up to make waterfalls. The largest hydroelectric power plant in the United States is at the Grand Coulee Dam across the Columbia River in the state of Washington. Although many states in the United States have hydroelectric power plants, the most productive ones are in Washington, Oregon, and California.

Figure 23.8 The water cycle

Hydroelectric power is related to solar energy. The sun heats water and causes it to evaporate from the surface of the earth. That water eventually falls back to the earth as rain or snow. If the water falls on a place that is above sea level, it runs downhill. It may eventually enter a river and flow to the ocean. While in a river, that water may be used to produce hydroelectric power. As long as there is sunlight to evaporate water so that it can fall back to earth and form rivers, the production of hydroelectric power will be possible.

3. Mark the strip of aluminum foil into 2.5-cm sections. At the first 2.5-cm mark fold the foil over to form a flap. At the next 2.5-cm mark, fold the foil under. Continue folding in this way to the end of the strip. The result should be a zigzag pattern.
4. Tape each zigzag section together to form the paddles of the water wheel. Then wind the paddles around the plastic straw, which will act as a bearing for the water wheel.
5. Shape one end of the coat hanger wire into a circular base so that the wire can stand upright. This upright portion should be about 6 cm high.
6. Bend the other end of the wire so it is L-shaped and parallel to the base. Thread this portion of the wire through the straw. With a pencil eraser, make a little dent in each blade of the wheel to catch the falling water. Blow on the wheel to be sure that it spins freely.
7. Cut a circle of black construction paper to fit the bottom of the jar. Place the circle in the jar. Into the jar pour water to a height of about 3 cm (about 1 in.).
8. Stand the water wheel in the jar. From the heavy-duty aluminum foil form a cone that will sit in the mouth of the jar but will not touch the blades of the wheel. Secure the foil cone by folding the foil over the edge of the jar. Fill the cone with ice. Cover the top of the jar to prevent evaporation.
9. Have each group set their jar in direct sunlight and observe it. How does the appearance of the cone change (becomes covered with water)? Where does the water come from (from moisture in the air inside the jar)? Does the amount of water in the jar increase (no, water from the funnel never enters the jar; the water in the jar evaporates and then condenses, falling into the jar again)? What happens to the wheel (it begins to spin as water falls on it)?
10. Have each student draw a diagram that shows what causes the wheel to spin (sun evaporates the water; the water vapor rises; the vapor condenses when it hits the cone; and water droplets fall on the wheel, causing it to spin). How would such a wheel be used to produce electricity (it would spin a generator)?

11. On your map or transparency of the United States, mark the location of several large hydroelectric plants (Hoover Dam on the Arizona-Nevada boundary, Grand Coulee Dam in northeast-central Washington, Robert Moses Dam at Niagara Falls). Ask what each plant has in common (proximity to a large river).
12. Ask the students whether waterpower is a renewable or a nonrenewable source of energy (renewable). Why then do we obtain only about four percent of our energy from hydroelectric power (dams cause changes in the environment, affecting plant and animal life)?

SCIENCE BACKGROUND

Experts estimate that if all the world's rivers had hydroelectric power plants, those plants could meet about one-third of the world's energy needs.

Figure 23.9 A hydroelectric power plant at the Bonneville Dam on the Columbia River in the state of Washington

Hydroelectric power is renewable and does not cause pollution. Why, then, is it not more widely used? Not all rivers are appropriate for the production of hydroelectric power. Also, the damming up of rivers causes drastic changes in the environment. Plant and animal life both in and along a river are threatened. In some areas, plans for hydroelectric power plants have been considered and finally canceled due to environmental concerns.

Many of the world's large rivers are in undeveloped areas of South America, Africa, and Asia. The potential for hydroelectric power in these areas is high. But, so far, the countries with that potential have not made use of it.

Another kind of energy that involves water is **tidal energy.** Gravitational forces involving the earth, the sun, and the moon cause ocean tides. Tidewaters can be used to turn turbines that generate electricity. The Rance tidal power plant in France has been producing electricity since 1966.

Tidal energy is renewable and does not cause pollution. However, the tidal areas that are involved may suffer some of the same changes in their environment as the river areas around hydroelectric power plants. This problem has limited the development of tidal power plants. Another limiting factor is the availability of coastline areas that are suitable for tidal power plants. A suitable coastline would be a narrow bay or inlet where there is a great difference in the water levels at high and low tides.

Figure 23.10 The Rance tidal power plant

Study Questions for 23.2

1. How is hydroelectric power produced?
2. What is the relationship between hydroelectric power and solar energy?
3. How do the dams associated with hydroelectric power plants affect the environment?
4. How is tidal energy used?

1. When falling water strikes the blades of a turbine, a rotational motion is produced. This motion is used to generate electricity.
2. The sun evaporates water. The water falls to Earth, forming rivers. The water in rivers is used to produce hydroelectric power.
3. The damming of rivers causes drastic changes in the environment that threaten plant and animal life both in and around the rivers.
4. Tidewaters are used to spin turbines to generate electricity.

13. Ask the students to identify the periodic rise and fall of ocean waters along a coastline (tides). What is the cause of tides (gravitational forces among the earth, moon, and sun)? Explain that because the tides are constantly rising and falling they can be used to produce electricity just as a moving river can.
14. Ask the students whether the tidal power used to produce electricity is renewable or nonrenewable (renewable). Can the students think of reasons why we don't have more tidal power plants (environmental damage, damage to recreational areas, unsuitable tides along some coastlines)?

FOLLOW-UP

Reinforcement: Depending on the Sun

Have each student create a poster to illustrate how hydroelectric and tidal powers depend on the sun.

Enrichment: Hydroelectric Power in the United States

Ask groups of students to select different hydroelectric dams in the United States. Each group should do research to determine when the dam was built, by whom, and how the electricity is presently used.

Tie-in: Water Power in United States History

Ask each student to create an illustrated time line to show how waterpower has been used by Americans since the Pilgrims landed in 1620.

23.3

OBJECTIVES: The student should be able to

- **list** some early uses of wind energy;
- **explain** how wind occurs;
- **explain** how wind energy is used;
- **explain** what is meant by geothermal energy;
- **describe** some uses of geothermal energy;
- **explain** what is meant by biomass;
- **describe** how biomass is used to produce energy.

SCIENCE TERMS

geothermal energy
biomass

LESSON PLAN

1. Begin the lesson by asking the students to name some early uses of wind energy (sailing boats, rotating wheels of windmills in order to grind grain and pump water). Explain that today windmills are used to produce electricity.
2. Ask the students to explain how wind is produced (the sun heats the air above the earth unevenly; as the warm air rises, cooler air rushes in to replace it). Ask the students whether wind energy is a renewable or a nonrenewable energy source (renewable). Why (as long as the sun exists, it will always heat air)?

SCIENCE BACKGROUND

Historians believe the Persians designed the first windmill. Wind was caught in triangular sails rigged between spokes radiating from a shaft. The Persian windmill was used to grind grain.

In the seventeenth century, the Netherlands was the world's most industrialized nation. It depended on wind power.

The most ambitious attempt to use wind energy on a commercial basis was made in 1941 on Grandpa's Knob, a hill outside Rutland, Vermont. The windmill ran successfully for sixteen months.

23.3 Wind and Other Sources of Energy

You will find out
- how wind energy is used;
- what is meant by geothermal energy;
- what is meant by biomass.

The wind is a source of energy that people have been using for centuries. Wind once filled the sails of ocean-going ships. It also turned windmills to grind grain and pump water.

Wind is the movement of masses of air from one place to another. This movement is the result of temperature differences in the atmosphere. The temperature differences come from uneven heating of the atmosphere by the sun. Wind energy, like solar energy, is a renewable energy source. Also like solar energy, the use of wind energy does not appear to threaten the environment.

Today, windmills are being used to generate electricity. The blades of a windmill act like a turbine. Wind strikes the blades, causing them to turn. This rotational motion is used to generate electricity.

Figure 23.11 *left:* Wind results when warmer, less dense air rises and cooler, denser air moves in to replace it. *right:* Using wind energy to generate electricity

In some areas, one windmill provides the electricity for a single home or business. In other areas, a large number of windmills generate enough electricity to service many homes or businesses. There is, of course, one obvious limitation to the large-scale use of windmills to generate electricity. Many areas do not receive enough wind on a regular basis to make windmills practical.

ACTIVITY

How Can You Measure Wind Speed?

Materials

- protractor
- long needle
- thread
- Ping-Pong ball
- wood (30 cm by 2 cm)
- glue

Procedure

1. Make a handle by gluing the wood to the corner of the protractor at the 180° mark.

2. Thread the needle and push it through the Ping-Pong ball. Pull the thread through the ball and tie a knot at one end.

3. Glue the free end of the thread to the center of the straight side of the protractor. When the protractor is held level, the thread and ball should hang at exactly 90°.

4. Use your device to measure the wind speed in five different areas near your school. Compare the protractor reading in degrees with those on the chart to determine approximate wind speed.

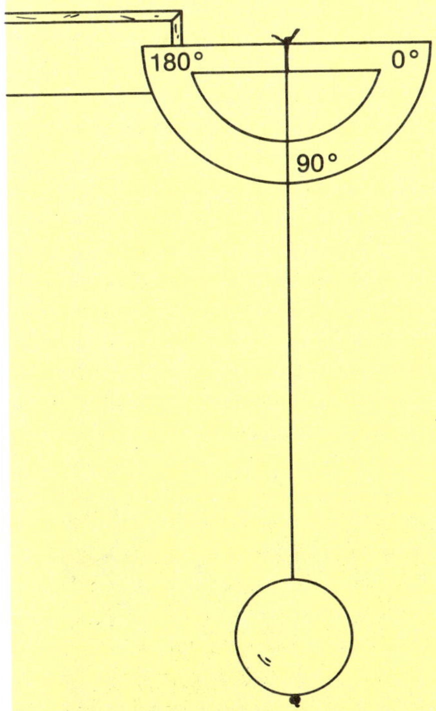

Angle	km/h
90°	0
80°	13
70°	19
60°	24
50°	29
40°	34
30°	42
20°	52

Questions

1. Make a bar graph to show the wind speed in the areas you visited.

2. To be useful in generating electricity, wind must have a constant speed of at least 13 km/h. Could the wind in any of the areas you visited be used to generate electricity?

1. Bar graphs should show the five different areas across the bottom. The height of the bars should indicate wind speed.
2. If students answer yes, the area they refer to should have a wind speed of 13 km/h or more.

ACTIVITY: How Can You Measure Wind Speed?

You may want to have the students make their wind speed measuring devices at home. A home economics teacher may be able to supply the needles.

The Ping-Pong ball should not hang very far below the protractor. Emphasize that the devices must be held level in order to obtain an accurate reading.

Be sure the students are adequately supervised when they are outside. You may want to have them measure the wind speeds in areas around their homes as a homework assignment.

◇ **Caution:** Students should be careful while using the needle.

- **Laboratory Skills: Manipulation of Materials Experimentation/Demonstration**
- **Mathematics Skill: Graphing**

3. Write the terms *geothermal energy* and *biomass* on the board. Divide the term *geothermal* between *geo* and *thermal*. Ask the students what each part of the word means (geo—earth, thermal—heat). Divide the term *biomass* between *bio* and *mass*. Ask the students what each part of the word means (bio—life, mass—quantity or amount).

Word Skill: Structure

4. Ask the students how we know the interior of the earth is hot (evidence from activity of geysers, volcanoes). Explain that the earth's heat, or geothermal energy, is used in Iceland and in areas of Italy and the United States. Ask how geothermal energy is used (to spin turbines for generating electricity, to heat buildings, to cook food).

5. Ask the students what biomass includes (all the materials in living things, their wastes, and their remains). How do the students think the wastes and remains of living things can be used to produce energy (by burning them)?

6. Ask the students whether they would classify geothermal energy and biomass as renewable or nonrenewable energy sources (renewable)? How would they justify their answer (geothermal energy—reactions in the earth will continue to produce heat; biomass—as long as there are living things, biomass will be a renewable energy source)?

SCIENCE BACKGROUND

The heat in the earth is thought to be the result of the natural, spontaneous decay of radioactive materials over vast periods of time.

Geothermal energy varies with depth. The temperature increases an average of about one degree for every 33 meters of depth.

The best areas for geothermal power are Iceland, Russia, Japan, New Zealand, Turkey, and El Salvador.

The interior of the earth is very hot. This fact may surprise you since people generally find basements and underground caves cool or even cold. However, this situation exists only in areas very near the earth's surface. If, for example, you were to go way down in the shaft of a deep coal mine, you would find it hot.

The high temperatures inside the earth can heat underground water supplies. This hot water and the steam it produces may come to or near the surface. It may erupt forcefully in geysers or flow into pools known as hot springs. The heat energy that comes from inside the earth is called **geothermal** [JEE-oh-thur-muhl] **energy.** Steam from geothermal energy can be used to spin turbines to generate electricity. Geothermal energy can also heat buildings and even cook food.

Geothermal energy is renewable and does not appear to seriously threaten the environment. However, the waters in geysers and hot springs do contain many minerals. In significant amounts, the distribution of these minerals could affect the environment.

Only certain locations are appropriate for the use of geothermal energy. Those areas have geysers or hot springs at or near the earth's surface. The Lardello area

Do You Know?
The steam wells tapped by The Geysers power plant are 2.5 km deep.

Figure 23.12 The Geysers power plant uses geothermal energy to generate electricity.

Figure 23.13 Wood, a form of biomass, is a renewable energy source.

of Italy has been using geothermal energy since 1904. Many areas of Iceland also rely on geothermal energy. In the United States, The Geysers power plant in California has been using geothermal energy since 1960.

Living things provide an energy source called **biomass** [BY-oh-mas]. Biomass includes all the materials in living things, their wastes, and their remains. Probably the most obvious use of energy from biomass is the burning of wood for heat. Other kinds of plant material can also be burned. In addition, efforts have been made to use garbage and other wastes for fuel.

Biomass is a renewable energy source. As long as there are living things there will be biomass. However, the burning of biomass does contribute to air pollution. Also, the careless cutting down of thousands of trees can drastically change the environment. Even if new trees are planted, it takes many years for them to grow. In the meantime, the area may lose topsoil due to the action of wind and running water. Food and shelter for animals are also lost when trees are cut down.

Study Questions for 23.3

1. How is wind energy used?
2. What is geothermal energy?
3. What is biomass?

1. Wind turns the blades of a windmill, creating a rotational motion that is used to generate electricity.
2. heat energy from inside the earth
3. the materials in living things, their wastes, and their remains

7. Have the students list some advantages and disadvantages of wind, geothermal energy, and biomass (advantages: renewable, inexpensive, available; disadvantages: cannot depend on strength and direction of wind, minerals can ruin pipes and machinery or make water too acidic or alkaline, burning biomass can cause air pollution, uncontrolled and unplanned cutting of trees can harm the environment).

8. Have the students form teams of two or three students. Each team should select one alternative energy source and prepare a display illustrating why it is the best energy source of the future. The display should also include how any disadvantages would be overcome.

FOLLOW-UP

Reinforcement: Wind, Geothermal Energy, and Biomass
Ask each student to write a definition of *wind, geothermal energy*, and *biomass* that could be used in the *New Energy Dictionary*.

Enrichment: Prevailing Winds
Ask interested students to research the types of wind patterns that exist around the world. Students should name the types of winds, the place of their origin, and the direction in which they are blowing. Which winds blow in the region where the students live? Are these winds reliable sources of energy for producing electricity for home use?

Tie-in: Bioconversion
Bioconversion is a new field of energy research. Ask the students to do research to find out about the two general schemes proposed by bioconversion scientists to relieve the energy crisis.

SCIENCE BACKGROUND

Nashville, Tennessee, uses the heat from burning garbage to produce electricity.

The city of Los Angeles uses as fuel the gas produced by the action of bacteria on garbage.

What Did You Learn?

23.1 • Solar energy is renewable and does not appear to threaten the environment.

23.1 • Solar collectors collect and change solar energy into heat energy.

23.1 • Solar cells collect and change solar energy into electric energy.

23.1 • There are buildings today that use solar energy for heating and electric power.

23.2 • The electricity generated by the use of moving water is called hydroelectric power.

23.2 • To produce hydroelectric power, rivers are dammed up to make waterfalls.

23.2 • Hydroelectric power is related to solar energy.

23.2 • The dams involved in the production of hydroelectric power cause drastic environmental changes.

23.2 • Tidal energy can be used to generate electricity.

23.3 • Wind energy can be used to generate electricity.

23.3 • Geothermal energy is heat energy that comes from inside the earth.

23.3 • Geothermal energy can be used to generate electricity and to heat buildings.

23.3 • Biomass includes all the materials in living things, their wastes, and their remains.

Key Terms

solar energy
renewable
solar collectors
solar cells
hydroelectric power
tidal energy
geothermal energy
biomass

Career

Environmental Engineer

An environmental engineer studies the effects of human activity on the environment and develops methods to protect the environment. An environmental engineer may work in one or more of three general areas: air quality management, water quality management, or land reclamation. Land reclamation involves the restoration of an area that has been damaged by things such as strip mining, floods, or fire.

An example of the work of environmental engineers in the area of land reclamation can be seen at the Big Sky Mine in Montana. The Big Sky Mine was a strip mining operation. The land that was scraped away to gain access to the coal below has been shaped into rolling hills and seeded with oats, alfalfa, and a mixture of native grasses.

Some environmental engineers who work for industry design systems that reduce air and water pollution. For example, they may work to produce automobile engines that are less polluting. Or they may design ways to dispose of industrial wastes without contaminating water supplies.

Environmental engineers working for the government or other institutions may study the pollutants produced by industry or transportation. Or they may be called upon to study the expected harm to the environment of some new energy system, such as a large dam or mining operation.

The solution to environmental problems requires the skills and experience of people who know about health, sanitation, physics, chemistry, biology, meteorology, engineering, and many other fields. The energy systems you may take for granted have limitations and disadvantages as well as advantages. Many environmental engineers will be needed in the future to cope with the problems and maximize the advantages of all energy systems.

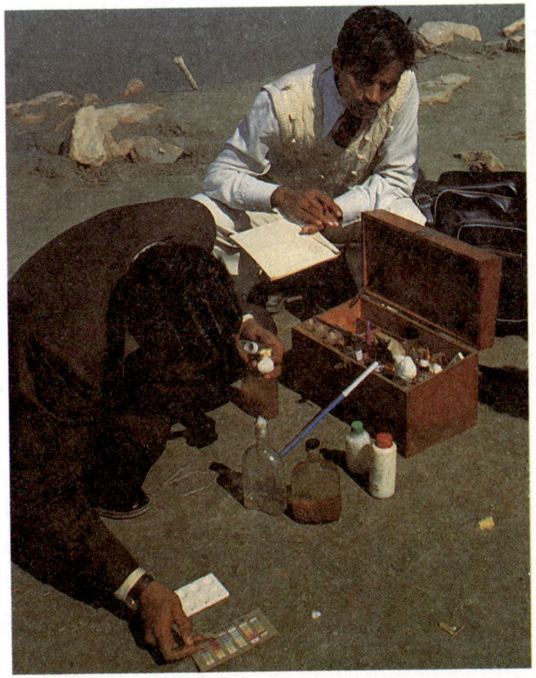

Vocabulary

1. solar cells
2. hydroelectric power
3. hydroelectric
4. true
5. geothermal, solar, or wind
6. true
7. wood
8. solar collectors
9. true
10. true

● Word Skill: Vocabulary

What Do You Remember?

1. Fossil and nuclear fuels are nonrenewable, create some serious environmental problems, and are becoming more and more expensive.
2. use solar cells to produce electricity, use a solar collector to produce hot water
3. Dark colors absorb solar energy better than light colors do.
4. A renewable energy source can be replaced. A nonrenewable energy source cannot be replaced and may eventually be used up.
5. solar energy, wind energy, geothermal energy
6. tidal energy, wind energy, moving water or hydroelectric energy, geothermal energy
7. Dams cause serious damage to the environment.

TO THINK ABOUT AND DO

Vocabulary ●

Number your paper from 1 to 10. Read each of the following sentences. If the sentence is true, write *true*. If it is false, make it true by changing the underlined term.

1. <u>Solar collectors</u> collect and change solar energy to electric energy.
2. The electricity generated with the use of moving water is called <u>turbine power</u>.
3. The generation of <u>geothermal</u> power requires the building of dams.
4. Wind energy and hydroelectric power are related to <u>solar energy</u>.
5. The use of <u>tidal</u> energy does not appear to threaten the environment.
6. Large land areas must be used in order to collect large amounts of <u>solar energy</u>.
7. The most obvious example of the use of energy from biomass is the burning of <u>coal</u>.
8. <u>Solar cells</u> collect and change solar energy to heat energy.
9. Geysers and hot springs contain <u>geothermal energy</u>.
10. The power plant at the Grand Coulee Dam produces <u>hydroelectric power</u>.

What Do You Remember?

1. Why are people interested in energy alternatives to fossil and nuclear fuels?
2. Name two ways a house can use solar energy.
3. Why are solar collectors usually painted black?
4. What is the difference between a renewable and a nonrenewable energy source?
5. Name three energy sources that do not appear to threaten the environment.
6. Name three energy sources that can be used to generate electricity.
7. What has limited the production of hydroelectricity?

Applying What You Have Learned ●

1. With all the sources of energy available on Earth, why do you think there is an energy crisis?
2. What kind of energy would you associate with lava from a volcano?
3. How is solar energy related to energy from biomass?
4. What source of energy do you associate with each of the following?
   (a) Hoover Dam in Nevada
   (b) a forest fire
   (c) Old Faithful in Yellowstone National Park
   (d) keeping a room cool by closing the drapes during the day
   (e) Skylab
   (f) a kite
5. How are fossil fuels and biomass related?

Research and Investigation ●

1. Make a map of the world. Indicate on the map geothermal hot spots and large rivers that could be used to produce hydroelectric power.
2. Two modern wind machines are the Savonius Rotor and the Helix Rotor. How are these machines different from ordinary windmills?

Applying What You Have Learned

1. Answers will vary. Some possibilities include uneven distribution of energy sources, lack of technology, and careless waste of energy.
2. geothermal
3. Biomass comes from living things. Living things need solar energy to live. Green plants need solar energy to make food. Animals need plants for food.
4. (a) hydroelectric power; (b) biomass; (c) geothermal energy) (d) solar energy; (e) solar cells, (f) wind energy
5. Both come from living things. Fossil fuels, however, take millions of years to form. Biomass can form in less than a year.

● Comprehension Skill: Applied

Research and Investigation

1. Students should have no trouble finding a map showing the world's rivers. Geothermal hot spots are in Russia, Iceland, Japan, New Zealand, Hungary, Turkey, El Salvador, the western United States, the Rift Valley in Africa, the Pacific Far East and Alberta, Canada.
2. Neither of these machines has blades like a windmill. The Savonius has two hollow cylinder-shaped cups. The Helix has six twisted strips.

● Reference/Research Skills

CHAPTER BACKGROUND

The purpose of this chapter is to point out the application of the scientific method and principles of physical science to astronomy, probably the most ancient of all the sciences.

The first lesson, "Investigating the Universe", shows how astronomers use the scientific method and principles of science to learn about the universe. Students become familiar with some of the instruments of astronomy such as the telescope and the spectroscope. You may also want to discuss some other tools that astronomers use. Cameras set for time-lapse photographs can be used to study the movement of stars. Maps of the heavens can be used to locate stars at any time during the year. An astrolabe can be used to determine the altitude of the sun or other celestial bodies. A sextant is a navigational instrument that can also be used to measure the altitudes of celestial bodies.

In the second lesson students learn that there is great variation among the stars, which may be in any one of several stages of their life cycle. Stars are luminous bodies; they give off their own light. Some stars are bigger and some are smaller than the sun. They vary from the size of some of the planets to over a thousand times larger than the sun. Some stars appear bright; others are faint. Some bright stars show colors, but most appear whitish.

Many of the brighter stars have been grouped into patterns called constellations. Some of the constellations such as Cygnus (the Swan) and Orion (the Hunter) were described thousands of years ago. Since these star patterns still fit the ancients' descriptions, astronomers have concluded that the relative positions of the stars change very little over the centuries.

Stars are formed by the condensation of dust or gas particles. As gravity pulls the particles closer, their temperature rises to the point where matter is converted to energy. The result is tremendous heat and light.

The last lesson, "The Space Frontier," provides a look at the future. It suggests some of the dreams that scientists have about space exploration and may spark student interest in participating in a space colony.

The Universe 24

Before class begins, get a filmstrip projector and cut several squares of aluminum foil the size of a frame of film. Use a pin to poke holes in the foil squares. The holes should outline the shape of some major constellations.

Introduce the lesson by showing the students your constellations. To do this, turn on the filmstrip projector and place the squares in the slot where the filmstrips go. Ask if anyone can name the constellations. Tell the students that in this chapter they will learn about stars. They will find out how scientists study stars and other objects in the universe. They will also learn about the ideas some scientists and engineers have for the future.

What is the structure of the universe? How big is it? Are all the objects in the universe moving? Where are they going? These and many other related questions have been asked again and again over hundreds of years. The ancient Egyptians believed the sun was carried across the sky in a boat. An ancient Hindu myth explains that the universe came from an egg. One half of the egg was gold. It became the heavens. The other half of the egg was silver. It became the earth.

As time went by, some of the questions about the universe were answered, or partly answered. Others are still mysteries. In this chapter, you will find out how the scientific method and many of the principles of physical science have helped to answer some of the questions about the universe. You will also learn of questions that remain unanswered. Some of these questions may be answered as people venture out into one of the few remaining frontiers, the space frontier.

Lesson Organizer

24.1 Investigating the Universe

Pages 496–499
Advanced Planning: Page 496
Activity: What Is the Diameter of the Sun? Page 497
Follow-up: Page 499

- **R** Tools of the Astronomer
- **E** Jocelyn Bell Burnell
- **T** Astronomy Versus Astrology

24.2 Stars

Pages 500–505
Advanced Planning: Page 500
Activity: How Far Away Are the Stars? Page 502
Follow-up: Page 505

- **R** The Formation of Stars
- **E** Constellations
- **T** Planetarium Visit

24.3 The Space Frontier

Pages 506–509
Advanced Planning: Page 506
Follow-up: Page 509

- **R** Satellites in Our Lives
- **E** Satellites in Orbit
- **T** Seeing the Future

See ACTIVITIES book pages 137–140 for laboratory, skill, and review exercises for Chapter 24.

24.1

OBJECTIVES: The student should be able to

- **describe** how astronomers study the universe;
- **name** some tools astronomers use;
- **state** some uses of the telescope;
- **state** some uses of the spectroscope.

SCIENCE TERMS

astronomers
spectroscope

ADVANCED PLANNING

1. To make a spectroscope, obtain a shoe box with cover, some black paint, a paint brush, some tape, a pair of scissors, and a piece of diffraction grating.

2. You may want to order the film series *The Search for Solutions*, which presents some interesting studies of how the scientific method has been used to advance astronomy.

LESSON PLAN

1. Introduce the lesson by asking the students how they would study the top of Mt. Everest if they could not go there. List their answers on the board.
2. Ask what the study of the stars and other objects in the universe is called (astronomy). Explain that astronomers, like people who study deep canyons or high mountains, cannot always visit the places they are studying.

SCIENCE BACKGROUND

Astronomers measure huge distances in space in parsecs. One parsec equals 3.26 light-years, or 3.09×10^{13} kilometers.

24.1 Investigating the Universe

You will find out
- how astronomers study the universe;
- what tools astronomers use.

Imagine that you want to study a very deep canyon. But the sides are so steep and the canyon is so deep that you cannot possibly climb down into it. How could you find out about the canyon? You could use instruments to make observations. You could send devices down to collect samples of the materials in the canyon and study the samples. You could compare your observations of the canyon with information about other canyons that are more accessible. You would have to piece together all these kinds of information to form an accurate picture of the canyon.

Scientists who study the universe, or **astronomers,** are faced with a similar challenge. They are studying objects and occurrences that they cannot visit or witness personally. They must study the universe the same way you would have to study the bottom of your imaginary canyon. To do this, astronomers use information gathered from many different sources.

Figure 24.2 *left:* Stonehenge, an ancient structure related to astronomical studies and observations; *right:* The modern complex at the Kitt Peak National Observatory in Arizona

You have already heard of one of the oldest and most important tools of astronomy. That tool is the telescope. Modern telescopes allow astronomers to carefully observe objects in the universe and monitor their motion. Some telescopes can pick up radio waves and other forms of radiation given off by objects in space. Information from telescopes continues to be very useful to astronomers.

ACTIVITY

What Is the Diameter of the Sun? ●

Materials

2 index cards
pin
meterstick
metric ruler
tape

Procedure

1. Label the index cards *A* and *B*. Use a pin to poke a hole in the center of card A.

2. Draw a square in the center of card B. The sides of the square should be 0.5 cm long.

3. Tape card A at the 20 cm mark on the meterstick as in the diagram.

4. Stand with your back to the sun. Hold the meterstick in one hand so that the sunlight shines directly on card A. Hold card B about 10 cm behind card A. Move card B slowly back and forth until a circle of light appears in the square on it. Measure the distance in centimeters between cards A and B when the circle appears in the square.

Questions

1. How far was card B from card A when the sun's image appeared in the square?

2. To find the diameter of the sun in kilometers, divide your answer from question 1 into 74,800,000. ●

3. The diameter of the earth is 12,640 km. How much larger in diameter is the sun?

1. The distance should be about 54 cm.
2. about 1,392,924 km
3. about 1,380,284 km larger

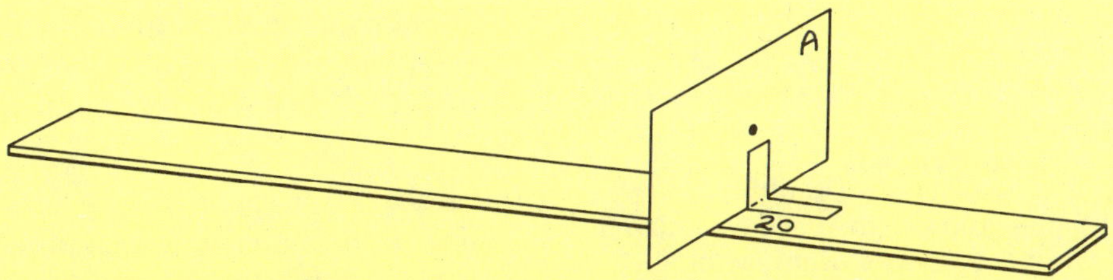

3. Ask the students to name one instrument that astronomers use to study the stars (they should all be familiar with the telescope). Review the principle of the reflecting telescope presented in Chapter 21, Lenses.

4. Have the students prepare a list of ways in which astronomers use the telescope (to see faraway objects, to monitor the motion of objects in the universe, to study radiation given off by objects in the universe).

ACTIVITY What Is the Diameter of the Sun?

Tell the students that in this activity they will be acting as astronomers by observing and measuring a very distant object from Earth. The number they will use to calculate the sun's diameter was developed by astronomers who applied some of the principles of physical science.

Card *A* must be taped to the meterstick no farther than 20 cm (8 in.) from the end. There must be ample space on the stick to move card *B* to 60 cm (20 to 24 in.) behind card *A*.

◇ **Caution:** Students should never look directly at the sun.

● Laboratory Skill: Manipulation of Materials

● Mathematics Skill: Computation

5. On the board write the terms *spectroscope*, *telescope*, and *microscope*. Underline the syllable *scope* in each. What does it mean (an instrument for observing or detecting, from the Latin word *scopium*, which means *to view*)?
6. Ask the students to recall what the spectrum is (colors found in white light). What is a spectroscope (instrument used to observe all the wavelengths of light given off by an object)?
7. Have students work individually or in small groups as they use the following directions to construct a spectroscope. Paint the inside of a shoe box and its cover black. Then make holes in both ends of the shoe box. Cover one hole by taping a piece of diffraction grating across it. Use the other hole for viewing. Tape the cover on the box. Before using the spectroscope, review the use of diffraction gratings in Lesson 4 of Chapter 20.

Figure 24.3 Astronauts and satellites have gathered large amounts of new information about the universe. *left:* An astronaut walks on the moon; *right:* A photograph of Jupiter taken from *Voyager*

Do You Know?
The first person from the United States to travel in space was Alan B. Shepard, Jr. His flight lasted 15 minutes and reached an altitude of 187.5 km.

Another important tool of astronomy is the **spectroscope** [SPEHK-truh-skohp]. A spectroscope is used to study the different wavelengths of light given off by a glowing object. It is known that different elements give off different wavelengths of light. By studying the light given off by glowing objects in space, astronomers can tell what elements are present in them.

A spectroscope can also be used to determine the surface temperature of an object that is giving off light. For example, a star that gives off more blue light than red light is a relatively hot star. Cooler stars give off more red light.

Astronomers can use a spectroscope to determine whether a glowing object is moving away from or toward the earth and how fast it is moving. For example, the light given off by a moving star may shift to include more of the longer red wavelengths. This is an indication that the star is moving away from the earth. The greater this shift to more red wavelengths, the faster the star is moving.

All the information astronomers have to work with has not been gathered from observations made only from Earth. The voyages of satellites, space probes, and the space shuttle have provided large amounts of new information. This information has been gathered by astronauts, cameras, and many specialized instruments.

Astronomers do not arrive at all their conclusions through direct observation and measurement alone. They apply the laws of gravitation, electricity, magnetism, nuclear reactions, and other principles of physical science to those observations and measurements. Then, following the steps of the scientific method, they develop theories to explain their observations and measurements. Thanks to recent advances in technology, astronomers can now call on computers to help them in their analysis of information about the universe.

Table 24-1 Milestones in Space Travel

Spacecraft	Launch Date	Length of Flight	Milestone
Vostok 1	April 12, 1961	1.8 h	First person in space
Freedom 7	May 5, 1961	15 min	First person from the U.S. in space
Friendship 7	Feb. 20, 1962	4.9 h	First person from the U.S. to orbit the earth
Vostok 6	June 16, 1963	70.8 h	First woman in space
Gemini 4	June 3, 1965	97.9 h	First person from the U.S. to walk in space
Apollo 11	July 16, 1969	195 h	First persons to walk on the moon
Skylab 1	May 25, 1973	672.7 h	First persons to travel in an earth-orbiting space station
Apollo-Soyuz Test Project	July 15, 1975	*Apollo*—216 h *Soyuz*—144 h	First cooperative space flight between the U.S. and the U.S.S.R.
Space Shuttle *Columbia*	April 12, 1981	54.4 h	First flight of a reusable spacecraft

Study Questions for 24.1

1. How do astronomers study the universe?
2. Name two important tools of astronomy.

1. Since they cannot visit or witness objects and occurrences personally, astronomers must piece together information from many different sources. Some of those sources are data from instruments, comparable objects or events that can be observed, and data from satellites, space probes, and the space shuttle.
2. telescope, spectroscope

8. Summarize all the methods astronomers use to make observations. Remind the students that observing is only one step of the scientific method. What are the other steps of the scientific method (experiment, analyze, synthesize)?

9. On the board write the following list of astronomers: Tycho Brahe, Kepler, Copernicus, Galileo, Ptolemy, and Herschel. Have each student select one astronomer from the list and do research to determine how that astronomer used the scientific method. This is a good time for students to read the career at the end of the chapter.

FOLLOW-UP

Reinforcement: Tools of the Astronomer

Ask each student to create an astronomer's dictionary. It should contain the names of some of the tools of an astronomer, some simple illustrations that identify each tool, and a brief description of ways in which the tool is used.

Enrichment: Jocelyn Bell Burnell

Ask students to do research on Jocelyn Bell Burnell. They should find out how she proved that one radio source in space did not consist of little green men signaling the earth.

Tie-in: Astronomy Versus Astrology

Ask students to determine the difference between astronomy and astrology. Why isn't astrology a true science? Have the students bring in the astrology predictions from the newspaper. How could they use the scientific method to prove that the predictions are not correct?

SCIENCE BACKGROUND

The Doppler shift can be used to calculate the direction and speed of stars. The shifting of spectral lines indicates the direction. If the shift is toward the red end of the spectrum, the stars are moving away from the earth. The Doppler shift is called the red shift.

24.2

OBJECTIVES: The student should be able to

- **explain** how scientists study the life cycle of stars;
- **name** the stages in the life cycle of a star;
- **explain** what scientists think black holes and quasars are.

SCIENCE TERMS

stars	neutron star
red giant	black hole
white dwarf	quasars
supernova	

ADVANCED PLANNING

1. For the dioramas collect scraps of material, boxes, and spools of thread.

2. Have available the materials you used to introduce the chapter.

LESSON PLAN

1. Introduce the lesson by showing the students some constellations, as you did in the introduction to the chapter. What is each member of a constellation called (a star)? What is a star (a glowing ball of gases that gives off energy)? What is the star closest to Earth (the sun)?

SCIENCE BACKGROUND

Stars are luminescent, or glowing, objects. Planets are not. Planets appear bright because they are reflecting light.

24.2 Stars

You will find out
- how astronomers have determined the life cycle of stars;
- what some of the stages are in the life cycle of stars;
- what astronomers believe about black holes and quasars.

Stars are large masses of gases that give off energy. The closest star to the earth is the sun. Even so, the sun is 150 million kilometers away! The sun and all other stars pass through different stages. These stages are often referred to as the life cycle of stars.

Unlike the relatively short life cycles of living things on Earth, the life cycles of stars cover millions or even billions of years! Certainly no astronomer has been alive long enough to observe the life cycle of a particular star. How, then, can astronomers understand the sequence of stages in the life cycle of a star?

Figure 24.4 A painting of the laboratory of Tycho Brahe, a 16th-century Danish astronomer who analyzed the positions of more than 777 stars without the aid of a telescope

Figure 24.5 The McMath Solar Telescope at Kitt Peak is 30 meters high and extends 90 meters into the ground. It is specially designed to study the star nearest to Earth, the sun.

To understand the methods astronomers use to piece together a life cycle that covers millions of years, imagine that you are on a brief visit to Earth from another planet. During your visit, you explore a forest. You see oak trees in all stages of development. You see acorns on the ground, young sprouts coming out of the ground, and small trees. You also see tall, fully grown trees, dying trees, and dead trees. Your visit isn't long enough to allow you to watch a single oak tree grow from an acorn to a fully grown tree and then die. But you see many trees in many different stages. Using the scientific method, you develop a theory about the life cycle of trees. In much the same way, astronomers have studied many stars and developed a theory about the life cycle of stars.

The life cycle of a star begins when it forms from a cloud of dust and gases. The cloud is made up mostly of hydrogen. When the cloud gets very dense, all its atoms are pulled together by gravitational forces. The cloud shrinks, or collapses. Its center becomes very hot. Its temperature reaches several million degrees Celsius. Individual atoms are separated into electrons and nuclei, forming a plasma. Nuclear fusion reactions begin. Hydrogen nuclei join together, forming helium nuclei. These reactions release enormous amounts of energy. The cloud is now a star.

2. Have a student compare the method of studying the life cycle of a tree to the method of studying the life cycle of a star. Ask the students to name any other objects that could be studied in this manner (any objects that show definite changes throughout their existance).

● **Comprehension Skill: Interpretative**

ACTIVITY: How Far Away Are the Stars?

You may want to have more advanced students convert the star distances from light-years to kilometers.

● Mathematics Skill: Graphing

ACTIVITY

How Far Away Are the Stars?

Materials

colored pencils
graph paper

Procedure

1. Study the chart and arrange the stars in order of their distance from the earth. (A light-year is the distance that light travels in one year. That distance is 9,461,000,000,000 km.)

2. Make a bar graph to illustrate how far each star is from the earth. The names of the stars should go across the bottom of the graph. The distance from the earth in light-years should go up the side of the graph. ●

The 20 Brightest Stars	Distance from Earth in Light-Years
Sirius	8.8
Canopus	98
Alpha Centauri	4.3
Arcturus	36
Vega	26
Capella	46
Rigel	900
Procyon	11
Betelgeuse	490
Achernar	114
Beta Centauri	290
Altair	16
Alpha Crucis	390
Aldebaran	68
Spica	300
Antares	250
Pollux	35
Fomalhaut	23
Deneb	1,630
Beta Crucis	490

Questions

1. Which star is closest to the earth?

2. Which star is farthest from the earth?

3. Are any stars equally distant from the earth?

4. Which star is about twice as far away as Alpha Centauri?

5. Which star is three times as far away as Spica?

6. Which star is about four times as far away as Sirius?

1. Alpha Centauri
2. Deneb
3. Betelgeuse and Beta Crucis (490 light-years)
4. Sirius
5. Rigel
6. Pollux

Fusion reactions continue in the star. As the hydrogen is used up in fusion, the center of the star collapses. At the same time, its outer layers expand. The overall diameter of the star increases a hundred or more times. At this stage, the star gives off mostly red light and is, therefore, called a **red giant.** The red-giant stage lasts several million years.

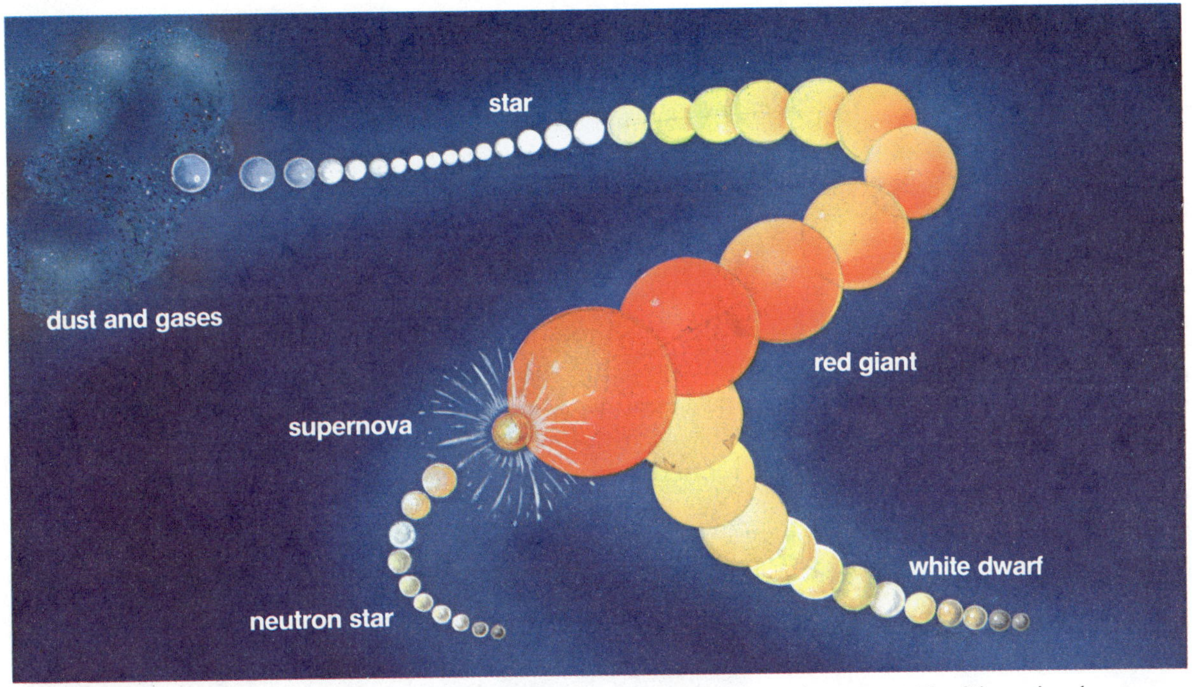

Figure 24.6 The life cycle of a star

Astronomers estimate that the sun will become a red giant in about five billion years. As a red giant, the sun's diameter would be so great that its outer surface would almost touch the earth! This situation would cause the earth's oceans to boil away!

What happens after the red-giant stage depends on the mass of the star. A star with a mass about equal to that of the sun will collapse to about the size of the earth. The result of this collapse is a very dense star called a **white dwarf.** The available hydrogen in a white dwarf is used up, so fusion reactions stop. No nuclear energy is released. However, white dwarfs are so hot that they glow for billions of years.

3. Have three or four students work together to prepare dioramas that illustrate the life cycle of a star. Have the students label the red giant, white dwarf, supernova, and neutron star in their dioramas.

● **Word Skill: Vocabulary**

4. Ask the following questions to reinforce an understanding of the differences between each stage in the life cycle of a star: Which stage is the densest (neutron star)? When do nuclear fusion reactions begin in a star (during stage in which star collapses, temperature rises to several million degrees Celsius, and plasma forms)? Which star gives off mostly red light (red giant)? What is the explosion of a red giant called (supernova)? What is the name of a red giant that collapses to a very dense star (white dwarf)?
5. Ask what force is exerted by objects that attract other objects to them (gravitational force). Point out that some regions in space have a gravitational force of such strength that nothing can escape. Ask what such regions are called (black holes). Of what are black holes made (stars that have collapsed into an extremely dense material)? What is the name of one black hole (Cygnus X-1)? Why can't black holes be seen (they allow no light to escape)?

● **Comprehension Skill: Literal**

Do You Know?
If you compacted all the material in 1,000 trailer trucks into one tiny dot, you would still not have a material as dense as the material in a neutron star.

The events after the red-giant stage are different for a star that is much more massive than the sun. Such a star will have a violent explosion. The explosion of a massive red giant is called a **supernova.** In a supernova, the star's outer layers are blown into space.

The collapsed star left behind after a supernova is a **neutron star.** A neutron star is extremely dense. Its atoms are completely crushed. It is difficult to imagine just how dense neutron stars are. A ball of neutron-star material the size of a period on this page would have a mass of 40,000,000 kilograms!

Like the density of neutron stars, many other things in the universe are hard to imagine. For example, in the early 1970's astronomers found something they call Cygnus X-1. In Cygnus X-1, there seems to be a region in which gravity is so strong that nothing, not even light, can escape. Cygnus X-1 is called a **black hole.** Black holes are thought to be stars that have collapsed into a material that is so dense that its gravity allows nothing ● to escape. Since black holes allow no light to escape, they cannot be seen. Astronomers can only detect their existence by observing their gravitational influence on other objects.

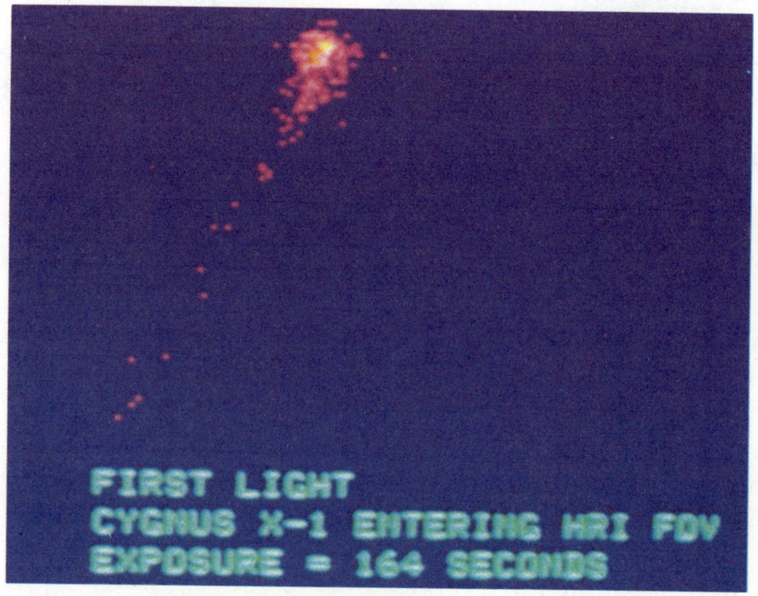

Figure 24.7 This is an X-ray image of Cygnus X-1. The X rays are thought to be produced by hot gases falling into the black hole.

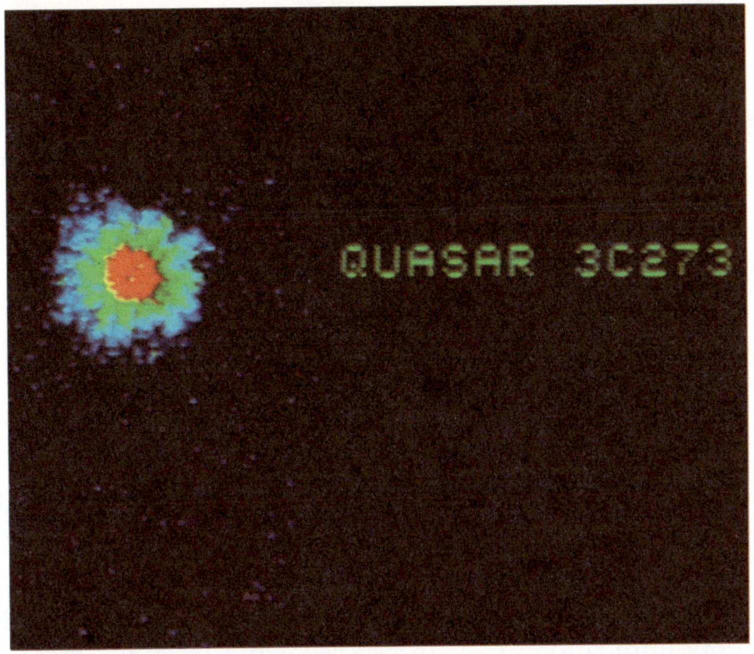

Figure 24.8 An X-ray image of a quasar

Astronomers discovered another unusual kind of object in space back in the early 1960's. These objects are called **quasars** [KWAY-zahrz]. Quasars appear to be extremely distant objects emitting radio waves, visible light waves, and X rays from a small area. This area may be associated with a very massive black hole. Quasars appear to be moving away from the earth at very high speeds. They also give off an amazingly large amount of energy. Astronomers are still puzzling over the source of this energy. Nuclear fusion reactions do not release enough energy to account for it. So far, this mystery remains one of the many unanswered questions about the universe.

Do You Know?
The term *quasar* is short for quasi-stellar radio source.

Study Questions for 24.2

1. How have astronomers pieced together the life cycle of stars?
2. Put the following stages in the life cycle of a star in order: red giant, neutron star, supernova, cloud of dust and gases.
3. What is Cygnus X-1?

1. They studied stars in many different stages and then, using the scientific method, developed a theory about the life cycle of stars.
2. cloud of dust and gases, red giant, supernova, neutron star
3. an object having a region in which gravity is so strong that not even light can escape, a black hole

6. Write the term *quasar* on the board. Ask a student to define the term (objects emitting radio waves, visible light waves, and xrays from a small area).
7. Stress that astronomers do not know where the quasars get their energy. Ask students to hypothesize about the source of energy of the quasars.

FOLLOW-UP

Reinforcement: The Formation of Stars

Ask the students to pretend that they are news reporters. Assign each student to write a magazine article describing how stars were formed. Have the students hypothesize about whether stars are still being formed.

Enrichment: Constellations

Have each student select a constellation and make a drawing of it. Each student should then do research to find out about the ancient myth behind the constellation.

Tie-in: Planetarium Visit

Take your students on a field trip to a nearby planetarium. Have one of the astronomers there describe some of the methods that are used to study the universe.

SCIENCE BACKGROUND

In its lifetime, a typical quasar gives out as much radiation as it would if it completely converted the mass of 10 million suns into energy.

24.3

OBJECTIVES: The student should be able to

- **describe** how satellites are being used;
- **list** the advantages of living in space;
- **describe** what space colonies might be like.

SCIENCE TERM

space colonies

ADVANCED PLANNING

1. Ask students who build models to bring in models of satellites. You may want to suspend the models from the ceiling for the duration of the lesson.

2. Ask students to read and bring in magazine and newspaper articles about satellites and Sky Lab.

3. You will need a piece of poster paper, a magic marker, and some yarn. Attach the poster paper to your bulletin board.

LESSON PLAN

1. Introduce the lesson by asking the students who brought in satellite models to describe the purpose of their satellites. Have several students summarize the articles that they brought in.
2. Based on the presentations in Step 1, ask the students to name some uses of satellites (telephone and television communications; information about weather patterns, air pollution, water, and movements of the earth's outer layer).

SCIENCE BACKGROUND

The heliosphere is a magnetic bubble that surrounds the solar system, expanding and contracting in response to solar winds. Pioneer 10 is studying this mysterious shroud.

24.3 The Space Frontier

You will find out

- how satellites are being used;
- what advantages there are to working in space;
- what space colonies might be like.

You are probably aware that the universe contains billions of objects and groups of objects such as stars, galaxies, comets, planets, and moons. But you are probably not consciously aware of most of these objects in your daily life. Your life centers mainly around one planet, the earth, and one star, the sun. However, many people believe that life could be very different in the future.

Things have been changing fast in recent years. People are now seriously thinking about venturing into space. Here on Earth, people are already benefiting from the use of satellites. Telephone and television communications can be relayed from one side of the earth to the other by satellites. Satellites equipped with telescopes and cameras gather information and relay it back to Earth. Such satellites are very useful in the study of weather patterns, air pollution, water and other natural resources, and movements of the earth's outer layer.

Figure 24.9 Information from satellites is used in making weather forecasts.

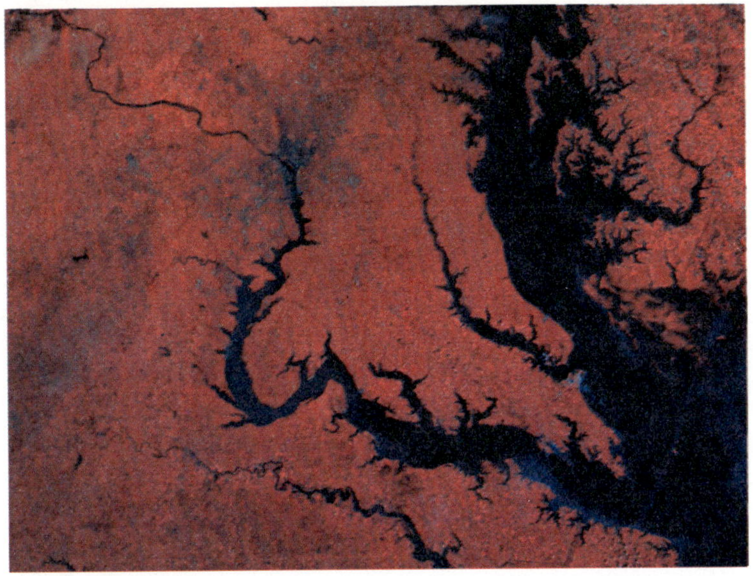

Figure 24.10 A satellite photograph of the Chesapeake Bay area of the United States

Already people have traveled in space. They have walked on the moon. So far, the main goal of such ventures has been to explore, experiment, and observe. However, many scientists believe that in the future people may venture into space not just to explore, but to actually live and work there. Groups of people living and working in space would form **space colonies.**

The limitations of working and living in space are obvious. A long list of essentials, including water and oxygen, would have to be supplied from Earth. Adjustments would have to be made for the absence of gravity. These and other practical aspects of life in a space colony pose many problems. Also, the funding of a space colony would require enormous amounts of money. Why, then, is anyone interested in space colonies? Many people are interested in them because there would be advantages to doing industrial and manufacturing work in space.

One advantage to doing work in space is the huge supply of solar energy available. In the United States, industry and manufacturing currently account for over 40 percent of the country's total energy consumption. The advantage to these businesses of a plentiful and relatively inexpensive energy source is obvious.

3. List the uses of satellites on the poster paper attached to the bulletin board. Have students put their articles on the bulletin board. A piece of yarn can be used to connect the article to the use of the satellite described in the article.

4. Draw a diagram of the solar system on the board. Draw a spaceship some distance from Earth. Ask the students to pretend that they have been asked to start a space colony at the point in space where the spaceship appears in the drawing. How would life be different from life on Earth (no gravity, probably no water, different amount of heat and light received from the sun, no air, no pollution)?

5. Ask the students how each of the differences between life on Earth and life in space could be used to benefit people (temperature—higher and lower temperatures would be available for use in manufacturing; no gravity—new and better mixtures could be made; different amount of heat and light from the sun—huge supply of solar energy would be available; no air—vacuums would be helpful in the manufacture of electronic devices).

● Comprehension Skill: Interpretative

Many industrial and manufacturing processes require very high or very low temperatures. These temperatures are readily available in space. With the use of lenses and mirrors, solar energy could be concentrated to provide temperatures of thousands of degrees Celsius. Very cold temperatures are easy to come by in a space colony. The outside temperatures in places shielded from the sun would be lower than −200°C.

The materials used to make products such as electronic devices and drugs must be highly pure. Molecules and tiny particles in the air can affect that purity. To protect the purity of the materials involved, the products are made in a vacuum. It is expensive and difficult to produce a vacuum on Earth. A better vacuum exists naturally in space. And it is free.

The absence of gravity would pose problems to people living and working in space colonies. However, at the same time, it would be an advantage to the production of certain materials. It is easier to form many kinds of ● mixtures in the absence of gravity. The different substances in the mixture would stay evenly distributed. No substances would settle out.

Figure 24.11 An artist's conception of a space colony

Studies have shown that a new kind of steel, ten to a hundred times stronger than any made on Earth, could be made in space. One company has estimated that at least 400 new mixtures of metals could be produced in space. Mixtures of metals are used to make many kinds of machinery and equipment. The new and improved mixtures of metals would be very useful.

The absence of gravity and the vacuum that exists in space would make it possible to grow large, pure crystals. Such crystals could be used in computers, lasers, solar cells, and many other devices.

The advantages of a space environment to industry and manufacturing are significant. These advantages and a spirit of curiosity and adventure may eventually bring people into space colonies. Some scientists and engineers have even drawn up plans for space colonies. One popular design is a huge cylinder about 20 kilometers long and 10 kilometers in diameter. Another design details a doughnut-shaped structure of about the same size. An effective substitute for gravity could be produced by spinning the colony.

The planners suggest that the interiors of the space colonies could be landscaped with hills, valleys, water, whatever was desired. Weather could be controlled. The number of daylight hours could be controlled.

It is difficult to say whether people will ever live in space colonies. And, if they do, exactly what those colonies will be like is an unanswered question. But it is fun and useful to think about the frontier of space and the possibilities that exist. What may seem impossible or unbelievable now may some day be taken for granted. Remember, there was a time when people believed that the earth was flat and that no one would ever walk on the moon.

Do You Know?
In the absence of gravity, you would be two or three centimeters taller.

Study Questions for 24.3

1. How are satellites being used?
2. Give three advantages to doing industrial and manufacturing work in space.
3. How do some people think space colonies will be designed?

1. to relay telephone and television communications, to gather information and relay it back to Earth
2. availability of very high and very low temperatures, availability of a vacuum, ability to make new and improved mixtures in the absence of gravity
3. a huge cylinder or doughnut-shaped structure that spins to provide a substitute for gravity

6. Ask the students whether they would be willing to live in a space colony. What would be some of the advantages (planned landscape, control of weather and number of hours of daylight)?
7. Have each student draw a diagram of an ideal space colony. Encourage the students to use their imaginations.

FOLLOW-UP

Reinforcement: Satellites in Our Lives
Assign each student to research one of the NASA missions. Each student should present an oral report to the class explaining the importance of the mission.

Enrichment: Satellites in Orbit
Have students research the meaning and importance of centripetal and centrifugal forces to orbiting bodies in space.

Tie-in: Seeing the Future
Ask students to write short stories that describe their adventures and lives as members of a space colony.

What Did You Learn?

24.1 • Astronomers use telescopes to observe objects and monitor their motion and to pick up radio waves and other forms of radiation given off by objects in space.
24.1 • A spectroscope can be used to study the surface temperature and the motion of objects in space.
24.1 • Astronomers develop theories by following the scientific method and applying the principles of physical science to the information they have gathered.
24.2 • Stars are large masses of gases that give off energy.
24.2 • Astronomers have studied many stars in many different stages and developed a theory about the life cycle of stars.
24.2 • Stars form from a cloud of dust and gases.
24.2 • Nuclear fusion reactions in stars release energy.
24.2 • A star with expanded outer layers that gives off mostly red light is called a red giant.
24.2 • A white dwarf is a collapsed red giant that glows for billions of years.
24.2 • The explosion of a massive red giant is a supernova.
24.2 • A neutron star is the collapsed star left behind after a supernova.
24.2 • A black hole is an object in which gravity is so strong that not even light can escape from it.
24.2 • Quasars appear to be extremely distant objects emitting radio waves, visible light waves, and X rays from a small area.
24.3 • Satellites are used to relay telephone and television communications and to gather information.
24.3 • There would be advantages to doing industrial and manufacturing work in space.
24.3 • Some scientists and engineers have drawn up plans for space colonies.

Key Terms

astronomers
spectroscope
stars
red giant
white dwarf
supernova
neutron star
black hole
quasars
space colonies

Biography

Beatrice Tinsley (1941–1981)

Beatrice Tinsley was an astronomer. In particular, she studied galaxies. She was interested in finding out how galaxies change as the stars in them change. Through her studies, Dr. Tinsley hoped to gain an understanding of what the future of the universe will be.

Early in her career, Dr. Tinsley had problems finding a permanent job. She had problems because it was difficult for a woman to get a permanent position at an astronomical observatory. She found a position at Yale University and remained there until her death.

While at Yale, Dr. Tinsley became a leader in her field. Just as important, she had a positive effect on the people she worked with. She took great joy in her own work and, at the same time, was interested in the work of others. She was happy and eager to help others with their work.

Dr. Tinsley became the Director of Graduate Studies at Yale. In this role she encouraged women to study science. Perhaps as a result of her work, Yale graduated several talented women scientists during the time that Dr. Tinsley was there.

Vocabulary

1. spectroscope
2. Stars
3. black hole
4. supernova
5. astronomers
6. red giant
7. Quasars
8. space colonies
9. white dwarf
10. neutron star

● Word Skill: Vocabulary

What Do You Remember?

1. They use a spectroscope to study the light given off by the glowing object. It is known that different elements give off different wavelengths of light.
2. When the density of a cloud of dust and gases gets very high, the cloud collapses. Its center becomes very hot and nuclear fusion reactions begin.
3. white dwarf
4. Astronomers observe the gravitational influence of black holes on other objects.
5. information on weather patterns, air pollution, water and other natural resources, and movements of the earth's outer layer
6. There would be advantages to doing industrial and manufacturing work in space.
7. Substances in the mixture would stay evenly distributed. No substances would settle out.

TO THINK ABOUT AND DO

Vocabulary ●

Number your paper from 1 to 10. Beside each number write the word or words from the list that best complete each sentence.

1. Astronomers use a(an) ________ to study the different wavelengths of light given off by a glowing object.
2. ________ are large masses of gases that give off energy.
3. Cygnus X-1 is a(an) ________.
4. A(An) ________ is the explosion of a massive red giant.
5. Spectroscopes and telescopes are tools used by ________.
6. A large star that gives off mostly red light is called a(an) ________.
7. ________ appear to be extremely distant objects emitting radio waves, visible light waves, and X rays.
8. Many people believe that in the future people may live and work in ________.
9. A red giant with a mass equal to that of the sun will eventually collapse to form a(an) ________.
10. The collapsed star left behind after a supernova is called a(an) ________.

white dwarf
black hole
spectroscope
astronomers
neutron star
space colonies
supernova
red giant
stars
quasars

What Do You Remember?

1. How can astronomers tell what elements are present in a glowing object in space?
2. How does a star form?
3. What stage will follow the red-giant stage for a star the size of the sun?
4. How do astronomers detect black holes?
5. What information can be gathered by satellites?
6. Why are many people interested in space colonies?
7. Why would it be easier to make mixtures in the absence of gravity?

8. Why is it better to produce electronic devices in a vacuum?
9. What is a white dwarf?
10. How do astronomers develop theories about the universe?

Applying What You Have Learned

1. Why is the sun the only star that is visible during the day?
2. Imagine that you have discovered a new star. How will you study it?
3. The sun is the most important star in your life. What would the earth be like if the sun did not exist?
4. Do you think you would like living in a space colony? Why or why not?

Research and Investigation

1. People have always found comets interesting. Perhaps the most well-known comet is Halley's Comet. Write a report on Halley's Comet.
2. Report on the use of satellites in weather forecasting.

8. In a vacuum, there are no tiny particles or molecules present to affect the purity of the materials used to make the devices.
9. a collapsed red giant that glows for billions of years
10. Following the steps of the scientific method, they apply the principles of physical science to their observations and measurements.

Applying What You Have Learned

1. Since the sun is the nearest star, when it is in view (during the day) its brightness obscures the light from other stars.
2. look at it through a telescope, try to pick up radio waves and forms of radiation from it with a telescope, use a spectroscope to identify the elements in the star, its surface temperature, and its motion
3. Answers will vary. Students may mention that without light and heat from the sun, life on Earth as it now exists would not be possible. They may also mention that since Earth is one of nine planets that revolve around the sun, if the sun did not exist, the location and motions of the planets would change.
4. Answers will vary. Stress that since the question calls for a feeling or opinion, no answer given can be wrong.

- Comprehension Skill: Application

Research and Investigation

1. Halley's Comet, like all comets, is an icy object that orbits the sun. It's orbit is ellipse shaped. It has a long tail that points away from the sun. As the comet moves closer to the sun, its tail becomes longer. Astronomers predict Halley's Comet will be visible again in 1986.
2. Weather satellites have been used by the United States since April 1, 1960. The first weather satellite was Tiros 1. It was used to take pictures of clouds. Satellites today take pictures of all weather systems and record temperatures. Satellites that are used to make long-range predictions move at the same speed as the earth. Thus, they remain over the same spot. These satellites take photographs every 30 minutes. Meteorologists study the photographs and use them to make long-range predictions.

- Reference/Research Skills

Using the Appendices

OBJECTIVES: The student should be able to

- **name** the basic metric units for measuring mass, length, and volume;
- **convert** from one metric unit to another metric unit;
- **estimate** the size in metric units of some common objects;
- **explain** how to light a Bunsen burner;
- **explain** how to use a laboratory balance.

LESSON PLAN

1. Introduce the lesson by having the students locate the appendices in their books and explain their purposes. Point out the different features of each appendix and how each feature can help the students.
2. Review the basic metric terms for measuring mass, length, and volume.
3. List the metric prefixes *kilo-*, *centi-*, and *milli-* on the board. Have the students explain the meaning of each.
4. Explain to the students that because the metric system is a decimal system, it is easy to convert from one metric unit to another metric unit. Draw the prefix line from page 514 on the board. Use it to go through the examples in the text.
5. Give the students several problems like the examples on page 514. Go over the problems with the students. You may wish to assign some additional problems as homework.

APPENDICES

SOME COMMONLY USED METRIC UNITS

LENGTH	meter (m) base unit	kilometer (km) 1,000 m = 1 km	centimeter (cm) 100 cm = 1 m	millimeter (mm) 1,000 mm = 1 m
MASS	gram (g) base unit	kilogram (kg) 1,000 g = 1 kg	centigram (cg) 100 cg = 1 g	milligram (mg) 1,000 mg = 1 g
VOLUME	liter (L) base unit	kiloliter (kL) 1,000 L = 1 kL	centiliter (cL) 100 cL = 1 L	milliliter (mL) 1,000 mL = 1 L cubic centimeter (cm^3) 1,000 cm^3 = 1 L

You can easily convert from one metric unit to another by moving the decimal point to the right or the left a certain number of spaces. You can figure out how many spaces and which way to move the decimal point by studying the diagram and examples below. The diagram shows units of length. If you are working with units of mass or volume, simply substitute g or L for the m in each unit.

Example

How many km are in 837.0 m?

1. How many spaces from m is km and in what direction? three spaces to the left
2. Move the decimal point in 837.0 three spaces to the left.
3. 837.0 m = 0.8370 km

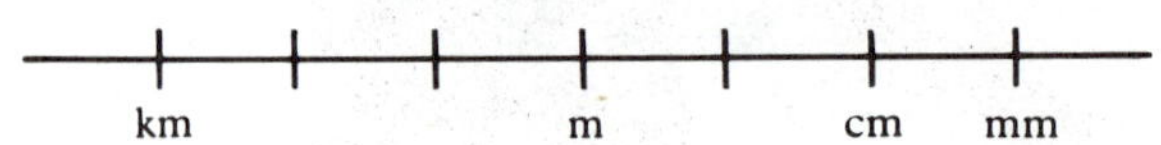

Example

How many mm are in 35.0 m?

1. How many spaces from m is mm and in what direction? three spaces to the right
2. Move the decimal point in 35.0 three spaces to the right.
3. 35.0 m = 35,000.0 mm

Example

How many mm are in 64.2 cm?

1. How many spaces from cm is mm and in what direction? one space to the right
2. Move the decimal point in 64.2 one space to the right.
3. 64.2 cm = 642.0 mm

THINKING IN METRIC TERMS

The following approximations will help you to get an idea of the sizes of some metric measurements.

LENGTH
1 km = the total length of 5 average city blocks
1 m = the length of a softball bat
1 cm = the width of the fingernail on your little finger
1 mm = the thickness of a dime

VOLUME
1 L = the total volume of about 4 single-serving milk cartons
1 mL = the total volume of 20 drops of water

MASS
1 kg = the mass of an average pineapple
1 g = the mass of a wire paper clip
1 cg = the mass of a flea

TEMPERATURE
Water freezes at 0°C.
Water boils at 100°C.
Normal human body temperature is about 37°C.
Room temperature is about 20°C.

THE LABORATORY BURNER

The Bunsen burner is the most common kind of laboratory burner. Some simple but important steps should be taken to light and use the burner safely and successfully. Remember, when near an open flame you should always roll up long sleeves, tie back long hair, and wear safety glasses.

1. Using an appropriate hose, connect the burner to the gas supply.
2. If you are using matches, light a match and then turn on the gas. Bring the match up along the side of the barrel to the top.
3. If you are using a sparker, turn on the gas and strike the sparker next to the top of the barrel.
4. Adjust the air supply so that the burner burns steadily with a light blue cone and without orange flames.

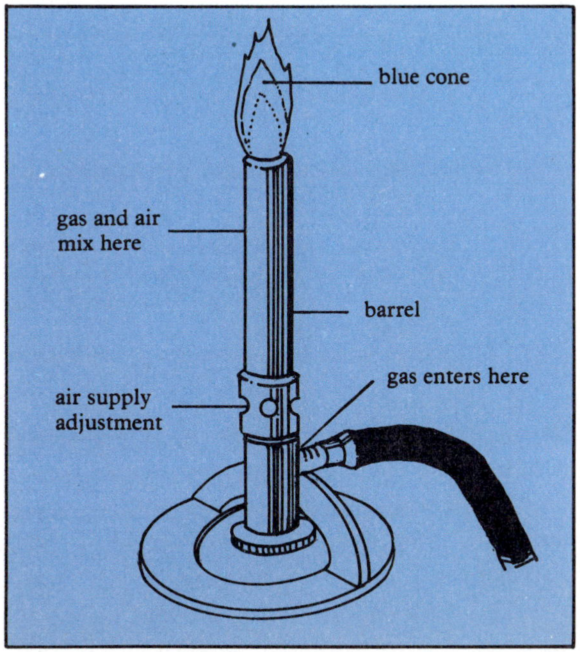

6. Ask the students to estimate in metric units the measurements of some familiar items. Then have the students measure the items to see how close their estimates are. Some examples are the following:

thickness of a slice of bread	1 cm
height of a stair step	17 cm
distance from the floor to a doorknob	1 m
distance from Chicago to Miami	2,000 km
volume of a shoe box	4,500 mL
1 gallon of milk	4 L
1 teardrop	1 mL
1 new pencil	5 g

7. Review the directions for using a Bunsen burner with the students. Stress safety as you demonstrate how to light the burner correctly. Have each student demonstrate that he or she can light a burner correctly.

8. Demonstrate for the students how to measure the mass of solids and liquids on the balances you have available. Stress the importance of accuracy. Remind the students that if a substance is in a container or on a paper while its mass is being measured, the mass of the container or paper alone must be subtracted from the mass of both together.

9. Provide the students with a variety of objects. Have them estimate the mass of each object and then check their estimates using a balance. Be sure the students are using the balance properly.

THE LABORATORY BALANCE

You use a balance to measure mass. There are many different kinds of balances. Your school probably has either equal-arm balances or triple-beam balances. No matter which kind of balance you use, the same general rules apply.

1. Always hold the balance with two hands at the base while carrying it.

2. Always place the balance on a level surface.

3. Always be sure the balance is zeroed before using it.
   If the pointer is still, the balance is zeroed when the pointer points to the zero mark.
   If the pointer is moving, the balance is zeroed when the pointer moves the same distance to each side of the zero mark.
   If the balance is not zeroed, ask your teacher to adjust it.

4. Never attempt to adjust or fix the balance yourself. Refer any problems to your teacher.

5. Never place chemicals directly on the balance.
   Place them on a paper. Then place the paper on the balance.

Using an Equal-Arm Balance

1. Place the object of unknown mass on the left pan.
2. Use the riders and the known masses to balance the object. The known masses should be placed on the right pan.
3. When the pointer moves the same distance to each side of the zero mark, the object is balanced.
4. The mass of the object is the total of the rider readings and the known masses on the right pan.

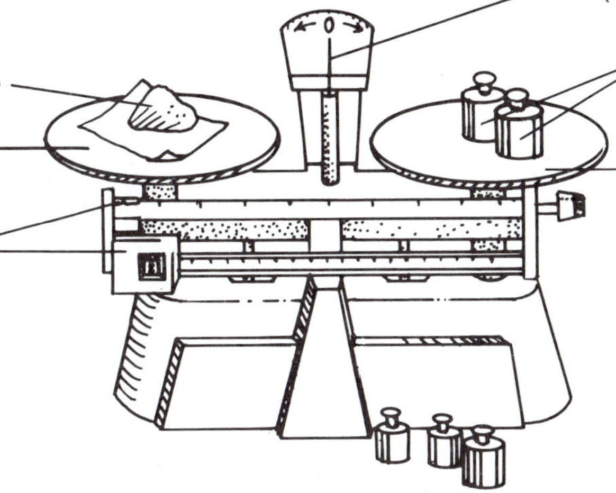

Using a Triple-Beam Balance

1. Place the object of unknown mass on the pan.
2. Use the riders to balance the object.
3. When the pointer moves the same distance to each side of the zero mark, the object is balanced.
4. The mass of the object is the total of the rider readings.

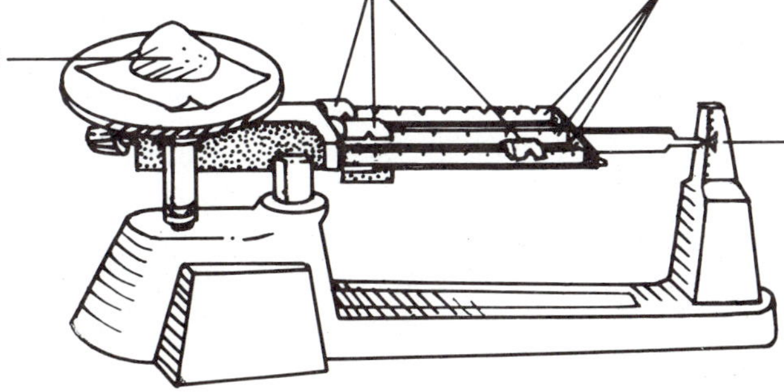

FOLLOW-UP

Reinforcement: Measure by Measure

Have the students form small groups. Each group should prepare a display or a series of posters that show how to use metric measuring tools.

Enrichment: Estimating

Organize a contest to see who is the best metric estimator in the class. For length, ask each contestant to draw a line a certain metric length and then measure to see who is the closest. For volume, have each contestant put a specified number of milliliters of water into a cup. Use graduated cylinders to determine whose estimate is the closest. For mass, have each contestant put a specified number of grams of sand on a piece of paper. Use balances to determine whose estimates is the closest. You may want to suggest to the contestants that they consult the section titled "Thinking in Metric Terms" on page 515. Examples in this section will give the students an idea of the size of some metric units

Tie-in: Early Methods of Measurement

Have the students do research to find out how measurements were made by early civilizations. How were things measured in ancient Egypt and Babylonia? How did the Romans measure things during the time of the Roman Empire?

GLOSSARY

PRONUNCIATION KEY

The glossary contains all the Key Terms in *Heath Physical Science* and their definitions. Some of these Key Terms are followed by their phonetic spelling, which shows you how to pronounce them. The words are divided into syllables and respelled according to the way each syllable sounds. The syllable that has the emphasis when the word is spoken appears in capital letters. The guide below is based on the Key to Pronunciation in *The World Book Encyclopedia.*

Sound	Respelling of Sound	Example	
hat, map	a	animal	AN-uh-muhl
age, face	ay	space	spays
care, air	ai	hair	hair
father, far	ah	charge	chahrj
child, much	ch	chocolate	CHAWK-luht
let, best	eh	energy	EHN-uhr-jee
equal, see,	ee	leaf	leef
machine, city		dream	dreem
term, learn	ur	earthworms	URTH-wurmz
sir, work		turkey	TUR-kee
it, pin	ih	system	SIHS-tuhm
ice, five	eye, y	iodine	EYE-uh-dyn
coat, hook	k	cloud	klowd
hot, rock	ah	cotton	KAHT-n
open, go,	oh	program	PROH-gram
grow		row	roh
order, all	aw	corn	kawrn
		fall	fawl
oil, voice	oy	poison	POY-zn
house, out	ow	fountain	FOWN-tuhn
say, nice	s	soil	soyl
she	sh	motion	MOH-shuhn
cup, butter,	uh	bulb	buhlb
flood		blood	bluhd
full, put,	u	pulley	PUL-ee
wood		wool	wul
rule, move,	oo	dew	doo
food		shoe	shoo
pleasure	zh	treasure	TREHZH-uhr
		vision	VIHZH-uhn
about	uh	sofa	SOH-fuh
taken	uh	paper	PAY-puhr
pencil	uh	fossil	FAHS-uhl
lemon	uh	carbon	KAHR-buhn
circus	uh	syrup	SIHR-uhp
curtain	uh	mountain	MOWN-tuhn
section	uh	action	AK-shuhn

A

absolute zero: −273° C; the lowest possible temperature

acceleration [ak-sehl-uh-RAY-shuhn]: the change in velocity divided by the time during which the change occurs

acid rain: rain containing the oxides of sulfur and nitrogen

acids: substances that produce positive hydrogen ions in liquid solution

actinide [AK-tih-nyd] **family:** the radioactive elements below the main body of the periodic table

activation energy: the energy that must be supplied to start a chemical reaction

alchemists: persons who established the laboratory and experimental approach of taking matter apart and analyzing it

alcohol: the compound formed when an OH group replaces a single hydrogen atom in a hydrocarbon

alkali [AL-kuh-ly] **metal family:** the family of very reactive elements with one valence electron

alkaline [AL-kuh-lyn] **earth metal family:** the family of elements that have two valence electrons

alpha particles: the form of nuclear radiation having two protons and two neutrons

alternating [awl-tur-NAY-tihng] **current:** current that is constantly switching directions

alternator: a generator that produces alternating current

amino acids: organic acids containing nitrogen

amperes [AM-pihrz]: the amount of electric current that passes a point in a conductor in one second

amplifier [AM-pluh-fy-uhr]: an electronic device that strengthens electronic signals

amplitude [AM-pluh-tood]: the height of a wave

analyze: to organize observations

Archimedes' principle: the principle that states that the buoyancy of any object is equal to the weight of the liquid that the object displaces

astronomers: scientists who study the universe

atmospheric pressure: the pressure exerted by the weight of the earth's atmosphere

atomic mass number: the sum of protons and neutrons in a nucleus

atomic number: the number of protons in a given nucleus

atoms: the smallest particles of individual elements

attract: to cause two oppositely charged objects to be pulled toward each other

B

bases: substances that produce a negative ion in liquid solution, usually the OH ion

beta particles: high-speed electrons that are a form of nuclear radiation

biomass [BY-oh-mas]: an energy source provided by all the materials in living things, their wastes, and their remains

black hole: an object thought to be a star that has collapsed into a material so dense that its gravity allows nothing to escape

boiling point: the temperature at which a liquid begins to boil

boron family: the family of elements with three valence electrons

breeder reactor: a nuclear reactor in which the fission of uranium produces plutonium 239, which can also be used as a fuel

buoyancy: the force that pushes an object up and makes it seem to lose weight in a fluid

C

calorie: the amount of heat required to raise the temperature of one gram of water one degree Celsius

carbon family: the family of elements that have four valence electrons

carburetor [KAHR-buh-rayt-uhr]: the part of a gasoline engine in which gasoline mixes with air

catalysts [KAT-uh-lihstz]: substances that affect the rate of a chemical reaction without being changed or used up in the reaction

cathode [KATH-ohd] **ray tube:** a vacuum tube in which a stream of electrons is directed onto a screen coated with a fluorescent material

center of gravity: the point around which an object's weight is evenly placed

chemical bond: the bond formed when atoms fit together, making a molecule

chemical changes: changes that affect the chemical properties of one or more substances

chemical equation [ih-KWAY-shuhn]: formulas and symbols that show what happens in a chemical reaction

chemical reaction: a process that results in chemical change

circuit breaker: a device that breaks or opens a circuit to stop electric current

coefficient [koh-uh-FIHSH-uhnt]: a number placed in front of a formula or symbol, indicating how many atoms or molecules of that substance are present

compounds: substances made of two or more elements chemically combined

compressional [kuhm-PRESH-uhn-ul] **wave:** a wave in which the disturbance occurs back and forth along the direction of wave motion

computer: a machine that can carry out a detailed set of instructions to perform a task

computer program: a set of instructions for a computer

concave [kahn-KAYV] **lens:** a lens with curved surfaces that dip at the middle and spread light waves apart

condensation [kahn-duhn-SAY-shuhn]: the change of a gas to a liquid

conduction [kuhn-DUHK-shuhn]: the process whereby heat energy is transferred from a hot object to a cooler one by collisions between particles

control rods: steel rods that control the rate of the chain reaction in a nuclear reactor

control unit: the unit that regulates the flow of information in a computer

convection [kuhn-VEHK-shuhn]: the spreading of heat energy by large movements of a liquid or a gas

convex [kahn-VEHKS] **lens:** a lens with curved surfaces that bulge out in the middle and bring light waves together

covalent [koh-VAY-luhnt] **bond:** a bond in which electrons are shared

crankshaft: the device connected to the rod of a heat engine and used to turn one or more wheels

crest: the highest point of a wave

crystals: solids in which the atoms are arranged in a regular pattern

cylinder: the part of a heat engine that confines a heated gas

D

deceleration: a change in velocity to a decreasing speed

decibels [DEHS-uh-behlz]: units that measure the intensity of sound

density: the mass of a substance in grams divided by its volume in cubic centimeters

deuterium [doo-TIHR-ee-uhm]: an isotope of hydrogen that has one proton and one neutron

diatomic [dy-uh-TAHM-ihk] **molecule:** a molecule in which only two atoms join together

diffraction [dih-FRAK-shuhn]: the bending of waves around the sides of a barrier

diffraction grating: a thin piece of glass with thousands of parallel slits close together

direct current: current that always flows in the same direction

distillation: the process by which a liquid is changed to a gas, then collected and cooled, returning the gas to a liquid state

Doppler effect: a change in the frequency of a wave as measured by an observer due to the motion of the wave source or the observer

double bonds: covalent bonds in which two pairs of electrons are shared

drag: the friction created when objects move in air

dry cell: a device that produces voltage as a result of chemical reactions between a moist paste and a metal

E

efficiency: a measure of the amount of work a machine puts out compared to the amount of work put into the machine

effort distance: the distance over which an effort force acts

effort force (F_E): the force that acts on a machine

electric charge: a characteristic of an object affected by electric forces

electric circuit [SUR-kiht]: the path of an electric current

electric current: the flow of electric charges from one point to another

electric discharge: the flow of an electric charge through a gas from a charged object to a conductor

electric forces: forces causing attraction or repulsion between two objects

electric generator: a machine that changes kinetic energy into electric energy

electric motor: a machine that changes electric energy into kinetic energy

electric power: the amount of electric energy that flows through a circuit at a given time

electrical ground: a very large conductor

electrically neutral objects: objects with equal numbers of positive and negative charges

electromagnetic induction: the creation of a current by a changing magnetic field

electromagnetic spectrum [SPEHK-truhm]: the range of all the kinds of electromagnetic waves

electromagnetic [ih-lehk-troh-mag-NEHT-ihk] **waves:** transverse waves that result from the action of electric and magnetic forces

electromagnetism [ih-lehk-troh-MAG-nuh-tihz-uhm]: the branch of physical science that involves the combined effects of electricity and magnetism

electromagnets: magnets in which magnetism is produced by an electric current

electron cloud: a three-dimensional area formed by the electron shells around the nucleus of an atom

electron dot diagram: a diagram in which the valence electrons are shown as dots

electronic [ih-lehk-TRAN-ihk] **devices:** devices that have rapid and accurate control of the flow of electrons

electrons: negatively charged particles surrounding the nucleus of an atom

electroscope: a device that detects the presence of an electric charge on an object

electrostatic induction [ih-lehk-truh-STAT-ihk in-DUHK-shuhn]: the production of an electric charge on an object by bringing it near a charged object

elements: substances that cannot be chemically broken down into simpler basic substances

endoscope [EHN-duh-skohp]: a device with optical fibers used by doctors to examine internal organs

endothermic [ehn-doh-THUHR-mihk] **reaction:** a chemical reaction in which energy is absorbed

energy: the capacity to move matter from one place to another, or to change matter from one substance to another

enzymes [EHN-zymz]: catalysts in living things

equilibrium [ee-kwuh-LIHB-ree-uhm]: a condition during which the net force is equal to zero

evaporation: the change from the liquid state to the gas state

exothermic [ehk-soh-THUHR-mihk] **reaction:** a chemical reaction that releases energy

experiments: measurements and tests under controlled conditions

F

facts: the building blocks of science

families: groups of elements with similar properties that appear in vertical columns in the periodic table

first law of motion: the law that states that every object stays in its equilibrium state unless it is acted on by outside forces

fluid: any substance that flows

fluorescent [floo-REHS-ehnt]: glowing when struck by electrons

focal length: the distance from the center of a lens to the focal point

focal [FOH-kuhl] **point:** the point at which light waves passing through a lens come together

force: a push or pull that is exerted on matter

fossil [FAHS-uhl] **fuels:** fuels formed from the remains of plants and animals that lived millions of years ago

fractional distillation: the process of distillation by which a substance is separated into many parts

freezing: the change of a liquid to a solid

frequency [FREE-kwuhn-see]: the number of crests of a wave that pass a point in a second

friction: a force that acts on a moving object in a direction opposite to the motion of the object

fulcrum: a fixed point of a lever

fuse: device that breaks or opens a circuit

G

gamma radiation: nuclear radiation in the form of energy without mass

gas: matter that has no definite shape or volume

geothermal [JEE-oh-thur-muhl] **energy:** heat that comes from inside the earth

gravitational force: the attraction between the earth and all objects on it

H

half-life: the time it takes for half the nuclei in a sample of a radioactive isotope to undergo radioactive decay

halogen [HAL-oh-jehn] **family:** the family of elements that have seven valence electrons and are extremely reactive

heat energy: the total kinetic energy of all the molecules or atoms in a substance

heat engine: a device that uses heat energy to do work

heat of fusion: the amount of heat energy used to melt a substance without changing its temperature

heat of vaporization [vay-puhr-ih-ZAY-shuhn]: the amount of heat energy used to evaporate a liquid

heat pump: a machine that takes heat from outdoors and carries it indoors

hertz [hurts]: a unit of measurement for wave frequency

holograms [HOH-luh-gramz]: three-dimensional photographs

hydrocarbons [hy-droh-KAHR-buhnz]: compounds made of only carbon and hydrogen

hydroelectric power: the electricity generated by the use of moving water

hypothesis [hy-PAHTH-uh-sis]: an educated guess that explains observations

I

image: a picture produced by light waves passing through a lens

impurities: elements that are added to semiconductor crystals to increase conductivity

inclined plane: a simple machine that increases the effort distance through which an object is moved and decreases the effort force needed to move it

indicator: a special type of organic compound that changes color as the pH of a solution changes

induced [ihn-DOOST] **magnetism:** the magnetizing of a material by a magnet

inertia [ihn-UR-shuh]: the tendency for objects to continue in their state of motion or rest

input unit: the part of a computer that takes in information

insulators: poor conductors of heat and electricity

integrated [ihn-tuh-GRAYT-ehd] **circuits:** circuits that are put on a semiconductor chip by placing impurities in a very precise arrangement

intensity [ihn-TEHN-suh-tee]: the loudness of a sound

interference [ihn-tuhr-FIHR-uhns]: the combined effects of two or more waves coming together

ionic [eye-AHN-ihk] **bond:** the bond formed when one atom shifts one or more electrons to another atom

isomers [EYE-suh-muhrz]: compounds with the same molecular formulas but different structural formulas

isotopes: atoms of the same element that differ in the number of neutrons in their nuclei

J

joule (J): the unit of work equal to a newton-meter

K

kilogram: a unit of mass

kilowatt: a unit of measure of large amounts of electric power

kilowatt-hour: the amount of energy produced by one kilowatt of power in one hour

kinetic energy: the energy of motion

kinetic theory: the theory that states that particles of matter move to varying degrees in solids, liquids, and gases

L

laboratory: the workshop of science

lanthanide [LAN-than-yd] **family:** the family of elements below the main body of the periodic table

laser: a device that produces laser light

laser light: an intense narrow beam of light made up of light waves all of the same wavelength and traveling in the same direction

law of conservation of mass: the law that states that in a chemical reaction, no atoms disappear and no new atoms are formed

lenses: curved pieces of glass or plastic that refract light waves

lever: a bar or board that is free to pivot about a fixed point

lift: the upward force of air moving around an object

liquids: matter that has a definite volume but does not have a definite shape

liter: a unit of volume

live wire: any wire carrying an electric current

M

machine: a device that helps a person to do work

magnet: a magnetized object

magnetic domains [doh-MAYNZ]: clusters of many atoms present in materials that can be magnetized

magnetic field: the area around a magnet where its magnetic force can be experienced

magnetic poles: areas of greatest magnetic force

magnetism: a force produced by the motion of electric charges in a material

mass: the amount of material in an object

matter: anything that has mass and takes up space

measurements: observations expressed in numbers

mechanical advantage: the amount of force gained by using a machine

medium: the matter or space through which a wave travels

melting: the change of a solid to a liquid

melting point: the temperature at which a solid changes to a liquid

memory unit: the part of a computer that stores all programs and information that have been put into it

meniscus [men-NIS-kuhs]: the curve at the surface of a liquid in a narrow cylinder

metallic [muh-TAL-ihk] **bond:** a bond in which many electrons are shared by many metal atoms

metals: materials that are good conductors of heat and electricity and that can be pounded into different shapes

meter: a unit of length

metric system: a decimal system of measurement based on the meter, the kilogram, and the liter

microelectronics: a branch of electronics involving integrated circuits

mirage: an illusion caused by the refraction of light waves, making objects that are far away appear to be nearby, floating in air, or upside down

mixture: two or more elements or compounds that are not chemically combined

molecular formulas: representations of molecules using symbols

molecule [MAHL-uh-kyool]: a combination of two or more atoms

momentum [moh-MEHN-tuhm]: the mass of an object multiplied by its velocity

N

natural elements: elements found in nature

negative ion: an atom with more electrons than protons

negative terminal: the place on the zinc jacket of a dry cell where electric contact is made

net force: the sum of opposing forces

neutralization [nu-truh-luh-ZAY-shuhn] **reaction:** a reaction in which an acid and base combine to produce water and a salt

neutron: a particle with no charge in the nucleus of an atom

neutron star: a collapsed star left behind after a supernova

newton (N): a unit of force or weight

nitrogen family: the family of elements that have five valence electrons

noble gas family: the family of elements that mark the end of each period on the periodic table and have eight valence electrons, except for helium

noise pollution: an excess of unpleasant sounds

nonmetals: materials that break instead of bend and are poor conductors of heat and electricity

nonrenewable: not able to be replaced

north geographic pole: the northernmost point of the earth

north magnetic pole: the north pole of the imaginary magnet stuck in the earth

nuclear chain reaction: a reaction in which each fission reaction causes more fission reactions

nuclear fission [FIHSH-uhn]: a reaction in which a heavy nucleus is split apart into nuclei of lighter elements and neutrons

nuclear force: the force holding neutrons and protons together in the nucleus

nuclear fusion [FYOO-zhuhn]: a reaction in which the nuclei of light atoms join together to form a heavier nucleus

nuclear radiation: the products of radioactive decay (alpha particles, beta particles, and gamma radiation)

nuclear reaction: a reaction in which the nucleus of one kind of atom changes into the nucleus of a different kind of atom

nuclear reactor: the part of a nuclear power plant where uranium fuel is bombarded with neutrons to produce a chain reaction that releases energy

nucleus: the small, dense core of an atom

O

observe: to record mentally or in writing things that are noticed

octet [ahk-TEHT] **rule:** the rule that states that atoms form chemical bonds to reach a more stable condition with eight valence electrons

ohms: units used to measure resistance

Ohm's law: the law that states that the greater the voltage, the higher the resistance

oil shale [shayl]: rock containing oil

optical fibers: thin strands of glass that use light waves to carry signals and messages

organic acids: hydrocarbon chains with a COOH group

organic [ohr-GAN-ihk] **compounds:** compounds that contain carbon

oscillator [AHS-uh-layt-uhr]: an electronic device that produces electronic signals from direct current

output unit: the part of a computer that changes the results of the processing unit from computer language into words, numbers, and symbols, and then prints them

overtones: notes of multiple frequencies produced by musical instruments

oxygen family: the family of elements having six valence electrons

P

parallel circuits: circuits in which the current divides and flows in two or more separate paths, or branches

periodic table: a table of the elements arranged according to properties

periods: horizontal rows of elements with increasing atomic numbers in the periodic table

photons: tiny bundles of energy present in electromagnetic waves

pH scale: scale that measures the strength of an acid or base

physical changes: changes that affect the size, shape, or physical state of a substance but not its chemical properties and formula

physical science: the branch of science that searches for answers to questions about matter and energy

piston: the movable top of a cylinder in a heat engine

pitch: the highness or lowness of a sound that the ear hears

plasma: the state of matter in which the particles of a gas break apart into electrically charged pieces

polarizing [POH-luh-ry-zing]: allowing only light waves with crests and troughs at a certain angle to pass through a material

polymers [PAHL-uh-muhrs]: organic molecules containing a large number of atoms

positive ion: an atom with fewer electrons than protons

positive terminal: the place on the carbon rod of a dry cell where electric contact is made

potential energy: stored energy that can be released when matter moves

power: the measurement of the rate at which work is done

pressure: force that is applied over a certain area

prism: a triangular piece of glass that bends light waves

processing: the carrying out of instructions and performance of operations using information that has been put into a computer

processing unit: the unit that carries out a computer program

products: substances that result from a chemical or nuclear reaction

property: a quality or characteristic of a substance

proteins: amino acids joined together in long chains

proton: positively charged particle in the nucleus of an atom

pulley: a lever that rotates around a fixed point

Q

quarks: small particles that are believed to make up protons and neutrons

quasars [KWAY-zahrz]: extremely distant objects that emit radio waves, visible light waves, and X rays from a small area

R

radiation [ray-dee-AY-shuhn]: a form of energy transfer that occurs through matter and through empty space

radioactive: giving off high-energy particles

radioactive decay: a nuclear reaction in which the nucleus of one kind of atom spontaneously changes into the nucleus of another kind of atom

radioactivity: the energy given off when a nuclear change occurs

reactants [ree-AK-tuhntz]: substances that undergo a chemical reaction

real image: an image that appears upside down and can be focused on a screen

rectifier [REHK-tuh-fy-uhr]: an electronic device that changes alternating current to direct current

red giant: a large star that gives off mostly red light

reflecting telescopes: devices that use a curved mirror and a convex lens to reflect and refract light waves from distant objects to produce a magnified image

reflection [rih-FLEHK-shuhn]: the bouncing back of a wave when it hits a boundary

refracting telescope: a device that uses two convex lenses to concentrate and bend light waves from distant objects to produce a magnified image

refraction [rih-FRAK-shuhn]: the bending of waves due to a change in the medium

renewable: able to be replaced

repel: to cause two similarly charged objects to push each other apart

resistance [rih-ZIHS-tuhns]: the opposition a material offers to the flow of electric charges

resistance distance: the distance an object on which work is being done is moved

resistance force (F_R): the weight of an object to be moved

resonance [REHZ-uh-nuhns]: the amplification of sound by musical instruments

retina [REHT-uhn-uh]: the surface in the eye upon which a real image is formed

reverberations [rih-vur-buh-RAY-shuhnz]: mixtures of reflected sounds

S

salt: the compound produced in a neutralization reaction when the hydrogen in the acid is replaced with another element from the base

science: a method of organizing curiosity

scientific method: investigation using observation, analysis, synthesis, and testing

screw: a combination of an inclined plane and a cylinder

second: a unit of time

second law of motion: the law that states that an object acted on by a constant force will move with a constant acceleration in the direction in which the force is acting

semiconductors: solid materials that have a conductivity in the range between that of an insulator and that of a good conductor

series circuit: a circuit with only one path for the current

shaft mining: the removal of coal deposits by cutting tunnels or shafts in the earth

shells: specific energy levels around the nucleus where electrons can be found

short circuit: a situation that occurs when current flows over an unintended path that has very low resistance

signals: short bursts of electric energy to which electronic devices respond

simple machines: devices that increase forces

solar cells: devices that collect solar energy and change it into electric energy

solar collectors: devices that collect and change solar energy into heat energy

solar energy: energy from the sun

solid: matter that has a definite shape and volume

solid-state devices: transistors or equipment utilizing transistors

solutions: mixtures in which substances are uniformly distributed

sound waves: compressional waves produced by a back-and-forth motion

space colonies: structures in space where people could live and work

spark plug: the part of a gasoline engine that ignites a mixture of gasoline and air

specific heat: the amount of heat needed to raise the temperature of one gram of a substance one degree Celsius

spectroscope [SPEHK-truh-skohp]: a device used to study different wavelengths of light given off by a glowing object

speed: the measure of how fast an object is moving

stars: large masses of gases that give off energy

static electricity: an electric charge that builds up on an object

steam turbine [TUHR-byn]: an engine in which steam under pressure is used to produce a rotational motion

straight chain hydrocarbons: compounds in which carbon atoms are bonded together in a single line

strip mining: the exposure of coal deposits with explosives and removal of the coal with giant shovels

strokes: movements of the piston in a gasoline engine

structural formula: a chemical formula using dashes to represent covalent bonds

supernova: the explosion of a massive red giant

synthesize [SIN-thuh-syz]: to put observations together in a new way

synthetic [sihn-THEHT-ihk]: word describing substances produced in the laboratory

T

technology: a way of applying scientific discoveries to the real world

temperature: the measure of the average kinetic energy of the atoms or molecules that make up a substance

terminal velocity: the velocity at which a falling object will not accelerate anymore

test: the check of a theory's predictions against new observations

theories: ideas that explain groups of facts

third law of motion: the law that states that for every action by one object there must be an equal and opposite reaction on the other object

tidal energy: energy in tidewaters that can be used to generate electricity

transformer [trans-FAWR-muhr]: a device used to increase or decrease voltage

transistor [tran-ZIHS-tuhr]: a device that uses semiconductors to control the flow of electrons

transition metals: elements in which the next to the last shell is filled as the number of electrons increases

transverse [trans-VURS] **waves:** waves in which the disturbance occurs at right angles to the direction of the wave motion

triple bonds: covalent bonds in which three pairs of electrons are shared

trough [trawf]: the lowest point of a wave

U

ultrasonic: sound waves with very short wavelengths and frequencies above 20,000 Hz

universal law of gravitation: the law that states that the force of gravitation is present between any two objects in the universe

V

vacuum tubes: electronic devices made of glass tubes from which nearly all air has been removed

valence [VAY-luhns] **electrons:** electrons in the shell farthest from the nucleus

velocity [vuh-LAHS-uh-tee]: a measure of both the speed of a body and its direction of motion

vibration: a back-and-forth motion

virtual [VUR-choo-wuhl] **image:** an image that appears right side up and cannot be focused on a screen

visible light: light waves that can be seen

voltage: the difference in electric potential that causes current to flow

volts: units for measuring voltage

volume: the measurement of the amount of space an object takes up

W

watt: a unit of power equal to one joule of work done in one second

wave: a disturbance that travels through matter or space

wavelength: the distance from one crest of a wave to the next

wedge: an inclined plane that moves

weight: the amount of gravitational pull on an object

wheel and axle: a lever that can move around in a circle

white dwarf: a verv dense, hot star formed as a result of the collapse of a red giant

white light: a mixture of all the colors of visible light

work: the application of a force moving an object a distance

INDEX

NOTE: Page number in **boldfaced** type indicates a definition.

S

T

U

V

W

X

ACKNOWLEDGEMENTS

Illustration

Michael Carroll 4, 127
Glenna Collett 138–139
James Conahan 109, 211, 235, 244–245
Genigraphics 17
David Hannum 463, 478
Walter Hortens 354, 416–417, 483, 503
Leonard Morgan 104, 143 bottom, 171, 208, 209, 233, 258, 284, 300, 344
Mike Prendergast 16, 21, 30, 56, 79, 83, 107, 108, 121, 142, 143 top, 145, 146, 147, 148, 149, 151, 153 top, 161, 162, 182, 185, 192, 213, 215 top, 237, 239, 240, 246, 259, 260, 261, 262, 265, 274, 275, 278, 279, 281, 285, 317, 322, 327, 338, 341, 353, 377, 379, 382, 384, 403, 420, 428, 429, 431, 442, 446, 447, 479, 497
Vytas Sakalas 16, 27, 45, 62, 76, 80, 86, 96, 120, 126, 129, 140, 144, 153 bottom, 157, 165, 167, 168, 170, 172, 184, 186, 212, 236, 238, 241, 271, 296, 297, 298, 309, 357, 364, 380, 416 top, 436, 437, 439, 444, 467, 487
Jeff Stock 33, 35, 95 bottom, 188
Will Tenney 14
Gary Torrisi 50, 51, 97, 99 bottom, 124, 169 top, 175, 440, 441
George Ulrich 2, 8, 13, 37, 38, 40, 41, 53, 58, 61, 78, 85, 94, 95 top, 99 top, 125, 141, 169 bottom, 190, 215 bottom, 223, 224, 240, 242, 266, 301, 304, 316, 318, 319, 320, 328, 336, 342, 345, 386, 394, 401, 417 center, 427, 438, 468, 486
Lily Yamamoto 359, 366, 399

Photography

Alpha: Jim Howard 227, 411; Richard Laird 462; D. C. Lowe 484; Alan McGee 407 left; James M. Mejuto 405; Blair Seitz 145; Al Zalon 406 top
Erik Anderson viii, 9, 10, 12, 180, 256 bottom row, 516
Animals Animals: 283; Margot Conte 32; Robert Mitchell 28 top
Peter Arnold, Inc.: Steve Allen 477; E. F. Bernstein 187 bottom; Cecile Brunswick 397 top; Richard Choy 259, 474; Jacques Jangoux 476; James H. Karales 269; Stephen J. Krasemann 496 right; Hans Pfletschinger 123; Mike Phillips 494; Clyde Smith 49
Atoz Images, Inc.: George Mars Cassidy 214; Peter Fronk 334; Fred Leavitt 230; Carol Lee 321; William Mean 407 right; L. L. T. Rhodes 114; Leonard Lee Rue 329, 398 bottom; L. Starz 308
Bell Laboratories 6 top and bottom right, 177, 361
The Bettman Archive 119 top
Biblioteca Marucelliana 19
Dr. Jason Birnholz, Brigham and Women's Hospital 6 top left
Black Star: Andy Levin 435; John Lopino 415; Debra & Flip Schulke 452; Fred Ward 447
Lee Boltin 146
David Brownell 48 right
The Burndy Library 43
James N. Butler 185 bottom
California Institute of Technology 142
Camerique 26, 204; Ron Dahlquist 378; J. Pickerell 68;
A. James Casner 52 bottom, 55, 63, 74, 75, 163 bottom
Stuart Cohen 91

Bruce Coleman, Inc.: A. Avis 216 top; D. Brewster 82 right; Robert P. Carr 71 left; Bob & Clara Calhoun 398 top; E. R. Degginger 173 bottom; John S. Flannery 81 top; K. Hisham 272; David Madison 57; M. Timothy O'Keefe 113; Bradley Olman 31; Vince Streano 216 bottom; Jonathan Wright 48 left, 188, 194, 263
Culver Pictures 150 left, 331, 353
E. R. Degginger 11, 28 center, 38, 63, 116 top and bottom right, 118, 122 bottom, 128, 134, 137 bottom, 155, 171, 189, 201, 371, 381, 392
Phil Degginger 28 right, 54, 60, 158
Leo de Wys, Inc.: 355, 429, 449, 460 bottom, 491; David Burnett 423; Calabri Photo 456; Everett C. Johnson 376
DPI: 101, 372; Don Askins 466; Mira Atkeson 414; Ken Biggs 282; Robert Isaer 111; J. Alex Langley 219, 351, 391; Delmar Lipp 489; T. Nebbia 173 top; Mickey Palmer 136
DRK Photo: Stephen J. Krasemann 28 left
E. I. DuPont de Nemours & Co. 199 top
Earth Images: Steve Kaminoff 70 left
Editorial Photocolor Archives, Inc. / Sotheby Parke-Bernet / Detail from Hokusai: *The Wave* 374
Esto Photographics, Inc. 314
William E. Ferguson 220, 229
Focus On Sports: 24, 36, 58, 255 left; Alistair Black 92
FPG: Herb Dallinger 418 top; Tom Meyers 461; Charles Orrico 237; J. Zimmerman 217
Agence Gamma: Christian Gallissian 187 top
General Electric Co. 277
Global Focus: Ted Cordingley 424; Lew Hedberg 363; Raya Jupjan 425 bottom; Rudi Langdon 326 left, 343; Joe Ofria 427; Brian W. Smith 367; Marsha Fall Stuart 433
Lee Gordon 105
The Granger Collection 65, 89, 140, 174, 234, 243, 249, 289, 294
Marcus Halevi 443, 452
Harvard Smithsonian Center for Astrophysics 504
Grant Heilman 87 right, 122 top, 147, 418 bottom
Ellis Herwig 286
Historical Picture Service, Chicago 276
Bohdan Hrynewych 106
Bruce Iverson: 148, 183; Malcolm Hewitt 207 top
Thomas Ives 52 top
Japanese National Railway 337
Jet Propulsion Laboratory 6 bottom left, 498 right
Paul E. Johnson 481
Breck P. Kent 267
Margaret Kissinger 195, 385
Jerg Kroener 440 left
Liaison Agency: Michael Abramson 67; David Hume Kennerley 222
Tom Magno 15, 29 top, 34 bottom, 70 right, 71 right, 72, 73, 77 bottom, 78, 81 bottom, 82 left, 98, 119 bottom, 131, 137 top, 161, 163 top, 184, 185 top, 191, 197, 207 bottom, 221, 225, 256 top, 257, 260, 264, 265, 287, 297, 298, 299, 301, 302, 305, 306, 326 right, 340, 426, 437, 440 right
Magnum Photos: Paul Fusco 3 top
Ralph Mercer 59, 77 top, 79, 84
MIT News Office: Calvin Campbell 107, 311
Monkmeyer Press: Bill Anderson 369; Bendick Associates 160; Paul Conklin 94, 480; Frank E. Gunnell 501; Freda Leinwand 425 top; Pro Pix 206, 458, 473; Hugh Rogers 339
Tim Morse 22
NASA 3 bottom, 5 left, 104, 232, 498 left, 507, 513
National Oceanic and Atmospheric Administration 506 left

Dave Olson 202
Tom Pantages 409, 445 right, 506 right
Photofile International, Ltd. 175, 210, 252
Photo Researchers, Inc.: Farrell Grehan 404 top
Phototake: Yoav Levy 116 top and bottom left
Photri: Don Davis 508
The Picture Cube: Ed Hof 412; Eric Roth 493; Steven M. Stone 261; R. Terry Walker 247; Richard Wood 347
Picture Group: Fredrik D. Bodin 29 bottom
Princeton Plasma Physics Laboratory 465
Rainbow: Coco McCoy 406; Dan McCoy 5 right, 149, 384, 419; Hank Morgan 291, 360
Shostal Associates 389, 485; Tony Linck 445 left; William Hamilton 387
Smithsonian Institution 505
Tom Stack and Associates: 193; Gerald A. Corsi 179, 486; William & Genny Garst 482; Bob McKeever 313; Gary Milburn 279
Stock, Boston: 421; Fredrik D. Bodin 255 right, 404 bottom; Daniel Brody 464, 469; Anestis Diakopolous 164, 397 bottom; Donald Dietz 87 left; Owen Franken 274, 356; Bill Gillett 457; Edith G. Haun 352, 358; James R. Holland 255 center; Bohdan Hrynewych 459; Litchfield 39; Mike Malyszko 254; Tony Mendoza 34 top; Peter Menzel 46, 488; Eric Neurath 333; Stacy Pick 325; Jonathan Rawle 328; John Running 218, 349, 402; Lincoln Russell 307; Frank Siteman 199 bottom; Peter Southwick 103; Peter Vandermark 37; Harry Wilks 292; Cary Wolinsky 182, 280
Taurus Photos: Mrs. T. W. Bennett 496 left; L. L. Rhodes 451; Frank E. Toman 151
U.S. Air Force 150 right
University of California 152, 500
University of Pennsylvania 362
Woods Hole Oceanographic Institution: Rod Catanach 100
Yale University: J. D. Levine 511
Dick Zimmerman 365

3 4 5 6 7 8 9 0

CHAPTER TESTS / UNIT TESTS

The following pages contain the annotated Chapter Tests and Unit Tests in a reduced format. The Chapter and Unit Tests are available both in a booklet of *Duplicating Masters* and in a booklet of *Copy Masters*.

In addition, Heath has provided Chapter Reviews that can be found in the *Heath Physical Science Activities* book. The Chapter Reviews are designed to help students prepare for the Chapter Tests .

CHAPTER 1 TEST

NAME ____
CLASS ____
DATE ____

Physical Science at First Glance

I. Select the best answer to complete each statement. Place its letter in the space provided.

1. In order to be scientific, when you gather information to answer a question, (a) you must have a hypothesis, (b) you must report your observations, (c) other people must be able to gather similar information, (d) other people must be able to form the same theories. **1.** c *1.1 1.3*

2. A good scientific theory (a) cannot be changed, (b) is correct under all conditions, (c) is agreed upon by all scientists, (d) makes predictions. **2.** d *1.1*

3. In this book on physical science, you will study (a) matter and energy, (b) the stars, the moon, and the universe, (c) plants and animals, (d) how the earth originated. **3.** a *1.1*

4. Building a pocket computer, going to the moon, and saving lives with the use of machines are all examples of (a) theories, (b) technology, (c) hypotheses, (d) all of the above. **4.** b *1.2*

5. While he was studying how the earth's gravity affects the way things move, Galileo developed (a) the pendulum, (b) some laws of space travel, (c) the scientific method, (d) a new theory of mathematics. **5.** c *1.3*

6. Galileo noted that as he rolled balls down a slanted track, they rolled faster if (a) they were heavier, (b) the track was tilted at a steeper angle, (c) they were painted a bright, shiny color, (d) they were all made of glass rather than wood. **6.** b *1.3*

7. A laboratory is a place where (a) information is gathered, (b) hypotheses are tested, (c) experiments are carried out, (d) all of these. **7.** d *1.4*

8. When Galileo dropped a heavy cannonball and a lighter musket ball at the same time from the Leaning Tower of Pisa (a) the balls both shattered when they hit the ground, (b) the heavier cannonball landed first, (c) the lighter musket ball landed first, (d) they both landed at the same time. **8.** d *1.3*

9. The meter is used to measure (a) distance, (b) mass, (c) volume, (d) time. **9.** a *1.5*

10. The instrument used to measure mass is called a (a) thermometer, (b) balance, (c) beaker, (d) graduated cylinder. **10.** b *1.5*

CHAPTER 1 TEST (Continued)

NAME ____

II. Fill in the missing word or phrase in the space provided.

11. Observations that are expressed in numbers are ____. **11.** measurements *1.5*

12. Temperature measurements in the metric scale are expressed in degrees ____. **12.** Celsius *1.5*

13. An educated guess is a ____. **13.** hypothesis *1.3*

14. A liter is ____ milliliters. **14.** 1,000 *1.5*

15. ____ is the practical application of science. **15.** Technology *1.2*

III. Use the information given to answer questions 16 and 17. Answer questions in the space provided.

Jimmy decided he would try to repeat the experiment in which Galileo dropped balls of different masses from the same height and measured the time it took for them to reach the ground. Jimmy built a series of platforms so that he could reach several levels. He used three different balls – one made of wood, one of shatterproof glass, and one of cast iron. He had Suzy help him time the balls' falls. They got the following results:

	Wooden Ball	Shatterproof Glass	Cast-iron Ball
5 ft	2 sec	2 sec	2 sec
10 ft	5.1 sec	5.1 sec	5.1 sec
15 ft	6.6 sec	6.6 sec	6.6 sec
20 ft	7.3 sec	7.3 sec	7.3 sec
25 ft	7.7 sec	7.7 sec	7.7 sec

16. Using the measurements in the chart, write two observations that Jimmy and Suzy were able to make about the relationship between height and falling objects. *1.3*

(1) The three different balls fell at the same rate.

(2) The greater the height, the greater the time it takes for the balls to fall.

17. State the four parts of the scientific method and describe how each one was used in Jimmy and Suzy's experiment. *1.3*

Answers will vary, but students should discuss:
observation (set up experiment)
analysis (organized experiment in a new way)
synthesis (based experiment on Galileo's theory)
testing (experimented under new conditions)

CHAPTER 2 TEST

NAME ____________
CLASS ____________
DATE ____________

Matter, Energy, and Motion

I. Match the definition in column A with the term in column B. Place the letter of the matching term in the space provided.

A	B		
1. ability to do work	**A.** kinetic energy	**1.** B	2.1
2. characteristic of a substance	**B.** energy	**2.** J	2.1
3. water behind a dam	**C.** matter	**3.** F	2.1
4. has mass and takes up space	**D.** vacuum	**4.** C	2.1
5. how fast an object is moving	**E.** acceleration	**5.** I	2.2
6. a speeding train	**F.** potential energy	**6.** A	2.1
7. layer of gas around the sun	**G.** velocity	**7.** H	2.1
8. change in velocity over a given time	**H.** plasma	**8.** E	2.3
9. absence of air	**I.** speed	**9.** D	2.3
10. speed in a given direction	**J.** property	**10.** G	2.2

II. Select the best answer to complete each statement. Place its letter in the space provided.

11. If you know the velocity and speed of an object, you know (a) how heavy it is, (b) how fast it is moving, (c) what effect mass has on the object, (d) the distance that the object is from the center of the earth. **11.** b 2.2

12. A substance that has a definite volume but no definite shape is called (a) a solid, (b) a liquid, (c) a gas, (d) plasma. **12.** b 2.1

13. If a baseball, a cardboard box, and a book are dropped from a tall building, the one that will most likely reach its terminal velocity first will be the (a) baseball, (b) cardboard box, (c) book, (d) all will reach terminal velocity at the same time. **13.** a 2.3

14. The characteristic *solid* is an example of (a) an energy, (b) a plasma, (c) a property, (d) a measurement. **14.** c 2.1

15. If three moving objects have the same mass, the one with the most momentum has (a) the greatest amount of matter, (b) the greatest velocity, (c) the greatest potential energy, (d) the greatest distance to go. **15.** b 2.4

CHAPTER 2 TEST (Continued)

NAME ____________

III. Answer the following. If calculations are required, show them under the question. Place your answer in the space provided.

16. A southbound train is traveling at 40 km/hour. A waiter with a tray is walking in the opposite direction on the train at 3 km/hour. On the tray is a fly walking south at 1 km/hour. What is the velocity of the fly relative to the ground on which the train is traveling? 2.2

40 km/hour − 3 km/hour + 1 km/hour = 38 km/hour

17. What is the momentum of a book that has a mass of 9 grams falling from a window at 2 m/sec? 2.4

Momentum = mass × velocity ; momentum = 9 g × 2 m/sec ;

Momentum = 18 (g)(m)/sec

18. A taxicab stops for a red light. When the light changes to green, the taxicab's velocity changes to 40 meters/second in 10 seconds. What was the taxicab's rate of acceleration at the light?

$$\text{Acceleration} = \frac{\text{change in velocity}}{\text{time}}$$

$$\text{Acceleration} = \frac{40 \text{ m/sec}}{10 \text{ sec}}$$

Acceleration = 4 meters/second/second

19. A lead weight is dropped from the top of a tower 50 m tall. It takes 20 seconds to reach the ground. What will its velocity be when it reaches the ground? 2.3

20 sec × 9.8 m/sec/sec

196 m/sec

(It will probably not reach its terminal velocity.)

20. Your teacher puts 20 g of crushed oyster shell in a container. He pours 30 mL of hydrochloric acid into the container and quickly closes the top. Your teacher finds that the total mass of the container, the acid, and the shell is 120 g. The mixture in the container foams and bubbles, indicating that a gas has formed. He then opens the container to allow the gas to escape. The total mass of the container and its contents is now 116 g. How do you account for the change in mass from the beginning of the experiment to the end? 2.1

The gas had a mass of 4 g. When it was released, the 4 g was lost from the total amount.

CHAPTER 3 TEST

NAME ____________
CLASS ____________
DATE ____________

Forces

I. Select the best answer to complete each statement. Place its letter in the space provided.

1. The state of an object when all forces are zero is called (a) gravity, (b) inertia, (c) equilibrium, (d) momentum. **1.** c 3.3

2. The center of gravity of an object is located (a) at the point where stable equilibrium exists, (b) along any surface, (c) at the point where the object is just touching a surface, (d) where the effect of gravity is the least. **2.** a 3.3

3. When all the net forces on an object are no longer zero, the (a) state of the object changes, (b) mass of the object changes, (c) weight of the object changes, (d) position of the object changes. **3.** d 3.1

4. Based on Newton's laws of motion, a hockey puck sliding across the ice slows down because (a) frictional forces between the puck and the ice are acting on the two surfaces, (b) the ice is too cold to keep the puck moving forever, (c) you must apply a force to keep the puck in motion, (d) the puck weighs too much. **4.** a 3.4

5. An object accelerates as it falls toward Earth because (a) no friction is present, (b) no force is acting on it, (c) the object is lighter than air, (d) the force of gravity is pulling on the object. **5.** d 3.1

6. Most forces that are not due to gravity are due to (a) electrical forces, (b) the laws of motion, (c) friction, (d) inertia. **6.** a 3.2

7. The amount of force acting on a moving object can be measured by its mass and its (a) inertia, (b) gravity, (c) acceleration, (d) momentum. **7.** c 3.5

8. A man standing on top of a mountain will weigh ________ than at sea level. (a) a little less (b) a little more (c) a lot less (d) a lot more **8.** a 3.1

9. An object that has an electrical charge will cause another charged object to (a) be repelled, (b) be attracted, (c) be either attracted or repelled, (d) remain the same. **9.** c 3.2

10. The amount of gravitational force one body exerts on another depends on the masses of the two bodies and (a) their state of equilibrium, (b) the electrical charge of the two bodies, (c) their acceleration, (d) the distance between them. **10.** d 3.1

CHAPTER 3 TEST (Continued)

NAME ____________

II. Read each of the following sentences. If the sentence is true, write *T*. If it is false, change the underlined expression to make it true. Place your answer in the space provided.

11. Electric forces keep the moon and the earth in orbit. **11.** Gravitational 3.1

12. A force is a push or a pull that is exerted on matter. **12.** T 3.1

13. Your weight is not changed by changes in gravitational force. **13.** mass 3.1

14. Objects with an equal number of positive and negative charges are said to be electrically neutral. **14.** T 3.2

15. The sum of opposing forces is called equilibrium. **15.** net force 3.3

III. Describe what law(s) of motion is (are) operating in the following situations. Explain. Place your answers in the spaces provided.

16. A person jumps from a boat to the dock. The boat moves away from the dock. 3.6

3rd law; the foot pushes the boat away with the same force that allows the person to reach the dock.

17. Unless a person is wearing a seatbelt, the person's head will hit the windshield if the car stops suddenly. 3.4

1st law; the person will continue to move forward unless acted upon by an outside force, the seatbelt.

18. It takes more force to move a fully loaded truck than it does to move an empty truck. 3.5

2nd law; the fully loaded truck has a greater mass and therefore a greater force, according to the formula F = m × a.

19. The moon continues to orbit the earth. 3.4

1st law; the moon was set in motion and will continue to orbit unless acted upon by an outside force.

20. You strike a ball with a bat. The ball goes flying into the air. A friend strikes the same ball with the same bat, but the ball goes farther. 3.5 3.6

2nd and 3rd laws; your friend hits the ball with a greater force so that it will have a greater acceleration—the ball will go as far as the force upon it (equal and opposite reaction).

CHAPTER 4 TEST

NAME ______
CLASS ______
DATE ______

Work, Machines, and Power

I. Match the definition in column A with the term in column B. Place the letter of the matching term in the space provided.

A	B	
1. simplest of all machines	A. mechanical advantage	1. H 4.3
2. lever that rotates around a fixed point	B. machine	2. J 4.3
3. measures how much force is gained by using a machine	C. watt	3. A 4.2
	D. joule	
4. measures rate at which work is done	E. power	4. E 4.4
5. done when a force moves an object over a distance	F. work	5. F 4.1
6. unit of work	G. efficiency	6. D 4.1
7. unit of power	H. inclined plane	7. C 4.4
8. a device that helps you do work	I. fulcrum	8. B 4.2
9. work put out compared to work put in	J. pulley	9. G 4.4
10. fixed point of a lever		10. I 4.3

II. Select the best answer to complete each statement. Place its letter in the space provided.

11. An example of a simple machine is a (a) light bulb, (b) can opener, (c) desk drawer, (d) pencil. — 11. b 4.1

12. The amount of work output from a machine is never as large as the work put into a machine. Some of this work is changed to (a) gravity, (b) weight, (c) magnetism, (d) friction. — 12. d 4.4

13. When using an inclined plane, a person pushes a heavy carton up the ramp. The person is using (a) effort distance, (b) effort force, (c) mechanical advantage, (d) all of these. — 13. d 4.2 4.3

14. When turning a screw, the more turns you use, the greater the ______ will be. (a) effort distance (b) effort force (c) mechanical advantage (d) all of these — 14. a 4.2 4.3

15. Work is being done when you (a) study for a test, (b) talk on the telephone, (c) turn on the television, (d) none of these. — 15. c 4.1

CHAPTER 4 TEST (Continued)

NAME ______

III. Answer the following. If calculations are required, show them under the question. Place your answer in the space provided. Be sure to include the proper units.

16. A book weighing 8.2 newtons is lifted 2 meters off the desk. How much work is done on the book? 4.1

Work = force × distance
= 8.2 N × 2 m
= 16.4 N.m or 16.4 J

17. A bag of sugar weighs 25 N. Mary is able to move the sugar 2 m in 10 sec. How much power has Mary used to move the sugar? 4.1 4.4

Work = force × distance
= 25 N × 2 m
= 50 J

Power = $\frac{\text{work}}{\text{time}}$
Power = $\frac{50\ J}{10\ sec}$
Power = 5 W

18. Sue and Sam are trying to balance a seesaw. Sue weighs 40 N and sits 3 m from the center. Sam weighs 60 N. How far should he sit from the center if the seesaw is going to balance? 4.3

$\frac{40}{3} = \frac{60}{x}$
$x = 4.5$ m

IV. Use the information in the diagram below to answer questions 19 through 21. If calculations are required, show them under the question. Place your answers in the spaces provided.

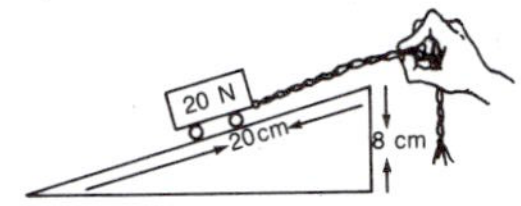

19. How much work would be done to lift the cart the entire distance up the ramp? 4.1

Work = force × distance
= 20 N × 8 cm
= 160 J

20. What is the mechanical advantage of using the inclined plane? 4.2

M.A. = $\frac{D_E}{D_R} = \frac{20\ cm}{8\ cm} = 2.5$

21. What force (F_E) needs to be exerted to lift the cart up the ramp? 4.2

work in = work out
$F_E \times D_E = F_R \times D_R$
$F_E \times 20$ cm = 20 N × 8 cm
$F_E = 8$ N

CHAPTER 5 TEST

NAME ______
CLASS ______
DATE ______

Forces in Liquids and Gases

I. Select the best answer to complete each statement. Place its letter in the space provided.

1. The force that makes it possible for you to float in salt water is (a) atmospheric pressure, (b) salinity, (c) buoyancy, (d) mass. — 1. c 5.2

2. If the buoyant force of an object is equal to the weight of the object, the object will (a) dissolve, (b) float, (c) sink, (d) drag. — 2. b 5.2

3. The person who first described buoyant force was (a) Archimedes, (b) Aristotle, (c) Newton, (d) Bernoulli. — 3. a 5.2

4. Air enters your lungs because when your lungs expand, (a) the air in your lungs is more dense than atmospheric pressure, (b) the air in your lungs is less buoyant, (c) low pressure inside your lungs creates a suction, (d) high pressure inside your lungs creates a suction. — 4. c 5.2

5. Atmospheric pressure refers to (a) the weight of air on Earth, (b) the pressure inside a space suit, (c) the pressure inside a bicycle pump, (d) the mass of a given volume of air. — 5. a 5.3

6. Propellers on helicopters operate on the same principle as ______ in order to lift the vehicle. (a) pointed noses on supersonic transports (b) jet engines (c) airplane propellers (d) airplane wings — 6. c 5.4

7. The easiest way to get a piece of clay to float in water is to (a) roll it as thin as possible, (b) form it into a tight ball, (c) dissolve it, (d) punch holes in it. — 7. a 5.2

8. Drag depends on (a) the density of the object, (b) the buoyancy of the object, (c) the shape of the object, (d) the weight of the object. — 8. c 5.4

9. The upward force of air on the wings of an airplane is called (a) drag, (b) lift, (c) atmospheric pressure, (d) air density. — 9. b 5.4

10. The faster a fluid moves, (a) the greater its density, (b) the colder it becomes, (c) the greater the pressure it exerts, (d) the less the pressure it exerts. — 10. d 5.1

CHAPTER 5 TEST (Continued)

NAME ______

II. Read each of the following sentences. If the sentence is true, write *T*. If it is false, change the underlined expression to make it true. Place your answer in the space provided.

11. If you change the shape of a fish tank to make the bottom, wider and the sides lower (but keep the volume constant), you will decrease the pressure of the water inside the fish tank. — 11. T 5.1

12. Density involves the mass of an object, and buoyancy involves its weight. — 12. T 5.2

13. As the temperature of air increases, its density increases. — 13. decreases 5.3

14. The pressure in a tank of water with an area of 520 cm^2 and a force of 260 N is 0.5 N/cm^2. — 14. T 5.1

15. The force exerted by a tank of water with a pressure of 3 N/cm^2 and an area of 100 cm^2 is 30 N. — 15. 300 N 5.1

III. Use the information below and the chart on the left to answer questions 16 through 19. Place answers in the spaces provided.

A scientist picked up a piece of metal on a field trip. She brought it back to her laboratory to make several tests on the metal in order to determine what it was. She first determined that the piece contained two kinds of metal. After melting the metal down, she found that one kind of metal floated to the top. She poured this off and put it aside. After both pieces solidified, she massed them and determined their volumes. The piece of metal she poured off weighed 13.25 grams and was 5 cm^3 in volume. The other piece was 32 grams and 4 cm^3 in volume.

Material	Density (g/cm^3)
magnesium	1.75
aluminum	2.7
iron	7.9
lead	11.3
mercury	13.6
gold	19.3

16. What was the first metal she poured off? 5.2

D = $\frac{\text{mass}}{\text{volume}} = \frac{13.25\ g}{5\ cm^3} = 2.65$

aluminum

17. What was the other metal? 5.2

$\frac{32\ g}{4\ cm^3} = 8.0$

iron

18. What property of metals did she use to determine their identity? 5.2

density

19. Which metal was more buoyant? 5.2

aluminum

CHAPTER 6 TEST

NAME ______
CLASS ______
DATE ______

The Atom

I. Match the definition in column A with the term in column B. Place the letter of the matching term in the space provided.

A	B	
1. smallest particle of an element	A. proton	1. C 6.3
2. pure substance	B. electron	2. J 6.1
3. positively charged particle	C. atom	3. A 6.3
4. particle with no charge	D. atomic mass number	4. I 6.3
5. number of protons in nucleus	E. atomic number	5. E 6.3
6. particle with a negative charge	F. nuclear force	6. B 6.2
7. combined number of protons and neutrons in the nucleus	G. electric force	7. D 6.3
	H. compound	
8. holds protons and neutrons together in the nucleus	I. neutron	8. F 6.4
9. holds electrons in the atom	J. element	9. G 6.4
10. combination of two or more elements		10. H 6.1

II. Select the best answer to complete each statement. Place its letter in the space provided.

11. Electrons move around the nucleus of an atom in paths called (a) systems, (b) clouds, (c) fields, (d) charges. **11.** b 6.4

12. An isotope differs from its parent element only by the number of ______ in its nucleus. (a) neutrons (b) electrons (c) protons (d) subatomic particles **12.** a 6.3

13. The particle that has been found in both protons and neutrons is called the (a) quark, (b) electron, (c) isotope, (d) nucleus. **13.** a 6.3

14. The name and behavior of a particular element is determined by the number of ______ it contains. (a) electrons (b) protons (c) neutrons (d) radioactive particles **14.** b 6.3

15. What makes an element radioactive is (a) how long it exists, (b) the size of its nucleus, (c) the fact that it is a natural element, (d) the fact that it gives off high-energy particles. **15.** d 6.2

CHAPTER 6 TEST (Continued)

NAME ______

III. Use the diagrams below to answer questions 16 through 22.

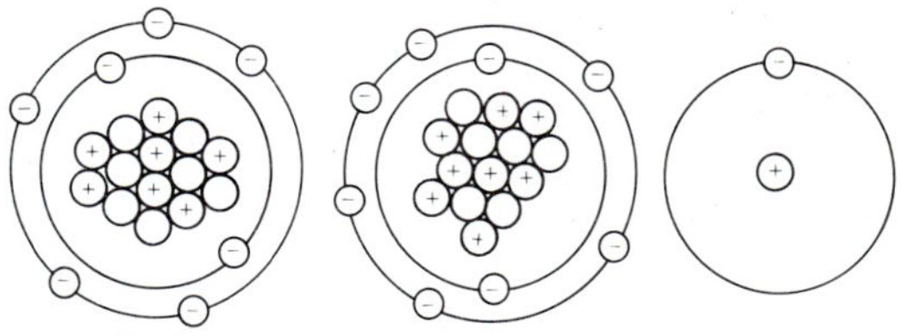

16. What does each diagram above represent? **16.** an atom 6.1

17. What does each circle represent? **17.** a shell 6.4

18. What is the atomic mass number of oxygen? **18.** 16 6.3

19. What is the atomic number of hydrogen? **19.** 1 6.3

20. What would nitrogen be called if it had another proton? **20.** oxygen 6.3

21. What would nitrogen be called if it had another neutron? **21.** isotope 6.3

22. If oxygen and hydrogen combined, what would they form? **22.** a compound 6.1

IV. Answer the following questions in the spaces provided.

23. How does the model of the solar system differ from the orbital model of electrons? 6.4

Planets travel in any orbit around the sun; electrons travel in well-defined orbits around the nucleus.

24. Why were alchemists unsuccessful in turning lead into gold? 6.1

Alchemists thought there was a "primary matter." Both lead and gold are elements that cannot be broken down further into more basic substances.

25. Explain the difference between a natural element and an element such as einsteinium. 6.1

Einsteinium only exists in a laboratory and then only for a fraction of a second. Natural elements occur in nature and last for a long time.

CHAPTER 7 TEST

NAME ______
CLASS ______
DATE ______

The Elements

I. Select the best answer to complete each statement. Place its letter in the space provided.

1. Actinides all have the property of being (a) a gas at room temperature, (b) radioactive, (c) only made in the laboratory, (d) nonmetals. **1.** b 7.5

2. The first person to group the elements according to their properties was (a) Rutherford, (b) Einstein, (c) Curie, (d) Mendeleev. **2.** d 7.1

3. Those elements that have a complete outer shell are naturally found as a (a) liquid, (b) solid, (c) gas, (d) plasma. **3.** c 7.4

4. An element of the alkali metal family (a) is not found alone in nature, (b) has 6 valence electrons, (c) is a nonmetal, (d) has electrons orbiting in 3 shells. **4.** a 7.3

5. The most familiar metals such as copper, iron, gold and silver, are in the family of (a) transition elements, (b) lanthanides, (c) alkali earth metals, (d) none of these. **5.** a 7.2

6. When referring to chemicals, the word *synthetic* means (a) found naturally, (b) coming from the sun, (c) made in the laboratory, (d) nonmetallic. **6.** c 7.5

7. The element that combines with more elements than any other is (a) hydrogen, (b) oxygen, (c) sulfur, (d) helium. **7.** b 7.3

8. Elements with similar properties are arranged in the periodic table (a) horizontally, (b) vertically, (c) scattered, (d) diagonally. **8.** b 7.1

9. One characteristic of all metals is that they (a) conduct electricity, (b) dissolve in water, (c) burn in air, (d) are highly reactive. **9.** a 7.1

10. The two most common elements on the earth's surface are (a) chlorine and iodine, (b) gold and silver, (c) hydrogen and carbon, (d) oxygen and silicon. **10.** d 7.3

CHAPTER 7 TEST (Continued)

NAME ______

II. Read each of the following sentences. If the sentence is true, write *T*. If it is false, change the underlined expression to make it true. Place your answers in the spaces provided.

11. Hydrogen is the most common element in the universe. **11.** T 7.2

12. Halogen compounds are used largely as disinfectants. **12.** T 7.3

13. More than half of all known elements are nonmetals. **13.** metals 7.1

14. Most of the material that makes up the earth is metals. **14.** nonmetals 7.1

15. The element carbon forms many compounds that are necessary for life. **15.** T 7.3

III. Use the diagram below that represents part of the periodic table, to answer questions 16 through 22.

III	IV	V	VI	VII	VIII
					4 He Helium 2
11 B Boron 5	12 C Carbon 6	14 N Nitrogen 7	16 O Oxygen 8	19 F Fluorine 9	20 Ne Neon 10
27 Al Aluminum 13	28 Si Silicon 14	31 P Phosphorus 15	32 S Sulfur 16	35.5 Cl Chlorine 17	40 Ar Argon 18
70 Ga Gallium 31	73 Ge Germanium 32	75 As Arsenic 33	79 Se Selenium 34	80 Br Bromine 35	84 Kr Krypton 36

16. VIII represents what family of elements? **16.** noble gas 7.4

17. What do the elements above the thick black line have in common? **17.** nonmetals 7.1

18. What do Al, Si, P, S, Cl, and Ar have in common? **18.** same period 7.1

19. How many valence electrons do the elements have in VI? **19.** 6 7.1 7.3

20. How many valence electrons do the elements have in VIII? **20.** 8 7.1 7.4

21. Knowing about what elements is most likely to give you information about Br? **21.** F, Cl 7.1

22. When Se is exposed to light, what strange property does it have for the family it is in? **22.** metallic 7.3

CHAPTER 8 TEST

NAME ____________
CLASS ____________
DATE ____________

Combining the Elements

I. Select the best answer to complete each statement. Place its letter in the space provided.

1. When two elements combine to form a compound, the elements (a) are not chemically combined, (b) combine in definite proportions, (c) keep their properties, (d) form a uniform mixture. **1.** b 8.1

2. Seawater is (a) an element, (b) a compound, (c) a mixture, (d) none of these. **2.** c 8.1

3. Sulfur and oxygen will combine to form a(n) ______ bond. (a) covalent (b) ionic (c) metallic (d) diatomic **3.** a 8.4

4. In the structure of atoms, valence electrons are found (a) in the nucleus, (b) in the shell closest to the nucleus, (c) in the shell farthest from the nucleus, (d) bound to the protons. **4.** c 8.2

5. Atoms whose electrons are shared among many other atoms are said to form (a) metallic bonds, (b) ionic bonds, (c) covalent bonds, (d) strong bonds. **5.** a 8.5

6. Metals can be identified because they (a) dissolve in water, (b) do not conduct heat, (c) conduct electricity, (d) form very hard crystals. **6.** c 8.5

7. Ions (a) have equal numbers of positive and negative charges, (b) conduct electricity, (c) have either gained or lost a proton, (d) are found in the nucleus of the atom. **7.** b 8.3

8. In the formula for sugar, $C_6H_{12}O_6$, the 6 refers to (a) the number of atoms in the molecule, (b) the number of molecules that are present, (c) the number of times carbon and oxygen will combine with hydrogen, (d) the number of carbon and oxygen atoms present in the molecule. **8.** d 8.2

9. The element carbon usually forms bonds that are (a) ionic, (b) covalent, (c) metallic, (d) easily broken. **9.** b 8.4

10. A crystal forms when (a) atoms are arranged in a regular pattern, (b) covalent bonding occurs, (c) strong bonds are formed, (d) freely moving electrons wander through metal. **10.** a 8.3

CHAPTER 8 TEST (Continued)

NAME ____________

II. List the name of the element and the number of atoms of each in the compounds below. Place your answer in the space provided.

11. CH_4 — **11.** 1 carbon, 4 hydrogen 8.2
12. NaOH — **12.** 1 sodium, 1 oxygen, 1 hydrogen 8.2
13. C_2H_5OH — **13.** 2 carbon, 6 hydrogen, 1 oxygen 8.2
14. H_2O — **14.** 2 hydrogen, 1 oxygen 8.2
15. $Ca(NO_3)_2$ — **15.** 1 calcium, 2 nitrogen, 6 oxygen 8.2

III. Use the information below to answer questions 16 through 21. Place your answers in the spaces provided.

A teacher opens a bottle labeled *ammonia* in front of the classroom. After five minutes, students in the front row detect an irritating smell and their eyes begin to water a little. After 15 minutes, students all over the room detect the odor. After smelling the bottle of ammonia, the students realize that the irritating smell came from the bottle. Several days later the liquid in the bottle no longer smells so strong.

16. What was the state of matter of the substance the students smelled after five minutes? **16.** gas 8.1

17. What kind of substance was formed with the substance from question 16 and the air? **17.** a mixture 8.1

18. What is the molecular formula for ammonia? **18.** NH_3 8.2

19. What kind of substance was the fluid inside the bottle? Explain your answer. 8.1
A solution; the ammonia gas escaped leaving water in which the gas was dissolved.

20. Why can hydrogen be said to *not* follow the octet rule while nitrogen *is* following the rule? 8.2
Hydrogen needs one electron to become stable; nitrogen needs 3 electrons to get a complete set of 8.

21. What kind of bonds do hydrogen and nitrogen form? How do you know? 8.4
Covalent bonds are formed when electrons are shared; also, hydrogen and nitrogen are nonmetals and tend to form covalent bonds.

CHAPTER 9 TEST

NAME ____________
CLASS ____________
DATE ____________

Carbon Compounds

I. Match the definition in column A with the term in column B. Place the letter of the matching term in the space provided.

A	B	Answer
1. These have an OH group.	A. polymers	1. H 9.3
2. compounds made of only carbon and hydrogen	B. organic acids	2. D 9.2
3. carbon compounds	C. isomers	3. F 9.1
4. These have the same molecular formula but different structural formulas.	D. hydrocarbons	4. C 9.2
5. process whereby a liquid is changed to a gas, then cooled to a liquid again	E. amino acids	5. I 9.2
6. These have a COOH group.	F. organic compounds	6. B 9.3
7. organic molecules with a large number of atoms	G. proteins	7. A 9.4
8. These have an NH_2 group.	H. alcohols	8. E 9.3
9. process whereby a liquid is separated into many parts	I. distillation	9. J 9.2
10. These have chains of amino acids.	J. fractional distillation	10. G 9.3

II. Complete the following by writing *true* if the statement is correct. If the statement is false, change the underlined word(s) to make it correct. Place your answers in the spaces on the right.

11. All polymers are synthetic. **11.** Some 9.4
12. Heavier long chain hydrocarbons are liquid or solid at room temperature. **12.** true 9.2
13. Short chain hydrocarbons are often used as fuels. **13.** true 9.2
14. Carbon atoms are unique because they have four valence electrons and many shells. **14.** few shells 9.1
15. The process by which crude oil is separated into gasoline and other products is called isomerization. **15.** fractional distillation 9.2

CHAPTER 9 TEST (Continued)

NAME ____________

III. Use the structural formulas below to answer questions 16 through 25. Place your answers in the spaces provided.

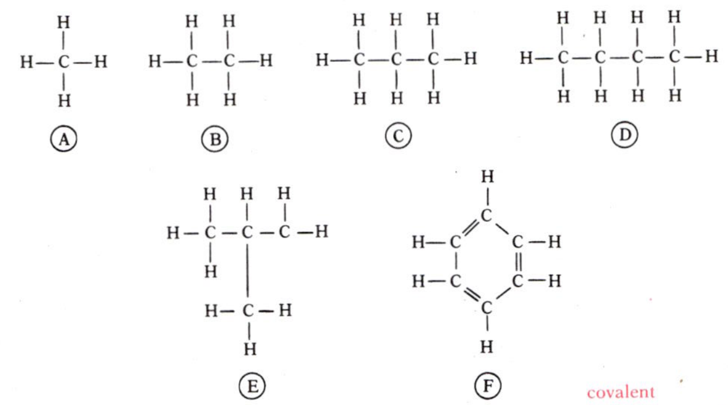

16. What does each dash represent? **16.** covalent bond 9.2
17. Which hydrocarbon(s) represent(s) a straight chain? **17.** B, C, D 9.2
18. Which hydrocarbon(s) represent(s) a branched chain? **18.** E 9.2
19. What is the molecular formula for hydrocarbon C? **19.** C_3H_8 9.2
20. What is the name of hydrocarbon B? **20.** ethane 9.2
21. If one of the H's in hydrocarbon A was replaced by an OH group, what would it be called? **21.** an alcohol 9.3 9.2
22. Which two hydrocarbons are isomers? **22.** D, E 9.2
23. Which hydrocarbon(s) has(have) double bonds? **23.** F 9.1
24. Which hydrocarbon(s) has(have) triple bonds? **24.** none 9.1
25. Which is the simplest hydrocarbon? **25.** A 9.4

IV. Answer the following in the space provided.

26. Name five types of organic compounds in the human body. 9.3
Answers will vary but should include five of these: proteins, amino acids, organic acids, fats, carbohydrates, hydrocarbons.

27. Describe how gasoline, diesel fuel, wax, and petroleum jelly can all be obtained from crude oil. 9.2
The process of fractional distillation separates the different substances out as they reach their differing boiling points.

CHAPTER 10 TEST

NAME ____
CLASS ____
DATE ____

Chemical Reactions

I. Select the best answer to complete each statement. Place its letter in the space provided.

1. A substance that produces hydrogen ions is called (a) a base, (b) an indicator, (c) an acid, (d) a reactant. — 1. c 10.5

2. In a chemical equation, the products are found on (a) the right side of the arrow, (b) the left side of the arrow, (c) both sides in certain reactions, (d) neither side. — 2. a 10.1

3. When an acid and a base combine, the products are (a) a metal and a nonmetal, (b) water and a salt, (c) water and a metal, (d) salts. — 3. b 10.5

4. Rusting of iron is an example of (a) physical change, (b) a chemical change, (c) a rapid reaction, (d) the use of a catalyst. — 4. b 10.1

5. When two or more elements combine to form a new product, the reaction is (a) displacement, (b) double displacement, (c) decomposition, (d) synthesis. — 5. d 10.1

6. A neutral substance has a pH of (a) 5, (b) 9, (c) 7, (d) 0. — 6. c 10.5

7. When an acid and a base combine, the reaction is called (a) synthesis, (b) a physical change, (c) neutralization, (d) composition. — 7. c 10.5

8. A chemical reaction in which energy is released is called (a) exothermic, (b) endothermic, (c) synthetic, (d) complete. — 8. a 10.4

9. Chemical equations must be balanced because (a) the formulas must be the same on both sides, (b) the reactants must be the same as the products, (c) the atoms must be the same as the molecules, (d) the mass on both sides must be the same. — 9. d 10.2

10. Chemical reactions take place because (a) catalysts are used, (b) bonds are broken in reactant molecules, (c) heat is absorbed by the reactants, (d) molecules collide. — 10. b 10.3

11. Catalysts found in living things are (a) acids, (b) bases, (c) enzymes, (d) salts. — 11. c 10.3

12. A substance with a pH of 14 is a (a) weak acid, (b) weak base, (c) strong acid, (d) strong base. — 12. d 10.5

13. An example of a physical change is (a) wood burning, (b) water boiling, (c) paper yellowing, (d) an explosion. — 13. b 10.1

14. In $2MgSO_4$, the 2 is the (a) coefficient, (b) valence, (c) pH, (d) atomic number. — 14. a 10.2

15. All of the following will speed up a reaction except (a) a catalyst, (b) greater concentration of reactant, (c) greater concentration of product, (d) increased temperature. — 15. c 10.4

CHAPTER 10 TEST (Continued)

NAME ____

II. Use the following equations to answer questions 16 through 25.

(A) $I_2 + KCl \rightarrow KI + Cl_2$

(B) $Zn + HCl \rightarrow ZnCl_2 + H_2$

(C) $Ba + H_2O \rightarrow Ba(OH)_2 + H_2$

(D) $KOH + HCl \rightarrow KCl + H_2O$

(E) $C + O_2 \rightarrow CO_2$

16. Which equation(s) represent(s) single displacement? — 16. A, B 10.1

17. Which equation(s) represent(s) double displacement? — 17. D 10.1

18. Which equation(s) represent(s) synthesis? — 18. E 10.1

19. Which equation(s) represent(s) decomposition? — 19. none 10.1

20. Balance equation A in the space below. 10.2

$I_2 + 2KCl \rightarrow 2KI + Cl_2$

21. Balance equation B in the space below. 10.2

$Zn + 2HCl \rightarrow ZnCl_2 + H_2$

22. Balance equation C in the space below. 10.2

$Ba + 2H_2O \rightarrow Ba(OH)_2 + H_2$

23. Balance equation D in the space below. 10.2

no change

24. If $Ba(OH)_2$ in equation C is in liquid solution, what will be the reaction of litmus paper to it? — 24. will turn blue 10.5

25. What is the pH of the H_2O formed in equation D? — 25. 7 10.5

III. Answer the following in the space provided.

26. What is the difference between an exothermic and an endothermic reaction? Give an example of each. 10.4

An exothermic reaction is one in which the final temperature is greater than the starting temperature. In an endothermic reaction, the starting temperature is greater than the final temperature.

Examples: exothermic—an explosion, iron rusting, fuel burning
endothermic—cake baking

27. How would you find out whether a liquid is a base or an acid? 10.5

Use an indicator: phenolphthalein or litmus paper.

CHAPTER 11 TEST

NAME ____
CLASS ____
DATE ____

Nuclear Reactions

I. Select the best answer to complete each statement. Place its letter in the space provided.

1. Henri Becquerel made his discoveries while he was storing the ore of (a) carbon, (b) nitrogen, (c) helium, (d) uranium. — 1. d 11.2

2. Unbalanced isotopes change or decay until they become (a) stable isotopes, (b) radioactive, (c) positively charged, (d) tracers. — 2. a 11.2

3. A balanced nucleus means (a) the atom is stable, (b) the number of protons equals the number of electrons, (c) the number of protons is greater than the number of neutrons, (d) the atom has more than 20 protons. — 3. a 11.2

4. The difference between chemical reactions and nuclear reactions is that in nuclear reactions (a) chemical bonds are broken, (b) the electrons of the product atom are affected, (c) the nucleus of an atom is changed, (d) new compounds are formed. — 4. c 11.1

5. C14 is used in finding the age of bones because (a) it has a long half-life, (b) it is found in dead organic matter, (c) it is being formed all the time, (d) all of these. — 5. d 11.3

6. All of the following statements are true of nuclear fission except (a) reactions can be multiplied in nuclear chain reactions, (b) reactions are caused by nuclear fusion reactions, (c) atoms with lighter nuclei are products, (d) the energy released is about 100 times greater than the energy released in radioactive decay. — 6. b 11.4

7. Fusion cannot be used as an energy source until scientists learn how to (a) cope with the high temperatures required, (b) join hydrogen nuclei, (c) control fission reactions, (d) split nuclei. — 7. a 11.5

8. The Curies discovered (a) uranium, (b) radioactivity, (c) fusion, (d) photographic plates. — 8. b 11.2

9. It is important to handle radioactive materials with care because (a) there is no protection from radioactivity, (b) overexposure can cause illness and even death, (c) the materials may contain an isotope, (d) all of these. — 9. b 11.2

10. Protons and neutrons are held tightly in the nucleus by (a) the nuclear force, (b) the electrons, (c) beta particles, (d) electric forces. — 10. a 11.1

CHAPTER 11 TEST (Continued)

NAME ____

II. Complete the following by writing *true* if the statement is correct. If the statement is false change the underlined word(s) to make it correct. Place your answers in the spaces on the right.

11. The length of time it takes for half the nuclei of a radioactive element to change is called its half-life. — 11. true 11.3

12. A positively charged particle most like a helium nucleus is an alpha particle. — 12. true 11.2

13. The relationship between plants and their environment is studied through the use of the element nitrogen. — 13. carbon 11.3

14. Isotopes are always radioactive. — 14. usually 11.3

15. A Geiger counter detects nuclear radiation. — 15. true 11.2

III. Use the graph on the left to answer questions 16 through 20. Fill in the missing information from the uranium decay series on the right in the spaces provided.

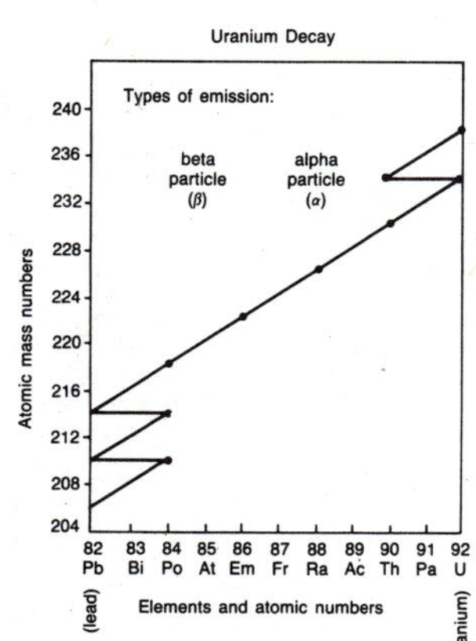

Uranium Decay Series

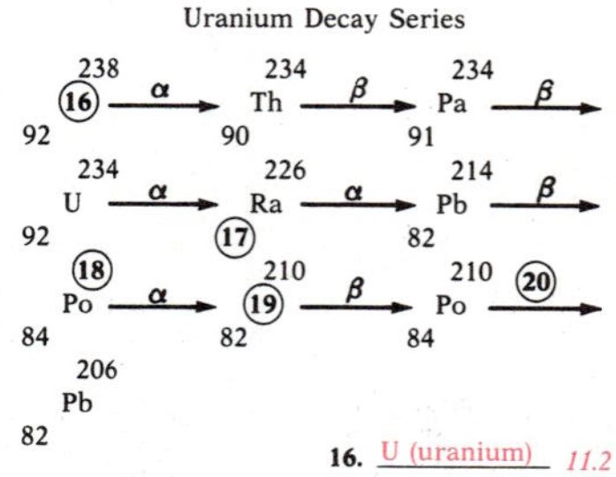

16. U (uranium) 11.2 11.3

17. 88 11.2 11.3

18. 214 11.2 11.3

19. Pb (lead) 11.2 11.3

20. alpha particle 11.2 11.3

CHAPTER 12 TEST

NAME ____________
CLASS ____________
DATE ____________

Heat

I. Match the definition in column A with the term in column B. Place the letter of the matching term in the space provided.

A	B		
1. temperature at which substance changes from solid to liquid	A. heat of fusion	1. C	12.3
2. energy transfer that can occur through empty space	B. heat of vaporization	2. G	12.4
3. amount of heat needed to change solid to liquid	C. melting point	3. A	12.3
4. amount of heat needed to raise 1 gram of water 1°C	D. specific heat	4. F	12.2
5. amount of heat needed to change liquid to gas	E. degree Celsius	5. B	12.3
6. escape of particles from surface of a liquid	F. calorie	6. J	12.3
7. amount of heat needed to raise 1 gram of a substance 1°C	G. radiation	7. D	12.2
8. measure of the average kinetic energy of a substance	H. convection	8. I	12.2
9. movement of large amounts of gas	I. temperature	9. H	12.4
10. unit for measuring temperature	J. evaporation	10. E	12.2

II. Select the best answer to complete each statement. Place its letter in the space provided.

11. Two units for measuring heat are (a) newton and joule, (b) meter and joule, (c) gram and meter, (d) joule and calorie. — 11. d 12.2

12. When heat is added to a substance, its particles (a) contract, (b) take up more space, (c) lose energy, (d) lose calories. — 12. b 12.1

13. The total kinetic energy of all the molecules or atoms of a substance is its (a) heat of fusion, (b) heat energy, (c) calorie count, (d) specific heat. — 13. b 12.1

14. A rock and a pebble with the same temperature do not have the same amount of heat because (a) the rock has a greater specific heat, (b) the particles are moving faster in the rock, (c) the rock has more moving particles, (d) the rock is expanding. — 14. c 12.2

15. Heat is transferred fastest by conduction through (a) gases, (b) liquids, (c) solids, (d) empty space. — 15. c 12.4

CHAPTER 12 TEST (Continued)

NAME ____________

III. While a pan of ice is heated, the temperature is measured every 30 seconds. The graph at the left shows the results. Use the graph to answer questions 16 through 20.

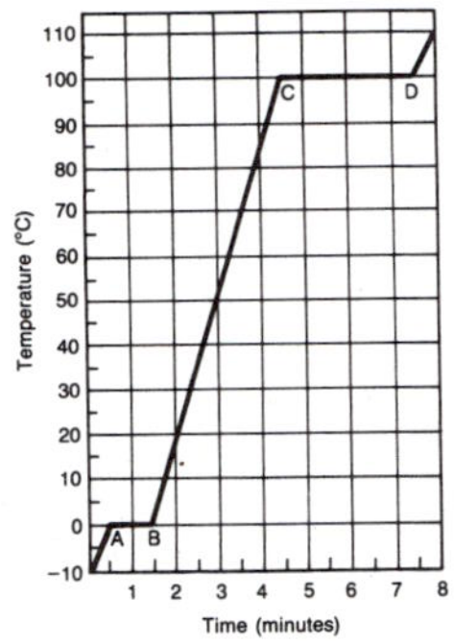

16. What process is taking place between point A and point B? — 16. melting 12.3

17. What explains why the graph levels off at point A? — 17. heat of fusion 12.3

18. What process is taking place between point C and point D? — 18. boiling 12.3

19. What explains why the graph levels off at point C? — 19. heat of vaporization 12.3

20. What is the state of matter of the H_2O at point D? — 20. gas (steam) 12.3

IV. Use the chart on the left to answer questions 21 and 22. Place your answers in the spaces provided.

Specific Heat (cal/g × °C)	
iron	0.11
glass	0.20
wood	0.4
water	1.00
mercury	0.033
air	0.25

21. Which substance has the highest specific heat? — 21. water 12.2

22. Which substance will require the least amount of heat to raise its temperature 1°C? — 22. mercury 12.2

V. Answer the following question in the space provided.

23. Based on what you have learned in Chapter 12, explain how evaporation is a cooling process. 12.3

As a substance is heated, some particles with a lot of energy escape from the surface of the substance into the surrounding air. Since energy (heat) is lost, the substance feels cooler.

CHAPTER 13 TEST

NAME ____________
CLASS ____________
DATE ____________

Heat Energy in Your Life

I. Select the best answer to complete each statement. Place its letter in the space provided.

1. A machine that takes heat from the outside air and uses it to heat a room is called a (a) turbine, (b) heat pump, (c) diesel, (d) compressor. — 1. b 13.4

2. One advantage of gasoline engines over diesel engines is that gasoline engines can be built (a) more efficiently, (b) more cheaply, (c) in larger quantities, (d) smaller and lighter. — 2. d 13.3

3. A machine that operates at 37% efficiency has a heat loss of (a) 37%, (b) 63%, (c) 73%, (d) 100%. — 3. b 13.3

4. The part of an engine that produces rotary motion is the (a) cylinder, (b) disc, (c) spark plug, (d) crankshaft. — 4. d 13.1

5. Heat engines work on the principle that (a) heat causes expansion of gases, (b) containers can expand, (c) work is force exerted over a distance, (d) heat changes work into energy. — 5. a 13.1

6. The first true car was (a) an internal combustion engine, (b) steam-driven, (c) gasoline-driven, (d) the Stanley steamer. — 6. b 13.3

7. Steam turbines work in a way similar to (a) refrigerators, (b) heat pumps, (c) water wheels, (d) car engines. — 7. c 13.2

8. Internal combustion engines proved to be better than steam engines for powering cars because internal combustion engines (a) are more efficient, (b) produce fewer wastes, (c) are larger, (d) are smaller. — 8. d 13.2

II. Complete the following by writing *true* if the statement is correct. If the statement is false, change the underlined word(s) to make it correct. Place your answers in the spaces on the right.

9. If burning takes place outside the cylinder, the engine is an internal combustion engine. — 9. external 13.1

10. In an internal combustion engine, gasoline and air mix in the carburetor. — 10. true 13.3

11. In refrigerator systems, Freon acts as a heat carrier. — 11. true 13.4

12. DeWitt Clinton developed the first efficient steam engine. — 12. James Watt 13.2

13. The part of a heat engine that corresponds to the cover of a pan with boiling water is the cylinder. — 13. piston 13.1

14. Heat pumps work in the opposite way in winter and summer. — 14. true 13.4

15. Air conditioners work in the same way as refrigerators. — 15. true 13.4

CHAPTER 13 TEST (Continued)

NAME ____________

III. Use the diagram below to answer questions 16 through 20.

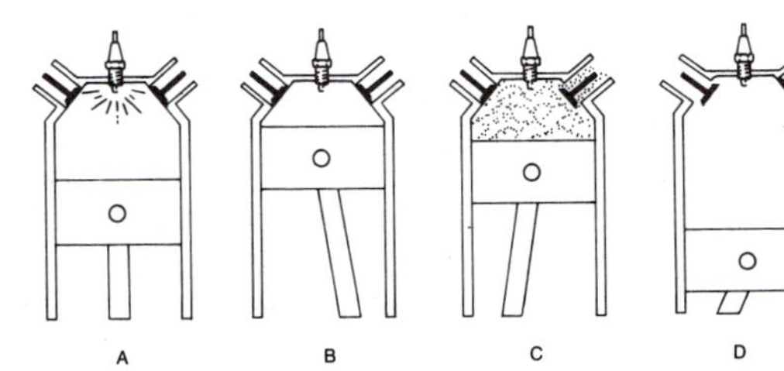

16. What letter represents the intake stroke? — 16. D 13.3

17. What letter represents the compression stroke? — 17. B 13.3

18. What letter represents the power stroke? — 18. A 13.3

19. What letter represents the exhaust stroke? — 19. C 13.3

20. What letter indicates when the crankshaft is turned? — 20. A 13.3

IV. Match the statement in column B with the kind(s) of machine(s) it describes in column A. Place the letter of your answer in the space provided. If more than one answer is correct, give the letters for all correct answers.

A Machines

A. internal combustion engine
B. external combustion engine
C. steam turbine
D. heat pump
E. air conditioner

B Statements

21. burns coal, gas, and oil as fuel — 21. A 13.1

22. B, C 13.2

22. A forceful stream of steam is directed at blades mounted on a wheel. — 23. D, E 13.3 13.4

23. absorbs heat through evaporation and releases it through condensation — 24. A, B, C 13.1

24. uses an increase in pressure to produce energy of motion — 25. A 13.3

25. A four-stroke system moves a crankshaft.

CHAPTER 14 TEST

NAME ____________
CLASS ____________
DATE ____________

Magnetic and Electric Forces

I. Select the best answer to complete each statement. Place its letter in the space provided.

1. A magnetic ore that is found in nature is called (a) alnico, (b) permatite, (c) magnetite, (d) insulation. **1.** c 14.1

2. The electric charge that builds up on an object is called (a) induced magnetism, (b) static magnetism, (c) induced electricity, (d) static electricity. **2.** d 14.3

3. Magnetism is a (a) form of energy, (b) force, (c) power, (d) form of electricity. **3.** b 14.1

4. The earth's magnetism is thought to be due to (a) lodestones, (b) circulation of the atmosphere, (c) rotation of the earth's liquid-iron center, (d) the orbit of the earth around the sun. **4.** c 14.2

5. Combing your hair with a plastic comb will produce (a) magnetic attraction, (b) static electricity, (c) an induced field, (d) neutral atoms. **5.** b 14.3

6. The safest place to be in a thunderstorm is (a) in a car, (b) under a tree, (c) in a field, (d) standing by a wire fence. **6.** a 14.4

7. Of the following, the best conductor of electricity is (a) helium, (b) sulfur, (c) glass, (d) copper. **7.** d 14.4

8. Permanent magnets can lose their magnetism if (a) they are heated to high temperatures, (b) their atomic motion is decreased, (c) the magnetic domains are aligned, (d) they are cut in half. **8.** a 14.4

9. All are true of the north magnetic pole except (a) it is located in northern Canada, (b) it is the same as the north geographic pole, (c) a compass will point to it, (d) it slowly wanders. **9.** b 14.2

10. Electric discharges occur when (a) there is a buildup of charges, (b) a charge is induced, (c) an object becomes grounded, (d) there is a flow of charges. **10.** d 14.4

II. Complete the following by writing *true* if the statement is correct. If the statement is false, change the underlined word(s) to make it correct. Place your answers in the spaces on the right.

11. Most materials can be magnetized. **11.** cannot 14.1

12. The earliest use of a lodestone was in navigation. **12.** true 14.1

13. An electroscope is used to detect a positive charge. **13.** static 14.3

14. Lightning rods are examples of electrical grounds. **14.** grounded conductors 14.4

15. The area around a magnet is a magnetic field. **15.** true 14.1

CHAPTER 14 TEST (Continued)

NAME ____________

III. Use the diagrams of magnets below to answer questions 16 through 18.

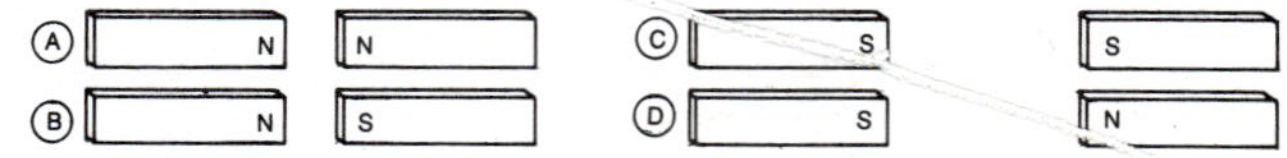

16. What is the letter with magnets having the greatest force of attraction? **16.** B 14.1

17. What is the letter with magnets having a lesser force of attraction? **17.** D 14.1

18. What is the letter with magnets having the greatest force of repulsion? **18.** A 14.1

IV. Use the sequence of events below to answer questions 19 through 23.

An electrically neutral metal rod is standing on a rubber base (1). A positively charged glass rod is brought near point A, inducing a charge on the metal (2). A negatively charged metal ball is brought close to the metal rod (3). The ball swings toward B (4).

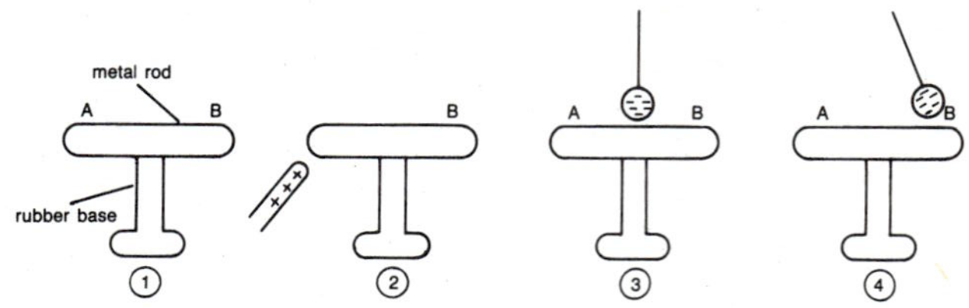

19. What is the charge induced by the glass rod on the rod at A? **19.** negative 14.3

20. What is the charge on the rod at B? **20.** positive 14.3

21. Why was rubber used for the base? **21.** for insulation 14.3

22. Why did the metal ball swing to B? **22.** attraction 14.3

23. What is the letter of the end of the rod that gained electrons? **23.** A 14.3

V. Answer question 24 in the space provided.

24. Why is a negatively charged comb able to pick up bits of paper? 14.3

The comb has excess electrons. They cause electrons in the paper to be repelled. The surface of the paper gains a positive charge and is attracted to the comb.

CHAPTER 15 TEST

NAME ____________
CLASS ____________
DATE ____________

Electricity

I. Select the best answer to complete each statement. Place its letter in the space provided.

1. In a dry cell, the carbon rod serves as the (a) negative terminal, (b) positive terminal, (c) resistor, (d) electron source. **1.** b 15.1

2. A wire gets hot because of its (a) voltage, (b) current, (c) resistance, (d) power. **2.** c 15.4

3. The purpose of a fuse or circuit breaker is to (a) direct current from one wire to another, (b) provide a source of electrons, (c) measure the current going through a wire, (d) shut off current when too much heat is produced. **3.** d 15.4

4. When current is flowing, a circuit is (a) closed, (b) open, (c) short, (d) live. **4.** a 15.4

5. A circuit in which all the lights go out if one light bulb burns out is called (a) a series circuit, (b) insulated, (c) a weak circuit, (d) a parallel circuit. **5.** a 15.2

6. A circuit in which the current branches into two or more separate paths is called (a) a series circuit, (b) a parallel circuit, (c) a divided circuit, (d) a separate circuit. **6.** b 15.2

7. The zinc jacket in a dry cell represents the (a) positive terminal, (b) insulator, (c) negative terminal, (d) resistor. **7.** c 15.1

8. As more charges flow through a conductor, the current (a) alternates, (b) reverses, (c) weakens, (d) gets stronger. **8.** d 15.2

9. Current is measured in (a) volts, (b) watts, (c) amperes, (d) ohms. **9.** c 15.2

10. The ohm is the unit for measuring (a) resistance, (b) power, (c) current, (d) voltage. **10.** a 15.2

II. Complete the following by writing *true* if the statement is correct. If the statement is false, change the underlined word(s) to make it correct. Place your answers in the spaces on the right.

11. The difference in electric potential that causes current to flow is called voltage. **11.** true 15.1

12. A live wire occurs when current flows over an unintended path that has low resistance. **12.** short circuit 15.4

13. The greater the voltage, the lower the resistance. **13.** higher 15.3

14. Edison found that voltage determines the amount of current that flows in a circuit. **14.** Ohm 15.2

15. The current-carrying properties of a material are summed up in the resistance of that material. **15.** true 15.2

CHAPTER 15 TEST (Continued)

NAME ____________

III. Use the chart below to answer questions 16 through 20.

Appliance	Average wattage	Use in average kWh/year	Cost at 13¢/kWh
refrigerator	615	1,829	55.7¢/day
dishwasher	1,200	363	7.8¢/use
iron	1,008	144	7.4¢/hour
color TV	200	660	4.3¢/hour
vacuum cleaner	1,000	46	8.6¢/hour
clothes dryer	5,000	993	38.9¢/load
washing machine	500	103	2.4¢/load
toaster	1,150	39	

16. Which appliance uses the most electric energy? **16.** refrigerator 15.3

17. Which appliance uses the least electric energy? **17.** toaster 15.3

18. Which appliance uses the most power? **18.** dryer 15.3

19. Which appliance uses the least power? **19.** color TV 15.3

20. How much power does a dishwasher use? **20.** 1,200 watts 15.3

IV. Answer the following. If calculations are required, show them under the question.

21. What is the resistance of heating coils in an electric toaster that uses 6 amperes of current in a 120-volt circuit? 15.2

$$\text{Resistance} = \frac{\text{voltage}}{\text{current}} = \frac{120 \text{ volts}}{6 \text{ amperes}} = 20 \text{ ohms}$$

22. How much power does a 100-watt light bulb use? 15.3

100 watts of power

23. Why should you be careful when using electric appliances? 15.4

If you provide a path to the ground for a current, the current will run through you and harm or even kill you.

CHAPTER 16 TEST

NAME ____________
CLASS ____________
DATE ____________

Electromagnetism

I. Match the definition in column A with the term in column B. Place the letter of the matching term in the space provided.

A	B	
1. device in which magnetism is produced by an electric current	**A.** kinetic energy	**1.** B *16.1*
	B. electromagnet	
2. detects a magnetic field	**C.** generator	**2.** J *16.1*
3. changes kinetic energy into electric energy	**D.** transformer	**3.** C *16.2*
4. changes electric energy into kinetic energy	**E.** alternator	**4.** I *16.3*
	F. electromagnetism	
5. increases or decreases voltage	**G.** electromagnetic induction	**5.** D *16.3*
6. produces alternating current		**6.** E *16.2*
7. energy of motion	**H.** voltage	**7.** A *16.2*
8. moves current along power lines	**I.** electric motor	**8.** H *16.3*
9. creates a current by changing a magnetic field	**J.** compass	**9.** G *16.2*
10. branch of physical science		**10.** F *16.1*

II. Select the best answer to complete each statement. Place its letter in the space provided.

11. The Danish scientist who discovered the relationship between electricity and magnetism was (a) Oersted, (b) Henry, (c) Faraday, (d) Edison. **11.** a *16.1*

12. The magnetic field in a coil can be reversed by (a) heating the coil, (b) reversing the current, (c) winding the coil tighter, (d) insulating the coil. **12.** b *16.1*

13. In order for a current to flow, (a) the magnetic field must change in relation to the coil, (b) the magnetic field must remain stationary, (c) the alternator must be disconnected, (d) the switch must be open. **13.** a *16.2*

14. In order for electric current to travel over very long distances, (a) the current must alternate, (b) the voltage must be increased, (c) the voltage must be decreased, (d) the current must be reduced. **14.** b *16.3*

15. Electric generators depend on the effects of (a) insulated wire, (b) alternating current, (c) electromagnetic induction, (d) high-voltage generators. **15.** c *16.2*

CHAPTER 16 TEST (Continued)

NAME ____________

III. Use the diagram of a generator below to answer questions 16 through 20.

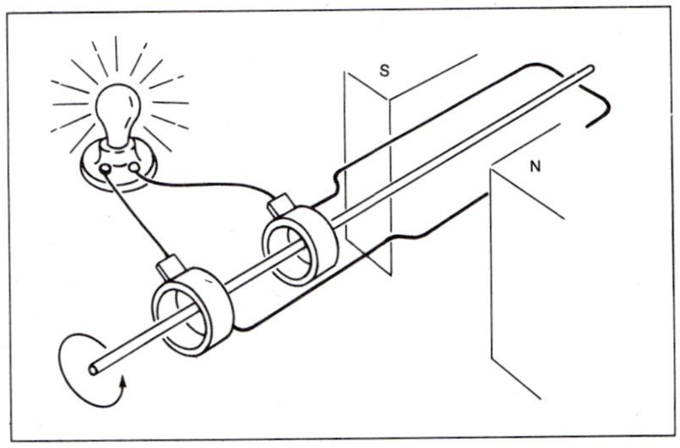

16. What kind of energy causes the light bulb to light? **16.** electric *16.2*

17. What kind of energy is used by the generator? **17.** kinetic *16.2*

18. What is the part of the generator that moves? **18.** coil *16.2*

19. What kind of current is produced? **19.** alternating *16.2*

20. What would make the above diagram a motor instead of a generator? *16.3*

If an electromagnet used electricity to rotate between the fixed magnets in order to turn a shaft.

IV. Answer the following in the space provided.

21. What purpose do transformers serve in moving electric currents? *16.3*

They increase voltage over long distances and decrease voltage for short distances.

22. Why is iron used in making electromagnets? *16.1*

Iron is magnetized when a current flows. When no current flows, iron loses its magnetism. If the current is reversed, iron becomes magnetized in the opposite direction.

CHAPTER 17 TEST

NAME ____________
CLASS ____________
DATE ____________

Electronics

I. Match the definition in column A with the term in column B. Place the letter of the matching term in the space provided.

A	B	
1. a device that changes alternating current to direct current	**A.** integrated circuit	**1.** C *17.2*
	B. amplifier	
2. the first electronic device	**C.** rectifier	**2.** E *17.1*
3. the instructions given to a computer	**D.** oscillator	**3.** G *17.4*
4. the part of a computer that stores information	**E.** vacuum tube	**4.** F *17.4*
5. short bursts of electric energy	**F.** memory unit	**5.** J *17.1*
6. a chip in which different areas conduct current differently	**G.** program	**6.** A *17.3*
	H. transistor	
7. a device that replaced vacuum tubes in electronic equipment	**I.** impurities	**7.** H *17.1*
	J. signals	
8. substances that increase the conductivity of semiconductors		**8.** I *17.2*
9. a device that produces electronic signals from direct current		**9.** D *17.2*
10. a device that strengthens an electronic signal		**10.** B *17.2*

II. Select the best answer to complete each statement. Place its letter in the space provided.

11. Crystals of germanium and silicon are the most commonly used (a) vacuum tubes, (b) semiconductors, (c) computers, (d) oscillators. **11.** b *17.1*

12. Electronic devices work by (a) providing bursts of electric energy, (b) controlling the flow of electrons, (c) creating a vacuum, (d) controlling the release of heat energy. **12.** b *17.1*

13. A material that glows when struck by electrons is (a) fluorescent, (b) charged, (c) silicon, (d) crystalline. **13.** a *17.1*

14. The earliest electronic device used (a) insulated wire, (b) batteries, (c) vacuum tubes, (d) integrated circuits. **14.** c *17.1*

15. The computer unit that usually involves a keyboard is the (a) memory unit, (b) processing unit, (c) input unit, (d) control unit. **15.** c *17.4*

CHAPTER 17 TEST (Continued)

NAME ____________

III. Use the diagrams below to answer questions 16 through 20. Place your answers in the spaces provided.

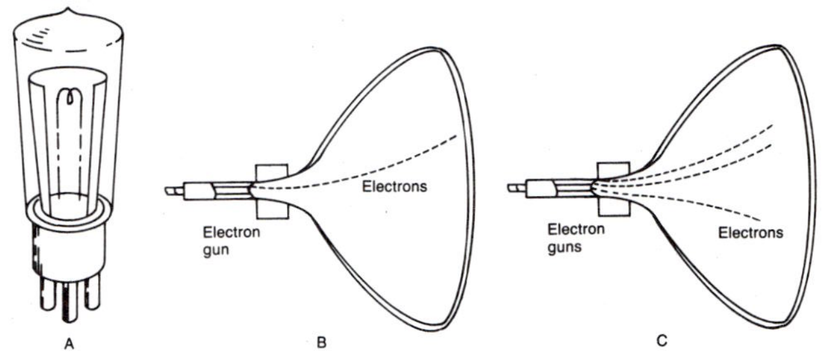

16. What is (are) the letter(s) of the diagram(s) of a cathode-ray tube? **16.** B, C *17.1*

17. What is (are) the letter(s) of the diagram(s) of a picture tube in a color television? **17.** C *17.1*

18. What is (are) the letter(s) of the diagram(s) of a picture tube in a black-and-white television? **18.** B *17.1*

19. What is (are) the letter(s) of the diagram(s) of a tube with a fluorescent screen? **19.** B, C *17.1*

20. Describe how the tube in diagram A works. *17.1*

Electrons flow through a vacuum from the negatively charged wire to the positively charged plate.

IV. Which of the following terms apply to the statements in 21 through 26? Place your answers in the spaces provided. Terms: *input, input unit, output, output unit, processing, processing unit, memory unit, control unit.*

21. A computer performs three basic functions. **21.** input, output, processing *17.4*

22. A computer prints new information on a screen. **22.** output *17.4*

23. Part of a computer regulates the flow of information. **23.** control unit *17.4*

24. An airline agent gets ticket prices and flight times. **24.** output unit *17.4*

25. A computer interprets weather patterns. **25.** processing *17.4*

26. Part of a computer scans a list of phone numbers to find the one you requested from Directory Assistance. **26.** processing unit *17.4*

CHAPTER 18 TEST

NAME ______ CLASS ______ DATE ______

Waves

I. Select the best answer to complete each statement. Place its letter in the space provided.

1. The angle at which a wave hits a barrier is called (a) the angle of reflection, (b) the angle of refraction, (c) the angle of incidence, (d) the angle of disturbance. — **1.** c *18.2*
2. A wave in which the disturbance moves back and forth in the direction of the wave motion is called (a) compressional, (b) transverse, (c) incidental, (d) reflected. — **2.** a *18.1*
3. Sound waves are (a) compressional waves, (b) transverse waves, (c) incidental waves, (d) reflected waves. — **3.** a *18.1*
4. Light waves are (a) compressional waves, (b) transverse waves, (c) refracted waves, (d) bent waves. — **4.** b *18.1*
5. Wavelength is measured in (a) hertz, (b) meters, (c) seconds, (d) joules. — **5.** b *18.1*
6. When light waves are refracted, the result may be (a) thunder, (b) clouds, (c) a rainbow, (d) an eclipse of the sun. — **6.** c *18.2*
7. When two waves meet, their combined effect is called (a) deflection, (b) interference, (c) the Doppler effect, (d) radar. — **7.** b *18.2*
8. When a wave travels through a medium, the medium (a) travels with the wave, (b) travels with the disturbance, (c) does not travel, (d) is not disturbed. — **8.** c *18.1*
9. As a train approaches you, the sound waves from it (a) appear to have a lower sound, (b) appear to have a lower frequency, (c) appear to have a higher frequency, (d) do not change. — **9.** c *18.3*
10. Destructive tidal waves have very great (a) frequencies, (b) amplitudes, (c) interferences, (d) hertz. — **10.** b *18.1*

II. Complete the following by writing *true* if the statement is correct. If the statement is false, change the underlined word(s) to make it correct. Place your answers in the spaces on the right.

11. The frequency of a wave compared to its motion is an effect that was studied by the scientist Hertz. — **11.** Doppler *18.3*
12. The reflection of sound waves produces echoes. — **12.** true *18.2*
13. Police radar equipment measures the frequency of waves. — **13.** true *18.3*
14. The bending of waves around a barrier is refraction. — **14.** diffraction *18.2*
15. The number of crests passing a point in a second is wavelength. — **15.** frequency *18.1*

CHAPTER 18 TEST (Continued)

NAME ______

III. Use the diagrams of waves below to answer questions 16 through 19. Place your answers in the spaces provided.

16. What is the letter of the wave with the greatest amplitude? — **16.** B *18.1*
17. What is the letter of the wave with the greatest frequency? — **17.** C *18.1*
18. What is the letter of the wave having the longest wavelength? — **18.** A *18.1*
19. What is the letter of the wave with the greatest energy? — **19.** B *18.1*

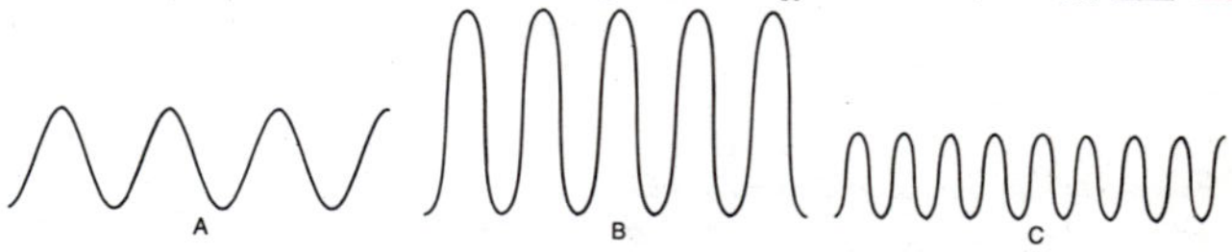

IV. Use the diagrams below to answer questions 20 through 25. In both diagrams, the arrows indicate the path of light waves. Place your answers in the spaces provided.

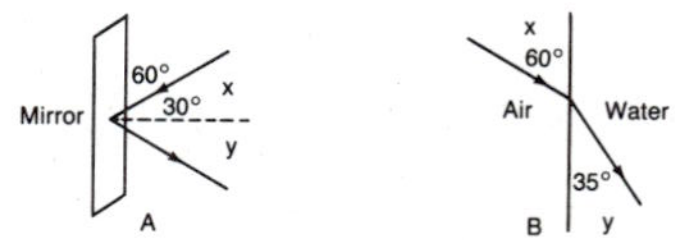

20. What is the letter of the diagram that shows reflection? — **20.** A *18.2*
21. What is the letter of the diagram that shows refraction? — **21.** B *18.2*
22. What is (are) the medium(s) light is traveling through in diagram B? — **22.** air, water *18.2*
23. What is the letter of the angle of incidence in diagram A? — **23.** x *18.2*
24. How many degrees is the angle indicated by the letter y in diagram A? — **24.** 30° *18.2*
25. If angle x in diagram B were larger, would angle y become smaller or larger? — **25.** larger *18.2*

V. Answer the following. If calculations are required, show them under the question.

26. What is the speed of a wave with a frequency of 50 Hz and a wavelength of 0.2 m? *18.1*

speed = wavelength × frequency = 0.2 m × 50 Hz = 10 m/s

27. Find the frequency of a wave that has a speed of 10 m/s and a wavelength of 0.5 m. *18.1*

$$\text{frequency} = \frac{\text{speed}}{\text{wavelength}} = \frac{10 \text{ m/s}}{0.5 \text{ m}} = 20 \text{ Hz}$$

CHAPTER 19 TEST

NAME ______ CLASS ______ DATE ______

Sound

I. Match the definition in column A with the term in column B. Place the letter of the matching term in the space provided.

A	B	Answer
1. a quality of sound that depends on the amplitude of the sound waves	A. vibration	1. B *19.2*
2. a rapid back-and-forth motion	B. intensity	2. A *19.1*
3. mixtures of reflected sounds	C. decibels	3. H *19.4*
4. notes that are multiple frequencies of simple notes	D. pitch	4. G *19.3*
5. that which describes the highness or lowness of a sound	E. ultrasonic	5. D *19.2*
6. the amplification of sound in a musical instrument	F. resonance	6. F *19.3*
7. unpleasant sounds	G. overtones	7. J *19.4*
8. units that measure loudness	H. reverberations	8. C *19.2*
9. a kind of sound heard by bats	I. echo	9. E *19.2*
10. a reflected sound	J. noise	10. I *19.4*

II. Select the best answer to complete each statement. Place its letter in the space provided.

11. Sounds cannot be heard in a total vacuum because (a) no matter is present to transmit the waves, (b) the temperature is too low for sound waves to travel, (c) the frequency is too high inside a vacuum, (d) the waves cannot move fast enough in a vacuum. — **11.** a *19.1*
12. The frequency of a sound wave depends on (a) its intensity, (b) how fast the source of the sound wave is vibrating, (c) the temperature, (d) the energy of the vibration. — **12.** b *19.2*
13. A sound wave moves faster when the medium contains (a) molecules that are close together, (b) molecules that are far apart, (c) air, (d) a vacuum. — **13.** a *19.1*
14. Sound waves travel (a) faster than light waves, (b) slower than light waves, (c) at the same speed as light waves, (d) faster at lower temperatures. — **14.** b *19.2*
15. An excess of unpleasant sounds is called (a) decibels, (b) reverberations, (c) echoes, (d) noise pollution. — **15.** d *19.4*

CHAPTER 19 TEST (Continued)

NAME ______

III. Use the diagrams below to answer questions 16 through 20. Place your answers in the spaces provided. The diagrams on the left show how a resultant wave form is created by a musical instrument from a note and the instrument's overtones. The diagrams on the right show the wave forms of three different musical instruments.

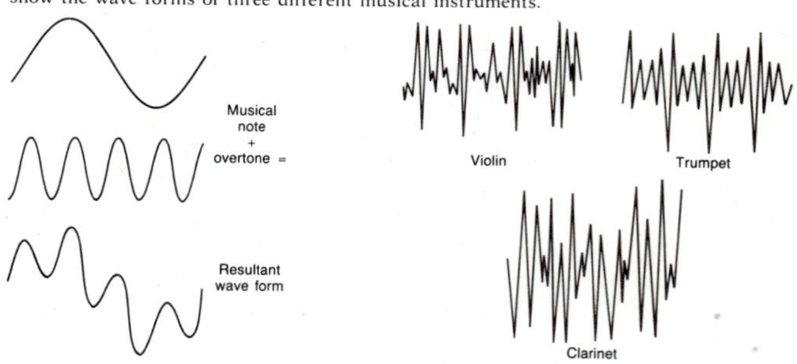

16. What instrument produced the loudest note? — **16.** clarinet *19.2*
17. What instrument produced the highest-pitched note? — **17.** violin *19.2*
18. What instrument produced the note with the highest frequency? — **18.** violin *19.2*
19. What instrument has tubes of different lengths through which sound resonates? — **19.** trumpet *19.3*
20. What do the body of the guitar and the column of the clarinet have in common? — **20.** Both resonate. *19.3*

IV. Answer the following in the space provided.

21. How could you use a row of bottles to produce a melody? *19.3*

Fill them with varying amounts of water and blow across the top to produce notes.

22. List at least three ways in which you could make your science classroom less noisy. *19.4*

Carpet the floors (and walls); upholster the seats; put drapes on the windows; install ceiling tiles; etc.

23. What would happen if you stuffed the body of a violin with cloth and then played it? *19.3*

The violin would not resonate.

CHAPTER 20 TEST

NAME ______
CLASS ______
DATE ______

Light

I. Select the best answer to complete each statement. Place its letter in the space provided.

1. Light waves are (a) transverse, (b) compressional, (c) longer than radio waves, (d) not present in a vacuum. — 1. a 20.1
2. Doctors destroy cancer cells with (a) ultraviolet light, (b) infrared light, (c) gamma rays, (d) microwaves. — 2. c 20.1
3. You know that all colors of light have been absorbed when you see (a) black, (b) white, (c) red, (d) blue. — 3. a 20.2
4. In white light, a red apple is red because (a) it reflects all colors except red, (b) it refracts all colors except red, (c) it absorbs all colors except red, (d) it bends red light waves. — 4. c 20.2
5. The scattering of shorter wavelengths makes the sky appear (a) red, (b) blue, (c) cloudy, (d) orange. — 5. b 20.2
6. Light is most accurately reflected from (a) a dull surface, (b) an uneven glossy surface, (c) a fun-house mirror, (d) a smooth shiny surface. — 6. d 20.3
7. A mirage occurs when light travels through (a) hot air, (b) cold air, (c) air of different densities, (d) water. — 7. c 20.3
8. You see lightning before you hear thunder because (a) sound is slowed down in a storm, (b) clouds permit light to travel very fast, (c) light waves travel faster than sound waves, (d) sound waves travel faster than light waves. — 8. c 20.1
9. Electromagnetic waves involve the production of (a) photons, (b) electrons, (c) neutrons, (d) quarks. — 9. a 20.1
10. All is true of diffraction gratings *except* (a) they cause bright bands of light to appear, (b) they have many parallel slits, (c) they can separate the wavelengths of light given off by a substance, (d) they produce mirages. — 10. d 20.4

II. Complete the following by writing *true* if the statement is correct. If the statement is false, change the underlined word(s) to make it correct. Place your answers in the spaces on the right.

11. Sound waves can travel through a vacuum. — 11. Light 20.1
12. The longest wavelength of visible light is the color blue. — 12. red 20.2
13. A triangular piece of glass that bends light is a prism. — 13. true 20.2
14. Reflecting sunglasses reduce glare. — 14. Polarizing 20.4
15. Both microwaves and infrared light can heat food. — 15. true 20.1

CHAPTER 20 TEST (Continued)

NAME ______

III. Use the diagram on the left to answer questions 16 through 20. The diagram represents light from the sun forming a rainbow. Place your answers in the spaces provided.

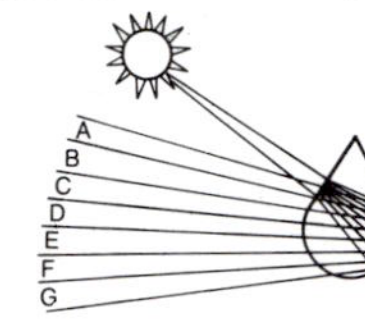

16. What color is represented by the letter A? — 16. red 20.2
17. What color is represented by the letter C? — 17. yellow 20.2
18. What letter represents the color with the longest wavelength? — 18. A 20.2
19. What letter represents the color with the shortest wavelength? — 19. G 20.2
20. What process creates a rainbow? — 20. refraction 20.2

IV. Use the diagram below to answer questions 21 through 25. Place your answers in the spaces provided.

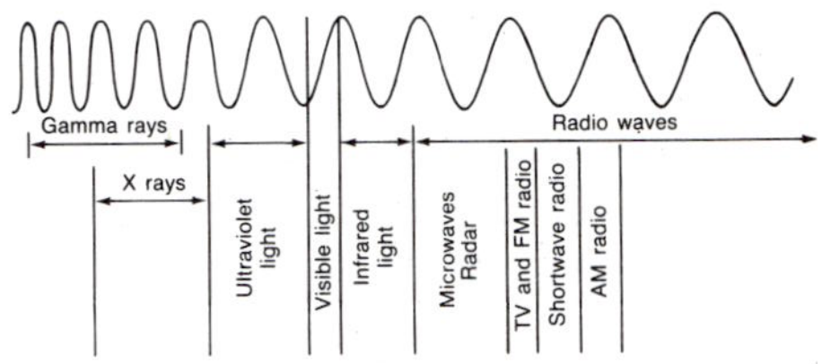

21. What type of electromagnetic wave has the longest wavelength? — 21. radio waves 20.1
22. What type of electromagnetic wave has the highest frequency? — 22. gamma rays 20.1
23. What type of electromagnetic wave has the greatest penetrating ability? — 23. gamma rays 20.1
24. Name three types of electromagnetic waves that can be used to take photographs. — 24. infrared light, X rays, visible light 20.1
25. Where do you think the names *ultraviolet* and *infrared* came from? 20.1
Ultraviolet light has wavelengths just shorter than the violet waves of visible light. Infrared light has wavelengths just longer than the red waves of visible light.

CHAPTER 21 TEST

NAME ______
CLASS ______
DATE ______

The Uses of Light

I. Match the definition in column A with the term in column B. Place the letter of the matching term in the space provided.

A	B	
1. a device used by doctors to examine internal organs	A. laser	1. G 21.3
2. curved pieces of glass used to refract light waves	B. laser light	2. I 21.1
3. light waves of the same wavelength traveling in the same direction	C. refracting telescope	3. B 21.3
4. an instrument that uses two or more lenses to magnify small objects	D. reflecting telescope	4. E 21.1
5. an instrument that uses a mirror and lenses to magnify distant objects	E. microscope	5. D 21.1
6. an instrument that uses two lenses to magnify distant objects	F. holograms	6. C 21.2
7. thin strands of glass transmitting signals carried by light waves	G. endoscope	7. H 21.3
8. a device used by doctors to reattach torn retinas	H. optical fibers	8. A 21.3
9. three-dimensional photographs	I. lenses	9. F 21.3
10. a picture formed by light waves that have passed through a lens	J. image	10. J 21.1

II. Select the best answer to complete each statement. Place its letter in the space provided.

11. The image formed on the retina is a (a) virtual image, (b) real image, (c) focal image, (d) defined image. — 11. b 21.1
12. The eye focuses images through the use of the (a) retina, (b) nerve, (c) lens, (d) pupil. — 12. c 21.2
13. Nearsightedness is corrected by (a) bifocals, (b) convex lenses, (c) concave lenses, (d) the retina. — 13. c 21.2
14. Convex lenses (a) bring light waves together, (b) are thinner in the middle, (c) correct nearsightedness, (d) spread light waves apart. — 14. a 21.1
15. What part of a camera corresponds to the retina of the eye? (a) lens, (b) shutter, (c) light meter, (d) film. — 15. d 21.2

CHAPTER 21 TEST (Continued)

NAME ______

III. Use the diagrams below to answer questions 16 through 22. The diagrams show light as it goes through two kinds of lenses. Place your answers in the spaces provided.

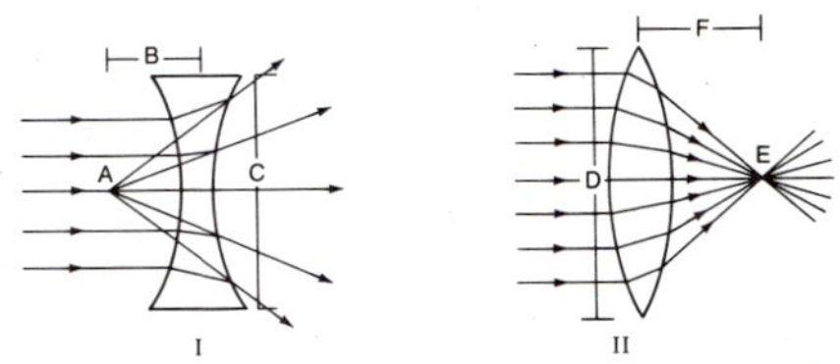

16. What is the letter representing the focal point in diagram II? — 16. E 21.1
17. What kind of lens is represented in diagram I? — 17. concave 21.1
18. What kind of lens is represented in diagram II? — 18. convex 21.1
19. What is the letter representing the focal distance in diagram II? — 19. F 21.1
20. What is the right-side-up image produced in diagram I called? — 20. virtual image 21.1
21. What kind of image is produced in diagram II? — 21. real 21.1
22. Which of the images produced can be focused on a screen? — 22. (II) real 21.1

IV. Answer the following in the space provided.

23. What are the three basic components of a laser? 21.3
(1) substance or medium that can produce laser light
(2) form of energy to excite the atoms of the medium
(3) control of direction of the laser light

24. How does the lens in your eye help you to see objects that are nearby as well as objects that are far away? 21.2
The lens of the eye is changeable. Muscles around the lens make it thicker so that you can see nearby objects. When you want to look at distant objects, muscles make the lens thinner.

CHAPTER 22 TEST

NAME ____________
CLASS ____________
DATE ____________

Energy and Society

I. Select the best answer to complete each statement. Place its letter in the space provided.

1. Energy sources that cannot be replaced are called (a) extinct, (b) fissionable, (c) renewable, (d) nonrenewable. — **1.** d 22.3
2. Control rods in a nuclear reactor (a) protect workers from radiation, (b) indicate when a reaction is completed, (c) regulate the rate of reaction, (d) monitor the temperature. — **2.** c 22.2
3. The most plentiful fossil fuel is (a) crude oil, (b) liquified natural gas, (c) petroleum, (d) coal. — **3.** d 22.1
4. A refinery (a) breaks down coal so it will burn easily, (b) separates crude oil into usable products, (c) makes LNG, (d) prevents crude oil from contaminating the environment. — **4.** b 22.1
5. Natural gas is found (a) in strip mines, (b) in oil shale, (c) in deep underground deposits, (d) mixed with tar sands. — **5.** c 22.1
6. Most of the energy losses from electric power plants are in the form of (a) heat, (b) incomplete burning of fuel, (c) transporting power, (d) radiation. — **6.** a 22.3
7. All is true of strip mining *except* (a) relatively inexpensive, (b) safe, (c) preserves the environment, (d) exposes coal with explosives. — **7.** c 22.1
8. Waste products of nuclear reactors (a) are not harmful, (b) are radioactive, (c) quickly form harmless products in waste dumps, (d) cause thermal pollution. — **8.** b 22.2

II. Complete the following by writing *true* if the statement is correct. If the statement is false, change the underlined word(s) to make it correct. Place your answers in the spaces on the right.

9. Deuterium is an isotope of hydrogen. — **9.** true 22.2
10. In a breeder reactor, hydrogen produces plutonium. — **10.** uranium 22.2
11. The oil that comes directly from a well is called shale oil. — **11.** crude 22.1
12. One of the best ways to cut fuel costs in your home is to install insulation. — **12.** true 22.3
13. In fission, energy is released when two nuclei join together to form a heavier nucleus. — **13.** fusion 22.2
14. The formation of fossil fuels takes hundreds of years. — **14.** millions 22.1
15. A synthetic fuel used to power cars is gasohol. — **15.** true 22.1

CHAPTER 22 TEST (Continued)

NAME ____________

III. Use the chart below to answer questions 16 through 20. Place your answers in the spaces provided.

Pollutants Emitted in a Given Year in the United States (in millions of metric tons)

Pollutant source	Particles	Sulfur oxides	Nitrogen oxides	Carbon monoxide	Hydrocarbons
Power and heating	8.1	22.1	9.1	1.7	0.6
Vehicles	1.1	0.7	7.3	57.9	15.1
Refuse disposal	0.9	0.1	0.5	7.1	1.5
Industry	6.8	6.6	0.2	8.8	4.2
Total	16.9	29.5	17.1	75.5	21.4

16. What is the main source of energy being used by all pollutant sources? — **16.** fossil fuels 22.1
17. What is the greatest pollutant source in general? — **17.** vehicles 22.1
18. What pollutant is emitted in the greatest amount? — **18.** carbon monoxide 22.1
19. Controlling the emission from what pollutant source would most reduce acid rain? — **19.** power and heating 22.1
20. Controlling the emission from what pollutant source would most reduce emissions of a deadly poison that prevents your body from getting the oxygen it needs? — **20.** vehicles 22.1

IV. Answer the following in the space provided.

21. Describe the negative and positive aspects of a breeder reactor. 22.2

The supply of uranium fuel will last longer, but plutonium fuel is especially dangerous to work with.

22. How does the design of the Alaskan pipeline protect the environment? 22.1

Most of the pipeline is on stilts. If the pipeline were buried, the heat from the oil flowing through the pipe would endanger plants and animals on the surface. If it were on the surface, it would interfere with the migration of animals.

CHAPTER 23 TEST

NAME ____________
CLASS ____________
DATE ____________

Energy Alternatives

I. Select the best answer to complete each statement. Place its letter in the space provided.

1. One problem with wind power is that it is not (a) plentiful, (b) cheap, (c) easy to use, (d) pollution-free. — **1.** a 23.3
2. One problem with solar power is that it is not (a) plentiful, (b) cheap, (c) practical everywhere, (d) pollution-free. — **2.** c 23.1
3. Solar cells change solar energy into (a) light energy, (b) heat energy, (c) electric energy, (d) kinetic energy. — **3.** c 23.1
4. Solar collectors change solar energy into (a) light energy, (b) heat energy, (c) electrical energy, (d) kinetic energy. — **4.** b 23.1
5. The use of which of the following energy sources causes pollution? (a) fossil fuels (b) solar energy (c) tidal energy (d) geothermal energy — **5.** a 23.1
6. The use of all of the following threatens the environment except (a) nuclear energy, (b) hydroelectric power, (c) wind energy, (d) geothermal energy. — **6.** c 23.3
7. All of the following are renewable energy sources except (a) nuclear energy, (b) solar energy, (c) tidal energy, (d) geothermal energy. — **7.** a 23.1
8. Hydroelectric power is generated using (a) heat from the earth, (b) the uneven heating of the land, (c) decaying plants and animals, (d) moving water. — **8.** d 23.3
9. The adobe used by the Pueblo Indians served as a (a) solar collector, (b) solar cell, (c) generator, (d) source of geothermal energy. — **9.** a 23.1
10. One of the first uses of wind power was to (a) cook food, (b) heat homes, (c) grind grain, (d) generate electricity. — **10.** c 23.3

II. Complete the following by writing *true* if the statement is correct. If the statement is false, change the underlined word(s) to make it correct. Place your answers in the spaces on the right.

11. A suitable coastline for the development of tidal energy has a small difference in water level at high and low tide. — **11.** large 23.2
12. Today it is possible to use solar energy to heat your home. — **12.** true 23.1
13. To produce hydroelectric power, a dam must be built. — **13.** true 23.2
14. Solar collectors are often painted white for better absorption. — **14.** black 23.1
15. Firewood is a form of biomass. — **15.** true 23.3

CHAPTER 23 TEST (Continued)

NAME ____________

III. Match the item in column A with the term in column B. The terms in column B can be used more than once. Place the letter of the matching term in the space provided.

A	B	
16. Iceland	A. tidal power	**16.** E 23.3
17. the desert regions of the United States	B. solar power	**17.** B 23.1
18. Grand Coulee Dam	C. hydroelectric power	**18.** C 23.2
19. a narrow bay or inlet	D. wind power	**19.** A 23.2
20. spacecraft	E. geothermal power	**20.** B 23.1
21. The Geysers		**21.** E 23.3
22. the uneven heating of the atmosphere		**22.** D 23.3

IV. Use the diagram below to answer questions 23 through 27. The diagram illustrates energy flow from two sources, the sun (solar) and the earth (terrestrial). The numbers indicate the relative amount of power available from each source. Place your answers in the spaces provided.

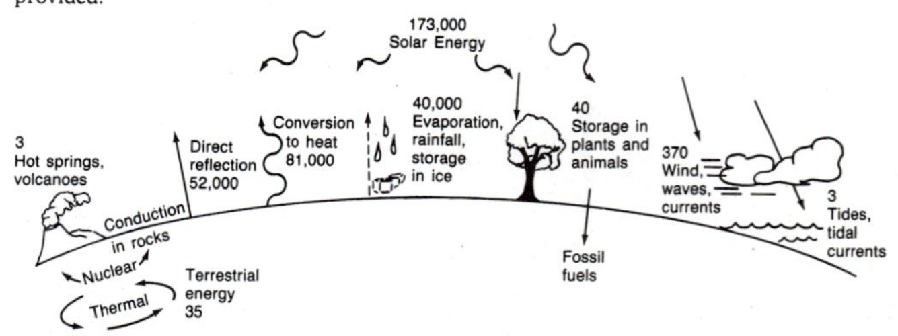

23. What is the relative amount of power available from the tides? — **23.** 3 23.2
24. What is the relative amount of power available from geothermal energy? — **24.** 35 23.3
25. What is the relative amount of power available from biomass? — **25.** 40 23.3
26. List the kinds of energy sources that are related to solar energy. 23.1 23.2 23.3

wind energy, tidal energy, solar energy, hydroelectric power, biomass, fossil fuels

27. Not all of the available energy can be used. Why? 23.1 23.2 23.3

All the energy from these energy sources needs to be collected. For each source, there is at least one factor that restricts our ability to collect all of the energy available.

NAME ______
CLASS ______
DATE ______

The Universe

I. Match the definition in column A with the term in column B. Place the letter of the matching term in the space provided.

A	B	
1. a very large, cool star	**A.** telescope	**1.** C 24.2
2. an extremely dense, collapsed star	**B.** spectroscope	**2.** E 24.2
3. a device that can pick up many forms of radiation	**C.** red giant	**3.** A 24.1
	D. supernova	
4. large masses of gas that give off energy	**E.** neutron star	**4.** H 24.1
5. extremely distant objects emitting radio waves	**F.** black hole	**5.** G 24.2
6. an object that can only be detected by observing its effect on other objects	**G.** quasars	**6.** F 24.2
	H. stars	
7. a device that detects different wavelengths of light	**I.** satellite	**7.** B 24.1
	J. computer	
8. the explosion of a red giant		**8.** D 24.2
9. a device used to gather information about Earth		**9.** I 24.1
10. a device that helps scientists analyze information		**10.** J 24.1

II. Select the best answer to complete each statement. Place its letter in the space provided.

11. To overcome the absence of gravity in space colonies, scientists plan to (a) limit working space, (b) tilt them, (c) launch them very slowly, (d) spin them. **11.** d 24.3

12. All of the energy needed in a space colony could easily come from (a) radio waves, (b) the sun, (c) storage batteries, (d) the heat generated as the colony moved through space. **12.** b 24.3

13. Items that would have to be supplied to space colonies are (a) communication lines, (b) facilities for providing heat, (c) vacuum pumps, (d) water and oxygen. **13.** d 24.3

14. A red giant may collapse to become a (a) quasar, (b) neutron star, (c) white dwarf, (d) white giant. **14.** c 24.2

15. Cygnus X-1 is called a (a) supernova, (b) black hole, (c) red giant, (d) quasar. **15.** b 24.2

NAME ______

III. Read the following paragraph. Then, using your knowledge of the principles of physics and the scientific method, select statement a, b, or c for questions 16 through 20. Place your answers in the spaces provided.

It is the year 2000. A scientist on Space Colony 5 is using a spectroscope supplied to her by the Department of International Science. Her job is to explore areas of space to determine whether black holes are present. On the morning of September 18, Earthtime, she observes light waves from an object. The light waves are shifting toward the red end of the spectrum. She uses another instrument and discovers that the object is emitting radio waves. The object itself does not appear to be large enough to give off the amount of energy detected.

(a) The paragraph supports the statement.
(b) The paragraph does not support the statement.
(c) The paragraph does not give enough information to support the statement.

16. The scientist is an astronomer. **16.** a 24.1

17. The object is moving toward her. **17.** b 24.1

18. The other instrument she used is a radio telescope. **18.** a 24.1

19. The object is probably a quasar. **19.** a 24.2

20. She has found a black hole. **20.** c 24.2

IV. Answer the following in the space provided.

21. How can astronomers learn about the life cycle of a star if that life cycle takes millions of years to complete? 24.2

Astronomers study many stars at different stages and use these observations to develop theories about how stars begin and how they change.

22. How would your life be different if you were living in a space colony? 24.3

Answers will vary. Students might mention that weather and daylight hours would be controlled, many essentials would have to be brought from Earth, and waste disposal and transportation systems would be very different from those on Earth.

Unit One Test
Motion, Force, and Energy

NAME ____________ CLASS ____________ DATE ____________

Select the best answer to complete each statement.

1. A book, milk, and air all have mass and occupy space. They are all examples of (a) solids, (b) properties, (c) matter, (d) plasma. 2.1
2. The state of matter with a definite volume but no definite shape is a (a) solid, (b) liquid, (c) gas, (d) plasma. 2.1
3. ________ moves matter from one place to another or changes matter from one substance to another. (a) Energy (b) Speed (c) Velocity (d) Momentum 2.1
4. If a car travels 600 kilometers north in six hours, its velocity is ________ kilometers per hour north. (a) 3,600 (b) 50 (c) 100 (d) 606 2.2
5. A boat is moving down a river at 50 kilometers per hour. If the velocity of the river is 30 kilometers per hour, what is the true velocity of the boat in kilometers per hour? (a) 20 (b) 80 (c) 1.6 (d) 6 2.2
6. A change in velocity to a lower speed is called (a) acceleration, (b) momentum, (c) terminal velocity, (d) deceleration. 2.3
7. Every object that falls, if it falls long enough, reaches a(an) ________. Once this is reached, the falling object does not accelerate any more; it falls at a constant rate. (a) momentum (b) potential energy (c) acceleration state (d) terminal velocity 2.3
8. The mass of a moving object multiplied by its velocity is called (a) lift, (b) drag, (c) momentum, (d) terminal velocity. 2.4
9. Any push or pull that affects matter is called a(an) (a) force, (b) mass, (c) inertia, (d) gravity. 3.1
10. On the earth, a man has a mass of 72 kilograms. On the moon, his mass would be ________ kilograms. (a) 432 (b) 12 (c) 72 (d) 36 3.1
11. If two objects with negative electric charges are placed near each other, they will (a) attract each other, (b) attract objects around them, (c) repel each other, (d) repel objects around them. 3.2
12. When all the opposing forces on an object are equal, a state of ________ has been reached. (a) inertia (b) equilibrium (c) gravitation (d) force 3.3
13. The tendency of every object to stay in its equilibrium state unless it is acted upon by outside forces is called (a) momentum, (b) friction, (c) terminal velocity, (d) inertia. 3.4
14. When an object is not in a state of equilibrium, it will tend to (a) vibrate, (b) decelerate, (c) remain motionless, (d) accelerate. 3.5
15. When you jump off a boat onto a dock, the boat moves away from the dock. This action can be explained by Newton's ________ law of motion. (a) first (b) second (c) third (d) fourth 3.6

Unit One Test (Continued)

NAME ____________

16. How much work is done when a force of 25 newtons moves a box 100 meters? (a) 2.5 newton-meters (b) 25 newton-meters (c) 75 newton-meters (d) 2,500 newton-meters 4.1
17. The newton-meter is usually called the (a) watt, (b) joule, (c) mechanical advantage, (d) work in. 4.1
18. Simple machines help people do work by (a) decreasing the force needed, (b) increasing the force needed, (c) decreasing the distance, (d) increasing the distance. 4.2
19. What is the mechanical advantage of a machine that uses a 200-newton force to lift a 500-newton weight? (a) 700 (b) 300 (c) 2.5 (d) 3.7 4.2
20. Of the following, the one that is NOT a simple machine is a (a) lever, (b) pulley, (c) wheel and axle, (d) lawn mower. 4.3
21. Of the following, the one that is NOT a lever is a (a) bottle opener, (b) nutcracker, (c) baseball bat, (d) flight of stairs. 4.3
22. How much power is used to do 100 newton-meters of work in 4 seconds? (a) 25 watts (b) 400 watts (c) 125 watts (d) 4,000 watts 4.2
23. The measure of how much work a machine puts out compared to the amount of work put into the machine is its (a) watts, (b) efficiency, (c) power, (d) mechanical advantage. 4.4
24. An aquarium has a bottom surface area of 600 cm^2. The weight of the aquarium exerts a force of 300 newtons on the table supporting it. The aquarium exerts a pressure of ________ N/cm^2. (a) 0.5 (b) 0.3 (c) 1,000 (d) 900 5.1
25. As a diver goes deeper into the water, the amount of pressure the water exerts on the diver (a) increases, (b) stabilizes, (c) decreases, (d) approaches zero. 5.1
26. The idea that the buoyancy of an object is equal to the weight of liquid that it displaces is called (a) Galileo's principle, (b) the law of buoyancy, (c) the displacement law, (d) Archimedes' principle. 5.2
27. What is the density of a piece of lead with a mass of 33.9 grams and a volume of 3 cubic centimeters? (a) 30.9 g/cm^3 (b) 36.9 g/cm^3 (c) 11.3 g/cm^3 (d) 101.7 g/cm^3 5.2
28. Astronauts wear pressurized suits because the farther you travel from Earth, the ________ the air pressure. (a) lower (b) higher (c) more variable (d) more stable 5.3
29. The friction created when objects move in air is called (a) turbulence, (b) lift, (c) drag, (d) reduction force. 5.4
30. When you blow across the top of a thin strip of paper, it rises. This action was explained by (a) Watt, (b) Bernoulli, (c) Lee, (d) Joule. 5.4

Unit Two Test
The Structure of Matter

NAME ____________ CLASS ____________ DATE ____________

Select the best answer to complete each statement.

1. The basic substances that make up all matter are called (a) elements, (b) atoms, (c) nuclei, (d) electrons. 6.1
2. The smallest particle of an individual element is a(an) (a) molecule, (b) compound, (c) nucleus, (d) atom. 6.1
3. The positively charged center of the atom is called the (a) nucleus, (b) electron, (c) neutron, (d) proton. 6.2
4. Elements that give off high-energy particles are said to be (a) energized, (b) electrical, (c) magnetized, (d) radioactive. 6.2
5. The positively charged particle in the atom is called a(an) (a) electron, (b) quark, (c) proton, (d) neutron. 6.3
6. The total number of particles in the nucleus is called the (a) isotope, (b) atomic number, (c) nuclear force, (d) atomic mass number. 6.3
7. The negatively charged particles in an atom are found in a three-dimensional area called the (a) nucleus, (b) electron cloud, (c) atomic area, (d) space. 6.4
8. Elements with similar properties and the same number of valence electrons are grouped in vertical rows on the periodic table. The groups are called (a) metals, (b) families, (c) nonmetals, (d) periods. 7.1
9. Of the following, the one that is NOT a property of metals is (a) are shiny, (b) can be drawn into fine wires, (c) are breakable, (d) are good conductors of heat and electricity. 7.1
10. Hydrogen and potassium are in family IA. This family is the most reactive family of metals. They are called the ________ metals. (a) alkali (b) transition (c) actinide (d) alkaline earth 7.2
11. The elements that are filling the next-to-the-last electron shell are the ________ metals. Most of these elements have two valence electrons. (a) alkaline earth (b) actinide (c) transition (d) alkali 7.2
12. Atoms of elements in the nitrogen family have ________ valence electrons, while atoms of elements in the halogen family have ________ valence electrons. (a) 7, 5 (b) 1, 4 (c) 5, 1 (d) 5, 7 7.3
13. The most common element in the earth's outer layer is ________. It has six valence electrons and easily combines with most metals. (a) lithium (b) oxygen (c) nitrogen (d) sulfur 7.3
14. The ________ differ from other elements because their outer electron shells are filled. As a result, they rarely join with other elements. (a) noble gases (b) halogens (c) actinides (d) lanthanides 7.4
15. Elements that are made in the laboratory by bombarding uranium with neutrons or nuclei of lighter atoms are called (a) actinides, (b) alkali metals, (c) lanthanides, (d) synthetics. 7.5

Unit Two Test (Continued)

NAME ____________

16. Substances made of two or more elements chemically combined are called (a) mixtures, (b) synthetics, (c) compounds, (d) transitionals. 8.1
17. Mixtures in which the substances are uniformly mixed are called (a) solutions, (b) molecules, (c) compounds, (d) crystals. 8.1
18. The formula H_2SO_4 means that there are ________ in one molecule of sulfuric acid.
    (a) 2 atoms of H, 4 atoms of S, 4 atoms of O
    (b) 8 atoms of H, 4 atoms of S, 4 atoms of O
    (c) 2 atoms of H, 1 atom of S, 4 atoms of O
    (d) 1 atom of H, 2 atoms of S, 4 atoms of H 8.2
19. When a compound forms due to a shifting of electrons from one atom to another, a(an) ________ bond is formed. (a) positive (b) ionic (c) covalent (d) negative 8.3
20. A solid in which the atoms are arranged in a regular pattern is called a (a) crystal, (b) molecule, (c) compound, (d) mixture. 8.3
21. When a compound forms due to a sharing of pairs of electrons between two atoms, the bond formed is a(an) ________ bond. (a) ionic (b) crystalline (c) negative (d) covalent 8.4
22. Of the following, the one that is NOT an example of a covalent molecule is (a) ammonia, (b) carbon dioxide, (c) salt, (d) a diamond. 8.4
23. Bonds such as the ones holding together the copper atoms in a penny are called (a) covalent, (b) electron, (c) ionic, (d) metallic. 8.5
24. The element that forms more compounds than any other element is ________. A member of family IVA, it has four valence electrons. (a) uranium (b) germanium (c) silicon (d) carbon 9.1
25. A compound that contains carbon, whether or not it is found in living matter, is called (a) ionic, (b) metallic, (c) organic, (d) acidic. 9.1
26. Compounds with the same molecular formula but different structural formulas are called (a) isomers, (b) organics, (c) alcohols, (d) hydrocarbons. 9.2
27. The process of separating oil into parts by heating it to different temperatures, collecting the gas produced at each temperature, and cooling each gas to a liquid state is called (a) polymerization, (b) fractional distillation, (c) triple bonding, (d) condensation. 9.2
28. When a hydrogen atom in a hydrocarbon is replaced by an OH group, a(an) ________ is formed. (a) organic acid (b) protein (c) carbohydrate (d) alcohol 9.3
29. Organic compounds that are an important part of muscles, hair, bones, skin, blood, and other body organs are (a) carbohydrates, (b) fats, (c) proteins, (d) alcohol. 9.3
30. To make a ________ in the laboratory, small molecules must be linked in long chains. (a) fat (b) polymer (c) amino acid (d) organic acid 9.4

Unit Three Test

Matter, Energy, and Change

NAME ______________ CLASS ______________ DATE ______________

Select the best answer to complete each statement.

1. Of the following, the one that is an example of a chemical change is (a) tearing paper, (b) burning wood, (c) dissolving salt in water, (d) melting ice. 10.1

2. Which of the following is a single displacement reaction?
(a) $Na + Cl_2 \rightarrow 2NaCl$ (b) $ZnCl_2 \rightarrow Zn + Cl_2$
(c) $BeF_2 + Mg \rightarrow MgF_2 + Be$ (d) $AgNO_3 + NaCl \rightarrow AgCl + NaNO_3$ 10.1

3. Which of the following illustrates the law of conservation of mass?
(a) $H_2O \rightarrow H_2 + O_2$ (b) $H_2O \rightarrow 2H_2 + O_2$
(c) $H_2O \rightarrow H + O$ (d) $2H_2O \rightarrow 2H_2 + O_2$ 10.2

4. Generally, increasing the temperature ________ the rate of a chemical reaction. (a) does not affect (b) increases (c) stabilizes (d) decreases 10.3

5. Substances that affect the rate of a chemical reaction without being changed or used up in the reaction are called (a) reactants, (b) products, (c) salts, (d) catalysts. 10.3

6. Which of the following is NOT an exothermic reaction? (a) a fire (b) baking a cake (c) rusting iron (d) burning oil 10.4

7. All of the following are properties of bases EXCEPT (a) react with metals, (b) feel slippery, (c) produce negative ions, (d) taste bitter. 10.5

8. The products of a neutralization reaction are (a) salt and water, (b) hydrogen and salt, (c) water and sugar, (d) hydrogen and water. 10.5

9. Which of the following indicates the presence of an acid? (a) Blue litmus paper turns red. (b) Phenolphthalein turns bright pink. (c) Red litmus paper turns blue. (d) The pH reading is 8. 10.5

10. When the nucleus of one kind of atom changes into the nucleus of another kind of atom, a ________ reaction occurs. (a) chemical (b) stabilizing (c) nuclear (d) physical 11.1

11. Which of the following did not contribute to the discovery of radioactivity? (a) A. H. Becquerel (b) M. Curie (c) Count Rumford (d) P. Curie 11.2

12. When an atom undergoes radioactive decay, it produces (a) water, (b) oxygen, (c) hydrogen, (d) energy. 11.2

13. The half-life of carbon 14 is 5,700 years. If you have 10 grams of carbon 14, how many grams will there be in 11,400 years? (a) 20 grams (b) 10 grams (c) 5 grams (d) 2.5 grams 11.3

Unit Three Test (Continued)

NAME ______________

14. Isotopes are radioactive when their ________ are unbalanced. (a) electrons (b) nuclei (c) protons (d) neutrons 11.3

15. When the nucleus of one kind of atom spontaneously changes into the nucleus of another kind of atom, the process is called (a) fusion, (b) radiation, (c) fission, (d) radioactive decay. 11.4

16. The sun and all the stars shine due to the energy produced by nuclear ________ reactions. (a) chemical (b) physical (c) fission (d) fusion 11.5

17. The idea that atoms and molecules are in constant motion is called the ________ theory. (a) potential (b) molecular (c) kinetic (d) nuclear 12.1

18. The unit that is used to measure the amount of heat energy that a substance contains is the (a) calorie, (b) newton, (c) degree, (d) watt. 12.2

19. When a solid becomes a liquid, the process is called (a) condensation, (b) freezing, (c) boiling, (d) melting. 12.3

20. The amount of heat energy required to turn the water in a pan into steam is called the heat of (a) vaporization, (b) fusion, (c) condensation, (d) fission. 12.3

21. Cracks are put between the blocks of cement in sidewalks to allow for (a) expansion, (b) conduction, (c) convection, (d) contraction. 12.3

22. A radiator placed on one side of a room heats the whole room by the process of (a) conduction, (b) insulation, (c) convection, (d) vaporization. 12.4

23. Heat energy from the sun reaches us through the process of (a) conduction, (b) radiation, (c) fusion, (d) convection. 12.4

24. The part of a heat engine that confines the heated gas is called the (a) crankshaft, (b) generator, (c) piston, (d) cylinder. 13.1

25. An engine in which steam under pressure is used to produce a rotational motion is the (a) car engine, (b) steam turbine, (c) jet engine, (d) crankshaft. 13.2

26. The first successful steam engine was built by (a) Thomas Edison, (b) Nicolas-Joseph Cugnot, (c) Benjamin Franklin, (d) James Watt. 13.2

27. The fuel of an internal combustion engine is ignited by the (a) carburetor, (b) spark plug, (c) piston, (d) cylinder. 13.3

28. The downward movement of the crankshaft to do work occurs in the ________ stroke of an internal combustion engine. (a) intake (b) compression (c) power (d) exhaust 13.3

29. When liquids evaporate, they ________ heat. (a) absorb (b) radiate (c) release (d) conduct 13.4

30. Machines that take heat from outdoors and use it to heat a house are called (a) air systems, (b) heat engines, (c) refrigeration units, (d) heat pumps. 13.4

Unit Four Test

Magnetism and Electricity

NAME ______________ CLASS ______________ DATE ______________

Select the best answer to complete each statement.

1. A force produced by the motion of electric charges in a material is called (a) electricity, (b) gravity, (c) magnetism, (d) inertia. 14.1

2. Clusters of atoms in a magnet that are lined up in a certain direction are called (a) induced magnets, (b) magnetic fields, (c) magnetic poles, (d) magnetic domains. 14.1

3. Of the following, a way in which magnetism and electricity are NOT similar is (a) like charges and like poles repel, (b) matter can have a single charge or a single magnetic pole, (c) as distance increases force decreases, (d) opposite charges and opposite poles attract. 14.1

4. The north magnetic pole is located to the ________ of the north geographic pole. (a) north (b) south (c) east (d) west 14.2

5. Scientists believe that the earth's magnetism is related to (a) the pull of the moon on the earth, (b) the rotation of its liquid iron core, (c) deposits of iron ore near the North Pole, (d) lava in the earth. 14.2

6. An electric charge that builds up on an object but does not flow is called (a) circuitry, (b) current electricity, (c) electric discharge, (d) static electricity. 14.3

7. An instrument that can be used to detect an electric charge is a(an) (a) telescope, (b) microscope, (c) electroscope, (d) periscope. 14.3

8. The conductivity of an object depends on the ability of the ________ in it to move. (a) neutrons (b) quarks (c) protons (d) electrons 14.4

9. Lightning rods protect buildings by allowing electric charges to pass through the rods into an electrical ground. Lightning rods are (a) electric dischargers, (b) grounded conductors, (c) grounded insulators, (d) insulators. 14.4

10. The difference in electric potential that causes current to flow is called (a) voltage, (b) amperes, (c) resistance, (d) power. 15.1

11. A circuit having only one path for the current to flow through is called a(an) ________ circuit. (a) single (b) series (c) parallel (d) branchless 15.2

12. ________ current constantly switches the direction in which it flows. (a) Reversible (b) Direct (c) Alternating (d) Movable 15.2

13. According to Ohm's law, when the voltage increases, the ________ increases. (a) resistance (b) amperes (c) watts (d) power 15.2

14. If one bulb in your circuit goes out but the others remain lighted, your bulbs are connected in a ________ circuit. (a) branched (b) series (c) single (d) parallel 15.2

Unit Four Test (Continued)

NAME ______________

15. The unit used to measure large amounts of electric power is the (a) volt, (b) ampere, (c) kilowatt, (d) ohm. 15.3

16. Devices that open a circuit to stop the flow of current are called (a) fuses, (b) transformers, (c) resistances, (d) short circuits. 15.4

17. Which of the following should you NEVER do? (a) stay clear of live wires (b) keep water away from electricity (c) allow insulation on wires to fray (d) replace burned-out fuses immediately 15.4

18. The scientist who discovered that electricity can produce magnetism is (a) Galileo, (b) Franklin, (c) Oersted, (d) Faraday. 16.1

19. To make a strong electromagnet, you would do all of the following EXCEPT (a) use a metallic core, (b) use a large current, (c) use a tight coil of wire, (d) use a nonmetallic core. 16.1

20. The creation of an electric current by a changing magnetic field is called (a) electromagnetic induction, (b) electrostatic induction, (c) direct current, (d) alternating current. 16.2

21. Machines that change kinetic energy into electric energy are called (a) motors, (b) transformers, (c) pulleys, (d) generators. 16.2

22. Devices that increase or decrease voltage are called (a) transformers, (b) generators, (c) turbines, (d) motors. 16.3

23. A short burst of electric energy to which an electronic device responds is called a(an) (a) transistor, (b) oscillation, (c) input, (d) signal. 17.1

24. The elements most commonly used in semiconductors are (a) carbon and neon, (b) germanium and silicon, (c) nitrogen and argon, (d) aluminum and cobalt. 17.1

25. Electronic devices that convert alternating current to direct current are called (a) oscillators, (b) rectifiers, (c) transistors, (d) amplifiers. 17.2

26. Electronic devices that are used to strengthen electronic signals are called (a) transistors, (b) oscillators, (c) amplifiers, (d) rectifiers. 17.2

27. A circuit put on a semiconductor by placing impurities in a precise arrangement is called a(an) ________ circuit. (a) micro (b) parallel (c) integrated (d) series 17.3

28. Of the following, the one that is NOT a function of a computer is to (a) take in information, (b) give out new information, (c) do something with the information, (d) operate independently. 17.4

29. The unit of a computer that regulates the flow of information is the ________ unit. (a) control (b) input (c) output (d) memory 17.4

30. The unit of a computer that carries out a program by following its instructions is the ________ unit. (a) memory (b) processing (c) input (d) control 17.4

Unit Five Test
The Action of Waves

NAME ________ CLASS ________ DATE ________

Select the best answer to complete each statement.

1. A disturbance that travels through matter or space is called a (a) wave, (b) crest, (c) compression, (d) trough. 18.1
2. The height of a wave is called its ________. It can indicate how much energy the wave has. (a) amplitude (b) frequency (c) crest (d) trough 18.1
3. The number of crests that pass a point in a second is called the wave (a) length, (b) amplitude, (c) speed, (d) frequency. 18.1
4. The bouncing back of a wave when it hits a boundary is called (a) refraction (b) reflection, (c) diffraction, (d) interference. 18.2
5. A pencil in a glass of water looks broken because light waves bend when they pass from one medium to another. This behavior of waves is called (a) interference, (b) refraction, (c) reflection, (d) diffraction. 18.2
6. The bending of light waves around the sides of a barrier is called (a) reflection, (b) interference, (c) diffraction, (d) refraction. 18.2
7. As a train moves toward you, its whistle sounds higher. As it moves away, the whistle sounds lower. This observation can be explained by (a) radar, (b) reflection, (c) interference, (d) the Doppler effect. 18.3
8. A ________ wave is a compressional wave produced by a back-and-forth motion. (a) light (b) sound (c) vibrational (d) radio 19.1
9. Two factors that affect the speed of sound waves are (a) temperature and the matter present, (b) temperature and loudness, (c) intensity and the matter present, (d) intensity and amplitude. 19.1
10. The ________ of a sound depends on the amplitude of the sound waves involved. (a) resonance (b) intensity (c) vibration (d) frequency 19.2
11. The highness or lowness of a sound is called its (a) loudness, (b) resonance, (c) intensity (d) pitch. 19.2
12. The sound waves that bats use to find their way in total darkness are called (a) resonance, (b) overtone, (c) ultrasonic, (d) sonic. 19.2
13. A guitar amplifies the sounds that its vibrating strings make when the body of the guitar vibrates. This amplification is an example of (a) echoing, (b) overtones, (c) resonance, (d) reverberations. 19.3
14. The same note played on a piano and on a flute sounds different because each instrument produces different multiple frequencies of the same note called (a) overtones, (b) pitch variations, (c) echoes, (d) ultrasonic waves. 19.3

Unit Five Test (Continued)

NAME ________

15. Mixtures of reflected sounds are called (a) overtones, (b) reverberations, (c) resonance, (d) echoes. 19.4
16. Of the following, the phrase that does NOT apply to electromagnetic waves is (a) act like particles and like waves, (b) are compressional, (c) travel faster than sound waves, (d) are transverse. 20.1
17. Bundles of energy given off when excited electrons return to their original energy levels are called (a) photons, (b) excitrons, (c) protons, (d) neutrons. 20.1
18. Electromagnetic waves that are used to sterilize objects are (a) gamma rays, (b) infrared, (c) X rays, (d) ultraviolet. 20.1
19. A blue shirt appears blue because (a) it reflects green and yellow light, (b) it reflects blue light, (c) it absorbs green and yellow light, (d) it absorbs blue light. 20.2
20. Raindrops and prisms separate white light into colors by (a) refraction, (b) reflection, (c) interference, (d) diffraction. 20.2
21. An illusion caused by the refraction of light is called a (a) real image, (b) mirage, (c) virtual image, (d) dream. 20.3
22. When you are fishing, you should not cast your line directly at the fish you see. ________ of light makes the fish look as if it is in a different place from where it really is. (a) Reflection (b) Refraction (c) Diffraction (d) Interference 20.3
23. A thin piece of glass with thousands of parallel slits is called a (a) prism, (b) diffraction grating, (c) microscope, (d) polarized lens. 20.4
24. Materials are described as ________ if they only allow light waves with crests and troughs at a certain angle to pass through them. (a) diffracting (b) refracting (c) polarizing (d) interfering 20.4
25. A lens that is thicker in the center and thinner at the edges is a ________ lens. (a) convex (b) focal (c) virtual (d) concave 21.1
26. An image that appears right side up and cannot be focused on a screen is called a ________ image. (a) convex (b) real (c) concave (d) virtual 21.1
27. The lens of your eye focuses a real image on the (a) eyeball, (b) pupil, (c) retina, (d) optic nerve. 21.2
28. Telescopes that use two convex lenses to magnify distant objects are ________ telescopes. (a) diffracting (b) reflecting (c) diffusing (d) refracting 21.2
29. Light involving waves of the same wavelength traveling in the same direction is called ________ light. (a) ruby (b) laser (c) polarized (d) diffracted 21.3
30. Thin strands of glass that use light waves to carry messages are called (a) holograms, (b) optical fibers, (c) lasers, (d) filaments. 21.3

Unit Six Test
Energy and the Future

NAME ________ CLASS ________ DATE ________

Select the best answer to complete each statement.

1. Fuels formed from the remains of plants and animals that lived millions of years ago are (a) tar sands, (b) oil shale, (c) fossil fuels, (d) managed resources. 22.1
2. Of the following, the one that is NOT a source of oil is (a) tar sands, (b) ocean sediments, (c) oil shale, (d) underground pools. 22.1
3. Liquified natural gas is formed by ________ natural gas. (a) cooling and compressing (b) cooling and expanding (c) heating and compressing (d) heating and expanding 22.1
4. Acid rain forms when oxides of ________ combine with water in the atmosphere. (a) nitrogen and phosphorus (b) nitrogen and sulfur (c) mercury and cadmium (d) lead and sulfur 22.1
5. Nuclear power plants produce ________ to generate electricity. (a) radiation (b) gamma rays (c) ions (d) steam 22.2
6. The rate of the reaction in a nuclear reactor is controlled by rods of steel and cadmium called ________ rods. (a) reactor (b) fission (c) fusion (d) control 22.2
7. In a(an) ________ reactor, uranium produces other nuclear fuels. (a) enhanced (b) multiplier (c) increased (d) breeder 22.2
8. Two nonrenewable energy sources are (a) wood and the sun, (b) biomass and the wind, (c) fossil fuels and nuclear fuels, (d) geysers and the tides. 22.3
9. All of the following are good ways to conserve energy EXCEPT (a) insulating homes, (b) driving faster, (c) using weather stripping, (d) turning off lights. 22.3
10. In the production of electricity, much energy in the form of ________ is lost. (a) heat (b) steam (c) light (d) sound 22.3
11. Solar energy is collected and changed to heat energy in solar (a) cells, (b) collectors, (c) traps, (d) storehouses. 23.1
12. Solar energy is collected and changed to electric energy in solar (a) generators, (b) traps, (c) turbines, (d) cells. 23.1
13. A major problem in using solar energy is that it is (a) nonrenewable, (b) spread over a very large area, (c) unavailable at night, (d) renewable. 23.1
14. It is possible to collect solar energy without taking over large land areas by using solar collectors in (a) ocean areas and satellites, (b) mountain ranges and polar areas, (c) polar areas and tropical islands, (d) tropical islands and mountain ranges. 23.1

Unit Six Test (Continued)

NAME ________

15. The source of hydroelectric power is (a) hydrogen, (b) moving water, (c) rain-filled clouds, (d) water vapor in the air. 23.2
16. Hydroelectric power (a) is difficult to transport, (b) is nonrenewable, (c) involves the building of dams, (d) is not related to solar energy. 23.2
17. Gravitational forces involving the earth, the sun, and the moon can be used to produce ________ energy. (a) wind (b) solar (c) geothermal (d) tidal 23.2
18. The movement of air masses from one place to another produces ________ energy that can be used to generate electricity. (a) wind (b) solar (c) biomass (d) geothermal 23.3
19. Geysers and hot springs involve ________ energy. (a) biomass (b) geothermal (c) solar (d) tidal 23.3
20. All materials in living things, their wastes, and their remains make up a source of energy called (a) fossil fuels, (b) biomass, (c) solar, (d) tar sands. 23.3
21. Astronomers use ________ to observe objects and monitor their motion. (a) microscopes (b) periscopes (c) oscilloscopes (d) telescopes 24.1
22. A ________ is used to study the surface temperature and the motion of objects in space. (a) spectroscope (b) weather balloon (c) thermometer (d) map of the heavens 24.1
23. Astronomers develop theories by using the scientific method and applying the principles of ________ science to the information they have gathered. (a) geological (b) biological (c) physical (d) meteorological 24.1
24. Large masses of gases that release energy are called (a) meteors, (b) moons, (c) planets, (d) stars. 24.2
25. A star with expanded outer layers that gives off mostly red light is called a (a) yellow image, (b) supernova, (c) red giant, (d) white dwarf. 24.2
26. The dense collapsed star that remains after a supernova is called a (a) neutron star, (b) red giant, (c) white dwarf, (d) nova. 24.2
27. Objects, like Cygnus X-1, that are so dense that their gravity allows nothing to escape are called (a) red giants, (b) supernovas, (c) black holes, (d) white dwarfs. 24.2
28. Satellites are used to do all of the following EXCEPT (a) study weather patterns, (b) study population changes, (c) study air pollution, (d) relay telephone and television communications. 24.2
29. If people were to live in space, they would have to take all of the following EXCEPT ________ with them. (a) food (b) light (c) water (d) air 24.3
30. Industries in space would be able to produce many new kinds of ________. (a) metals (b) gases (c) foods (d) elements 24.3

3 4 5 6 7 8 9 0